The Tumor Microenvironment of High Grade Serous Ovarian Cancer

The Tumor Microenvironment of High Grade Serous Ovarian Cancer

Special Issue Editors

M. Sharon Stack
Kenneth P. Nephew
Joanna E. Burdette
Anirban K. Mitra

MDPI • Basel • Beijing • Wuhan • Barcelona • Belgrade

Special Issue Editors

M. Sharon Stack
University of Notre Dame
USA

Kenneth P. Nephew
Indiana University School of Medicine
USA

Joanna E. Burdette
University of Illinois
USA

Anirban K. Mitra
Indiana University School of Medicine
USA

Editorial Office
MDPI
St. Alban-Anlage 66
4052 Basel, Switzerland

This is a reprint of articles from the Special Issue published online in the open access journal *Cancers* (ISSN 2072-6694) from 2018 to 2019 (available at: https://www.mdpi.com/journal/cancers/special_issues/Ovarian_Cancer)

For citation purposes, cite each article independently as indicated on the article page online and as indicated below:

LastName, A.A.; LastName, B.B.; LastName, C.C. Article Title. *Journal Name* **Year**, *Article Number*, Page Range.

ISBN 978-3-03897-554-0 (Pbk)
ISBN 978-3-03897-555-7 (PDF)

Cover image courtesy of Anirban K. Mitra and Yiming Fang.

Contents

About the Special Issue Editors

M. Sharon Stack, PhD, is Professor of Chemistry and Biochemistry and Director of the Harper Cancer Research Institute (HCRI) at the University of Notre Dame. She received her PhD degree in Biochemistry from the University of Louisville School of Medicine and post-doctoral fellowship training at Duke University Medical Center. In 1994, she joined the faculty in the Department of Cell & Molecular Biology at Northwestern University Feinberg Medical School, where she rose through the ranks to tenured Professor. In 2011, she was recruited to the University of Notre Dame to lead the HCRI. She has 30 years of experience in mechanistic cancer biology with focus on cell:cell and cell:matrix interactions, matrix biology, and mechanisms of metastasis. Dr. Stack was elected a Fellow of the American Association for the Advancement of Science in 2012.

Kenneth P Nephew (PhD) is a Professor of Cellular and Integrative Physiology at Indiana University School of Medicine, leader of the Tumor Microenvironment and Metastasis Program and Co-Director of Ovarian Cancer Research Group both of the Indiana University Simon Cancer Center. He received his PhD from the Ohio State University and postdoctoral fellowship training at University of Kansas School of Medicine and University of Cincinnati College of Medicine. He has studied ovarian cancer biology for over 20 years. His main areas of research are epigenetics (DNA methylation, histone modifications, non-coding RNAs), women's cancers, cancer stem cells, drug resistance and bench-bedside studies. He has published over 200 original research articles, is an Editorial Board Member of Cancer Research and Cancers, a member of Van Andel Research Institute Stand Up to Cancer Epigenetics Dream Team, and serves on scientific advisory boards for Ovarian Cancer Research Alliance and Ovarian Cancer Action.

Joanna E. Burdette (Dr.) earned her B.S. from Emory University and her Ph.D. at the University of Illinois at Chicago. She was a postdoctoral fellow at Northwestern under the direction of Dr. Teresa K. Woodruff. She is currently a Professor at the University of Illinois at Chicago in the Department of Medicinal Chemistry and Pharmacognosy. She has served as the Associate Dean for Research in the College of Pharmacy since 2014. Her research has helped to demonstrate that high grade serous ovarian cancer can arise from the fallopian tube.

Anirban K. Mitra (Dr.) is an Assistant Professor of Medical and Molecular Genetics in the Medical Sciences Program of Indiana University School of Medicine (IUSM) and a full member of the IU Simon Cancer Center (IUSCC). Dr. Mitra earned his master's degree in Biotechnology from DAVV University, Indore, India and PhD in Biochemistry from the University of Mumbai, India. His postdoctoral training was in ovarian cancer biology under the mentorship of Dr. Ernst Lengyel at the University of Chicago. Dr. Mitra studies the regulation of ovarian cancer metastatic colonization by microRNAs and transcription factors and epigenetic regulators. His focus is on reciprocal interactions between ovarian cancer cells with the metastatic and tumor microenvironment and understanding reprogramming of resident normal fibroblasts into cancer associated fibroblasts through the altered microRNA expression. He demonstrated epigenetic regulation of microRNAs and transcription factors by paracrine/juxtacrine signals from the metastatic microenvironment was essential for metastatic colonization.

cancers

MDPI

Editorial

The Tumor Microenvironment of High Grade Serous Ovarian Cancer

M. Sharon Stack [1,2], Kenneth P. Nephew [3,4,5,*], Joanna E. Burdette [6] and Anirban K. Mitra [7,8,9]

1. Department of Chemistry and Biochemistry, Harper Cancer Research Institute, University of Notre Dame, Notre Dame, IN 46617, USA; Sharon.Stack.11@nd.edu
2. Harper Cancer Research Institute, A200 Harper Hall, University of Notre Dame, South Bend, IN 46617, USA
3. Cell, Molecular and Cancer Biology Program, Medical Sciences, Indiana University School of Medicine, Bloomington, IN 47405, USA
4. Department of Cellular and Integrative Physiology and Obstetrics and Gynecology, Indiana University School of Medicine, Indianapolis, IN 46202, USA
5. Indiana University Simon Cancer Center, Indianapolis, IN 46202, USA
6. Department of Medicinal Chemistry and Pharmacognosy, University of Illinois, Chicago, IL 60607, USA; joannab@uic.edu
7. Medical Sciences Program, Indiana University School of Medicine, Bloomington, IN 47401, USA; anmitra@indiana.edu
8. Indiana University Melvin and Bren Simon Cancer Center, Indianapolis, IN 46202, USA
9. Department of Medical and Molecular Genetics, Indiana University School of Medicine, Indianapolis, IN 46202, USA

* Correspondence: knephew@indiana.edu

Received: 24 December 2018; Accepted: 25 December 2018; Published: 26 December 2018

The Special Issue on high grade serous ovarian cancer (HGSOC) and the contribution of the tumor micro-environment (TME) consisted of reviews contributed by leaders in the ovarian cancer (OC) field. As HGSOC metastases have a highly complex TME, there is an urgent need to better understand the TME in general, its distinct components in particular, and the role of the TME in the context of disease recurrence and development of chemoresistance. The Special Issue integrated the current understanding of the TME components, including malignant cells, surrounding host stromal cells, and infiltrating (recruited) immune cells. In addition to cellular contributors to the TME, the role of ascites fluid components including soluble factors such as cytokines, chemokines, and growth factors; cell–cell and cell–matrix adhesion molecules; extracellular matrix remodeling; and abnormal vascular and lymphatic networks were the subject of reviews. Reviews covered the relationship between the molecular mechanisms of HGSOC progression, including genomic, epigenomic, and transcriptomic changes, and alterations of the immune cell landscape, as these may provide attractive new molecular targets for HGSOC therapy.

Prof. Dr. Stack et al. [1] illustrated how the aging process has been shown to modulate the TME in ways that are beneficial to the spread and survival of ovarian cancer. Aging hosts have been shown to better facilitate cancer associated inflammation, invasion, and adhesion of cancer cells, while also putting forth a weakened immune response. More research into the unique features of an aging peritoneum is needed to better treat aging HGSOC patients.

Prof. Dr. Nephew and Prof. Dr. Klemenko illustrated how cells constituting the TME are also involved in epigenetic crosstalk with ovarian cancer cells [2]. Ovarian cancer cells have been shown to epigenetically reprogram a wide variety of cell types in their microenvironment to promote tumor growth, survival, and metastasis. There is also growing evidence to suggest that cells from the tumor microenvironment are capable of epigenetically modifying cancer cells. Prof. Dr. Mitra et al. discussed cancer associated fibroblasts; fibroblasts that have been reprogrammed by cancer cells to support tumor growth, survival, and spread, through the secretion of cancer promoting factors. Further research into their origin and identifying markers is needed in order to better characterize their function within the TME [3].

There is still a great deal of research to be done regarding the roles of individual proteins in ovarian cancer. As illustrated by Prof. Dr. Burdette et al. [4], the paired box protein PAX8 is overexpressed in HGSOCs and confers advantages in growth, survival, and migration. While PAX2 has been shown to impart similar growth advantages, expression of it is lost early on in carcinogenesis. Prof. Dr. Hilliard's review focused on mesothelin which is believed to play a role in survival, proliferation, tumor progression, and adherence. Though its native biological function is poorly understood, it is known to bind to the ovarian cancer biomarker CA125, through which it plays a role in metastasis [5]. Prof. Dr. Hudson and colleagues discussed how many of the signaling pathways implicated in HGSOC converge on the small GTPase Rac1, which is associated not only with actin remodeling, adhesion, and migration, but also EMT, stemness, angiogenesis, and chemoresistance [6]. Rac1's role in such a high number of cancer associated signaling pathways makes it an appealing target for anticancer therapies.

HGSOC presents unique challenges in the development of effective immunotherapies to combat spread and progression. Prof. Dr. Vanderhyden et al. [7,8] showed how the low mutational burden, recruitment of T-regs, upregulation of immune checkpoint proteins, and heterogeneity associated with epithelial ovarian cancer have served as road blocks in the development of ovarian cancer immune therapies. Prof. Dr. Drakes and Prof. Dr. Stiff illustrated the importance of understanding the factors in the TME that contribute to the immunogenicity of HGSOC in the development of immune therapies and more accurate prognosis of patients [9]. Improvements in immune therapies that result from better characterizing immune modulating TME factors, combined with treatments targeting other areas of the malignancy are important efforts to increase the survival of patients. In their chapter, Prof. Dr. Khabele et al. [10,11] illustrated how macrophages in the TME represent cancer promoting and antitumor forces in ovarian cancer. Cancer promoting M2 tumor associated macrophages (TAM) represent an attractive target for anticancer therapies. Reprogramming of these M2 cells to M1 tumoricidal macrophages constitutes a promising means of manipulating the TME to be less amenable to the malignancy.

HGSOC is a malignancy once thought to originate exclusively from the ovarian surface epithelium. Prof. Dr. Kim et al. [12] reviewed current evidence that suggests HGSOC likely also originates from serous tubal intraepithelial carcinomas (STICs) from the fallopian tube epithelium. Though STIC has been shown to correlate with an increased risk of HGSOC, it is still important to show causation. Additionally, it is important to elucidate the differences between STIC lesions likely to remain benign vs those that are likely to develop into HGSOC.

Epithelial ovarian cancer is commonly associated with metastasis to the peritoneum, though it has been shown to colonize a wide range of other tissues. Prof. Dr. Barbolina's review focused on the mechanisms of transcoelomic, hematogenous, and lymphatic metastasis [13]. Though most patients typically succumb to transcoelomic, the presence of distant metastasis is associated with worse prognosis. As treatment of transcoelomic metastasis improves, it is likely that more research will have to be devoted to hematogenous and lymphatic spread in order to further improve patient outcomes.

Understanding the interactions between ovarian cancer and metabolites is critical to understanding and treating the disease. In his review, Prof. Dr. Xu outlined how supportive cells have been shown to produce cancer-promoting oncolipids [14]. Improvements to existing detection methods will be valuable in the use of oncolipids as a diagnostic marker in gynecological cancers. Additionally, HGSOC exhibits a reliance on oxidative phosphorylation for its energy needs, making inhibition of the OXPHOS pathway an intriguing target for novel therapies. Prof. Dr. Patankar et al. [15] discussed how OXPHOS inhibition slows proliferation through energy depletion and increases oxidative damage, and through it, cell death. Development of targeted delivery systems for inhibitors of this pathway are needed.

Given the complex roles that the TME plays in supporting tumors, it follows that more sophisticated in vitro models are needed to recapitulate the conditions in which cancer cells exist. Prof. Dr. Kenny et al. [16] described some of the recent developments in 3D modeling of ovarian cancer.

Models approximating in situ carcinoma in the fallopian tube, dissemination into the peritoneal cavity, early metastatic attachment to the mesothelial-lined surfaces of the omentum, bowel, and abdominal wall, and late chemoresistant metastases are needed.

The TME of ovarian cancer has numerous unique features that need to be considered in the study and treatment of the disease. According to Prof. Dr. O'Hagan and colleagues [17], recent studies have shown an association between inflammation and an increase in ovarian cancer risk. Though this phenomenon has been well characterized in colon and pancreatic cancers, the mechanisms through which inflammation contributes to ovarian cancer risk need further study. Prof. Dr. Hawkins et al. [18] explained how endometriosis increases the risk of developing endometrioid, clear cell carcinoma, and low grade serous ovarian cancers. The unique tumor microenvironment created by endometriosis facilitates tumorigenesis through upregulation of many gene products associated with ovarian cancers.

A number of other therapeutic avenues are being explored in ovarian cancer treatment. Efficient targeting of a chemoresistant sub-population of cancer cells known as cancer stem cells (CSC) is a rapidly growing field in the study of high grade serous ovarian cancer. As shown by Prof. Dr. Dahl and Prof. Dr. Roy, the PI3K/PTEN/AKT, Jak/STAT, NFkB, Notch, Wnt, and Hedgehog pathways have all been implicated in the maintenance of cancer stem cells, many of which have therapeutics targeting them currently undergoing clinical trials. Further study is needed to better identify cancer stem cell populations and the mechanisms through which these pathways are involved in their maintenance.

Another treatment option, known as heated intraperitoneal chemotherapy was discussed by Prof. Dr. Jewell et al. [19] The higher drug levels delivered through HIPEC combined with hyperthermia are thought to increase the efficacy of chemotherapy. Studies showing the benefit of HIPEC along with CRS have been promising, however, use of HIPEC in recurrent cancers warrants further study. The relative safety of this treatment also warrants further investigation. Strategies for treatments targeting supportive components of the TME outlined by Prof. Dr. Matei et al. [20] represent a promising avenue in improving clinical outcomes in patients. Inhibition of angiogenesis, immune therapies, and therapies targeting supportive stromal cells are being explored as therapies to improve survival in HGSOC patients, though more research is needed to increase the efficacy of these approaches.

The planning of this special issue was motivated by the Albert Trust Midwest Ovarian Cancer Coalition (MWOCC) biannual meeting, held in June 2018. We would like to express our deepest gratitude to the conference sponsor, The Leo and Anne Albert Charitable Trust, Mr. Gene Pranzo (Trustee), and Susan Brogan (organizer). We sincerely thank all the attendees and participants who kindly contributed manuscripts of the highest quality to this issue.

References

1. Harper, E.; Sheedy, E.; Stack, M. With Great Age Comes Great Metastatic Ability: Ovarian Cancer and the Appeal of the Aging Peritoneal Microenvironment. *Cancers* **2018**, *10*, 230. [CrossRef] [PubMed]
2. Klymenko, Y.; Nephew, K. Epigenetic Crosstalk between the Tumor Microenvironment and Ovarian Cancer Cells: A Therapeutic Road Less Traveled. *Cancers* **2018**, *10*, 295. [CrossRef] [PubMed]
3. Dasari, S.; Fang, Y.; K. Mitra, A. Cancer Associated Fibroblasts: Naughty Neighbors That Drive Ovarian Cancer Progression. *Cancers* **2018**, *10*, 406. [CrossRef] [PubMed]
4. Hardy, L.; Salvi, A.; Burdette, J. UnPAXing the Divergent Roles of PAX2 and PAX8 in High-Grade Serous Ovarian Cancer. *Cancers* **2018**, *10*, 262. [CrossRef] [PubMed]
5. Hilliard, T. The Impact of Mesothelin in the Ovarian Cancer Tumor Microenvironment. *Cancers* **2018**, *10*, 277. [CrossRef] [PubMed]
6. Hudson, L.; Gillette, J.; Kang, H.; Rivera, M.; Wandinger-Ness, A. Ovarian Tumor Microenvironment Signaling: Convergence on the Rac1 GTPase. *Cancers* **2018**, *10*, 358. [CrossRef] [PubMed]
7. Rodriguez, G.; Galpin, K.; McCloskey, C.; Vanderhyden, B. The Tumor Microenvironment of Epithelial Ovarian Cancer and Its Influence on Response to Immunotherapy. *Cancers* **2018**, *10*, 242. [CrossRef] [PubMed]
8. McCloskey, C.; Rodriguez, G.; Galpin, K.; Vanderhyden, B. Ovarian Cancer Immunotherapy: Preclinical Models and Emerging Therapeutics. *Cancers* **2018**, *10*, 244. [CrossRef] [PubMed]

9. Drakes, M.; Stiff, P. Regulation of Ovarian Cancer Prognosis by Immune Cells in the Tumor Microenvironment. *Cancers* **2018**, *10*, 302. [CrossRef] [PubMed]
10. Jewell, A.; McMahon, M.; Khabele, D. Heated Intraperitoneal Chemotherapy in the Management of Advanced Ovarian Cancer. *Cancers* **2018**, *10*, 296. [CrossRef] [PubMed]
11. Gupta, V.; Yull, F.; Khabele, D. Bipolar Tumor-Associated Macrophages in Ovarian Cancer as Targets for Therapy. *Cancers* **2018**, *10*, 366. [CrossRef] [PubMed]
12. Kim, J.; Park, E.; Kim, O.; Schilder, J.; Coffey, D.; Cho, C.; Bast, R. Cell Origins of High-Grade Serous Ovarian Cancer. *Cancers* **2018**, *10*, 433. [CrossRef] [PubMed]
13. Barbolina, M. Molecular Mechanisms Regulating Organ-Specific Metastases in Epithelial Ovarian Carcinoma. *Cancers* **2018**, *10*, 444. [CrossRef] [PubMed]
14. Xu, Y. Lysophospholipid Signaling in the Epithelial Ovarian Cancer Tumor Microenvironment. *Cancers* **2018**, *10*, 227. [CrossRef] [PubMed]
15. Nayak, A.; Kapur, A.; Barroilhet, L.; Patankar, M. Oxidative Phosphorylation: A Target for Novel Therapeutic Strategies Against Ovarian Cancer. *Cancers* **2018**, *10*, 337. [CrossRef] [PubMed]
16. Watters, K.; Bajwa, P.; Kenny, H. Organotypic 3D Models of the Ovarian Cancer Tumor Microenvironment. *Cancers* **2018**, *10*, 265. [CrossRef] [PubMed]
17. Savant, S.; Sriramkumar, S.; O'Hagan, H. The Role of Inflammation and Inflammatory Mediators in the Development, Progression, Metastasis, and Chemoresistance of Epithelial Ovarian Cancer. *Cancers* **2018**, *10*, 251. [CrossRef] [PubMed]
18. Wendel, J.; Wang, X.; Hawkins, S. The Endometriotic Tumor Microenvironment in Ovarian Cancer. *Cancers* **2018**, *10*, 261. [CrossRef] [PubMed]
19. Roy, L.; Cowden Dahl, K. Can Stemness and Chemoresistance Be Therapeutically Targeted via Signaling Pathways in Ovarian Cancer? *Cancers* **2018**, *10*, 241. [CrossRef] [PubMed]
20. Nwani, N.; Sima, L.; Nieves-Neira, W.; Matei, D. Targeting the Microenvironment in High Grade Serous Ovarian Cancer. *Cancers* **2018**, *10*, 266. [CrossRef] [PubMed]

cancers

MDPI

Review

With Great Age Comes Great Metastatic Ability: Ovarian Cancer and the Appeal of the Aging Peritoneal Microenvironment

Elizabeth I. Harper [1,2,3], Emma F. Sheedy [2,4] and M. Sharon Stack [1,2,*]

1 Department of Chemistry and Biochemistry, University of Notre Dame, South Bend, IN 46617, USA; eharper1@nd.edu
2 Harper Cancer Research Institute, University of Notre Dame, South Bend, IN 46617, USA; esheedy1@nd.edu
3 Integrated Biomedical Sciences Program, University of Notre Dame, South Bend, IN 46617, USA
4 Department of Mathematics, University of Notre Dame, South Bend, IN 46617, USA
* Correspondence: sstack@nd.edu; Tel.: +1-574-631-4100

Received: 8 June 2018; Accepted: 4 July 2018; Published: 10 July 2018

Abstract: Age is one of the biggest risk factors for ovarian cancer. Older women have higher rates of diagnosis and death associated with the disease. In mouse models, it was shown that aged mice had greater tumor burden than their younger counterparts when intraperitoneally injected with ovarian tumor cells. While very few papers have been published looking at the direct link between ovarian cancer metastasis and age, there is a wealth of information on how age affects metastatic microenvironments. Mesothelial cells, the peritoneal extracellular matrix (ECM), fibroblasts, adipocytes and immune cells all exhibit distinct changes with age. The aged peritoneum hosts a higher number of senescent cells than its younger counterpart, in both the mesothelium and the stroma. These senescent cells promote an inflammatory profile and overexpress Matrix Metalloproteinases (MMPs), which remodel the ECM. The aged ECM is also modified by dysregulated collagen and laminin synthesis, increases in age-related crosslinking and increasing ovarian cancer invasion into the matrix. These changes contribute to a vastly different microenvironment in young and aged models for circulating ovarian cancer cells, creating a more welcoming "soil".

Keywords: ovarian cancer; age; tumor microenvironment; extracellular matrix; mesothelial cells; immune; fibroblast; adipocytes; peritoneum

1. Introduction

Ovarian cancer (OvCa) is the deadliest gynecological cancer, with a survival rate under 50% [1]. One of the biggest risk factors for OvCa is age, where the median age of diagnosis is 63 and median age of death is 70 [1]. Aging, as defined in the Hallmarks of Aging, is "the time-dependent functional decline that affects most living organisms" [2]. A call for research investigating the relationship between OvCa and aging was voiced in 1993 by Yancik after a review of epidemiologic data, showing older women were not only more likely to be diagnosed with OvCa but were more likely to die from their disease [3]. Yancik raised the question that has propelled the research in this field: why is there is a difference in survival between young and aged patients? Is there a difference in treatment, or does the cancer behave differently in older women? In 2013, epidemiological data were reviewed again by Trillsch et al. and their data suggest that older patients often receive less radical treatment, contributing to this disparity [4]. However, it is likely that there is more contributing to this disparity than physician partiality alone. A separate epidemiological study showed that older OvCa patients have a two-fold increase in peritoneal metastases relative to younger patients at time of diagnosis, suggesting that there is more to be discovered in the relationship between OvCa and aging [5]. Here we review the aging studies in the OvCa field, as well as aging studies involving distinct components of the peritoneal metastatic microenvironment.

OvCa metastasizes in a very unique fashion, where cells are exfoliated from the primary tumor as either single cells or multicellular aggregates and circulate through the peritoneal cavity via diffusion in the peritoneal fluid [6]. The circulating cells adhere to secondary sites, such as the omentum and parietal peritoneum, via interactions with mesothelial cells [6]. The OvCa cells induce mesothelial cell retraction, then invade into and anchor in the collagen-rich submesothelial matrix [6]. The OvCa cells can then proliferate and form a metastatic lesion [6]. Aging can affect nearly every step of this process.

The peritoneum is a vast, serous membrane covering the interior of the abdomen and the visceral organs. The parietal peritoneum covers the interior of the abdominal wall, then folds to form the omentum, which lies between the parietal peritoneum and the anterior surface of the abdominal organs. The omentum is an organ rich in adipocytes and immune cells. Both the omentum and the parietal peritoneum are composed of a collagen-rich matrix covered by the mesothelium, separated by a thin basement membrane. The mesothelium is a single monolayer of simple squamous epithelial-like cells, or mesothelial cells, that cover the surface of the peritoneum. The basement membrane is a thin layer (<100 nm) composed mostly of collagen IV and laminin that separates the mesothelium from the elastic matrix below [7]. This matrix is comprised mostly of collagens I and III but contains other entities such as fibroblasts, immune cells, adipocytes, lymphatics and limited cardiovasculature [7]. Each of the components has the potential to react differently to OvCa cells through age-related changes (Figure 1).

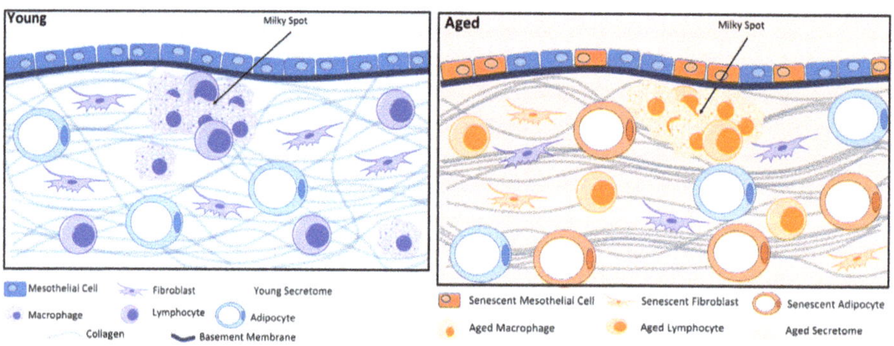

Figure 1. Changes in the Aged Microenvironment. Young: In the young metastatic microenvironment, collagens I and III form a directionally random meshwork that supports the tissue. In addition, there are low rates of senescence in mesothelial cells, fibroblasts and adipocytes, all of which secrete factors into the extracellular environment, forming the young secretome. The young secretome is characterized by decreased Matrix Metalloproteinase (MMP) expression, increased cytokine expression by immune cells, decreased cytokine expression by fibroblasts and decreased adipocyte-associated inflammatory factors. Milky spot immune cell aggregates exist in both young and aged metastatic environments, providing the tumor with abundant vascularization. Aged: In the aged metastatic microenvironment, there are lower levels of collagens I and III, which are remodeled to form more aligned, linear structures. In addition, higher levels of senescence alter the secretome, increasing inflammation and other factors that can promote ovarian cancer (OvCa) metastasis.

The aging peritoneal microenvironment is defined in large part by two processes: extracellular matrix (ECM) remodeling and cellular senescence. Changes in collagen, laminin and fibronectin have the potential to alter how the metastatic OvCa cells invade into the peritoneum [8–10]. Senescence-induced changes in fibroblasts, mesothelial cells and immune cells drastically alter the secretome of the microenvironment, causing an increase in the transcription of factors that are associated with inflammation and angiogenesis [11,12]. Senescence is a cell's permanent exit from the cell cycle and was first attributed to telomere attrition [13]. More recently, a number of factors have been identified that contribute to cellular senescence, including DNA damage [14,15], oxidative stress [16,17], high levels of glucose [18,19], transforming growth factor-β (TGF-β), [18,20–22] and the tumor suppressors p16^{INK4a} and p53 [2]. While senescence within the tumor itself suppresses tumor growth [2,23], senescence in the microenvironment has been shown to increase tumor growth [24,25].

Interestingly, the role of p53 also varies greatly between OvCa cells and microenvironment. It was reported that in a C57Bl/6 model, ID8 cells with a p53 deletion showed greater tumor growth than the ID8 parental cells [26]. However, p53 is overexpressed in the aging OvCa tumor microenvironment as a result of oxidative stress, oncogenic stress and DNA damage [27,28]. In response to severe damage, p53 determines cell fate, inducing either senescence or apoptosis. In epithelial and stromal cell lines, p53 more frequently induces senescence [28]. This leads to increased Matrix Metalloproteinase (MMP) secretion, a remodeling of the ECM and disruption of normal epithelial cell differentiation [28,29]. For reasons to be addressed, these effects contribute to increased OvCa metastasis and occur more dramatically in aged individuals.

2. Aging Modifies the Metastatic Microenvironment

In vivo models of intraperitoneal (IP) metastasis have been utilized to demonstrate an age-related difference in tumor burden in mice injected with ovarian tumor cells. When IP injected with syngeneic tumor cell lines, both C57Bl/6 and FVB mice exhibited a dramatic difference in disease progression between the young (3–6 months) and aged (20–23 months) cohort, with the aged mice harboring greater tumor burden than their younger counterparts [26]. Transcriptome analysis of gonadal adipose tissue from young and aged mice points to a difference in immune response in the aged mice but it is likely that the immune system is only one of the components of the microenvironment that is contributing to the age-related disparity in metastasis [26].

2.1. Mesothelial Cells

The mesothelium, a cobblestone monolayer of cells that exhibit characteristics of both epithelial and mesenchymal cells, lines the surface of the peritoneum. Its function in normal tissue is to create a barrier and limit the permeability of the peritoneum, as well as secretion of factors that are involved in peritoneal homeostasis and launching appropriate immune responses to pathogens [30]. These cells are very important in the adhesion of OvCa cells to secondary metastatic site. The senescent mesothelial population increases as the host ages, due to both increased rates of senescence as well as the resistance of senescent cells to pro-apoptotic signaling [5,31].

Senescent mesothelial cells change the cellular signaling in the tumor microenvironment, expressing factors such as fibronectin [16,32], intercellular adhesion molecule-1 (ICAM-1) [33], beta-galactosidase [31,34] and thymosin beta-10 [35]. Fibronectin, a mediator of cell-extracellular matrix interaction, has been shown to be increased in aging tissues [36]. This increase has been linked with increased OvCa cell adhesion [16] and increases tissue stiffness (which will be discussed in more detail in Section 2.3.1) [37]. The increase in OvCa cell adhesion is partially mediated by mesothelial ICAM-1, an adhesion molecule expressed by mesothelial cells that has been shown to be important in other abdominal cancers that metastasize to the peritoneum [33]. In addition, profiles of human peritoneal mesothelial cells isolated from young (mature adults under the age of 65) and aged (over the age of 65) patients showed an increase in inflammation-associated factors, suggesting increased inflammation in the aged mesothelium [38]. It was shown that age was associated with an increase in both the cyclooxygenase (COX) and nitric oxide synthase (NOX) pro-inflammatory systems, an upregulation of nuclear factor-κB (NF-κB) and inflammatory cytokines and an increase in reactive oxygen species (ROS) in mesothelial cells [38]. ROS have been shown to be a mediator of senescence; increased ROS results in increased cellular senescence [39]. Additional information on inflammation and the role of the immune system is included in Section 2.4.

Senescent mesothelial cells have been shown to interact with metastasizing OvCa cells, altering the OvCa secretome to express angiogenic agents such as chemokine CXC ligand 2 (CXCL1), chemokine CXC ligand 8 (CXCL8), hepatocyte growth factor (HGF) and vascular endothelial growth factor (VEGF) [40]. Mikuła-Pietrasik et al. saw increased angiogenesis in mouse models when OvCa was coinjected with senescent human peritoneal mesothelial cells (HPMCs) [40]. This process is mediated by TGF-β1 and IL-6, which are overexpressed in aged mesothelial cells [38,40]. When OvCa cells were incubated with senescent mesothelial cell conditioned media, they experienced higher levels of proliferation than those incubated with conditioned media from young cells, suggesting soluble factors released by senescent mesothelial cells promote the proliferation of OvCa cells [40]. OvCa cells with conditioned media from senescent mesothelial cells also showed greater migration and invasion *in vitro* [40]. In addition, histological analysis of patient tumors showed the presence of senescent mesothelial cells in cancerous tissues [40]. It is likely that an accumulation of senescent mesothelial cells, as seen in tissue from aged patients, provides a more welcoming metastatic niche for circulating OvCa cells [5].

Hyaluronic acid, or hyaluronan (HA), is a glycosaminoglycan secreted by cells with mesenchymal characteristics, such as mesothelial cells. It acts as a mediator of ECM organization as well as a lubricant on the mesothelial surface [41,42]. HA is also an FDA-approved treatment for osteoarthritis and is a popular treatment used by plastic surgeons to reverse the signs of aging of the skin [43,44]. Relevant studies have shown two divergent lines of research: HA increasing [45–47] or decreasing [48–50] cell adhesion. However, certain OvCa cells lines have been shown to bind directly to HA, suggesting HA increases OvCa cell adhesion [46]. This likely contributes to the observation in ovarian and prostate cancer patients, where overexpression of HA generally results in a poorer prognosis [11,51]. In addition, HA has been shown to impact cell migration [52–54] and cell proliferation [54–56], to activate skin fibroblasts [57] and to be upregulated in response to inflammation [58]. There is not much information available on the effects of age on HA other than an observed decrease in aged tissue, likely due to the decreased synthetic capacity of aged cells [41,59]. However, the use of HA in the treatment of age-related diseases suggests that the role of HA in the aging microenvironment warrants further investigation.

2.2. Extracellular Matrix

The peritoneal ECM is a complex system that supports the cells of the peritoneum. Made up of collagen, laminin and fibronection, the ECM plays an integral role in both normal peritoneal structure as well as the metastatic success of OvCa. Directly beneath the mesothelial layer is a thin basement membrane composed of collagen IV and laminin, covering an elastic matrix of collagens I and III, laminin and fibronectin [60]. The ECM changes drastically with age, which can change how integrins and syndecans bind to the ECM, thus altering the interaction between the metastasizing OvCa cells and the tumor microenvironment [60], including increased adhesion of macrophages [61] and increased cancer cell invasion [62].

2.2.1. Collagen

Collagen is one of the most abundant proteins in the body and forms a large portion of the peritoneal extracellular matrix. There are multiple types of collagens; in the context of the peritoneum, collagens I and III are the most notable, both of which are fibrous collagens [60]. On a molecular level, both I and III have a similar amino acid structure distinct from other proteins, with glycine repeating every third amino acid and a high percentage of prolines, which are often post-translationally modified to become hydroxyprolines [63]. These amino acids chains come together to form the characteristic triple helix, which are banded together in an overlapping manner to form fibrils with the distinct D-banding pattern [63].

While little research has been done on the effects of aging on peritoneal collagen, there is a wealth of information on skin collagen. As far back as 1975, scientists noted a significant decrease in the amount of collagen in aged skin [64]. An immunohistochemical analysis showed amounts of collagens I and III change as an individual ages [65]. Both collagens decrease in aged tissue but the ratio of collagen I/collagen III increases, suggesting that collagen I is decreasing at a slower rate than collagen III [65]. The structure of collagen is disrupted with age, resulting in disorganization of the fibers (Figure 2) [11]. In addition to skin, collagen extracted from human arteries, mouse tails and mouse prostates showed alterations not only in structure but also in mRNA and protein expression, pointing to a decrease in collagen synthesis as the culprit behind the decreasing amounts of collagen [11,66,67]. Later research showed that this decrease is likely due not only to decreased synthesis but increased degradation as well [11,68].

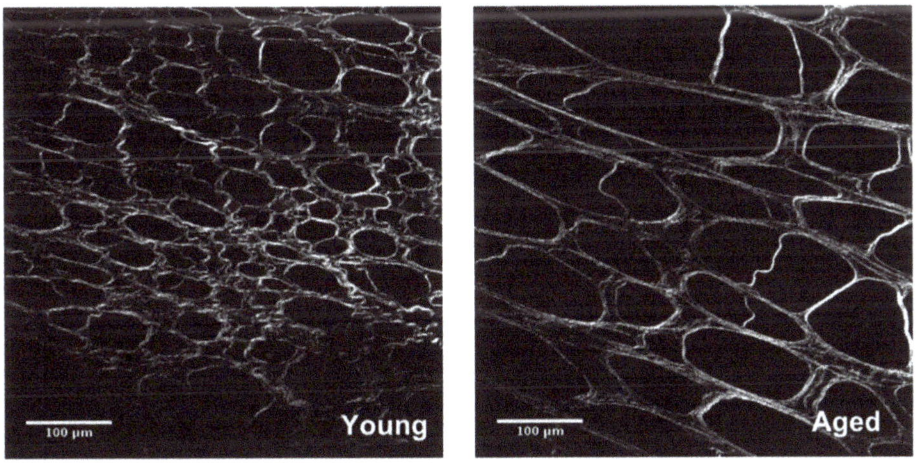

Figure 2. Age-related changes in omental collagen structure. Second harmonic generation imaging of omental tissue isolated from young and aged mice shows a distinct difference in structure. Aged collagen forms crosslinks that result in the loss of meshwork, formation of tendon-like structures and increased anisotropy. This causes a disruption of tissue structure that can affect how metastasizing OvCa cells interact with the tumor microenvironment.

Matrix Metalloproteinases, or MMPs, are the main source of ECM degradation [69]. MMPs are a family of 23 zinc-dependent enzymes that are divided into distinct groups: collagenases, gelatinases, matrilysins and membrane-type MMPs [69]. Outside of gene regulation, MMPs are regulated in two major ways: they require activation from the zymogen form in order to be active and active MMPs are regulated by tissue inhibitors of metalloproteinases, or TIMPs [69,70]. MMPs are secreted from numerous cell types within the microenvironment, such as fibroblasts and immune cells, as well as the OvCa cells themselves [69,71]. Three MMPs have been shown to be upregulated in OvCa: MMP2, MMP9 and MMP14 (also referred to as MT-MMP1) [13]. MMP14, a membrane-type, is present at high levels in the tumor cells themselves, while MMP9, a gelatinase specific to collagens IV and V, is more often upregulated in the stroma [71]. In addition, senescent cells have been shown to have an increased expression of MMPs and addition of an MMP inhibitor reverses some senescent-specific tissue phenotypes [72]. An upregulation of either MMP9 or MMP14 in the stroma around the tumor cells is correlated with a more invasive phenotype, pointing to a critical role of MMPs in the tumor microenvironment [71].

Due to its long half-life, post-translational modifications accumulate in collagenous tissue over time [73]. Of particular interest here are modifications that create covalent crosslinks between collagen molecules. Lysyl oxidase, or LOX, is a family of enzymes that modify lysine sidechains to form desmosine through a Schiff base intermediate [74]. Recently, increased LOX crosslinks have been shown to play a role in chemoresistance [75]. Advanced Glycation End-products, or AGEs, are formed non-enzymatically as a result of glycosylation over time. These crosslinks have been shown to change the structure and mechanical properties of collagen-rich tissues, such as the peritoneum [76]. Crosslinked collagen has higher fiber alignment, resulting in more tendon-like structures, causing the tissue to lose elasticity and become stiffer than non-crosslinked collagen (Figure 2) [76,77]. An increase in AGEs has been correlated with increased peritoneal permeability, which could contribute to increased OvCa invasion [8]. In addition, stiffer matrices have been shown to increase cell motility, proliferation and adhesion [9,10].

When AGEs occur in serum albumin, they can bind to the AGE receptor (RAGE) on monocytes and trigger the release of tumor necrosis factor-α (TNF-α), leading to insulin resistance [78]. When bound to RAGE on adipocytes, AGEs can induce the formation of ROS [78]. As mentioned in Section 2.1, ROS have been shown to be a mediator of cellular senescence, where high levels cause enough cellular damage for the cells to leave the cell cycle permanently [39]. Additionally, ROS can activate p53, which is another pathway leading to cellular senescence [39]. In addition to their role in changing the structure of the ECM, AGEs can also induce senescence in numerous cell types in the microenvironment through formation of ROS and subsequent pathways [39,78].

2.2.2. Fibronectin

In contrast to the helical nature of collagen, fibronectin is a structural glycoprotein that forms repeating beta-sheets in its folded form [37,79]. One of the main roles of fibronectin is mediating cell-cell interactions [36]. Not only does the amount of fibronectin increase in aged tissues but aging fibronectin, like collagen, shows an increase in anisotropy with age [36,37,79]. Fibronectin has also been shown to stretch with age, resulting in increased stiffness [37]. In addition, fibroblasts interacting with aged fibronectin responded differently than when interacting with young fibronectin [37]. The fibroblasts interacting with aged fibronectin were shown to have longer β1 integrin adhesions as well as more actin stress fibers [37]. In addition, as mentioned in Section 2.1, senescent mesothelial cells express more fibronection, contributing to increased OvCa cell adhesion mediated by the α5β1 integrin [16].

2.2.3. Basement Membrane

The basement membrane (BM) is a component of the extracellular matrix that separates epithelial cells from underlying connective tissue. It is primarily composed of collagen IV, intertwined with laminin polymers [80,81]. The BM exhibits structural changes as it ages, most notably with aged cells synthesizing less collagen IV than young cells [82,83]. While the basement membrane is understood to thicken with age, the declined synthesis of collagen IV indicates that the thickness is due to decreased turnover of aged tissues [82,83].

In primary ovarian tumors, collagen IV is absent on the ovarian surface [84]. This indicates that OvCa cells must firstly degrade the ovarian BM (specifically, degrading collagen IV) to detach from the ovary and shed into the intraperitoneal space [84]. Following this migration, cells then alter the mesothelial BM to anchor and proliferate [85]. The mesothelial BM also has high collagen IV and laminin content [85], so OvCa cells must again degrade collagen IV to gain entry into the underlying ECM.

Disabled-2 (Dab2) is a signal transduction protein and tumor suppressor that also functions in positional organization of ovarian surface cells. In OvCa, genetic and epigenetic changes to Dab2 enable tumor cells to escape ovarian BM control and proliferate in a disorganized fashion, resulting in diffusion into the peritoneal cavity and metastasis [84]. Hypermethylation of the Dab2 promoter results in epigenetic silencing of the gene, which is correlated with a loss of expression of collagen IV [86]. Methylation patterns are known to change with age [87] and the effects of aging on methylation can vary from inducing DNA hypomethylation to inducing hypermethylation. Such age-associated deviation in methylation leads to advanced epigenetic damage in aged individuals [88]. It is possible that DNA hyper-methylation of the Dab2 promoter may be affected by age, thereby impeding collagen IV expression—increasing BM susceptibility to degradation.

OvCa cells first bind to mesothelial cells to gain entry into the underlying matrix [85]. This adhesion is facilitated by ovarian cancer antigen CA125 and mesothelin interaction [89], and/or by integrins such as β_1-integrin and cell surface receptors such as CD44 (the receptor for HA) [85]. Upon attachment to the mesothelium, OvCa cells upregulate MMP production, including that of MMP2 [85]. MMP2 preferentially interacts with collagen IV, resulting in the loss of basement membrane [90]. As aged cells are downregulated in their expression of collagen IV [83], this may lead to more efficient BM degradation in the aged host. Additionally, in many cancers, Dab2 downregulation leads to increased transcription of the ribonucleoprotein hnRNPK, which then enhances MMP2 transcription by the metastatic cells [91]. Thus, downregulation of Dab2, as observed in OvCa metastatic cell lines, may be correlated with increased MMP2 expression.

In addition to collagen IV, laminin provides structural support in the basement membrane [11,92]. Laminin is a trimeric protein with high homology between the alpha, beta and gamma trimers [11,92]. It is highly regulated in adults; the biggest changes observed in aging studies are the replacement of fetal laminin with adult laminin [11,93]. However, in carcinogenesis, it was observed that prostate tissues experience a loss of adult laminin, which results in disorganization of epithelial cells [11,93]. In addition, some tumor cells have been shown to increase expression of laminins, increasing cell adhesion and invasion [92]. In the context of aging, it has been shown that there are decreased levels of laminin in aged basement membranes [94,95]. In addition, laminins can also be AGE-modified, leading to decreased laminin-collagen IV binding, which may make it easier for the OvCa cells to invade through the basement membrane [94]. AGE modifications have also been shown to increase laminin synthesis, however they also impair laminin assembly, likely contributing to the described decrease in total laminin [94].

2.3. Fibroblasts

2.3.1. Senescent Fibroblasts

Fibroblasts are a stromal cell type, functioning in upkeep of the connective tissue environment and ECM [96]. This upkeep is greatly altered with age in ways that promote tumorigenesis, such as increasing angiogenesis and stimulating OvCa cell growth [97,98]. Aged fibroblasts secrete less collagen and other proteins than their younger counterparts [96]. Furthermore, fibroblasts isolated from older individuals had far higher rates of senescence than fibroblasts from younger individuals with age, senescent fibroblasts accrue and replace presenescent cells (Figure 3) [24,25,96,99]—greatly altering the function of the tissue in the process. Notably, accumulation of senescent fibroblasts in the OvCa microenvironment is associated with increased cell proliferation and metastatic potential due to interactions with the cancer cells [24].

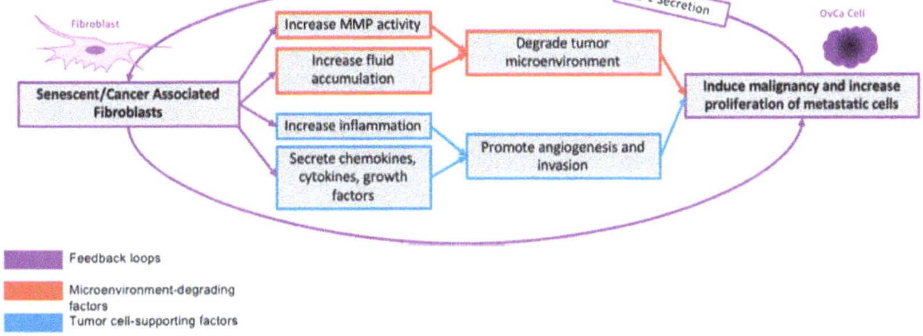

Figure 3. Stromal-Epithelial Crosstalk. Active crosstalk occurs between senescent and activated fibroblasts and OvCa cells. This induces activated fibroblasts, while concurrently inducing proliferation and malignancy of the invading tumor.

In a murine model, senescent fibroblasts partake in significant stromal-epithelial crosstalk (Figure 3) [100], inducing premalignant epithelial cells to lose differentiation capacity, increase invasiveness and eventually become fully malignant cells [99]. This can be attributed to many factors secreted by senescent fibroblasts that alter the tissue microenvironment and stimulate growth of epithelial cells expressing oncogenic mutations [12,101,102]. Increased fibroblast senescence results in greater secretion of vascular endothelial growth factor (VEGF), which increases angiogenesis of the region [97,98]. As tumors necessitate a vascular supply for efficient growth [103], increased angiogenesis supports epithelial tumor growth. Senescent fibroblasts also secrete more MMPs [102], which degrade collagen and the basement membrane [69,90]. These effects have been widely shown to be correlated with increased cancer cell growth (Figure 3) [71,84]. Secretion of MMPs by senescent fibroblasts also results in heightened microvascular permeability leading to a buildup of extracellular fluid, which increases inflammation and damages the surrounding tissue matrix, possibly altering the natural anti-tumorigenic nature of the presenescent microenvironment [72].

2.3.2. Fibroblast Activation

Epithelial tumor cells activate fibroblasts in the tumor stroma, stimulating a phenotypic switch from normal fibroblasts to cancer-associated fibroblasts (CAFs) [104]. Epithelial OvCa cells secrete factors such as chemokine growth-regulated oncogene 1 (Gro-1) [100]. Gro-1 induces the CAF phenotype and, as Gro-1 is overexpressed in OvCa patients, there is significant evidence that Gro-1 alters the stromal environment to induce senescence in fibroblasts. This epithelial-stromal interaction is critical in tumor initiation and proliferation. Ovarian CAFs promote tumor growth by secreting cytokines and chemokines into the microenvironment [98,104], while non-recruited, presenescent fibroblasts do not enhance tumorigenesis [100].

CAFs exhibit many of the same general characteristics as senescent fibroblasts [25,102]. Thus, as OvCa cells recruit CAFs, they also induce pro-tumorigenic microenvironment changes as caused by senescent fibroblasts described above (Figure 3). Both senescent fibroblasts and CAFs secrete CXCL12 [105], among other pertinent factors such as IL6, IL8 and MMPs [106]. These increase inflammation and promote angiogenesis, invasiveness and metastasis [105,106]. Secretion of chemokines—as observed in both CAFs and senescent fibroblasts—is likely a key cancer-promoting function of fibroblasts [105]. Thus, aging and increased senescence of fibroblasts alter the microenvironment and oncogenic cells themselves in a way highly conducive to tumor growth (Figure 3).

2.4. *Immune Cells*

2.4.1. Tumor Cells Preferentially Adhere to Immune Cell Clusters

Ovarian cancer cells shed from the primary tumor and adhere preferentially to the peritoneum or omentum in the abdominal cavity. The omentum, a visceral adipose tissue, is known to have a large influence on peritoneal immunity due to its high quantity of lymphoid aggregates (Figure 1), often called milky spots [107]. Within the omentum, initial attachment and growth of tumors were observed to be most prevalent surrounding organized aggregates of immune cells [108]. Omental stem cells exhibit a large capacity to produce angiogenic growth factors, resulting in high vascularization of the region, particularly surrounding immune cells [103]. Avascular tumors are severely limited in growth due to a lack of blood supply. Tumors must make an "angiogenic switch" to proliferate, where the initial metastatic tumor initiates the formation of new vessels for increased blood supply [109]. However, the tumor must anchor to a membrane before it can make the angiogenic switch. Studies show that tumor cells preferentially bind to mesothelial cells directly above the omental immune cell cluster, where the initial tumor is provided with an abundant blood supply from the existing vasculature of the immune cell cluster. This contributes to the high survival rate of metastatic cells in the omentum [108].

Intraperitoneal injection of green fluorescent protein (GFP)-expressing tumor cells showed localization to milky spots in the omentum [108]. This supports prior conclusions that migration and attachment of tumor cells to the omentum and specifically to immune aggregates, occurs from migration from the peritoneal cavity and does not necessitate intravascular transportation. As tumor cells metastasize, they disturb the structure of the immune cell aggregate and eventually displace all immune cells from the metastatic tumor mass [108]. It is important to note, however, that while hematogenous metastasis of OvCa to the peritoneal and omentum is not critical for cancer spread, intravascular transportation of the tumor does occur with significant metastatic results [110,111].

2.4.2. Aging Affects Antitumor Macrophage Function in Peritoneum

Milky spot aggregates are comprised primarily of macrophages [112]. Studies exploring the effect of aging on macrophage function prior to tumor exposure have yielded conflicting results, although such discrepancies could be due to differences in sex, strain, species, or site of tissue in macrophage isolation [113]. Isolated macrophages specifically from the peritoneum indicate lower levels of inflammatory cytokine production with age (Table 1) [113,114]. Particularly, replicated *in vitro* results indicate that macrophages of young mice produce higher amounts of tumor necrosis factor-α (TNF-α), MMPs and have a higher phagocytic capacity than aged mice (Table 1) [115–117].

Studying the effect of decreased cytokine secretion on cancer in models of aging yields highly conflicting results (Table 1, Figure 4). Firstly, TNF-α has both pro- and anti-tumorigenic effects. On one hand, TNF-α could promote cancer due to its activation of cancer-promoting pathways such as NF-κB and its correlation with increased angiogenesis, cell growth and metastasis [106,118]. On the other hand, TNF-α also has inherent anti-tumor effects: the cytokine activates tumor-infiltrating dendritic cells and promotes tumor stroma destruction [106,119]. Phagocytic efficiency and general cytotoxic capabilities also decrease in aged models [120]. This could lead to increased cell proliferation in the aged host, a hallmark of cancer [121]. However, increased cytokine secretion can lead to increased inflammation in the tissue. This results in a mutagenic microenvironment abundant in growth factors and cytokines that sustain angiogenesis, proliferation and invasion [122], which are three other hallmarks [121]. It is difficult to conclude whether the anti-tumor killing abilities of the aged macrophage outweigh its inability to provide support to the tumor (Table 1).

Notably, the innate immune response of macrophages is affected by their environment [113,115]. Peritoneal macrophage function, including cytokine secretion, was observed to be altered with age only due to changes in the aged microenvironment, not inherent age-related dysfunction of the macrophage itself [115]. Thus, it is possible to restore the macrophage to its full secretory phenotype by changing its environment [113,115]. Epithelial cancer cells and stromal cells do just this–they secrete growth factors and cytokines such as macrophage colony-stimulating factor 1 (CSF-1) to recruit macrophages, converting their phenotype into tumor-associated macrophages (TAMs) [121,123,124]. When activated, TAMs work similarly to cancer associated fibroblasts (CAFs). They promote metastasis by secreting growth factors and cytokines by increasing angiogenesis and participating in cross-talk with epithelial cells and stromal cells [121,124]. Increased CSF1 density and increased TAM occurrence are correlated with decreased survival rates [121,124]. However, it is not understood whether cytokine secretion is downregulated in aged TAMs, as occurs in pre-activated macrophages.

2.4.3. Tumor Infiltrating Lymphocytes: B and T cells

T-cell associated tumor infiltrating lymphocytes (TILs) are correlated with increased survival in OvCa patients (Table 1). CD4+ and CD8+ T-lymphocytes are two types of TILs which recognize cancer antigens and inhibit cancer proliferation. CD4+ TILs elicit dendritic cell responses, which then induce CD8+ cells to provide extended cytotoxicity, killing tumor cells (Figure 4). Thus, an increase in CD4+ and CD8+ T-lymphocytes is a survival advantage in OvCa patients [121,125]. One factor in this pathway is IL-2 secretion: increased IL-2 secretion results in activated macrophages and tumor lysis directly from CD8+ T-lymphocytes [125].

T-cell production and function is widely known to decrease with age. Notably, aged CD4+ T-cells experience higher degrees of apoptosis and decreased function when compared to young T-cells in an aged murine model. Aged CD4+ T-cells showed less expression of CD4 and a lower mitochondrial mass [126]. Furthermore, aged CD4+ T-cells secreted less IL-2 than young phenotypes [127] and have decreased memory capabilities [128]. These factors indicate that aged CD4+ TILs are inherently less active than young TILs and therefore express less antitumorigenic capacity (Table 1).

The effect of B-cell TIL function on OvCa presents more difficult data (Table 1). In some studies, B-cell TILs, such as CD20+, are also understood to bear a tumor survival advantage in OvCa patients [125,129]. Studies showed that using anti-CD20+ antibodies in B-cells result in decreased CD8+ antitumor functionality, which links B-cell advantage to that of CD8+ T-cells. A lack of CD20+ secretion results in decreased CD8+ cytotoxic capabilities, promoting cancer development [130]. While CD8+ T-cells function in antitumor activity on their own, effectiveness is shown to increase in the presence of CD20+ [129]. Similar to T-cells, aged B-cells exhibit decreased antibody affinity and memory responses [131]. Consequently, aging downregulates the CD20+ and CD8+ association, resulting in decreased tumor lysis and poorer OvCa prognosis (Table 1). However, reports of certain aged B-cells such as B1a lose many immunosuppressive functions with age but notably gain the capacity to stimulate T-cell CD8+ tumor-killing activity [132]. Other reports on OvCa, also present data that increased B-cell inflammatory activity in ovarian tumors is associated with poorer prognoses [133,134]. Certain populations of B-cells, such as CD138+, instead increase angiogenesis and disrupt the T-cell lymphocyte antitumor response. Reports show reduced survival of individuals with ovarian tumors presenting high CD138+ B-cell counts, possibly due to tumor-induced alterations of B-cell phenotype [133]. Studies of OvCa patients also conclude that higher numbers of CD19+ B-cells are correlated with increased tumor severity [134]. High B-cell activity is a trait generally attributed to a younger individual [131] and thus the conflicting results of B-cell TIL contribution to OvCa proliferation cannot be fully resolved by literature results.

Table 1. Summary of aging-related immune changes.

Immune Cell Component	Effect of Aging	Effect on OvCa Metastasis
T-cell Tumor Infiltrating Lymphocytes	• Decreased cytokine secretion • Increased apoptosis • Decreased lymphocyte association	Decreased tumor lysis leads to increased proliferation
B-Cell tumor Infiltrating Lymphocytes	• Decreased cytokine secretion • Decreased T-cell association	Possibly increased angiogenesis, possibly decreased tumor lysis
Pre-Activated Macrophages	• Decreased cytokine secretion • Decreased phagocytic activity	Unknown, possibly mixed effects.

2.5. Adipocytes

Adipocytes make up the majority of the omentum and are present throughout the peritoneum [6]. Adipocytes are a complicated cell group that play a very important role in metabolism. In addition, adipocytes fuel OvCa metastatic success by providing energy in the form of fatty acids and lipids [135]. In addition to this role, adipocytes have been shown to secrete IL-8 and adipokines, which help guide OvCa cells to metastatic sites [135,136]. It has been shown that body fat percentage increases with age, as well as the capability of adipocytes to migrate out of their normal adipose tissues and into other sites of the body, causing site-specific alterations [136–138]. Specifically, aged adipocytes migrate to the viscera in the abdominal cavity, which is linked with higher disease rates than fat depots in other areas [137,139]. In fact, surgical removal of visceral fat in rats alleviated obesity-related symptoms, such as metabolic disease and insulin resistance and lengthened the lifespan of the rats [139–141]. Epidemiologic data show that obesity is a risk factor for worse disease in women, notably age-related diseases such as OvCa [136,142]. An *in vivo* pre-clinical study showed that obese mice intraperitoneally injected with OvCa cells (either diet-induced obesity or leptin-mutant) have an increased tumor burden over their lean counterparts [143]. Recently, there has been a surge of research on aging adipocytes; it has even been suggested that obesity accelerates aging, or that aging- and obesity-related processes mirror each other [136].

Aging adipocytes have been correlated with chronic inflammation [136]. Adipose tissue macrophages, or ATMs, have been shown to increase with age [136]. These immune cells secrete IL-6, promoting inflammation [136]. In addition, aged adipose tissue has higher rates of cellular senescence, as seen in the other cell types mentioned in previous sections [136]. These senescent cells also promote inflammation through the secretion of factors such as chemokines, cytokines, growth factors and MMPs [136]. In addition, the amount of differentiated and mature adipocytes formed from preadipocytes decrease with age, increasing the percentage of preadipocytes in aged tissue [137]. These preadipocytes secrete a proinflammatory profile similar to senescent cells, with factors such as PAI, IL-6 and proinflammatory cytokines and chemokines [137,144].

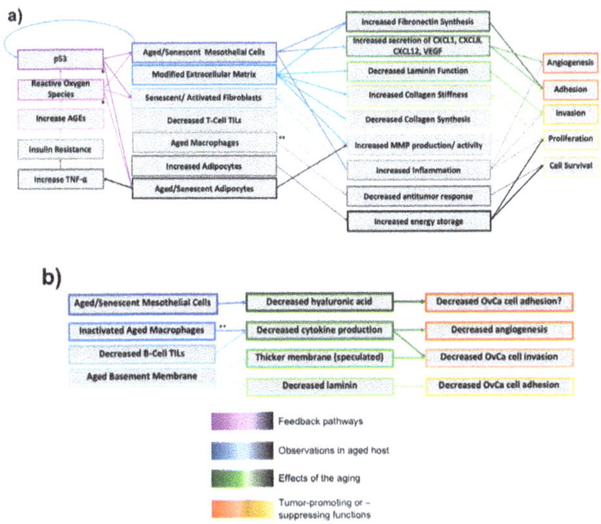

Figure 4. The effects of aging on the peritoneal microenvironment. (**a**) Tumor-Inducing Effects: Aging of the host stimulates a vast and interconnected network of alterations to the peritoneal microenvironment. These changes are often correlated with increased tumor burden due to heightened angiogenesis of the region and OvCa cell adhesion, invasion, proliferation and survival. As depicted, the multifactorial causes and results of aging present significant challenges for analysis; (**b**) Tumor-Suppressing Effects: While our review generally concludes that aging alters the microenvironment in a way conducive to tumor growth, in contrast certain aspects of aging seem to impair tumorigenesis. Aged and senescent mesothelial cells secrete less hyaluronic acid, which is hypothesized to decrease OvCa adhesion to the extracellular matrix (ECM). Inactivated aged macrophages are less capable of cytokine secretion, which thereby decreases angiogenesis potential and cell invasion. The aged BM thickens due to less collagen IV turnover, which we speculate could in theory decrease OvCa invasion (however, to our knowledge no conclusions have been drawn regarding this). The aged basement membrane (BM) also has a decreased laminin content, which may decrease cell adhesion. * While not shown to be a causative link, in aged adipose tissue there is an increase in reactive oxygen species (ROS) that is correlated with adipocytes presenting a senescent phenotype, suggesting that ROS plays the same role in adipocytes that it does in other cell types [145]. p53 has been shown to have numerous effects on adipose tissue and is likely also contributing to the senescent phenotype [145]. ** Aged macrophages paradigm: aged macrophages have been shown to both induce tumorigenesis and inhibit it., we depict both pathways. Note: Color gradients intended to help viewer differentiate between different effects of each component of the aging microenvironment.

In addition to inflammation, aging adipocytes have been correlated with insulin resistance [136,145]. AGE modifications on serum albumin have been shown to cause an increase in ROS in adipocytes, which blocks cell differentiation and leads to insulin resistance [78]. AGEs prevent cellular uptake of glucose, which can raise glucose levels, potentially contributing to AGE-mediated collagen crosslinks (see Section 2.2.1) [78,146]. Serum-AGE levels were shown to be higher in aged mice versus young, contributing to more ROS and less glucose uptake [78]. In addition, serum AGEs have been shown to stimulate TNF-α in monocytes, which causes insulin resistance [145,147].

3. Conclusions

While this review has divided the peritoneal microenvironment into distinct cellular or functional units, in reality there is complex crosstalk between all components of the microenvironment that is just beginning to be uncovered and understood. The end result is a vastly different metastatic microenvironment in aged patients relative to that seen in young patients (Figure 1), reminiscent of one of the first big debates in the field: the seed-and-soil hypothesis. Based on the research discussed above, it is clear that the aging peritoneum provides a better "soil" for metastasizing OvCa cells. Each component of the microenvironment has the potential to affect OvCa metastasis in a variety of ways (Figure 4).

At every step of the establishment of metastases, we see differences in aged hosts. OvCa cells first adhere to mesothelial cells; aged hosts have higher numbers of senescent mesothelial cells, which increase inflammation and also increase factors such as fibronectin and ICAM-1 that mediate cell-cell adhesion [16,33,38]. Once the OvCa cells adhere to and disrupt the mesothelial cells, they next invade into the collagen-rich matrix below. Aged hosts have an increase in MMP activity and lower rates of collagen synthesis, resulting in a less dense matrix that facilitates invasion. In addition, aged collagen accumulates crosslinks, which make the tissue stiffer and more aligned, allowing OvCa cells to adhere more readily [8,75]. The other cells present in the microenvironment, including fibroblasts, immune cells and adipocytes, also play a large role in changing the metastatic microenvironment. Aged fibroblasts secrete less collagen than their younger counterparts and senescent fibroblasts share many of the characteristics of CAFs, promoting OvCa metastasis [24,96]. In the immune landscape, it is unclear whether the effect of age on macrophages promotes or obstructs tumor growth. However, it can be concluded T-cell lymphocytes and certain B-cell lymphocytes experience a loss of function with age, resulting in less regulated tumor growth and increased proliferation [125]. Aged individuals have been shown to have increased adipocyte deposits, which provide energy for the OvCa metastases [137]. Aged adipose tissue also has a chronic inflammation response, resulting in immune stimulation as well as secretion of elements such as growth factors and MMPs, that can contribute to OvCa invasion and proliferation [136]. These molecular processes may also represent targets for therapeutic intervention in the aged host.

There are not many therapeutic interventions that target aging. Recent studies of senescence and the senescent-associated secretory phenotype (SASP) illuminate the field of senolytics as a promising anti-cancer treatment [148,149]. Many senolytic drugs have been discovered and tested in murine models, working to selectively target the senescent cells' anti-apoptotic pathways to induce cell death [148]. In murine models, this decreases the SASP to decrease cancer spread [148]. Notably, this is a selective treatment [148,149]—not every senescent cell has to be eliminated. Much work remains to bring this field to clinical trial stages but this review supports the observation that senolytic treatments are a propitious focus for age-associated cancers.

The studies performed in this field to date have shown that aging has multi-faceted effects on the tumor microenvironment. However, many questions remain. Much of the work reviewed here is not specific to the peritoneal tumor microenvironment and many studies were performed outside the context of OvCa metastasis. Just as Yancik voiced in 1993, there is still a need for aging research in the OvCa field. As the field progresses, integrating research on the molecular mechanisms of aging may reveal new targets for anti-metastatic therapies for OvCa patients.

Funding: This work was supported by Grants RO1CA109545 (MSS) and RO1CA086984 (MSS) from the National Institutes of Health/National Cancer Institute; from the Leo and Anne Albert Charitable Trust (MSS); University of Notre Dame College of Science Summer Undergraduate Research Fellowship (EFS).

Acknowledgments: We would like to acknowledge Elizabeth Loughran for her help with editing this manuscript.

Conflicts of Interest: The authors declare no conflict of interest.

References

1. Howlader, N.; Noore, A.M.; Krapcho, M.; Miller, D.; Bishop, K.; Kosary, C.L.; Yu, M.; Rahl, J.; Tatalovich, Z.; Mariotto, A.; et al. *Seer Cancer Statistics Review, 1975–2014*; National Cancer Institute: Bethesda, MD, USA, 2017.
2. López-Otín, C.; Blasco, M.A.; Partridge, L.; Serrano, M.; Kroemer, G. The hallmarks of aging. *Cell* **2013**, *153*, 1194–1217. [CrossRef] [PubMed]
3. Yancik, R. Ovarian cancer: Age contrasts in incidence, histology, disease stage at diagnosis, and mortality. *Cancer* **1993**, *71*, 517–523. [CrossRef] [PubMed]
4. Trillsch, F.; Woelber, L.; Eulenburg, C.; Braicu, I.; Lambrechts, S.; Chekerov, R.; van Nieuwenhuysen, E.; Speiser, P.; Zeimet, A.; Castillo-Tong, D.C.; et al. Treatment reality in elderly patients with advanced ovarian cancer: A prospective analysis of the ovcad consortium. *J. Ovarian Res.* **2013**, *6*, 42. [CrossRef] [PubMed]
5. Mikuła-Pietrasik, J.; Uruski, P.; Sosińska, P.; Maksin, K.; Piotrowska-Kempisty, H.; Kucińska, M.; Murias, M.; Szubert, S.; Woźniak, A.; Szpurek, D.; et al. Senescent peritoneal mesothelium creates a niche for ovarian cancer metastases. *Cell Death Dis.* **2016**, *7*, e2565. [CrossRef] [PubMed]
6. Kenny, H.A.; Krausz, T.; Yamada, S.D.; Lengyel, E. Use of a novel 3d culture model to elucidate the role of mesothelial cells, fibroblasts and extra-cellular matrices on adhesion and invasion of ovarian cancer cells to the omentum. *Int. J. Cancer* **2007**, *121*, 1463–1472. [CrossRef] [PubMed]
7. Van Baal, J.O.A.M.; Van de Vijver, K.K.; Nieuwland, R.; van Noorden, C.J.F.; van Driel, W.J.; Sturk, A.; Kenter, G.G.; Rikkert, L.G.; Lok, C.A.R. The histophysiology and pathophysiology of the peritoneum. *Tissue Cell* **2017**, *49*, 95–105. [CrossRef] [PubMed]
8. Park, M.S.; Lee, H.A.; Chu, W.S.; Yang, D.H.; Hwang, S.D. Peritoneal accumulation of age and peritoneal membrane permeability. *Perit. Dial. Int.* **2000**, *20*, 452–460. [PubMed]
9. Ulrich, T.A.; Pardo, E.M.D.J.; Kumar, S. The mechanical rigidity of the extracellular matrix regulates the structure, motility, and proliferation of glioma cells. *Cancer Res.* **2009**, *69*, 4167–4174. [CrossRef] [PubMed]
10. Wells, R.G. The role of matrix stiffness in regulating cell behavior. *Hepatology* **2008**, *47*, 1394–1400. [CrossRef] [PubMed]
11. Sprenger, C.C.; Plymate, S.R.; Reed, M.J. Aging-related alterations in the extracellular matrix modulate the microenvironment and influence tumor progression. *Int. J. Cancer* **2010**, *127*, 2739–2748. [CrossRef] [PubMed]
12. Eyman, D.; Damodarasamy, M.; Plymate, S.R.; Reed, M.J. Ccl5 secreted by senescent aged fibroblasts induces proliferation of prostate epithelial cells and expression of genes that modulate angiogenesis. *J. Cell. Physiol.* **2009**, *220*, 376–381. [CrossRef] [PubMed]
13. Hayflick, L. Mortality and immortality at the cellular level. A review. *Biochem. Mosc.* **1997**, *62*, 1180–1190.
14. Ksiazek, K.; Passos, J.F.; Olijslagers, S.; Saretzki, G.; Martin-Ruiz, C.; von Zglinicki, T. Premature senescence of mesothelial cells is associated with non-telomeric DNA damage. *Biochem. Biophys. Res. Commun.* **2007**, *362*, 707–711. [CrossRef] [PubMed]
15. Di Fagagna, F.D.A.; Reaper, P.M.; Clay-Farrace, L.; Fiegler, H.; Carr, P.; von Zglinicki, T.; Saretzki, G.; Carter, N.P.; Jackson, S.P. A DNA damage checkpoint response in telomere-initiated senescence. *Nature* **2003**, *426*, 194–198. [CrossRef] [PubMed]
16. Ksiazek, K.; Mikula-Pietrasik, J.; Korybalska, K.; Dworacki, G.; Jörres, A.; Witowski, J. Senescent peritoneal mesothelial cells promote ovarian cancer cell adhesion: The role of oxidative stress-induced fibronectin. *Am. J. Pathol.* **2009**, *174*, 1230–1240. [CrossRef] [PubMed]
17. Ksiazek, K.; Passos, J.F.; Olijslagers, S.; von Zglinicki, T. Mitochondrial dysfunction is a possible cause of accelerated senescence of mesothelial cells exposed to high glucose. *Biochem. Biophys. Res. Commun.* **2008**, *366*, 793–799. [CrossRef] [PubMed]
18. Książek, K.; Korybalska, K.; Jörres, A.; Witowski, J. Accelerated senescence of human peritoneal mesothelial cells exposed to high glucose: The role of tgf-β1. *Lab. Investig.* **2007**, *87*, 345–356. [CrossRef] [PubMed]

19. Blazer, S.; Khankin, E.; Segev, Y.; Ofir, R.; Yalon-Hacohen, M.; Kra-Oz, Z.; Gottfried, Y.; Larisch, S.; Skorecki, K.L. High glucose-induced replicative senescence: Point of no return and effect of telomerase. *Biochem. Biophys. Res. Commun.* **2002**, *296*, 93–101. [CrossRef]
20. Margetts, P.J.; Bonniaud, P.; Liu, L.; Hoff, C.M.; Holmes, C.J.; West-Mays, J.A.; Kelly, M.M. Transient overexpression of tgf-β1 induces epithelial mesenchymal transition in the rodent peritoneum. *J. Am. Soc. Nephrol.* **2005**, *16*, 425–436. [CrossRef] [PubMed]
21. Debacq-Chainiaux, F.; Borlon, C.; Pascal, T.; Royer, V.; Eliaers, F.; Ninane, N.; Carrard, G.; Friguet, B.; de Longueville, F.; Boffe, S.; et al. Repeated exposure of human skin fibroblasts to uvb at subcytotoxic level triggers premature senescence through the tgf-β1 signaling pathway. *J. Cell Sci.* **2005**, *118*, 743–758. [CrossRef] [PubMed]
22. Kim, K.-H.; Park, G.-T.; Lim, Y.-B.; Rue, S.-W.; Jung, J.-C.; Sonn, J.-K.; Bae, Y.-S.; Park, J.-W.; Lee, Y.-S. Expression of connective tissue growth factor, a biomarker in senescence of human diploid fibroblasts, is up-regulated by a transforming growth factor-β-mediated signaling pathway. *Biochem. Biophys. Res. Commun.* **2004**, *318*, 819–825. [CrossRef] [PubMed]
23. Collado, M.; Gil, J.; Efeyan, A.; Guerra, C.; Schuhmacher, A.J.; Barradas, M.; Benguría, A.; Zaballos, A.; Flores, J.M.; Barbacid, M.; et al. Tumour biology: Senescence in premalignant tumours. *Nature* **2005**, *436*, 642. [CrossRef] [PubMed]
24. Lawrenson, K.; Grun, B.; Benjamin, E.; Jacobs, I.J.; Dafou, D.; Gayther, S.A. Senescent fibroblasts promote neoplastic transformation of partially transformed ovarian epithelial cells in a three-dimensional model of early stage ovarian cancer. *Neoplasia* **2010**, *12*, 317–325. [CrossRef] [PubMed]
25. Krtolica, A.; Campisi, J. Cancer and aging: A model for the cancer promoting effects of the aging stroma. *Int. J. Biochem. Biol.* **2002**, *34*, 1401–1414. [CrossRef]
26. Loughran, E.A.; Leonard, A.K.; Hilliard, T.S.; Phan, R.C.; Yemc, M.G.; Harper, E.; Sheedy, E.; Klymenko, Y.; Asem, M.; Liu, Y.; et al. Aging increases susceptibility to ovarian cancer metastasis in murine allograft models and alters immune composition of peritoneal adipose tissue. *Neoplasia* **2018**, *20*, 621–631. [CrossRef] [PubMed]
27. Sastre, J. The role of mitochondrial oxidative stress in aging. *Free Radical Biol. Med.* **2003**, *35*, 1–8. [CrossRef]
28. Rodier, F.; Campisi, J.; Bhaumik, D. Two faces of p53: Aging and tumor suppression. *Nucleic Acids Res.* **2007**, *35*, 7475–7484. [CrossRef] [PubMed]
29. Donehower, L.A. Using mice to examine p53 functions in cancer, aging, and longevity. *Cold Spring Harb. Perspect. Biol.* **2009**, *1*, a001081. [CrossRef] [PubMed]
30. Yáñez-Mó, M.; Lara-Pezzi, E.; Selgas, R.; Ramírez-Huesca, M.; Domínguez-Jiménez, C.; Jiménez-Heffernan, J.A.; Aguilera, A.; Sánchez-Tomero, J.A.; Bajo, M.A.; Álvarez, V.; et al. Peritoneal dialysis and epithelial-to-mesenchymal transition of mesothelial cells. *N. Engl. J. Med.* **2003**, *348*, 403–413. [CrossRef] [PubMed]
31. Dimri, G.P.; Lee, X.; Basile, G.; Acosta, M.; Scott, G.; Roskelley, C.; Medrano, E.E.; Linskens, M.; Rubelj, I.; Pereira-Smith, O. A biomarker that identifies senescent human cells in culture and in aging skin in vivo. *Proc. Natl. Acad. Sci. USA* **1995**, *92*, 9363–9367. [CrossRef] [PubMed]
32. Fodil-Bourahla, I.; Drubaix, I.; Robert, L. Effect of in vitro aging on the biosynthesis of glycosaminoglycans by human skin fibroblasts. Modulation by the elastin-laminin receptor. *Mech. Ageing Dev.* **1999**, *106*, 241–260. [CrossRef]
33. Ksiażek, K.; Mikuła-Pietrasik, J.; Catar, R.; Dworacki, G.; Winckiewicz, M.; Frydrychowicz, M.; Dragun, D.; Staniszewski, R.; Jörres, A.; Witowski, J. Oxidative stress-dependent increase in icam-1 expression promotes adhesion of colorectal and pancreatic cancers to the senescent peritoneal mesothelium. *Int. J. Cancer* **2010**, *127*, 293–303. [PubMed]
34. Debacq-Chainiaux, F.; Erusalimsky, J.D.; Campisi, J.; Toussaint, O. Protocols to detect senescence-associated beta-galactosidase (sa-βgal) activity, a biomarker of senescent cells in culture and *in vivo*. *Nat. Protoc.* **2009**, *4*, 1798–1806. [CrossRef] [PubMed]
35. Vasile, E.; Tomita, Y.; Brown, L.F.; Kocher, O.; Dvorak, H.F. Differential expression of thymosin β-10 by early passage and senescent vascular endothelium is modulated by vpf/vegf: Evidence for senescent endothelial cells *in vivo* at sites of atherosclerosis. *FASEB J.* **2001**, *15*, 458–466. [CrossRef] [PubMed]
36. Labat-Robert, J. Fibronectin in malignancy. *Semin. Cancer Biol.* **2002**, *12*, 187–195. [CrossRef]
37. Antia, M.; Baneyx, G.E.; Kubow, K.; Vogel, V. Fibronectin in aging extracellular matrix fibrils is progressively unfolded by cells and elicits an enhanced rigidity response. *Faraday Discuss.* **2008**, *139*, 229–249. [CrossRef] [PubMed]

38. Nevado, J.; Vallejo, S.; El-Assar, M.; Peiró, C.; Sánchez-Ferrer, C.F.; Rodríguez-Mañas, L. Changes in the human peritoneal mesothelial cells during aging. *Kidney Int.* **2006**, *69*, 313–322. [CrossRef] [PubMed]
39. Colavitti, R.; Finkel, T. Reactive oxygen species as mediators of cellular senescence. *IUBMB Life* **2008**, *57*, 277–281. [CrossRef] [PubMed]
40. Mikuła-Pietrasik, J.; Sosińska, P.; Naumowicz, E.; Maksin, K.; Piotrowska, H.; Woźniak, A.; Szpurek, D.; Książek, K. Senescent peritoneal mesothelium induces a pro-angiogenic phenotype in ovarian cancer cells in vitro and in a mouse xenograft model in vivo. *Clin. Exp. Metastasis* **2016**, *33*, 15–27. [CrossRef] [PubMed]
41. Robert, L.; Robert, A.M.; Renard, G. Biological effects of hyaluronan in connective tissues, eye, skin, venous wall. Role in aging. *Pathol. Biol.* **2010**, *58*, 187–198. [CrossRef] [PubMed]
42. Young, V.J.; Brown, J.K.; Saunders, P.T.; Horne, A.W. The role of the peritoneum in the pathogenesis of endometriosis. *Hum. Reprod. Update* **2013**, *19*, 558–569. [CrossRef] [PubMed]
43. Gallorini, M.; Berardi, A.C.; Berardocco, M.; Gissi, C.; Maffulli, N.; Cataldi, A.; Oliva, F. Hyaluronic acid increases tendon derived cell viability and proliferation in vitro: Comparative study of two different hyaluronic acid preparations by molecular weight. *Muscles Ligaments Tendons J.* **2017**, *7*, 208–214. [PubMed]
44. Landau, M.; Fagien, S. Science of hyaluronic acid beyond filling: Fibroblasts and their response to the extracellular matrix. *Plast. Reconstr. Surg.* **2015**, *136*, 188S–195S. [CrossRef] [PubMed]
45. Simpson, M.A.; Reiland, J.; Burger, S.R.; Furcht, L.T.; Spicer, A.P.; Oegema, T.R.; McCarthy, J.B. Hyaluronan synthase elevation in metastatic prostate carcinoma cells correlates with hyaluronan surface retention, a prerequisite for rapid adhesion to bone marrow endothelial cells. *J. Biol. Chem.* **2001**, *276*, 17949–17957. [CrossRef] [PubMed]
46. Gardner, M.J.; Catterall, J.B.; Jones, L.M.H.; Turner, G.A. Human ovarian tumour cells can bind hyaluronic acid via membrane cd44: A possible step in peritoneal metastasis. *Clin. Exp. Metastasis* **1996**, *14*, 325–334. [CrossRef] [PubMed]
47. Dechaud, H.; Witz, C.A.; Montoya-Rodriguez, I.A.; Degraffenreid, L.A.; Schenken, R.S. Mesothelial cell–associated hyaluronic acid promotes adhesion of endometrial cells to mesothelium. *Fert. Steril.* **2001**, *76*, 1012–1018. [CrossRef]
48. Jones, L.M.H.; Gardner, M.J.; Catterall, J.B.; Turner, G.A. Hyaluronic acid secreted by mesothelial cells: A natural barrier to ovarian cancer cell adhesion. *Clin. Exp. Metastasis* **1995**, *13*, 373–380. [CrossRef] [PubMed]
49. Wiig, M.; Abrahamsson, S.O.; Lundborg, G. Effects of hyaluronan on cell proliferation and collagen synthesis: A study of rabbit flexor tendons in vitro. *J. Hand Surg. Am.* **1996**, *21*, 599–604. [CrossRef]
50. Klein, E.S.; Asculai, S.S.; Ben-Ari, G.Y. Effects of hyaluronic acid on fibroblast behavior in peritoneal injury. *J. Surg. Res.* **1996**, *61*, 473–476. [CrossRef] [PubMed]
51. Anttila, M.A.; Tammi, R.H.; Tammi, M.I.; Syrjänen, K.J.; Saarikoski, S.V.; Kosma, V.-M. High levels of stromal hyaluronan predict poor disease outcome in epithelial ovarian cancer. *Cancer Res.* **2000**, *60*, 150–155. [PubMed]
52. Hall, C.L.; Wang, C.; Lange, L.A.; Turley, E.A. Hyaluronan and the hyaluronan receptor rhamm promote focal adhesion turnover and transient tyrosine kinase activity. *J. Cell Biol.* **1994**, *126*, 575–588. [CrossRef] [PubMed]
53. Chen, W.Y.; Grant, M.E.; Schor, A.M.; Schor, S.L. Differences between adult and foetal fibroblasts in the regulation of hyaluronate synthesis: Correlation with migratory activity. *J. Cell Sci.* **1989**, *94 Pt 3*, 577–584. [PubMed]
54. Melrose, J.; Numata, Y.; Ghosh, P. Biotinylated hyaluronan: A versatile and highly sensitive probe capable of detecting nanogram levels of hyaluronan binding proteins (hyaladherins) on electroblots by a novel affinity detection procedure. *Electrophoresis* **1996**, *17*, 205–212. [CrossRef] [PubMed]
55. Matuoka, K.; Namba, M.; Mitsui, Y. Hyaluronate synthetase inhibition by normal and transformed human fibroblasts during growth reduction. *J. Cell Biol.* **1987**, *104*, 1105–1115. [CrossRef] [PubMed]
56. Yeo, T.K.; Nagy, J.A.; Yeo, K.T.; Dvorak, H.F.; Toole, B.P. Increased hyaluronan at sites of attachment to mesentery by cd44-positive mouse ovarian and breast tumor cells. *Am. J. Pathol.* **1996**, *148*, 1733–1740. [PubMed]
57. Bernard, E. Effect of hyaluronan on the elastase-type activity of human skin fibroblasts. *Cell Biol. Int.* **1994**, *18*, 967–972. [CrossRef] [PubMed]
58. Breborowicz, A.; Polubinska, A.; Moberly, J.; Ogle, K.; Martis, L.; Oreopoulos, D. Hyaluronan modifies inflammatory response and peritoneal permeability during peritonitis in rats. *Am. J. Kidney Dis.* **2001**, *37*, 594–600. [CrossRef] [PubMed]

59. Terazawa, S.; Nakajima, H.; Tobita, K.; Imokawa, G. The decreased secretion of hyaluronan by older human fibroblasts under physiological conditions is mainly associated with the down-regulated expression of hyaluronan synthases but not with the expression levels of hyaluronidases. *Cytotechnology* **2015**, *67*, 609–620. [CrossRef] [PubMed]
60. Witz, C.A.; Montoya-Rodriguez, I.A.; Cho, S.; Centonze, V.E.; Bonewald, L.F.; Schenken, R.S. Composition of the extracellular matrix of the peritoneum. *J. Soc. Gynecol. Investig.* **2001**, *8*, 299–304. [CrossRef] [PubMed]
61. Iiyama, M.; Shimada, Y.; Kita, T.; Ito, H. Effect of aging on macrophage adherence to extracellular matrix proteins. *Mech. Ageing Dev.* **1992**, *66*, 149–158. [CrossRef]
62. Sharaf, H.; Matou-Nasri, S.; Wang, Q.; Rabhan, Z.; Al-Eidi, H.; Al Abdulrahman, A.; Ahmed, N. Advanced glycation endproducts increase proliferation, migration and invasion of the breast cancer cell line mda-mb-231. *BBA-Mol. Basis Dis.* **2015**, *1852*, 429–441. [CrossRef] [PubMed]
63. Lodish, H.; Berk, A.; Zipursky, S.L.; Matsudaira, P.; Baltimore, D.; Darnell, J. Collagen: The fibrous proteins of the matrix. *Mol. Cell Biol.* **2000**, *4*.
64. Shuster, S.; Black, M.M.; McVitie, E. The influence of age and sex on skin thickness, skin collagen and density. *Br. J. Dermatol.* **1975**, *93*, 639–643. [CrossRef] [PubMed]
65. Cheng, W.; Rrong, Y.; Ning, F.; Zhang, G. The content and ratio of type i and iii collagen in skin differ with age and injury. *Afr. J. Biotechnol.* **2011**, *10*, 2524–2529.
66. Damodarasamy, M.; Vernon, R.B.; Karres, N.; Chang, C.H.; Bianchi-Frias, D.; Nelson, P.S.; Reed, M.J. Collagen extracts derived from young and aged mice demonstrate different structural properties and cellular effects in three-dimensional gels. *J. Gerontol. A Biol. Sci. Med. Sci.* **2010**, *65A*, 209–218. [CrossRef] [PubMed]
67. Bianchi-Frias, D.; Vakar-Lopez, F.; Coleman, I.M.; Plymate, S.R.; Reed, M.J.; Nelson, P.S. The effects of aging on the molecular and cellular composition of the prostate microenvironment. *PLoS ONE* **2010**, *5*, e12501. [CrossRef] [PubMed]
68. Qin, Z.; Balimunkwe, R.M.; Quan, T. Age-related reduction of dermal fibroblast size upregulates multiple matrix metalloproteinases as observed in aged human skin in vivo. *Brit. J. Dermatol.* **2017**, *177*, 1337–1348. [CrossRef] [PubMed]
69. Jabłońska-Trypuć, A.; Matejczyk, M.; Rosochacki, S. Matrix metalloproteinases (mmps), the main extracellular matrix (ecm) enzymes in collagen degradation, as a target for anticancer drugs. *J. Enzym. Inhib. Med. Chem.* **2016**, *31*, 177–183. [CrossRef] [PubMed]
70. Jacob, M.P. Extracellular matrix remodeling and matrix metalloproteinases in the vascular wall during aging and in pathological conditions. *Biomed. Pharmacother.* **2003**, *57*, 195–202. [CrossRef]
71. Kamat, A.A.; Fletcher, M.; Gruman, L.M.; Mueller, P.; Lopez, A.; Landen, C.N.; Han, L.; Gershenson, D.M.; Sood, A.K. The clinical relevance of stromal matrix metalloproteinase expression in ovarian cancer. *Clin. Cancer Res.* **2006**, *12*, 1707–1714. [CrossRef] [PubMed]
72. Liu, D.; Hornsby, P.J. Senescent human fibroblasts increase the early growth of xenograft tumors via matrix metalloproteinase secretion. *Cancer Res.* **2007**, *67*, 3117–3126. [CrossRef] [PubMed]
73. Verzijl, N.; DeGroot, J.; Thorpe, S.R.; Bank, R.A.; Shaw, J.N.; Lyons, T.J.; Bijlsma, J.W.J.; Lafeber, F.P.J.G.; Baynes, J.W.; TeKoppele, J.M. Effect of collagen turnover on the accumulation of advanced glycation end products. *J. Biol. Chem.* **2000**, *275*, 39027–39031. [CrossRef] [PubMed]
74. Kagan, H.M.; Trackman, P.C. Properties and function of lysyl oxidase. *Am. J. Respir. Cell Mol. Biol.* **1991**, *5*, 206–210. [CrossRef] [PubMed]
75. Rossow, L.; Veitl, S.; Vorlová, S.; Wax, J.K.; Kuhn, A.E.; Maltzahn, V.; Upcin, B.; Karl, F.; Hoffmann, H.; Gätzner, S.; et al. Lox-catalyzed collagen stabilization is a proximal cause for intrinsic resistance to chemotherapy. *Oncogene* **2018**, *1*. [CrossRef] [PubMed]
76. Panwar, P.; Lamour, G.; Mackenzie, N.C.W.; Yang, H.; Ko, F.; Li, H.; Brömme, D. Changes in structural-mechanical properties and degradability of collagen during aging-associated modifications. *J. Biol. Chem.* **2015**, *290*, 23291–23306. [CrossRef] [PubMed]
77. Bailey, A.J.; Paul, R.G.; Knott, L. Mechanisms of maturation and ageing of collagen. *Mech. Ageing Dev.* **1998**, *106*, 1–56. [CrossRef]
78. Unoki, H.; Bujo, H.; Yamagishi, S.-I.; Takeuchi, M.; Imaizumi, T.; Saito, Y. Advanced glycation end products attenuate cellular insulin sensitivity by increasing the generation of intracellular reactive oxygen species in adipocytes. *Diabetes Res. Clin. Pract.* **2007**, *76*, 236–244. [CrossRef] [PubMed]

79. Robert, L. Mechanisms of aging of the extracellular matrix: Role of the elastin-laminin receptor. *Gerontology* **1998**, *44*, 307–317. [CrossRef] [PubMed]
80. Liotta, L.A.; Tryggvason, K.; Garbisa, S.; Hart, I.; Foltz, C.M.; Shafie, S. Metastatic potential correlates with enzymatic degradation of basement membrane collagen. *Nature* **1980**, *284*, 67–68. [CrossRef] [PubMed]
81. Hotary, K.; Li, X.Y.; Allen, E.; Stevens, S.L.; Weiss, S.J. A cancer cell metalloprotease triad regulates the basement membrane transmigration program. *Genes Dev.* **2006**, *20*, 2673–2686. [CrossRef] [PubMed]
82. Richardson, L.L.; Kleinman, H.K.; Dym, M. The effects of aging on basement membrane in the testis. *J. Androl.* **1995**, *16*, 118–126. [PubMed]
83. Vázquez, F.; Palacios, S.; Alemañ, N.; Guerrero, F. Changes of the basement membrane and type iv collagen in human skin during aging. *Maturitas* **1996**, *25*, 209–215. [CrossRef]
84. Capo-Chichi, C.D.; Smith, E.R.; Yang, D.-H.; Roland, I.H.; Vanderveer, L.; Cohen, C.; Hamilton, T.C.; Godwin, A.K.; Xu, X.-X. Dynamic alterations of the extracellular environment of ovarian surface epithelial cells in premalignant transformation, tumorigenicity, and metastasis. *Cancer* **2002**, *95*, 1802–1815. [CrossRef] [PubMed]
85. Lengyel, E. Ovarian cancer development and metastasis. *Am. J. Pathol.* **2010**, *177*, 1053–1064. [CrossRef] [PubMed]
86. Bagadi, S.A.R.; Prasad, C.P.; Srivastava, A.; Prashad, R.; Gupta, S.D.; Ralhan, R. Frequent loss of dab2 protein and infrequent promoter hypermethylation in breast cancer. *Breast Cancer Res. Treat.* **2007**, *104*, 277–286. [CrossRef] [PubMed]
87. Johnson, A.A.; Akman, K.; Calimport, S.R.G.; Wuttke, D.; Stolzing, A.; de Magalhães, J.P. The role of DNA methylation in aging, rejuvenation, and age-related disease. *Rejuv. Res.* **2012**, *15*, 483–494. [CrossRef] [PubMed]
88. Zampieri, M.; Ciccarone, F.; Calabrese, R.; Franceschi, C.; Bürkle, A.; Caiafa, P. Reconfiguration of DNA methylation in aging. *Mech. Ageing Dev.* **2015**, *151*, 60–70. [CrossRef] [PubMed]
89. Rump, A.; Morikawa, Y.; Tanaka, M.; Minami, S.; Umesaki, N.; Takeuchi, M.; Miyajima, A. Binding of ovarian cancer antigen ca125/muc16 to mesothelin mediates cell adhesion. *J. Biol. Chem.* **2004**, *279*, 9190–9198. [CrossRef] [PubMed]
90. Monaco, S.; Sparano, V.; Gioia, M.; Sbardella, D.; Di Pierro, D.; Marini, S.; Coletta, M. Enzymatic processing of collagen iv by mmp-2 (gelatinase a) affects neutrophil migration and it is modulated by extracatalytic domains. *Protein Sci.* **2006**, *15*, 2805–2815. [CrossRef] [PubMed]
91. Zhu, X.H.; Wang, J.M.; Yang, S.S.; Wang, F.F.; Hu, J.L.; Xin, S.N.; Men, H.; Lu, G.F.; Lan, X.L.; Zhang, D.; et al. Down-regulation of dab2ip promotes colorectal cancer invasion and metastasis by translocating hnrnpk into nucleus to enhance the transcription of mmp2: Down-regulation of dab2ip. *Int. J. Cancer* **2017**, *141*, 172–183. [CrossRef] [PubMed]
92. Patarroyo, M.; Tryggvason, K.; Virtanen, I. Laminin isoforms in tumor invasion, angiogenesis and metastasis. *Semin. Cancer Biol.* **2002**, *12*, 197–207. [CrossRef]
93. Brar, P.K.; Dalkin, B.L.; Weyer, C.; Sallam, K.; Virtanen, I.; Nagle, R.B. Laminin alpha-1, alpha-3, and alpha-5 chain expression in human prepubetal benign prostate glands and adult benign and malignant prostate glands. *Prostate* **2003**, *55*, 65–70. [CrossRef] [PubMed]
94. Brownlee, M. Advanced protein glycosylation in diabetes and aging. *Annu. Rev. Med.* **1995**, *46*, 223–234. [CrossRef] [PubMed]
95. Karttunen, T.; Risteli, J.; Autio–Harmainen, H.; Risteli, L. Effect of age and diabetes on type iv collagen and laminin in human kidney cortex. *Kidney Int.* **1986**, *30*, 586–591. [CrossRef] [PubMed]
96. Johnson, B.D. Effects of donor age on protein and collagen synthesis in vitro by human diploid fibroblasts. *Lab. Investig.* **1986**, *55*, 490–496. [PubMed]
97. Coppé, J.-P.; Kauser, K.; Campisi, J.; Beauséjour, C.M. Secretion of vascular endothelial growth factor by primary human fibroblasts at senescence. *J. Biol. Chem.* **2006**, *281*, 29568–29574. [CrossRef] [PubMed]
98. Schosserer, M.; Grillari, J.; Breitenbach, M. The dual role of cellular senescence in developing tumors and their response to cancer therapy. *Front. Oncol.* **2017**, *7*, 278. [CrossRef] [PubMed]
99. Parrinello, S. Stromal-epithelial interactions in aging and cancer: Senescent fibroblasts alter epithelial cell differentiation. *J. Cell Sci.* **2005**, *118*, 485–496. [CrossRef] [PubMed]
100. Yang, G.; Rosen, D.G.; Zhang, Z.; Bast, R.C.; Mills, G.B.; Colacino, J.A.; Mercado-Uribe, I.; Liu, J. The chemokine growth-regulated oncogene 1 (gro-1) links ras signaling to the senescence of stromal fibroblasts and ovarian tumorigenesis. *Proc. Natl. Acad. Sci. USA* **2006**, *103*, 16472–16477. [CrossRef] [PubMed]

101. Krtolica, A.; Parrinello, S.; Lockett, S.; Desprez, P.Y.; Campisi, J. Senescent fibroblasts promote epithelial cell growth and tumorigenesis: A link between cancer and aging. *Proc. Natl. Acad. Sci. USA* **2001**, *98*, 12072–12077. [CrossRef] [PubMed]
102. Campisi, J. Senescent cells, tumor suppression, and organismal aging: Good citizens, bad neighbors. *Cell* **2005**, *120*, 513–522. [CrossRef] [PubMed]
103. García-Gómez, I.; Goldsmith, H.S.; Angulo, J.; Prados, A.; López-Hervás, P.; Cuevas, B.; Dujovny, M.; Cuevas, P. Angiogenic capacity of human omental stem cells. *Neurol. Res.* **2005**, *27*, 807–811. [CrossRef] [PubMed]
104. Tao, L.; Huang, G.; Song, H.; Chen, Y.; Chen, L. Cancer associated fibroblasts: An essential role in the tumor microenvironment. *Oncol. Lett.* **2017**, *14*, 2611–2620. [CrossRef] [PubMed]
105. Mishra, P.; Banerjee, D.; Ben-Baruch, A. Chemokines at the crossroads of tumor-fibroblast interactions that promote malignancy. *J. Leukoc. Biol.* **2011**, *89*, 31–39. [CrossRef] [PubMed]
106. Wang, T.; Notta, F.; Navab, R.; Joseph, J.; Ibrahimov, E.; Xu, J.; Zhu, C.-Q.; Borgida, A.; Gallinger, S.; Tsao, M.-S. Senescent carcinoma-associated fibroblasts upregulate il8 to enhance prometastatic phenotypes. *Mol. Cancer Res.* **2017**, *15*, 3–14. [CrossRef] [PubMed]
107. Meza-Perez, S.; Randall, T.D. Immunological functions of the omentum. *Trends Immunol.* **2017**, *38*, 526–536. [CrossRef] [PubMed]
108. Gerber, S.A.; Rybalko, V.Y.; Bigelow, C.E.; Lugade, A.A.; Foster, T.H.; Frelinger, J.G.; Lord, E.M. Preferential attachment of peritoneal tumor metastases to omental immune aggregates and possible role of a unique vascular microenvironment in metastatic survival and growth. *Am. J. Pathol.* **2006**, *169*, 1739–1752. [CrossRef] [PubMed]
109. Folkman, J. Role of angiogenesis in tumor growth and metastasis. *Semin. Oncol.* **2002**, *29*, 15–18. [CrossRef] [PubMed]
110. Pradeep, S.; Kim, S.W.; Wu, S.Y.; Nishimura, M.; Chaluvally-Raghavan, P.; Miyake, T.; Pecot, C.V.; Kim, S.-J.; Choi, H.J.; Bischoff, F.Z.; et al. Hematogenous metastasis of ovarian cancer: Rethinking mode of spread. *Cancer Cell* **2014**, *26*, 77–91. [CrossRef] [PubMed]
111. Coffman, L.G.; Burgos-Ojeda, D.; Wu, R.; Cho, K.; Bai, S.; Buckanovich, R.J. New models of hematogenous ovarian cancer metastasis demonstrate preferential spread to the ovary and a requirement for the ovary for abdominal dissemination. *Transl. Res.* **2016**, *175*, 92–102. [CrossRef] [PubMed]
112. Krist, L.F.G.; Eestermans, I.L.; Steenbergen, J.J.E.; Hoefsmit, E.C.M.; Cuesta, M.A.; Meyer, S.; Beelen, R.H.J. Cellular composition of milky spots in the human greater omentum: An immunochemical and ultrastructural study. *Anat. Rec.* **1995**, *241*, 163–174. [CrossRef] [PubMed]
113. Stout, R.D.; Suttles, J. Functional plasticity of macrophages: Reversible adaptation to changing microenvironments. *J. Leukoc. Biol.* **2004**, *76*, 509–513. [CrossRef] [PubMed]
114. Sebastián, C.; Espia, M.; Serra, M.; Celada, A.; Lloberas, J. Macrophaging: A cellular and molecular review. *Immunobiology* **2005**, *210*, 121–126. [CrossRef] [PubMed]
115. Linehan, E.; Fitzgerald, D. Ageing and the immune system: Focus on macrophages. *Eur. J. Microbiol. Immunol.* **2015**, *5*, 14–24. [CrossRef] [PubMed]
116. Kohut, M.L.; Senchina, D.S.; Madden, K.S.; Martin, A.E.; Felten, D.L.; Moynihan, J.A. Age effects on macrophage function vary by tissue site, nature of stimulant, and exercise behavior. *Exp. Gerontol.* **2004**, *39*, 1347–1360. [CrossRef] [PubMed]
117. De la Fuente, M.; Del Río, M.; Medina, S. Changes with aging in the modulation by neuropeptide y of murine peritoneal macrophage functions. *J. Neuroimmunol.* **2001**, *116*, 156–167. [CrossRef]
118. Kulbe, H.; Thompson, R.; Wilson, J.L.; Robinson, S.; Hagemann, T.; Fatah, R.; Gould, D.; Ayhan, A.; Balkwill, F. The inflammatory cytokine tumor necrosis factor- generates an autocrine tumor-promoting network in epithelial ovarian cancer cells. *Cancer Res.* **2007**, *67*, 585–592. [CrossRef] [PubMed]
119. Wallace, P.K.; Eisenstein, T.K.; Meissler, J.J.; Morahan, P.S. Decreases in macrophage mediated antitumor activity with aging. *Mech. Ageing Dev.* **1995**, *77*, 169–184. [CrossRef]
120. Linehan, E.; Dombrowski, Y.; Snoddy, R.; Fallon, P.G.; Kissenpfennig, A.; Fitzgerald, D.C. Aging impairs peritoneal but not bone marrow-derived macrophage phagocytosis. *Aging Cell* **2014**, *13*, 699–708. [CrossRef] [PubMed]
121. Hanahan, D.; Weinberg, R.A. Hallmarks of cancer: The next generation. *Cell* **2011**, *144*, 646–674. [CrossRef] [PubMed]

122. Condeelis, J.; Pollard, J.W. Macrophages: Obligate partners for tumor cell migration, invasion, and metastasis. *Cell* **2006**, *124*, 263–266. [CrossRef] [PubMed]
123. Dhabekar, G.; Dandekar, R.; Kingaonkar, A. Role of macrophages in malignancy. *Ann. Maxillofac. Surg.* **2011**, *1*, 150. [CrossRef] [PubMed]
124. Pollard, J.W. Tumour-educated macrophages promote tumour progression and metastasis. *Nat. Rev. Cancer* **2004**, *4*, 71–78. [CrossRef] [PubMed]
125. Santoiemma, P.P.; Powell, D.J. Tumor infiltrating lymphocytes in ovarian cancer. *Cancer Biol. Ther.* **2015**, *16*, 807–820. [CrossRef] [PubMed]
126. Salam, N.; Rane, S.; Das, R.; Faulkner, M.; Gund, R.; Kandpal, U.; Lewis, V.; Mattoo, H.; Prabhu, S.; Ranganathan, V.; et al. T cell ageing: Effects of age on development, survival & function. *Indian J. Med. Res.* **2013**, *138*, 595–608. [PubMed]
127. Haynes, L.; Eaton, S.M.; Burns, E.M.; Randall, T.D.; Swain, S.L. Cd4 t cell memory derived from young naive cells functions well into old age, but memory generated from aged naive cells functions poorly. *Proc. Natl. Acad. Sci. USA* **2003**, *100*, 15053–15058. [CrossRef] [PubMed]
128. Linton, P.J. Antigen-independent changes in naive cd4 t cells with aging. *J. Exp. Med.* **1996**, *184*, 1891–1900. [CrossRef] [PubMed]
129. Nielsen, J.S.; Nelson, B.H. Tumor-infiltrating b cells and t cells: Working together to promote patient survival. *OncoImmunology* **2012**, *1*, 1623–1625. [CrossRef] [PubMed]
130. Montfort, A.; Pearce, O.; Maniati, E.; Vincent, B.G.; Bixby, L.; Böhm, S.; Dowe, T.; Wilkes, E.H.; Chakravarty, P.; Thompson, R.; et al. A strong b-cell response is part of the immune landscape in human high-grade serous ovarian metastases. *Clin. Cancer Res.* **2017**, *23*, 250–262. [CrossRef] [PubMed]
131. Frasca, D.; Blomberg, B.B. Effects of aging on b cell function. *Curr. Opin. Immunol.* **2009**, *21*, 425–430. [CrossRef] [PubMed]
132. Lee-Chang, C.; Bodogai, M.; Moritoh, K.; Chen, X.; Wersto, R.; Sen, R.; Young, H.A.; Croft, M.; Ferrucci, L.; Biragyn, A. Aging converts innate b1a cells into potent cd8$^+$ t cell inducers. *J. Immunol.* **2016**, *196*, 3385–3397. [CrossRef] [PubMed]
133. Lundgren, S.; Berntsson, J.; Nodin, B.; Micke, P.; Jirström, K. Prognostic impact of tumour-associated b cells and plasma cells in epithelial ovarian cancer. *J. Ovarian Res.* **2016**, *9*, 21. [CrossRef] [PubMed]
134. Dong, H.P.; Elstrand, M.B.; Holth, A.; Silins, I.; Berner, A.; Trope, C.G.; Davidson, B.; Risberg, B. Nk- and b-cell infiltration correlates with worse outcome in metastatic ovarian carcinoma. *Am. J. Clin. Pathol.* **2006**, *125*, 451–458. [CrossRef] [PubMed]
135. Nieman, K.M.; Kenny, H.A.; Penicka, C.V.; Ladanyi, A.; Buell-Gutbrod, R.; Zillhardt, M.R.; Romero, I.L.; Carey, M.S.; Mills, G.B.; Hotamisligil, G.S.; et al. Adipocytes promote ovarian cancer metastasis and provide energy for rapid tumor growth. *Nat. Med.* **2011**, *17*, 1498–1503. [CrossRef] [PubMed]
136. Palmer, A.K.; Kirkland, J.L. Aging and adipose tissue: Potential interventions for diabetes and regenerative medicine. *Exp. Gerontol.* **2016**, *86*, 97–105. [CrossRef] [PubMed]
137. Cartwright, M.; Tchkonia, T.; Kirkland, J. Aging in adipocytes: Potential impact of inherent, depot-specific mechanisms. *Exp. Gerontol.* **2007**, *42*, 463–471. [CrossRef] [PubMed]
138. Guo, T.; Marmol, P.; Moliner, A.; Björnholm, M.; Zhang, C.; Shokat, K.M.; Ibanez, C.F. Adipocyte alk7 links nutrient overload to catecholamine resistance in obesity. *eLife* **2014**, *3*, e03245. [CrossRef] [PubMed]
139. Huffman, D.M.; Barzilai, N. Role of visceral adipose tissue in aging. *BBA-Gen. Subj.* **2009**, *1790*, 1117–1123. [CrossRef] [PubMed]
140. Gabriely, I.; Ma, X.H.; Yang, X.M.; Atzmon, G.; Rajala, M.W.; Berg, A.H.; Scherer, P.; Rossetti, L.; Barzilai, N. Removal of visceral fat prevents insulin resistance and glucose intolerance of aging: An adipokine-mediated process? *Diabetes* **2002**, *51*, 2951–2958. [CrossRef] [PubMed]
141. Muzumdar, R.; Allison, D.B.; Huffman, D.M.; Ma, X.; Atzmon, G.; Einstein, F.H.; Fishman, S.; Poduval, A.D.; McVei, T.; Keith, S.W.; et al. Visceral adipose tissue modulates mammalian longevity. *Aging Cell* **2008**, *7*, 438–440. [CrossRef] [PubMed]
142. Foong, K.W.; Bolton, H. Obesity and ovarian cancer risk: A systematic review. *Post Reprod. Health* **2017**, *23*, 183–198. [CrossRef] [PubMed]

143. Liu, Y.; Metzinger, M.N.; Lewellen, K.A.; Cripps, S.N.; Carey, K.D.; Harper, E.I.; Shi, Z.; Tarwater, L.; Grisoli, A.; Lee, E.; et al. Obesity contributes to ovarian cancer metastatic success through increased lipogenesis, enhanced vascularity, and decreased infiltration of m1 macrophages. *Cancer Res.* **2015**, *75*, 5046–5057. [CrossRef] [PubMed]
144. Tchkonia, T.; Morbeck, D.E.; Zglinicki, T.V.; Deursen, J.V.; Lustgarten, J.; Scrable, H.; Khosla, S.; Jensen, M.D.; Kirkland, J.L. Fat tissue, aging, and cellular senescence. *Aging Cell* **2010**, *9*, 667–684. [CrossRef] [PubMed]
145. Ahima, R.S. Connecting obesity, aging and diabetes. *Nat. Med.* **2009**, *15*, 996–997. [CrossRef] [PubMed]
146. Gkogkolou, P.; Böhm, M. Advanced glycation end products. *Dermatoendocrinol* **2012**, *4*, 259–270. [CrossRef] [PubMed]
147. Naitoh, T.; Kitahara, M.; Tsuruzoe, N. Tumor necrosis factor-alpha is induced through phorbol ester–and glycated human albumin-dependent pathway in thp-1 cells. *Cell. Signal.* **2001**, *13*, 331–334. [CrossRef]
148. Kirkland, J.L.; Tchkonia, T. Cellular senescence: A translational perspective. *eBioMedicine* **2017**, *21*, 21–28. [CrossRef] [PubMed]
149. Zhu, Y.; Tchkonia, T.; Pirtskhalava, T.; Gower, A.C.; Ding, H.; Giorgadze, N.; Palmer, A.K.; Ikeno, Y.; Hubbard, G.B.; Lenburg, M.; et al. The achilles' heel of senescent cells: From transcriptome to senolytic drugs. *Aging Cell* **2015**, *14*, 644–658. [CrossRef] [PubMed]

cancers

MDPI

Review

Epigenetic Crosstalk between the Tumor Microenvironment and Ovarian Cancer Cells: A Therapeutic Road Less Traveled

Yuliya Klymenko [1,2,*] and Kenneth P. Nephew [1,3,4,*]

1 Cell, Molecular and Cancer Biology Program, Medical Sciences, Indiana University School of Medicine, Bloomington, IN 47405, USA

2 Department of Chemistry and Biochemistry, Harper Cancer Research Institute, University of Notre Dame, South Bend, IN 46617, USA

3 Department of Cellular and Integrative Physiology and Obstetrics and Gynecology, Indiana University School of Medicine, Indianapolis, IN 46202, USA

4 Indiana University Simon Cancer Center, Indianapolis, IN 46202, USA

* Correspondence: yklymenk@iu.edu (Y.K.); knephew@indiana.edu (K.P.N.); Tel.: +1-812-855-2118 (Y.K.); +1-812-855-9445 (K.P.N.)

Received: 22 July 2018; Accepted: 28 August 2018; Published: 30 August 2018

Abstract: Metastatic dissemination of epithelial ovarian cancer (EOC) predominantly occurs through direct cell shedding from the primary tumor into the intra-abdominal cavity that is filled with malignant ascitic effusions. Facilitated by the fluid flow, cells distribute throughout the cavity, broadly seed and invade through peritoneal lining, and resume secondary tumor growth in abdominal and pelvic organs. At all steps of this unique metastatic process, cancer cells exist within a multidimensional tumor microenvironment consisting of intraperitoneally residing cancer-reprogramed fibroblasts, adipose, immune, mesenchymal stem, mesothelial, and vascular cells that exert miscellaneous bioactive molecules into malignant ascites and contribute to EOC progression and metastasis via distinct molecular mechanisms and epigenetic dysregulation. This review outlines basic epigenetic mechanisms, including DNA methylation, histone modifications, chromatin remodeling, and non-coding RNA regulators, and summarizes current knowledge on reciprocal interactions between each participant of the EOC cellular milieu and tumor cells in the context of aberrant epigenetic crosstalk. Promising research directions and potential therapeutic strategies that may encompass epigenetic tailoring as a component of complex EOC treatment are discussed.

Keywords: ovarian cancer; epigenetics; tumor microenvironment; DNA methylation; histone modifications; chromatin remodeling; non-coding RNAs

1. Introduction

Epithelial ovarian cancer (EOC), a histopathologically, morphologically, and molecularly heterogeneous group of neoplasms [1], is the leading cause of gynecological malignancy-related deaths in women, with >14,000 deaths in the United States (US) and ~152,000 deaths worldwide yearly [2–4]. Most women have vastly disseminated intraperitoneal disease at the time of diagnosis contributing to a five-year survival rate of only 30% [5]. Development of multidrug resistant and essentially incurable tumor recurrence in the majority of patients after initial good response to standard platinum/taxane-based chemotherapy are also significant factors contributing to this deadly disease [6,7].

1.1. Tumor Microenvironment (TME) Associated with Ovarian Neoplasms

EOC initiation results from accumulation of genetic mutations and epigenetic changes resulting in malicious transformation of epithelial cells, stem cells, or transient metaplastic regions at the primary site, either ovary or the fallopian tube fimbriae [8–18]. While lymph node and hematogenous metastasis of ovarian cancer have been reported in human EOC cancer and/or model systems [19,20], the current consensus is that expansion of ovarian neoplastic masses occurs primarily via transcoelomic route, including the direct exfoliation of anoikis-resistant cancer cells and multi-cellular clusters from the original tumor, ascitic fluid-facilitated intraperitoneal dissemination, subsequent mesothelial adhesion and retraction, submesothelial extracellular matrix invasion, and ultimate establishment of secondary lesions in peritoneum-sheathed surfaces and organs [18,21–23]. During this metastasis process, ovarian cancer cells are confined to and nurtured by the complex host intraperitoneal cellular milieu, encompassing cells co-existing within the tumor bulk, freely available in ascitic effusions, and residing in peritoneal and adipose tissues—fibroblasts, mesothelial cells, adipocytes, infiltrating lymphocytes, macrophages, plasmacytoid dendritic cells, mesenchymal stem cells, and others (Figure 1) [24–29]. Both EOC and host non-cancerous cells secrete a plethora of bioactive soluble constituents—proteins, growth factors, phospholipids, hormones, cytokines—into the extracellular space and malignant ascites [23,27,30–44], collectively generating a dynamic intraperitoneal TME that mediates ovarian cancer development, metastatic progression, and therapeutic response through receptor-ligand (autocrine, paracrine, endocrine) signaling, contact-dependent (juxtacrine) cell signaling, as well as epigenetic regulation (Figure 1B).

1.2. Basic Epigenetic Mechanisms at a Glance

Epigenetic modifications are heritable alterations in gene expression (activation or suppression) that occur as a result of perturbed chromatin organization and altered gene accessibility for transcriptional machinery in the absence of changes to the DNA itself [45]. Additionally, epigenetic mediation encompasses the modulation of gene expression at the posttranscriptional level via altered mRNA translation into protein (Figure 2). Fundamental epigenetic regulatory mechanisms include:

1. DNA methylation—addition of methyl groups to DNA CpG sites without altering DNA nucleotide sequence. Methylation occurs by means of enzymes called DNA methyltransferases (DNMTs), which place methyl groups on symmetric cytosine residues in double-stranded CpG sites [46,47]. Hypermethylation of CpG islands (nucleotide sequences enriched for CpG sites) in the promoter regions of tumor suppressor genes (TSGs) and growth regulatory genes prompts gene silencing [46,47] as attached methyl groups physically block binding of transcription factors to the gene promoters. Alternatively, dense DNA methylation interferes with the proper nucleosome positioning [48]. Within the DNMT family (including three active enzymes, DNMT1, DNMT3a, and DNMT3b), DNMT1 exhibits high preference for hemimethylated DNA (in which one of two complimentary DNA strands already possess attached methyl groups), and is therefore responsible for so called "maintenance methylation" [49,50]. DNMT3a and DNMT3b are primarily responsible for the "de novo" methylation of previously unmethylated CpG regions [51,52], but both of these methyltransferases have been shown to carry out maintenance methylation as well [53]. Importantly, in human neoplastic cells, it has been shown that DNMT1 provides both de novo and maintenance DNA methylation of TSGs [54–56]. The demethylating agents (or hypomethylating agents (HMAs) that inhibit these enzymes (azacitidine or AZA; decitabine or DAC; SGI-110 or guadecitabine) are discussed below).
2. Histone modifications—various posttranslational modifications (PTMs) at histone protein N-terminal tails, which impair proper interactions between adjacent nucleosomes to affect the compact packing of the chromatin and impede the binding ability of other factors/enzymes that are involved in gene transcription [57,58]. The most common and well-characterized PTM, histone acetylation, is a dynamic, reversible process in which positively charged histone lysine

residues are neutralized via the addition of acetyl groups by histone acetyltransferases (HATs), resulting in the attenuation of bonds between negatively charged DNA string and a histone complex. In the reverse reaction, deacetylation, enzymes histone deacetylases (HDACs) remove acetyl groups, and reinforce positive charge of the lysines, securing compact wrapping of DNA around histones [59,60]. Similarly, histone (de)phosphorylation utilizes protein kinases and phosphatases to attach or remove negatively charged phosphate groups, respectively, influencing chemical attraction between DNA and histone tails (reviewed in [60,61]). Histone (de)methylation—addition/removal of methyl groups by histone-specific methyltransferases and demethylases—can either activate or silence gene transcription. Remarkably, the functional consequences of each histone (de)methylation event depend on the histone, amino acid and residue methylated, degree of modification (mono-, di- or tri-methylation), and attraction of additional function-specific protein cofactors to the site, as well as existence of other methyl or acetyl groups in close proximity (reviewed in [62]). Comprehensive analyses of currently known histone PTMs, including those less common (ubiquitylation, sumoylation, deamination, etc.), their functional outcomes and complex interplay between the DNA methylation and histone modifications have been recently published [63,64].

3. Chromatin remodeling—rearrangement of chromatin organization through complete or partial nucleosome repositioning and altering gene access for transcription. Chromatin remodeling can occur via nucleosome sliding (movement of the core histone octamer nexus across DNA segment with no evident disintegration of the octamer itself), nucleosome ejection (nucleosome segregation from the chromatin chain), or histone eviction (removal of histone H2A/H2B dimers from the DNA-associated nucleosome, sometimes with an alternative histone replacement) [65–67]. These processes are mediated by a number of ATP-dependent chromatin remodelers with high binding affinity to modified core histone tails, as well as transcriptional enzymes, which are extensively described in [65–67]. In particular, ARID1A and SMARCA4 are prominent chromatin remodeler examples in ovarian cancer. ARID1A is frequently mutated in ovarian clear cell (~50%) and low grade ovarian endometrioid (30%) carcinomas [68,69]. Most interestingly, tumors with ARID1A mutations acquire sensitivity to pan-HDAC inhibitors, thus making ARID1A-bearing cancers attractive for HDAC-based therapy [70]. SMARCA4 is frequently (over 90%) mutated in ovarian small cell carcinomas of the hypercalcemic type [71,72], however, to our knowledge, the first case of a germline SMARCA4 mutation in a patient with HGSOC was recently reported [73]. Further investigation on the role and clinical applicability of SMARCA4 and ARID1A in HGSOC is warranted [74,75]. Altogether, the three epigenetic mechanisms that are described above work closely to mediate DNA (un)coiling around the core histones and ensure dynamic chromatin reassembly between heterochromatin (condensed or closed, silent) and euchromatin (loose or open, transcription-permissive) states.
4. Non-coding RNA interference—a group of epigenetic regulatory mechanisms that involves microRNAs (miRNAs; miR) and long non-coding RNAs (lncRNAs). MiRNAs are short (~22 nucleotides) non-messenger RNAs that act primarily at a posttranscriptional level by base pairing with their complimentary mRNA targets to alter mRNA translation into protein [76]. Remarkably, one miRNA may complement a variety of mRNAs, whereas the same mRNA transcript might be a target of multiple miRNAs. Additionally, miRNAs may act as mRNA destabilizers causing poly-A-tail shortening [77] or interfere at the gene transcriptional level by means of PTMs (e.g., initiation of histone H3 lysine^{9} methylation with RNA interference machinery, followed by DNA methylation and gene transcription repression) and heterochromatic silencing [78]. LncRNAs are long (>200 for up to a hundred thousand nucleotides) non-messenger RNA that execute epigenetic regulation via several mechanisms: engage in post-translational histone modifications through association with chromatin-modifying proteins as an obligatory active player in the complex or as a scaffold that brings different protein complexes in close vicinity for proper functioning; serve as endogenous competitors to mRNA by base pairing with

miRNAs and uncovering mRNAs for effective protein translation; or, serve as precursor RNAs for miRNAs (all mechanisms are detailed in [79–81]).

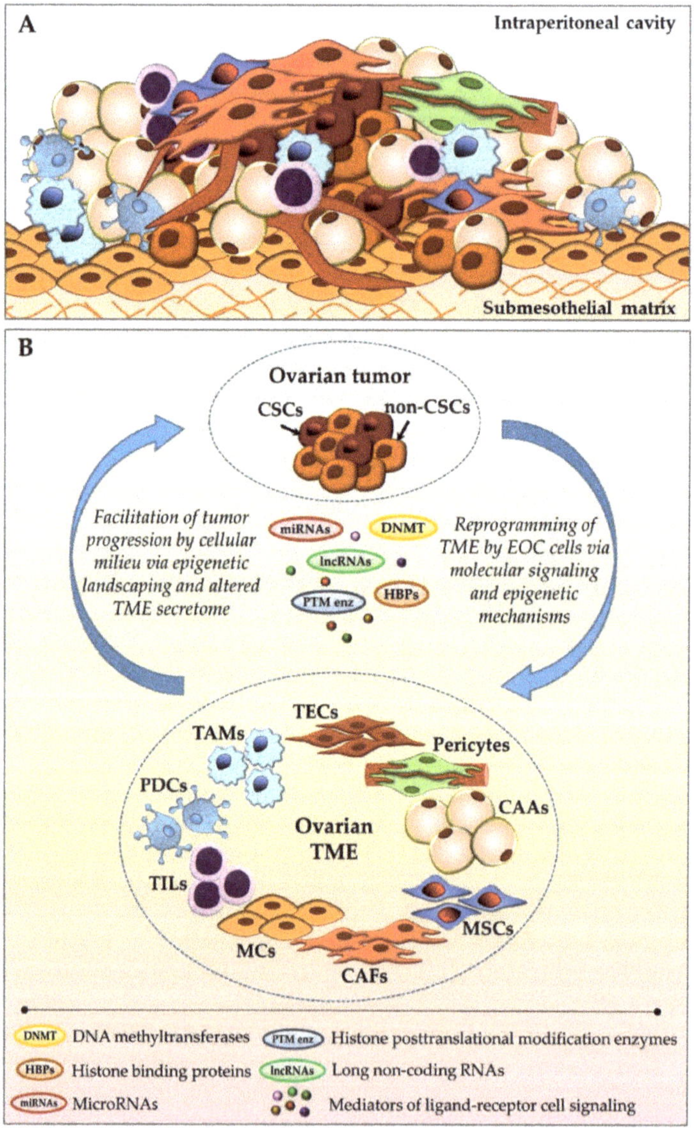

Figure 1. Ovarian tumor-stroma bidirectional crosstalk. (**A**) Schematic representation of cellular diversity within the complex ovarian tumor bulk; and, (**B**) Reciprocal communication between ovarian cancer cells and intraperitoneally residing cancer-associated cellular milieu components via molecular signaling pathways and epigenetic regulation. CAAs—cancer-associated adipocytes; CAFs—cancer-associated fibroblasts; CSCs—cancer stem cells; EOC—epithelial ovarian cancer; MCs—mesothelial cells; MSCs—mesenchymal stem cells; PDCs—plasmacytoid dendritic cells; TAMs—tumor-associated macrophages; TECs—tumor-associated endothelial cells; TILs—tumor-infiltrating lymphocytes; TME—tumor microenvironment (see main text for details).

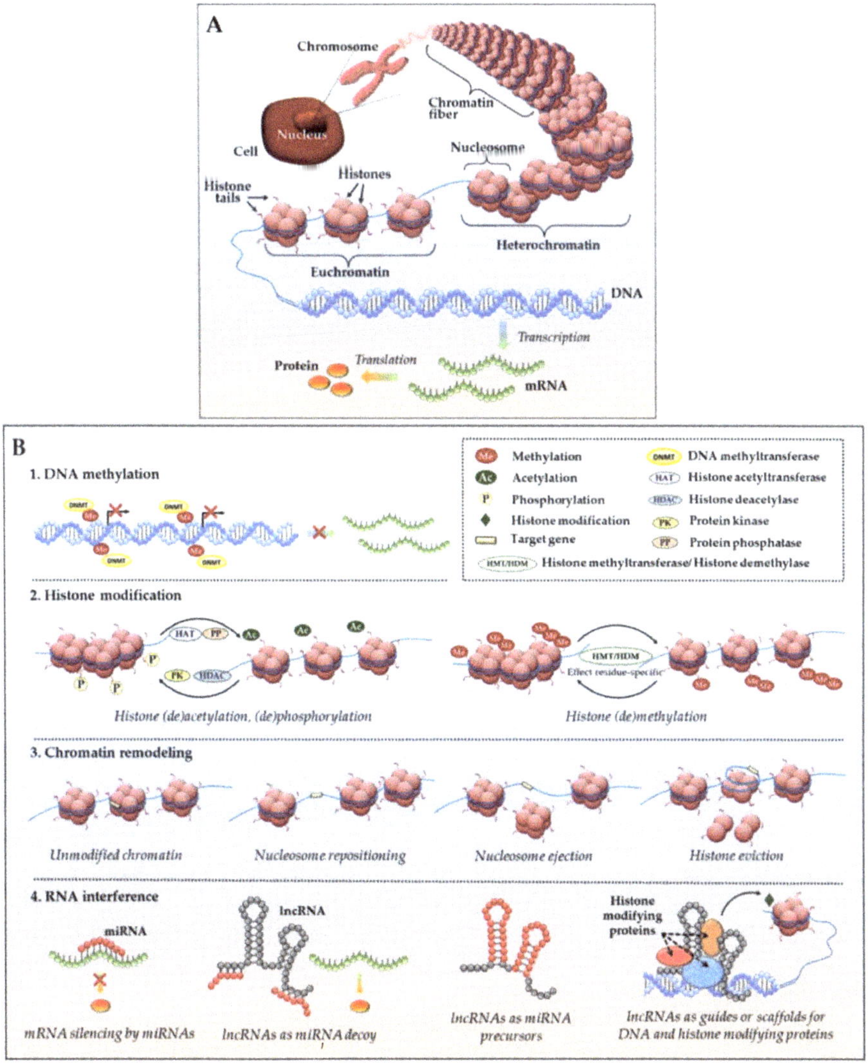

Figure 2. Epigenetic regulation of gene expression. (**A**) DNA packing in a eukaryotic cell: a DNA molecule (chromosome) located inside the cell nucleus is composed of chromatin fibers which are made of nucleosomes—histone octamers wrapped by a DNA helix. Condensed chromatin (heterochromatin) is transcriptionally silent; loosely packed euchromatin allows access to DNA and active transcription into mRNA, followed by translation into protein; (**B**) Major epigenetic mechanisms of gene expression include: (1) DNA methylation; (2) histone modifications, such as histone acetylation, methylation, phosphorylation, etc.; (3) chromatin remodeling, including nucleosome sliding, nucleosome ejection and histone eviction; and, (4) mRNA interference with miRNAs and lncRNAs (see main text for details).

The importance of epigenetic dysregulation for cancer progression cannot be understated, as various aberrant epigenetic modifications trigger activation of oncogenes, repression of tumor suppressor genes, and altered transcription of protein-encoding genes that are collectively responsible for proper expression of signaling proteins that are necessary for physiological cell life span and functions. In the light of significant advances in technical tools and the rapid emergence of large "omics" data, the current knowledge in the field of cancer epigenetics and understanding their translational potential have vastly expanded. Multiple elegant reviews illustrate a wealth of currently available data on cancer epigenomics [46,82–87]. In the context of EOC specifically, several recent works highlight major epigenetic players—DNA methylation [88–90], histone PTMs [88,89], miRNAs [88,89,91,92], lncRNAs [93,94], and discuss current and potential therapeutic implication strategies and challenges [85,95–98].

2. Epigenetic Crosstalk between EOC Cells and TME Cellular Components

The biological importance of TME as a reactive platform orchestrating diverse aspects of tumor initiation, evolvement, metastatic progression, altered immune response, development of therapeutic resilience, and cancer recurrence is unquestionable and continuously reaffirmed, as highlighted in a multitude of studies [99–104]. Given the unique ovarian cancer intraperitoneal TME and the emerging evidence of benefits from epigenetic-targeted therapeutics, we systematize epigenetic dysregulations linked to EOC progression from the perspective of complex reciprocal relationship between ovarian tumor cells and tumor-associated microenvironmental non-malignant cell setting. These interactions will allow for researchers to evaluate, comprehensively, the potential for exploiting specific epigenetic vulnerabilities in both cancerous and TME cells as molecular biomarkers, prognostic indicators, and complementation to conventional chemotherapy and TME-targeted interventions.

2.1. Cancer-Associated Fibroblasts (CAFs)

CAFs are traditionally abundantly present within the tumor stroma (Figure 1A). They serve a tumor-supporting role via remodeling of extracellular matrix, a scaffold for the tumor bulk. CAFs perform endocrine/paracrine communication with the surrounding tumor and other stromal components via the excretion of a variety of growth factors and chemokines, hence contributing to tumor growth, immune response, angiogenesis, chemoresistance, and cell stemness; furthermore, CAFs differentiate into other cell types, as comprehensively discussed elsewhere [105,106]. CAFs exhibit substantial heterogeneity, which is attributed to existence of multiple proposed CAF origins, including reprogramming of the normal resident fibroblasts, conversion from adipocytes, endothelial or epithelial cells, or differentiation of the bone marrow derived mesenchymal or hematopoietic stem cells (summarized in [107]). Gene expression analysis of CAFs purified from different types of cancers, including ovarian carcinomas, revealed normal genotype, and acquisition of somatic mutations is considered to be an extremely infrequent event [108–110]. Hence, epigenetic changes likely play an essential role in CAF regulation.

Studies have reported altered DNA methylation status of genes (either at the gene promoter regions or global) in stromal fibroblasts dissected from various cancer tissues that corresponded to methylation profiles that are found in adjacent malignant cells [111,112]. In turn, as was shown by Mathot et al. for breast cancer [113], malignant cells maintained in the presence of CAF soluble mediators exhibit a wide gene upregulation pattern (372 genes upregulated total in the study) epigenetically modulated via DNA hypermethylation that could be reversed by the DNMT inhibitor decitabine. Pistone and team [114] demonstrated through global transcriptome and DNA methylation analysis that CAF-secreted factors trigger combinatorial DNA hyper/hypomethylation changes, collectively responsible for epithelial-to-mesenchymal transition (EMT) and stemness phenotype in prostate cancer cells. Albrengues and colleagues [115] reported the transformation of fibroblasts into pro-invasive CAFs in response to a cytokine termed leukemia inhibitory factor (LIF), which triggers the continued activation of JAK1/STAT3 signaling pathway via the histone acetylation of STAT3, followed by the activation of DNMT3b, methylation and abrogation of the Src homology region 2 domain-containing phosphatase-1 (SHP-1), and ultimate sustained phosphorylation of JAK1. The process is chaperoned by DNMT1, whereas the inhibition of DNMTs and JAK signaling abolishes the CAF phenotype [115].

Noteworthy, even brief transitory communication between cancer cells and normal fibroblasts leads to an elevated release of transforming growth factor beta 1 (TGF-β1) by activated fibroblasts in response to cancer cell presence [116]. This may be of particular relevance in the context of ovarian cancer epigenetic regulation, as TGF-β is known to catalyze global DNA hypermethylation alterations in EOC cells, promoting EMT and metastasis [117]. Moreover, Cardenas and co-workers found that TGF-β enhances both expression and enzymatic activity of DNMT-1, -3a, and -3b in EOC cells [117]. Importantly, the hypermethylation effect and EMT may be abrogated by treatment with DNMT inhibitors [117]. On the other hand, TGF-β assists in the reprogramming of CAFs from other cell types [118] and it favors further aggressive invasion of HGSOC through TGF-β-induced secretion of prometastatic mediators by CAFs [118,119]. Collectively, these data suggest formation of a constitutively active positive feedback loop in EOC, where initial fibroblast-malignant cell interaction triggers secretion of TGF-β by stromal cells, followed by EMT-associated global epigenetic alterations in EOC cells, as well as promotion of CAF phenotype and secretory activity, which closes the cycle and further aggravates EOC. In this scenario, the combined usage of demethylating agents and drugs targeting TGF-β signaling requires further investigation.

To our knowledge, posttranscriptional histone modifications in EOC-associated CAFs have not been directly addressed, whereas similar studies in the context of other cancer types are very limited. The histone mark H3K27me3 in breast CAFs was shown to be reduced, along with decreased expression of enhancer of zeste homolog 2 (EZH2), a methyltransferase that is responsible for the H3K27me3 histone mark [120], resulting in promotion of cancer cell invasion by CAFs via upregulated thrombospondin type 1 motif 1 [121]. In EOC, however, EZH2 is commonly upregulated [122], and thereby, such a mechanism is not likely. Undoubtedly, studies are needed in elucidating histone PTMs possibly taking place during malignant cell-CAF interplay.

Malignant TME-driven implication of miRNAs in CAF phenotype and functioning has been dissected for various cancer types (breast [123], cervical [124], gastric [125], prostate [126], colorectal [127,128], and bladder [129]), including EOC. In an elegant gain/loss-of-function study employing the co-transfection of multiple miRNA mimics and inhibitors, Mitra et al. [130] discovered a combination of three dysregulated miRNAs (upregulation of miR-155 plus repression of miR-31 and miR-214) that are capable of converting quiescent fibroblasts into ovarian CAFs with extensive expression of chemokines, in particular, direct target of miR-214 CCL-5 (C-C motif ligand 5), CXCL-10 (C-X-C motif ligand 10), CCL-7 and CCL-8, among others, accompanied by the substantial augmentation of ovarian tumor growth. The respective reverse experiments have shown the restoration of wild-type fibroblast phenotype and alleviation of ovarian tumor growth in co-culture [130], suggesting these miRNAs as novel therapeutic targets for halting EOC progression by means of manipulating stromal signals.

Zhao and coworkers [131] have recently discerned prognostically unfavorable upregulation of lncRNA LINC00092 in EOC cells which is induced by high CXCL-14 (C-X-C motif ligand 14) chemokine expression in ovarian CAFs. Expressed LINC00092 binds to a glycolytic enzyme and it facilitates glycolysis in EOC cells, boosts metastatic activity, and reciprocally supports pro-active CAF phenotype [131]. A differential expression analysis of CAFs purified from 67 high grade serous ovarian carcinomas (HGSOC) and 10 normal ovarian fibroblast samples identified 39 divergently expressed lncRNAs (out of 1970 lncRNAs total analyzed) in HGSOC CAFs in comparison with normal fibroblasts [132]. Subsequent context-specific regulatory network construction and pathway analyses have linked the upregulation of seven lncRNAs (FLJ39739, GAS5, H19, LOC100499466, MALAT1, NEAT1, and TUG1) and downregulation of four lncRNAs (CASC2, DLEU2, HCG18, and LOC100133669) to the furtherance of metastasis-associated pathways in HGSOC [132]. Mechanistic insight on the interaction of these differentially expressed lncRNAs in CAFs with proteins/pathways is essential for the further translation of these findings into diagnostic and/or therapeutic strategies.

2.2. Cancer-Associated Adipocytes (CAAs)

One of the distinguishing characteristics of the ovarian cancer TME is tumor cells that are in close vicinity to adipose tissue at all stages of development and metastasis. Adipose tissue exists at the site of the primary tumor (ovarian fat pad, mesosalpingeal/mesoovarian adipose layers) and around the intra-abdominal cavity in the form of substantial omental adipose nodes, large adipose bundles in the mesentery (both serve as prevalent EOC metastasis locations), as well as smaller fat depots in the parietal and diaphragmatic peritoneum (Figure 1A). It is now commonly accepted that adipose tissue is a highly communicative metabolic and secretory organ. Aside from functioning as energy (lipid) storage, it produces adipokines, metabolic substrates, growth factors, hormones, and immune mediators, and it contains other stromal components, such as fibroblasts, stromal vascular fraction, macrophages and other immune cells, nerve tissue, and extracellular matrix [133]. In presence of malignant setting, normal adipocytes rapidly acquire a highly active phenotype (CAAs) and respond by metabolic and secretory profile changes, causing pro-inflammatory, pro-invasive, proliferative, and radio- and chemoresistance effect on cancer cells [28,134–139].

There is accumulating evidence for tumor-adipocyte epigenetic interactions. During maturation, murine preadipocytes demonstrate substantial upregulation of miR-17-92 cluster, which further stimulates adipocyte differentiation via negative targeting of tumor-suppressor Rb2/p130 and is known to boost cell proliferation in various cancers [140]. In breast cancer (another type of adipose tissue-rich neoplasm), the transition of adipocytes into inflammatory CAAs in the vicinity of malignant milieu is dependent on miRNA regulatory mechanism. In particular, in the presence of breast cancer cells, reprogramming CAA increase expression of miR-5112, which suppresses the translation of Cpeb1, a negative regulator of interleukin (IL)-6. As a result, the pro-inflammatory CAAs exhibit an increased expression of IL-6 and a proliferation-promoting effect on breast cancer cells [141]. During ovarian cancer metastatic progression, omental CAAs and CAFs deliver miR-21-containing exosomes to cancer cells. Mir-21 targets apoptotic protease activating factor 1, inhibits ovarian cancer apoptosis, and confers paclitaxel chemoresistance [142], suggesting stromal-derived miR-21 blockade as a potential therapeutic strategy for metastatic and refractory ovarian cancer. Profiling of miRNAs in tumor interstitial fluid of human breast tumor tissues identified a list of 23 miRNAs that were associated with the presence of adipocytes and immune cells in tissues, suggesting epigenetic tumor-stroma crosstalk [143].

A genome-wide expression profiling of peri-prostate adipose tissue samples taken from prostate cancer patients revealed shifted expression of genes collectively accounting for increased proliferative and anti-apoptotic activity and mitigated immunosurveillance, suggesting that the peri-prostate adipose tissue cultivates prostate cancer development [144]. A related epigenome-wide DNA methylation pilot study of peri-prostate adipose tissue in normal weight and obese prostate cancer patients, performed by the same research group [145], revealed abundant DNA hypermethylation in cancer patients with excessive adiposity, with the epigenetically altered genes contributing to altered fatty acid metabolism, immune perturbations (including those providing tumor immune evasion), pluripotency of stem cells, and other pathways that are advantageous for cancer support. Concordantly, analysis of DNA methylation in omental tissue of obese women revealed significant DNA hypermethylation, whereas subsequent weight loss had a substantial hypomethylating effect [146]. Interestingly, obesity-induced proinflammatory cytokines triggered expression of DNMT1 and methylation of adipokine gene, whereas treatment with the DNMT inhibitor reverted the process in adipocytes [147]. Recent work by Tang et al. [148] demonstrated that treatment of adipocytes alone with the DNMT inhibitor resulted in the re-expression of tumor suppressor genes (e.g., *SUSD2*, *TFP12*, *GREM1*, *TRIM29*), altered expression of EMT mediators (e.g., *CDH1*, *CDH2*, *FN1*, and *SLUG*), and diminished migrative and invasive properties of co-cultured EOC cells. These data provide clear evidence that therapeutic tailoring of epigenetic aberrations in CAAs may, in turn, have anti-metastatic effect on ovarian malignant cell behavior [148].

Taking into consideration that discovery of lncRNAs as a class occurred very recently, there is an ample gap in evidence of their regulatory role in tumor-stroma interplay. However, recent studies report that lncRNAs are capable of mediating the expression of genes that are associated with lipid adipogenesis and metabolism via RNA, DNA or miRNA complementation, or recruitment of proteins involved in modulation of histone markers [149–151]. A recent study including transcriptome profiling of primary brown and white adipocytes, preadipocytes, and cultured adipocytes reported a list of 175 lncRNAs that are specifically regulated during adipogenesis [152]. Given a constantly increasing number of lncRNAs with documented significance in the development of a variety of cancers, including ovarian [94,153,154], investigation of lncRNA-associated epigenetic modifications in the context of interrelationship between EOC and CAAs will undoubtedly be a very fruitful and promising direction for further basic/translational research.

2.3. Mesenchymal Stem Cells (MSCs)

MSCs encompass a diverse multipotent cell subgroup that was recruited to the tumor stroma from such sources as bone marrow, adipose tissue, umbilical cord, endometrium (menstrual blood), pericytes, as well as other organ-specific locations. The existence of MSCs has been reported for the majority of organs and tissues, including ovary (Figure 1A) [155]. Due to their capacity to differentiate into other active pro-cancerous stromal components (such as CAFs and CAAs) as well as sustain a cancer stem cell (CSC) population, MSCs are strongly associated with cancer progression yet retain a normal (non-malignant) genotype. In particular, ovarian carcinoma-associated MSCs (CA-MSCs) exhibit multipotent potential and strong EOC growth-permissive and stemness-promoting properties, increasing resistance to platinum-based chemotherapy, providing tumor stromal support and neovascularization [156–161]. Contradictory to those reports are studies demonstrating tumor-restricting effect of human MSCs on EOC cells via stimulation of EOC cell cycle arrest, enhanced apoptosis, altered mitochondria membrane potential and suppressed neoangiogenesis [162], and the inhibition of cisplatin-resistant ovarian carcinoma xenograft growth [163]. Given a wealth of controversial data on the assisting vs restricting role of MSCs on EOC development and the extensive interest in therapeutic applications of MSCs as vehicles for EOC-targeted drug delivery due to their high tumor site tropism [164–168], further investigations to determine the underlying mechanisms of MSC-EOC interactions are clearly necessary.

While the overwhelming majority of research work is focused on the elucidation of cell signaling pathway implications, the epigenetic interplay between MSCs and EOC remains largely unexplored. While considering the well accepted role of epigenetic aberrations in CSC reprogramming (comprehensively outlined in [169]), emerging evidence on efficient epigenetic transformation of MSCs into CSCs via tailoring chromatin remodeling (imbalanced DNA methylation of cancer-implicated genes, application/loss of histone marks, etc.) [170,171], and epigenetic modulation of MSC differentiation into diverse stromal cell lineages [172–174], the involvement of EOC-mediated epigenetic factors on MSC phenotype and functioning, and the subsequent reciprocal effect of reprogrammed MSCs on ovarian carcinoma, is highly plausible. Toward these possibilities, Reza et al. [175] reported the anti-proliferative and pro-apoptotic effect of adipose-derived MSC exosomes on ovarian cancer A2780 and SKOV3 cells; subsequent exosomal RNA sequencing identified a list of enriched miRNAs targeting EOC cell survival pathway genes [175]. MiRNA expression profiling in aging MSCs revealed altered expression of two miRNAs, miR-638 and miR-572, both of which have been reported to be upregulated in ovarian carcinoma; however, distinct studies display controversial findings on the impact of these miRNAs on EOC [176–179]. Among other non-coding RNAs, abundant in MSC-derived extracellular vesicles and associated with ovarian cancer development are miR-21, miR-92a and miR-221 [180]. Alternatively, Ho and colleagues [181] demonstrated, that ovarian cancer stromal progenitor cells isolated from tumor tissues and ascitic effusions of EOC patients displayed 40 hypermethylated tumor suppressor genes (TSGs) (with DLC1, RASS382, CDH13, BRCA1, TIMP3, HIN-1, ESR1, CDKN2A, CCND2, CDKN2B, as most frequently hypermethylated and correlating with validated mRNA expression decrease in DLC1, RASSF1A, CCND2, and CDKN2B) in comparison with matching patient ovarian cancer cells and were capable of promoting tumor growth in vivo when co-injected with SKOV3 cells. Most importantly, treatment with the hypomethylating agent (HMA) 5-aza-2-deoxycytidine (decitabine or DAC, a desoxyribonucloside that exclusively incorporates into DNA and inhibits DNMTs by irreversible binding of the latter [182]) resulted in efficient demethylation of TSGs in stromal progenitor cells, repressed their self-renewal and growth and mitigated proliferation of ovarian tumor cells [181]. Collectively, these studies suggest a direct or indirect epigenetic relationship between MSCs and ovarian tumor cells and warrant additional research that can potentially lead to the identification of novel therapeutic targets for ovarian carcinomas.

2.4. Tumor-Associated Endothelial Cells (TECs)

TECs refer to endothelial cells that line the inner walls of newly formed blood vessels in tumors. TECs support blood flow, tumor tissue trophics, and accelerate tumor progression. TECs are genetically non-malignant cells, however, they differ from normal endothelial cells by exhibiting cytogenetic abnormalities [183], distorted morphology [184], and altered molecular profiles [185], as well as improper functional characteristics. TECs form a disorganized, excessively sprouted, branched, fragile, and gap-prone endothelial network, which allows for chaotic non-laminar blood flow, increased vascular permeability and escape of primary tumor cells into blood circulation. TECs help circulating cancer cells overcome anoikis and escort their movement to metastatic niches [186]. Finally, TECs are able to develop taxol resistance in the presence of surrounding malignant milieu [187].

TECs exhibit considerable heterogeneity (reviewed in [188]), and diverse features of TECs in various tumors are attributed to the specific malignant settings [189]. In the context of epigenetic regulation, Maishi et al. [190] implanted TECs that were isolated from the high metastatic (HM) melanomas into low metastatic (LM) melanomas and achieved metastatic enhancement as a result of elevated levels of proteoglycan biglycan, attracting tumor cells to intravasate. Strikingly, upregulation of biglycan was due to DNA demethylation of its promoter region in HM-TECs as opposed to normal endothelial cells, LM-TECs, and tumor cells [190]. Furthermore, exposure of LM-TECs to conditioned medium from HM-tumors and treatment with the HMA decitabine proved the upregulation of biglycan in LM-TECs via DNA demethylation triggered in the presence of HM-tumor [190]. The study elegantly reaffirmed epigenetically-mediated bidirectional communication between tumor cells and TECs, and further suggested epigenetic perturbations as attractive targets to manipulate neoangiogenesis and tumor metastasis.

Epigenetic regulation of TECs has been reported. DNMT and HDAC inhibitors efficiently mitigated TEC growth in vitro and in vivo [191,192]. Moreover, epigenetic modifications were shown to regulate TEC-mediated immune response [193]. Downregulation of intercellular adhesion molecule-1 (ICAM-1) in TECs due to promoter histone H3 deacetylation and the loss of histone H3 Lysine^4 methylation was reported, and re-expression of ICAM-1 in TECs by DNMT and HDAC inhibitors successfully restored leukocyte-TEC adhesion and leukocyte infiltration of vessel walls in vitro and in vivo, respectively [193]. In a separate study, the same group identified additional anti-angiogenesis genes (including clusterin, fibrillin 1, and quiescin Q6) downregulated in normal endothelial cells subjected to a tumor-mimicking setting (conditioned medium) via epigenetic silencing by promoter histone H3 deacetylation and loss of histone H3 Lysine^4 methylation [194]. Furthermore, subsequent treatment with DNMT and HDAC inhibitors reversed the gene silencing effects [194].

An influential study assessing the role of EOC on TECs was conducted by Sood group [195], who performed a comparative genome-wide gene expression analysis of endothelial cells from five different normal ovarian tissues and TECs isolated from 10 invasive ovarian carcinomas and revealed a list of 400 differentially expressed genes, among which EZH2 was significantly upregulated. EZH2, a histone-lysine N-methyltransferase, is frequently overexpressed in ovarian tumor tissues and is a well-established epigenetic mediator (stimulator) of EOC tumor growth, invasion, metastasis, neoangiogenesis, platinum-resistance and ovarian cancer stem cell renewal through transcriptional repression of signaling pathways, as well as direct/indirect interaction with multiple miRNAs and lncRNAs (reviewed in [196]). Overexpression of EZH2 in TECs by the surrounding EOC setting occurred in a VEGF-mediated paracrine manner, which in turn promotes EZH2 direct binding, methylation, and transcriptional repression of vasohibin-1 (a selective negative regulator of TEC migration, proliferation, and vessel tube formation), stimulating further angiogenesis [197]. Remarkably, the silencing of the EZH2 gene in vitro and in vivo (via systemic siRNA nanoparticle delivery in both EOC cells and TECs in mouse EOC xenografts) substantially increased vasohibin-1 expression and reduced neoangiogenesis and tumor growth [197], underlining the additional therapeutic potential of concomitant EZH2 targeting in TECs and EOCs. Noteworthy, the expression of EZH2 in ovarian cancers has been shown to be regulated by different non-coding RNAs, for example miR-298 [198] and lncRNA HOTAIR (HOX transcript antisense RNA) [199], providing additional druggable epigenetic targets in EOC cells and TECs in order to suppress neoangiogenesis and ovarian tumor progression.

2.5. Pericytes

Another type of vessel-associated cells, pericytes (Figure 1), play a key role in normal vasculature development, including neoangiogenesis, and in cancer contribute to tumor development and metastasis. These perivascular cells reside in the microvessel walls, immediately beyond the basement membrane, and provide physical support to endothelial cells, control integrity of the vessel endothelial layer, vascular permeability, and blood flow [200], serve as an origin of mesenchymal stem cells (MSCs) for some human tissues [201] and as multipotent precursors for other types of cells [202], and exert immune mediators [203]. During cancer progression, pericytes essentially contribute to rapid neoangiogenesis and tumor growth, expansion of the CSC population, on one hand, but also strengthen immune defense against tumor invading and restrict metastatic seeding, on the other hand, as comprehensively outlined in [204,205]. In EOC patients, high pericyte score (carrying a pericyte gene signature) was shown to be an unequivocal predictor of earlier relapse and poor prognosis [206]. Dual targeting of pericytes in combination with endothelial cells [207], as well as the disruption of connections between pericytes and endothelial cells via insertion of N-cadherin blocking peptides [208], are potential anti-angiogenesis therapeutic strategies in EOC.

Intriguingly, in a HGSOC xenograft model, injection of pericytes concomitantly with EOC cells amplified tumor growth and metastasis rate without altering tumor vasculature, highlighting a direct impact of perycites on EOC cells, independent of neoangiogenesis [206]. No studies have adequately described epigenetic tumor-perycite crosstalk in ovarian carcinomas. However, proximity of the malignant (glioblastoma) setting [209], hypoxic conditions [210] and inflammatory stimulation [211] trigger an epigenetically mediated (through altered non-coding RNA) pericyte response. Though these findings are quite limited, they raise the possibility of dynamic epigenetic cooperation between pericytes and EOC cells (and other cancer-associated stromal cells) within a highly reactive, proinfammatory, and hypoxia-prone ascitic TME. In addition, a very interesting preclinical study by Kratzsch et al. [212] assessed the effect of 5-azacitidine (AZA; a ribonucleoside that is capable to incorporate into cellular RNA and DNA and acts as a HMA interfering with the DNMT activity [182,213]), valproic acid (HDAC inhibitor), temozolomide (a standard DNA alkylating chemotherapeutic positive control), and a bevacizumab (an angiogenesis inhibitor targeted therapy positive control) on the tumor growth and neovasculature status in the murine glioblastoma multiforme models. Strikingly, besides the suppressive effect on glioma tumor growth, the HMA also resulted in a notable antiangiogenic effect via the substantial diminishing of endothelial cells and the decreasing number of pericytes [212]. Contrarily, HDAC inhibitor valproic acid had no mitigating effect either on tumor growth or on the vasculature [212]. These results suggest concordant epigenetic expression changes taking place within the malignant tissue and the adjacent TME cells as a result of their mutual communication (as both tumorous and angiogenesis cells selectively responded to the same epigenetic-tailoring drug, but not to the other). In dissonance with that is work by Karén et al. [214], who reported successful inhibition of pericyte differentiation, migration, and proliferation in response to HDAC inhibitors in vitro, and HDAC suppression stimulated expression of genes regulating vessel stabilization and maturation in pericytes. Collectively, these data highlight translational potential of cancer-prone epigenetic signatures not only in the neoplastic component *per se*, but in the surrounding stroma as well.

2.6. Tumor-Associated Macrophages (TAMs)

TAMs within malignant stroma (Figure 1A) are considered to be a fundamental immune cell subpopulation responsible for cancer-associated inflammation, matrix remodeling, tumor immune escape, growth, invasion, angiogenesis, metastasis, cancer cell stemness, and drug resistance. Macrophage diversity and ability to transfer between M1 (classic, host immune defense activating, tumoricidal) and M2 (alternative, immunosuppressive, pro-tumorigenic) phenotypes are defined by the distinct tumor microenvironments [215]. Several recent reviews [216,217] elegantly describe bidirectional TAM interactions with cancer cells, other immune cell subpopulations and non-malignant stromal components, such as CAFs, TECs, B cells, eosinophils, basophils, dendritic cells, natural killer cells, and others.

In addition to ample evidence on receptor-ligand signaling interrelationship between EOC and TAMs [218–223], a better understanding of their epigenetic cooperation has begun to accrue. Ying et al. [224] reported initiation of M2 polarization and tumor-promoting capabilities in ovarian TAMs by EOC-released exosomal miR-222-3p through SOCS3/STAT3 pathway. Similarly, macrophage M2 phenotype shift was induced by exosomal miR-940 released from hypoxic epithelial ovarian tumors [225]. Importantly, TAM-secreted exosomes suppress endothelial cell migration by targeting miR-146b-5p/TRAF6/NF-κB/MMP2 pathway, whereas the EOC-released exosomal delivery of lncRNAs to the site efficiently restores the endothelial cell movement [226]. Alternatively, Hu et al. [227] observed altered expression of 19 miRNAs in TAMs exposed to tumor necrosis factor-related inducer of apoptosis (TWEAK; commonly expressed by immune cells, such as dendritic cells and natural killer cells) and demonstrated that the exosomal transportation of overexpressed miRNAs, and in particular, miR-7, from TAMs to EOC cells significantly repressed their metastatic activity in vitro and in vivo via repression of EGFR/AKT/ERK1/2 pathway. Furthermore, the insertion of antagomiR-7 into TAMs repressed levels of miR-7 in TAMs, in released exosome vesicles and in the recipient ovarian malignant cells, and stimulated EOC metastasis [227].

Chromatin remodeling-related epigenetic modifications resulting from TAM-EOC interaction remain largely unknown. However, as reported for gastric cancer, TAMs are capable of silencing TSG gelsolin in cancer cells by increased DNMT1 expression and DNA methylation of gelsolin promoter via activation of CCL5/CCR5/STAT3 signaling [228]. Most importantly, either suppressing DNMT enzyme activity, or treatment with demethylation agent, or interfering with CCL5/CCR5/STAT3 pathway led to decreased in vivo tumor growth, suggesting several possible routes of epigenetic inhibition of gastric cancer development [228]. In oral cancer, interferon-γ mRNA expression in TAMs was substantially decreased in comparison with normal or benign oral tumor specimens as a result of promoter region methylation, with the level of methylation strongly correlating with the clinical stage, histopathology grade, and primary tumor scale [229]. Promoter hypermethylation and downregulation of the follistatin like-1 (FSTL-1) glycoprotein was associated with metastatic activity of nasopharyngeal carcinomas and dysfunction of macrophages, whereas treatment with recombinant human FSTL-1 protein elevated IL-1β and tumor necrosis factor-α in TAMs and repressed cancer cell immune evasion [230]. Using a panel of distinct cancer type cells, including cervical, hepatocellular, epidermoid carcinomas, glioblastoma, rabdomyosarcoma, and murine melanoma, Osawa and co-authors [231] demonstrated that cancer cell hypoxia and nutrient starvation lead to activation of histone demethylase JMJD1A (Jumonji domain-containing 1A), followed by increased AKT phosphorylation, cancer cell metastatic properties, increased angiogenesis, and infiltration of macrophages into cervical cancer and muscle sarcoma tissues in vivo. Remarkably, JMJD1A repression mitigated tumor progression through decreased neovascularization and alleviated TAM infiltration, and importantly, enhanced the anti-tumor effect of anti-angiogenesis agents bevacizumab and sunitinib [231]. Ishii et al. [232] described M2 macrophage polarization via reciprocal epigenetic changes in histone H3 lysine^{4} and histone H3 lysine^{27} methylation through STAT6 mediation, which collectively lead to transcriptional activation of specific M2 maker genes. Altogether, these findings underline the immediate importance of continued research accessing TAM-modulated epigenetic changes in EOC, as they may reveal novel diagnostic and therapeutic approaches to mitigate TAM-associated pro-tumoral inflammation and cancer progression, and boost host immune defense mechanisms.

2.7. Tumor-Infiltrating Lymphocytes (TILs)

TILs are collectively represented by varying amounts of T and B lymphocytes recruited from the circulatory system to the tumor site to fulfill the host immune response function. In EOC, TILs may be present in primary tumors and advanced metastatic lesions, both within malignant (intratumoral) bulk and stromal compartment, and differential representation of certain TIL subsets (helpers, killers, regulatory/suppressors, effectors, memory cells, and such) depends on the disease stage, therapeutic management, chemotherapy response status, and possesses prognostic significance (Figure 1A) [29,233–238].

The extensive evidence on signaling molecules and pathways implicated in TIL functioning and ovarian cancer prognosis, as well as current attempts to differentially manipulate TIL subsets and signaling networks towards boosting adequate anti-tumor immunity, have been comprehensively summarized by Santoiemma and Powell in a recent review [239]. The epigenetic regulation of TIL functioning in ovarian cancer and epigenetic-based immunotherapeutic strategies are emerging areas of interest. Sehouli et al. [240] proposed the concept of "epigenetic immunophenotyping" (identification of certain epigenetic marks) of both overall and specialized TIL populations. By using matching healthy and cancer tissues, including ovarian, they established a strong correlation between epigenetics and cancer prognosis [240]. In support of this concept, treatment with DNMT inhibitor 5-azacitidine led to the substantial enrichment for immunomodulatory pathways (interferon signaling, antigen processing and presentation, and cytokines/chemokines) in ovarian and other cancers [241]. Similarly, with the use of global gene expression profiling of EOC treated with DNMT inhibitor decitabine, Wang et al. [242] discovered prominent upregulation of immunoregulatory genes cohort in decitabine-exposed malignant cells. The group further showed that decitabine treatment stimulated TIL infiltration and anti-tumor function in both subcutaneous and intraperitoneal syngeneic murine ovarian cancer models [242]. Moreover, combining a DNMT inhibitor with the standard immune checkpoint blockade antibody stimulated conversion of naïve T cells into T effectors and contributed to better mouse survival [242]. Based on this work, Stone et al. [243] further demonstrated that the activation of type I interferon signaling in response to DNMT inhibitor 5-azacytidine was a key requirement for efficient stimulation of CD45^{+} (leukocyte common antigen) immune cells, CD8^{+} cytotoxic T cells, natural killer cells, restriction of macrophages, and myeloid-derived suppressor cells in the ovarian TME in vivo, prompting the inhibition of tumor growth and increased survival. In support, Adair and Hogan [244] demonstrated the enhanced expression of cancer-testis antigens and class I major histocompatibility complex (MHC)-encoded molecules in EOC cells that were treated with DNMT inhibitors and subsequent infiltration of antigen-reactive CD8^{+} cytotoxic T cells to EOC. Consistent with that, seminal work by Chiappinelli and co-authors [245] established that DNMT inhibitors activate interferon signaling, TIL infiltration, and EOC cell death via the upregulation of viral defense pathway, as hypomethylation of endogenous retrovirus (ERV) genes leads to upregulated viral defense gene expression and boosts immune response.

In the context of non-coding RNA interference, two prognostic miRNA signatures—malignancy signature and immunological signature—have been recently identified in advanced EOC [246]. Briefly, using integrative mRNA/miRNA co-expression analysis of primary tumor tissues from advanced

EOC patients, two modules were established: a malignancy module (let-7f, let-7g, miR-106a, miR-17, miR200c, miR-26a, miR-26b, and miR-328), which was associated with the more aggressive EOC growth and an immunological module (miR-197, miR-22, miR-22#, miR-28, miR-339–5p, miR-340#, miR-628–5p, miR-629, miR-661, and miR-98), which strongly correlated with intratumoral infiltration by T cells, natural killer cells, cytotoxic lymphocytes, and macrophages [246]. These microRNA signatures may serve as prognostic and treatment efficacy biomarkers as well as potential targets for epigenetic-based immunotherapy of advanced ovarian cancer [246].

2.8. Plasmacytoid Dendritic Cells (PDCs)

PDCs constitute a rare, yet critically important and highly specialized immune cell subpopulation whose main role in immune surveillance is rapid recognition of foreign pathogens via selectively expressed toll-like receptors and the immediate activation of both innate and acquired immune systems (Figure 1A) [247]. Incessant stimulation of PDCs by self-DNA (a situation when PDCs, which do not normally react to inert DNA of organism's cells, become continually activated by the altered DNA of transformed cells) is a characteristic of a variety of neoplasms, including ovarian carcinomas [248–250]. Accumulation of PDCs within the epithelial ovarian tumor bulk promotes vasculogenesis [251,252] and immune tolerance [253], and it is associated with unfavorable disease prognosis [254]. Importantly, the elimination of immature PDCs in mice bearing ovarian tumors via the targeting of specific markers led to vascular ablation, substantially reduced tumor growth, and triggered anti-tumor immune response and tumor re-sensitization to chemotherapeutic agents [255].

Epigenetic regulation of hematopoietic stem cell lineage commitment and subsequent differentiation of dendritic cell precursors into specific dendritic subtypes (including PDCs) and interferon response in the context of chromatin remodeling, miRNA and lncRNAs interference are well documented [256–260] and not focused on herein. Meanwhile to our knowledge, the epigenetic impact on and by PDCs in relation to EOC remains largely undefined. A single study demonstrated that exogenous supplementation of miR-155 (which is downregulated in tumor-associated PDCs) via targeted nanoparticle delivery to EOC-associated PDCs results in the genome-wide silencing of multiple immunosuppressive mediators, following stimulation of immune defense mechanisms and abolishment of ovarian carcinoma progression in vivo [261]. Importantly, a comparison of methylome profiles of dendritic cells and myeloid-derived suppressor cells (MDSCs, an immune cell subtype bearing same progenitor with PDCs) grown under tumor-associated conditions (exposure to prostaglandin E2 or malignant conditioned medium) revealed MDSC-specific DNMT3A enzyme activation, DNA methylation signature, and silencing of immunogenic-associated genes analogous to changes that were observed in primary MDSCs dissected from ovarian cancer patients [262]. Moreover, the suppression of DNMT3A activity abrogates MDSC-specific hypermethylation and MDSC immunosuppressive properties [262]. These findings support the epigenetic tuning of immune surveillance by ovarian TME and underline the necessity for further investigations in this direction.

2.9. Mesothelial Cells (MCs)

MCs are simple squamous epithelial cells that line the intra-abdominal cavity as an upper single layer (mesothelium) of the peritoneum, immediately supported by the basal membrane, underneath which is the collagen-rich extracellular matrix (Figure 1A). MCs function as an active physical barrier against the invasion of ovarian neoplastic cells into submesothelial matrix and metastasis formation, and mesothelial disruption (clearance) imposes a higher level of EOC peritoneal adhesion and dispersal [263,264]. Besides, MCs serve a secretory role by releasing bioactive soluble factors into malignant ascites, such as phospholipid lysophosphatidic acid [265] and VEGF [266], which are known to enhance EOC tumor growth, metastasis, apoptotic resistance, and angiogenesis [265,267–269]. MCs also produce large amounts of tumor-associated immunostimulatory protein K90 [270], associated with the poor prognosis in ovarian and breast cancers [271,272] and implicated in drug resistance [273].

The bidirectional interplay between EOC cells and MCs has been recently assessed in a study by Matte et al. [274], who reported global gene expression changes (649 altered genes primarily related to cell growth and proliferation, apoptosis, cell cycle, cell assembly, and organization) in MCs exposed to EOC ascitic effusions [274]; reciprocally, EOC cells exposed to EOC-associated MC setting exhibited an enhanced resistance to induced apoptosis [274]. However, epigenetic regulation underlying these gene expression alterations are unknown. Clearly, DNA methylation, histone modification profiling, and non-coding RNA profiling are further key steps necessary to distinguish specific epigenetic mechanisms responsible for gene expression switches occurring in EOC cells and cancer-affected mesothelium because of their interdependence. Gaining better understanding of these events may provide valuable insight on cellular adjustment during tumor-mesothelial adhesion step and suggest novel approaches to block metastatic seeding of the peritoneum.

3. Ovarian TME: Potential Epigenomic-Based Therapeutic Strategies

The growing body of information on epigenetic control of ovarian cancer metastatic advancement and acquisition of resistance to standard-of-care chemotherapy make epigenomic alterations attractive prognostic markers and druggable targets to improve therapeutic outcomes in EOC patients especially with platinum-resistant and recurrent disease. Our group and others are actively exploiting DNA hypermethylation, modifications of histone marks, and aberrant expression of non-coding RNAs in malignant EOC cells in an attempt to treat EOC, re-sensitize ovarian tumors to conventional chemotherapy, and prevent/delay disease recurrence (Figure 3A). We and others aim to discern global DNA methylation patterns and methylation states of specific candidate genes to prognosticate and improve patient response to chemotherapies with hypomethylating agents (HMAs) [15,88,275–277]. We recently demonstrated successful in vitro and in vivo re-sensitization of multiple cisplatin-resistant ovarian cancer cell lines using a novel small-molecule DNMT inhibitor guadecitabine (SGI-110) [275]. Efficient epigenetic targeting of ovarian CSC population by the HMA with induction of differentiation, restriction of tumor-initiating capacity, and re-sensitization to platinum was also reported [278]. In a recently completed phase I clinical trial, guadecitabine "priming" in combination with carboplatin induced clinical response in patients with platinum-resistant, recurrent HGSOC [279]. Similarly, HDAC inhibitors are being examined in advanced, refractory, and recurrent ovarian cancer [280–282]. However, despite some promising results, cell adaptation to HDAC inhibitor-mediated epigenomic disruption has been frequently observed [283–285]. Finally, the recent identification of large miRNA and lncRNA classes as epigenetic regulators of gene expression granted new opportunities for miRNA- and lncRNA-employed prognostic evaluation and targeting of EOC [92–94,153,286].

Continuous accumulation of knowledge on epigenomic perturbations in ovarian TME cellular compartments in the context of their communication with EOC cells holds great translational promise. The fact that non-malignant stromal cells lack genetic mutations and acquire potentially reversible tumor-specific molecular traits and cancer-indulgent functions via epigenetic reprogramming by EOC cells suggests the epigenetic tailoring of ovarian TME as a promising strategy in ovarian cancer management. Given the varying stroma representation in ovarian malignancies (stromal compartment may range from 7 to 83% in different ovarian tumors [287]) and high level of TME cell heterogeneity, it is pivotal to correctly evaluate TME composition within the tumor bulk in each disease case. Prevalence of one or another type of TME cells may dictate activation of certain mechanistic pathways, help to dissect "first choice" epigenetic therapy candidates, and possibly lead to the identification of epigenetic alterations in the TME that are specific for each patient

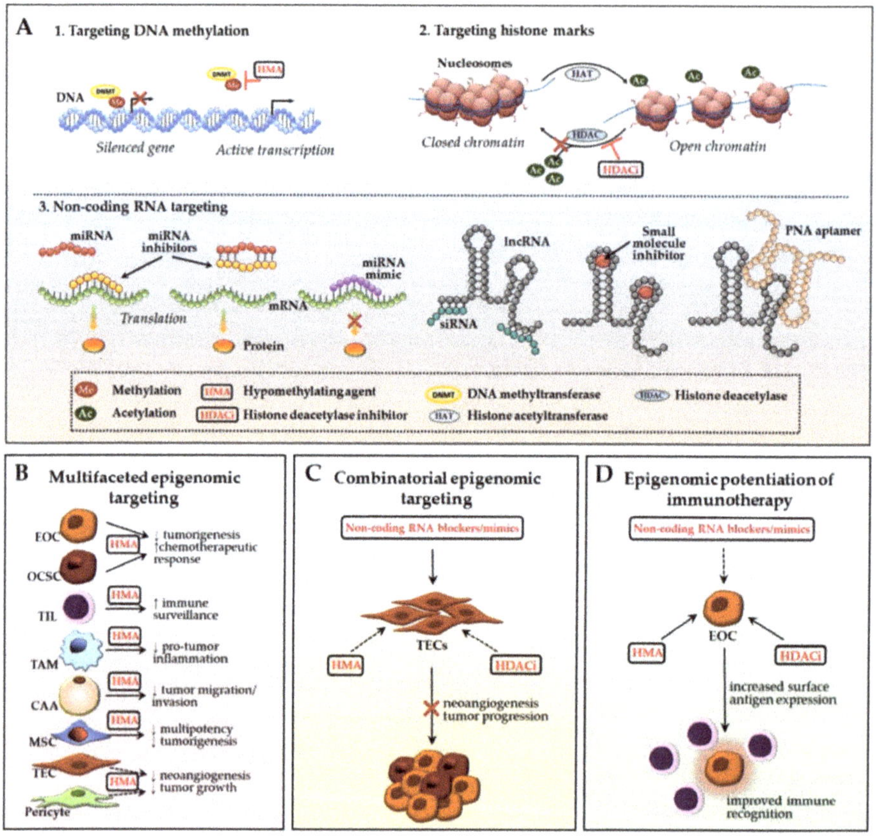

Figure 3. Ovarian TME: potential epigenomic-based therapeutic strategies. (**A**) Main epigenetic therapy mechanisms include: (1) DNA demethylation with hypomethylating agents (HMAs) to re-activate transcription of silenced genes; (2) restoration of open, transcription-permissive euchromatin state with histone deacetylase (HDAC) inhibitors preventing removal of acetyl groups from the histone tails; (3) targeting non-coding RNA-mediated epigenetic perturbations via delivery of exogenous miRNA inhibitors (antagomirs/blockmirs/sponges) or mimics, lncRNA siRNAs, small molecules or peptide nucleic acid aptamers; (**B**) Multifaceted epigenomic targeting approach suggests simultaneous targeting of several malignant and TME cell types using one epigenetic drug; (**C**) Combinatorial epigenomic targeting involves employment of multiple epigenetic mechanisms to revert cancer-associated phenotype in TME cells and inhibit their tumor-promoting effect on EOC cells; (**D**) Epigenomic potentiation of immunotherapy involves epigenetic stimulation of cancer cell immune gene/pathway representation which allows for increased immune surveillance efficacy. CAA—cancer-associated adipocyte; HMA—hypomethylating agent; HDACi—histone deacetylase inhibitor; EOC—epithelial ovarian cancer cell; MSC—mesenchymal stem cell; OCSC—ovarian cancer stem cell; PNA—peptide nucleic acid; TAM—tumor-associated macrophage; TEC—tumor-associated endothelial cell; TIL—tumor-infiltrating lymphocyte. Black arrows represent interactions reported in EOC; dashed arrows designate patterns that were observed in other cancer types and may potentially be applicable towards ovarian cancer (see main text for details).

Treatment of different stromal components with HMAs demonstrated tumor-restrictive potential. In particular, the exposure of adipocytes to guadecitabine attenuated HGSOC cells metastasis-associated behaviors [148]. HMA decitabine effectively inhibited the multipotent and tumor-promoting capabilities of ovarian cancer-associated MSCs and led to decrease in cancer cell proliferation [181]. While not specified for EOC, treatment with the DNMT inhibitor azacitidine showed tumor-inhibiting and anti-angiogenesis effect with a decreased amount of TECs and pericytes in other cancer types [212]. HMA therapy also showed change in immunoregulatory cell response within the tumor bulk with increased infiltration of tumor-restrictive TIL subsets and suppression of TAM-mediated tumor progression in EOC and other cancers [228,242–244]. Such multifaceted epigenomic targeting of multiple stromal cell types in parallel with primary EOC cell targeting through assessing the same epigenetic mechanism (Figure 3B) may improve efficacy, while also diminishing the toxic effects that are associated with multi-drug treatment regimens.

Alternatively, similarly to an improved anticancer effect of a DNMT/HDAC inhibitor combination on ovarian tumors [288] and other malignancies [284,289], the combinatorial epigenomic targeting of ovarian TME components using several epigenomic approaches is plausible (Figure 3C). For example, anti-angiogenesis and anti-proliferative effect of simultaneous utilization of an HDAC inhibitor to target pro-tumorigenic histone marks and an HMA to demethylate DNA promoter CpG islands in colon tumor-conditioned TECs was reported [194]. Analogous drug combination may have therapeutic prospective to block ovarian cancer neovascularization. Moreover, systemic nanoparticle delivery of siRNAs or antagomirs targeting tumor-assisting non-coding RNAs (such as EZH2, HOTAIR, and miR-298) into ovarian TECs may provide an additional option for combinatorial blocking of EOC tumor growth and angiogenesis [197–199]. Current data provide strong rationale for combining HDAC and DNMT inhibitors to affect EOC-associated pericytes [212,214].

Given that CAF reprogramming is shown to repress ovarian tumor advancement [290], prevention or reversion of CAF phenotype may potentially be accomplished through combined treatment with a HMA [113,115] and a CAF-delivered mixture of miRNA miR-31/miR-214 mimics plus miR-155 antagonist to interfere with the EOC-associated CAF miRNA signature [130]. The recent discovery of a number of altered lncRNAs in HGSOC CAFs [132] may suggest novel options for restricting CAF-promoted EOC growth via targeted siRNA injections. Similarly, non-coding RNA signatures recently identified in EOC-related immune cell populations (discussed earlier in this review) warrant further siRNA/miRNAs targeted conveyance studies in this direction. Importantly, in contrast to CAFs, miR-155 was downregulated in EOC-associated PDCs and its exogenous delivery to these cells, in turn, boosted anti-tumor immune defense response and EOC suppression [261]. Taking into consideration such drastically opposite effects of the same epigenomic regulators on different ovarian TME cell compartments, the development of methodologies for highly specific nanoparticle transportation and precise cell type targeting is pivotal. Several studies investigating the potential of exosome-facilitated epigenomic (miRNA) cargo delivery for EOC treatment are also in progress [291].

Exploiting host immune system in managing ovarian cancer growth and metastatic spread has emerged as another key research and therapeutic direction [292–296]. A number of comprehensive analyses of epigenomic strategies employed to advance cancer immunotherapies [297–299] provide detailed illustration on how currently available epigenetic drugs may potentially contribute to modulation of cancer immunosurveillance and anti-tumor responses (Figure 3D). In ovarian cancer in particular, Stone et al. [243] demonstrated while using preclinical in vivo models that combining DNMT inhibitor and HDAC inhibitor with immune checkpoint blockade resulted in the most notable anti-cancer effect and survival due to activation of type I interferon signaling, more efficient recruitment of anti-tumor TIL subsets and the restriction of tumor-indulgent TAMs and myeloid-derived suppressor cells to the ovarian TME. Alternatively, Song and coworkers [300] reported HDAC inhibitor-stimulated increase of NKG2D ligand expression on the surface of ovarian cancer cells, which allowed for better recognition and killing of the latter by engineered NGK2DL-specific chimeric antigen receptor T cells. Finally, Wargo and colleagues [301] reported improved recognition of tumor cells by engineered peripheral blood lymphocytes in response to increased cancer-testis antigen NY-ESO-1 expression (typically expressed in the majority of epithelial tumors, including ovarian [302]), catalyzed by DNMT inhibitor (alone or combined with HDAC inhibitor). Furthermore, the results of a recent phase I clinical trial by Odunsi et al. [303] showed improved efficacy of NY-ESO-1 vaccine therapy when combined with escalated doses of HMA decitabine in patients with EOC relapse. Current clinical trials combining immunotherapeutic approach with epigenetic drug(s) are summarized in Table 1. In addition to epigenomic potentiation of immunotherapy by chromatin remodelers, boosting regulatory T cell-mediated immune response may be achieved via targeting non-coding RNAs [304] and it requires further testing for applicability in EOC (Figure 3D).

Table 1. Clinical trials evaluating safety and efficacy of immunotherapeutic agents in combination with epigenetic drugs in patients with ovarian cancer (https://clinicaltrials.gov).

Study Name	Phase	Status	ClinicalTrials.gov Identifier
Decitabine, Vaccine Therapy, and Pegylated Liposomal Doxorubicin Hydrochloride in Treating Patients With Recurrent Ovarian Epithelial Cancer, Fallopian Tube Cancer, or Peritoneal Cancer	I	Completed Completion: June 2013	NCT01673217
Atezolizumab, Guadecitabine, and CDX-1401 Vaccine in Treating Patients With Recurrent Ovarian, Fallopian Tube, or Primary Peritoneal Cancer	I/IIb	Ongoing Patient recruitment suspended Estimated completion: March 2020	NCT03206047
Guadecitabine and Pembrolizumab in Treating Patients With Recurrent Ovarian, Primary Peritoneal, or Fallopian Tube Cancer	II	Ongoing Recruiting patients Estimated completion: March 2020	NCT02901899
Genetically Modified T Cells and Decitabine in Treating Patients With Recurrent or Refractory Ovarian, Primary Peritoneal, or Fallopian Tube Cancer	I	Ongoing Recruiting patients Estimated completion: August 2020	NCT03017131
Study of Azacitidine and Durvalumab in Advanced Solid Tumors (METADUR)	II	Ongoing Recruiting patients Estimated completion: January 2022	NCT02811497

4. Conclusions

To summarize, a deeper understanding of epigenomic involvement in reciprocal EOC tumor-stroma interrelationship is essential and it will help to determine pharmacological routes to alter the TME, which in turn could inhibit EOC metastatic progression and the development of chemoresistance and tumor recurrence. Furthermore, we believe epigenetically-mediated pharmacological engagement of certain TME players, such as immune cells, to boost immune responses towards complete malignant cell elimination has tremendous potential.

Author Contributions: Y.K. contributed to the conceptualization and writing (original draft and figure preparation, review and editing). K.P.N. contributed to conceptualization and writing (review and editing).

Funding: This work was supported in part by the Ovarian Cancer Research Fund Alliance [Epigenetic Vulnerabilities of Ovarian Cancer Stem Cells, Grant Number 458788] (K.P.N.); the Leo and Anne Albert Charitable Trust (K.P.N.); and the Indiana Clinical and Translational Sciences Institute Grant Award (Y.K.).

Conflicts of Interest: The authors declare no conflict of interest.

References

1. Vaughan, S.; Coward, J.I.; Bast, R.C.; Berchuck, A.; Berek, J.S.; Brenton, J.D.; Coukos, G.; Crum, C.C.; Drapkin, R.; Etemadmoghadam, D. Rethinking Ovarian Cancer: Recommendations for Improving Outcomes. *Nat. Rev. Cancer* **2011**, *11*, 719–725. [CrossRef] [PubMed]
2. American Cancer Society. *Cancer Facts & Figures 2018*; American Cancer Society: Atlanta, GA, USA, 2018.

3. Siegel, R.L.; Miller, K.D.; Ahmedin, J. Cancer Statistics, 2018. *CA Cancer J. Clin.* **2018**, *68*, 7–30. [CrossRef] [PubMed]
4. Jacques, F.; Isabelle, S.; Rajesh, D.; Sultan, E.; Colin, M.; Marise, R.; Parkin, D.M.; David, F.; Freddie, B. Cancer Incidence and Mortality Worldwide: Sources, Methods and Major Patterns in GLOBOCAN 2012. *Int. J. Cancer* **2015**, *136*, E359–E386.
5. Howlader, N.; Noone, A.M.; Krapcho, M.; Garshell, J.; Miller, D.; Altekruse, S.F.; Kosary, C.L.; Yu, M.; Ruhl, J.; Tatalovich, Z.; et al. (Eds.) *SEER Cancer Statistics Review, 1975–2011*; National Cancer Institute: Bethesda, MD, USA, 2013. Available online: http://Seer.Cancer.Gov/Csr/1975_2011/ (accessed on 20 April 2014).
6. Marcus, C.S.; Maxwell, G.L.; Darcy, K.M.; Hamilton, C.A.; McGuire, W.P. Current Approaches and Challenges in Managing and Monitoring Treatment Response in Ovarian Cancer. *J. Cancer* **2014**, *5*, 25–30. [CrossRef] [PubMed]
7. Bowtell, D.D.; Böhm, S.; Ahmed, A.A.; Aspuria, P.; Bast, R.C., Jr.; Beral, V.; Berek, J.S.; Birrer, M.J.; Blagden, S.; Bookman, M.A. Rethinking Ovarian Cancer II: Reducing Mortality from High-Grade Serous Ovarian Cancer. *Nat. Rev. Cancer* **2015**, *15*, 668–679. [CrossRef] [PubMed]
8. Auersperg, N.; Wong, A.S.T.; Choi, K.; Kang, S.K.; Leung, P.C.K. Ovarian Surface Epithelium: Biology, Endocrinology, and Pathology. *Endocr. Rev.* **2001**, *22*, 255–288. [CrossRef] [PubMed]
9. Auersperg, N. The Origin of Ovarian Cancers—Hypotheses and Controversies. *Front. Biosci.* **2013**, *5*, 709–719. [CrossRef]
10. Bowen, N.J.; Walker, L.D.; Matyunina, L.V.; Logani, S.; Totten, K.A.; Benigno, B.B.; McDonald, J.F. Gene Expression Profiling Supports the Hypothesis that Human Ovarian Surface Epithelia are Multipotent and Capable of Serving as Ovarian Cancer Initiating Cells. *BMC Med. Genom.* **2009**, *2*, 71. [CrossRef] [PubMed]
11. Chene, G.; Dauplat, J.; Radosevic-Robin, N.; Cayre, A.; Penault-Llorca, F. Tu-be Or Not Tu-be: That is the Question … About Serous Ovarian Carcinogenesis. *Crit. Rev. Oncol.* **2013**, *88*, 134–143. [CrossRef] [PubMed]
12. Flesken-Nikitin, A.; Hwang, C.; Cheng, C.; Michurina, T.V.; Enikolopov, G.; Nikitin, A.Y. Ovarian Surface Epithelium at the Junction Area Contains a Cancer-Prone Stem Cell Niche. *Nature* **2013**, *495*, 241–245. [CrossRef] [PubMed]
13. Seidman, J.D.; Yemelyanova, A.; Zaino, R.J.; Kurman, R.J. The Fallopian Tube-Peritoneal Junction: A Potential Site of Carcinogenesis. *Int. J. Gynecol. Pathol.* **2011**, *30*, 4–11. [CrossRef] [PubMed]
14. Earp, M.A.; Cunningham, J.M. DNA Methylation Changes in Epithelial Ovarian Cancer Histotypes. *Genomics* **2015**, *106*, 311–321. [CrossRef] [PubMed]
15. Barton, C.A.; Hacker, N.F.; Clark, S.J.; O'Brien, P.M. DNA Methylation Changes in Ovarian Cancer: Implications for Early Diagnosis, Prognosis and Treatment. *Gynecol. Oncol.* **2008**, *109*, 129–139. [CrossRef] [PubMed]
16. Montavon, C.; Gloss, B.S.; Warton, K.; Barton, C.A.; Statham, A.L.; Scurry, J.P.; Tabor, B.; Nguyen, T.V.; Qu, W.; Samimi, G. Prognostic and Diagnostic Significance of DNA Methylation Patterns in High Grade Serous Ovarian Cancer. *Gynecol. Oncol.* **2012**, *124*, 582–588. [CrossRef] [PubMed]
17. Robles, A.I.; Jen, J.; Harris, C.C. Clinical Outcomes of TP53 Mutations in Cancers. *Cold Spring Harb. Perspect. Med.* **2016**, *6*. [CrossRef] [PubMed]
18. Klymenko, Y.; Kim, O.; Stack, M.S. Complex Determinants of Epithelial: Mesenchymal Phenotypic Plasticity in Ovarian Cancer. *Cancers* **2017**, *9*, 104. [CrossRef] [PubMed]
19. Pu, T.; Xiong, L.; Liu, Q.; Zhang, M.; Cai, Q.; Liu, H.; Sood, A.K.; Li, G.; Kang, Y.; Xu, C. Delineation of Retroperitoneal Metastatic Lymph Nodes in Ovarian Cancer with Near-infrared Fluorescence Imaging. *Oncol. Lett.* **2017**, *14*, 2869–2877. [CrossRef] [PubMed]
20. Pradeep, S.; Kim, S.W.; Wu, S.Y.; Nishimura, M.; Chaluvally-Raghavan, P.; Miyake, T.; Pecot, C.V.; Kim, S.; Choi, H.J.; Bischoff, F.Z. Hematogenous Metastasis of Ovarian Cancer: Rethinking Mode of Spread. *Cancer Cell* **2014**, *26*, 77–91. [CrossRef] [PubMed]
21. Lengyel, E. Ovarian Cancer Development and Metastasis. *Am. J. Pathol.* **2010**, *177*, 1053–1064. [CrossRef] [PubMed]
22. Shield, K.; Ackland, M.L.; Ahmed, N.; Rice, G.E. Multicellular Spheroids in Ovarian Cancer Metastases: Biology and Pathology. *Gynecol. Oncol.* **2009**, *113*, 143–148. [CrossRef] [PubMed]
23. Hudson, L.G.; Zeineldin, R.; Stack, M.S. Phenotypic Plasticity of Neoplastic Ovarian Epithelium: Unique Cadherin Profiles in Tumor Progression. *Clin. Exp. Metastasis* **2008**, *25*, 643–655. [CrossRef] [PubMed]

24. Davidson, B. Ovarian Carcinoma and Serous Effusions. Changing Views Regarding Tumor Progression and Review of Current Literature1. *Anal. Cell. Pathol.* **2001**, *23*, 107–128. [CrossRef] [PubMed]
25. Burleson, K.M.; Casey, R.C.; Skubitz, K.M.; Pambuccian, S.E.; Oegema, T.R., Jr.; Skubitz, A.P.N. Ovarian Carcinoma Ascites Spheroids Adhere to Extracellular Matrix Components and Mesothelial Cell Monolayers. *Gynecol. Oncol.* **2004**, *93*, 170–181. [CrossRef] [PubMed]
26. Ho, C.; Chang, S.; Hsiao, C.; Chien, T.; Shih, D.T. Isolation and Characterization of Stromal Progenitor Cells from Ascites of Patients with Epithelial Ovarian Adenocarcinoma. *J. Biomed. Sci.* **2012**, *19*, 23. [CrossRef] [PubMed]
27. Ahmed, N.; Stenvers, K. Getting to Know Ovarian Cancer Ascites: Opportunities for Targeted Therapy-Based Translational Research. *Front. Oncol.* **2013**, *3*, 256. [CrossRef] [PubMed]
28. Nieman, K.M.; Romero, I.L.; Van Houten, B.; Lengyel, E. Adipose Tissue and Adipocytes Support Tumorigenesis and Metastasis. *Biochim. Biophys. Acta* **2013**, *1831*, 1533–1541. [CrossRef] [PubMed]
29. Zhang, L.; Conejo-Garcia, J.R.; Katsaros, D.; Gimotty, P.A.; Massobrio, M.; Regnani, G.; Makrigiannakis, A.; Gray, H.; Schlienger, K.; Liebman, M.N. Intratumoral T Cells, Recurrence, and Survival in Epithelial Ovarian Cancer. *N. Engl. J. Med.* **2003**, *348*, 203–213. [CrossRef] [PubMed]
30. Becker, G.; Galandi, D.; Blum, H.E. Malignant Ascites: Systematic Review and Guideline for Treatment. *Eur. J. Cancer* **2006**, *42*, 589–597. [CrossRef] [PubMed]
31. Byrne, A.T.; Ross, L.; Holash, J.; Nakanishi, M.; Hu, L.; Hofmann, J.I.; Yancopoulos, G.D.; Jaffe, R.B. Vascular Endothelial Growth Factor-Trap Decreases Tumor Burden, Inhibits Ascites, and Causes Dramatic Vascular Remodeling in an Ovarian Cancer Model. *Clin. Cancer Res.* **2003**, *9*, 5721–5728. [PubMed]
32. Fang, X.; Schummer, M.; Mao, M.; Yu, S.; Tabassam, F.H.; Swaby, R.; Hasegawa, Y.; Tanyi, J.L.; LaPushin, R.; Eder, A. Lysophosphatidic Acid is a Bioactive Mediator in Ovarian Cancer. *Biochim. Biophys. Acta* **2002**, *1582*, 257–264. [CrossRef]
33. Garrison, R.N.; Galloway, R.H.; Heuser, L.S. Mechanisms of Malignant Ascites Production. *J. Surg. Res.* **1987**, *42*, 126–132. [CrossRef]
34. Gotlieb, W.H.; Abrams, J.S.; Watson, J.M.; Velu, T.J.; Berek, J.S.; Martínez-Maza, O. Presence of Interleukin 10 (IL-10) in the Ascites of Patients with Ovarian and Other Intra-Abdominal Cancers. *Cytokine* **1992**, *4*, 385–390. [CrossRef]
35. Hu, L.; Hofmann, J.; Zaloudek, C.; Ferrara, N.; Hamilton, T.; Jaffe, R.B. Vascular Endothelial Growth Factor Immunoneutralization Plus Paclitaxel Markedly Reduces Tumor Burden and Ascites in Athymic Mouse Model of Ovarian Cancer. *Am. J. Pathol.* **2002**, *161*, 1917–1924. [CrossRef]
36. Mesiano, S.; Ferrara, N.; Jaffe, R.B. Role of Vascular Endothelial Growth Factor in Ovarian Cancer: Inhibition of Ascites Formation by Immunoneutralization. *Am. J. Pathol.* **1998**, *153*, 1249–1256. [CrossRef]
37. Moradi, M.M.; Carson, L.F.; Twiggs, L.B.; Weinberg, J.B.; Haney, A.; Ramakrishnan, S. Serum and Ascitic Fluid Levels of Interleukin-1, Interleukin-6, and Tumor Necrosis Factor-alpha in Patients with Ovarian Epithelial Cancer. *Cancer* **1993**, *72*, 2433–2440. [CrossRef]
38. Pedersen, N.; Schmitt, M.; Ronne, E.; Nicoletti, M.I.; Hoyer-Hansen, G.; Conese, M.; Giavazzi, R.; Dano, K.; Kuhn, W.; Janicke, F. A Ligand-Free, Soluble Urokinase Receptor is Present in the Ascitic Fluid from Patients with Ovarian Cancer. *J. Clin. Investig.* **1993**, *92*, 2160–2167. [CrossRef] [PubMed]
39. Santin, A.D.; Hermonat, P.L.; Ravaggi, A.; Cannon, M.J.; Pecorelli, S.; Parham, G.P. Secretion of Vascular Endothelial Growth Factor in Ovarian Cancer. *Eur. J. Gynaecol. Oncol.* **1999**, *20*, 177–181. [PubMed]
40. Topalak, O.; Saygili, U.; Soyturk, M.; Karaca, N.; Batur, Y.; Uslu, T.; Erten, O. Serum, Pleural Effusion, and Ascites CA-125 Levels in Ovarian Cancer and Nonovarian Benign and Malignant Diseases: A Comparative Study. *Gynecol. Oncol.* **2002**, *85*, 108–113. [CrossRef] [PubMed]
41. Wilhelm, O.; Hafter, R.; Coppenrath, E.; Pflanz, M.A.; Schmitt, M.; Babic, R.; Linke, R.; Gossner, W.; Graeff, H. Fibrin-Fibronectin Compounds in Human Ovarian Tumor Ascites and their Possible Relation to the Tumor Stroma. *Cancer Res.* **1988**, *48*, 3507–3514. [PubMed]
42. Xu, Y.; Shen, Z.; Wiper, D.W.; Wu, M.; Morton, R.E.; Elson, P.; Kennedy, A.W.; Belinson, J.; Markman, M.; Casey, G. Lysophosphatidic Acid as a Potential Biomarker for Ovarian and Other Gynecologic Cancers. *JAMA* **1998**, *280*, 719–723. [CrossRef] [PubMed]
43. Yabushita, H.; Shimazu, M.; Noguchi, M.; Kishida, T.; Narumiya, H.; Sawaguchi, K.; Noguchi, M. Vascular Endothelial Growth Factor Activating Matrix Metalloproteinase in Ascitic Fluid during Peritoneal Dissemination of Ovarian Cancer. *Oncol. Rep.* **2003**, *10*, 89–95. [CrossRef] [PubMed]

44. Carduner, L.; Leroy-Dudal, J.; Picot, C.R.; Gallet, O.; Carreiras, F.; Kellouche, S. Ascites-induced shift along epithelial-mesenchymal spectrum in ovarian cancer cells: Enhancement of their invasive behavior partly dependant on αv integrins. *Clin. Exp. Metastasis* **2014**, *31*, 675–688. [CrossRef] [PubMed]
45. Waddington, C. Embryology, Epigenetics and Biogenetics. *Nature* **1956**, *177*, 1241. [CrossRef]
46. Esteller, M. Cancer Epigenomics: DNA Methylomes and Histone-Modification Maps. *Nat. Rev. Genet.* **2007**, *8*, 286–298. [CrossRef] [PubMed]
47. Lyko, F. The DNA Methyltransferase Family: A Versatile Toolkit for Epigenetic Regulation. *Nat. Rev. Genet.* **2018**, *19*, 81–92. [CrossRef] [PubMed]
48. Huff, J.; Zilberman, D. Dnmt1-Independent CG Methylation Contributes to Nucleosome Positioning in Diverse Eukaryotes. *Cell* **2014**, *156*, 1286–1297. [CrossRef] [PubMed]
49. Bestor, T.; Laudano, A.; Mattaliano, R.; Ingram, V. Cloning and Sequencing of a cDNA Encoding DNA Methyltransferase of Mouse Cells: The Carboxyl-Terminal Domain of the Mammalian Enzymes is Related to Bacterial Restriction Methyltransferases. *J. Mol. Biol.* **1988**, *203*, 971–983. [CrossRef]
50. Hermann, A.; Goyal, R.; Jeltsch, A. The Dnmt1 DNA-(Cytosine-C5)-Methyltransferase Methylates DNA Processively with High Preference for Hemimethylated Target Sites. *J. Biol. Chem.* **2004**, *279*, 48350–48359. [CrossRef] [PubMed]
51. Lyko, F.; Ramsahoye, B.H.; Kashevsky, H.; Tudor, M.; Mastrangelo, M.; Orr-Weaver, T.L.; Jaenisch, R. Mammalian (Cytosine-5) Methyltransferases Cause Genomic DNA Methylation and Lethality in Drosophila. *Nat. Genet.* **1999**, *23*, 363–366. [CrossRef] [PubMed]
52. Bird, A. DNA Methylation De Novo. *Science* **1999**, *286*, 2287–2288. [CrossRef] [PubMed]
53. Rhee, I.; Jair, K.; Yen, R.C.; Lengauer, C.; Herman, J.G.; Kinzler, K.W.; Vogelstein, B.; Baylin, S.B.; Schuebel, K.E. CpG Methylation is Maintained in Human Cancer Cells Lacking DNMT1. *Nature* **2000**, *404*, 1003–1007. [CrossRef] [PubMed]
54. Jair, K.W.; Bachman, K.E.; Suzuki, H.; Ting, A.H.; Rhee, I.; Yen, R.W.; Baylin, S.B.; Schuebel, K.E. De Novo CpG Island Methylation in Human Cancer Cells. *Cancer Res.* **2006**, *66*, 682–692. [CrossRef] [PubMed]
55. Ting, A.H.; Jair, K.W.; Schuebel, K.E.; Baylin, S.B. Differential Requirement for DNA Methyltransferase 1 in Maintaining Human Cancer Cell Gene Promoter Hypermethylation. *Cancer Res.* **2006**, *66*, 729–735. [CrossRef] [PubMed]
56. Robert, M.; Morin, S.; Beaulieu, N.; Gauthier, F.; Chute, I.C.; Barsalou, A.; MacLeod, A.R. DNMT1 is Required to Maintain CpG Methylation and Aberrant Gene Silencing in Human Cancer Cells. *Nat. Genet.* **2003**, *33*, 61–65. [CrossRef] [PubMed]
57. Luger, K.; Mäder, A.W.; Richmond, R.K.; Sargent, D.F.; Richmond, T.J. Crystal Structure of the Nucleosome Core Particle at 2.8 Å. Resolution. *Nature* **1997**, *389*, 251–260. [CrossRef] [PubMed]
58. Ellis, L.; Atadja, P.W.; Johnstone, R.W. Epigenetics in Cancer: Targeting Chromatin Modifications. *Mol. Cancer. Ther.* **2009**, *8*, 1409–1420. [CrossRef] [PubMed]
59. Yang, X.; Seto, E. HATs and HDACs: From Structure, Function and Regulation to Novel Strategies for Therapy and Prevention. *Oncogene* **2007**, *26*, 5310–5318. [CrossRef] [PubMed]
60. Zentner, G.E.; Henikoff, S. Regulation of Nucleosome Dynamics by Histone Modifications. *Nat. Struct. Mol. Biol.* **2013**, *20*, 259–266. [CrossRef] [PubMed]
61. Oki, M.; Aihara, H.; Ito, T. Role of histone phosphorylation in chromatin dynamics and its implications in diseases. In *Chromatin and Disease*; Springer: Berlin, Germany, 2007; pp. 323–340.
62. Greer, E.L.; Shi, Y. Histone Methylation: A Dynamic Mark in Health, Disease and Inheritance. *Nat. Rev. Genet.* **2012**, *13*, 343–357. [CrossRef] [PubMed]
63. Bannister, A.J.; Kouzarides, T. Regulation of Chromatin by Histone Modifications. *Cell Res.* **2011**, *21*, 381–395. [CrossRef] [PubMed]
64. Cedar, H.; Bergman, Y. Linking DNA Methylation and Histone Modification: Patterns and Paradigms. *Nat. Rev. Genet.* **2009**, *10*, 295–304. [CrossRef] [PubMed]
65. Hamiche, A.; Sandaltzopoulos, R.; Gdula, D.A.; Wu, C. ATP-Dependent Histone Octamer Sliding Mediated by the Chromatin Remodeling Complex NURF. *Cell* **1999**, *97*, 833–842. [CrossRef]
66. Clapier, C.R.; Cairns, B.R. Chromatin remodeling complexes. In *Fundamentals of Chromatin*; Springer: Berlin, Germany, 2014; pp. 69–146.
67. Clapier, C.R.; Iwasa, J.; Cairns, B.R.; Peterson, C.L. Mechanisms of Action and Regulation of ATP-Dependent Chromatin-Remodelling Complexes. *Nat. Rev. Mol. Cell Biol.* **2017**, *18*, 407–422. [CrossRef] [PubMed]

68. Wiegand, K.C.; Shah, S.P.; Al-Agha, O.M.; Zhao, Y.; Tse, K.; Zeng, T.; Senz, J.; McConechy, M.K.; Anglesio, M.S.; Kalloger, S.E. ARID1A Mutations in Endometriosis-Associated Ovarian Carcinomas. *N. Engl. J. Med.* **2010**, *363*, 1532–1543. [CrossRef] [PubMed]
69. Jones, S.; Wang, T.L.; Shih, I.; Mao, T.L.; Nakayama, K.; Roden, R.; Glas, R.; Slamon, D.; Diaz, L.A., Jr.; Vogelstein, B.; et al. Frequent Mutations of Chromatin Remodeling Gene ARID1A in Ovarian Clear Cell Carcinoma. *Science* **2010**, *330*, 228–231. [CrossRef] [PubMed]
70. Fukumoto, T.; Park, P.H.; Wu, S.; Fatkhutdinov, N.; Karakashev, S.; Nacarelli, T.; Kossenkov, A.V.; Speicher, D.W.; Jean, S.; Zhang, L. Repurposing Pan-HDAC Inhibitors for ARID1A-Mutated Ovarian Cancer. *Cell Rep.* **2018**, *22*, 3393–3400. [CrossRef] [PubMed]
71. Jelinic, P.; Mueller, J.J.; Olvera, N.; Dao, F.; Scott, S.N.; Shah, R.; Gao, J.; Schultz, N.; Gonen, M.; Soslow, R.A.; et al. Recurrent SMARCA4 Mutations in Small Cell Carcinoma of the Ovary. *Nat. Genet.* **2014**, *46*, 424–426. [CrossRef] [PubMed]
72. Witkowski, L.; Carrot-Zhang, J.; Albrecht, S.; Fahiminiya, S.; Hamel, N.; Tomiak, E.; Grynspan, D.; Saloustros, E.; Nadaf, J.; Rivera, B. Germline and Somatic SMARCA4 Mutations Characterize Small Cell Carcinoma of the Ovary, Hypercalcemic Type. *Nat. Genet.* **2014**, *46*, 438–443. [CrossRef] [PubMed]
73. Muppala, R.; Donenberg, T.; Huang, M.S.; Schlumbrecht, M.P. SMARCA4 Germline Gene Mutation in a Patient with Epithelial Ovarian: A Case Report. *Gynecol. Oncol. Rep.* **2017**, *22*, 45–47. [CrossRef] [PubMed]
74. Zhang, R.; Fukumoto, T.; Magno, E. SWI/SNF Complexes in Ovarian Cancer: Mechanistic Insights and Therapeutic Implications. *Mol. Cancer Res.* **2018**. [CrossRef] [PubMed]
75. Bitler, B.G.; Aird, K.M.; Garipov, A.; Li, H.; Amatangelo, M.; Kossenkov, A.V.; Schultz, D.C.; Liu, Q.; Shih, I.; Conejo-Garcia, J.R. Synthetic Lethality by Targeting EZH2 Methyltransferase Activity in ARID1A-Mutated Cancers. *Nat. Med.* **2015**, *21*, 231–238. [CrossRef] [PubMed]
76. Bartel, D.P. MicroRNAs: Genomics, Biogenesis, Mechanism, and Function. *Cell* **2004**, *116*, 281–297. [CrossRef]
77. Eichhorn, S.W.; Guo, H.; McGeary, S.E.; Rodriguez-Mias, R.A.; Shin, C.; Baek, D.; Hsu, S.; Ghoshal, K.; Villén, J.; Bartel, D.P. mRNA Destabilization is the Dominant Effect of Mammalian microRNAs by the Time Substantial Repression Ensues. *Mol. Cell* **2014**, *56*, 104–115. [CrossRef] [PubMed]
78. Volpe, T.A.; Kidner, C.; Hall, I.M.; Teng, G.; Grewal, S.I.; Martienssen, R.A. Regulation of Heterochromatic Silencing and Histone H3 Lysine-9 Methylation by RNAi. *Science* **2002**, *297*, 1833–1837. [CrossRef] [PubMed]
79. Wang, K.C.; Chang, H.Y. Molecular Mechanisms of Long Noncoding RNAs. *Mol. Cell* **2011**, *43*, 904–914. [CrossRef] [PubMed]
80. Engreitz, J.M.; Ollikainen, N.; Guttman, M. Long Non-Coding RNAs: Spatial Amplifiers that Control Nuclear Structure and Gene Expression. *Nat. Rev. Mol. Cell Biol.* **2016**, *17*, 756–770. [CrossRef] [PubMed]
81. Yamamura, S.; Imai-Sumida, M.; Tanaka, Y.; Dahiya, R. Interaction and Cross-Talk between Non-Coding RNAs. *Cell. Mol. Life Sci.* **2018**, *75*, 467–484. [CrossRef] [PubMed]
82. Jones, P.A.; Baylin, S.B. The Epigenomics of Cancer. *Cell* **2007**, *128*, 683–692. [CrossRef] [PubMed]
83. You, J.; Jones, P. Cancer Genetics and Epigenetics: Two Sides of the Same Coin? *Cancer Cell* **2012**, *22*, 9–20. [CrossRef] [PubMed]
84. Feinberg, A.P. The Epigenetics of Cancer Etiology. *Semin. Cancer Biol.* **2004**, *14*, 427–432. [CrossRef] [PubMed]
85. Waldmann, T.; Schneider, R. Targeting Histone Modifications—Epigenetics in Cancer. *Curr. Opin. Cell Biol.* **2013**, *25*, 184–189. [CrossRef] [PubMed]
86. Sawan, C.; Herceg, Z. Histone modifications and cancer. In *Advances in Genetics*; Elsevier: Amsterdam, The Netherlands, 2010; pp. 57–85.
87. Audia, J.E.; Campbell, R.M. Histone Modifications and Cancer. *Cold Spring Harb. Perspect. Biol.* **2016**, *8*, a019521. [CrossRef] [PubMed]
88. Balch, C.; Fang, F.; Matei, D.E.; Huang, T.H.; Nephew, K.P. Minireview: Epigenetic Changes in Ovarian Cancer. *Endocrinology* **2009**, *150*, 4003–4011. [CrossRef] [PubMed]
89. Balch, C.; Matei, D.E.; Huang, T.H.; Nephew, K.P. Role of Epigenomics in Ovarian and Endometrial Cancers. *Epigenomics* **2010**, *2*, 419–447. [CrossRef] [PubMed]
90. Natanzon, Y.; Goode, E.L.; Cunningham, J.M. Epigenetics in Ovarian Cancer. *Semin. Cancer Biol.* **2018**, *51*, 160–169. [CrossRef] [PubMed]
91. Zhang, L.; Volinia, S.; Bonome, T.; Calin, G.A.; Greshock, J.; Yang, N.; Liu, C.; Giannakakis, A.; Alexiou, P.; Hasegawa, K. Genomic and Epigenetic Alterations Deregulate microRNA Expression in Human Epithelial Ovarian Cancer. *Proc. Natl. Acad. Sci. USA* **2008**, *105*, 7004–7009. [CrossRef] [PubMed]

92. Kinose, Y.; Sawada, K.; Nakamura, K.; Kimura, T. The Role of microRNAs in Ovarian Cancer. *Biomed. Res. Int.* **2014**, *2014*, 249393. [CrossRef] [PubMed]
93. Meryet-Figuiere, M.; Lambert, B.; Gauduchon, P.; Vigneron, N.; Brotin, E.; Poulain, L.; Denoyelle, C. An Overview of Long Non-Coding RNAs in Ovarian Cancers. *Oncotarget* **2016**, *7*, 44719–44734. [CrossRef] [PubMed]
94. Worku, T.; Bhattarai, D.; Ayers, D.; Wang, K.; Wang, C.; Rehman, Z.U.; Talpur, H.S.; Yang, L. Long Non-Coding RNAs: The New Horizon of Gene Regulation in Ovarian Cancer. *Cell. Physiol. Biochem.* **2017**, *44*, 948–966. [CrossRef] [PubMed]
95. Asadollahi, R.; Hyde, C.A.; Zhong, X.Y. Epigenetics of Ovarian Cancer: From the Lab to the Clinic. *Gynecol. Oncol.* **2010**, *118*, 81–87. [CrossRef] [PubMed]
96. Balch, C.; Huang, T.H.; Brown, R.; Nephew, K.P. The Epigenetics of Ovarian Cancer Drug Resistance and Resensitization. *Am. J. Obstet. Gynecol.* **2004**, *191*, 1552–1572. [CrossRef] [PubMed]
97. Bennett, R.L.; Licht, J.D. Targeting Epigenetics in Cancer. *Annu. Rev. Pharmacol. Toxicol.* **2018**, *58*, 187–207. [CrossRef] [PubMed]
98. Pfister, S.X.; Ashworth, A. Marked for Death: Targeting Epigenetic Changes in Cancer. *Nat. Rev. Drug Discov.* **2017**, *16*, 241–263. [CrossRef] [PubMed]
99. Balkwill, F.R.; Capasso, M.; Hagemann, T. The Tumor Microenvironment at a Glance. *J. Cell. Sci.* **2012**, *125*, 5591–5596. [CrossRef] [PubMed]
100. Quail, D.F.; Joyce, J.A. Microenvironmental Regulation of Tumor Progression and Metastasis. *Nat. Med.* **2013**, *19*, 1423–1437. [CrossRef] [PubMed]
101. Chen, F.; Zhuang, X.; Lin, L.; Yu, P.; Wang, Y.; Shi, Y.; Hu, G.; Sun, Y. New Horizons in Tumor Microenvironment Biology: Challenges and Opportunities. *BMC Med.* **2015**, *13*, 45. [CrossRef] [PubMed]
102. Wang, M.; Zhao, J.; Zhang, L.; Wei, F.; Lian, Y.; Wu, Y.; Gong, Z.; Zhang, S.; Zhou, J.; Cao, K.; et al. Role of Tumor Microenvironment in Tumorigenesis. *J. Cancer* **2017**, *8*, 761–773. [CrossRef] [PubMed]
103. Lyssiotis, C.A.; Kimmelman, A.C. Metabolic Interactions in the Tumor Microenvironment. *Trends Cell Biol.* **2017**, *27*, 863–875. [CrossRef] [PubMed]
104. Fearon, D.T. Immune-Suppressing Cellular Elements of the Tumor Microenvironment. *Annu. Rev. Cancer Biol.* **2017**, *1*, 241–255. [CrossRef]
105. Kalluri, R. The Biology and Function of Fibroblasts in Cancer. *Nat. Rev. Cancer* **2016**, *16*, 582–598. [CrossRef] [PubMed]
106. Lin, H.; Lin, J. Seed-in-Soil: Pancreatic Cancer Influenced by Tumor Microenvironment. *Cancers* **2017**, *9*, 93. [CrossRef] [PubMed]
107. Shiga, K.; Hara, M.; Nagasaki, T.; Sato, T.; Takahashi, H.; Takeyama, H. Cancer-Associated Fibroblasts: Their Characteristics and Their Roles in Tumor Growth. *Cancers* **2015**, *7*, 2443–2458. [CrossRef] [PubMed]
108. Allinen, M.; Beroukhim, R.; Cai, L.; Brennan, C.; Lahti-Domenici, J.; Huang, H.; Porter, D.; Hu, M.; Chin, L.; Richardson, A. Molecular Characterization of the Tumor Microenvironment in Breast Cancer. *Cancer Cell* **2004**, *6*, 17–32. [CrossRef] [PubMed]
109. Walter, K.; Omura, N.; Hong, S.; Griffith, M.; Goggins, M. Pancreatic Cancer Associated Fibroblasts Display Normal Allelotypes. *Cancer Biol. Ther.* **2008**, *7*, 882–888. [CrossRef] [PubMed]
110. Qiu, W.; Hu, M.; Sridhar, A.; Opeskin, K.; Fox, S.; Shipitsin, M.; Trivett, M.; Thompson, E.R.; Ramakrishna, M.; Gorringe, K.L. No Evidence of Clonal Somatic Genetic Alterations in Cancer-Associated Fibroblasts from Human Breast and Ovarian Carcinomas. *Nat. Genet.* **2008**, *40*, 650–655. [CrossRef] [PubMed]
111. Hanson, J.A.; Gillespie, J.W.; Grover, A.; Tangrea, M.A.; Chuaqui, R.F.; Emmert-Buck, M.R.; Tangrea, J.A.; Libutti, S.K.; Linehan, W.M.; Woodson, K.G. Gene Promoter Methylation in Prostate Tumor–Associated Stromal Cells. *J. Natl. Cancer Inst.* **2006**, *98*, 255–261. [CrossRef] [PubMed]
112. Hu, M.; Yao, J.; Cai, L.; Bachman, K.E.; van den Brûle, F.; Velculescu, V.; Polyak, K. Distinct Epigenetic Changes in the Stromal Cells of Breast Cancers. *Nat. Genet.* **2005**, *37*, 899–905. [CrossRef] [PubMed]
113. Mathot, P.; Grandin, M.; Devailly, G.; Souaze, F.; Cahais, V.; Moran, S.; Campone, M.; Herceg, Z.; Esteller, M.; Juin, P. DNA Methylation Signal Has a Major Role in the Response of Human Breast Cancer Cells to the Microenvironment. *Oncogenesis* **2017**, *6*, e390. [CrossRef] [PubMed]
114. Pistore, C.; Giannoni, E.; Colangelo, T.; Rizzo, F.; Magnani, E.; Muccillo, L.; Giurato, G.; Mancini, M.; Rizzo, S.; Riccardi, M. DNA Methylation Variations are Required for Epithelial-to-Mesenchymal Transition Induced by Cancer-Associated Fibroblasts in Prostate Cancer Cells. *Oncogene* **2017**, *36*, 5551–5566. [CrossRef] [PubMed]

115. Albrengues, J.; Bertero, T.; Grasset, E.; Bonan, S.; Maiel, M.; Bourget, I.; Philippe, C.; Serrano, C.H.; Benamar, S.; Croce, O. Epigenetic Switch Drives the Conversion of Fibroblasts into Proinvasive Cancer-Associated Fibroblasts. *Nat. Commun.* **2015**, *6*, 10204. [CrossRef] [PubMed]
116. Stuelten, C.H.; Busch, J.I.; Tang, B.; Flanders, K.C.; Oshima, A.; Sutton, E.; Karpova, T.S.; Roberts, A.B.; Wakefield, L.M.; Niederhuber, J.E. Transient Tumor-Fibroblast Interactions Increase Tumor Cell Malignancy by a TGF-B Mediated Mechanism in a Mouse Xenograft Model of Breast Cancer. *PLoS ONE* **2010**, *5*, e9832. [CrossRef] [PubMed]
117. Cardenas, H.; Vieth, E.; Lee, J.; Segar, M.; Liu, Y.; Nephew, K.P.; Matei, D. TGF-B Induces Global Changes in DNA Methylation during the Epithelial-to-Mesenchymal Transition in Ovarian Cancer Cells. *Epigenetics* **2014**, *9*, 1461–1472. [CrossRef] [PubMed]
118. Calon, A.; Tauriello, D.V.F.; Batlle, E. TGF-Beta in CAF-Mediated Tumor Growth and Metastasis. *Semin. Cancer Biol.* **2014**, *25*, 15–22. [CrossRef] [PubMed]
119. Yeung, T.L.; Leung, C.S.; Wong, K.K.; Samimi, G.; Thompson, M.S.; Liu, J.; Zaid, T.M.; Ghosh, S.; Birrer, M.J.; Mok, S.C. TGF-Beta Modulates Ovarian Cancer Invasion by Upregulating CAF-Derived Versican in the Tumor Microenvironment. *Cancer Res.* **2013**, *73*, 5016–5028. [CrossRef] [PubMed]
120. Bracken, A.P.; Dietrich, N.; Pasini, D.; Hansen, K.H.; Helin, K. Genome-Wide Mapping of Polycomb Target Genes Unravels their Roles in Cell Fate Transitions. *Genes Dev.* **2006**, *20*, 1123–1136. [CrossRef] [PubMed]
121. Tyan, S.; Hsu, C.; Peng, K.; Chen, C.; Kuo, W.; Eva, Y.L.; Shew, J.; Chang, K.; Juan, L.; Lee, W. Breast Cancer Cells Induce Stromal Fibroblasts to Secrete ADAMTS1 for Cancer Invasion through an Epigenetic Change. *PLoS ONE* **2012**, *7*, e35128. [CrossRef] [PubMed]
122. Li, H.; Zhang, R. Role of EZH2 in Epithelial Ovarian Cancer: From Biological Insights to Therapeutic Target. *Front. Oncol.* **2013**, *3*, 47. [CrossRef] [PubMed]
123. Zhao, L.; Sun, Y.; Hou, Y.; Peng, Q.; Wang, L.; Luo, H.; Tang, X.; Zeng, Z.; Liu, M. MiRNA Expression Analysis of Cancer-Associated Fibroblasts and Normal Fibroblasts in Breast Cancer. *Int. J. Biochem. Cell Biol.* **2012**, *44*, 2051–2059. [CrossRef] [PubMed]
124. Wang, X.; Tang, S.; Le, S.; Lu, R.; Rader, J.S.; Meyers, C.; Zheng, Z. Aberrant Expression of Oncogenic and Tumor-Suppressive microRNAs in Cervical Cancer is Required for Cancer Cell Growth. *PLoS ONE* **2008**, *3*, e2557. [CrossRef] [PubMed]
125. Naito, Y.; Sakamoto, N.; Oue, N.; Yashiro, M.; Sentani, K.; Yanagihara, K.; Hirakawa, K.; Yasui, W. MicroRNA-143 Regulates Collagen Type III Expression in Stromal Fibroblasts of Scirrhous Type Gastric Cancer. *Cancer Sci.* **2014**, *105*, 228–235. [CrossRef] [PubMed]
126. Musumeci, M.; Coppola, V.; Addario, A.; Patrizii, M.; Maugeri-Sacca, M.; Memeo, L.; Colarossi, C.; Francescangeli, F.; Biffoni, M.; Collura, D. Control of Tumor and Microenvironment Cross-Talk by miR-15a and miR-16 in Prostate Cancer. *Oncogene* **2011**, *30*, 4231–4242. [CrossRef] [PubMed]
127. Yamamichi, N.; Shimomura, R.; Inada, K.; Sakurai, K.; Haraguchi, T.; Ozaki, Y.; Fujita, S.; Mizutani, T.; Furukawa, C.; Fujishiro, M.; et al. Locked Nucleic Acid in Situ Hybridization Analysis of miR-21 Expression during Colorectal Cancer Development. Clin. *Cancer Res.* **2009**, *15*, 4009–4016.
128. Schepeler, T.; Reinert, J.T.; Ostenfeld, M.S.; Christensen, L.L.; Silahtaroglu, A.N.; Dyrskjot, L.; Wiuf, C.; Sorensen, F.J.; Kruhoffer, M.; Laurberg, S.; et al. Diagnostic and Prognostic microRNAs in Stage II Colon Cancer. *Cancer Res.* **2008**, *68*, 6416–6424. [CrossRef] [PubMed]
129. Enkelmann, A.; Heinzelmann, J.; von Eggeling, F.; Walter, M.; Berndt, A.; Wunderlich, H.; Junker, K. Specific Protein and miRNA Patterns Characterise Tumour-Associated Fibroblasts in Bladder Cancer. *J. Cancer Res. Clin. Oncol.* **2011**, *137*, 751–759. [CrossRef] [PubMed]
130. Mitra, A.K.; Zillhardt, M.; Hua, Y.; Tiwari, P.; Murmann, A.E.; Peter, M.E.; Lengyel, E. MicroRNAs Reprogram Normal Fibroblasts into Cancer-Associated Fibroblasts in Ovarian Cancer. *Cancer Discov.* **2012**, *2*, 1100–1108. [CrossRef] [PubMed]
131. Zhao, L.; Ji, G.; Le, X.; Wang, C.; Xu, L.; Feng, M.; Zhang, Y.; Yang, H.; Xuan, Y.; Yang, Y.; et al. Long Noncoding RNA LINC00092 Acts in Cancer-Associated Fibroblasts to Drive Glycolysis and Progression of Ovarian Cancer. *Cancer Res.* **2017**, *77*, 1369–1382. [CrossRef] [PubMed]
132. Vafaee, F.; Colvin, E.K.; Mok, S.C.; Howell, V.M.; Samimi, G. Functional Prediction of Long Non-Coding RNAs in Ovarian Cancer-Associated Fibroblasts Indicate a Potential Role in Metastasis. *Sci. Rep.* **2017**, *7*, 10374. [CrossRef] [PubMed]

133. Kershaw, E.E.; Flier, J.S. Adipose Tissue as an Endocrine Organ. *J. Clin. Endocrinol. Metab.* **2004**, *89*, 2548–2556. [CrossRef] [PubMed]
134. Rio, M. The role of cancer-associated adipocytes (CAA) in the dynamic interaction between the tumor and the host. In *Tumor-Associated Fibroblasts and Their Matrix*; Springer: Berlin, Germany, 2011; pp. 111–123.
135. Hoy, A.J.; Balaban, S.; Saunders, D.N. Adipocyte–Tumor Cell Metabolic Crosstalk in Breast Cancer. *Trends Mol. Med.* **2017**, *23*, 381–392. [CrossRef] [PubMed]
136. Toren, P.; Venkateswaran, V. Periprostatic Adipose Tissue and Prostate Cancer Progression: New Insights into the Tumor Microenvironment. *Clin. Genitourin. Cancer* **2014**, *12*, 21–26. [CrossRef] [PubMed]
137. Dirat, B.; Bochet, L.; Dabek, M.; Daviaud, D.; Dauvillier, S.; Majed, B.; Wang, Y.Y.; Meulle, A.; Salles, B.; Le Gonidec, S.; et al. Cancer-Associated Adipocytes Exhibit an Activated Phenotype and Contribute to Breast Cancer Invasion. *Cancer Res.* **2011**, *71*, 2455–2465. [CrossRef] [PubMed]
138. Bochet, L.; Meulle, A.; Imbert, S.; Salles, B.; Valet, P.; Muller, C. Cancer-Associated Adipocytes Promotes Breast Tumor Radioresistance. Biochem. Biophys. *Res. Commun.* **2011**, *411*, 102–106.
139. Tabuso, M.; Homer-Vanniasinkam, S.; Adya, R.; Arasaradnam, R.P. Role of Tissue Microenvironment Resident Adipocytes in Colon Cancer. *World J. Gastroenterol.* **2017**, *23*, 5829–5835. [CrossRef] [PubMed]
140. Wang, Q.; Li, Y.C.; Wang, J.; Kong, J.; Qi, Y.; Quigg, R.J.; Li, X. miR-17-92 Cluster Accelerates Adipocyte Differentiation by Negatively Regulating Tumor-Suppressor Rb2/p130. *Proc. Natl. Acad. Sci. USA* **2008**, *105*, 2889–2894. [CrossRef] [PubMed]
141. Lee, J.; Hong, B.S.; Ryu, H.S.; Lee, H.; Lee, M.; Park, I.A.; Kim, J.; Han, W.; Noh, D.; Moon, H. Transition into Inflammatory Cancer-Associated Adipocytes in Breast Cancer Microenvironment Requires microRNA Regulatory Mechanism. *PLoS ONE* **2017**, *12*, e0174126. [CrossRef] [PubMed]
142. Au Yeung, C.L.; Co, N.N.; Tsuruga, T.; Yeung, T.L.; Kwan, S.Y.; Leung, C.S.; Li, Y.; Lu, E.S.; Kwan, K.; Wong, K.K.; et al. Exosomal Transfer of Stroma-Derived miR21 Confers Paclitaxel Resistance in Ovarian Cancer Cells through Targeting APAF1. *Nat. Commun.* **2016**, *7*, 11150. [CrossRef] [PubMed]
143. Halvorsen, A.R.; Helland, Å.; Gromov, P.; Wielenga, V.T.; Talman, M.M.; Brunner, N.; Sandhu, V.; Børresen-Dale, A.; Gromova, I.; Haakensen, V.D. Profiling of microRNAs in Tumor Interstitial Fluid of Breast Tumors—A Novel Resource to Identify Biomarkers for Prognostic Classification and Detection of Cancer. *Mol. Oncol.* **2017**, *11*, 220–234. [CrossRef] [PubMed]
144. Ribeiro, R.; Monteiro, C.; Catalán, V.; Hu, P.; Cunha, V.; Rodríguez, A.; Gómez-Ambrosi, J.; Fraga, A.; Príncipe, P.; Lobato, C. Obesity and Prostate Cancer: Gene Expression Signature of Human Periprostatic Adipose Tissue. *BMC Med.* **2012**, *10*, 108. [CrossRef] [PubMed]
145. Cheng, Y.; Monteiro, C.; Matos, A.; You, J.; Fraga, A.; Pereira, C.; Catalán, V.; Rodríguez, A.; Gómez-Ambrosi, J.; Frühbeck, G. Epigenome-Wide DNA Methylation Profiling of Periprostatic Adipose Tissue in Prostate Cancer Patients with Excess Adiposity—A Pilot Study. *Clin. Epigenet.* **2018**, *10*, 54. [CrossRef] [PubMed]
146. Benton, M.C.; Johnstone, A.; Eccles, D.; Harmon, B.; Hayes, M.T.; Lea, R.A.; Griffiths, L.; Hoffman, E.P.; Stubbs, R.S.; Macartney-Coxson, D. An Analysis of DNA Methylation in Human Adipose Tissue Reveals Differential Modification of Obesity Genes before and After Gastric Bypass and Weight Loss. *Genome Biol.* **2015**, *16*, 8. [CrossRef] [PubMed]
147. Kim, A.Y.; Park, Y.J.; Pan, X.; Shin, K.C.; Kwak, S.; Bassas, A.F.; Sallam, R.M.; Park, K.S.; Alfadda, A.A.; Xu, A. Obesity-Induced DNA Hypermethylation of the Adiponectin Gene Mediates Insulin Resistance. *Nat. Commun.* **2015**, *6*, 7585. [CrossRef] [PubMed]
148. Tang, J.; Pulliam, N.; Ozes, A.; Buechlein, A.; Ding, N.; Rusch, D.; Keer, H.; O'Hagan, H.; Stack, M.S.; Nephew, K.P. Epigenetic Targeting of Adipocytes Inhibits High-Grade Serous Ovarian Cancer Cell Migration and Invasion. *Mol. Cancer. Res.* **2018**, *16*, 1226–1240. [CrossRef] [PubMed]
149. Chen, Z. Progress and Prospects of Long Noncoding RNAs in Lipid Homeostasis. *Mol. Metab.* **2016**, *5*, 164–170. [CrossRef] [PubMed]
150. Xiao, T.; Liu, L.; Li, H.; Sun, Y.; Luo, H.; Li, T.; Wang, S.; Dalton, S.; Zhao, R.C.; Chen, R. Long Noncoding RNA ADINR Regulates Adipogenesis by Transcriptionally Activating C/EBPα. *Stem Cell Rep.* **2015**, *5*, 856–865. [CrossRef] [PubMed]
151. Huang, Y.; Jin, C.; Zheng, Y.; Li, X.; Zhang, S.; Zhang, Y.; Jia, L.; Li, W. Knockdown of lncRNA MIR31HG Inhibits Adipocyte Differentiation of Human Adipose-Derived Stem Cells Via Histone Modification of FABP4. *Sci. Rep.* **2017**, *7*, 8080. [CrossRef] [PubMed]

152. Sun, L.; Goff, L.A.; Trapnell, C.; Alexander, R.; Lo, K.A.; Hacisuleyman, E.; Sauvageau, M.; Tazon-Vega, B.; Kelley, D.R.; Hendrickson, D.G.; et al. Long Noncoding RNAs Regulate Adipogenesis. *Proc. Natl. Acad. Sci. USA* **2013**, *110*, 3387–3392. [CrossRef] [PubMed]
153. Nikpayam, E.; Tasharrofi, B.; Sarrafzadeh, S.; Ghafouri-Fard, S. The Role of Long Non-Coding RNAs in Ovarian Cancer. *Iran. Biomed. J.* **2017**, *21*, 3–15. [CrossRef] [PubMed]
154. Hosseini, E.S.; Meryet-Figuiere, M.; Sabzalipoor, H.; Kashani, H.H.; Nikzad, H.; Asemi, Z. Dysregulated Expression of Long Noncoding RNAs in Gynecologic Cancers. *Mol. Cancer* **2017**, *16*, 107. [CrossRef] [PubMed]
155. Da Silva Meirelles, L.; Chagastelles, P.C.; Nardi, N.B. Mesenchymal Stem Cells Reside in Virtually all Post-Natal Organs and Tissues. *J. Cell Sci.* **2006**, *119*, 2204–2213. [CrossRef] [PubMed]
156. McLean, K.; Gong, Y.; Choi, Y.; Deng, N.; Yang, K.; Bai, S.; Cabrera, L.; Keller, E.; McCauley, L.; Cho, K.R.; et al. Human Ovarian Carcinoma-Associated Mesenchymal Stem Cells Regulate Cancer Stem Cells and Tumorigenesis Via Altered BMP Production. *J. Clin. Investig.* **2011**, *121*, 3206–3219. [CrossRef] [PubMed]
157. Coffman, L.G.; Choi, Y.J.; McLean, K.; Allen, B.L.; di Magliano, M.P.; Buckanovich, R.J. Human Carcinoma-Associated Mesenchymal Stem Cells Promote Ovarian Cancer Chemotherapy Resistance Via a BMP4/HH Signaling Loop. *Oncotarget* **2016**, *7*, 6916–6932. [CrossRef] [PubMed]
158. Touboul, C.; Lis, R.; Al Farsi, H.; Raynaud, C.M.; Warfa, M.; Althawadi, H.; Mery, E.; Mirshahi, M.; Rafii, A. Mesenchymal Stem Cells Enhance Ovarian Cancer Cell Infiltration through IL6 Secretion in an Amniochorionic Membrane Based 3D Model. *J. Transl. Med.* **2013**, *11*, 28. [CrossRef] [PubMed]
159. Lis, R.; Touboul, C.; Mirshahi, P.; Ali, F.; Mathew, S.; Nolan, D.J.; Maleki, M.; Abdalla, S.A.; Raynaud, C.M.; Querleu, D. Tumor Associated Mesenchymal Stem Cells Protects Ovarian Cancer Cells from Hyperthermia through CXCL12. *Int. J. Cancer* **2011**, *128*, 715–725. [CrossRef] [PubMed]
160. Spaeth, E.L.; Dembinski, J.L.; Sasser, A.K.; Watson, K.; Klopp, A.; Hall, B.; Andreeff, M.; Marini, F. Mesenchymal Stem Cell Transition to Tumor-Associated Fibroblasts Contributes to Fibrovascular Network Expansion and Tumor Progression. *PLoS ONE* **2009**, *4*, e4992. [CrossRef] [PubMed]
161. Cho, J.A.; Park, H.; Lim, E.H.; Kim, K.H.; Choi, J.S.; Lee, J.H.; Shin, J.W.; Lee, K.W. Exosomes from Ovarian Cancer Cells Induce Adipose Tissue-Derived Mesenchymal Stem Cells to Acquire the Physical and Functional Characteristics of Tumor-Supporting Myofibroblasts. *Gynecol. Oncol.* **2011**, *123*, 379–386. [CrossRef] [PubMed]
162. Bu, S.; Wang, Q.; Zhang, Q.; Sun, J.; He, B.; Xiang, C.; Liu, Z.; Lai, D. Human Endometrial Mesenchymal Stem Cells Exhibit Intrinsic Anti-Tumor Properties on Human Epithelial Ovarian Cancer Cells. *Sci. Rep.* **2016**, *6*, 37019. [CrossRef] [PubMed]
163. Zhu, P.; Chen, M.; Wang, L.; Ning, Y.; Liang, J.; Zhang, H.; Xu, C.; Chen, S.; Yao, L. Systemic Mesenchymal Stem Cells Reduce Growth Rate of Cisplatin-Resistant Ovarian Cancer. *Int. J. Clin. Exp. Pathol.* **2013**, *6*, 2506–2514. [PubMed]
164. Zhang, Y.; Wang, J.; Ren, M.; Li, M.; Chen, D.; Chen, J.; Shi, F.; Wang, X.; Dou, J. Gene Therapy of Ovarian Cancer Using IL-21-Secreting Human Umbilical Cord Mesenchymal Stem Cells in Nude Mice. *J. Ovar. Res.* **2014**, *7*, 8. [CrossRef] [PubMed]
165. Jiang, J.; Chen, W.; Zhuang, R.; Song, T.; Li, P. The Effect of Endostatin Mediated by Human Mesenchymal Stem Cells on Ovarian Cancer Cells in Vitro. *J. Cancer Res. Clin. Oncol.* **2010**, *136*, 873–881. [CrossRef] [PubMed]
166. Dembinski, J.L.; Wilson, S.M.; Spaeth, E.L.; Studeny, M.; Zompetta, C.; Samudio, I.; Roby, K.; Andreeff, M.; Marini, F.C. Tumor Stroma Engraftment of Gene-Modified Mesenchymal Stem Cells as Anti-Tumor Therapy Against Ovarian Cancer. *Cytotherapy* **2013**, *15*, 20–32. [CrossRef] [PubMed]
167. Jiang, J.; Wei, D.; Sun, L.; Wang, Y.; Wu, X.; Li, Y.; Fang, Z.; Shang, H.; Wei, Z. A Preliminary Study on the Construction of Double Suicide Gene Delivery Vectors by Mesenchymal Stem Cells and the in Vitro Inhibitory Effects on SKOV3 Cells. *Oncol. Rep.* **2014**, *31*, 781–787. [CrossRef] [PubMed]
168. Amara, I.; Touati, W.; Beaune, P.; de Waziers, I. Mesenchymal Stem Cells as Cellular Vehicles for Prodrug Gene Therapy Against Tumors. *Biochimie* **2014**, *105*, 4–11. [CrossRef] [PubMed]
169. Muñoz, P.; Iliou, M.S.; Esteller, M. Epigenetic Alterations Involved in Cancer Stem Cell Reprogramming. *Mol. Oncol.* **2012**, *6*, 620–636. [CrossRef] [PubMed]
170. Leu, Y.; Huang, T.H.; Hsiao, S. Epigenetic reprogramming of mesenchymal stem cells. In *Epigenetic Alterations in Oncogenesis*; Springer: Berlin, Germany, 2013; pp. 195–211.

171. Teven, C.M.; Liu, X.; Hu, N.; Tang, N.; Kim, S.H.; Huang, E.; Yang, K.; Li, M.; Gao, J.L.; Liu, H.; et al. Epigenetic Regulation of Mesenchymal Stem Cells: A Focus on Osteogenic and Adipogenic Differentiation. *Stem Cells Int.* **2011**, *2011*, 201371. [CrossRef] [PubMed]
172. Mortada, I.; Mortada, R. Epigenetic Changes in Mesenchymal Stem Cells Differentiation. *Eur. J. Med. Genet.* **2017**, *61*, 114–118. [CrossRef] [PubMed]
173. Meyer, M.B.; Benkusky, N.A.; Sen, B.; Rubin, J.; Pike, J.W. Epigenetic Plasticity Drives Adipogenic and Osteogenic Differentiation of Marrow-Derived Mesenchymal Stem Cells. *J. Biol. Chem.* **2016**, *291*, 17829–17847. [CrossRef] [PubMed]
174. Collino, F.; Bruno, S.; Deregibus, M.C.; Tetta, C.; Camussi, G. MicroRNAs and mesenchymal stem cells. In *Vitamins & Hormones*; Elsevier: Amsterdam, The Netherlands, 2011; pp. 291–320.
175. Reza, A.M.M.T.; Choi, Y.-J.; Yasuda, H.; Kim, J.-H. Human Adipose Mesenchymal Stem Cell-Derived Exosomal-miRNAs are Critical Factors for Inducing Anti-Proliferation Signalling to A2780 and SKOV-3 Ovarian Cancer Cells. *Sci. Rep.* **2016**, *6*, 38498. [CrossRef] [PubMed]
176. Bellayr, I.H.; Kumar, A.; Puri, R.K. MicroRNA Expression in Bone Marrow-Derived Human Multipotent Stromal Cells. *BMC Genom.* **2017**, *18*, 605. [CrossRef] [PubMed]
177. Lu, X.; Chen, L.; Chen, Y.; Shao, Q.; Qin, W. Bafilomycin A1 Inhibits the Growth and Metastatic Potential of the BEL-7402 Liver Cancer and HO-8910 Ovarian Cancer Cell Lines and Induces Alterations in their microRNA Expression. *Exp. Ther. Med.* **2015**, *10*, 1829–1834. [CrossRef] [PubMed]
178. Zhang, X.; Liu, J.; Zang, D.; Wu, S.; Liu, A.; Zhu, J.; Wu, G.; Li, J.; Jiang, L. Upregulation of miR-572 Transcriptionally Suppresses SOCS1 and p21 and Contributes to Human Ovarian Cancer Progression. *Oncotarget* **2015**, *6*, 15180–15193. [CrossRef] [PubMed]
179. Wu, A.; Huang, Y.; Zhang, L.; Tian, G.; Liao, Q.; Chen, S. MiR-572 Prompted Cell Proliferation of Human Ovarian Cancer Cells by Suppressing PPP2R2C Expression. *Biomed. Pharmacother.* **2016**, *77*, 92–97. [CrossRef] [PubMed]
180. Lopatina, T.; Gai, C.; Deregibus, M.C.; Kholia, S.; Camussi, G. Cross Talk between Cancer and Mesenchymal Stem Cells through Extracellular Vesicles Carrying Nucleic Acids. *Front. Oncol.* **2016**, *6*, 125. [CrossRef] [PubMed]
181. Ho, C.; Shih, D.T.; Hsiao, C.; Huang, S.; Chang, S.; Cheng, W. Gene Methylation of Human Ovarian Carcinoma Stromal Progenitor Cells Promotes Tumorigenesis. *J. Transl. Med.* **2015**, *13*, 367. [CrossRef] [PubMed]
182. Jones, P.A.; Taylor, S.M. Cellular Differentiation, Cytidine Analogs and DNA Methylation. *Cell* **1980**, *20*, 85–93. [CrossRef]
183. Akino, T.; Hida, K.; Hida, Y.; Tsuchiya, K.; Freedman, D.; Muraki, C.; Ohga, N.; Matsuda, K.; Akiyama, K.; Harabayashi, T.; et al. Cytogenetic Abnormalities of Tumor-Associated Endothelial Cells in Human Malignant Tumors. *Am. J. Pathol.* **2009**, *175*, 2657–2667. [CrossRef] [PubMed]
184. Konerding, M.; Malkusch, W.; Klapthor, B.; Van Ackern, C.; Fait, E.; Hill, S.; Parkins, C.; Chaplin, D.; Presta, M.; Denekamp, J. Evidence for Characteristic Vascular Patterns in Solid Tumours: Quantitative Studies using Corrosion Casts. *Br. J. Cancer* **1999**, *80*, 724–732. [CrossRef] [PubMed]
185. Seaman, S.; Stevens, J.; Yang, M.Y.; Logsdon, D.; Graff-Cherry, C.; Croix, B.S. Genes that Distinguish Physiological and Pathological Angiogenesis. *Cancer Cell* **2007**, *11*, 539–554. [CrossRef] [PubMed]
186. Yadav, A.; Kumar, B.; Yu, J.; Old, M.; Teknos, T.N.; Kumar, P. Tumor-Associated Endothelial Cells Promote Tumor Metastasis by Chaperoning Circulating Tumor Cells and Protecting them from Anoikis. *PLoS ONE* **2015**, *10*, e0141602. [CrossRef] [PubMed]
187. Akiyama, K.; Ohga, N.; Hida, Y.; Kawamoto, T.; Sadamoto, Y.; Ishikawa, S.; Maishi, N.; Akino, T.; Kondoh, M.; Matsuda, A. Tumor Endothelial Cells Acquire Drug Resistance by MDR1 Up-Regulation Via VEGF Signaling in Tumor Microenvironment. *Am. J. Pathol.* **2012**, *180*, 1283–1293. [CrossRef] [PubMed]
188. Hida, K.; Ohga, N.; Akiyama, K.; Maishi, N.; Hida, Y. Heterogeneity of Tumor Endothelial Cells. *Cancer Sci.* **2013**, *104*, 1391–1395. [CrossRef] [PubMed]
189. Ohga, N.; Ishikawa, S.; Maishi, N.; Akiyama, K.; Hida, Y.; Kawamoto, T.; Sadamoto, Y.; Osawa, T.; Yamamoto, K.; Kondoh, M. Heterogeneity of Tumor Endothelial Cells: Comparison between Tumor Endothelial Cells Isolated from High-and Low-Metastatic Tumors. *Am. J. Pathol.* **2012**, *180*, 1294–1307. [CrossRef] [PubMed]

190. Maishi, N.; Ohba, Y.; Akiyama, K.; Ohga, N.; Hamada, J.; Nagao-Kitamoto, H.; Alam, M.T.; Yamamoto, K.; Kawamoto, T.; Inoue, N. Tumour Endothelial Cells in High Metastatic Tumours Promote Metastasis Via Epigenetic Dysregulation of Biglycan. *Sci. Rep.* **2016**, *6*, 28039. [CrossRef] [PubMed]
191. Hellebrekers, D.M.; Jair, K.W.; Vire, E.; Eguchi, S.; Hoebers, N.T.; Fraga, M.F.; Esteller, M.; Fuks, F.; Baylin, S.B.; van Engeland, M.; et al. Angiostatic Activity of DNA Methyltransferase Inhibitors. *Mol. Cancer Ther.* **2006**, *5*, 467–475. [CrossRef] [PubMed]
192. Kim, M.S.; Kwon, H.J.; Lee, Y.M.; Baek, J.H.; Jang, J.; Lee, S.; Moon, E.; Kim, H.; Lee, S.; Chung, H.Y. Histone Deacetylases Induce Angiogenesis by Negative Regulation of Tumor Suppressor Genes. *Nat. Med.* **2001**, *7*, 437–443. [CrossRef] [PubMed]
193. Hellebrekers, D.M.; Castermans, K.; Vire, E.; Dings, R.P.; Hoebers, N.T.; Mayo, K.H.; Oude Egbrink, M.G.; Molema, G.; Fuks, F.; van Engeland, M.; et al. Epigenetic Regulation of Tumor Endothelial Cell Anergy: Silencing of Intercellular Adhesion Molecule-1 by Histone Modifications. *Cancer Res.* **2006**, *66*, 10770–10777. [CrossRef] [PubMed]
194. Hellebrekers, D.M.; Melotte, V.; Vire, E.; Langenkamp, E.; Molema, G.; Fuks, F.; Herman, J.G.; Van Criekinge, W.; Griffioen, A.W.; van Engeland, M. Identification of Epigenetically Silenced Genes in Tumor Endothelial Cells. *Cancer Res.* **2007**, *67*, 4138–4148. [CrossRef] [PubMed]
195. Lu, C.; Bonome, T.; Li, Y.; Kamat, A.A.; Han, L.Y.; Schmandt, R.; Coleman, R.L.; Gershenson, D.M.; Jaffe, R.B.; Birrer, M.J.; et al. Gene Alterations Identified by Expression Profiling in Tumor-Associated Endothelial Cells from Invasive Ovarian Carcinoma. *Cancer Res.* **2007**, *67*, 1757–1768. [CrossRef] [PubMed]
196. Jones, B.A.; Varambally, S.; Arend, R.C. Histone Methyltransferase EZH2: A Therapeutic Target for Ovarian Cancer. *Mol. Cancer Ther.* **2018**, *17*, 591–602. [CrossRef] [PubMed]
197. Lu, C.; Han, H.D.; Mangala, L.S.; Ali-Fehmi, R.; Newton, C.S.; Ozbun, L.; Armaiz-Pena, G.N.; Hu, W.; Stone, R.L.; Munkarah, A. Regulation of Tumor Angiogenesis by EZH2. *Cancer Cell* **2010**, *18*, 185–197. [CrossRef] [PubMed]
198. Zhou, F.; Chen, J.; Wang, H. MicroRNA-298 Inhibits Malignant Phenotypes of Epithelial Ovarian Cancer by Regulating the Expression of EZH2. *Oncol. Lett.* **2016**, *12*, 3926–3932. [CrossRef] [PubMed]
199. Özeş, A.R.; Wang, Y.; Zong, X.; Fang, F.; Pilrose, J.; Nephew, K.P. Therapeutic Targeting using Tumor Specific Peptides Inhibits Long Non-Coding RNA HOTAIR Activity in Ovarian and Breast Cancer. *Sci. Rep.* **2017**, *7*, 894. [CrossRef] [PubMed]
200. Bergers, G.; Song, S. The Role of Pericytes in Blood-Vessel Formation and Maintenance. *Neuro-Oncology* **2005**, *7*, 452–464. [CrossRef] [PubMed]
201. Crisan, M.; Yap, S.; Casteilla, L.; Chen, C.; Corselli, M.; Park, T.S.; Andriolo, G.; Sun, B.; Zheng, B.; Zhang, L. A Perivascular Origin for Mesenchymal Stem Cells in Multiple Human Organs. *Cell Stem Cell* **2008**, *3*, 301–313. [CrossRef] [PubMed]
202. Farrington-Rock, C.; Crofts, N.J.; Doherty, M.J.; Ashton, B.A.; Griffin-Jones, C.; Canfield, A.E. Chondrogenic and Adipogenic Potential of Microvascular Pericytes. *Circulation* **2004**, *110*, 2226–2232. [CrossRef] [PubMed]
203. Balabanov, R.; Washington, R.; Wagnerova, J.; Dore-Duffy, P. CNS Microvascular Pericytes Express Macrophage-Like Function, Cell Surface Integrin αM, and Macrophage Marker ED-2. *Microvasc. Res.* **1996**, *52*, 127–142. [CrossRef] [PubMed]
204. Ribeiro, A.L.; Okamoto, O.K. Combined Effects of Pericytes in the Tumor Microenvironment. *Stem Cells Int.* **2015**, *2015*, 868475. [CrossRef] [PubMed]
205. Ferland-McCollough, D.; Slater, S.; Richard, J.; Reni, C.; Mangialardi, G. Pericytes, an Overlooked Player in Vascular Pathobiology. *Pharmacol. Ther.* **2017**, *171*, 30–42. [CrossRef] [PubMed]
206. Sinha, D.; Chong, L.; George, J.; Schluter, H.; Monchgesang, S.; Mills, S.; Li, J.; Parish, C.; Bowtell, D.; Kaur, P.; et al. Pericytes Promote Malignant Ovarian Cancer Progression in Mice and Predict Poor Prognosis in Serous Ovarian Cancer Patients. *Clin. Cancer Res.* **2016**, *22*, 1813–1824. [CrossRef] [PubMed]
207. Lu, C.; Shahzad, M.M.; Moreno-Smith, M.; Lin, Y.; Jennings, N.B.; Allen, J.K.; Landen, C.N.; Mangala, L.S.; Armaiz-Pena, G.N.; Schmandt, R. Targeting Pericytes with a PDGF-B Aptamer in Human Ovarian Carcinoma Models. *Cancer Biol. Ther.* **2010**, *9*, 176–182. [CrossRef] [PubMed]
208. Perotti, A.; Sessa, C.; Mancuso, A.; Noberasco, C.; Cresta, S.; Locatelli, A.; Carcangiu, M.L.; Passera, K.; Braghetti, A.; Scaramuzza, D.; et al. Clinical and Pharmacological Phase I Evaluation of Exherin (ADH-1), a Selective Anti-N-Cadherin Peptide in Patients with N-Cadherin-Expressing Solid Tumours. *Ann. Oncol.* **2009**, *20*, 741–745. [CrossRef] [PubMed]

209. Yi, D.; Xiang, W.; Zhang, Q.; Cen, Y.; Su, Q.; Zhang, F.; Lu, Y.; Zhao, H.; Fu, P. Human Glioblastoma-Derived Mesenchymal Stem Cell to Pericytes Transition and Angiogenic Capacity in Glioblastoma Microenvironment. *Cell Physiol. Biochem.* **2018**, *46*, 279–290. [CrossRef] [PubMed]

210. Truettner, J.S.; Katyshev, V.; Esen-Bilgin, N.; Dietrich, W.D.; Dore-Duffy, P. Hypoxia Alters MicroRNA Expression in Rat Cortical Pericytes. *Microrna* **2013**, *2*, 32–45. [CrossRef] [PubMed]

211. Yamamoto, S.; Niida, S.; Azuma, E.; Yanagibashi, T.; Muramatsu, M.; Huang, T.T.; Sagara, H.; Higaki, S.; Ikutani, M.; Nagai, Y. Inflammation-Induced Endothelial Cell-Derived Extracellular Vesicles Modulate the Cellular Status of Pericytes. *Sci. Rep.* **2015**, *5*, 8505. [CrossRef] [PubMed]

212. Kratzsch, T.; Kuhn, S.A.; Joedicke, A.; Hanisch, U.K.; Vajkoczy, P.; Hoffmann, J.; Fichtner, I. Treatment with 5-Azacitidine Delay Growth of Glioblastoma Xenografts: A Potential New Treatment Approach for Glioblastomas. *J. Cancer Res. Clin. Oncol.* **2018**, *144*, 809–819. [CrossRef] [PubMed]

213. Li, L.; Olin, E.; Buskirk, H.; Reineke, L. Cytotoxicity and Mode of Action of 5-Azacytidine on L1210 Leukemia. *Cancer Res.* **1970**, *30*, 2760–2769. [PubMed]

214. Karén, J.; Rodriguez, A.; Friman, T.; Dencker, L.; Sundberg, C.; Scholz, B. Effects of the Histone Deacetylase Inhibitor Valproic Acid on Human Pericytes in Vitro. *PLoS ONE* **2011**, *6*, e24954. [CrossRef] [PubMed]

215. Lewis, C.E.; Pollard, J.W. Distinct Role of Macrophages in Different Tumor Microenvironments. *Cancer Res.* **2006**, *66*, 605–612. [CrossRef] [PubMed]

216. Zheng, X.; Turkowski, K.; Mora, J.; Brune, B.; Seeger, W.; Weigert, A.; Savai, R. Redirecting Tumor-Associated Macrophages to Become Tumoricidal Effectors as a Novel Strategy for Cancer Therapy. *Oncotarget* **2017**, *8*, 48436–48452. [CrossRef] [PubMed]

217. Mantovani, A.; Marchesi, F.; Malesci, A.; Laghi, L.; Allavena, P. Tumour-Associated Macrophages as Treatment Targets in Oncology. *Nat. Rev. Clin. Oncol.* **2017**, *14*, 399–416. [CrossRef] [PubMed]

218. Hagemann, T.; Wilson, J.; Burke, F.; Kulbe, H.; Li, N.F.; Pluddemann, A.; Charles, K.; Gordon, S.; Balkwill, F.R. Ovarian Cancer Cells Polarize Macrophages Toward a Tumor-Associated Phenotype. *J. Immunol.* **2006**, *176*, 5023–5032. [CrossRef] [PubMed]

219. Kawamura, K.; Komohara, Y.; Takaishi, K.; Katabuchi, H.; Takeya, M. Detection of M2 Macrophages and Colony-stimulating Factor 1 Expression in Serous and Mucinous Ovarian Epithelial Tumors. *Pathol. Int.* **2009**, *59*, 300–305. [CrossRef] [PubMed]

220. Colvin, E.K. Tumor-Associated Macrophages Contribute to Tumor Progression in Ovarian Cancer. *Front. Oncol.* **2014**, *4*, 137. [CrossRef] [PubMed]

221. Reinartz, S.; Finkernagel, F.; Adhikary, T.; Rohnalter, V.; Schumann, T.; Schober, Y.; Nockher, W.A.; Nist, A.; Stiewe, T.; Jansen, J.M. A Transcriptome-Based Global Map of Signaling Pathways in the Ovarian Cancer Microenvironment Associated with Clinical Outcome. *Genome Biol.* **2016**, *17*, 108. [CrossRef] [PubMed]

222. Worzfeld, T.; Finkernagel, F.; Reinartz, S.; Konzer, A.; Adhikary, T.; Nist, A.; Stiewe, T.; Wagner, U.; Looso, M.; Graumann, J.; et al. Proteotranscriptomics Reveal Signaling Networks in the Ovarian Cancer Microenvironment. *Mol. Cell Proteom.* **2018**, *17*, 270–289. [CrossRef] [PubMed]

223. Adhikary, A.; Chakraborty, S.; Mazumdar, M.; Ghosh, S.; Mukherjee, S.; Manna, A.; Mohanty, S.; Nakka, K.K.; Joshi, S.; De, A.; et al. Inhibition of Epithelial to Mesenchymal Transition by E-Cadherin Up-Regulation Via Repression of Slug Transcription and Inhibition of E-Cadherin Degradation: Dual Role of Scaffold/Matrix Attachment Region-Binding Protein 1 (SMAR1) in Breast Cancer Cells. *J. Biol. Chem.* **2014**, *289*, 25431–25444. [CrossRef] [PubMed]

224. Ying, X.; Wu, Q.; Wu, X.; Zhu, Q.; Wang, X.; Jiang, L.; Chen, X.; Wang, X. Epithelial Ovarian Cancer-Secreted Exosomal miR-222-3p Induces Polarization of Tumor-Associated Macrophages. *Oncotarget* **2016**, *7*, 43076–43087. [CrossRef] [PubMed]

225. Chen, X.; Ying, X.; Wang, X.; Wu, X.; Zhu, Q.; Wang, X. Exosomes Derived from Hypoxic Epithelial Ovarian Cancer Deliver microRNA-940 to Induce Macrophage M2 Polarization. *Oncol. Rep.* **2017**, *38*, 522–528. [CrossRef] [PubMed]

226. Wu, Q.; Wu, X.; Ying, X.; Zhu, Q.; Wang, X.; Jiang, L.; Chen, X.; Wu, Y.; Wang, X. Suppression of Endothelial Cell Migration by Tumor Associated Macrophage-Derived Exosomes is Reversed by Epithelial Ovarian Cancer Exosomal lncRNA. *Cancer Cell Int.* **2017**, *17*, 62. [CrossRef] [PubMed]

227. Hu, Y.; Li, D.; Wu, A.; Qiu, X.; Di, W.; Huang, L.; Qiu, L. TWEAK-Stimulated Macrophages Inhibit Metastasis of Epithelial Ovarian Cancer Via Exosomal Shuttling of microRNA. *Cancer Lett.* **2017**, *393*, 60–67. [CrossRef] [PubMed]

228. Wang, H.; Chen, C.; Yang, C.; Tsai, I.; Hou, Y.; Chen, C.; Shan, Y. Tumor-Associated Macrophages Promote Epigenetic Silencing of Gelsolin through DNA Methyltransferase 1 in Gastric Cancer Cells. *Cancer Immunol. Res.* **2017**, *5*, 885–897. [CrossRef] [PubMed]
229. Tian, S.; Jiang, C.; Liu, X.; Xu, S.; Zhang, Z.; Chen, H.; Zhang, Y.; Liu, Y.; Ma, D. Hypermethylation of IFN-Γ in Oral Cancer Tissues. *Clin. Oral Investig.* **2017**, *21*, 2535–2542. [CrossRef] [PubMed]
230. Zhou, X.; Xiao, X.; Huang, T.; Du, C.; Wang, S.; Mo, Y.; Ma, N.; Murata, M.; Li, B.; Wen, W.; et al. Epigenetic Inactivation of Follistatin-Like 1 Mediates Tumor Immune Evasion in Nasopharyngeal Carcinoma. *Oncotarget* **2016**, *7*, 16433–16444. [CrossRef] [PubMed]
231. Osawa, T.; Tsuchida, R.; Muramatsu, M.; Shimamura, T.; Wang, F.; Suehiro, J.; Kanki, Y.; Wada, Y.; Yuasa, Y.; Aburatani, H.; et al. Inhibition of Histone Demethylase JMJD1A Improves Anti-Angiogenic Therapy and Reduces Tumor-Associated Macrophages. *Cancer Res.* **2013**, *73*, 3019–3028. [CrossRef] [PubMed]
232. Ishii, M.; Wen, H.; Corsa, C.A.; Liu, T.; Coelho, A.L.; Allen, R.M.; Carson, W.F., 4th; Cavassani, K.A.; Li, X.; Lukacs, N.W.; et al. Epigenetic Regulation of the Alternatively Activated Macrophage Phenotype. *Blood* **2009**, *114*, 3244–3254. [CrossRef] [PubMed]
233. Leffers, N.; Gooden, M.J.; de Jong, R.A.; Hoogeboom, B.; Klaske, A.; Hollema, H.; Boezen, H.M.; van der Zee, A.G.J.; Daemen, T.; Nijman, H.W. Prognostic Significance of Tumor-Infiltrating T-Lymphocytes in Primary and Metastatic Lesions of Advanced Stage Ovarian Cancer. *Cancer Immunol. Immunother.* **2009**, *58*, 449–459. [CrossRef] [PubMed]
234. James, F.R.; Jiminez-Linan, M.; Alsop, J.; Mack, M.; Song, H.; Brenton, J.D.; Pharoah, P.D.; Ali, H.R. Association between Tumour Infiltrating Lymphocytes, Histotype and Clinical Outcome in Epithelial Ovarian Cancer. *BMC Cancer* **2017**, *17*, 657. [CrossRef] [PubMed]
235. Barnett, J.C.; Bean, S.M.; Whitaker, R.S.; Kondoh, E.; Baba, T.; Fujii, S.; Marks, J.R.; Dressman, H.K.; Murphy, S.K.; Berchuck, A. Ovarian Cancer Tumor Infiltrating T-Regulatory (T(Reg)) Cells are Associated with a Metastatic Phenotype. *Gynecol. Oncol.* **2010**, *116*, 556–562. [CrossRef] [PubMed]
236. Tsiatas, M.L.; Gyftaki, R.; Liacos, C.; Politi, E.; Rodolakis, A.; Dimopoulos, M.A.; Bamias, A. Study of T Lymphocytes Infiltrating Peritoneal Metastases in Advanced Ovarian Cancer: Associations with Vascular Endothelial Growth Factor Levels and Prognosis in Patients Receiving Platinum-Based Chemotherapy. *Int. J. Gynecol. Cancer* **2009**, *19*, 1329–1334. [CrossRef] [PubMed]
237. Napoletano, C.; Bellati, F.; Landi, R.; Pauselli, S.; Marchetti, C.; Visconti, V.; Sale, P.; Liberati, M.; Rughetti, A.; Frati, L. Ovarian Cancer Cytoreduction Induces Changes in T Cell Population Subsets Reducing Immunosuppression. *J. Cell. Mol. Med.* **2010**, *14*, 2748–2759. [CrossRef] [PubMed]
238. Curiel, T.J.; Coukos, G.; Zou, L.; Alvarez, X.; Cheng, P.; Mottram, P.; Evdemon-Hogan, M.; Conejo-Garcia, J.R.; Zhang, L.; Burow, M. Specific Recruitment of Regulatory T Cells in Ovarian Carcinoma Fosters Immune Privilege and Predicts Reduced Survival. *Nat. Med.* **2004**, *10*, 942–949. [CrossRef] [PubMed]
239. Santoiemma, P.P.; Powell, D.J., Jr. Tumor Infiltrating Lymphocytes in Ovarian Cancer. *Cancer Biol. Ther.* **2015**, *16*, 807–820. [CrossRef] [PubMed]
240. Sehouli, J.; Loddenkemper, C.; Cornu, T.; Schwachula, T.; Hoffmüller, U.; Grützkau, A.; Lohneis, P.; Dickhaus, T.; Gröne, J.; Kruschewski, M. Epigenetic Quantification of Tumor-Infiltrating T-Lymphocytes. *Epigenetics* **2011**, *6*, 236–246. [CrossRef] [PubMed]
241. Li, H.; Chiappinelli, K.B.; Guzzetta, A.A.; Easwaran, H.; Yen, R.W.; Vatapalli, R.; Topper, M.J.; Luo, J.; Connolly, R.M.; Azad, N.S.; et al. Immune Regulation by Low Doses of the DNA Methyltransferase Inhibitor 5-Azacitidine in Common Human Epithelial Cancers. *Oncotarget* **2014**, *5*, 587–598. [CrossRef] [PubMed]
242. Wang, L.; Amoozgar, Z.; Huang, J.; Saleh, M.H.; Xing, D.; Orsulic, S.; Goldberg, M.S. Decitabine Enhances Lymphocyte Migration and Function and Synergizes with CTLA-4 Blockade in a Murine Ovarian Cancer Model. *Cancer. Immunol. Res.* **2015**, *3*, 1030–1041. [CrossRef] [PubMed]
243. Stone, M.L.; Chiappinelli, K.B.; Li, H.; Murphy, L.M.; Travers, M.E.; Topper, M.J.; Mathios, D.; Lim, M.; Shih, I.M.; Wang, T.L.; et al. Epigenetic Therapy Activates Type I Interferon Signaling in Murine Ovarian Cancer to Reduce Immunosuppression and Tumor Burden. *Proc. Natl. Acad. Sci. USA* **2017**, *114*, E10981–E10990. [CrossRef] [PubMed]
244. Adair, S.J.; Hogan, K.T. Treatment of Ovarian Cancer Cell Lines with 5-Aza-2′-Deoxycytidine Upregulates the Expression of Cancer-Testis Antigens and Class I Major Histocompatibility Complex-Encoded Molecules. *Cancer Immunol. Immunother.* **2009**, *58*, 589–601. [CrossRef] [PubMed]

245. Chiappinelli, K.B.; Strissel, P.L.; Desrichard, A.; Li, H.; Henke, C.; Akman, B.; Hein, A.; Rote, N.S.; Cope, L.M.; Snyder, A. Inhibiting DNA Methylation Causes an Interferon Response in Cancer Via dsRNA Including Endogenous Retroviruses. *Cell* **2015**, *162*, 974–986. [CrossRef] [PubMed]
246. Korsunsky, I.; Parameswaran, J.; Shapira, I.; Lovecchio, J.; Menzin, A.; Whyte, J.; Dos Santos, L.; Liang, S.; Bhuiya, T.; Keogh, M.; et al. Two microRNA Signatures for Malignancy and Immune Infiltration Predict overall Survival in Advanced Epithelial Ovarian Cancer. *J. Investig. Med.* **2017**, *65*, 1068–1076. [CrossRef] [PubMed]
247. McKenna, K.; Beignon, A.S.; Bhardwaj, N. Plasmacytoid Dendritic Cells: Linking Innate and Adaptive Immunity. *J. Virol.* **2005**, *79*, 17–27. [CrossRef] [PubMed]
248. Vermi, W.; Soncini, M.; Melocchi, L.; Sozzani, S.; Facchetti, F. Plasmacytoid Dendritic Cells and Cancer. *J. Leukoc. Biol.* **2011**, *90*, 681–690. [CrossRef] [PubMed]
249. Li, S.; Wu, J.; Zhu, S.; Liu, Y.; Chen, J. Disease-Associated Plasmacytoid Dendritic Cells. *Front. Immunol.* **2017**, *8*, 1268. [CrossRef] [PubMed]
250. Wertel, I.; Polak, G.; Bednarek, W.; Barczyński, B.; Roliński, J.; Kotarski, J. Dendritic Cell Subsets in the Peritoneal Fluid and Peripheral Blood of Women Suffering from Ovarian Cancer. *Cytom. Part B Clin. Cytom.* **2008**, *74*, 251–258. [CrossRef] [PubMed]
251. Curiel, T.J.; Cheng, P.; Mottram, P.; Alvarez, X.; Moons, L.; Evdemon-Hogan, M.; Wei, S.; Zou, L.; Kryczek, I.; Hoyle, G.; et al. Dendritic Cell Subsets Differentially Regulate Angiogenesis in Human Ovarian Cancer. *Cancer Res.* **2004**, *64*, 5535–5538. [CrossRef] [PubMed]
252. Conejo-Garcia, J.R.; Benencia, F.; Courreges, M.; Kang, E.; Mohamed-Hadley, A.; Buckanovich, R.J.; Holtz, D.O.; Jenkins, A.; Na, H.; Zhang, L. Tumor-Infiltrating Dendritic Cell Precursors Recruited by a B-Defensin Contribute to Vasculogenesis Under the Influence of Vegf-A. *Nat. Med.* **2004**, *10*, 950–958. [CrossRef] [PubMed]
253. Wei, S.; Kryczek, I.; Zou, L.; Daniel, B.; Cheng, P.; Mottram, P.; Curiel, T.; Lange, A.; Zou, W. Plasmacytoid Dendritic Cells Induce CD8+ Regulatory T Cells in Human Ovarian Carcinoma. *Cancer Res.* **2005**, *65*, 5020–5026. [CrossRef] [PubMed]
254. Labidi-Galy, S.I.; Treilleux, I.; Goddard-Leon, S.; Combes, J.; Blay, J.; Ray-Coquard, I.; Caux, C.; Bendriss-Vermare, N. Plasmacytoid Dendritic Cells Infiltrating Ovarian Cancer are Associated with Poor Prognosis. *Oncoimmunology* **2012**, *1*, 380–382. [CrossRef] [PubMed]
255. Huarte, E.; Cubillos-Ruiz, J.R.; Nesbeth, Y.C.; Scarlett, U.K.; Martinez, D.G.; Buckanovich, R.J.; Benencia, F.; Stan, R.V.; Keler, T.; Sarobe, P.; et al. Depletion of Dendritic Cells Delays Ovarian Cancer Progression by Boosting Antitumor Immunity. *Cancer Res.* **2008**, *68*, 7684–7691. [CrossRef] [PubMed]
256. Tian, Y.; Meng, L.; Zhang, Y. Epigenetic Regulation of Dendritic Cell Development and Function. *Cancer J.* **2017**, *23*, 302–307. [CrossRef] [PubMed]
257. Paul, F.; Amit, I. Plasticity in the Transcriptional and Epigenetic Circuits Regulating Dendritic Cell Lineage Specification and Function. *Curr. Opin. Immunol.* **2014**, *30*, 1–8. [CrossRef] [PubMed]
258. Kim, H.; Lee, Y.; Park, J.H.; Kim, Y. Transcriptional and Epigenetic Networks in the Development and Maturation of Dendritic Cells. *Epigenomics* **2013**, *5*, 195–204. [CrossRef] [PubMed]
259. Smyth, L.A.; Boardman, D.A.; Tung, S.L.; Lechler, R.; Lombardi, G. MicroRNAs Affect Dendritic Cell Function and Phenotype. *Immunology* **2015**, *144*, 197–205. [CrossRef] [PubMed]
260. Valadkhan, S.; Gunawardane, L.S. lncRNA-Mediated Regulation of the Interferon Response. *Virus Res.* **2016**, *212*, 127–136. [CrossRef] [PubMed]
261. Cubillos-Ruiz, J.R.; Baird, J.R.; Tesone, A.J.; Rutkowski, M.R.; Scarlett, U.K.; Camposeco-Jacobs, A.L.; Anadon-Arnillas, J.; Harwood, N.M.; Korc, M.; Fiering, S.N.; et al. Reprogramming Tumor-Associated Dendritic Cells in Vivo using miRNA Mimetics Triggers Protective Immunity Against Ovarian Cancer. *Cancer Res.* **2012**, *72*, 1683–1693. [CrossRef] [PubMed]
262. Rodríguez-Ubreva, J.; Català-Moll, F.; Obermajer, N.; Álvarez-Errico, D.; Ramirez, R.N.; Vento-Tormo, R.; Moreno-Bueno, G.; Edwards, R.P.; Mortazavi, A.; Kalinski, P. Prostaglandin E2 Leads to the Acquisition of DNMT3A-Dependent Tolerogenic Functions in Human Myeloid-Derived Suppressor Cells. *Cell Rep.* **2017**, *21*, 154–167. [CrossRef] [PubMed]
263. Davidowitz, R.A.; Iwanicki, M.P.; Brugge, J.S. In Vitro Mesothelial Clearance Assay that Models the Early Steps of Ovarian Cancer Metastasis. *J. Vis. Exp.* **2012**, *60*, 3888. [CrossRef] [PubMed]

264. Klymenko, Y.; Kim, O.; Loughran, E.; Yang, J.; Lombard, R.; Alber, M.; Stack, M. Cadherin Composition and Multicellular Aggregate Invasion in Organotypic Models of Epithelial Ovarian Cancer Intraperitoneal Metastasis. *Oncogene* **2017**, *19*, 5840–5851. [CrossRef] [PubMed]
265. Ren, J.; Xiao, Y.J.; Singh, L.S.; Zhao, X.; Zhao, Z.; Feng, L.; Rose, T.M.; Prestwich, G.D.; Xu, Y. Lysophosphatidic Acid is Constitutively Produced by Human Peritoneal Mesothelial Cells and Enhances Adhesion, Migration, and Invasion of Ovarian Cancer Cells. *Cancer Res.* **2006**, *66*, 3006–3014. [CrossRef] [PubMed]
266. Stadlmann, S.; Amberger, A.; Pollheimer, J.; Gastl, G.; Offner, F.A.; Margreiter, R.; Zeimet, A.G. Ovarian Carcinoma Cells and IL-1beta-Activated Human Peritoneal Mesothelial Cells are Possible Sources of Vascular Endothelial Growth Factor in Inflammatory and Malignant Peritoneal Effusions. *Gynecol. Oncol.* **2005**, *97*, 784–789. [CrossRef] [PubMed]
267. Kassim, S.K.; El-Salahy, E.M.; Fayed, S.T.; Helal, S.A.; Helal, T.; Azzam, E.E.; Khalifa, A. Vascular Endothelial Growth Factor and Interleukin-8 are Associated with Poor Prognosis in Epithelial Ovarian Cancer Patients. *Clin. Biochem.* **2004**, *37*, 363–369. [CrossRef] [PubMed]
268. Moghaddam, S.M.; Amini, A.; Morris, D.L.; Pourgholami, M.H. Significance of Vascular Endothelial Growth Factor in Growth and Peritoneal Dissemination of Ovarian Cancer. *Cancer Metastasis Rev.* **2012**, *31*, 143–162. [CrossRef] [PubMed]
269. Li, H.; Wang, D.; Zhang, H.; Kirmani, K.; Zhao, Z.; Steinmetz, R.; Xu, Y. Lysophosphatidic Acid Stimulates Cell Migration, Invasion, and Colony Formation as Well as Tumorigenesis/Metastasis of Mouse Ovarian Cancer in Immunocompetent Mice. *Mol. Cancer Ther.* **2009**, *8*, 1692–1701. [CrossRef] [PubMed]
270. Zeimet, A.G.; Stadlmann, S.; Natoli, C.; Widschwendter, M.; Hermann, M.; Abendstein, B.; Daxenbichler, G.; Offner, F.A.; Iacobelli, S.; Marth, C. Peritoneal Mesothelial Cells as a Significant Source of Ascitic Immunostimulatory Protein 90K. *Anticancer Res.* **2000**, *20*, 4507–4511. [PubMed]
271. Zeimet, A.G.; Natoli, C.; Herold, M.; Fuchs, D.; Windbichler, G.; Daxenbichler, G.; Iacobelli, S.; Dapunt, O.; Marth, C. Circulating Immunostimulatory Protein 90K and Soluble Interleukin-2-receptor in Human Ovarian Cancer. *Int. J. Cancer* **1996**, *68*, 34–38. [CrossRef]
272. Iacobelli, S.; Sismondi, P.; Giai, M.; d'Egidio, M.; Tinari, N.; Amatetti, C.; Di Stefano, P.; Natoli, C. Prognostic Value of a Novel Circulating Serum 90K Antigen in Breast Cancer. *Br. J. Cancer* **1994**, *69*, 172–176. [CrossRef] [PubMed]
273. Fornarini, B.; D'Ambrosio, C.; Natoli, C.; Tinari, N.; Silingardi, V.; Iacobelli, S. Adhesion to 90K (Mac-2 BP) as a Mechanism for Lymphoma Drug Resistance in Vivo. *Blood* **2000**, *96*, 3282–3285. [PubMed]
274. Matte, I.; Lane, D.; Bachvarov, D.; Rancourt, C.; Piché, A. Role of Malignant Ascites on Human Mesothelial Cells and their Gene Expression Profiles. *BMC Cancer* **2014**, *14*, 288. [CrossRef] [PubMed]
275. Fang, F.; Munck, J.; Tang, J.; Taverna, P.; Wang, Y.; Miller, D.F.; Pilrose, J.; Choy, G.; Azab, M.; Pawelczak, K.S.; et al. The Novel, Small-Molecule DNA Methylation Inhibitor SGI-110 as an Ovarian Cancer Chemosensitizer. Clin. *Cancer Res.* **2014**, *20*, 6504–6516.
276. Fang, F.; Cardenas, H.; Huang, H.; Jiang, G.; Perkins, S.M.; Zhang, C.; Keer, H.N.; Liu, Y.; Nephew, K.P.; Matei, D. Genomic and Epigenomic Signatures in Ovarian Cancer Associated with Resensitization to Platinum Drugs. *Cancer Res.* **2018**, *78*, 631–644. [CrossRef] [PubMed]
277. Fang, F.; Balch, C.; Schilder, J.; Breen, T.; Zhang, S.; Shen, C.; Li, L.; Kulesavage, C.; Snyder, A.J.; Nephew, K.P. A Phase 1 and Pharmacodynamic Study of Decitabine in Combination with Carboplatin in Patients with Recurrent, Platinum-resistant, Epithelial Ovarian Cancer. *Cancer* **2010**, *116*, 4043–4053. [CrossRef] [PubMed]
278. Wang, Y.; Cardenas, H.; Fang, F.; Condello, S.; Taverna, P.; Segar, M.; Liu, Y.; Nephew, K.P.; Matei, D. Epigenetic Targeting of Ovarian Cancer Stem Cells. *Cancer Res.* **2014**, *74*, 4922–4936. [CrossRef] [PubMed]
279. Matei, D.; Ghamande, S.; Roman, L.; Alvarez Secord, A.; Nemunaitis, J.; Markham, M.J.; Nephew, K.P.; Jueliger, S.; Oganesian, A.; Naim, S.; et al. A Phase I Clinical Trial of Guadecitabine and Carboplatin in Platinum-Resistant, Recurrent Ovarian Cancer: Clinical, Pharmacokinetic, and Pharmacodynamic Analyses. *Clin. Cancer Res.* **2018**, *24*, 2285–2293. [CrossRef] [PubMed]
280. Modesitt, S.C.; Sill, M.; Hoffman, J.S.; Bender, D.P. A Phase II Study of Vorinostat in the Treatment of Persistent or Recurrent Epithelial Ovarian Or Primary Peritoneal Carcinoma: A Gynecologic Oncology Group Study. *Gynecol. Oncol.* **2008**, *109*, 182–186. [CrossRef] [PubMed]
281. Qian, X.; LaRochelle, W.J.; Ara, G.; Wu, F.; Petersen, K.D.; Thougaard, A.; Sehested, M.; Lichenstein, H.S.; Jeffers, M. Activity of PXD101, a Histone Deacetylase Inhibitor, in Preclinical Ovarian Cancer Studies. *Mol. Cancer. Ther.* **2006**, *5*, 2086–2095. [CrossRef] [PubMed]

282. Yang, Y.T.; Balch, C.; Kulp, S.K.; Mand, M.R.; Nephew, K.P.; Chen, C.S. A Rationally Designed Histone Deacetylase Inhibitor with Distinct Antitumor Activity Against Ovarian Cancer. *Neoplasia* **2009**, *11*, 552–563, 3 p following 563. [CrossRef] [PubMed]
283. Halsall, J.A.; Turan, N.; Wiersma, M.; Turner, B.M. Cells Adapt to the Epigenomic Disruption Caused by Histone Deacetylase Inhibitors through a Coordinated, Chromatin-Mediated Transcriptional Response. *Epigenet. Chromatin* **2015**, *8*, 29. [CrossRef] [PubMed]
284. Adler, J.T.; Hottinger, D.G.; Kunnimalaiyaan, M.; Chen, H. Combination Therapy with Histone Deacetylase Inhibitors and Lithium Chloride: A Novel Treatment for Carcinoid Tumors. *Ann. Surg. Oncol.* **2009**, *16*, 481–486. [CrossRef] [PubMed]
285. Halsall, J.A.; Turner, B.M. Histone Deacetylase Inhibitors for Cancer Therapy: An Evolutionarily Ancient Resistance Response May Explain Their Limited Success. *Bioessays* **2016**, *38*, 1102–1110. [CrossRef] [PubMed]
286. Meseure, D.; Drak Alsibai, K.; Nicolas, A.; Bieche, I.; Morillon, A. Long Noncoding RNAs as New Architects in Cancer Epigenetics, Prognostic Biomarkers, and Potential Therapeutic Targets. *BioMed Res. Int.* **2015**, *2015*, 320214. [CrossRef] [PubMed]
287. Labiche, A.; Heutte, N.; Herlin, P.; Chasle, J.; Gauduchon, P.; Elie, N. Stromal Compartment as a Survival Prognostic Factor in Advanced Ovarian Carcinoma. *Int. J. Gynecol. Cancer* **2010**, *20*, 28–33. [CrossRef] [PubMed]
288. Steele, N.; Finn, P.; Brown, R.; Plumb, J. Combined Inhibition of DNA Methylation and Histone Acetylation Enhances Gene Re-Expression and Drug Sensitivity in Vivo. *Br. J. Cancer* **2009**, *100*, 758–763. [CrossRef] [PubMed]
289. Ganesan, A.; Nolan, L.; Crabb, S.; Packham, G. Epigenetic Therapy: Histone Acetylation, DNA Methylation and Anti-Cancer Drug Discovery. *Curr. Cancer Drug Targets* **2009**, *9*, 963–981. [CrossRef] [PubMed]
290. Yeung, T.; Leung, C.S.; Mok, S.C. CAF Reprogramming Inhibits Ovarian Cancer Progression. *Cell Cycle* **2014**, *13*, 3783–3784. [CrossRef] [PubMed]
291. Cheng, L.; Wu, S.; Zhang, K.; Xu, T. A Comprehensive Overview of Exosomes in Ovarian Cancer: Emerging Biomarkers and Therapeutic Strategies. *J. Ovar. Res.* **2017**, *10*, 73. [CrossRef] [PubMed]
292. Chiriva-Internati, M.; Mirandola, L.; Kast, W.M.; Jenkins, M.R.; Cobos, E.; Cannon, M.J. Understanding the Cross-Talk between Ovarian Tumors and Immune Cells: Mechanisms for Effective Immunotherapies. *Int. Rev. Immunol.* **2011**, *30*, 71–86. [CrossRef] [PubMed]
293. Chae, C.; Teran-Cabanillas, E.; Cubillos-Ruiz, J.R. Dendritic Cell Rehab: New Strategies to Unleash Therapeutic Immunity in Ovarian Cancer. *Cancer Immunol. Immunother.* **2017**, *66*, 969–977. [CrossRef] [PubMed]
294. Drakes, M.L.; Stiff, P.J. Understanding Dendritic Cell Immunotherapy in Ovarian Cancer. *Expert Rev. Anticancer Ther.* **2016**, *16*, 643–652. [CrossRef] [PubMed]
295. Wefers, C.; Lambert, L.J.; Torensma, R.; Hato, S.V. Cellular Immunotherapy in Ovarian Cancer: Targeting the Stem of Recurrence. *Gynecol. Oncol.* **2015**, *137*, 335–342. [CrossRef] [PubMed]
296. Odunsi, K. Immunotherapy in Ovarian Cancer. *Ann. Oncol.* **2017**, *28*, viii1–viii7. [CrossRef] [PubMed]
297. Barrero, M.J. Epigenetic Strategies to Boost Cancer Immunotherapies. *Int. J. Mol. Sci.* **2017**, *18*, 1108. [CrossRef] [PubMed]
298. Terranova-Barberio, M.; Thomas, S.; Munster, P.N. Epigenetic Modifiers in Immunotherapy: A Focus on Checkpoint Inhibitors. *Immunotherapy* **2016**, *8*, 705–719. [CrossRef] [PubMed]
299. Strick, R.; Strissel, P.L.; Baylin, S.B.; Chiappinelli, K.B. Unraveling the Molecular Pathways of DNA-Methylation Inhibitors: Human Endogenous Retroviruses Induce the Innate Immune Response in Tumors. *Oncoimmunology* **2016**, *5*, e1122160. [CrossRef] [PubMed]
300. Song, D.; Ye, Q.; Santoro, S.; Fang, C.; Best, A.; Powell, D.J., Jr. Chimeric NKG2D CAR-Expressing T Cell-Mediated Attack of Human Ovarian Cancer is Enhanced by Histone Deacetylase Inhibition. *Hum. Gene Ther.* **2013**, *24*, 295–305. [CrossRef] [PubMed]
301. Wargo, J.A.; Robbins, P.F.; Li, Y.; Zhao, Y.; El-Gamil, M.; Caragacianu, D.; Zheng, Z.; Hong, J.A.; Downey, S.; Schrump, D.S. Recognition of NY-ESO-1 Tumor Cells by Engineered Lymphocytes is Enhanced by Improved Vector Design and Epigenetic Modulation of Tumor Antigen Expression. *Cancer Immunol. Immunother.* **2009**, *58*, 383–394. [CrossRef] [PubMed]
302. Scanlan, M.J.; Gure, A.O.; Jungbluth, A.A.; Old, L.J.; Chen, Y. Cancer/Testis Antigens: An Expanding Family of Targets for Cancer Immunotherapy. *Immunol. Rev.* **2002**, *188*, 22–32. [CrossRef] [PubMed]

303. Odunsi, K.; Matsuzaki, J.; James, S.R.; Mhawech-Fauceglia, P.; Tsuji, T.; Miller, A.; Zhang, W.; Akers, S.N.; Griffiths, E.A.; Miliotto, A.; et al. Epigenetic Potentiation of NY-ESO-1 Vaccine Therapy in Human Ovarian Cancer. *Cancer. Immunol. Res.* **2014**, *2*, 37–49. [CrossRef] [PubMed]
304. Wang, D.; Quiros, J.; Mahuron, K.; Pai, C.; Ranzani, V.; Young, A.; Silveria, S.; Harwin, T.; Abnousian, A.; Pagani, M. Targeting EZH2 Reprograms Intratumoral Regulatory T Cells to Enhance Cancer Immunity. *Cell Rep.* **2018**, *23*, 3262–3274. [CrossRef] [PubMed]

cancers

MDPI

Review

Cancer Associated Fibroblasts: Naughty Neighbors That Drive Ovarian Cancer Progression

Subramanyam Dasari [1,†], Yiming Fang [1,†] and Anirban K. Mitra [1,2,3,*]

1 Medical Sciences Program, Indiana University School of Medicine, Bloomington, IN 47401, USA; sudasari@iu.edu (S.D.); yimfang@iu.edu (Y.F.)
2 Indiana University Melvin and Bren Simon Cancer Center, Indianapolis, IN 46202, USA
3 Department of Medical and Molecular Genetics, Indiana University School of Medicine, Indianapolis, IN 46202, USA
* Correspondence: anmitra@indiana.edu; Tel.: +1-812-856-1207
† These authors contributed equally to this work.

Received: 12 September 2018; Accepted: 24 October 2018; Published: 29 October 2018

Abstract: Ovarian cancer is the most lethal gynecologic malignancy, and patient prognosis has not improved significantly over the last several decades. In order to improve therapeutic approaches and patient outcomes, there is a critical need for focused research towards better understanding of the disease. Recent findings have revealed that the tumor microenvironment plays an essential role in promoting cancer progression and metastasis. The tumor microenvironment consists of cancer cells and several different types of normal cells recruited and reprogrammed by the cancer cells to produce factors beneficial to tumor growth and spread. These normal cells present within the tumor, along with the various extracellular matrix proteins and secreted factors, constitute the tumor stroma and can compose 10–60% of the tumor volume. Cancer associated fibroblasts (CAFs) are a major constituent of the tumor microenvironment, and play a critical role in promoting many aspects of tumor function. This review will describe the various hypotheses about the origin of CAFs, their major functions in the tumor microenvironment in ovarian cancer, and will discuss the potential of targeting CAFs as a possible therapeutic approach.

Keywords: ovarian cancer; tumor microenvironment; cancer associated fibroblasts; fibroblast; cross-talk; invasion; angiogenesis; ECM; chemoresistance; therapy

1. Introduction

Ovarian cancer is the deadliest of all the gynecologic malignancies, and is the fifth leading cause of cancer related deaths among women in the United States. There has been only a modest improvement in ovarian cancer patient prognosis over the last several decades [1,2]. Therefore, there is a critical need for focused research to improve our understanding of the disease and develop novel therapies that are more effective. In the past, most of the research efforts were focused on the cancer cells in isolation. However, recent research has identified the tumor microenvironment as a key factor in promoting cancer progression [3–7]. The cancer cells recruit various normal cells and reprogram them to produce factors beneficial to tumor growth and spread. These normal cells present within the tumor constitute the tumor stroma and can compose 10–60% of the tumor volume [3].

In order to survive and proliferate, the cancer cells productively interact with their microenvironment. The tumor microenvironment is complex and contains a variety of cells constituting the tumor stroma. Tumors however, can only grow if their complex tissue environment provides them with a milieu of factors and conditions that can sustain their growth and spread [8]. A complicated bidirectional interaction is therefore happening at the interface between the genetically unstable malignant cells and the genetically stable stroma, a process that will determine the degree of tumor promotion and proliferation, invasiveness, potential for spread, and even patient prognosis [9].

The tumor stroma consists of cellular components like the cancer associated fibroblasts (CAFs), immune cells, endothelial cells, pericytes, adipocytes, and so forth, as well as acellular components like the extra cellular matrix proteins (ECMs) [3–7]. Each of these tumor microenvironmental elements has been shown to play important roles in tumor growth and progression in various cancers, including ovarian cancer. CAFs are an important constituent of the tumor stroma, and this review will focus on providing an overview of the origin, function, and potential targeting of CAFs in ovarian cancer therapy.

2. Origin of CAFs

There are several hypotheses about the origin of CAFs, which include the reprogramming of the resident normal fibroblasts by the cancer cells, and differentiation of mesenchymal stem cells (Figure 1). A less widely accepted theory is that they are already present as a small subpopulation of normal fibroblasts, which are selected for by the cancer cells [10]. These subpopulations may have acquired a mutation, or epigenetic alterations, independent of the tumor cells, which transform them into activated fibroblasts. A proinflammatory microenvironment resulting in the generation of reactive oxygen species may promote acquisition of genetic lesions [11]. As the tumor develops in their vicinity, these subpopulations might be selected for by the cancer cells for their ability to support tumor growth [10]. Since mutations are not commonly found in the tumor stroma, and CAFs are not believed to have clonal populations with distinct genetic changes, this hypothesis has not gained much traction [12]. Fibroblasts are mesenchymal cells that are generally present in the basement membrane and serve as a scaffold, secreting ECMs, and growth and trophic factors [13,14]. They are generally in a quiescent "inactive" state, but retain some plasticity, and are capable of getting "activated" by various physiological stimuli. During development, dermal fibroblasts play a role in tissue patterning, and can differentiate into multiple kind of cells, including hair follicle cells, papillary cells, reticular cells, and pre-adipocytes [15]. The fibroblasts get activated at the site of wound healing by factors such as insulin-like growth factors (IGFs), transforming growth factor beta 1,2,3 (TGF-β 1,2,3), and platelet-derived growth factor (PDGF), among others. These activated fibroblasts express α-smooth muscle actin (α-SMA), which makes them contractile, and helps in wound closure. These α-SMA expressing fibroblasts are called myofibroblasts. They secrete various ECMs and extracellular proteases, which help in the initial wound healing and development of the scar. They also secrete factors like TGF-β to stimulate epithelial to mesenchymal transition in the epithelial cells around the wound. This enables the epithelial cells to move and close the wound. Thereafter, as the wound heals, epithelialization is promoted by epidermal growth factors (EGFs) and the keratinocyte growth factor (KGF) produced by the fibroblasts [16].

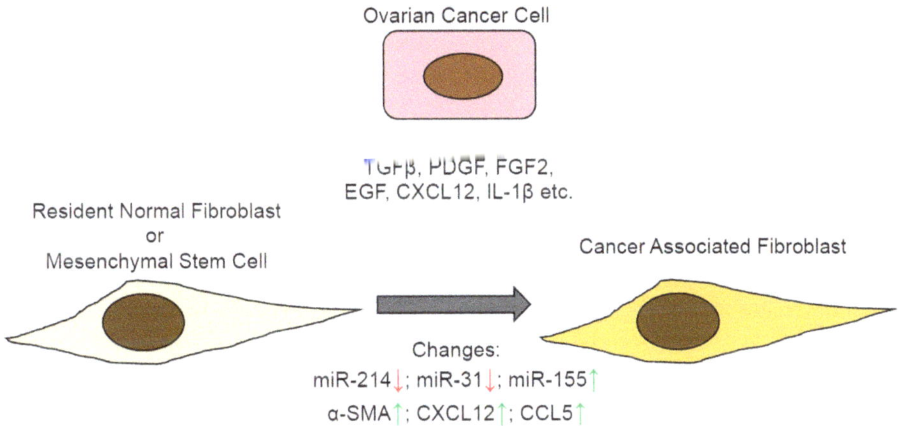

Figure 1. Formation of cancer associated fibroblasts (CAFs) through the reprogramming of resident normal fibroblasts or mesenchymal stem cells by ovarian cancer cells. TGF-β: transforming growth factor beta; PDGF: platelet-derived growth factor; FGF: fibroblast growth factor; EGF: epidermal growth factors; CXCL12: C-X-C Motif Chemokine Ligand 12; CCL5: C-C Motif Chemokine Ligand 5; ↑: upregulated;↓: downregulated.

CAFs display several traits of the activated fibroblasts found in healing wounds, including upregulation of TGF-β, increased secretion of ECMs, extra cellular proteases, and expression of α-SMA. It is believed that cancer cells can recruit the resident normal fibroblasts and reprogram them into CAFs. Several reports have demonstrated evidence in support of this hypothesis in ovarian cancer [7]. Ovarian cancer cells produce factors including TGF-β and PDGF, that can change normal fibroblasts into "activated" CAFs. We have previously shown that ovarian cancer cells can induce a change in expression of a set of 3 microRNAs in the resident normal omental fibroblasts, which reprograms them into CAFs [7]. miR-214 and miR-31 were found to decrease, while miR-155 expression increased, in the normal fibroblasts because of their interaction with the metastasizing ovarian cancer cells. This resulted in their reprogramming into CAFs (Figure 1). It was the first report of a set of microRNAs reprogramming normal fibroblasts into CAFs. Simultaneous inhibition of miR-214 and miR-31, along with overexpression of miR-155, could convert normal fibroblasts into CAFs. These results supported previous findings in other cancers, which demonstrated the absence of mutations in CAFs. Moreover, CAFs can be isolated from tumors and cultured in vitro for several passages, and yet retain their phenotype and their ability to support cancer cell functions. This suggested a potential role of epigenetic regulation [17]. The role of microRNAs in reprogramming of normal fibroblasts into CAFs further revealed a potential mechanism. Interestingly, overexpressing miR-214 and miR-31 and inhibiting miR-155 simultaneously in CAFs could revert them back into normal fibroblasts. This offers a potential opportunity to normalize a key component of the tumor microenvironment. Research on targeting the tumor microenvironment has revealed that the normalization of the tumor microenvironment is a more effective approach as compared to attempts at obliterating it altogether. The latter typically leads to the cancer cells becoming more aggressive. Depleting α-SMA positive CAFs in a transgenic mouse model of pancreatic ductal adenocarcinoma through induction of thymidine kinase by ganciclovir administration, either early in the tumor precursor stage or late carcinoma stage, led to the development of undifferentiated tumors, which were highly invasive and resulted in decreased survival [18]. Since this was observed when the CAFs were ablated in the precursor lesions or in the late carcinomas, it indicated that irrespective of tumor stage, in the absence of the microenvironmental support, the more aggressive cancer cell clones are selected. These findings are similar to the increased metastasis observed upon pericyte depletion [19].

Other hypotheses about the origin of CAFs include the recruitment of mesenchymal stem cells by the cancer cells to the tumor [10,20]. Human pancreatic cancer cells were shown to recruit bone marrow derived progenitors when injected in mice [21]. The resulting tumors had about 40% myofibroblasts derived from bone marrow cells. Similarly, CAFs were shown to be derived from mesenchymal stem cells in ovarian cancer and supported tumor growth through the secretion of the paracrine factor IL-6 [22]. Ovarian cancer cells have been reported to secrete IL-1β that leads to the decreased expression of p53 protein in the ovarian fibroblasts, converting them into CAFs [23]. The decreased p53 resulted in increased secretion of IL-8, growth regulated oncogene-alpha (GRO-α), IL-6, IL-1β, and vascular endothelial growth factor (VEGF) by the CAFs. Mesenchymal stem cell derived CAFs were also shown to regulate ovarian cancer stem cells through bone morphogenetic protein secretion, which resulted in resistance to chemotherapy [24,25]. The mesothelial cells lining the peritoneum and the omentum have also been reported as a source of CAFs in ovarian cancer peritoneal and omental metastasis [26]. The mesothelial cells have been shown to undergo mesothelial to mesenchymal transition under the influence of ovarian cancer cell secreted TGF-β, which can form a subpopulation of the CAFs in ovarian cancer metastatic tumors [27]. Others have demonstrated that ovarian cancer cells interact with the mesothelial cells in a β1-integrin-dependent manner to induce mesothelial to mesenchymal transition and convert them into CAFs [28].

3. CAF Markers

Since CAFs in the tumor microenvironment are functionally very similar to the activated fibroblasts in healing wounds, they both share several markers. CAFs express α-SMA, which is also a marker of myofibroblasts in wound healing [10]. Ovarian cancer CAFs also express α-SMA, while normal fibroblasts do not [29]. However, the expression of α-SMA is only one of many changes that occur in activated fibroblasts [30]. The levels of expression of α-SMA may also vary between CAFs. As evidenced by the findings of Mhawech-Fauceglia et al., the CAFs in ovarian tumors are predominantly α-SMA positive, but not all of them stain for the protein [29]. This indicates the existence of a certain level heterogeneity within the CAF population.

In addition to α-SMA, many other markers have been reported to distinguish CAFs from normal fibroblasts. They include fibroblast activated protein (FAP), S100A4, and platelet derived growth factor receptor, among others [7,29,31]. FAP, a cell surface serine protease, has emerged as a specific marker of CAFs in ovarian cancer [32]. While each of them has been shown to be an effective marker by different groups, there is no clear consensus about a universal marker for CAFs. The probable reason for this is that CAFs are a heterogeneous population, with some expressing one marker and others expressing other markers. α-SMA and PDGF positive CAFs do not overlap with S100A4 positive CAFs in pancreatic cancer [33]. The mutual exclusivity and heterogeneity in CAF marker expression may impart unique functions; for example, FAP and podoplanin positive CAFs were found to be immunosuppressive through a nitric oxide-dependent mechanism, while FAP positive and podoplanin negative CAFs were not immunosuppressive in lung adenocarcinoma [34]. For prostate cancer, CAFs expressing high CD90 had greater tumor promoting capacity than CAFs expressing low CD90 [35]. Pancreatic ductal adenocarcinomas have a subpopulation of CAFs that are distinct from those expressing α-SMA. These CAFs express proinflammatory mediators like IL-6, and mediate a paracrine interaction with the carcinoma cells [36]. In ovarian cancer, the expression levels of different CAF markers, CD10, podoplanin, FAP, Platelet-derived growth factor receptor alpha (PDGFRα), Platelet-derived growth factor receptor beta (PDGFRβ), S100 calcium binding protein A4 (S100A4), α-SMA, snail family transcriptional repressor 2 (SNAI2, commonly known as Slug), Zinc finger E-box-binding homeobox 1 (ZEB1), and twist family bHLH transcription factor 1 (TWIST1), clustered the CAFs into different subgroups showing different protein expression patterns [31].

Due to the continuous reciprocal interactions of CAFs with cancer cells, it is quite possible that the CAFs can undergo dynamic changes in their marker expression and functions depending on the heterogeneity of the cancer cells within the tumors. A recent study identified a unique subset of CAFs expressing the metallo-endopeptidase CD10 and the complement anaphylatoxin receptor GPR77 [37]. These CAFs were enriched following neoadjuvant chemotherapy, and were shown to promote cancer stem cell self-renewal through the secretion of IL-6 and IL-8. Therefore, their abundance in the tumors of breast cancer patients predicted a poor prognosis. Similarly, the evolving ovarian cancer metastatic tumors are metabolically reprogrammed by CAFs through the secretion of C-C Motif Chemokine Ligand 5 (CCL5), C-X-C motif chemokine ligand 10 (CXCL10), and IL-6 to utilize glycogen [38]. The CAFs with activated p38 MAP kinase signaling were capable of inducing such reprogramming. Therefore, considering the heterogeneity of CAFs, it is important to take into account their functional effects in promoting tumor progression as well as the potential of dynamic changes in them.

4. CAF Functions

CAFs have multiple functions in the tumor microenvironment, which directly or indirectly promote tumor progression. Most of these functions are mediated through the secretion of paracrine factors, ECMs, and proteases, as well as through cell surface receptors and direct contact with cancer cells. These functions and their underlying mechanisms are detailed below, outlined in Figure 2, and listed in Table 1.

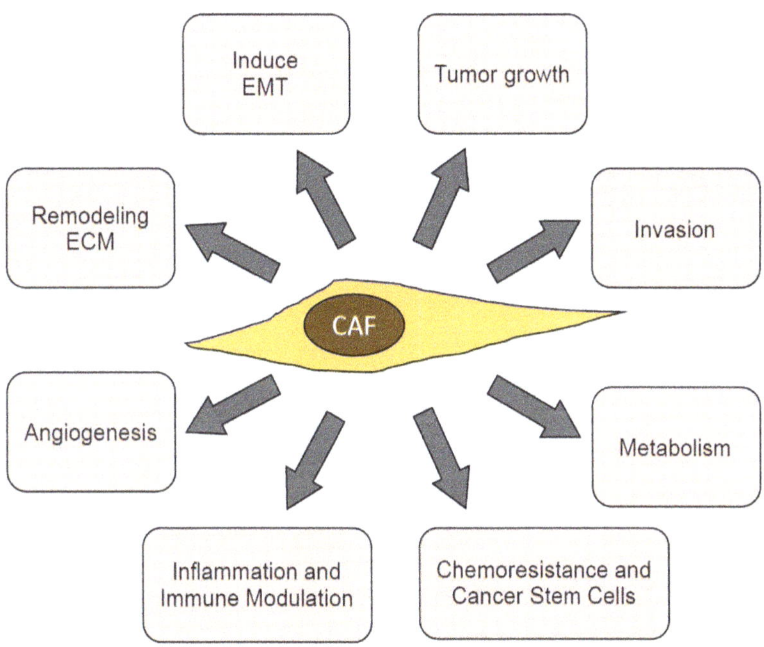

Figure 2. Functions of CAFs contributing towards tumor progression. ECM: extra cellular matrix; EMT: epithelial–mesenchymal transition.

Table 1. Functional roles of cancer associated fibroblasts (CAFs) in tumor progression.

No.	Functional Role of CAF	References
1	Promoting of tumor growth	[39]
2	Promoting tumor invasion	[3,40]
3	Inducing EMT in cancer cells	[41,42]
4	Remodeling the ECM	[43,44]
5	Inducing angiogenesis	[45,46]
6	Inflammation and immune modulation	[7,47]
7	Promoting chemoresistance and cancer stem cells	[48,49]
8	Reprogramming cancer metabolism	[3,4,38]

ECM: extra cellular matrix; EMT: epithelial–mesenchymal transition.

4.1. Promoting Tumor Growth

CAFs have predominantly been demonstrated to have tumor-promoting functions. They stimulate cancer cell survival, growth, and invasion, enhance the stiffness of the extracellular matrix, contribute to angiogenesis by releasing pro-angiogenic factors, contribute to a pro-inflammatory milieu, and impact on the activation state of various immune cells [50]. CAFs are crucial in tumor-stroma communication through modulation of the extracellular matrix, fibrogenesis, and chemoattraction of other stromal cells. Several studies have demonstrated that tumor–CAF crosstalk promotes growth and invasion of the particular cancer cells [39]. CAFs produce autocrine and paracrine cytokines that promote the growth and biological characteristics of tumors [7]. In addition to classical growth factors, including EGF and hepatocyte growth factor (HGF), novel CAF secreted proteins (secreted frizzled related protein 1, and IGF like family member (IGF) 1 and 2) and membrane molecules (integrin α11 and syndecan 1) have also been identified to possess cancer cell supporting roles. These factors directly or indirectly stimulate tumor growth and survival, or enhance their migratory and invasive properties [51]. Several secreted molecular regulators of CAFs have a pro-tumorigenic role, such as the TGF-β superfamily and bone morphogenic proteins (BMPs), PDGFs, EGFs, fibroblast growth factors (FGFs), and sonic hedgehog (SHH) [52].

Initial experiments with co-injection of CAFs with simian virus 40 (SV40)-transformed prostate epithelial cells in mice resulted in tumors resembling prostatic intraepithelial neoplasia, whereas normal fibroblasts did not. Similarly, co-transplantation of myofibroblasts with Ras-transformed hepatocytes strongly enhances tumor growth through the TGF-β/PDGF axis [39]. In addition, CAFs induce forkhead box Q1 (FOXQ1) expression; as a result, N-myc downstream-regulated gene 1 (NDRG1) is trans-activated to enhance hepatocellular carcinoma (HCC) initiation. Interestingly, pSTAT6/CCL26 signaling is induced by the FOXQ1/NDRG1 axis, thus recruiting hepatic stellate cells (HSCs), the main cellular source of CAFs, to the tumor microenvironment. Thereby, tumor-initiating properties are enhanced at least partly through a positive feedback loop between CAFs and HCC cells [53]. Taken together, these indicate that CAFs can provide growth-promoting signals to epithelial cells, and support epithelial transformation [54].

Tumor-derived TGF-β1 has been reported to activate tumor stroma, and thereby facilitate tumor growth and progression. Inhibition of the TGF-β pathway in mouse fibroblasts through conditional inactivation of the *TGFBR2* gene is associated with increased oncogenic potential of the adjacent epithelia [55]. An invasive breast cancer cohort study, using a randomized tamoxifen trial, demonstrated that TGF-β receptor type-2 expression in cancer-associated fibroblasts regulates breast cancer cell growth and survival, and is a prognostic marker in pre-menopausal breast cancer [56]. Mesenchymal stem cell derived CAFs recruited to the stroma of the dysplastic stomach express IL-6, Wnt5a, and bone morphogenetic protein 4, which promote tumor growth through DNA hypomethylation [57]. In oral squamous cell carcinoma (OCC), CAFs promote the production of endogenous reactive oxygen species (ROS) through CCL2 expression, which induces the cell cycle regulatory proteins, and promotes OCC proliferation, migration, and invasion [58]. CAFs have also been reported to promote Th2 polarization of the tumor microenvironment, and stimulate tumor growth and metastasis by recruiting tumor-associated macrophages (TAMs), myeloid derived suppressor cells (MDSCs), and T regulatory cells (T_{regs}) [8,59].

In ovarian cancer, CAFs promote tumor invasion and growth through the secretion of a number of chemokines, cytokines, and growth factors like CCL5, IL-6, IL-8, HB-EGF, and TGF-α, among others [7]. These secreted factors were regulated by the decreased expression of miR-214 and miR-31, and an increased expression of miR-155, in CAFs induced by ovarian cancer cells. CCL5 was a target of miR-214 and miR-31, and was responsible for homing of the ovarian cancer cells onto plugs of CAFs in vitro [7]. Inhibiting CCL5 with a neutralizing antibody was sufficient to reduce tumor growth of co-injected CAFs and ovarian cancer cells in mice [7].

4.2. Promoting Tumor Invasion

Tumor invasion is a key hallmark of cancer and is essential for successful dissemination of the cancer cells. Myofibroblasts have the inherent ability to invade through the ECM in the basement membrane during wound healing. Similarly, CAFs have the ability to invade through matrix, and have been widely reported to promote invasiveness of cancer cells [3]. There are several potential mechanisms by which CAFs can directly or indirectly promote cancer cell invasiveness. These include secretion of factors and proteases that help in the invasion. Zhu et al. (2016) [40] reported that Gal-1-regulated CAF activation promotes breast cancer cell metastasis by upregulating MMP-9 expression in breast cancer. Recent studies have shown that breast CAFs overexpress the chemokine CXCL1, a key regulator of tumor invasion and chemo-resistance. TGF-β negatively regulates CXCL1 expression in CAFs through Smad2/3 binding to the promoter, and through suppression of HGF/c-Met autocrine signaling [60]. CAFs can also induce changes in the cancer cells, which helps in their invasiveness. They have been reported to promote the metastatic activity of breast cancer cells by activating the transcription of HOTAIR via TGF-β1 secretion [61].

CAFs can serve as engines for collective invasion of directly interacting cancer cells through heterotypic interactions between the N-cadherin expressed on CAFs with the E-cadherin on cancer cells [62]. Interestingly, a dual mechanism is involved. CAFs favor invasion of cancer cells by pulling them away from the tumor, while cancer cells further enhance their spread by polarizing CAF migration away from the tumor. Along similar lines, vimentin is reported to be necessary for lung adenocarcinoma metastasis by maintaining heterotypic tumor cell–CAF interactions during collective invasion [63]. Cdc42EP3—a member of the BORG family of Cdc42 effectors—is highly expressed in CAFs, and regulates the actin and septin fibrillar networks. Coordination between the actin and the septin networks in CAFs is required for force-mediated matrix remodeling, promoting cancer cell invasion, angiogenesis, and tumor growth [64].

In ovarian cancer we have previously shown that CAFs can promote coordinated invasion of the cancer cells, which promotes metastasis [7]. Using a novel 3D live confocal imaging-based co-invasion assay, we observed that the CAFs derived from ovarian cancer omental metastasis are able to closely interact with ovarian cancer cells and invade through matrigel by forming distinct networks of CAFs and cancer cells, which invaded together. Cancer cells alone invaded at a slower rate and failed to form the network consisting of cellular associations. The mechanism of CAF–cancer cell interactions reported by Labernadie et al. [62] involving heterophilic E-cadherin/N-cadherin junctions could potentially be playing a role in these interactions between the CAFs and ovarian cancer cells. A comparison of gene expression profiles of CAFs from omental metastasis with normal omental fibroblasts revealed that the CAFs secrete multiple chemokines and cytokines that can potentially regulate invasion and motility of cancer cells [7]. Among them, CCL5 was identified as a key CAF derived mediator of metastasis, which was itself regulated by miR-214 and miR-31. Both these microRNAs are downregulated in CAFs during the reprogramming of normal fibroblasts by metastasizing ovarian cancer cells [7]. Zhao et al.

(2017) identified STAT4 as a key regulator of ovarian cancer metastasis via Wnt7a-induced activation of CAFs [65]. The concomitant increased production of CXCL12, IL6, and VEGFA by CAFs within the tumor microenvironment could enable peritoneal metastasis of ovarian cancer via induction of the EMT program. They also established a model of promotion of ovarian cancer metastasis by STAT4 via tumor-derived Wnt7a-induced activation of CAFs [65]. CAFs promote ovarian cancer cell proliferation, migration, and invasion through the paracrine FGF-1 factor. The FGF-1/FGFR-4 signaling axis regulates the stromal microenvironment in ovarian carcinomas. CAFs also activate the expression of Snail1 and MMP3, as well as reduce the expression of E-cadherin [66].

4.3. Inducing EMT in Cancer Cells

An epithelial-mesenchymal transition (EMT) is a biological process that allows a polarized epithelial cell, which normally interacts with the basement membrane via its basal surface, to undergo multiple biochemical changes that enable it to assume a mesenchymal cell phenotype, which includes enhanced migratory capacity, invasiveness, elevated resistance to apoptosis, and greatly increased production of ECM components. The completion of an EMT is signaled by the degradation of the underlying basement membrane and the formation of a mesenchymal cell that can migrate away from the epithelial layer in which it originated [67]. The idea that epithelial cells can downregulate epithelial characteristics and acquire mesenchymal characteristics arose in the early 1980s from observations made by Elizabeth Hay [68]. Over the subsequent years, the importance of EMT in cancer progression has been well established. The heterotypic interactions of cancer cells with the microenvironment, including CAFs, has been shown to be a key inducer of EMT.

Coculturing CAFs with lung cancer cells can induce miR-33b downregulation and promote epithelial cells EMT. miR-33b overexpression in lung cancer cells can counteract CAF-induced EMT. Interestingly, Snail1 expression in fibroblasts activates the inductive effects of CAFs on lung cancer cells. Snail1-expressing cancer-associated fibroblasts induce lung cancer cell epithelial-mesenchymal transition through miR-33b [41]. CAF conditioned medium induced epithelial-mesenchymal transition (EMT) by regulating the expression of EMT-associated markers E-cadherin and vimentin, and modulated metastasis-related genes MMP-2 and VEGF, both in vitro and in vivo. Further studies demonstrated that CAFs enhanced the metastatic potential of lung cancer cells by secreting IL-6 and subsequently activating the JAK2/STAT3 signaling pathway [42]. TGF-β1 secreted by CAFs can induce EMT in the interacting cancer cells and promote metastasis [69,70]. In ovarian tumors, CAF derived exosomes contain higher levels of TGF-β1 compared to those derived from normal omental fibroblasts [71]. Theses exosomes, upon uptake by ovarian cancer cells, induce EMT through the activation of SMADs. Activation of STAT3 by microenvironmental IL-6 can also induce EMT in ovarian cancer cells [72]. CAFs were reported to be the major source of IL-6 in the tumor microenvironment of ovarian tumors [72]. Therefore, CAFs can make cancer cells more aggressive by inducing EMT in them through various paracrine mechanisms.

4.4. Remodeling the ECM

Every organ has an ECM with unique composition, providing physical support for tissue integrity and elasticity. It is a dynamic structure that is constantly remodeled to control tissue homeostasis [73]. Dysregulation of ECM composition, structure, stiffness, and abundance contributes to several pathological conditions, such as fibrosis and invasive cancer. Typically, tumors have much stiffer ECMs, causing altered dynamics of the biophysical and biochemical interactions of the cancer cells with their microenvironment [74]. The increased stiffness of the matrix promotes invasiveness and motility of the cancer cells through improved invadosome and lamella formation [75]. Matrix stiffness drives EMT and metastasis through the TWIST1–G3BP2 mechanotransduction pathway in breast cancer [76]. A 9-gene matrisome gene signature has been identified through the analysis of available databases, and is associated with tumor progression through promotion of EMT, angiogenesis, hypoxia, inflammation, and altered metabolism in several cancers, including ovarian cancer [43]. Altered ECMs and ECM remodeling enzymes, like matrix metalloproteinases (MMPs), tissue inhibitors of metalloproteinases (TIMPs), and lysyl oxidases (LOXs), work to create a stiffer microenvironment permissive for tumor cell growth, migration, and invasion [44]. The increased secretion of fibronectin and LOXs by breast cancer CAFs contribute towards the remodeling of the ECMs in these tumors, promoting invasion and metastasis [77]. TGF-β activates the secretion of versican by CAFs in high-grade serous ovarian tumors, which induces the expression of MMP9 and CD44 by the cancer cells, resulting in ECM remodeling and invasion [78]. A recent study has demonstrated the role of ovarian cancer cell derived inhibin βA in effectively inducing CAFs, which then secrete increased amounts of collagens and other ECMs [79]. Therefore, CAFs serve the important function of remodeling the ECMs through altered secretion of the matrisome components, and provide the ideal microenvironmental stiffness for tumor progression.

4.5. Inducing Angiogenesis

As the tumor grows, the cancer cells are further removed from the existing blood vessels, and as a result, experience depleted levels of oxygen and nutrients. This typically places a limit to the tumor size, as cell proliferation in the regions well supplied by the blood vessels is balanced by cell death in the regions deprived of oxygen and nutrients. Therefore, in order to progress, the tumors must induce angiogenesis. It is the formation of a new vascular network to supply nutrients and oxygen, and remove waste products. Multiple factors, like VEGF, basic fibroblast growth factor (bFGF), interleukin-8 (IL-8), placenta-like growth factor (PLGF), TGF-β, platelet-derived endothelial growth factor (PD-EGF), pleiotrophin, activated hypoxia-inducible factor-1α (HIF-1α), and so forth, have been shown to trigger angiogenesis [80,81]. Many of these pro-angiogenic factors are contributed by the tumor microenvironment [82]. CAFs induce angiogenesis directly, as well as indirectly, through VEGFA, PDGFC, FGF2, CXCL12, osteopontin, and CSF3 secretion, ECM production, and recruitment of myeloid cells [82]. SDF-1 secreted by breast cancer CAFs has been involved in mobilization of endothelial precursor cells from bone marrow, thereby inducing de novo angiogenesis, as well as in tumor growth through a paracrine effect on CXCR4 expressing cancer cells [45]. Similarly, fibroblast-derived SDF-1 synergized with IL-8 in the promotion of a complete angiogenic response in recruited endothelial cells in pancreatic cancer [46]. SDF-1 is induced in breast cancer CAFs by oxidative stress-mediated activation of HIF-1 [83]. Chloride intracellular channel protein 3 (CLIC3) secreted by breast cancer CAFs induces angiogenesis through an active transglutaminase-2 (TGM2) mediated mechanism [84]. MMP-13 secreted by CAFs promotes tumor angiogenesis by releasing VEGF entrapped in the ECM, thereby leading to increased invasion of endothelial cells in squamous cell carcinoma and melanoma [85]. Ovarian cancer CAFs can indirectly induce angiogenesis through increased secretion of pro-inflammatory factors IL-6, COX-2, and CXCL1 [86]. Ovarian cancer cell expression HOXA9 induces CAFs to secrete CXCL12, IL-6, and VEGF-A expression, which promotes angiogenesis [87]. Ovarian cancer CAFs have also been shown to induce angiogenesis by secreting VEGF-C as a result of induction by Sonic Hedgehog (SHH) from ovarian cancer cells [88]. While angiogenesis can be induced by cancer cells as well as the tumor microenvironment, CAFs are important direct or indirect contributors to the process, and hence towards cancer progression.

4.6. Inflammation and Immune Modulation

Inflammation is a normal physiological response that is initiated in injured tissue and helps in its healing. In clinical settings, chronic inflammation and cancer are closely related, and cancer is referred to as "wounds that never heal". During tissue injury associated with wounding, cell proliferation is enhanced while the tissue regenerates; proliferation and inflammation subside after the assaulting agent is removed or the repair completed. In contrast, proliferating neoplastic cells continue to proliferate in microenvironments rich in inflammatory cells, and growth and survival factors, that support their growth [89]. Pro-inflammatory cytokines are secreted by cancer cells and CAFs to attract immune cells to the tumor. Macrophages actively attracted into tumor regions along defined chemotactic gradients start to differentiate into tumor-associated macrophages (TAMs), which further enhance the growth and metastasis of cancer cells [52]. CAFs are functionally required for mediating inflammation during squamous cell carcinogenesis, starting at the earliest pre-neoplastic stages [90]. A recent paper demonstrated that CAFs associated to incipient neoplasia exhibit a pro-inflammatory signature, leading them to mainly overexpress SDF-1, IL-6, and IL-1β, as well as to recruit proangiogenic macrophages. This gene set is under the transcriptional control of nuclear factor-κB (NF-κB) and cyclooxygenase 2 (COX-2), thereby strengthening the link between CAFs and inflammatory mediators in tumor progression [47]. We have demonstrated that ovarian cancer CAFs produce an array of chemokines and cytokines, which can potentially induce a proinflammatory response [7]. These chemokines and cytokines were directly or indirectly regulated by miR-214, miR-31, and miR-155 in CAFs [7]. Several other groups have also shown many chemokines and cytokines, including IL-6, COX-2, and CXCL1, to be involved in tumor-related inflammation in epithelial ovarian cancer [50,91].

CAFs in the tumor microenvironment exert immunomodulatory effects through secretion of immunomodulatory factors that polarize responsive immune populations, such as macrophages [92]. In order for the tumor to survive, any immune response directed toward the tumor cells needs to be suppressed [52]. CAFs play important roles in shaping the tumor immunosuppressive microenvironment in oral squamous cell carcinoma by inducing the protumor M2 macrophages [93]. Immunosuppressive activity of CAFs has been reported in head and neck squamous cell carcinoma through increased expression levels of IL-6, CXCL8, and TGF1 [94]. Genetic ablation of Chitinase 3-like 1 (Chi3L1) in fibroblasts in vivo attenuated tumor growth, macrophage recruitment, and reprogramming to an M2-like phenotype, enhanced tumor infiltration by CD8+ and CD4+ T cells, and promoted a Th1 phenotype. These results indicate that CAF-derived Chi3L1 promotes tumor growth and shifts the balance of the immune milieu towards type 2 immunity [95]. Activation of the PD1/PDL1 signaling pathway in T-cells leads to T-cell exhaustion and immune suppression. IL6 secreted by CAFs in hepatocellular carcinoma activates neutrophils in the tumor microenvironment, and induces PDL1 expression in them through the JAK-STAT3 pathway. The PDL1 expressing neutrophils inhibit T-cell mediated immunity and create a protumor microenvironment [96]. Being the most abundant cellular component of the stroma, CAFs can exert their effects indirectly on tumor progression through secreted factors that help evolve a proinflammatory and immunosuppressive microenvironment for the cancer cells to thrive in.

4.7. Promoting Chemoresistance and Cancer Stem Cells

The eventual development of chemoresistance is the cause of most ovarian cancer related mortalities. The role of the tumor microenvironment in this process has generated great interest in recent years. Glutathione and cysteine released by fibroblasts in ovarian tumors contribute towards the depletion of platinum in the nuclei of the adjacent ovarian cancer cells, and thus impart resistance to platinum based chemotherapies [97]. CAFs can also induce therapy resistance by reducing the bioavailability of the drugs, by causing tumor microvessel leakiness. CAFs induce the *lipoma-preferred partner* (*LPP*) gene in microvascular endothelial cells through a calcium-dependent FAK/ERK/MLC2/CREB signaling pathway [48]. In one study, the increased *LPP* enhanced the endothelial cell motility and permeability through increased focal adhesions and stress fiber formation [48]. CAFs can also induce chemoresistance in cancer cells by inducing apoptosis resistance. CAFs activate STAT3 signaling in ovarian cancer cells resulting in the development of chemoresistance through the increased expression of the antiapoptotic survivin and Bcl-2 [49]. CAFs in pancreatic ductal adenocarcinoma can similarly decrease apoptosis and increase the chemoresistance of the cancer cells by the induction of transcriptional downregulation of caspases by promoter hypermethylation [98]. Ovarian cancer apoptosis was also inhibited by CAF derived exosomes that transfer miR-21 to the cancer cells. The miR-21 then downregulated its direct target apoptotic protease activating factor 1 (APAF1), conferring chemoresistance [99].

Cancer stem cells are known to be resistant to cytotoxic chemotherapy and can give rise to tumor relapse. CD10 and GPR77 expressing CAFs induce cancer stem cells in breast and lung cancer through the consistent secretion of IL-6 and IL-8 [37]. These CAFs also increase the take rate of patient derived xenograft tumors, and inhibition of GPR77 abolishes this effect [37]. In one study, autophagic CAFs in luminal breast cancer induced stemness in the cancer cells through the secretion of high mobility group box 1 (HMGB1), resulting in the activation of toll-like receptor 4 in the cancer cells [100]. In endocrine resistant metastatic breast cancer, the transfer of CAF derived exosomes containing miR-221 activated an ER^{lo}/$Notch^{hi}$ feed-forward loop that generated $CD133^{hi}$ cancer stem cells [101]. In ovarian cancer, the evidence of induction of cancer stem cells is limited, with a few reports indicating the role of CAF derived fibroblast growth factor 4 and IL-6 in inducing cancer stem cells [102,103]. Targeting the FHF4/FGFR2 axis that mediates the CAF-cancer stem cell signaling prevented the CAFs from inducing cancer stem cells [102]. Insulin growth factor receptor activation in ovarian cancer cells by CAFs, and the resulting increased Nanog expression, is also reported to promote cancer stem cells in ovarian cancer [104]. Overall, the potential role of CAFs in providing a stem cell niche for cancer stem cells is an idea that needs to be systematically researched.

4.8. Reprogramming Cancer Cell Metabolism

Cancer cells have an altered metabolism to cope with their different growth rate, nutrient availability, and the hypoxia they experience as compared to normal cells. This altered metabolism is considered one of the hallmarks of cancer, and the tumor microenvironment is a major contributor towards this phenomenon [3,4,38,105]. CAFs have been reported to secrete vesicles, which created hypoxia mimicking conditions in the cancer cells, causing reductive carboxylation of glutamine in them, and decreased oxidative phosphorylation [106]. The CAF derived vesicles are also carriers of metabolites feeding into the tricarboxylic acid cycle in the cancer cells. This brings forth a novel mechanism by which CAFs can influence the cancer cells through the transfer of metabolites and pushing away from an oxygen based energy metabolism. Using stable isotope labeling of amino acids in cell culture (SILAC) in cocultures of CAFs with ovarian cancer cells, a recent study demonstrated how the CAFs can help ovarian cancer cells switch from utilizing lipids to using glycogen reserves for energy [38]. As the metastatic tumor grows and depletes the adipocytes in the omentum, the IL-6, CXCL10, and CCL5 secreted by the CAFs induce the ovarian cancer cells to start utilizing glycogen. The activation of p38 MAP kinase in the CAFs drives the cytokine secretion, which in turn activates glycogen phosphorylase in the ovarian cancer cells [38]. This demonstrates that the dynamic changes happening in the tumor microenvironment as the tumor progresses, forces the cancer cells to reset their energy sources. CAFs can orchestrate this switch by turning off glycogen synthesis and activating glycogen utilization for glycolysis. Therefore, targeting the key enzyme in this process, glycogen phosphorylase, would be a potential therapeutic option to treat ovarian cancer metastasis.

5. Targeting CAFs Clinically

Since CAFs contribute towards so many critical aspects essential for tumor progression, strategies targeting CAFs to treat ovarian cancer can be potentially effective. Moreover, since CAFs themselves are genetically stable and do not have the propensity to mutate, acquiring resistance against these therapies would be less likely. Since CAFs overexpress FAP, a humanized antibody (sibrotuzumab) directed against human FAP has been tested in phase 1 clinical trials to demonstrate that it is safe to administer to patients with high levels of FAP expression in their tumors [107]. However, it did not have any beneficial effect in a phase II trial for metastatic colorectal cancer [108]. A fusion protein consisting of an anti-FAP antibody fused with IL-2 (RO6874281) is presently under clinical trials as a combination therapy with atezolizumab—an anti-PDL-1 antibody—for advanced or metastatic solid tumors (ClinicalTrials.gov: NCT03386721). In addition, a phase I clinical trial is ongoing to test RO6874281 as a single agent, or in combination with trastuzumab or cetuximab, for solid tumor, and breast, head, and neck tumors (ClinicalTrials.gov: NCT02627274). As TGF-β plays an essential role in stromal-epithelial interaction and CAF induction, targeting TGF-β is a potentially promising approach to target CAFs as well as cancer cells. Transgenic mice expressing a TGF-β antagonist were resistant to metastasis to multiple organs while not exhibiting the adverse pathological outcomes observed in TGF-β-null mice [109]. Transcription profiling of CAFs microdissected from ovarian cancer patient tumors identified a subpopulation that had activation of SMAD signaling [110]. These CAFs were markers of poor patient progression, and targeting SMAD signaling with calcitriol inhibited tumor progression in mice. At present there are as many as 60 active clinical trials on TGF-β in cancers (clinicaltrials.gov). However, it is very difficult to differentiate the role of CAF derived TGF-β from other stromal sources and cancer cell autocrine TGF-β signaling. The HGF-cMet pathway, involving the cross-talk between CAFs and cancer cells, plays a role in cancer metastasis, and is another potential target for blocking CAF–cancer cell interaction. Targeting c-Met or HGF has shown promising tumor growth inhibition and gemcitabine sensitization in vivo [111,112]. There are 69 active studies on cMet listed in clinicaltrials.gov. Targeting CAF can decrease the immunosuppressive microenvironment of the tumor, as well as lead to CD8+ T-cell activation, and enhance anti-tumor immunity [18,113].

While several strategies to target CAFs in tumors have been attempted, much remains to be studied before it can be effectively translated to the clinic. Strategies like targeting TGF-β may benefit from attacking both the cancer and stromal compartments. Importantly, the potential of combining such therapies with existing platinum and taxane based chemotherapies should be tested for ovarian cancer. However, previous experiences with targeting the tumor microenvironment have taught us that an approach towards normalization is preferable to an eradication of the tumor stroma. This is because the latter approach tends to give rise to more aggressive cancer cells. Therefore, targeting CAFs should aim for reverting them back to normal fibroblasts, rather than depleting them altogether.

6. Conclusions

CAFs are an important constituent of the ovarian cancer tumor microenvironment, and have been demonstrated to play an important role in tumor progression, metastasis, and chemoresistance. While a universal CAF marker has not been identified, several markers have been demonstrated in unique subpopulations, indicating that CAFs are heterogenous in this context, and this may also dictate their function. Continuous reciprocal interactions of CAFs with cancer cells and other components of the microenvironment shape their fate, marker expression, and function in the tumor. Continuing research towards a better understanding of their plasticity, regulation, function, and heterogeneity would greatly enhance the way we perceive tumors, and will determine how we treat them. Strategies to "normalize" CAFs and deprive the cancer cells of the gamut of factors provided by them may be an effective approach to complement existing therapies targeting the cancer cells.

Author Contributions: Conceptualization, manuscript preparation, figures, and funding acquisition, A.K.M.; Literature review and draft preparation, S.D., Y.F., and A.K.M.; Editing, S.D. and A.K.M.

Funding: This research was funded by a DoD OCRP Ovarian Cancer Academy Award (W81XWH-15-0253) (A.K.M.), a Colleen's Dream Foundation award (A.K.M.), and a Ralph W. and Grace M. Showalter Research Award (A.K.M.).

Conflicts of Interest: The authors declare that they have no conflict of interest.

References

1. Torre, L.A.; Trabert, B.; DeSantis, C.E.; Miller, K.D.; Samimi, G.; Runowicz, C.D.; Gaudet, M.M.; Jemal, A.; Siegel, R.L. Ovarian cancer statistics. *CA Cancer J. Clin.* **2018**, *68*, 284–296. [CrossRef] [PubMed]
2. Siegel, R.L.; Miller, K.D.; Jemal, A. Cancer statistics, 2016. *CA Cancer J. Clin.* **2016**, *66*, 7–30. [CrossRef] [PubMed]
3. Hanahan, D.; Coussens, L.M. Accessories to the crime: Functions of cells recruited to the tumor microenvironment. *Cancer Cell* **2012**, *21*, 309–322. [CrossRef] [PubMed]

4. Ladanyi, A.; Mukherjee, A.; Kenny, H.A.; Johnson, A.; Mitra, A.K.; Sundaresan, S.; Nieman, K.M.; Pascual, G.; Benitah, S.A.; Montag, A.; et al. Adipocyte-induced CD36 expression drives ovarian cancer progression and metastasis. *Oncogene* **2018**, *37*, 2285–2301. [CrossRef] [PubMed]
5. Tomar, S.; Plotnik, J.P.; Haley, J.; Scantland, J.; Dasari, S.; Sheikh, Z.; Emerson, R.; Lenz, D.; Hollenhorst, P.C.; Mitra, A.K. ETS1 induction by the microenvironment promotes ovarian cancer metastasis through focal adhesion kinase. *Cancer Lett.* **2018**, *411*, 190–204. [CrossRef] [PubMed]
6. Mitra, A.K.; Chiang, C.Y.; Tiwari, P.; Tomar, S.; Watters, K.M.; Peter, M.E.; Lengyel, E. Microenvironment-induced downregulation of miR-193b drives ovarian cancer metastasis. *Oncogene* **2015**, *34*, 5923–5932. [CrossRef] [PubMed]
7. Mitra, A.K.; Zillhardt, M.; Hua, Y.; Tiwari, P.; Murmann, A.E.; Peter, M.E.; Lengyel, E. MicroRNAs reprogram normal fibroblasts into cancer-associated fibroblasts in ovarian cancer. *Cancer Discov.* **2012**, *2*, 1100–1108. [CrossRef] [PubMed]
8. Goubran, H.A.; Kotb, R.R.; Stakiw, J.; Emara, M.E.; Burnouf, T. Regulation of Tumor Growth and Metastasis: The Role of Tumor Microenvironment. *Cancer Growth Metastasis* **2014**, *7*, 9–18. [CrossRef] [PubMed]
9. Joyce, J.A.; Pollard, J.W. Microenvironmental regulation of metastasis. *Nat. Rev. Cancer* **2009**, *9*, 239–252. [CrossRef] [PubMed]
10. Orimo, A.; Weinberg, R.A. Stromal fibroblasts in cancer: A novel tumor-promoting cell type. *Cell Cycle* **2006**, *5*, 1597–1601. [CrossRef] [PubMed]
11. Legrand, A.J.; Poletto, M.; Pankova, D.; Clementi, E.; Moore, J.; Castro-Giner, F.; Ryan, A.J.; O'Neill, E.; Markkanen, E.; Dianov, G.L. Persistent DNA strand breaks induce a CAF-like phenotype in normal fibroblasts. *Oncotarget* **2018**, *9*, 13666–13681. [CrossRef] [PubMed]
12. Qiu, W.; Hu, M.; Sridhar, A.; Opeskin, K.; Fox, S.; Shipitsin, M.; Trivett, M.; Thompson, E.R.; Ramakrishna, M.; Gorringe, K.L.; et al. No evidence of clonal somatic genetic alterations in cancer-associated fibroblasts from human breast and ovarian carcinomas. *Nat. Genet.* **2008**, *40*, 650–655. [CrossRef] [PubMed]
13. Li, H.; Fan, X.; Houghton, J. Tumor microenvironment: The role of the tumor stroma in cancer. *J. Cell Biochem.* **2007**, *101*, 805–815. [CrossRef] [PubMed]
14. Kalluri, R.; Zeisberg, M. Fibroblasts in cancer. *Nat. Rev. Cancer* **2006**, *6*, 392–401. [CrossRef] [PubMed]
15. Thulabandu, V.; Chen, D.; Atit, R.P. Dermal fibroblast in cutaneous development and healing. *Wiley Interdiscip. Rev. Dev. Biol.* **2018**, *7*, e307. [CrossRef] [PubMed]
16. Tomasek, J.J.; Gabbiani, G.; Hinz, B.; Chaponnier, C.; Brown, R.A. Myofibroblasts and mechano-regulation of connective tissue remodelling. *Nat. Rev. Mol. Cell Biol.* **2002**, *3*, 349–363. [CrossRef] [PubMed]
17. Hu, M.; Yao, J.; Cai, L.; Bachman, K.E.; van den Brule, F.; Velculescu, V.; Polyak, K. Distinct epigenetic changes in the stromal cells of breast cancers. *Nat. Genet.* **2005**, *37*, 899–905. [CrossRef] [PubMed]
18. Ozdemir, B.C.; Pentcheva-Hoang, T.; Carstens, J.L.; Zheng, X.; Wu, C.C.; Simpson, T.R.; Laklai, H.; Sugimoto, H.; Kahlert, C.; Novitskiy, S.V.; et al. Depletion of carcinoma-associated fibroblasts and fibrosis induces immunosuppression and accelerates pancreas cancer with reduced survival. *Cancer Cell* **2014**, *25*, 719–734. [CrossRef] [PubMed]
19. Cooke, V.G.; LeBleu, V.S.; Keskin, D.; Khan, Z.; O'Connell, J.T.; Teng, Y.; Duncan, M.B.; Xie, L.; Maeda, G.; Vong, S.; et al. Pericyte depletion results in hypoxia-associated epithelial-to-mesenchymal transition and metastasis mediated by met signaling pathway. *Cancer Cell* **2012**, *21*, 66–81. [CrossRef] [PubMed]
20. Direkze, N.C.; Hodivala-Dilke, K.; Jeffery, R.; Hunt, T.; Poulsom, R.; Oukrif, D.; Alison, M.R.; Wright, N.A. Bone marrow contribution to tumor-associated myofibroblasts and fibroblasts. *Cancer Res.* **2004**, *64*, 8492–8495. [CrossRef] [PubMed]
21. Ishii, G.; Sangai, T.; Oda, T.; Aoyagi, Y.; Hasebe, T.; Kanomata, N.; Endoh, Y.; Okumura, C.; Okuhara, Y.; Magae, J.; et al. Bone-marrow-derived myofibroblasts contribute to the cancer-induced stromal reaction. *Biochem. Biophys. Res. Commun.* **2003**, *309*, 232–240. [CrossRef]
22. Spaeth, E.L.; Dembinski, J.L.; Sasser, A.K.; Watson, K.; Klopp, A.; Hall, B.; Andreeff, M.; Marini, F. Mesenchymal Stem Cell Transition to Tumor-Associated Fibroblasts Contributes to Fibrovascular Network Expansion and Tumor Progression. *PLoS ONE* **2009**, *4*, e4992. [CrossRef] [PubMed]
23. Schauer, I.G.; Zhang, J.; Xing, Z.; Guo, X.; Mercado-Uribe, I.; Sood, A.K.; Huang, P.; Liu, J. Interleukin-1β promotes ovarian tumorigenesis through a p53/NF-κB-mediated inflammatory response in stromal fibroblasts. *Neoplasia* **2013**, *15*, 409–420. [CrossRef] [PubMed]

24. McLean, K.; Gong, Y.; Choi, Y.; Deng, N.; Yang, K.; Bai, S.; Cabrera, L.; Keller, E.; McCauley, L.; Cho, K.R.; et al. Human ovarian carcinoma-associated mesenchymal stem cells regulate cancer stem cells and tumorigenesis via altered BMP production. *J. Clin. Investig.* **2011**, *121*, 3206–3219. [CrossRef] [PubMed]
25. Coffman, L.G.; Choi, Y.J.; McLean, K.; Allen, B.L.; di Magliano, M.P.; Buckanovich, R.J. Human carcinoma-associated mesenchymal stem cells promote ovarian cancer chemotherapy resistance via a BMP4/HH signaling loop. *Oncotarget* **2016**, *7*, 6916–6932. [CrossRef] [PubMed]
26. Rynne-Vidal, A.; Jimenez-Heffernan, J.A.; Fernandez-Chacon, C.; Lopez-Cabrera, M.; Sandoval, P. The Mesothelial Origin of Carcinoma Associated-Fibroblasts in Peritoneal Metastasis. *Cancers* **2015**, *7*, 1994–2011. [CrossRef] [PubMed]
27. Kenny, H.A.; Chiang, C.Y.; White, E.A.; Schryver, E.M.; Habis, M.; Romero, I.L.; Ladanyi, A.; Penicka, C.V.; George, J.; Matlin, K.; et al. Mesothelial cells promote early ovarian cancer metastasis through fibronectin secretion. *J. Clin. Investig.* **2014**, *124*, 4614–4628. [CrossRef] [PubMed]
28. Sandoval, P.; Jimenez-Heffernan, J.A.; Rynne-Vidal, A.; Perez-Lozano, M.L.; Gilsanz, A.; Ruiz-Carpio, V.; Reyes, R.; Garcia-Bordas, J.; Stamatakis, K.; Dotor, J.; et al. Carcinoma-associated fibroblasts derive from mesothelial cells via mesothelial-to-mesenchymal transition in peritoneal metastasis. *J. Pathol.* **2013**, *231*, 517–531. [CrossRef] [PubMed]
29. Mhawech-Fauceglia, P.; Wang, D.; Samrao, D.; Kim, G.; Lawrenson, K.; Meneses, T.; Liu, S.; Yessaian, A.; Pejovic, T. Clinical Implications of Marker Expression of Carcinoma-Associated Fibroblasts (CAFs) in Patients with Epithelial Ovarian Carcinoma After Treatment with Neoadjuvant Chemotherapy. *Cancer Microenviron.* **2014**, *7*, 33–39. [CrossRef] [PubMed]
30. Strutz, F.; Zeisberg, M. Renal fibroblasts and myofibroblasts in chronic kidney disease. *J. Am. Soc. Nephrol.* **2006**, *17*, 2992–2998. [CrossRef] [PubMed]
31. Fukagawa, D.; Sugai, T.; Osakabe, M.; Suga, Y.; Nagasawa, T.; Itamochi, H.; Sugiyama, T. Protein expression patterns in cancer-associated fibroblasts and cells undergoing the epithelial-mesenchymal transition in ovarian cancers. *Oncotarget* **2018**, *9*, 27514–27524. [CrossRef] [PubMed]
32. Lai, D.; Ma, L.; Wang, F. Fibroblast activation protein regulates tumor-associated fibroblasts and epithelial ovarian cancer cells. *Int. J. Oncol.* **2012**, *41*, 541–550. [CrossRef] [PubMed]
33. Sugimoto, H.; Mundel, T.M.; Kieran, M.W.; Kalluri, R. Identification of fibroblast heterogeneity in the tumor microenvironment. *Cancer Biol. Ther.* **2006**, *5*, 1640–1646. [CrossRef] [PubMed]
34. Cremasco, V.; Astarita, J.L.; Grauel, A.L.; Keerthivasan, S.; MacIsaac, K.D.; Woodruff, M.C.; Wu, M.; Spel, L.; Santoro, S.; Amoozgar, Z.; et al. FAP delineates heterogeneous and functionally divergent stromal cells in immune-excluded breast tumors. *Cancer Immunol. Res.* **2018**, *18*, e0098. [CrossRef] [PubMed]
35. Zhao, H.; Peehl, D.M. Tumor-promoting phenotype of CD90hi prostate cancer-associated fibroblasts. *Prostate* **2009**, *69*, 991–1000. [CrossRef] [PubMed]
36. Ohlund, D.; Handly-Santana, A.; Biffi, G.; Elyada, E.; Almeida, A.S. Distinct populations of inflammatory fibroblasts and myofibroblasts in pancreatic cancer. *J. Exp. Med.* **2017**, *214*, 579–596. [PubMed]
37. Su, S.; Chen, J.; Yao, H.; Liu, J.; Yu, S.; Lao, L.; Wang, M.; Luo, M.; Xing, Y.; Chen, F.; et al. CD10(+) GPR77(+) Cancer-Associated Fibroblasts Promote Cancer Formation and Chemoresistance by Sustaining Cancer Stemness. *Cell* **2018**, *172*, 841–856. [CrossRef] [PubMed]
38. Curtis, M.; Kenny, H.A.; Ashcroft, B.; Mukherjee, A.; Johnson, A.; Zhang, Y.; Helou, Y.; Batlle, R.; Liu, X.; Gutierrez, N.; et al. Fibroblasts Mobilize Tumor Cell Glycogen to Promote Proliferation and Metastasis. *Cell Metab.* **2018**, in press. [CrossRef] [PubMed]
39. Van Zijl, F.; Mair, M.; Csiszar, A.; Schneller, D.; Zulehner, G.; Huber, H.; Eferl, R.; Beug, H.; Dolznig, H.; Mikulits, W. Hepatic tumor–stroma crosstalk guides epithelial to mesenchymal transition at the tumor edge. *Oncogene* **2009**, *28*, 4022–4033. [CrossRef] [PubMed]
40. Zhu, X.; Wang, K.; Zhang, K.; Xu, F.; Yin, Y.; Zhu, L.; Zhou, F. Galectin-1 knockdown in carcinoma-associated fibroblasts inhibits migration and invasion of human MDA-MB-231 breast cancer cells by modulating MMP-9 expression. *Acta Biochim. Biophys. Sin. (Shanghai)* **2016**, *48*, 462–467. [CrossRef] [PubMed]
41. You, J.; Li, M.; Tan, Y.; Cao, L.; Gu, Q.; Yang, H.; Hu, C. Snail1-expressing cancer-associated fibroblasts induce lung cancer cell epithelial-mesenchymal transition through miR-33b. *Oncotarget* **2017**, *8*, 114769–114786. [CrossRef] [PubMed]

42. Wang, L.; Cao, L.; Wang, H.; Liu, B.; Zhang, Q.; Meng, Z.; Wu, X.; Zhou, Q.; Xu, K. Cancer-associated fibroblasts enhance metastatic potential of lung cancer cells through IL-6/STAT3 signaling pathway. *Oncotarget* **2017**, *8*, 76116–76128. [CrossRef] [PubMed]
43. Yuzhalin, A.E.; Urbonas, T.; Silva, M.A.; Muschel, R.J.; Gordon-Weeks, A.N. A core matrisome gene signature predicts cancer outcome. *Br. J. Cancer* **2018**, *118*, 435–440. [CrossRef] [PubMed]
44. Choe, C.; Shin, Y.S.; Kim, C.; Choi, S.I.; Lee, J.; Kim, D.Y.; Cho, Y.B.; Kim, J. Crosstalk with cancer-associated fibroblasts induces resistance of non-small cell lung cancer cells to epidermal growth factor receptor tyrosine kinase inhibition. *OncoTargets Ther.* **2015**, *8*, 3665–3678. [CrossRef] [PubMed]
45. Orimo, A.; Gupta, P.B.; Sgroi, D.C.; Arenzana-Seisdedos, F.; Delaunay, T.; Naeem, R.; Carey, V.J.; Richardson, A.L.; Weinberg, R.A. Stromal fibroblasts present in invasive human breast carcinomas promote tumor growth and angiogenesis through elevated SDF-1/CXCL12 secretion. *Cell* **2005**, *121*, 335–348. [CrossRef] [PubMed]
46. Matsuo, Y.; Ochi, N.; Sawai, H.; Yasuda, A.; Takahashi, H.; Funahashi, H.; Takeyama, H.; Tong, Z.; Guha, S. CXCL8/IL-8 and CXCL12/SDF-1alpha co-operatively promote invasiveness and angiogenesis in pancreatic cancer. *Int. J. Cancer* **2009**, *124*, 853–861. [CrossRef] [PubMed]
47. Kojima, Y.; Acar, A.; Eaton, E.N.; Mellody, K.T.; Scheel, C.; Ben-Porath, I.; Onder, T.T.; Wang, Z.C.; Richardson, A.L.; Weinberg, R.A.; et al. Autocrine TGF-β and stromal cell-derived factor-1 (SDF-1) signaling drives the evolution of tumor-promoting mammary stromal myofibroblasts. *Proc. Nat. Acad. Sci. USA* **2010**, *107*, 20009–20014. [CrossRef] [PubMed]
48. Leung, C.S.; Yeung, T.L.; Yip, K.P.; Wong, K.K.; Ho, S.Y.; Mangala, L.S.; Sood, A.K.; Lopez-Berestein, G.; Sheng, J.; Wong, S.T.; et al. Cancer-associated fibroblasts regulate endothelial adhesion protein LPP to promote ovarian cancer chemoresistance. *J. Clin. Investig.* **2018**, *128*, 589–606. [CrossRef] [PubMed]
49. Yan, H.; Guo, B.Y.; Zhang, S. Cancer-associated fibroblasts attenuate Cisplatin-induced apoptosis in ovarian cancer cells by promoting STAT3 signaling. *Biochem. Biophys. Res. Commun.* **2016**, *470*, 947–954. [CrossRef] [PubMed]
50. Schauer, I.G.; Sood, A.K.; Mok, S.; Liu, J. Cancer-Associated Fibroblasts and Their Putative Role in Potentiating the Initiation and Development of Epithelial Ovarian Cancer. *Neoplasia* **2011**, *13*, 393–405. [CrossRef] [PubMed]
51. Chen, J.S.; Liang, L.L.; Xu, H.X.; Chen, F.; Shen, S.L.; Chen, W.; Chen, L.Z.; Su, Q.; Zhang, L.J.; Bi, J.; et al. miR-338-3p inhibits epithelial-mesenchymal transition and metastasis in hepatocellular carcinoma cells. *Oncotarget* **2017**, *8*, 71418–71429. [CrossRef] [PubMed]
52. Xing, F.; Saidou, J.; Watabe, K. Cancer associated fibroblasts (CAFs) in tumor microenvironment. *Front. Biosci.* **2010**, *15*, 166–179. [CrossRef]
53. Luo, Q.; Wang, C.Q.; Yang, L.Y.; Gao, X.M.; Sun, H.T.; Zhang, Y.; Zhang, K.L.; Zhu, Y.; Zheng, Y.; Sheng, Y.Y.; et al. FOXQ1/NDRG1 axis exacerbates hepatocellular carcinoma initiation via enhancing crosstalk between fibroblasts and tumor cells. *Cancer Lett.* **2018**, *417*, 21–34. [CrossRef] [PubMed]
54. Olumi, A.F.; Grossfeld, G.D.; Hayward, S.W.; Carroll, P.R.; Tlsty, T.D.; Cunha, G.R. Carcinoma-associated fibroblasts direct tumor progression of initiated human prostatic epithelium. *Cancer Res.* **1999**, *59*, 5002–5011. [PubMed]
55. Bhowmick, N.A.; Chytil, A.; Plieth, D.; Gorska, A.E.; Dumont, N.; Shappell, S.; Washington, M.K.; Neilson, E.G.; Moses, H.L. TGF-β signaling in fibroblasts modulates the oncogenic potential of adjacent epithelia. *Science* **2004**, *303*, 848–851. [CrossRef] [PubMed]
56. Busch, S.; Acar, A.; Magnusson, Y.; Gregersson, P.; Ryden, L.; Landberg, G. TGF-β receptor type-2 expression in cancer-associated fibroblasts regulates breast cancer cell growth and survival and is a prognostic marker in pre-menopausal breast cancer. *Oncogene* **2015**, *34*, 27–38. [CrossRef] [PubMed]
57. Quante, M.; Tu, S.P.; Tomita, H.; Gonda, T.; Wang, S.S.; Takashi, S.; Baik, G.H.; Shibata, W.; Diprete, B.; Betz, K.S.; et al. Bone marrow-derived myofibroblasts contribute to the mesenchymal stem cell niche and promote tumor growth. *Cancer Cell* **2011**, *19*, 257–272. [CrossRef] [PubMed]
58. Li, X.; Xu, Q.; Wu, Y.; Li, J.; Tang, D.; Han, L.; Fan, Q. A CCL2/ROS autoregulation loop is critical for cancer-associated fibroblasts-enhanced tumor growth of oral squamous cell carcinoma. *Carcinogenesis* **2014**, *35*, 1362–1370. [CrossRef] [PubMed]

59. Liao, D.; Luo, Y.; Markowitz, D.; Xiang, R.; Reisfeld, R.A. Cancer associated fibroblasts promote tumor growth and metastasis by modulating the tumor immune microenvironment in a 4T1 murine breast cancer model. *PLoS ONE* **2009**, *4*, e7965. [CrossRef] [PubMed]
60. Fang, W.B.; Mafuvadze, B.; Yao, M.; Zou, A.; Portsche, M.; Cheng, N. TGF-β Negatively Regulates CXCL1 Chemokine Expression in Mammary Fibroblasts through Enhancement of Smad2/3 and Suppression of HGF/c-Met Signaling Mechanisms. *PLoS ONE* **2015**, *10*, e0135063. [CrossRef] [PubMed]
61. Ren, Y.; Jia, H.H.; Xu, Y.Q.; Zhou, X.; Zhao, X.H.; Wang, Y.F.; Song, X.; Zhu, Z.Y.; Sun, T.; Dou, Y.; et al. Paracrine and epigenetic control of CAF-induced metastasis: The role of HOTAIR stimulated by TGF-ss1 secretion. *Mol. Cancer* **2018**, *17*, 5. [CrossRef] [PubMed]
62. Labernadie, A.; Kato, T.; Brugués, A.; Serra-Picamal, X.; Derzsi, S.; Arwert, E.; Weston, A.; González-Tarragó, V.; Elosegui-Artola, A.; Albertazzi, L.; et al. A mechanically active heterotypic E-cadherin/N-cadherin adhesion enables fibroblasts to drive cancer cell invasion. *Nat. Cell Biol.* **2017**, *19*, 224. [CrossRef] [PubMed]
63. Richardson, A.M.; Havel, L.S.; Koyen, A.E.; Konen, J.M.; Shupe, J.; Wiles, W.G.T.; Martin, W.D.; Grossniklaus, H.E.; Sica, G.; Gilbert-Ross, M.; et al. Vimentin Is Required for Lung Adenocarcinoma Metastasis via Heterotypic Tumor Cell-Cancer-Associated Fibroblast Interactions during Collective Invasion. *Clin. Cancer Res.* **2018**, *24*, 420–432. [CrossRef] [PubMed]
64. Calvo, F.; Ranftl, R.; Hooper, S.; Farrugia, A.J.; Moeendarbary, E.; Bruckbauer, A.; Batista, F.; Charras, G.; Sahai, E. Cdc42EP3/BORG2 and Septin Network Enables Mechano-transduction and the Emergence of Cancer-Associated Fibroblasts. *Cell Rep.* **2015**, *13*, 2699–2714. [CrossRef] [PubMed]
65. Zhao, L.; Ji, G.; Le, X.; Luo, Z.; Wang, C.; Feng, M.; Xu, L.; Zhang, Y.; Lau, W.B.; Lau, B.; et al. An integrated analysis identifies STAT4 as a key regulator of ovarian cancer metastasis. *Oncogene* **2017**, *36*, 3384–3396. [CrossRef] [PubMed]
66. Sun, Y.; Fan, X.; Zhang, Q.; Shi, X.; Xu, G.; Zou, C. Cancer-associated fibroblasts secrete FGF-1 to promote ovarian proliferation, migration, and invasion through the activation of FGF-1/FGFR4 signaling. *Tumour Biol.* **2017**, *39*, e1010428317712592. [CrossRef] [PubMed]
67. Kalluri, R.; Weinberg, R.A. The basics of epithelial-mesenchymal transition. *J. Clin. Investig.* **2009**, *119*, 1420–1428. [CrossRef] [PubMed]
68. Lamouille, S.; Xu, J.; Derynck, R. Molecular mechanisms of epithelial–mesenchymal transition. *Nat. Rev. Mol. Cell Biol.* **2014**, *15*, 178–196. [CrossRef] [PubMed]
69. Zhuang, J.; Lu, Q.; Shen, B.; Huang, X.; Shen, L.; Zheng, X.; Huang, R.; Yan, J.; Guo, H. TGFβ1 secreted by cancer-associated fibroblasts induces epithelial-mesenchymal transition of bladder cancer cells through lncRNA-ZEB2NAT. *Sci. Rep.* **2015**, *5*, e11924. [CrossRef] [PubMed]
70. Yu, Y.; Xiao, C.H.; Tan, L.D.; Wang, Q.S.; Li, X.Q.; Feng, Y.M. Cancer-associated fibroblasts induce epithelial-mesenchymal transition of breast cancer cells through paracrine TGF-β signalling. *Br. J. Cancer* **2014**, *110*, 724–732. [CrossRef] [PubMed]
71. Li, W.; Zhang, X.; Wang, J.; Li, M.; Cao, C.; Tan, J.; Ma, D.; Gao, Q. TGFβ1 in fibroblasts-derived exosomes promotes epithelial-mesenchymal transition of ovarian cancer cells. *Oncotarget* **2017**, *8*, 96035–96047. [PubMed]
72. Wang, L.; Zhang, F.; Cui, J.Y.; Chen, L.; Chen, Y.T.; Liu, B.W. CAFs enhance paclitaxel resistance by inducing EMT through the IL6/JAK2/STAT3 pathway. *Oncol. Rep.* **2018**, *39*, 2081–2090. [PubMed]
73. Bonnans, C.; Chou, J.; Werb, Z. Remodelling the extracellular matrix in development and disease. *Nat. Rev. Mol. Cell Biol.* **2014**, *15*, 786–801. [CrossRef] [PubMed]
74. DuFort, C.C.; Paszek, M.J.; Weaver, V.M. Balancing forces: Architectural control of mechanotransduction. *Nat. Rev. Mol. Cell Biol.* **2011**, *12*, 308–319. [CrossRef] [PubMed]
75. Kai, F.; Laklai, H.; Weaver, V.M. Force Matters: Biomechanical Regulation of Cell Invasion and Migration in Disease. *Trends Cell Biol.* **2016**, *26*, 486–497. [CrossRef] [PubMed]
76. Wei, S.C.; Fattet, L.; Tsai, J.H.; Guo, Y.; Pai, V.H.; Majeski, H.E.; Chen, A.C.; Sah, R.L.; Taylor, S.S.; Engler, A.J.; et al. Matrix stiffness drives epithelial-mesenchymal transition and tumour metastasis through a TWIST1-G3BP2 mechanotransduction pathway. *Nat. Cell Biol.* **2015**, *17*, 678–688. [CrossRef] [PubMed]
77. Tang, X.; Hou, Y.; Yang, G.; Wang, X.; Tang, S.; Du, Y.E.; Yang, L.; Yu, T.; Zhang, H.; Zhou, M.; et al. Stromal miR-200s contribute to breast cancer cell invasion through CAF activation and ECM remodeling. *Cell Death Differ.* **2016**, *23*, 132–145. [CrossRef] [PubMed]

78. Yeung, T.L.; Leung, C.S.; Wong, K.K.; Samimi, G.; Thompson, M.S.; Liu, J.; Zaid, T.M.; Ghosh, S.; Birrer, M.J.; Mok, S.C. TGF-β modulates ovarian cancer invasion by upregulating CAF-derived versican in the tumor microenvironment. *Cancer Res.* **2013**, *73*, 5016–5028. [CrossRef] [PubMed]
79. Nagaraja, A.S.; Dood, R.L.; Armaiz-Pena, G.; Kang, Y.; Wu, S.Y.; Allen, J.K.; Jennings, N.B.; Mangala, L.S.; Pradeep, S.; Lyons, Y.; et al. Adrenergic-mediated increases in INHBA drive CAF phenotype and collagens. *JCI Insight* **2017**, *2*, e93076. [CrossRef] [PubMed]
80. Kerbel, R.; Folkman, J. Clinical translation of angiogenesis inhibitors. *Nat. Rev. Cancer* **2002**, *2*, 727–739. [CrossRef] [PubMed]
81. Nishida, N.; Yano, H.; Nishida, T.; Kamura, T.; Kojiro, M. Angiogenesis in Cancer. *Vasc. Health Risk Manag.* **2006**, *2*, 213–219. [CrossRef] [PubMed]
82. De Palma, M.; Biziato, D.; Petrova, T.V. Microenvironmental regulation of tumour angiogenesis. *Nat. Rev. Cancer* **2017**, *17*, 457–474. [CrossRef] [PubMed]
83. Toullec, A.; Gerald, D.; Despouy, G.; Bourachot, B.; Cardon, M.; Lefort, S.; Richardson, M.; Rigaill, G.; Parrini, M.C.; Lucchesi, C.; et al. Oxidative stress promotes myofibroblast differentiation and tumour spreading. *EMBO Mol. Med.* **2010**, *2*, 211–230. [CrossRef] [PubMed]
84. Hernandez-Fernaud, J.R.; Ruengeler, E.; Casazza, A.; Neilson, L.J.; Pulleine, E.; Santi, A.; Ismail, S.; Lilla, S.; Dhayade, S.; MacPherson, I.R.; et al. Secreted CLIC3 drives cancer progression through its glutathione-dependent oxidoreductase activity. *Nat. Commun.* **2017**, *8*, e14206.
85. Lederle, W.; Hartenstein, B.; Meides, A.; Kunzelmann, H.; Werb, Z.; Angel, P.; Mueller, M.M. MMP13 as a stromal mediator in controlling persistent angiogenesis in skin carcinoma. *Carcinogenesis* **2010**, *31*, 1175–1184. [CrossRef] [PubMed]
86. Erez, N.; Glanz, S.; Raz, Y.; Avivi, C.; Barshack, I. Cancer associated fibroblasts express pro-inflammatory factors in human breast and ovarian tumors. *Biochem. Biophys. Res. Commun.* **2013**, *437*, 397–402. [CrossRef] [PubMed]
87. Ko, S.Y.; Barengo, N.; Ladanyi, A.; Lee, J.S.; Marini, F.; Lengyel, E.; Naora, H. HOXA9 promotes ovarian cancer growth by stimulating cancer-associated fibroblasts. *J. Clin. Investig.* **2012**, *122*, 3603–3617. [CrossRef] [PubMed]
88. Wei, R.; Lv, M.; Li, F.; Cheng, T.; Zhang, Z.; Jiang, G.; Zhou, Y.; Gao, R.; Wei, X.; Lou, J.; et al. Human CAFs promote lymphangiogenesis in ovarian cancer via the Hh-VEGF-C signaling axis. *Oncotarget* **2017**, *8*, 67315–67328. [CrossRef] [PubMed]
89. Arnold, K.M.; Opdenaker, L.M.; Flynn, D.; Sims-Mourtada, J. Wound Healing and Cancer Stem Cells: Inflammation as a Driver of Treatment Resistance in Breast Cancer. *Cancer Growth Metastasis* **2015**, *8*, 1–13. [CrossRef] [PubMed]
90. Erez, N.; Truitt, M.; Olson, P.; Arron, S.T.; Hanahan, D. Cancer-Associated Fibroblasts Are Activated in Incipient Neoplasia to Orchestrate Tumor-Promoting Inflammation in an NF-κB-Dependent Manner. *Cancer Cell* **2010**, *17*, 135–147. [CrossRef] [PubMed]
91. Yang, G.; Rosen, D.G.; Liu, G.; Yang, F.; Guo, X.; Xiao, X.; Xue, F.; Mercado-Uribe, I.; Huang, J.; Lin, S.H.; et al. CXCR2 promotes ovarian cancer growth through dysregulated cell cycle, diminished apoptosis, and enhanced angiogenesis. *Clin. Cancer Res.* **2010**, *16*, 3875–3886. [CrossRef] [PubMed]
92. Lakins, M.A.; Ghorani, E.; Munir, H.; Martins, C.P.; Shields, J.D. Cancer-associated fibroblasts induce antigen-specific deletion of CD8 + T Cells to protect tumour cells. *Nat. Commun.* **2018**, *9*, e948. [CrossRef] [PubMed]
93. Takahashi, H.; Sakakura, K.; Kudo, T.; Toyoda, M.; Kaira, K.; Oyama, T.; Chikamatsu, K. Cancer-associated fibroblasts promote an immunosuppressive microenvironment through the induction and accumulation of protumoral macrophages. *Oncotarget* **2017**, *8*, 8633–8647. [CrossRef] [PubMed]
94. Takahashi, H.; Sakakura, K.; Kawabata-Iwakawa, R.; Rokudai, S.; Toyoda, M.; Nishiyama, M.; Chikamatsu, K. Immunosuppressive activity of cancer-associated fibroblasts in head and neck squamous cell carcinoma. *Cancer Immunol. Immunother.* **2015**, *64*, 1407–1417. [CrossRef] [PubMed]
95. Cohen, N.; Shani, O.; Raz, Y.; Sharon, Y.; Hoffman, D.; Abramovitz, L.; Erez, N. Fibroblasts drive an immunosuppressive and growth-promoting microenvironment in breast cancer via secretion of Chitinase 3-like 1. *Oncogene* **2017**, *36*, 4457–4468. [CrossRef] [PubMed]
96. Cheng, Y.; Li, H.; Deng, Y.; Tai, Y.; Zeng, K.; Zhang, Y.; Liu, W.; Zhang, Q.; Yang, Y. Cancer-associated fibroblasts induce PDL1+ neutrophils through the IL6-STAT3 pathway that foster immune suppression in hepatocellular carcinoma. *Cell Death Dis.* **2018**, *9*, e422. [CrossRef] [PubMed]

97. Wang, W.; Kryczek, I.; Dostal, L.; Lin, H.; Tan, L.; Zhao, L.; Lu, F.; Wei, S.; Maj, T.; Peng, D.; et al. Effector T Cells Abrogate Stroma-Mediated Chemoresistance in Ovarian Cancer. *Cell* **2016**, *165*, 1092–1105. [CrossRef] [PubMed]
98. Muerkoster, S.S.; Werbing, V.; Koch, D.; Sipos, B.; Ammerpohl, O.; Kalthoff, H.; Tsao, M.S.; Folsch, U.R.; Schafer, H. Role of myofibroblasts in innate chemoresistance of pancreatic carcinoma—Epigenetic downregulation of caspases. *Int. J. Cancer* **2008**, *123*, 1751–1760. [CrossRef] [PubMed]
99. Au Yeung, C.L.; Co, N.N.; Tsuruga, T.; Yeung, T.L.; Kwan, S.Y.; Leung, C.S.; Li, Y.; Lu, E.S.; Kwan, K.; Wong, K.K. Exosomal transfer of stroma-derived miR21 confers paclitaxel resistance in ovarian cancer cells through targeting. *APAF1* **2016**, *7*, e11150. [CrossRef] [PubMed]
100. Zhao, X.L.; Lin, Y.; Jiang, J.; Tang, Z.; Yang, S.; Lu, L.; Liang, Y.; Liu, X.; Tan, J.; Hu, X.G.; et al. High-mobility group box 1 released by autophagic cancer-associated fibroblasts maintains the stemness of luminal breast cancer cells. *J. Pathol.* **2017**, *243*, 376–389. [CrossRef] [PubMed]
101. Sansone, P.; Berishaj, M.; Rajasekhar, V.K.; Ceccarelli, C.; Chang, Q.; Strillacci, A.; Savini, C.; Shapiro, L.; Bowman, R.L.; Mastroleo, C.; et al. Evolution of Cancer Stem-like Cells in Endocrine-Resistant Metastatic Breast Cancers Is Mediated by Stromal Microvesicles. *Cancer Res.* **2017**, *77*, 1927–1941. [CrossRef] [PubMed]
102. Yasuda, K.; Torigoe, T.; Mariya, T.; Asano, T.; Kuroda, T.; Matsuzaki, J.; Ikeda, K.; Yamauchi, M.; Emori, M.; Asanuma, H.; et al. Fibroblasts induce expression of FGF4 in ovarian cancer stem-like cells/cancer-initiating cells and upregulate their tumor initiation capacity. *Lab Investig.* **2014**, *94*, 1355–1369. [CrossRef] [PubMed]
103. Pasquier, J.; Rafii, A. Role of the microenvironment in ovarian cancer stem cell maintenance. *Biomed. Res. Int.* **2013**, *2013*, e630782. [CrossRef] [PubMed]
104. Kalluri, R. The biology and function of fibroblasts in cancer. *Nat. Rev. Cancer* **2016**, *16*, 582–598. [CrossRef] [PubMed]
105. Nieman, K.M.; Kenny, H.A.; Penicka, C.V.; Ladanyi, A.; Buell-Gutbrod, R.; Zillhardt, M.R.; Romero, I.L.; Carey, M.S.; Mills, G.B.; Hotamisligil, G.S.; et al. Adipocytes promote ovarian cancer metastasis and provide energy for rapid tumor growth. *Nat. Med.* **2011**, *17*, 1498–1503. [CrossRef] [PubMed]
106. Zhao, H.; Yang, L.; Baddour, J.; Achreja, A.; Bernard, V.; Moss, T.; Marini, J.C.; Tudawe, T.; Seviour, E.G.; San Lucas, F.A.; et al. Tumor microenvironment derived exosomes pleiotropically modulate cancer cell metabolism. *Elife* **2016**, *5*, e10250. [CrossRef] [PubMed]
107. Scott, A.M.; Wiseman, G.; Welt, S.; Adjei, A.; Lee, F.T.; Hopkins, W.; Divgi, C.R.; Hanson, L.H.; Mitchell, P.; Gansen, D.N.; et al. A Phase I dose-escalation study of sibrotuzumab in patients with advanced or metastatic fibroblast activation protein-positive cancer. *Clin. Cancer Res.* **2003**, *9*, 1639–1647. [PubMed]
108. Hofheinz, R.D.; Al-Batran, S.E.; Hartmann, F.; Hartung, G.; Jager, D.; Renner, C.; Tanswell, P.; Kunz, U.; Amelsberg, A.; Kuthan, H.; et al. Stromal antigen targeting by a humanised monoclonal antibody: An early phase II trial of sibrotuzumab in patients with metastatic colorectal cancer. *Onkologie* **2003**, *26*, 44–48. [CrossRef] [PubMed]
109. Yang, Y.A.; Dukhanina, O.; Tang, B.; Mamura, M.; Letterio, J.J.; MacGregor, J.; Patel, S.C.; Khozin, S.; Liu, Z.Y.; Green, J.; et al. Lifetime exposure to a soluble TGF-β antagonist protects mice against metastasis without adverse side effects. *J. Clin. Investig.* **2002**, *109*, 1607–1615. [CrossRef] [PubMed]
110. Yeung, T.-L.; Sheng, J.; Leung, C.S.; Li, F.; Kim, J.; Ho, S.Y.; Matzuk, M.M.; Lu, K.H.; Wong, S.T.C.; Mok, S.C. Systematic Identification of Druggable Epithelial–Stromal Crosstalk Signaling Networks in Ovarian Cancer. *J. Natl. Cancer Inst.* **2018**, *10*, e097. [CrossRef] [PubMed]
111. Brandes, F.; Schmidt, K.; Wagner, C.; Redekopf, J.; Schlitt, H.J.; Geissler, E.K.; Lang, S.A. Targeting cMET with INC280 impairs tumour growth and improves efficacy of gemcitabine in a pancreatic cancer model. *BMC Cancer* **2015**, *15*, e71. [CrossRef] [PubMed]
112. Chen, L.; Li, C.; Zhu, Y. The HGF inhibitory peptide HGP-1 displays promising in vitro and in vivo efficacy for targeted cancer therapy. *Oncotarget* **2015**, *6*, 30088–30101. [CrossRef] [PubMed]
113. Ohshio, Y.; Teramoto, K.; Hanaoka, J.; Tezuka, N.; Itoh, Y.; Asai, T.; Daigo, Y.; Ogasawara, K. Cancer-associated fibroblast-targeted strategy enhances antitumor immune responses in dendritic cell-based vaccine. *Cancer Sci.* **2015**, *106*, 134–142. [CrossRef] [PubMed]

cancers

MDPI

Review

UnPAXing the Divergent Roles of PAX2 and PAX8 in High-Grade Serous Ovarian Cancer

Laura R. Hardy, Amrita Salvi and Joanna E. Burdette *

Department of Medicinal Chemistry and Pharmacognosy, College of Pharmacy, University of Illinois at Chicago, Chicago, IL 60607, USA; lrodge4@uic.edu (L.R.H.); amritas@uic.edu (A.S.)

* Correspondence: joannab@uic.edu; Tel.: +1-312-996-6153

Received: 14 June 2018; Accepted: 4 August 2018; Published: 8 August 2018

Abstract: High-grade serous ovarian cancer is a deadly disease that can originate from the fallopian tube or the ovarian surface epithelium. The PAX (paired box) genes PAX2 and PAX8 are lineage-specific transcription factors required during development of the fallopian tube but not in the development of the ovary. PAX2 expression is lost early in serous cancer progression, while PAX8 is expressed ubiquitously. These proteins are implicated in migration, invasion, proliferation, cell survival, stem cell maintenance, and tumor growth. Hence, targeting PAX2 and PAX8 represents a promising drug strategy that could inhibit these pro-tumorigenic effects. In this review, we examine the implications of PAX2 and PAX8 expression in the cell of origin of serous cancer and their potential efficacy as drug targets by summarizing their role in the molecular pathogenesis of ovarian cancer.

Keywords: high-grade serous ovarian carcinoma (HGSC); PAX2; PAX8; cell of origin; ovary; fallopian tube

1. Introduction

In 2017, there were 22,440 new cases of ovarian cancer and 14,080 deaths [1]. Ovarian cancer is the fifth leading cause of cancer related death in women and the most lethal gynecological malignancy. High-grade serous carcinoma (HGSC) accounts for 80% of ovarian cancer cases and it is the deadliest histological subtype of epithelial ovarian cancer (EOC). This high mortality rate is due in part to the insidious nature of the disease, as the majority of cases are detected at an advanced stage with distant metastases. Symptoms of HGSC, such as abdominal pressure, bloating, and urinary frequency, are non-specific and do not present until after the tumor cells have metastasized and obstructed organs in the peritoneum. Current treatment strategies at this late stage include surgical debulking followed by chemotherapy with platinum and taxane drugs. While tumors are initially responsive to chemotherapy, the 5-year survival rate remains poor because of drug resistance and subsequent patient relapse. Patients with chemoresistant disease may receive chemotherapy in combination with targeted therapy against PARP (Olaparib) or VEGF-A (Bevacizumab) [2,3].

While it was originally believed that the ovary was the primary site of HGSC development, accumulating histologic, molecular, and animal model evidence suggests that the majority of cases originate from the fallopian tube epithelium [4–7]. The PAX (paired box) genes PAX2 and PAX8 are lineage-specific transcription factors that are involved in epithelial development of the fallopian tube but not the ovary [8,9]. PAX8 is expressed in HGSC tumors derived from both the fallopian tube and ovarian surface epithelium (OSE), at least in murine models where the source of the tumor is experimentally derived. In regard to the other histotypes of EOC, PAX8 shows high expression in clear cell and endometrioid tumors and reduced expression in mucinous tumors [10–12]. PAX2 is lost early in the molecular progression of fallopian tube derived cancer and is absent in ~85% of HGSC. PAX2 is detected in clear cell and mucinous tumors and absent in most endometrioid tumors [13–16].

Studying PAX2 and PAX8 in this context provides valuable insight into the site of origin of ovarian cancer and the tumorigenic properties that make the PAX proteins promising drug targets for treatment of HGSC.

2. Site of Origin of HGSC

The origin of HGSC has perplexed researchers for decades and it is now known that HGSC can originate from the fallopian tube epithelium as well as the OSE. Since PAX2 and PAX8 are expressed in the fallopian tube, and PAX8 expression is maintained in HGSC, the expression and regulation of PAX proteins may help to explain the source of ovarian cancer. The OSE was historically believed to be the site of origin of serous carcinoma based on the incessant ovulation hypothesis. This hypothesis suggests that during ovulation, fragments of the OSE get trapped within the wound created by follicle rupture, forming an ovarian cyst [17]. The epithelium trapped within the cyst has direct contact with the stroma and therefore has increased exposure to the stromal microenvironment, including growth factors and cytokines [18]. As a result, cells within an ovarian cyst have a higher likelihood of transforming into tubal-like cells that express markers of ovarian cancer, including PAX8, CA-125 and E-cadherin [18,19]. This hypothesis is supported by epidemiological data showing pregnancy and oral contraceptive use, both of which decrease the number of ovulatory cycles, are correlated with a decrease in ovarian cancer risk [20,21].

The OSE is unique to the female reproductive tract in that it is formed embryologically from the mesodermally derived colemic epithelium. In contrast, other components of the female reproductive tract, including the fallopian tube, cervix, and uterus, are Müllerian-derived structures. These Müllerian-derived structures express PAX8, while the OSE does not. This difference in embryonic origin has implications for adult cells. The adult OSE contains a mix of epithelial and mesenchymal-like cells that appear to be less differentiated than the rest of the female reproductive tract. These cells do not express molecular markers characteristic of epithelial cells, including CA-125 and E-Cadherin, but rather express mesenchymal markers, including keratin and vimentin [18]. Serous tumors that are derived from the OSE, however, obtain expression of these epithelial markers as well as phenotypic characteristics of the epithelium, including papillary serous structures [22]. Thus, in mouse models, HGSC can experimentally originate from the OSE.

The differentiated serous histology of HGSC is an interesting paradox since most cancers are less differentiated than the tissue of origin. Cheng et al. hypothesized that the OSE is an incompletely differentiated tissue type that can differentiate during oncogenic transformation through expression of *HOX* genes [23]. *HOX* genes are tightly controlled genes involved in developmental programming of the Müllerian duct, but they are not expressed in development of the OSE. This is similar to the *PAX8* gene, which is expressed in the fallopian tube and in serous tumors, but not in the OSE. By experimentally expressing *Hoxa9*, researchers observed the OSE formed serous papillary tumors. The OSE may also harbor a stem cell niche within the transitional zone of the ovarian hilum that has increased tumorigenic properties. Researchers experimentally demonstrated that cells within the ovarian hilum express stem cell markers that contribute to regeneration of the OSE [24]. Importantly, these stem cells had greater transformative ability after conditional inactivation of *p53* and *RB1*. It would be interesting to examine whether these stem cells also gained developmental markers, such as HOXA9 or PAX8 that would induce differentiation to a serous histotype.

Increasing evidence indicates that the fallopian tube epithelium serves as the main site of origin of HGSC. Under this scenario, serous tumors found on the ovarian surface are secondary metastasis from the fallopian tube, and thus resemble this lineage history. Piek et al. presented the first clinical evidence supporting this hypothesis by identifying pre-neoplastic lesions with increased staining for p53 and Ki67 in the fallopian tubes of BRCA-positive women who are predisposed to developing ovarian carcinoma [5]. Identical p53 mutations were identified in the precursor lesions of the fallopian tube and in concurrent ovarian carcinomas [6,25]. Molecular profiling of serous tumors identified a gene signature in HGSC tumors that more closely correlated with the normal fallopian tube epithelium than the normal OSE [26,27]. Clinically, bilateral salpingectomy reduced the risk of serous carcinoma by 61% and prophylactic salpingo-oophorectomy in BRCA-positive women reduced the risk of serous carcinoma by 80% [28,29]. Therefore, the current recommendation states that BRCA-positive women after child bearing age should undergo prophylactic salpingo-oophorectomy [30].

The fallopian tube origin for ovarian cancer is further supported by multiple animal models. Immortalized fallopian tube secretory epithelial cells are transformed into HGSC through *H-Ras*^{V12} mutation or *c-Myc* expression [31]. *Dicer-Pten* deletion from the reproductive tract resulted in HGSC formation, even after bilateral removal of the ovaries, demonstrating that these tumors originated in the fallopian tube [7]. *Pax8* promoter-driven deletion of *Brca*, *Tp53*, and *Pten* in the fallopian tube also led to HGSC development [32]. Since a common molecular alteration in these models is loss of *Pten*, Russo and colleagues examined the effects of *Pten* loss alone from the fallopian tube epithelium. Homozygous loss of *Pten* was sufficient to drive the development of borderline serous and endometrioid carcinoma that could metastasize to the ovary [33]. Interestingly, in a cell-based model, *Pten* loss in combination with *Kras* mutation formed highly aggressive tumors, while addition of constitutively active *Akt* attenuated this phenotype [34]. Research from the Cho laboratory demonstrated how serous carcinoma progresses from serous tubal intraepithelial carcinoma (STIC) to HGSC using various combinatorial deletions in *Rb1*, *Brca1*, *p53*, *Nf1* [35]. These tumor models derived from the fallopian tube epithelium provide researchers with the tools to study the molecular progression from pre-neoplastic lesion to aggressive serous carcinoma.

Careful examination and sequencing of patients with HGSC paints a more nuanced picture of the cell of origin debate. Laser-capture tumor microdissection of multiple anatomic sites in patients with HGSC showed an identical *p53* mutation at all sites [36]. The metastatic trajectory of HGSC was elucidated using phylogenetic clustering that compared tumor mutations to a patient's germline DNA. While the majority of patient tumors clustered in the "basal STIC" category, with the STIC showing the highest similarity to germline DNA, some tumors showed "STIC metastases". These findings call into question the assumption that the presence of STICs is always evidence for a fallopian tube origin for HGSC. A separate evolutionary analysis study that sequenced STICs, ovarian cancer, and metastases in nine patients found tumor-specific alterations in p53, BRCA1, BRCA2, or PTEN to be present in STICs [37]. This finding implies that in the majority of cases, mutations that drive HGSC occur early, before metastasis to the ovary. In a proteomic study of HGSC cell lines and patient tumor samples, 26 ovarian cancer cell lines and five HGSC tumors were grouped into three distinct categories: epithelial, clear cell, and mesenchymal [38]. While most cell lines and tumors in this study clustered in the epithelial group, suggesting a fallopian tube cell of origin, the authors identified a subset of cell lines and one HGSC tumor that grouped in the mesenchymal category, suggesting an ovarian cell of origin [38]. This demonstrates HGSC may arise from both the fallopian tube and OSE or that cells may acquire markers during tumorigenesis that resemble different tissues.

PAX2 and PAX8 are expressed in the fallopian tube epithelium, however, PAX2 is lost in ~85% HGSC and it has been shown that mutant p53 and loss of PTEN represses PAX2 expression in a fallopian tube-derived mode of ovarian cancer [39]. On the contrary, PAX8 is expressed in 85–90% of HGSC and is a widely used biomarker for HGSC [4,16,40]. PAX2 and PAX8 are differentially regulated in HGSC and it will be interesting to know whether loss of PAX2 during HGSC progression leads to dependence of HGSC on PAX8. Thus, studying the shared regulatory mechanisms of PAX2 and PAX8 expression between the fallopian tube and ovary will be essential to developing effective treatment therapies until the site of origin of a patient's tumor can be definitively identified.

3. Role of PAX2 and PAX8 in Development and Adult Tissues

The *PAX* genes are a set of developmental transcription factors that are key regulators for proper tissue formation and cellular differentiation [41]. This is convincingly supported by mouse models with *Pax* gene deletions. PAX2 is required for mesenchymal-to-epithelial transition of the intermediate mesoderm into the epithelial structures of the inner ear, kidneys, ureters, Wolffian and Müllerian ducts, including the oviducts, uterus, and vagina [42]. Mice with *Pax2* homozygous mutation do not develop these structures. Research has shown that *Pax2* is a tissue-specific epigenetic regulatory gene that ensures proper temporal and spatial development of these epithelial structures. The Hashino laboratory demonstrated that in progenitor cells of the inner ear, the histone demethylase LSD1 recruits the NuRD co-repressor complex to bind and repress PAX2 target genes. This inhibition ensures tight temporal control of PAX2-regulated genes. Once cells enter the differentiated state to become epithelial cells, LSD1 and the NuRD complex are released from the PAX2 binding site, and transcription can occur. This switch from progenitor intermediate mesoderm to differentiated epithelium is irreversible and is maintained over rapidly dividing cell populations through PAX2-regulated epigenetic modifications [43]. Research from the Dressler laboratory demonstrated that PAX2 promotes assembly of the histone H3K4 methylation complex by recruiting PTIP (PAX transcription activation domain interacting protein) at PAX2 binding elements [44]. This histone modification is associated with active promoters and increased transcription. PTIP deletion inhibits histone H3K4 methylation, even though PAX2 still binds to the chromosome. These data suggest that PTIP regulates epigenetic modifications required for activation of PAX2 targets that are essential for development and maintenance of epithelial structures.

PAX2 expression persists in adult reproductive tissues (epididymis, vas deferens, oviduct), ureters, bladder, kidneys, and mammary glands [45]. Cai et al. demonstrated that PAX2 levels are osmotically regulated [46]. Exposing medullary epithelial cells in vitro to high levels of NaCl increased PAX2 levels, while reducing in vivo renal inner-medullary interstitial NaCl levels decreased PAX2 levels. This increase in PAX2 appears to protect against cell death induced by osmotic stress. The stem cells of the mammary duct also express PAX2 where it may protect against apoptosis [47]. This is supported by research in *C. elegans* which demonstrates PAX2/5/8 can upregulate transcription of the anti-apoptotic *Bcl2* [48].

PAX8 is a closely related paralog to PAX2 that is expressed during embryogenesis in the thyroid, metanephros, central nervous system, and Müllerian duct. Inactivation of the *Pax8* gene in mice leads to complete loss of thyroid follicular cells, severe growth retardation, and death in the perinatal period [49]. Providing exogenous thyroid hormone to $Pax8^{-/-}$ mice rescued the hypothyroid phenotype, but these mice remained infertile due to nonfunctional uteri and closed vaginal openings [50].

PAX8 continues to be expressed in the adult kidneys, cervix, endometrium, fallopian tube, seminal vesicle, epididymis, thyroid, pancreas, and lymphoid cells [10,51]. There is also evidence that a subset of cells in the OSE express PAX8, but further research will need to examine the mechanism for this acquired expression [51,52]. The majority of our understanding of PAX8 function in adults is based on studies in the thyroid. Zannini and colleagues demonstrated that PAX8 is required for expression of the thyroid-specific genes: thyroglobulin, thyroperoxidase, and sodium/iodide symporter [53,54]. Interestingly, ChIP-Seq demonstrated PAX8 tends to bind in intronic regions (82%) over 5′-UTR regulatory regions (2%) [55]. This suggests PAX8 may bind alternative promoters or ncRNAs that regulate gene expression. Additionally, immunoprecipitation studies demonstrated that PAX8 binds CTCF and SP1, both of which are involved in chromatin remodeling [55]. These data suggest PAX8 functions both to directly increase transcription and to remodel the chromatin landscape.

4. Role of PAX2 and PAX8 in HGSC

Examining the histologic and molecular events that give rise to serous carcinoma is crucial to understanding the drivers of ovarian cancer. Secretory cell outgrowths (SCOUTs) are precursor lesions of serous carcinoma that can be found in the proximal and distal fallopian tube. Normal fallopian tube epithelial cells express high levels of PAX2 but approximately 90% of SCOUTs have lost PAX2 expression [16]. Almost all serous tumor cells also have mutation in the tumor suppressor p53, yet only 25% of SCOUTs have p53 mutation that can be detected histologically [16,56]. SCOUTs located at the fimbrial edge with p53 mutation are coined 'p53 signatures' [16]. Cells with the p53 signature have PAX2 loss, suggesting a step-wise progression from PAX2 loss to p53 signature to STIC to metastatic serous carcinoma. This progression has been extensively researched, and there are many excellent reviews detailing these findings [4,30,57,58].

Through molecular characterization of SCOUTs, Ning and colleagues demonstrated that PAX2 loss is associated with an increased stem cell phenotype [59]. They show through in vitro culture of SCOUTs that these cells can differentiate into both ciliated and basal cell histotypes. PAX2 knockdown in fallopian tube epithelial cell lines increased expression of the stem cell markers CD44 and SCA1 and decreased the capability of these cells to form differentiated epithelial luminal structures [60]. Modi et al. demonstrated in murine oviductal epithelial cells that *Pax2* loss and *p53* mutation increased proliferation and migration, but was insufficient to drive tumorigenesis [39]. This is consistent with human histological findings that p53 signatures are benign secretory outgrowths. ChIP analysis revealed wild type *p53* enhances *Pax2* transcription while mutant *p53* decreases *Pax2* transcription, suggesting a mechanism for sustained *Pax2* loss in neoplastic lesions [39]. Interestingly, cells lost *Pax2* expression in a fallopian tube model of ovarian cancer derived through loss of *Pten* [39]. Re-expression of *Pax2* inhibited the tumorigenic properties of these cells and prolonged survival (Figure 1). Alternatively, *Pax2* expression in a spontaneous OSE derived model of HGSC (called STOSE) reduced proliferation and metastasis by increasing COX2 and reducing HTRA1 expression [61]. Taken together, these findings suggest *Pax2* loss is an early molecular event in ovarian cancer progression that predisposes cells to further mutations that can drive tumorigenesis, regardless of cell of origin. Further research should examine the mechanistic requirement for *Pax2* loss in HGSC progression, especially considering that there is increased hypomethylation and activation of *Pax2* in endometrial and renal carcinoma, yet The Cancer Genome Atlas (TCGA) does not find increased methylation at this locus in HGSC tumor samples [47,62,63].

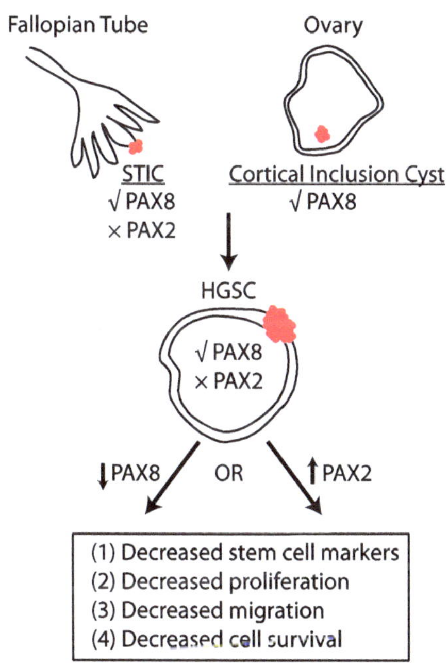

Figure 1. PAX2 and PAX8 regulate tumor formation in HGSC in an opposing manner. Serous tubal intraepithelial carcinomas (STICs) express PAX8, but not PAX2. Similarly, epithelial cells in cortical inclusion cysts express PAX8. HGSC tumor cells express PAX8 and it has been experimentally shown that PAX8 reduction decreases characteristics that enhance tumor formation. PAX2 is not expressed in HGSC and re-expression of PAX2 inhibited the tumorigenic properties of tumor cells.

Pathologists have used PAX8 for decades as a histologic marker to define HGSC, but a genome-wide RNA interference screen of cancer cell lines was the first to identify the importance of PAX8 in ovarian cancer [64]. PAX8 was the top-ranked differentially expressed gene in the screen between ovarian and non-ovarian cancer cell lines. *PAX8* knockdown reduced proliferation, migration and invasion and increased apoptosis in ovarian cancer cells [65]. *Pax8* was shown to directly bind and increase the transcription of *p53*, which then increased *p21* to induce proliferation [66]. *Pax8* also promoted tumor cell growth by increasing transcription of the cell cycle regulator *E2f1* through direct binding to the *E2f1* promoter in a complex with the RB protein [67]. In thyroid follicular carcinoma, a translocation event results in *PAX8-PPAR*γ*1* fusion [68], but this genetic event is not observed in HGSC (regulation of PAX2 and PAX8 in specific cancers is summarized in Table 1). To better understand the mechanism of PAX8 oncogenesis in HGSC despite its normal expression in the fallopian tube, several research groups have examined the role of PAX8 in the ovary and fallopian tube. Serial passaging of the normal OSE transforms cells into serous carcinoma with PAX8 expressed [22]. Loss of LKB1 and PTEN in the OSE also leads to a HGSC cell line with acquired PAX8 expression [69]. Rodgers and colleagues demonstrated that forced PAX8 expression in normal OSE increases proliferation, migration, and epithelial-mesenchymal transition through upregulation of the FOXM1 pathway [70]. Correspondingly, *PAX8* knockdown in three human HGSC cell lines decreased expression of FOXM1, decreased proliferation, and increased apoptosis [70]. Reducing PAX8 expression in the normal fallopian tube, however, did not produce noticeable phenotypic effects, suggesting that targeting PAX8 pharmacologically would not affect normal tissues. These phenotypic observations were corroborated by Elias and colleagues who performed an RNA sequencing experiment demonstrating few transcripts altered in the fallopian tube but increased transcript alterations in serous tumors after *PAX8* knockdown. The authors suggest alterations to the *PAX8* cistrome are responsible for changes in gene expression leading to HGSC derived from the fallopian tube. The *PAX8* consensus binding motif is altered between the fallopian tube and serous tumor cells that may affect downstream regulated genes. Elias et al. show differential association between PAX8 and Yes-associated protein (YAP1), a major downstream regulator of the evolutionarily conserved Hippo pathway that regulates organ size, cell proliferation, and apoptosis [71]. Interestingly, ChIP-Seq identified PAX8 mostly binds at non-promoter sites and is enriched at super-enhancers, where PAX8 can globally regulate genes involved in tumorigenesis [72]. Taken together, these findings suggest PAX8 could be targeted for drug development to reduce proliferation, migration and survival of tumor cells while leaving other organs unaffected (Figure 1).

Table 1. Mechanism of PAX2 and PAX8 regulation in specific cancer types.

Cancer Type	PAX2 Regulation	PAX8 Regulation	References
HGSC	Transcriptional downregulation	No change	[39]
Endometrial	Promoter hypomethylation	No change	[62]
Thyroid	No change	PAX8-PPARγ1 fusion	[68]
Renal	Promoter hypomethylation	Increased protein levels	[63,73]
Wilms tumor	Transcriptional upregulation	Transcriptional upregulation	[74,75]
Breast	Transcriptional upregulation	No change	[76]
Glioma	Transcriptional upregulation	Transcriptional upregulation	[77,78]

5. Clinical Strategies to Target PAX2 and PAX8

Ovarian cancer is a heterogeneous disease with few common molecular alterations [56]. Developing therapeutic strategies that target common molecular alterations, such as loss of PAX2 or gain of PAX8, may produce greater therapeutic benefits. A promoter activation screen identified luteolin as a small molecule that restores PAX2 expression in cells with wild type p53 [39]. Luteolin could be taken as a preventative supplement to decrease the occurrence of SCOUTs, but it would be ineffective in treating serous tumors with p53 mutation. Further screens or combination therapy studies should be performed in HGSC cells to identify molecules that increase PAX2 in tumors. The effect of these molecules on the homologue PAX8 should also be explored. Molecules that increase expression of PAX2 may also increase expression of PAX8, which could then increase the aggressive properties of a tumor cell. Therapies that reduce transcription of these PAX proteins, however, may significantly mediate the deleterious effect of PAX8 while maintaining the already decreased PAX2 levels.

PAX8 seems to have little functional effect in the fully differentiated adult fallopian tube, but mediates several tumorigenic effects in HGSC, including proliferation, migration, angiogenesis, and apoptosis [65,70–72,79]. Reducing PAX8 levels or disrupting the transcriptional activity of PAX8 may inhibit these pro-cancerous effects while leaving the normal fallopian tube epithelium unaffected. Using a virtual screen that modeled paired domain binding to DNA, Grimley and colleagues identified small molecules that disrupt binding of the paired domain of PAX2/5/8 to DNA [80]. Other potential drug targets include the adapter proteins that bind to the chromosome in a complex with PAX8. PAX8 requires interactions with YAP1, CTCF and SP1 to initiate transcription, as discussed earlier. Disrupting these interactions may mediate the deleterious effects of PAX8 in serous carcinoma.

6. Concluding Remarks

Proper temporal and spatial expression of the PAX protein family is essential for embryonic development. PAX2 and PAX8 are co-expressed during mesenchymal-to-epithelial transition of the Müllerian duct and they continue to be expressed in adult structures, such as the fallopian tube. These proteins maintain a regenerative stem cell population in adult tissues. In HGSC, PAX8 provides growth advantages by enhancing the proliferative, migratory, and survival capabilities of cancer cells from the fallopian tube and ovary. The OSE does not normally express PAX8, yet it acquires PAX8 expression during malignant transformation in certain mouse models. More work is required to tease apart the role of PAX8 in tumors derived from the fallopian tube or OSE. PAX2 is a homolog of PAX8 that has been shown to impart similar growth advantages, yet tumors derived from the fallopian tube epithelium lose PAX2 expression during malignant transformation. Further research is required to understand the importance and regulatory machinery that leads to PAX2 loss and PAX8 dependence in HGSC.

Identifying drug targets for novel cancer treatments in HGSC has been challenging because it is a heterogeneous disease with few shared mutations. The PAX proteins are promising because PAX8 is ubiquitously expressed in serous tumors and PAX2 loss is an early molecular event shared in the progression from benign to malignant carcinoma. Targeting these proteins may hold promise in reducing tumor growth and progression in a majority of patients and significantly improving patient survival.

Author Contributions: Literature search, L.R.H., A.S. and J.E.B.; Writing—Original Draft Preparation, L.R.H., A.S.; Writing—Review & Editing, A.S., J.E.B., Figures, L.R.H.; Funding Acquisition, L.R.H. and J.E.B.

Funding: This research was funded by NCI grant number F30CA224986, NIH R21 CA208610, and by DOD OCRP grant number OC160076.

Conflicts of Interest: The authors declare no conflict of interest.

References

1. Siegel, R.L.; Miller, K.D.; Jemal, A. Cancer statistics, 2017. *CA Cancer J. Clin.* **2017**, *67*, 7–30. [CrossRef] [PubMed]
2. Kim, G.; Ison, G.; McKee, A.E.; Zhang, H.; Tang, S.; Gwise, T.; Sridhara, R.; Lee, E.; Tzou, A.; Philip, R.; et al. FDA Approval Summary: Olaparib Monotherapy in Patients with Deleterious Germline BRCA-Mutated Advanced Ovarian Cancer Treated with Three or More Lines of Chemotherapy. *Clin. Cancer Res.* **2015**, *21*, 4257–4261. [CrossRef] [PubMed]
3. Pujade-Lauraine, E.; Hilpert, F.; Weber, B.; Reuss, A.; Poveda, A.; Kristensen, G.; Sorio, R.; Vergote, I.; Witteveen, P.; Bamias, A.; et al. Bevacizumab Combined With Chemotherapy for Platinum-Resistant Recurrent Ovarian Cancer: The AURELIA Open-Label Randomized Phase III Trial. *Obstet. Gynecol. Surv.* **2014**, *69*, 402. [CrossRef]

4. Kurman, R.J.; Shih, I.-M. The Origin and Pathogenesis of Epithelial Ovarian Cancer: A Proposed Unifying Theory. *Am. J. Surg. Pathol.* **2010**, *34*, 433–443. [CrossRef] [PubMed]
5. Piek, J.M.; van Diest, P.J.; Zweemer, R.P.; Jansen, J.W.; Poort-Keesom, R.J.; Menko, F.H.; Gille, J.J.; Jongsma, A.P.; Pals, G.; Kenemans, P.; et al. Dysplastic changes in prophylactically removed Fallopian tubes of women predisposed to developing ovarian cancer. *J. Pathol.* **2001**, *195*, 451–456. [CrossRef] [PubMed]
6. Kindelberger, D.W.; Lee, Y.; Miron, A.; Hirsch, M.S.; Feltmate, C.; Medeiros, F.; Callahan, M.J.; Garner, E.O.; Gordon, R.W.; Birch, C.; et al. Intraepithelial carcinoma of the fimbria and pelvic serous carcinoma: Evidence for a causal relationship. *Am. J. Surg. Pathol.* **2007**, *31*, 161–169. [CrossRef] [PubMed]
7. Kim, J.; Coffey, D.M.; Creighton, C.J.; Yu, Z.; Hawkins, S.M.; Matzuk, M.M. High-grade serous ovarian cancer arises from fallopian tube in a mouse model. *Proc. Natl. Acad. Sci. USA* **2012**, *109*, 3921–3926. [CrossRef] [PubMed]
8. Robson, E.J.D.; He, S.-J.; Eccles, M.R. A PANorama of PAX genes in cancer and development. *Nat. Rev. Cancer* **2006**, *6*, 52–62. [CrossRef] [PubMed]
9. Eccles, M.R.; Li, C.G. PAX Genes in Cancer; Friends or Foes? *Front. Genet.* **2012**, *3*. [CrossRef]
10. Tacha, D.; Zhou, D.; Cheng, L. Expression of PAX8 in normal and neoplastic tissues: A comprehensive immunohistochemical study. *Appl. Immunohistochem. Mol. Morphol.* **2011**, *19*, 293–299. [CrossRef] [PubMed]
11. Nonaka, D.; Chiriboga, L.; Soslow, R.A. Expression of pax8 as a useful marker in distinguishing ovarian carcinomas from mammary carcinomas. *Am. J. Surg. Pathol.* **2008**, *32*, 1566–1571. [CrossRef] [PubMed]
12. Laury, A.R.; Perets, R.; Piao, H.; Krane, J.F.; Barletta, J.A.; French, C.; Chirieac, L.R.; Lis, R.; Loda, M.; Hornick, J.L.; et al. A comprehensive analysis of PAX8 expression in human epithelial tumors. *Am. J. Surg. Pathol.* **2011**, *35*, 816–826. [CrossRef] [PubMed]
13. Song, H.; Kwan, S.-Y.; Izaguirre, D.I.; Zu, Z.; Tsang, Y.T.; Tung, C.S.; King, E.R.; Mok, S.C.; Gershenson, D.M.; Wong, K.-K. PAX2 Expression in Ovarian Cancer. *Int. J. Mol. Sci.* **2013**, *14*, 6090–6105. [CrossRef] [PubMed]
14. Tung, C.S.; Mok, S.C.; Tsang, Y.T.M.; Zu, Z.; Song, H.; Liu, J.; Deavers, M.T.; Malpica, A.; Wolf, J.K.; Lu, K.H.; et al. PAX2 expression in low malignant potential ovarian tumors and low-grade ovarian serous carcinomas. *Mod. Pathol.* **2009**, *22*, 1243–1250. [CrossRef] [PubMed]
15. Tong, G.-X.; Chiriboga, L.; Hamele-Bena, D.; Borczuk, A.C. Expression of PAX2 in papillary serous carcinoma of the ovary: Immunohistochemical evidence of fallopian tube or secondary Müllerian system origin? *Mod. Pathol.* **2007**, *20*, 856–863. [CrossRef] [PubMed]
16. Chen, E.Y.; Mehra, K.; Mehrad, M.; Ning, G.; Miron, A.; Mutter, G.L.; Monte, N.; Quade, B.J.; McKeon, F.D.; Yassin, Y.; et al. Secretory cell outgrowth, PAX2 and serous carcinogenesis in the Fallopian tube. *J. Pathol.* **2010**, *222*, 110–116. [CrossRef] [PubMed]
17. Murdoch, W.J.; McDonnel, A.C. Roles of the ovarian surface epithelium in ovulation and carcinogenesis. *Reproduction* **2002**, *123*, 743–750. [CrossRef] [PubMed]
18. Auersperg, N.; Wong, A.S.; Choi, K.C.; Kang, S.K.; Leung, P.C. Ovarian surface epithelium: Biology, endocrinology, and pathology. *Endocr. Rev.* **2001**, *22*, 255–288. [CrossRef] [PubMed]
19. Bowen, N.J.; Logani, S.; Dickerson, E.B.; Kapa, L.B.; Akhtar, M.; Benigno, B.B.; McDonald, J.F. Emerging roles for PAX8 in ovarian cancer and endosalpingeal development. *Gynecol. Oncol.* **2007**, *104*, 331–337. [CrossRef] [PubMed]
20. Hankinson, S.E.; Colditz, G.A.; Hunter, D.J.; Willett, W.C.; Stampfer, M.J.; Rosner, B.; Hennekens, C.H.; Speizer, F.E. A prospective study of reproductive factors and risk of epithelial ovarian cancer. *Cancer* **1995**, *76*, 284–290. [CrossRef]
21. La Vecchia, C.; Negri, E.; Franceschi, S.; Parazzini, F. Long-term impact of reproductive factors on cancer risk. *Int. J. Cancer* **1993**, *53*, 215–219. [CrossRef] [PubMed]
22. McCloskey, C.W.; Goldberg, R.L.; Carter, L.E.; Gamwell, L.F.; Al-Hujaily, E.M.; Collins, O.; Macdonald, E.A.; Garson, K.; Daneshmand, M.; Carmona, E.; et al. A new spontaneously transformed syngeneic model of high-grade serous ovarian cancer with a tumor-initiating cell population. *Front. Oncol* **2014**, *4*, 53. [CrossRef] [PubMed]
23. Cheng, W.; Liu, J.; Yoshida, H.; Rosen, D.; Naora, H. Lineage infidelity of epithelial ovarian cancers is controlled by HOX genes that specify regional identity in the reproductive tract. *Nat. Med.* **2005**, *11*, 531–537. [CrossRef] [PubMed]

24. Flesken-Nikitin, A.; Choi, K.-C.; Eng, J.P.; Shmidt, E.N.; Nikitin, A.Y. Induction of carcinogenesis by concurrent inactivation of p53 and Rb1 in the mouse ovarian surface epithelium. *Cancer Res.* **2003**, *63*, 3459–3463. [PubMed]
25. Kuhn, E.; Kurman, R.J.; Vang, R.; Sehdev, A.S.; Han, G.; Soslow, R.; Wang, T.L.; Shih, I.-M. *TP53* mutations in serous tubal intraepithelial carcinoma and concurrent pelvic high-grade serous carcinoma evidence supporting the clonal relationship of the two lesions. *J. Pathol.* **2012**, *226*, 421–426. [CrossRef] [PubMed]
26. Marquez, R.T. Patterns of Gene Expression in Different Histotypes of Epithelial Ovarian Cancer Correlate with Those in Normal Fallopian Tube, Endometrium, and Colon. *Clin. Cancer Res.* **2005**, *11*, 6116–6126. [CrossRef] [PubMed]
27. Merritt, M.A.; Bentink, S.; Schwede, M.; Iwanicki, M.P.; Quackenbush, J.; Woo, T.; Agoston, E.S.; Reinhardt, F.; Crum, C.P.; Berkowitz, R.S.; et al. Gene Expression Signature of Normal Cell-of-Origin Predicts Ovarian Tumor Outcomes. *PLoS ONE* **2013**, *8*, e80314-10. [CrossRef] [PubMed]
28. Rebbeck, T.R.; Lynch, H.T.; Neuhausen, S.L.; Narod, S.A.; van't Veer, L.; Garber, J.E.; Evans, G.; Isaacs, C.; Daly, M.B.; Matloff, E.; et al. Prophylactic Oophorectomy in Carriers of BRCA1 or BRCA2 Mutations. *Mass Med. Soc.* **2002**, *346*, 1616–1622. [CrossRef]
29. Falconer, H.; Yin, L.; Grönberg, H.; Altman, D. Ovarian cancer risk after salpingectomy: A nationwide population-based study. *J. Natl. Cancer Inst.* **2015**, *107*. [CrossRef] [PubMed]
30. Gross, A.L.; Kurman, R.J.; Vang, R.; Shih, I.-M.; Visvanathan, K. Precursor lesions of high-grade serous ovarian carcinoma: Morphological and molecular characteristics. *J. Oncol.* **2010**, *2010*, 126295. [CrossRef] [PubMed]
31. Karst, A.M.; Levanon, K.; Drapkin, R. Modeling high-grade serous ovarian carcinogenesis from the fallopian tube. *Proc. Natl. Acad. Sci. USA* **2011**, *108*, 7547–7552. [CrossRef] [PubMed]
32. Perets, R.; Wyant, G.A.; Muto, K.W.; Bijron, J.G.; Poole, B.B.; Chin, K.T.; Chen, J.Y.; Ohman, A.W.; Stepule, C.D.; Kwak, S.; et al. Transformation of the fallopian tube secretory epithelium leads to high-grade serous ovarian cancer in *Brca;Tp53;Pten* models. *Cancer Cell* **2013**, *24*, 751–765. [CrossRef] [PubMed]
33. Russo, A.; Czarnecki, A.A.; Dean, M.; Modi, D.A.; Lantvit, D.D.; Hardy, L.; Baligod, S.; Davis, D.A.; Wei, J.-J.; Burdette, J.E. PTEN loss in the fallopian tube induces hyperplasia and ovarian tumor formation. *Oncogene* **2018**, *67*, 7. [CrossRef] [PubMed]
34. Eddie, S.L.; Quartuccio, S.M.; Ó hAinmhir, E.; Moyle-Heyrman, G.; Lantvit, D.D.; Wei, J.-J.; Vanderhyden, B.C.; Burdette, J.E. Tumorigenesis and peritoneal colonization from fallopian tube epithelium. *Oncotarget* **2015**, *6*, 20500–20512. [CrossRef] [PubMed]
35. Zhai, Y.; Wu, R.; Kuick, R.; Sessine, M.S.; Schulman, S.; Green, M.; Fearon, E.R.; Cho, K.R. High-grade serous carcinomas arise in the mouse oviduct via defects linked to the human disease. *J. Pathol.* **2017**, *243*, 16–25. [CrossRef] [PubMed]
36. Eckert, M.A.; Pan, S.; Hernandez, K.M.; Loth, R.M.; Andrade, J.; Volchenboum, S.L.; Faber, P.; Montag, A.; Lastra, R.; Peter, M.E.; et al. Genomics of Ovarian Cancer Progression Reveals Diverse Metastatic Trajectories Including Intraepithelial Metastasis to the Fallopian Tube. *Cancer Discov.* **2016**, *6*, 1342–1351. [CrossRef] [PubMed]
37. Labidi-Galy, S.I.; Papp, E.; Hallberg, D.; Niknafs, N.; Adleff, V.; Noe, M.; Bhattacharya, R.; Novak, M.; Jones, S.; Phallen, J.; et al. High grade serous ovarian carcinomas originate in the fallopian tube. *Nat. Commun.* **2017**, *8*, 1093. [CrossRef] [PubMed]
38. Coscia, F.; Watters, K.M.; Curtis, M.; Eckert, M.A.; Chiang, C.Y.; Tyanova, S.; Montag, A.; Lastra, R.R.; Lengyel, E.; Mann, M. Integrative proteomic profiling of ovarian cancer cell lines reveals precursor cell associated proteins and functional status. *Nat. Commun.* **2016**, *7*, 1–14. [CrossRef] [PubMed]
39. Modi, D.A.; Tagare, R.D.; Karthikeyan, S.; Russo, A.; Dean, M.; Davis, D.A.; Lantvit, D.D.; Burdette, J.E. PAX2 function, regulation and targeting in fallopian tube-derived high-grade serous ovarian cancer. *Oncogene* **2017**, *36*, 3015–3024. [CrossRef] [PubMed]
40. Blake, J.A.; Ziman, M.R. Pax genes: Regulators of lineage specification and progenitor cell maintenance. *Development* **2014**, *141*, 737–751. [CrossRef] [PubMed]
41. Dahl, E.; Koseki, H.; Balling, R. Pax genes and organogenesis. *Bioessays* **1997**, *19*, 755–765. [CrossRef] [PubMed]
42. Torres, M.; Gómez-Pardo, E.; Dressler, G.R.; Gruss, P. Pax-2 controls multiple steps of urogenital development. *Development* **1995**, *121*, 4057–4065. [PubMed]

43. Patel, D.; Shimomura, A.; Majumdar, S.; Holley, M.C.; Hashino, E. The histone demethylase LSD1 regulates inner ear progenitor differentiation through interactions with Pax2 and the NuRD repressor complex. *PLoS ONE* **2018**, *13*, e0191689-20. [CrossRef] [PubMed]
44. Patel, S.R.; Kim, D.; Levitan, I.; Dressler, G.R. The BRCT-Domain Containing Protein PTIP Links PAX2 to a Histone H3, Lysine 4 Methyltransferase Complex. *Dev. Cell* **2007**, *13*, 580–592. [CrossRef] [PubMed]
45. Zhai, Q.J.; Ozcan, A.; Hamilton, C.; Shen, S.S.; Coffey, D.; Krishnan, B.; Truong, L.D. PAX-2 expression in non-neoplastic, primary neoplastic, and metastatic neoplastic tissue: A comprehensive immunohistochemical study. *Appl. Immunohistochem. Mol. Morphol.* **2010**, *18*, 323–332. [CrossRef] [PubMed]
46. Cai, Q.; Dmitrieva, N.I.; Ferraris, J.D.; Brooks, H.L.; van Balkom, B.W.M.; Burg, M. Pax2 expression occurs in renal medullary epithelial cells in vivo and in cell culture, is osmoregulated, and promotes osmotic tolerance. *Proc. Natl. Acad. Sci. USA* **2005**, *102*, 503–508. [CrossRef] [PubMed]
47. Silberstein, G.B.; Dressler, G.R.; Van Horn, K. Expression of the PAX2 oncogene in human breast cancer and its role in progesterone-dependent mammary growth. *Oncogene* **2002**, *21*, 1009–1016. [CrossRef] [PubMed]
48. Park, D.; Jia, H.; Rajakumar, V.; Chamberlin, H.M. Pax2/5/8 proteins promote cell survival in *C. elegans*. *Development* **2006**, *133*, 4193–4202. [CrossRef] [PubMed]
49. Mansouri, A.; Chowdhury, K.; Gruss, P. Follicular cells of the thyroid gland require Pax8 gene function. *Nat. Genet.* **1998**, *19*, 87–90. [CrossRef] [PubMed]
50. Mittag, J.; Winterhager, E.; Bauer, K.; Grümmer, R. Congenital Hypothyroid Female Pax8-Deficient Mice Are Infertile Despite Thyroid Hormone Replacement Therapy. *Endocrinology* **2007**, *148*, 719–725. [CrossRef] [PubMed]
51. Ozcan, A.; Shen, S.S.; Hamilton, C.; Anjana, K.; Coffey, D.; Krishnan, B.; Truong, L.D. PAX 8 expression in non-neoplastic tissues, primary tumors, and metastatic tumors: A comprehensive immunohistochemical study. *Mod. Pathol.* **2011**, *24*, 751–764. [CrossRef] [PubMed]
52. Adler, E.; Mhawech-Fauceglia, P.; Gayther, S.A.; Lawrenson, K. PAX8 expression in ovarian surface epithelial cells. *Hum. Pathol.* **2015**, *46*, 948–956. [CrossRef] [PubMed]
53. Zannini, M.; Francis-Lang, H.; Plachov, D.; Di Lauro, R. Pax-8, a paired domain-containing protein, binds to a sequence overlapping the recognition site of a homeodomain and activates transcription from two thyroid-specific promoters. *Mol. Cell. Biol.* **1992**, *12*, 4230–4241. [CrossRef] [PubMed]
54. Ohno, M.; Zannini, M.; Levy, O.; Carrasco, N.; Di Lauro, R. The paired-domain transcription factor Pax8 binds to the upstream enhancer of the rat sodium/iodide symporter gene and participates in both thyroid-specific and cyclic-AMP-dependent transcription. *Mol. Cell. Biol.* **1999**, *19*, 2051–2060. [CrossRef] [PubMed]
55. Ruiz-Llorente, S.; Carrillo Santa de Pau, E.; Sastre-Perona, A.; Montero-Conde, C.; Gómez-López, G.; Fagin, J.A.; Valencia, A.; Pisano, D.G.; Santisteban, P. Genome-wide analysis of Pax8 binding provides new insights into thyroid functions. *BMC Genom.* **2012**, *13*, 147. [CrossRef] [PubMed]
56. Cancer Genome Atlas Research Network Integrated genomic analyses of ovarian carcinoma. *Nature* **2011**, *474*, 609–615. [CrossRef] [PubMed]
57. Levanon, K.; Crum, C.; Drapkin, R. New Insights Into the Pathogenesis of Serous Ovarian Cancer and Its Clinical Impact. *JCO* **2008**, *26*, 5284–5293. [CrossRef] [PubMed]
58. Reade, C.J.; McVey, R.M.; Tone, A.A.; Finlayson, S.J.; McAlpine, J.N.; Fung-Kee-Fung, M.; Ferguson, S.E. The Fallopian Tube as the Origin of High Grade Serous Ovarian Cancer: Review of a Paradigm Shift. *J. Obstet. Gynaecol. Can.* **2014**, *36*, 133–140. [CrossRef]
59. Ning, G.; Bijron, J.G.; Yamamoto, Y.; Wang, X.; Howitt, B.E.; Herfs, M.; Yang, E.; Hong, Y.; Cornille, M.; Wu, L.; et al. The PAX2-null immunophenotype defines multiple lineages with common expression signatures in benign and neoplastic oviductal epithelium. *J. Pathol.* **2014**, *234*, 478–487. [CrossRef] [PubMed]
60. Alwosaibai, K.; Abedini, A.; Al-Hujaily, E.M.; Tang, Y.; Garson, K.; Collins, O.; Vanderhyden, B.C. PAX2 maintains the differentiation of mouse oviductal epithelium and inhibits the transition to a stem cell-like state. *Oncotarget* **2017**, *8*, 76881–76897. [CrossRef] [PubMed]
61. Al-Hujaily, E.M.; Tang, Y.; Yao, D.S.; Carmona, E.; Garson, K.; Vanderhyden, B.C. Divergent Roles of PAX2 in the Etiology and Progression of Ovarian Cancer. *Cancer Prev. Res.* **2015**, *8*, 1163–1173. [CrossRef] [PubMed]
62. Wu, H.; Chen, Y.; Liang, J.; Shi, B.; Wu, G.; Zhang, Y.; Wang, D.; Li, R.; Yi, X.; Zhang, H.; et al. Hypomethylation-linked activation of PAX2 mediates tamoxifen-stimulated endometrial carcinogenesis. *Nature* **2005**, *438*, 981–987. [CrossRef] [PubMed]

63. Metsuyanim, S.; Pode-Shakked, N.; Schmidt-Ott, K.M.; Keshet, G.; Rechavi, G.; Blumental, D.; Dekel, B. Accumulation of Malignant Renal Stem Cells Is Associated with Epigenetic Changes in Normal Renal Progenitor Genes. *Stem Cells* **2008**, *26*, 1808–1817. [CrossRef] [PubMed]
64. Cheung, H.W.; Cowley, G.S.; Weir, B.A.; Boehm, J.S.; Rusin, S.; Scott, J.A.; East, A.; Ali, L.D.; Lizotte, P.H.; Wong, T.C.; et al. Systematic investigation of genetic vulnerabilities across cancer cell lines reveals lineage-specific dependencies in ovarian cancer. *Proc. Natl. Acad. Sci. USA* **2011**, *108*, 12372–12377. [CrossRef] [PubMed]
65. Di Palma, T.; Lucci, V.; de Cristofaro, T.; Filippone, M.G.; Zannini, M. A role for PAX8 in the tumorigenic phenotype of ovarian cancer cells. *BMC Cancer* **2014**, *14*, 292. [CrossRef] [PubMed]
66. Ghannam-Shahbari, D.; Jacob, E.; Kakun, R.R.; Wasserman, T.; Korsensky, L.; Sternfeld, O.; Kagan, J.; Bublik, D.R.; Aviel-Ronen, S.; Levanon, K.; et al. PAX8 activates a p53-p21-dependent pro-proliferative effect in high grade serous ovarian carcinoma. *Oncogene* **2018**, 1–12. [CrossRef] [PubMed]
67. Li, C.G.; Nyman, J.E.; Braithwaite, A.W.; Eccles, M.R. PAX8 promotes tumor cell growth by transcriptionally regulating E2F1 and stabilizing RB protein. *Oncogene* **2011**, *30*, 4824–4834. [CrossRef] [PubMed]
68. Kroll, T.G.; Sarraf, P.; Pecciarini, L.; Chen, C.J.; Mueller, E.; Spiegelman, B.M.; Fletcher, J.A. PAX8-PPARgamma1 fusion oncogene in human thyroid carcinoma [corrected]. *Science* **2000**, *289*, 1357–1360. [CrossRef] [PubMed]
69. Tanwar, P.S.; Mohapatra, G.; Chiang, S.; Engler, D.A.; Zhang, L.; Kaneko-Tarui, T.; Ohguchi, Y.; Birrer, M.J.; Teixeira, J.M. Loss of LKB1 and PTEN tumor suppressor genes in the ovarian surface epithelium induces papillary serous ovarian cancer. *Carcinogenesis* **2014**, *35*, 546–553. [CrossRef] [PubMed]
70. Rodgers, L.H.; Ó hAinmhire, E.; Young, A.N.; Burdette, J.E. Loss of PAX8 in high-grade serous ovarian cancer reduces cell survival despite unique modes of action in the fallopian tube and ovarian surface epithelium. *Oncotarget* **2016**, *7*, 32785–32795. [CrossRef] [PubMed]
71. Elias, K.M.; Emori, M.M.; Westerling, T.; Long, H.; Budina-Kolomets, A.; Li, F.; MacDuffie, E.; Davis, M.R.; Holman, A.; Lawney, B.; et al. Epigenetic remodeling regulates transcriptional changes between ovarian cancer and benign precursors. *JCI Insight* **2016**, *1*. [CrossRef] [PubMed]
72. Adler, E.K.; Corona, R.I.; Lee, J.M.; Rodriguez-Malave, N.; Mhawech-Fauceglia, P.; Sowter, H.; Hazelett, D.J.; Lawrenson, K.; Gayther, S.A. The PAX8 cistrome in epithelial ovarian cancer. *Oncotarget* **2017**, *8*, 108316–108332. [CrossRef] [PubMed]
73. Ozcan, A.; de la Roza, G.; Ro, J.Y.; Shen, S.S.; Truong, L.D. PAX2 and PAX8 Expression in Primary and Metastatic Renal Tumors: A Comprehensive Comparison. *Arch. Pathol. Lab. Med.* **2012**, *136*, 1541–1551. [CrossRef] [PubMed]
74. Dressler, G.R.; Douglass, E.C. Pax-2 is a DNA-binding protein expressed in embryonic kidney and Wilms tumor. *Proc. Natl Acad. Sci. USA* **1992**, *89*, 1179–1183. [CrossRef] [PubMed]
75. Tagge, E.P.; Hanson, P.; Re, G.G.; Othersen, H.B.; Smith, C.D.; Garvin, A.J. Paired box gene expression in Wilms' tumor. *J. Pediatr. Surg.* **1994**, *29*, 134–141. [CrossRef]
76. Hurtado, A.; Holmes, K.A.; Geistlinger, T.R.; Hutcheson, I.R.; Nicholson, R.I.; Brown, M.; Jiang, J.; Howat, W.J.; Ali, S.; Carroll, J.S. Regulation of ERBB2 by oestrogen receptor–PAX2 determines response to tamoxifen. *Nature* **2008**, *456*, 663–666. [CrossRef] [PubMed]
77. Chen, Y.-J.; Campbell, H.G.; Wiles, A.K.; Eccles, M.R.; Reddel, R.R.; Braithwaite, A.W.; Royds, J.A. PAX8 Regulates Telomerase Reverse Transcriptase and Telomerase RNA Component in Glioma. *Cancer Res.* **2008**, *68*, 5724–5732. [CrossRef] [PubMed]
78. Burger, M.C.; Brucker, D.P.; Baumgarten, P.; Ronellenfitsch, M.W.; Wanka, C.; Hasselblatt, M.; Eccles, M.R.; Klingebiel, T.; Weller, M.; Rieger, J.; et al. PAX2 is an antiapoptotic molecule with deregulated expression in medulloblastoma. *Int. J. Oncol.* **2012**, *41*, 235–241. [CrossRef] [PubMed]
79. De Cristofaro, T.; Di Palma, T.; Soriano, A.A.; Monticelli, A.; Affinito, O.; Cocozza, S.; Zannini, M. Candidate genes and pathways downstream of PAX8 involved in ovarian high-grade serous carcinoma. *Oncotarget* **2016**, *7*, 41929–41947. [CrossRef] [PubMed]
80. Grimley, E.; Liao, C.; Ranghini, E.J.; Nikolovska-Coleska, Z.; Dressler, G.R. Inhibition of Pax2 Transcription Activation with a Small Molecule that Targets the DNA Binding Domain. *ACS Chem. Biol.* **2017**, *12*, 724–734. [CrossRef] [PubMed]

cancers

MDPI

Review

The Impact of Mesothelin in the Ovarian Cancer Tumor Microenvironment

Tyvette S. Hilliard

Department of Chemistry and Biochemistry, Harper Cancer Research Institute, University of Notre Dame, Notre Dame, IN 46617, USA; thilliar@nd.edu; Tel.: +1-574-631-2452

Received: 8 June 2018; Accepted: 18 August 2018; Published: 21 August 2018

Abstract: Ovarian cancer is the deadliest gynecological disease among U.S. women. Poor 5-year survival rates (<30%) are due to presentation of most women at diagnosis with advanced stage disease with widely disseminated intraperitoneal metastasis. However, when diagnosed before metastatic propagation the overall 5-year survival rate is >90%. Metastasizing tumor cells grow rapidly and aggressively attach to the mesothelium of all organs within the peritoneal cavity, including the parietal peritoneum and the omentum, producing secondary lesions. In this review, the involvement of mesothelin (MSLN) in the tumor microenvironment is discussed. MSLN, a 40kDa glycoprotein that is overexpressed in many cancers including ovarian and mesotheliomas is suggested to play a role in cell survival, proliferation, tumor progression, and adherence. However, the biological function of MSLN is not fully understood as MSLN knockout mice do not present with an abnormal phenotype. Conversely, MSLN has been shown to bind to the ovarian cancer antigen, CA-125, and thought to play a role in the peritoneal diffusion of ovarian tumor cells. Although the cancer-specific expression of MSLN makes it a potential therapeutic target, more studies are needed to validate the role of MSLN in tumor metastasis.

Keywords: ovarian cancer; mesothelin; CA125; tumor microenvironment

1. Introduction

Ovarian cancer is the fifth leading cause of cancer death in U.S. women, making it the most lethal gynecological malignancy. The American Cancer Society estimates that about 22,240 new cases of ovarian cancer will be diagnosed in the United States in 2018, of which 14,070 (>60%) women will die of the disease [1]. The overall 5-year survival rate of women diagnosed with ovarian cancer is 47% and for women diagnosed with advanced stage disease, presenting with intraperitoneal metastasis, the 5-year survival rate is only 29% [1,2]. Ovarian cancer is a heterogeneous disease composed of seven histological subtypes: high-grade serous, low-grade serous, mucinous, endometrioid, clear cell, carcinosarcoma, and Brenner tumors [3]. Approximately 90% of ovarian cancers are classified as malignant epithelial ovarian carcinomas (EOCs), of which high-grade serous carcinomas (HGSC) account for 70% of tumor types [4–7]. Early signs or symptoms of ovarian cancer are often subtle and nonspecific which are frequently ignored or treated with medicine to relieve discomfort. In 50–80% of high-grade serous carcinomas, the most frequent genetic change is a p53 mutation found in tumors of all stages [8–10]. Mutations in BRCA1 and BRCA2, tumor suppressor genes, are found in about 50% and 70% of ovarian cancer patients with a family history of ovarian cancer, but 95% of ovarian cancer cases are sporadic [11–13].

The major cause of death is due to therapy-resistant metastasis from the primary tumor to the peritoneum [14–18]. The lack of successful eradication of the disease can be owing to the various complex overlapping signaling networks, together with the peritoneal tumor microenvironment composed of mesothelial cells, the submesothelial matrix, and adipose. Unlike other cancers, ovarian cancer uniquely metastasizes by the detachment of tumor cells, either single or multicellular aggregates, from the primary ovarian/fallopian tube tumor instead of the classically studied pattern of hematogenous metastasis (Figure 1A,B) [15,16,18,19]. Recent studies have challenged this mode of metastasis, suggesting that hematogenous spread of ovarian cancer may play a larger role in ovarian cancer cell metastasis; however, for the purpose of this review, ovarian cancer metastasis will be discussed as direct shedding of tumor cells [20,21]. This distinctive process bypasses several steps of intra- and extravasation before metastasis to other organs [19]. These detached cells undergo epithelial to mesenchymal transition before detaching, resulting in the loss of E-cadherin, a glycoprotein located at cellular junctions, and an invasive phenotype [22]. The metastatic cells disseminate throughout the peritoneal cavity, facilitated by natural fluid flow and preferentially attach to the mesothelium that covers all the organs in the peritoneal cavity including the omentum, abdominal peritoneum and the contralateral ovary (Figure 1C,D) [14,23,24]. Proliferation of disseminated tumor cells on the omentum eventually results in the obstruction of the bowel and stomach [25,26]. It is unknown if the primary tumor prepares secondary metastatic sites, including the omentum and peritoneum, for colonization, a process that has been implicated in other cancers [19].

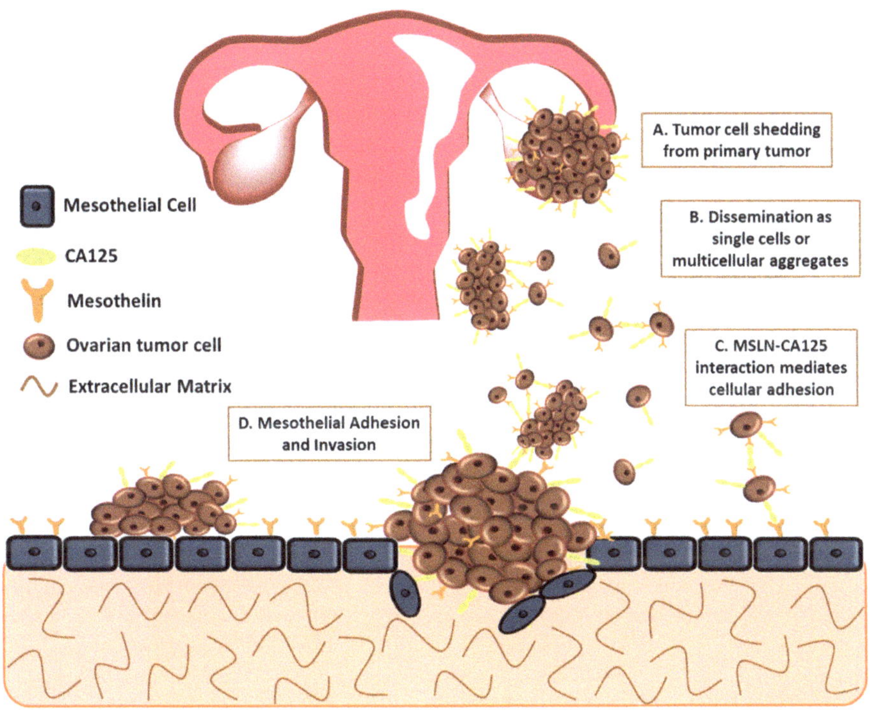

Figure 1. Model for peritoneal metastasis of ovarian tumors. Ovarian cancer metastasis is unique as tumor cells shed from the primary tumor and spread throughout the peritoneal cavity. MSLN:CA125 interaction mediates heterotypic and homotypic cellular adhesion.

Currently, there are no simple screening tests available to detect ovarian cancer. However, available diagnostic testing includes pelvic examinations, transvaginal ultrasonography and serum measurements of cancer antigen-125 (CA125) [27]. Identification of additional screening strategies to accurately diagnose patients in early stages are of great need. Moreover, mesothelin, a glycoprotein expressed in mesothelial cells and overexpressed in EOCs, could be useful as both a screening biomarker as well as a therapeutic target [28]. Understanding the interaction of the tumor and mesothelium and regulating the molecules that modify the metastatic tumor microenvironment is of great importance for the development of future therapeutics.

2. CA125

CA125, a repeating peptide epitope of the mucin 16 (MUC16), is a large membrane-bound cell surface mucin, discovered in 1981 by a monoclonal antibody OC125 developed from mice immunized with human ovarian cancer cells [29]. CA125 is a heavily glycosylated type I transmembrane protein belonging to the family of tethered mucins containing both *O*-linked and *N*-linked oligosaccharides [30]. CA125 is overexpressed in many tumors of epithelial origin suggesting that it plays an important role in tumorigenesis [30,31]. CA125 is commonly used as a biomarker to monitor ovarian cancer disease progression and relapse as it is highly expressed in ovarian carcinomas yet minimally expressed in normal ovarian tissues [32–34]. CA125/MUC16 has been shown to inhibit cytolytic responses of human natural killer cells in ovarian cancer, therefore acting as a suppressor of the immune response directed against the ovarian tumors [35,36]. CA125 has been shown to promote cancer cell proliferation [37]. Although the role of CA125 is mainly studied in ovarian cancer, recent studies have shown that CA125 is also highly expressed in other cancers including peritoneal mesotheliomas, pancreatic, and colorectal cancer, implicating a mesothelial cell interaction [38–40].

3. Mesothelial Cells

All organs of the abdominal cavity are covered by the mesothelium, a monolayer of mesothelial cells covering a basement membrane composed of fibronectin, collagen I and IV and laminin [41,42]. Mesothelial cells are flattened squamous-like cells derived from the mesoderm and possess both epithelial and mesenchymal characteristics [43,44]. Mesothelial cells have well-developed cell–cell junction complexes, including tight junctions, that are critical for cell surface polarity and the formation and maintenance of a semi-permeable diffusion barrier. The mesothelium functions to provide a protective barrier as well as a frictionless interface for the free movement of organs and tissues [45]. The mesothelium also plays an important role in contributing to the homeostasis of the peritoneal cavity, fluid and cell transport, tissue repair, initiation and resolution of inflammation and possibly tumor dissemination [46,47]. In the tumor microenvironment, mesothelial cells are preconditioned by the cancer cell secretome to induce the expression of multiple pro-inflammatory factors [48]. Mesothelial cells are implicated in both epithelial-to-mesenchymal transition (EMT) and mesothelial-to-mesenchymal transition (MMT), an EMT-like process [49–51]. EMT is the biological process by which epithelial cells lose cell–cell adhesion and gain migratory properties and MMT is a biologic process in which mesothelial cells of the peritoneal cavity acquire a fibroblast-like phenotype, with increased migratory capabilities [50,52]. Mesothelial cells, expressing mesothelin, line the peritoneal wall and all the organs of the peritoneal cavity that is susceptible to ovarian cancer metastasis.

4. Mesothelin

Mesothelin (MSLN), first identified in 1992 [53], is synthesized as a 70 kDa precursor that is proteolytically cleaved at Arg295, resulting in an approximately 30 kDa fragment called megakaryocyte potentiating factor (MPF) and the 40 kDa MSLN membrane-bound fragment (Figure 2) [54,55]. Both MSLN and MPF are biologically active; however, the exact function remains unknown [56]. MSLN is a glycosylphosphatidylinositol (GPI)-anchored membrane glycoprotein that is physiologically expressed at the cell surface of mesothelial cells lining the pleura, pericardium, and peritoneum [57,58]. Composed of 16 exons spanning 7733 bp, the human MSLN gene occupies approximately 8 kb located at chromosome 16 p 13.3. Alternative splicing results in the predominant variant 1 encoded by MSLN1, variant 2 (24 bp insert), and variant 3 (82 bp insert) [55,57,59,60].

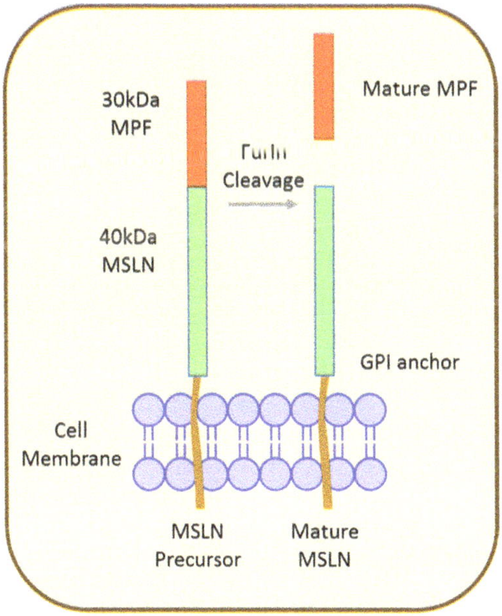

Figure 2. Structure of mesothelin (MSLN). The 70 kDa MSLN precursor protein is proteolytically cleaved to release the 30 kDa N-terminal megakaryocyte potentiating factor (MPF) and is displayed as mature MSLN on the cell surface.

Although many prediction programs have attempted to predict the three-dimensional structure of the MSLN precursor and mature MSLN, the structure still remains unknown [61]. MSLN1 was found by Hellstrom et al. to be primarily expressed at the cell surface and was also released into body fluids of patients of several tumor types. Soluble MSLN results from a cleavage of variants 1 at the C-terminal domain [60]. An 18-bp enhancer sequence, CanScript, located −65 to −46 bp 5′ of one of three transcriptional start sites in the promoter region of the *MSLN* gene, was identified in cancer cell lines with aberrant overexpression of MSLN. The CanScript sequence enhancer consists of two functionally putative binding motifs: the conventional MCAT element and a SP1-like element [62]. All eight nucleotides in the MCAT element were shown to be essential for its function; conversely, the SP1-like element was shown to have two mutations suggesting, that the cancer-specific expression of MSLN is thought to occur through the binding of an unknown transcription factor. Transcription factors such as KLF6 and YAP1 have been investigated but binding of these factors are not adequate for MSLN overexpression in certain cancer types [63]. Nonetheless, the essential transcriptional factor that regulates the MSLN overexpression in human cancers has not been identified.

MSLN is normally expressed in mesothelial cells in trace amounts. In contrast, MSLN is highly expressed in human cancers including 70% of ovarian cancers [54,64–66], mesotheliomas [54], and pancreatic adenocarcinoma [67,68] and therefore identified as a tumor-associated marker. The biological function of MSLN is not fully understood as MSLN knockout mice do not present with an abnormal phenotype, suggesting that MSLN is a non-essential protein [58]. Furthermore, MSLN is reported to play a role in cell adhesion [69], tumor progression [65,70–73], and chemoresistance [73–76]. Specifically, MSLN has been shown to have oncogenic properties by increasing ovarian cancer invasion by inducing MMP-7 through MAPK/ERK and JNK pathways and by inducing drug resistance through PI3K/AKT and MAPK/ERK signaling pathways [65,74]. Albeit, mechanisms that regulate MSLN cell-surface expression are not well understood.

5. MSLN and CA125

CA125, the ovarian cancer antigen/biomarker, has been identified as a MSLN ligand and could potentially mediate cell adhesion [69]. Rump et al. demonstrated MSLN–CA125 interaction mediates heterotypic cellular adhesion (Figure 1C) of the human ovarian cancer cell line, OVCAR3, expressing CA125 to a MSLN expressing endothelial-like cell line [69]. Additionally, Gubbels et al. established that MSLN binds to CA125 in a specific and N-linked glycan-dependent manner, thus CA125-expressing ovarian tumor cells could bind specifically to the mesothelin-expressing peritoneal lining (Figure 1D) [77]. The *N*-linked oligosaccharides of CA125 are necessary for the binding to MSLN with MSLN having a strong affinity to CA125 with an apparent dissociation constant (K_d) of 5 nM [77–79]. Consequently, MSLN:CA125-dependent cell attachment may play an important role in the peritoneal implantation of ovarian tumor cells [54,80]. The MSLN:CA125 role in cell attachment is supported by work from Bruney et al., demonstrating the overexpression of membrane type 1 matrix metalloproteinase (MT1-MMP) in human ovarian cancer cells (OVCA433-MT)-decreased cell surface expression of CA125/MUC16, subsequently increasing CA125/MUC16 ectodomain shedding, resulting in the release of CA125 from the cell surface. Additionally, there was decreased adhesion of OVCA433-MT to human mesothelial cells (LP9) and to intact peritoneal explants, suggesting the importance of MSLN:CA125 initial adhesion of [81]. After initial attachment of ovarian cancer cells to the peritoneal mesothelium, the co-overexpression of both MSLN and CA125 can lead to recruitment of other ovarian cancer cells being sloughed off from the primary site (Figure 1B,C) [82]. Therefore, the tumor load at secondary sites could be a combination of excessive proliferation and adhesion of circulating single or multicellular aggregates in peritoneal ascites fluid [77,83]. Conversely, the exact function of MSLN in tumor progression remains unclear [84]; however, understanding the importance of CA125:mesothelin binding may lead to novel therapies to control ovarian peritoneal metastasis.

6. Targeting MSLN

Clarifying the function of MSLN will enhance its clinical application in ovarian cancer, including early detection, chemo-response, prognosis and therapeutic targeting. Several features of MSLN make it a useful candidate for cancer therapy, including that it is well-internalized, enabling it to be a good target for immunotoxins [85]. Additionally, MSLN is actively shed from the cell surface generating a pool of antigens in ascites or blood circulation allowing for the quantification of circulating serum MSLN levels potentially used for diagnosis of ovarian cancer patients [28,86–88]. The use of MSLN as a plasma biomarker has been investigated by several groups using blood ELISA tests and demonstrated that serum MSLN levels decrease after surgical therapy and, therefore, may be useful in monitoring treatment response in MSLN expression cancers [86,89]. Pools of antigens, from shed MSLN, in the tumor interstitial space will unavoidably interact with a targeting agent during tumor dissemination [60,85,87]. The first identified sheddase, TNF-α converting enzyme (TACE) was shown to mediate MSLN shedding. TACE is a transmembrane glycoprotein, known for its role in releasing EGFR ligands from the cell surface, therefore regulating the activation of the EGFR pathway [60,90]. Tumor targeting is a complex process and, furthermore, modulation of MSLN shedding could have an influence on drug kinetics in both circulation and tumor tissue. However, shedding is not the only way MSLN could be modulated. The expression levels of MSLN could potentially be regulated similarly to other antigens by trogocytosis [91] or antigen masking [92]; however, the role of these antigens remains to be elucidated. Furthermore, MSLN is expressed in dispensable mesothelial cells so the risk of non-specific toxicity is decreased.

6.1. Molecular Imaging for the Detection of MSLN

Mesothelin has recently been investigated as a target for molecular imaging probes. These probes are designed to guide antibody-based treatments that can be used to assess tumor uptake, response to treatment and the distribution in primary tumors and secondary sites. Prantner et al. identified and characterized an antimesothelin nanobody (NbG3a) used for in vitro diagnostic applications [93]. Further studies from the same group established the potential use of NbG3a for a novel molecular imaging probe with promising results for human imaging and therapeutic applications [94]. Terwisscha van Scheltinga et al. investigated the use of an antibody–drug conjugate (anti-mesothelin antibody-monomethyl auristatin E) coupled to molecular imaging with ^{89}Zr immuno-positive emission tomography (PET). Using this technique, quantitative immuno-PET measurement of relative antibody uptake was determined to correlate with tumor growth inhibition [95]. Furthermore, non-antibody protein scaffolds have successfully been engineered to bind to mesothelin with high affinity [96]. Unlike antibodies that are large in size and have slow clearance from circulation, non-antibody protein scaffolds have demonstrated specific binding to identify tumors expressing the molecular target in murine models [97–99] and have demonstrated promising results in both preclinical and clinical evaluations [100]. The use of these techniques demonstrates the translational potential of MSLN.

6.2. Clinical Trials

There are many clinical trials testing MSLN-targeting agents using strategies such as antibody-based immunotoxins such as SS1P, consisting of an anti-MLSN Fv obtained from a phage display library of immunized mice with recombinant MSLN fused to a truncated form of the Pseudomonas Exotoxin PE38 that mediates cell death. The mechanism of action of an immunotoxin is threefold. First, the immunotoxin binds to cell-bound MSLN; second, this complex is internalized by endocytosis, undergoes retrograde transport to the endoplasmic reticulum and the PE portion is translocated to the cytosol; and third, the PE catalyzes ADP-ribosylation of the elongation factor-2, halting protein synthesis and activating apoptosis [85,101]. There have been two Phase I clinical trials with different modes of administration using either continuous infusion or as bolus intravenous infusions in mesotheliomas, ovarian, and pancreatic cancers. Continuous infusion was well tolerated and showed modest clinical activity; however, there was advantage seen over bolus dosing [102,103]. Additionally, there is a high affinity chimeric antibody, amatuximab (MORAb-009), with high affinity and specificity for mesothelin that is under investigation in clinical trials. Amatuximab works by inducing antibody-dependent cellular cytoxicity [104]. It was observed that upon treatment with amatuximab, patients had an increase in CA125 levels suggesting that amatuximab interferes with the MSLN:CA125 interaction [105]. A tumor vaccine CRS-207 utilizing a live attenuated strain of bacterium *Listeria monocytogenes* (*Lm*) expressing human MSLN has shown good tolerance and MSLN-specific T-cell response in a phase I study of safety clinical trial. This phase I study not only demonstrated that vaccines are safe and tolerable but also showed that a tumor antigen-modified *Lm* can induce tumor antigen-specific T-cell responses in patients with advanced cancer, suggesting that further evaluation of *Lm* vaccine as a candidate biomarker of improved clinical outcomes is needed [106]. A two-part phase I/II trial is underway using combination therapy with CRS-207, epacadostat, and pembrolizumab (keytruda) in patients with platinum-resistant ovarian, fallopian tube, and peritoneal cancers using different combinations of the three treatments (ClinicalTrials.gov Identifier NCT02575807). Antibody–drug conjugates is another strategy used to target MSLN. An ongoing phase I clinical trial with anetumab ravtansine (BAY94-9343) to determine the safety and maximum tolerated dose for patients with advanced solid tumors including ovarian carcinoma and mesothelioma opened in 2011 (ClinicalTrials.gov Identifier NCT01439152). Anetumab ravtansine consists of the fully human anti-MSLN IgG1 linked to a potent tubulin-binding drug, DM4. In preclinical trials, anetumab ravtansine inhibited both subcutaneous and orthotopic tumor growth in xenograft models of ovarian, pancreatic, and mesothelioma cancers [107]. In patients with recurrent MSLN-expressing platinum-resistant recurrent ovarian, fallopian tube or primary peritoneal cancer, a phase Ib clinical trial to determine the maximum tolerated dose of anetumab ravtansine that could be safely combined with pegylated liposomal doxorubicin is underway (ClinicalTrials.gov Identifier NCT02751918). Several ongoing clinical trials are utilizing anti-MSLN CAR-modified T cells as MSLN targeting agent. The T cells are obtained by apheresis and introduced to a temporary gene which will cause them to make a new type of antibody that will attach to MSLN. Once attached, the cells will become activated and stimulate the host immune system to attack the MSLN-expressing cells [108]. The above clinical trials have confirmed that targeting MSLN could be beneficial in improving existing therapeutic options for patients diagnosed with a MSLN-expressing cancer, including ovarian cancer.

7. Conclusions

Ovarian cancer is the deadliest gynecological malignancy among U.S. women and is often diagnosed at a late stage when the disease has metastasized into the peritoneal cavity. Mesothelin, a glycoprotein normally expressed in mesothelial cells, is highly expressed in several cancers including ovarian, pancreatic, and mesotheliomas. It has been shown that MSLN binds to the ovarian cancer biomarker CA125 and this interaction plays a role in the peritoneal metastasis of ovarian cancer. The differential expression of mesothelin in normal and cancer tissues makes it a promising candidate for targeted therapeutics. Several candidate immunotherapies targeting MSLN are in ongoing clinical trials. New strategies to disrupt the MSLN:CA125 interaction are emerging. Although MSLN is implicated in many cancers, the role of MSLN is still poorly understood warranting further investigation and clinical trial studies. Future advances in ovarian cancer therapy depend on novel treatment mechanisms in combination with current chemotherapeutic approaches that will result in cytotoxicity, inhibition of metastasis and angiogenesis, and increasing the immunological detection of tumors. Further mechanistic studies on MSLN are needed to validate the potential role of MSLN in tumor metastasis that possibly will provide insight for effective MSLN-targeting therapies for several cancers.

Funding: This work was supported by grant KO1CA218305 from the National Institutes of Health/National Cancer Institute.

Acknowledgments: I would like to thank members of the Stack lab at the University of Notre Dame for help with editing this manuscript.

Conflicts of Interest: The authors declare no conflict of interest.

References

1. Siegel, R.L.; Miller, K.D.; Jemal, A. Cancer statistics, 2018. *CA Cancer J. Clin.* **2018**, *68*, 7–30. [CrossRef] [PubMed]
2. Peres, L.C.; Cushing-Haugen, K.L.; Kobel, M.; Harris, H.R.; Berchuck, A.; Rossing, M.A.; Schildkraut, J.M.; Doherty, J.A. Invasive epithelial ovarian cancer survival by histotype and disease stage. *J. Natl. Cancer Inst.* **2018**. [CrossRef] [PubMed]
3. Meinhold-Heerlein, I.; Fotopoulou, C.; Harter, P.; Kurzeder, C.; Mustea, A.; Wimberger, P.; Hauptmann, S.; Sehouli, J. The new who classification of ovarian, fallopian tube, and primary peritoneal cancer and its clinical implications. *Arch. Gynecol. Obstet.* **2016**, *293*, 695–700. [CrossRef] [PubMed]
4. Auersperg, N.; Wong, A.S.; Choi, K.C.; Kang, S.K.; Leung, P.C. Ovarian surface epithelium: Biology, endocrinology, and pathology. *Endocr. Rev.* **2001**, *22*, 255–288. [CrossRef] [PubMed]
5. Choi, J. Ovarian epithelial cancer: Etiology and pathogenesis. *Biowave* **2008**, *10*, 32.
6. Auersperg, N. The origin of ovarian cancers–hypotheses and controversies. *Front. Biosci. (Sch. Ed.)* **2013**, *5*, 709–719. [CrossRef]
7. Mutch, D.G.; Prat, J. 2014 FIGO staging for ovarian, fallopian tube and peritoneal cancer. *Gynecol. Oncol.* **2014**, *133*, 401–404. [CrossRef] [PubMed]
8. Kohler, M.F.; Marks, J.R.; Wiseman, R.W.; Jacobs, I.J.; Davidoff, A.M.; Clarke-Pearson, D.L.; Soper, J.T.; Bast, R.C., Jr.; Berchuck, A. Spectrum of mutation and frequency of allelic deletion of the p53 gene in ovarian cancer. *J. Natl. Cancer Inst.* **1993**, *85*, 1513–1519. [CrossRef] [PubMed]

9. Singer, G.; Stohr, R.; Cope, L.; Dehari, R.; Hartmann, A.; Cao, D.F.; Wang, T.L.; Kurman, R.J.; Shih, I.M. Patterns of p53 mutations separate ovarian serous borderline tumors and low- and high-grade carcinomas and provide support for a new model of ovarian carcinogenesis: A mutational analysis with immunohistochemical correlation. *Am. J. Surg. Pathol.* **2005**, *29*, 218–224. [CrossRef] [PubMed]
10. Milner, B.J.; Allan, L.A.; Eccles, D.M.; Kitchener, H.C.; Leonard, R.C.; Kelly, K.F.; Parkin, D.E.; Haites, N.E. P53 mutation is a common genetic event in ovarian carcinoma. *Cancer Res.* **1993**, *53*, 2128–2132. [PubMed]
11. Struewing, J.P.; Hartge, P.; Wacholder, S.; Baker, S.M.; Berlin, M.; McAdams, M.; Timmerman, M.M.; Brody, L.C.; Tucker, M.A. The risk of cancer associated with specific mutations of brca1 and brca2 among ashkenazi jews. *N. Engl. J. Med.* **1997**, *336*, 1401–1408. [CrossRef] [PubMed]
12. Antoniou, A.; Pharoah, P.D.; Narod, S.; Risch, H.A.; Eyfjord, J.E.; Hopper, J.L.; Loman, N.; Olsson, H.; Johannsson, O.; Borg, A.; et al. Average risks of breast and ovarian cancer associated with BRCA1 or BRCA2 mutations detected in case series unselected for family history: A combined analysis of 22 studies. *Am. J. Hum. Genet.* **2003**, *72*, 1117–1130. [CrossRef] [PubMed]
13. Mavaddat, N.; Peock, S.; Frost, D.; Ellis, S.; Platte, R.; Fineberg, E.; Evans, D.G.; Izatt, L.; Eeles, R.A.; Adlard, J.; et al. Cancer risks for BRCA1 and BRCA2 mutation carriers: Results from prospective analysis of embrace. *J. Natl. Cancer Inst.* **2013**, *105*, 812–822. [CrossRef] [PubMed]
14. Tan, D.S.; Agarwal, R.; Kaye, S.B. Mechanisms of transcoelomic metastasis in ovarian cancer. *Lancet Oncol.* **2006**, *7*, 925–934. [CrossRef]
15. Burleson, K.M.; Casey, R.C.; Skubitz, K.M.; Pambuccian, S.E.; Oegema, T.R., Jr.; Skubitz, A.P. Ovarian carcinoma ascites spheroids adhere to extracellular matrix components and mesothelial cell monolayers. *Gynecol. Oncol.* **2004**, *93*, 170–181. [CrossRef] [PubMed]
16. Burleson, K.M.; Hansen, L.K.; Skubitz, A.P. Ovarian carcinoma spheroids disaggregate on type i collagen and invade live human mesothelial cell monolayers. *Clin. Exp. Metastasis* **2004**, *21*, 685–697. [CrossRef] [PubMed]
17. Carmignani, C.P.; Sugarbaker, T.A.; Bromley, C.M.; Sugarbaker, P.H. Intraperitoneal cancer dissemination: Mechanisms of the patterns of spread. *Cancer Metastasis Rev.* **2003**, *22*, 465–472. [CrossRef] [PubMed]
18. Buy, J.N.; Moss, A.A.; Ghossain, M.A.; Sciot, C.; Malbec, L.; Vadrot, D.; Paniel, B.J.; Decroix, Y. Peritoneal implants from ovarian tumors: Ct findings. *Radiology* **1988**, *169*, 691–694. [CrossRef] [PubMed]
19. Gupta, G.P.; Massague, J. Cancer metastasis: Building a framework. *Cell* **2006**, *127*, 679–695. [CrossRef] [PubMed]
20. Pradeep, S.; Kim, S.W.; Wu, S.Y.; Nishimura, M.; Chaluvally-Raghavan, P.; Miyake, T.; Pecot, C.V.; Kim, S.J.; Choi, H.J.; Bischoff, F.Z.; et al. Hematogenous metastasis of ovarian cancer: Rethinking mode of spread. *Cancer Cell* **2014**, *26*, 77–91. [CrossRef] [PubMed]
21. Coffman, L.G.; Burgos-Ojeda, D.; Wu, R.; Cho, K.; Bai, S.; Buckanovich, R.J. New models of hematogenous ovarian cancer metastasis demonstrate preferential spread to the ovary and a requirement for the ovary for abdominal dissemination. *Transl. Res.* **2016**, *175*, 92–102. [CrossRef] [PubMed]
22. Kalluri, R.; Weinberg, R.A. The basics of epithelial-mesenchymal transition. *J. Clin. Investig.* **2009**, *119*, 1420–1428. [CrossRef] [PubMed]
23. Meyers, M.A.; Oliphant, M.; Berne, A.S.; Feldberg, M.A. The peritoneal ligaments and mesenteries: Pathways of intraabdominal spread of disease. *Radiology* **1987**, *163*, 593–604. [CrossRef] [PubMed]
24. Coakley, F.V.; Hricak, H. Imaging of peritoneal and mesenteric disease: Key concepts for the clinical radiologist. *Clin. Radiol.* **1999**, *54*, 563–574. [CrossRef]
25. White, E.A.; Kenny, H.A.; Lengyel, E. Three-dimensional modeling of ovarian cancer. *Adv. Drug Deliv. Rev.* **2014**, *79–80*, 184–192. [CrossRef] [PubMed]
26. Lengyel, E. Ovarian cancer development and metastasis. *Am. J. Pathol.* **2010**, *177*, 1053–1064. [CrossRef] [PubMed]
27. Cannistra, S.A. Cancer of the ovary. *N. Engl. J. Med.* **2004**, *351*, 2519–2529. [CrossRef] [PubMed]
28. Scholler, N.; Fu, N.; Yang, Y.; Ye, Z.; Goodman, G.E.; Hellstrom, K.E.; Hellstrom, I. Soluble member(s) of the mesothelin/megakaryocyte potentiating factor family are detectable in sera from patients with ovarian carcinoma. *Proc. Natl. Acad. Sci. USA* **1999**, *96*, 11531–11536. [CrossRef] [PubMed]
29. Bast, R.C., Jr.; Feeney, M.; Lazarus, H.; Nadler, L.M.; Colvin, R.B.; Knapp, R.C. Reactivity of a monoclonal antibody with human ovarian carcinoma. *J. Clin. Investig.* **1981**, *68*, 1331–1337. [CrossRef] [PubMed]

30. O'Brien, T.J.; Beard, J.B.; Underwood, L.J.; Shigemasa, K. The ca 125 gene: A newly discovered extension of the glycosylated N-terminal domain doubles the size of this extracellular superstructure. *Tumour Biol.* **2002**, *23*, 154–169. [PubMed]
31. Yin, B.W.; Lloyd, K.O. Molecular cloning of the ca125 ovarian cancer antigen: Identification as a new mucin, muc16. *J. Biol. Chem.* **2001**, *276*, 27371–27375. [CrossRef] [PubMed]
32. Gubbels, J.A.; Claussen, N.; Kapur, A.K.; Connor, J.P.; Patankar, M.S. The detection, treatment, and biology of epithelial ovarian cancer. *J. Ovarian Res.* **2010**, *3*, 8. [CrossRef] [PubMed]
33. Bast, R.C., Jr.; Klug, T.L.; St John, E.; Jenison, E.; Niloff, J.M.; Lazarus, H.; Berkowitz, R.S.; Leavitt, T.; Griffiths, C.T.; Parker, L.; et al. A radioimmunoassay using a monoclonal antibody to monitor the course of epithelial ovarian cancer. *N. Engl. J. Med.* **1983**, *309*, 883–887. [CrossRef] [PubMed]
34. Yin, B.W.; Dnistrian, A.; Lloyd, K.O. Ovarian cancer antigen ca125 is encoded by the muc16 mucin gene. *Int. J. Cancer* **2002**, *98*, 737–740. [CrossRef] [PubMed]
35. Patankar, M.S.; Jing, Y.; Morrison, J.C.; Belisle, J.A.; Lattanzio, F.A.; Deng, Y.; Wong, N.K.; Morris, H.R.; Dell, A.; Clark, G.F. Potent suppression of natural killer cell response mediated by the ovarian tumor marker ca125. *Gynecol. Oncol.* **2005**, *99*, 704–713. [CrossRef] [PubMed]
36. Senapati, S.; Das, S.; Batra, S.K. Mucin-interacting proteins: From function to therapeutics. *Trends Biochem. Sci.* **2010**, *35*, 236–245. [CrossRef] [PubMed]
37. Reinartz, S.; Failer, S.; Schuell, T.; Wagner, U. Ca125 (muc16) gene silencing suppresses growth properties of ovarian and breast cancer cells. *Eur. J. Cancer* **2012**, *48*, 1558–1569. [CrossRef] [PubMed]
38. Streppel, M.M.; Vincent, A.; Mukherjee, R.; Campbell, N.R.; Chen, S.H.; Konstantopoulos, K.; Goggins, M.G.; Van Seuningen, I.; Maitra, A.; Montgomery, E.A. Mucin 16 (cancer antigen 125) expression in human tissues and cell lines and correlation with clinical outcome in adenocarcinomas of the pancreas, esophagus, stomach, and colon. *Hum. Pathol.* **2012**, *43*, 1755–1763. [CrossRef] [PubMed]
39. Shimizu, A.; Hirono, S.; Tani, M.; Kawai, M.; Okada, K.; Miyazawa, M.; Kitahata, Y.; Nakamura, Y.; Noda, T.; Yokoyama, S.; et al. Coexpression of muc16 and mesothelin is related to the invasion process in pancreatic ductal adenocarcinoma. *Cancer Sci.* **2012**, *103*, 739–746. [CrossRef] [PubMed]
40. Baratti, D.; Kusamura, S.; Martinetti, A.; Seregni, E.; Oliva, D.G.; Laterza, B.; Deraco, M. Circulating ca125 in patients with peritoneal mesothelioma treated with cytoreductive surgery and intraperitoneal hyperthermic perfusion. *Ann. Surg. Oncol.* **2007**, *14*, 500–508. [CrossRef] [PubMed]
41. Yeung, T.L.; Leung, C.S.; Yip, K.P.; Au Yeung, C.L.; Wong, S.T.; Mok, S.C. Cellular and molecular processes in ovarian cancer metastasis. *Am. J. Physiol. Cell Physiol.* **2015**, *309*, C444–C456. [CrossRef] [PubMed]
42. Kenny, H.A.; Chiang, C.Y.; White, E.A.; Schryver, E.M.; Habis, M.; Romero, I.L.; Ladanyi, A.; Penicka, C.V.; George, J.; Matlin, K.; et al. Mesothelial cells promote early ovarian cancer metastasis through fibronectin secretion. *J. Clin. Investig.* **2014**, *124*, 4614–4628. [CrossRef] [PubMed]
43. Van Baal, J.O.; Van de Vijver, K.K.; Nieuwland, R.; van Noorden, C.J.; van Driel, W.J.; Sturk, A.; Kenter, G.G.; Rikkert, L.G.; Lok, C.A. The histophysiology and pathophysiology of the peritoneum. *Tissue Cell* **2017**, *49*, 95–105. [CrossRef] [PubMed]
44. Ferrandez-Izquierdo, A.; Navarro-Fos, S.; Gonzalez-Devesa, M.; Gil-Benso, R.; Llombart-Bosch, A. Immunocytochemical typification of mesothelial cells in effusions: In vivo and in vitro models. *Diagn. Cytopathol.* **1994**, *10*, 256–262. [CrossRef] [PubMed]
45. Mutsaers, S.E. The mesothelial cell. *Int. J. Biochem. Cell Biol.* **2004**, *36*, 9–16. [CrossRef]
46. Roth, J. Ultrahistochemical demonstration of saccharide components of complex carbohydrates at the alveolar cell surface and at the mesothelial cell surface of the pleura visceralis of mice by means of concanavalin A. *Exp. Pathol. (Jena)* **1973**, *8*, 157–167. [PubMed]
47. Yung, S.; Chan, T.M. Pathophysiology of the peritoneal membrane during peritoneal dialysis: The role of hyaluronan. *J. Biomed. Biotechnol.* **2011**, *2011*, 180594. [CrossRef] [PubMed]
48. Ren, J.; Xiao, Y.J.; Singh, L.S.; Zhao, X.; Zhao, Z.; Feng, L.; Rose, T.M.; Prestwich, G.D.; Xu, Y. Lysophosphatidic acid is constitutively produced by human peritoneal mesothelial cells and enhances adhesion, migration, and invasion of ovarian cancer cells. *Cancer Res.* **2006**, *66*, 3006–3014. [CrossRef] [PubMed]
49. Du, J.; Sun, B.; Zhao, X.; Gu, Q.; Dong, X.; Mo, J.; Sun, T.; Wang, J.; Sun, R.; Liu, Y. Hypoxia promotes vasculogenic mimicry formation by inducing epithelial-mesenchymal transition in ovarian carcinoma. *Gynecol. Oncol.* **2014**, *133*, 575–583. [CrossRef] [PubMed]

50. Yanez-Mo, M.; Lara-Pezzi, E.; Selgas, R.; Ramirez-Huesca, M.; Dominguez-Jimenez, C.; Jimenez-Heffernan, J.A.; Aguilera, A.; Sanchez-Tomero, J.A.; Bajo, M.A.; Alvarez, V.; et al. Peritoneal dialysis and epithelial-to-mesenchymal transition of mesothelial cells. *N. Engl. J. Med.* **2003**, *348*, 403–413. [CrossRef] [PubMed]
51. Sandoval, P.; Jimenez-Heffernan, J.A.; Rynne-Vidal, A.; Perez-Lozano, M.L.; Gilsanz, A.; Ruiz-Carpio, V.; Reyes, R.; Garcia-Bordas, J.; Stamatakis, K.; Dotor, J.; et al. Carcinoma-associated fibroblasts derive from mesothelial cells via mesothelial-to-mesenchymal transition in peritoneal metastasis. *J. Pathol.* **2013**, *231*, 517–531. [CrossRef] [PubMed]
52. Rynne-Vidal, A.; Au-Yeung, C.L.; Jimenez-Heffernan, J.A.; Perez-Lozano, M.L.; Cremades-Jimeno, L.; Barcena, C.; Cristobal-Garcia, I.; Fernandez-Chacon, C.; Yeung, T.L.; Mok, S.C.; et al. Mesothelial-to-mesenchymal transition as a possible therapeutic target in peritoneal metastasis of ovarian cancer. *J. Pathol.* **2017**, *242*, 140–151. [CrossRef] [PubMed]
53. Chang, K.; Pastan, I.; Willingham, M.C. Isolation and characterization of a monoclonal antibody, k1, reactive with ovarian cancers and normal mesothelium. *Int. J. Cancer* **1992**, *50*, 373–381. [CrossRef] [PubMed]
54. Chang, K.; Pastan, I. Molecular cloning of mesothelin, a differentiation antigen present on mesothelium, mesotheliomas, and ovarian cancers. *Proc. Natl. Acad. Sci. USA* **1996**, *93*, 136–140. [CrossRef] [PubMed]
55. Hassan, R.; Bera, T.; Pastan, I. Mesothelin: A new target for immunotherapy. *Clin. Cancer Res. Off. J. Am. Assoc. Cancer Res.* **2004**, *10*, 3937–3942. [CrossRef] [PubMed]
56. Yamaguchi, N.; Yamamura, Y.; Konishi, E.; Ueda, K.; Kojima, T.; Hattori, K.; Oheda, M.; Imai, N.; Taniguchi, Y.; Tamura, M.; et al. Characterization, molecular cloning and expression of megakaryocyte potentiating factor. *Stem Cells* **1996**, *14*, 62–74. [CrossRef] [PubMed]
57. Scholler, N. Mesothelin. In *Encyclopedia of Cancer*; Schwab, M., Ed.; Springer: Berlin/Heidelberg, Germany, 2011; pp. 2241–2245.
58. Bera, T.K.; Pastan, I. Mesothelin is not required for normal mouse development or reproduction. *Mol. Cell. Biol.* **2000**, *20*, 2902–2906. [CrossRef] [PubMed]
59. Muminova, Z.E.; Strong, T.V.; Shaw, D.R. Characterization of human mesothelin transcripts in ovarian and pancreatic cancer. *BMC Cancer* **2004**, *4*, 19. [CrossRef] [PubMed]
60. Hellstrom, I.; Raycraft, J.; Kanan, S.; Sardesai, N.Y.; Verch, T.; Yang, Y.; Hellstrom, K.E. Mesothelin variant 1 is released from tumor cells as a diagnostic marker. *Cancer Epidemiol. Biomark. Prev.* **2006**, *15*, 1014–1020. [CrossRef] [PubMed]
61. Sathyanarayana, B.K.; Hahn, Y.; Patankar, M.S.; Pastan, I.; Lee, B. Mesothelin, stereocilin, and otoancorin are predicted to have superhelical structures with arm-type repeats. *BMC Struct. Biol.* **2009**, *9*, 1. [CrossRef] [PubMed]
62. Hucl, T.; Brody, J.R.; Gallmeier, E.; Iacobuzio-Donahue, C.A.; Farrance, I.K.; Kern, S.E. High cancer-specific expression of mesothelin (MSLN) is attributable to an upstream enhancer containing a transcription enhancer factor dependent mcat motif. *Cancer Res.* **2007**, *67*, 9055–9065. [CrossRef] [PubMed]
63. Ren, Y.R.; Patel, K.; Paun, B.C.; Kern, S.E. Structural analysis of the cancer-specific promoter in mesothelin and in other genes overexpressed in cancers. *J. Biol. Chem.* **2011**, *286*, 11960–11969. [CrossRef] [PubMed]
64. Hassan, R.; Kreitman, R.J.; Pastan, I.; Willingham, M.C. Localization of mesothelin in epithelial ovarian cancer. *Appl. Immunohistochem. Mol. Morphol.* **2005**, *13*, 243–247. [CrossRef] [PubMed]
65. Chang, M.C.; Chen, C.A.; Chen, P.J.; Chiang, Y.C.; Chen, Y.L.; Mao, T.L.; Lin, H.W.; Lin Chiang, W.H.; Cheng, W.F. Mesothelin enhances invasion of ovarian cancer by inducing mmp-7 through Mapk/Erk and Jnk pathways. *Biochem. J.* **2012**, *442*, 293–302. [CrossRef] [PubMed]
66. Yen, M.J.; Hsu, C.Y.; Mao, T.L.; Wu, T.C.; Roden, R.; Wang, T.L.; Shih Ie, M. Diffuse mesothelin expression correlates with prolonged patient survival in ovarian serous carcinoma. *Clin. Cancer Res. Off. J. Am. Assoc. Cancer Res.* **2006**, *12*, 827–831. [CrossRef] [PubMed]
67. Wang, K.; Bodempudi, V.; Liu, Z.; Borrego-Diaz, E.; Yamoutpoor, F.; Meyer, A.; Woo, R.A.; Pan, W.; Dudek, A.Z.; Olyaee, M.S.; et al. Inhibition of mesothelin as a novel strategy for targeting cancer cells. *PLoS ONE* **2012**, *7*, e33214. [CrossRef] [PubMed]
68. Argani, P.; Iacobuzio-Donahue, C.; Ryu, B.; Rosty, C.; Goggins, M.; Wilentz, R.E.; Murugesan, S.R.; Leach, S.D.; Jaffee, E.; Yeo, C.J.; et al. Mesothelin is overexpressed in the vast majority of ductal adenocarcinomas of the pancreas: Identification of a new pancreatic cancer marker by serial analysis of gene expression (sage). *Clin. Cancer Res. Off. J. Am. Assoc. Cancer Res.* **2001**, *7*, 3862–3868.

69. Rump, A.; Morikawa, Y.; Tanaka, M.; Minami, S.; Umesaki, N.; Takeuchi, M.; Miyajima, A. Binding of ovarian cancer antigen ca125/muc16 to mesothelin mediates cell adhesion. *J. Biol. Chem.* **2004**, *279*, 9190–9198. [CrossRef] [PubMed]
70. Li, M.; Bharadwaj, U.; Zhang, R.; Zhang, S.; Mu, H.; Fisher, W.E.; Brunicardi, F.C.; Chen, C.; Yao, Q. Mesothelin is a malignant factor and therapeutic vaccine target for pancreatic cancer. *Mol. Cancer Ther.* **2008**, *7*, 286–296. [CrossRef] [PubMed]
71. Bharadwaj, U.; Li, M.; Chen, C.; Yao, Q. Mesothelin-induced pancreatic cancer cell proliferation involves alteration of cyclin e via activation of signal transducer and activator of transcription protein 3. *Mol. Cancer Res.* **2008**, *6*, 1755–1765. [CrossRef] [PubMed]
72. Bharadwaj, U.; Marin-Muller, C.; Li, M.; Chen, C.; Yao, Q. Mesothelin overexpression promotes autocrine il-6/sil-6r trans-signaling to stimulate pancreatic cancer cell proliferation. *Carcinogenesis* **2011**, *32*, 1013–1024. [CrossRef] [PubMed]
73. Bharadwaj, U.; Marin-Muller, C.; Li, M.; Chen, C.; Yao, Q. Mesothelin confers pancreatic cancer cell resistance to tnf-alpha-induced apoptosis through akt/pi3k/nf-kappab activation and il-6/mcl-1 overexpression. *Mol. Cancer* **2011**, *10*, 106. [CrossRef] [PubMed]
74. Chang, M.C.; Chen, C.A.; Hsieh, C.Y.; Lee, C.N.; Su, Y.N.; Hu, Y.H.; Cheng, W.F. Mesothelin inhibits paclitaxel-induced apoptosis through the pi3k pathway. *Biochem. J.* **2009**, *424*, 449–458. [CrossRef] [PubMed]
75. Cheng, W.F.; Huang, C.Y.; Chang, M.C.; Hu, Y.H.; Chiang, Y.C.; Chen, Y.L.; Hsieh, C.Y.; Chen, C.A. High mesothelin correlates with chemoresistance and poor survival in epithelial ovarian carcinoma. *Br. J. Cancer* **2009**, *100*, 1144–1153. [CrossRef] [PubMed]
76. Uehara, N.; Matsuoka, Y.; Tsubura, A. Mesothelin promotes anchorage-independent growth and prevents anoikis via extracellular signal-regulated kinase signaling pathway in human breast cancer cells. *Mol. Cancer Res.* **2008**, *6*, 186–193. [CrossRef] [PubMed]
77. Gubbels, J.A.; Belisle, J.; Onda, M.; Rancourt, C.; Migneault, M.; Ho, M.; Bera, T.K.; Connor, J.; Sathyanarayana, B.K.; Lee, B.; et al. Mesothelin-muc16 binding is a high affinity, n-glycan dependent interaction that facilitates peritoneal metastasis of ovarian tumors. *Mol. Cancer* **2006**, *5*, 50. [CrossRef] [PubMed]
78. Kui Wong, N.; Easton, R.L.; Panico, M.; Sutton-Smith, M.; Morrison, J.C.; Lattanzio, F.A.; Morris, H.R.; Clark, G.F.; Dell, A.; Patankar, M.S. Characterization of the oligosaccharides associated with the human ovarian tumor marker ca125. *J. Biol. Chem.* **2003**, *278*, 28619–28634. [CrossRef] [PubMed]
79. Kaneko, O.; Gong, L.; Zhang, J.; Hansen, J.K.; Hassan, R.; Lee, B.; Ho, M. A binding domain on mesothelin for ca125/muc16. *J. Biol. Chem.* **2009**, *284*, 3739–3749. [CrossRef] [PubMed]
80. Scholler, N.; Garvik, B.; Hayden-Ledbetter, M.; Kline, T.; Urban, N. Development of a ca125-mesothelin cell adhesion assay as a screening tool for biologics discovery. *Cancer Lett.* **2007**, *247*, 130–136. [CrossRef] [PubMed]
81. Bruney, L.; Conley, K.C.; Moss, N.M.; Liu, Y.; Stack, M.S. Membrane-type i matrix metalloproteinase-dependent ectodomain shedding of mucin16/ca-125 on ovarian cancer cells modulates adhesion and invasion of peritoneal mesothelium. *Biol. Chem.* **2014**, *395*, 1221–1231. [CrossRef] [PubMed]
82. Tang, Z.; Qian, M.; Ho, M. The role of mesothelin in tumor progression and targeted therapy. *Anticancer Agents Med. Chem.* **2013**, *13*, 276–280. [CrossRef] [PubMed]
83. Bast, R.C., Jr.; Badgwell, D.; Lu, Z.; Marquez, R.; Rosen, D.; Liu, J.; Baggerly, K.A.; Atkinson, E.N.; Skates, S.; Zhang, Z.; et al. New tumor markers: Ca125 and beyond. *Int. J. Gynecol. Cancer* **2005**, *15*, 274–281. [CrossRef] [PubMed]
84. Hilliard, T.; Iwamoto, K.; Loughran, E.; Liu, Y.; Yang, J.; Asem, M.; Tarwater, L.; Klymenko, Y.; Johnson, J.; Shi, Z.; et al. Mesothelin expression increases ovarian cancer metastasis in the peritoneal microenvironment. *Cancer Lett.* **2018**, in press.
85. Pastan, I.; Zhang, Y. Modulating mesothelin shedding to improve therapy. *Oncotarget* **2012**, *3*, 114–115. [CrossRef] [PubMed]
86. Hassan, R.; Remaley, A.T.; Sampson, M.L.; Zhang, J.; Cox, D.D.; Pingpank, J.; Alexander, R.; Willingham, M.; Pastan, I.; Onda, M. Detection and quantitation of serum mesothelin, a tumor marker for patients with mesothelioma and ovarian cancer. *Clin. Cancer Res. Off. J. Am. Assoc. Cancer Res.* **2006**, *12*, 447–453. [CrossRef] [PubMed]

87. Ho, M.; Onda, M.; Wang, Q.C.; Hassan, R.; Pastan, I.; Lively, M.O. Mesothelin is shed from tumor cells. *Cancer Epidemiol. Biomark. Prev.* **2006**, *15*, 1751. [CrossRef] [PubMed]
88. Huang, C.Y.; Cheng, W.F.; Lee, C.N.; Su, Y.N.; Chien, S.C.; Tzeng, Y.L.; Hsieh, C.Y.; Chen, C.A. Serum mesothelin in epithelial ovarian carcinoma: A new screening marker and prognostic factor. *Anticancer Res.* **2006**, *26*, 4721–4728. [PubMed]
89. McIntosh, M.W.; Drescher, C.; Karlan, B.; Scholler, N.; Urban, N.; Hellstrom, K.E.; Hellstrom, I. Combining ca 125 and smr serum markers for diagnosis and early detection of ovarian carcinoma. *Gynecol. Oncol.* **2004**, *95*, 9–15. [CrossRef] [PubMed]
90. Zhang, Y.; Chertov, O.; Zhang, J.; Hassan, R.; Pastan, I. Cytotoxic activity of immunotoxin ss1p is modulated by tace-dependent mesothelin shedding. *Cancer Res.* **2011**, *71*, 5915–5922. [CrossRef] [PubMed]
91. Pham, T.; Mero, P.; Booth, J.W. Dynamics of macrophage trogocytosis of rituximab-coated b cells. *PLoS ONE* **2011**, *6*, e14498. [CrossRef] [PubMed]
92. Nagy, P.; Friedlander, E.; Tanner, M.; Kapanen, A.I.; Carraway, K.L.; Isola, J.; Jovin, T.M. Decreased accessibility and lack of activation of erbb2 in jimt-1, a herceptin-resistant, muc4-expressing breast cancer cell line. *Cancer Res.* **2005**, *65*, 473–482. [PubMed]
93. Prantner, A.M.; Turini, M.; Kerfelec, B.; Joshi, S.; Baty, D.; Chames, P.; Scholler, N. Anti-mesothelin nanobodies for both conventional and nanoparticle-based biomedical applications. *J. Biomed. Nanotechnol.* **2015**, *11*, 1201–1212. [CrossRef] [PubMed]
94. Prantner, A.M.; Yin, C.; Kamat, K.; Sharma, K.; Lowenthal, A.C.; Madrid, P.B.; Scholler, N. Molecular imaging of mesothelin-expressing ovarian cancer with a human and mouse cross-reactive nanobody. *Mol. Pharm.* **2018**, *15*, 1403–1411. [CrossRef] [PubMed]
95. Terwisscha van Scheltinga, A.G.; Ogasawara, A.; Pacheco, G.; Vanderbilt, A.N.; Tinianow, J.N.; Gupta, N.; Li, D.; Firestein, R.; Marik, J.; Scales, S.J.; et al. Preclinical efficacy of an antibody-drug conjugate targeting mesothelin correlates with quantitative 89zr-immunopet. *Mol. Cancer Ther.* **2017**, *16*, 134–142. [CrossRef] [PubMed]
96. Sirois, A.R.; Deny, D.A.; Baierl, S.R.; George, K.S.; Moore, S.J. Fn3 proteins engineered to recognize tumor biomarker mesothelin internalize upon binding. *PLoS ONE* **2018**, *13*, e0197029. [CrossRef] [PubMed]
97. Hackel, B.J.; Kimura, R.H.; Gambhir, S.S. Use of (64)cu-labeled fibronectin domain with egfr-overexpressing tumor xenograft: Molecular imaging. *Radiology* **2012**, *263*, 179–188. [CrossRef] [PubMed]
98. Abou-Elkacem, L.; Wilson, K.E.; Johnson, S.M.; Chowdhury, S.M.; Bachawal, S.; Hackel, B.J.; Tian, L.; Willmann, J.K. Ultrasound molecular imaging of the breast cancer neovasculature using engineered fibronectin scaffold ligands: A novel class of targeted contrast ultrasound agent. *Theranostics* **2016**, *6*, 1740–1752. [CrossRef] [PubMed]
99. Park, S.H.; Park, S.; Kim, D.Y.; Pyo, A.; Kimura, R.H.; Sathirachinda, A.; Choy, H.E.; Min, J.J.; Gambhir, S.S.; Hong, Y. Isolation and characterization of a monobody with a fibronectin domain iii scaffold that specifically binds epha2. *PLoS ONE* **2015**, *10*, e0132976. [CrossRef] [PubMed]
100. Miao, Z.; Levi, J.; Cheng, Z. Protein scaffold-based molecular probes for cancer molecular imaging. *Amino Acids* **2011**, *41*, 1037–1047. [CrossRef] [PubMed]
101. Alewine, C.; Hassan, R.; Pastan, I. Advances in anticancer immunotoxin therapy. *Oncologist* **2015**, *20*, 176–185. [CrossRef] [PubMed]
102. Kreitman, R.J.; Hassan, R.; Fitzgerald, D.J.; Pastan, I. Phase i trial of continuous infusion anti-mesothelin recombinant immunotoxin ss1p. *Clin. Cancer Res. Off. J. Am. Assoc. Cancer Res.* **2009**, *15*, 5274–5279. [CrossRef] [PubMed]
103. Hassan, R.; Bullock, S.; Premkumar, A.; Kreitman, R.J.; Kindler, H.; Willingham, M.C.; Pastan, I. Phase i study of ss1p, a recombinant anti-mesothelin immunotoxin given as a bolus i.V. Infusion to patients with mesothelin-expressing mesothelioma, ovarian, and pancreatic cancers. *Clin. Cancer Res. Off. J. Am. Assoc. Cancer Res.* **2007**, *13*, 5144–5149. [CrossRef] [PubMed]
104. Hassan, R.; Schweizer, C.; Lu, K.F.; Schuler, B.; Remaley, A.T.; Weil, S.C.; Pastan, I. Inhibition of mesothelin-ca-125 interaction in patients with mesothelioma by the anti-mesothelin monoclonal antibody morab-009: Implications for cancer therapy. *Lung Cancer* **2010**, *68*, 455–459. [CrossRef] [PubMed]

105. Hassan, R.; Cohen, S.J.; Phillips, M.; Pastan, I.; Sharon, E.; Kelly, R.J.; Schweizer, C.; Weil, S.; Laheru, D. Phase i clinical trial of the chimeric anti-mesothelin monoclonal antibody morab-009 in patients with mesothelin-expressing cancers. *Clin. Cancer Res. Off. J. Am. Assoc. Cancer Res.* **2010**, *16*, 6132–6138. [CrossRef] [PubMed]
106. Le, D.T.; Brockstedt, D.G.; Nir-Paz, R.; Hampl, J.; Mathur, S.; Nemunaitis, J.; Sterman, D.H.; Hassan, R.; Lutz, E.; Moyer, B.; et al. A live-attenuated listeria vaccine (anz-100) and a live-attenuated listeria vaccine expressing mesothelin (crs-207) for advanced cancers: Phase i studies of safety and immune induction. *Clin. Cancer Res. Off. J. Am. Assoc. Cancer Res.* **2012**, *18*, 858–868. [CrossRef] [PubMed]
107. Golfier, S.; Kopitz, C.; Kahnert, A.; Heisler, I.; Schatz, C.A.; Stelte-Ludwig, B.; Mayer-Bartschmid, A.; Unterschemmann, K.; Bruder, S.; Linden, L.; et al. Anetumab ravtansine: A novel mesothelin-targeting antibody-drug conjugate cures tumors with heterogeneous target expression favored by bystander effect. *Mol. Cancer Ther.* **2014**, *13*, 1537–1548. [CrossRef] [PubMed]
108. Beatty, G.L.; Haas, A.R.; Maus, M.V.; Torigian, D.A.; Soulen, M.C.; Plesa, G.; Chew, A.; Zhao, Y.; Levine, B.L.; Albelda, S.M.; et al. Mesothelin-specific chimeric antigen receptor mrna-engineered t cells induce anti-tumor activity in solid malignancies. *Cancer Immunol. Res.* **2014**, *2*, 112–120. [CrossRef] [PubMed]

cancers

MDPI

Review

Ovarian Tumor Microenvironment Signaling: Convergence on the Rac1 GTPase

Laurie G. Hudson [1,2,*,†], Jennifer M. Gillette [2,3,†], Huining Kang [2,4], Melanie R. Rivera [2,3] and Angela Wandinger-Ness [2,3,†]

1 Department of Pharmaceutical Sciences, University of New Mexico Health Sciences Center, Albuquerque, NM 87131, USA

2 Comprehensive Cancer Center, University of New Mexico Health Sciences Center, Albuquerque, NM 87131, USA; jgillette@salud.unm.edu (J.M.G.); HuKang@salud.unm.edu (H.K.); MelRivera@salud.unm.edu (M.R.R.); awandinger-ness@salud.unm.edu (A.W.-N.)

3 Department of Pathology, University of New Mexico Health Sciences Center, Albuquerque, NM 87131, USA

4 Department of Medicine, University of New Mexico Health Sciences Center, Albuquerque, NM 87131, USA

* Correspondence: lhudson@salud.unm.edu; Tel.: +1-505-272-2482

† These authors contributed equally to this manuscript.

Received: 1 September 2018; Accepted: 25 September 2018; Published: 27 September 2018

Abstract: The tumor microenvironment for epithelial ovarian cancer is complex and rich in bioactive molecules that modulate cell-cell interactions and stimulate numerous signal transduction cascades. These signals ultimately modulate all aspects of tumor behavior including progression, metastasis and therapeutic response. Many of the signaling pathways converge on the small GTPase Ras-related C3 botulinum toxin substrate (Rac)1. In addition to regulating actin cytoskeleton remodeling necessary for tumor cell adhesion, migration and invasion, Rac1 through its downstream effectors, regulates cancer cell survival, tumor angiogenesis, phenotypic plasticity, quiescence, and resistance to therapeutics. In this review we discuss evidence for Rac1 activation within the ovarian tumor microenvironment, mechanisms of Rac1 dysregulation as they apply to ovarian cancer, and the potential benefits of targeting aberrant Rac1 activity in this disease. The potential for Rac1 contribution to extraperitoneal dissemination of ovarian cancer is addressed.

Keywords: Rho-GTPase; Rac1; guanine nucleotide exchange factors (GEFs); GTPase activating proteins (GAPs); oncogene; oncoprotein; ovarian cancer; tumor microenvironment; bone niche; therapeutic targeting

1. Introduction

Despite advances in treatment, long-term outcomes for epithelial ovarian cancer (EOC) patients remain discouraging. Challenges to effective treatment include factors such as diagnosis after tumor dissemination, presence of residual disease after treatment, a limited number of identified targets for maintenance therapy, and acquired chemoresistance leading to relapse after initial clinical remission [1,2]. EOC displays a high degree of genomic heterogeneity [3,4] and it has been proposed that tumor microenvironmental factors may also contribute to tumor heterogeneity [5].

EOC dissemination occurs predominantly through tumor cell exfoliation into the peritoneal cavity thereby providing a unique environment for tumor growth and metastasis when compared to the majority of solid tumors [6–10]. There is heterogeneity of sites within the peritoneal cavity leading to diverse localized environments. For example, the omentum is rich in adipocytes and provides a distinct niche when compared to the mesothelium of the peritoneal wall [10–18]. Furthermore, the tissue underlying the mesothelial lining at various locations differs in architecture and local production of chemotactic factors thus promoting different adhesive and invasive behaviors [11]. It may be more accurate to consider the peritoneal cavity as home to multiple tumor microenvironments (TMEs) presenting additional challenges to effective treatment.

Tumor cells within the ovarian cancer TME are exposed to a variety of regulatory signals. Tumor cells interact with mesothelium, fibroblasts, endothelium, immune cells and other cells in the TME [6–10,19,20]. Invasive cells come into contact with the extracellular matrix (ECM) underlying the mesothelium. This leads to intracellular signaling due to integrin engagement and exposure to ECM-associated growth factors. Each cell type in the TME, as well as the tumor cells themselves, secrete bioactive molecules that accumulate in the peritoneal fluids and drive adverse tumor cell behaviors such as proliferation, invasion, and phenotypes promoting chemoresistance. These cell-cell interactions between tumor cells or other cells in the TME, cell-matrix interactions, and exposure to growth factors and cytokines present in peritoneal fluids all stimulate signaling cascades that dictate aspects of tumor cell function. Many of these diverse signals converge upon, and are integrated through, the small GTPase Ras-related C3 botulinum toxin substrate (Rac) 1 (Figure 1) [21–26].

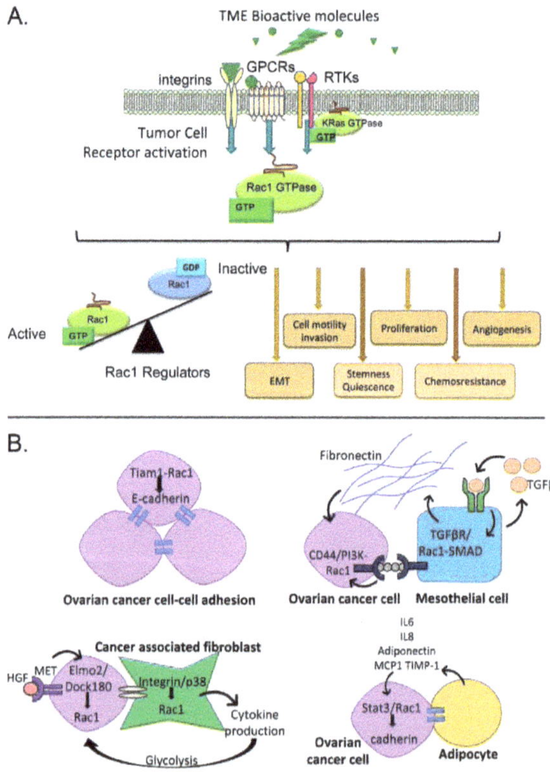

Figure 1. Bioactive molecules in the tumor microenvironment (TME) activate multiple receptors that converge on Rac1. (**A**) Examples of receptors that activate Rac1 in response to bioactive molecules in the TME are shown (GPCR = G protein-coupled receptor; RTK = receptor tyrosine kinase). Rac1 activity is balanced through multiple regulatory mechanisms discussed in this review that serve to control diverse physiological outcomes. (**B**) Cell-cell interactions between tumor cells themselves or with cells in the TME (adipocytes, cancer-associated fibroblasts, mesothelia) can also cause Rac1 activation and further modulate the TME. See Section 3.3 for further detail.

The Rac subfamily of Rho family small GTPases has three members. Rac1 is the best-characterized member of this subfamily with strong evidence for Rac1 dysregulation in cancer [21,23–25,27,28]. Rac2 expression is confined to hematopoietic cells [29] and Rac3 has not been studied in the context of ovarian cancer so these two proteins will not be discussed further in this review. Rac1 acts as a molecular switch by cycling between active and inactive states that depend upon nucleotide binding (Figure 1A) and other regulatory mechanisms discussed below. As a focal point for multiple signaling pathways, Rac1 is capable of shunting cells between proliferation, apoptosis or quiescence, altering cell differentiation and transcription, and modulating cell-environment interactions. Based on the known activities of Rac1, this protein can play important roles in multiple steps of tumor development, dissemination and disease recurrence.

2. Consequences of Rac1 Activation in Cancer

Rac1 cycles between an active GTP-bound state and an inactive GDP-bound conformation (Figure 1A) regulated by guanine nucleotide exchange factors (GEFs), GTPase-activating proteins (GAPs) and guanine nucleotide dissociation inhibitors (GDIs) [24]. Aberrant activation of Rac1 is implicated in numerous aspects of tumor development and progression and is the subject of several recent reviews [25,26,30,31]. Rac1 is best recognized for translating extracellular signaling into downstream changes in actin remodeling, cell adhesion, motility and invasion [23,26,32]. There is strong emerging evidence that Rac1 also contributes to the tumor stem cell phenotype, epithelial to mesenchymal transition (EMT), angiogenesis, and chemoresistance [33–38]. Elevated Rac1 activity is associated with enhanced stem cell characteristics in multiple cancers and its inhibition attenuates the stem cell phenotype [34,36,39]. Although the relationship between Rac1 expression and/or elevated activity and cancer stem cells has been reported for several cancer types, there is little information for ovarian cancer. However, a splice variant of Vav3, a GEF and enhancer of Rac1 activity, is overexpressed in multi-drug resistant stem cell-like fractions of ovarian cancer cells [40]. This finding suggests that elevated Rac1 activity may promote stem cell characteristics in ovarian cancer similar to the reports for other tumor types.

Ovarian cancer cells display phenotypic plasticity with gains and losses of epithelial characteristics during tumor development and peritoneal metastasis [41–43]. EMT is viewed as a critical aspect of tumor invasion and metastasis [44,45] and Rac1 is implicated in promoting EMT in a number of cancers [25]. Experimental evidence demonstrates that elevated Rac1 activity is sufficient to drive aspects of EMT in ovarian tumor cells. When a mutationally activated form of Rac1 (Rac1G12V) was introduced into ovarian tumor cells with an epithelial phenotype, cells displayed morphologic characteristics of EMT including down-regulation of the epithelial marker E-cadherin, up-regulation of the mesenchymal marker vimentin, and increased invasive capacity [46]. Inhibition of Rac1 activity or knockdown of Rac1 expression restored epithelial characteristics to ovarian tumor cells [13,46,47] and inhibited migration and invasion [47–49]. The significance of EMT in ovarian cancer is demonstrated by the presence of ovarian cancer cells in extraperitoneal sites [50–52] and the circulation [50,53–55] where the circulating tumor cells display mesenchymal characteristics [54,56]

Tumor angiogenesis supplies necessary nutrients and fosters tumor growth [22]. Angiogenesis is a critical aspect of ovarian cancer and this process is targeted by therapeutics in current care [57–59]. Rac1 is involved in angiogenesis and required for vascular integrity and blood vessel sprouting as demonstrated in a conditional Rac1 knockout mouse model [60]. In humans, Rac1 expression correlated with blood vessel invasion in a meta-analysis of multiple cancer studies [38]. Rac1 is activated by the angiogenic factors vascular endothelial growth factor (VEGF)-A, angiopoietin 1, basic fibroblast growth factor (FGF) and others [61]. Activation of Rac1 in endothelial cells regulates adhesion, filopodia, morphogenesis, cell proliferation and migration [33,62–65]. Two different Rac1 inhibitors displayed anti-angiogenic activity in breast cancer models in vivo [66,67] supporting a potential benefit of Rac1 inhibition as an alternate anti-angiogenic strategy in cancer, including ovarian cancer.

Lymphangiogenesis is driven by the VEGF-C ligand and its high affinity receptor VEGFR3 [68,69]. The well-established omental niche site has vessels that express high levels of the neoangiogenic VEGFR3, which serve to recruit ovarian tumor cells and offer a supportive environment for neovascularization [70]. High VEGF-C expression is associated with worse overall survival in ovarian cancer patients and tumor cell expression of VEGF-C is critical for lymphatic invasion and lymphangiogenesis [71,72]. Mechanistic studies show VEGF-C signaling to Rac1 requires VEGFR3 endocytosis mediated by EphrinB2 [73]. In colorectal and lung cancers, lymph node metastasis mediated by VEGF-C is linked to high expression of the Rac1 activating GEF Tiam1 [74,75]. Conversely, a chemical library screen identified statins as potent inhibitors of lymphangiogenesis by blocking Rac1 prenylation and plasma membrane recruitment [69]. In this regard it is worth noting that inhibition of VEGFR3 signaling in OVCAR8 cells, via Maz51, induced chemosensitization through downregulation of BRCA gene expression. This finding suggests that combined targeting of VEGFR3 and Rac1 may have benefit for dually blocking metastasis and enhancing tumor cell killing [76].

Rac1 is gaining substantial attention as a mediator of chemoresistance [37,77,78]. Rac1 is implicated in treatment resistance in multiple cancers [37] and Rac1 inhibition increases sensitivity to doxorubicin for squamous cell carcinoma cells, 5-fluorouracil and cisplatin in gastric adenocarcinoma spheroids, and fludarabine for chronic lymphocytic leukemia (reviewed in [37]). In addition to these conventional chemotherapies, Rac1 is suspected in resistance to a number of targeted therapies through regulation of compensatory mechanisms. These include therapies directed against the epidermal growth factor (EGF) receptor and human epidermal growth factor receptor (HER)-2 for lung and breast cancers, B-RAF protein inhibitors in melanoma, estrogen targeted therapies in breast cancer and VEGF/VEGFR targeted therapies in prostate cancer (reviewed in [30]). In many cases sensitivity to the targeted therapeutic is restored upon Rac1 inhibition. The contributions of Rac1 activation to chemoresistance is likely multifaceted based on specific mechanisms along distinct drug action pathways, as well as non-specific mechanisms related to Rac1 promotion of EMT and stem cell characteristics [42,79–81].

3. Pathways for Rac1 Activation by the Ovarian Tumor Microenvironment

Extracellular signals mediated by various cell surface receptors such as integrins, cadherins, cytokine receptors, G-protein-coupled receptors (GPCRs) and receptor tyrosine kinases (RTKs) activate GEFs and recruit Rac1 (sequestered with GDIs) from the cytosol to the plasma membrane or other cellular locations (Figure 3A [21,30,82,83]). Rac1 then activates effector molecules including proteins involved in actin remodeling, kinases, and adapter proteins that are responsible for propagating Rac1-dependent signals and subsequent biological responses. The specific stimulus can dictate distinct responses to Rac1 activation based on post-translational modifications of Rac1, GEFs or other Rac1 modulatory molecules or effectors [21]. Because Rac1 is responsive to an array of signals, Rac1 is capable of driving multiple steps of tumor development, dissemination and recurrence. A few examples of Rac1 activation by common components of ascitic fluids are described in more detail below.

3.1. Activators of G-Protein Coupled Receptors and Rac1 Activity

The bioactive lipids lysophosphatidic acid (LPA) and sphingosine-1-phosphate (S1P) are present in ascitic fluid of ovarian cancer patients and activate GPCRs upstream of Rac1. Elevated levels of LPA and S1P are both associated with ovarian tumor cell migration, invasion and metastasis [6,84] and these processes require Rac1-dependent actin remodeling. Pharmacologic inhibition of Rac1 decreased S1P-dependent ovarian tumor cell invasion [85]. When multiple ovarian tumor cell lines were studied, the ability of LPA to stimulate migration was highly correlated with LPA-dependent Rac1 activation [86]. Expression of a dominant negative form of Rac1 ablated LPA-stimulated cell migration and in vivo metastatic colonization in responsive cell lines. Conversely, expression of a constitutively active form of Rac1 conferred migration and in vivo implantation to cell lines non-responsive to LPA [86]. Knock-down strategies determined that a Rac1-activating SOS1/EPS8/ABI1 complex unique to metastatic cells was responsible for the LPA stimulated migration and invasive implantation in mice [86]. LPA activation of Rac1 has also been reported to be dependent on a Src/p130Cas pathway for ovarian cell migration [87] and the Rac1 GEF βPIX was necessary for LPA-induced invadopodia formation [88] although βPIX knock-down did not disrupt LPA-stimulated migration in certain ovarian tumor cell lines [86]. The reported observations indicate that distinct Rac1 regulatory mechanisms are responsible for different functional outputs and there may be cell-specific differences based on the expression or activity of Rac1 regulators.

3.2. Activators of Tyrosine Kinases and Rac1 Activity

Ligands for RTKs such as the EGF receptor and VEGF receptor are prevalent in ovarian cancer ascites and regulate Rac1 activation through multiple mechanisms. Signaling through RTKs activate phosphatidylinositol-3 kinase and phospholipase C-γ to modulate targeting of Rac1 regulatory proteins such as GEFs and GAPs and recruit GEFs to signaling complexes through post-translational modifications (reviewed in [21,89]). In certain cases, signaling receptors can modify Rac1 activity directly. For example, EGF receptor-stimulated ERK phosphorylation of Rac1 on T108 targets Rac1 for nuclear translocation [21]. Rac1 has been shown to be an essential component of EGF receptor signaling in different tumor types [90,91] and implicated in EGF receptor driven tumorigenesis [91]. Ligands present in the ovarian TME are likely to activate Rac1 by impinging on ErbB3, ErbB4 and MET receptors, which are expressed in 76–98% of ovarian tumors [92]. For example, heregulin stimulation of ErbB3 and ErbB4 causes upregulation of C-X-C chemokine receptor type 4 (CXCR4) and increases Rac1 activation through a stromal cell-derived factor (SDF)-1-CXCR4 mediated PREX1 GEF mechanism in breast cancer cells [93]. Hepatocyte growth factor (HGF) induces a MET-AXL-ELMO2-DOCK180 complex that activates Rac1-dependent cancer cell migration and invasion [94]. Pharmacologic inhibition of Rac1 inhibited EGF-stimulated p21-activating kinase (PAK) phosphorylation, filopodia formation and invadopodia [48,95] in ovarian tumor cell lines indicating contributions of Rac1 in cancer-relevant functions. Although specific mechanisms of Rac1 activation by VEGF have not been explored in ovarian cancer models, there is abundant evidence that Rac1 is a component of VEGF signaling to angiogenesis. Ablation of Rac1 in endothelial cells in development is embryonic lethal due to lack of neovascularization [96]. Studies show that Rac1 activation is critical for normal in vivo angiogenesis in adult mice due to junctional stabilization required for mature vessels [97]. More recent work indicates that lumen formation and stable cell:cell contacts are mediated through the GEF DOCK4 activation of Rac1 [62]. The combined data indicate that further study of Rac1 activation in ovarian cancer by tyrosine kinase receptors and their interfaces with G-protein coupled receptors is warranted.

3.3. Cell Interactions Leading to Rac1 Activation

An article in the present series and other recent reviews provide an in depth analysis of cell-cell interactions in the ovarian tumor microenvironment that drive ovarian cancer progression [9,57,98]. Here, we briefly highlight how some of these interactions may promote ovarian cancer metastasis through Rac1-dependent mechanisms (Figure 1B).

3.3.1. Tumor Cell-Cell Adhesion

Rac1 signaling is important for cell-cell adhesion. Ovarian cancer cells in the ascites fluid form multicellular aggregates (spheroids) that facilitate angiogenesis and invasion of various peritoneal organs [11]. Tumor cell-cell adhesions are mediated by E-cadherin maintenance of cell-cell junctions that depend on a Rac1-Tiam1 GEF-IQGAP1 effector complex and promote an anti-migratory phenotype [99]. Ovarian cancer spheroids with high E-cadherin expression are less sensitive to cisplatin treatment suggesting an important role for cell-cell adhesions in spheroid chemoresistance [100].

3.3.2. Mesothelial Cells

Ovarian cancer frequently metastasizes to the peritoneal wall, which is lined with mesothelial cells. Ovarian cancer cell interactions with mesothelial cells can stimulate mesothelial cell production of fibronectin through the autocrine secretion of transforming growth factor (TGF)-β1. This activates a TGF-βR1/Rac1/SMAD-dependent signaling pathway in mesothelial cells. The activated mesothelial cells and production of fibronectin contributes to metastasis by supporting tumor cell adhesion, invasion, and proliferation [13,57]. Co-culture of ovarian cancer cell lines with mesothelial cells led to upregulated expression of the hyaluronan receptor and stem cell marker CD44 and promoted tumorigenesis in a xenograft model [101]. CD44 promotes ovarian tumor cell-peritoneal cell adhesion through binding of its ligand hyaluronan in complex with versican [102] and is generally known to signal through multiple pathways downstream of Rac1 to promote tumor cell invasion [103].

3.3.3. Fibroblasts

Ovarian tumor cell-fibroblast interactions cause conversion of normal fibroblasts to cancer-associated fibroblasts (CAFs, distinguished by smooth muscle actin expression) and lead to increased tumor cell adhesion and overexpression of HGF and matrix metalloproteinase (MMP) [104]. MET receptor activation by HGF induced recruitment of the bipartite Rac1 GEF Elmo2/Dock180 and promoted Rac1-dependent migration and invasion of multiple cancer cell lines in vitro, though ovarian cell lines were not specifically tested [94]. Interactions between human omental CAF and ovarian tumor cells also result in an integrin/p38/Rac1-dependent activation of cytokine secretion by CAFs, which in turn promotes tumor cell proliferation and metastasis through activated glycogen breakdown and glycolysis [105].

3.3.4. Adipocytes

The omentum is a favored ovarian tumor cell niche based on initial chemoattraction by adipocyte secreted factors that can stimulate Stat3-mediated Rac1 activation [106]. In turn, the activation of these pathways can strengthen cadherin-dependent binding of tumor cells, provide tumor cells with an energy source through mutual changes in lipid metabolism, and promote invasion [14,106].

The selected illustrations do not capture the entire scope of potential ovarian cancer TME regulation of Rac1 activity. Inflammatory cytokines such as interleukins 6 and 8, tumor necrosis factor (TNF) α, and TGFβ are among the additional soluble factors in ascites fluids that are associated with worse prognosis and variously associated with proliferation, metastatic spread, angiogenesis, EMT and treatment resistance [6,107]. Each of these bioactive molecules is capable of stimulating signaling cascades leading to Rac1 activation through direct or indirect mechanisms [24]. In addition, integrin engagement and focal adhesion kinase activation recruits Rac1 to regulate spreading and adhesion on the extracellular matrix [26,89,108]. Immune cells are an integral part of the ovarian cancer TME and perform immune suppressive and activating functions that are pivotal in disease pathology [109,110] and these cells serve as important therapeutic targets [111,112]. The best-studied example of immune cell coupling to Rac1 activation in ovarian cancer is through cytokine activation of CXCR4 as detailed in Sections 4.2 and 6. A more complete understanding of the complexities of Rac1 regulation by the ovarian cancer TME will require further study.

4. Mechanisms of Rac1 Dysregulation and Evidence in Ovarian Cancer

We reported that Rac1 protein is overexpressed and hyperactivated in ovarian cancer patient samples [113]. Addressing the function of Rac1 hyperactivation in ovarian cancer is an important research area because of the known roles of Rac1 in cancer metastasis and recurrence. In cancer, Rac1 is frequently released from normal control mechanisms through mutation [114–118], aberrant regulation of nucleotide binding and hydrolysis [26,30,119], and altered splicing [120–130]. Insight into possible mechanisms leading to Rac1 overexpression and hyperactivation in ovarian cancer is garnered from analyses of the Catalogue of Somatic Mutations in Cancer (COSMIC) and The Cancer Genome Atlas (TCGA) databases as detailed below.

4.1. Rac1 Overexpression and Somatic Mutation

There are 239 pathogenic missense mutations across diverse cancer types affecting 46 of the 192 amino acids in RAC1 (COSMIC v86 database updated in August 2018, https://cancer.sanger.ac.uk/cosmic/download). The mutants are clustered in conserved residues relevant to GTPase activity or affect residues close in 3D space that are important to Rac1 function (Figure 2A,B [117,118,131–133]). Select point mutants are the primary cause of constitutive Rac1 activation in some cancer types (melanoma, lung and germ cell cancers) (Figure 2A [114–118,132,134]). The highest prevalence (9%) of the constitutively active, fast cycling P29S mutant is found in melanoma [115]. To date, the functionally characterized Rac1 missense mutants (P29S, A159V, C18S and G15) all increase Rac1 activation and possibly expression [118,135]. Rac1 is not found mutant in the 315 serous ovarian cancer patient samples in the TCGA. However, given the low frequency of Rac1 missense mutants (0.01–0.02% for G15, C18 [118]) such rare mutations would be undetectable in the sample size and should not be taken as lack of evidence for the importance of Rac1 in ovarian cancer. For example, an shRNA essentiality screen of 29 ovarian cancer cell lines showed SKOV3, COV362, JHM + OM1 and SNU840 to have significantly decreased growth fitness with the loss of Rac1 (Harmonizome Achilles [136,137]. As another case in point, Rac1 is overexpressed due to gene amplification or mRNA upregulation in 21% (66/316) of the primary tumors in TCGA [138]. Despite the low frequency of RAC1 gene mutations, RAC1 is similar to well-known oncogenes and tumor suppressors in being categorized as a Tier 1 cancer-causing gene in the COSMIC cancer gene census. Therefore, further systematic study of the 239 Rac1 missense mutations is warranted. In contrast to tumor suppressor genes, where truncating mutations are prevalent and cause loss of function, the Rac1 mutations are like those in the oncoprotein Ras. The mutations appear in hotspots and tend to be activating mutations [139]. Thus, RAC1 is a Tier 1 cancer-causing gene and the mutational patterns in Rac1 are similar to many well-known oncogenes which are positive drivers of cancer.

4.2. Rac1 Regulators

The activity of Rac1 is tightly controlled through a large network of GEF and GAP regulatory factors (Figure 3A,B [30,131,140–143]). This network is much greater than most other Ras-related GTPases. Rac1 GEF and GAP regulatory factors are mutant or exhibit altered expression in ovarian serous adenocarcinoma with a frequency of 0.3–1.6% based on our analyses of 28 relevant regulatory proteins in COSMIC v86 [144] and the cBioPortal platform for TCGA data viewing [19,131] (Figure 4 [20,83,118,138,144–147]).

Notably, the regulatory protein mutants show a high level of concurrent expression in tumor cells, suggesting that hitting multiple nodes releases key Rac1-regulated pathways from normal control. Even while the identified mutations often lie in known GEF and GAP regulatory domains, as well as in lipid or protein interaction domains, no systematic analyses have been completed to identify hotspot mutations or determine their pathogenicity in ovarian cancer. Nevertheless, some insights can be drawn from a handful of analyses of regulatory protein overexpression [109,148,149], truncation [150] or altered splice variants [40]; see also review [30].

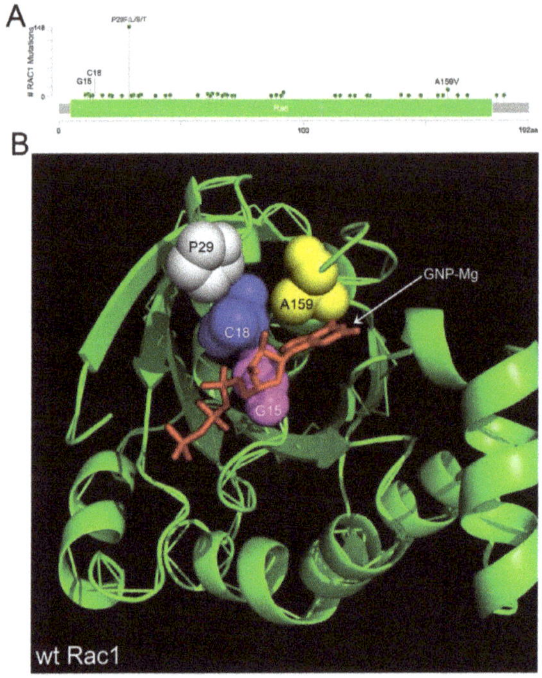

Figure 2. Pathogenic Cancer Mutations in Rac1. Rac1 gain of function mutations occur with low frequency (0.2–1%) in multiple cancer types, though as yet none have been found in serous ovarian cancer. (**A**) There are 239 pathogenic mutations in Rac1, resulting in missense substitutions at 46 amino acid residues. Melanoma has the highest frequency of Rac1 mutations, leading to substitutions at proline 29 and constitutive activation through GEF-independent fast nucleotide exchange. (**B**) Thirteen of the missense mutants are likely oncogenic (G12R/V/E, G15S, C18S/Y, P29F/L/S/T, Q61R/K, A159V) evidenced by recurrence at hotspots, paralogous with oncogenic mutations in Ras, or affecting residues that are clustered in the 3D structure close to the nucleotide binding site. Shown is the proximity of 4 point mutants in the crystal structure of wild-type Rac1 (PDB 3th5) rendered with MacPyMOL: PyMOL v1.5.0.5 (Schrödinger LLC).

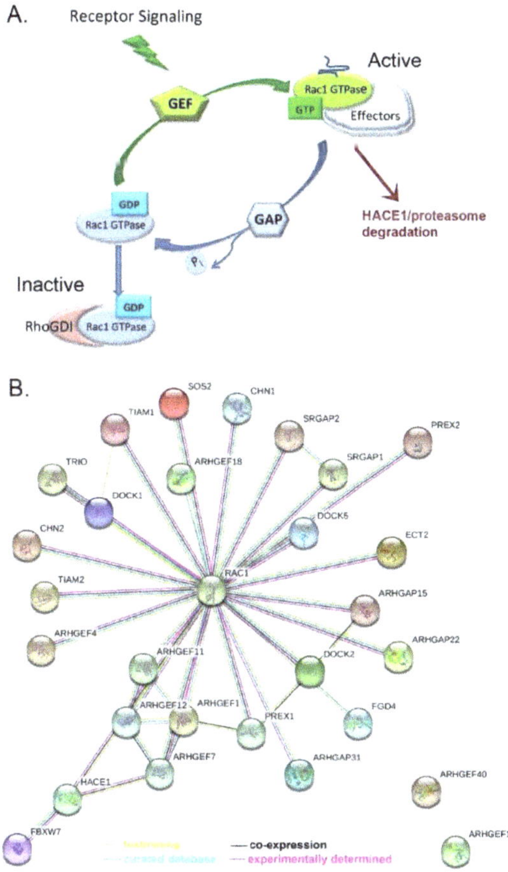

Figure 3. Rac1 Regulatory Network in Ovarian Cancer. Rac1 integrates signals downstream of tyrosine kinase receptors, adhesion molecules and G-protein coupled receptors and has nuclear functions. (**A**) There are over 40 GEF and GAP proteins involved in the regulation of Rac1 activity through the stimulation of nucleotide binding (GEFs) and hydrolysis (GAPs). Activated Rac1 binds to effectors that recognize the unique conformation of the GTP-bound GTPase and mediate downstream physiologic responses to receptor signaling. Activated Rac1 is ubiquitinated by HACE1 or FBXW7 and targeted for proteasomal degradation, further controlling protein levels and activity. (**B**) Twenty-six Rac1 GEF and GAP proteins, HACE1 and FBXW7 ubiquitin ligases have somatic mutations in serous ovarian cancer patient samples. String network analysis places Rac1 at the core with a large array of close functional associations with Rac1 regulatory proteins whose functions in ovarian cancer largely remain to be determined.

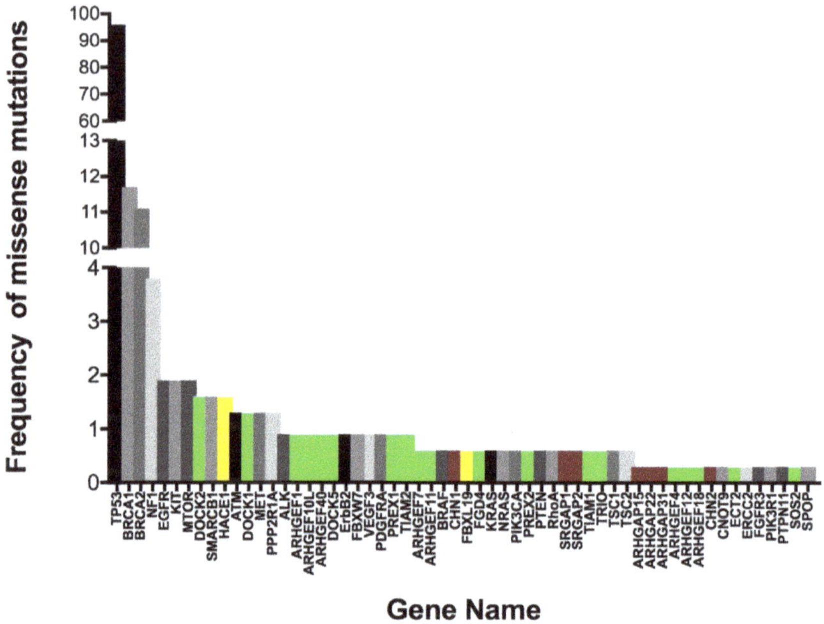

Figure 4. Rac1 regulators and effectors are part of the long tail of oncogenic drivers in ovarian cancer. A list of 54 genes with frequent missense mutations in cancers was derived from OncoKB: A precision oncology knowledge base and two recent publications on rare mutations. The cancer gene list was combined with a selected list of 28 Rac1 regulatory proteins (GEFs, GAPs, ubiquitin ligases). Plotted is the frequency of missense mutations in genes with mutation frequencies above 0 (58 of 82 analyzed) among 315 serous ovarian cancer patient samples in TCGA. The frequency for BRCA1 and BRCA2 gene mutations is the sum of somatic and germline missense mutations. For all other genes no germ line mutations are reported. Among the analyzed genes, 53 (9.2%) are Tier 1 of 574 reported in COSMIC v86; Tier 1 genes have "*documented activity relevant to cancer, plus evidence of mutations in cancer, which change the activity of the gene product in a way that promotes oncogenic transformation*". Among the Rac1 regulators only ARHGEF12 and ARHGEF10L are validated as Tier 1 and Tier 2 ("*strong indication of a role in cancer*"), respectively. Cancer genes (black and gray), Rac1 GEFs (green), Rac1 GAPs (red), ubiquitin ligases (yellow). The data show that even though missense mutations in individual Rac1 regulators occur with low frequency, there are at least 26 possible targets (10 with co-occurring alterations, $p < 0.05$) that might lead to Rac1 activation or inactivation in ovarian cancer.

Overexpression of the Rac1 GEF DOCK180 drives glioblastoma invasion through the activation of a Rac1-dependent kinase pathway [149]. A truncating mutant of PREX2 in melanoma has increased Rac1 GEF activity, and activates PI3K/AKT signaling, while abolishing binding to the PTEN tumor suppressor in melanoma [150]. An N-terminally truncated splice variant of the Vav3 GEF (Vav3.1) is a predictor of poor prognosis and platinum-response and highly expressed in ovarian cancer stem-like cell populations isolated from established cell lines [40]. These examples are supportive of a requirement for Rac1 activation in multiple cancers. Recent analyses of the metastatic TME using omental samples from patients with high grade serous ovarian cancer characterized secreted, matrix and cellular components [109]. Multivariate regression analyses of data were used to model the relationships between all TME components. Comprehensive RNA seq analysis of the TME identified 31 Rac1 GEFs, GAPS and ubiquitin ligases significantly associated with disease score by Pearson's and Spearman's tests; five GEFs and GAPs were significant based on Pearson's only (supplementary

Table 13 in [109]). Recent analyses of a large cohort of Canadian ovarian cancer patients identified variants in ARHGEF10L to be significantly associated with invasive disease [151] and three somatic missense mutations have been identified in ovarian cancer patient samples (COSMIC v86). The limited information on ARHGEF10L suggests in vitro GEF activity for RhoA, but not Rac1 or Cdc42 [152]. Since RhoA and Rac1 are often reciprocally active, connections between the two GTPases may need further analysis in ovarian cancer. Alterations in GAP expression in vivo have both activating and inhibitory effects on tumorigenesis and metastasis, likely due to dual roles as scaffolding proteins and GTP hydrolysis regulators [30]. When considering how to tackle prioritization of GEF and GAP proteins for study, categorizing potential tumor suppressive vs. promoting activity might be gained by using a ratiometric analysis of truncating/frameshift vs. missense mutations [139]. Additionally, functional analyses of select point mutants in key regulatory domains is an essential complementary effort that is necessary to understand effects on regulatory protein activity and pathway interconnections. The composite data are suggestive that Rac1 hyperactivation is an important driver in ovarian cancer and may result largely from the misregulation of GEF and GAP regulatory cascades rather than through activating mutations in Rac1 itself.

Emerging evidence suggests that Rac1 regulatory proteins function in spatially localized molecular assemblies. Such assemblies restrict Rac1 activity temporally and spatially to specific subcellular domains, which in turn restricts what downstream pathways are triggered by Rac1. In ovarian cancer, a recently described tripartite complex that includes the SOS1 GEF is essential for LPA-mediated Rac1 activation and metastasis [86]. Activation of Rac1 by the Tiam1 or PREX1 GEF proteins is spatially distinct in the cell and dictates anti- or pro-migratory responses in ovarian cancer cells [99]. The translocation of Rac1 in response to signaling and transient assembly of Rac1 GEFs at the plasma membrane can also occur through specific actin and protein based recruitment [82]. On the other hand, Rac1 forms a stable plasma membrane complex with CXCR4 independent of GTP-bound status, which is important for maintaining CXCR4 in a signaling competent conformation [153]. The PREX1 GEF is speculated to enable rapid response of Rac1 activation downstream of CXCR4 signaling. Therefore, functional studies of Rac1 and associated regulatory factors in the ovarian metastatic cascade will need to carefully consider spatiotemporal organization.

4.3. *Rac1b Splice Variant*

The constitutively active Rac1b splice variant mRNA level [113] and protein levels are moderate to high in the majority of serous papillary ovarian adenocarcinoma cells (Figure 5). Interestingly, Rac1b is also differentially expressed in underlying stromal cells in malignant serous papillary ovarian adenocarcinoma tissue as compared to normal ovary. The prognostic or diagnostic significance of overexpression of canonical Rac1 in ovarian cancer and/or the potential role(s) of the hyperactivated, fast cycling Rac1b isoform remain open questions. We analyzed RAC1 mRNA expression data for 298 Stage III primary serous ovarian cancer patient samples in TCGA using isoform analysis tools [154,155]. The results demonstrate that high total RAC1 mRNA expression is associated with worse outcomes (Figure 6A,B) and concur with a report that analyzed Rac1 as a risk factor in a cohort of 150 Chinese ovarian cancer patients [47]. High expression of the canonical RAC1 isoform also trended to worse outcomes but was not statistically significant (Figure 6C). The impact of RAC1b isoform expression on ovarian patient survival has not been reported and was of particular interest. Rac1b protein drives tumor cell proliferation and EMT and is upregulated by MMP3, a known survival risk factor in breast, lung, and pancreatic cancers [124,156–158]. High mRNA expression of the fast cycling and constitutively active RAC1b isoform does not predict ovarian cancer patient survival and trended toward higher survival probability (Figure 6D [113,120–122,129,130,155]); the finding was consistent irrespective of various groupings, treatment as a continuous variable or when expressed as a fraction of total RAC1 mRNA expression. The only other study assessing the significance of RAC1b isoform expression measured the prognostic value of RAC1b in progression free and overall survival [159]. Findings were based on quantitative RT-PCR analyses of 157 metastatic colorectal cancer patient samples following relapse after first line chemotherapy. In contrast to our findings in primary ovarian tumors, fractional RAC1b overexpression was significantly associated with poor progression free (HR 0.54, $p = 0.49$) and overall survival (HR 0.53, $p = 0.039$) in metastatic colorectal cancer patients. Similar to the ovarian cancer patients, RAC1b expression was not mutually exclusive and 152/157 (97%) of the metastatic colorectal patients had higher canonical RAC1 than RAC1b expression. To date there are no studies that have distinguished the functions of Rac1 and Rac1b overexpression or activity in the absence of endogenous protein, in part due to the essentiality of Rac1 function [128,160]. Together, these data indicate that overexpression and aberrant Rac1 and/or Rac1b activity are closely tied to malignant ovarian cancer and further dissection of their respective roles in tumor microenvironment responsiveness, metastasis and relapse is warranted.

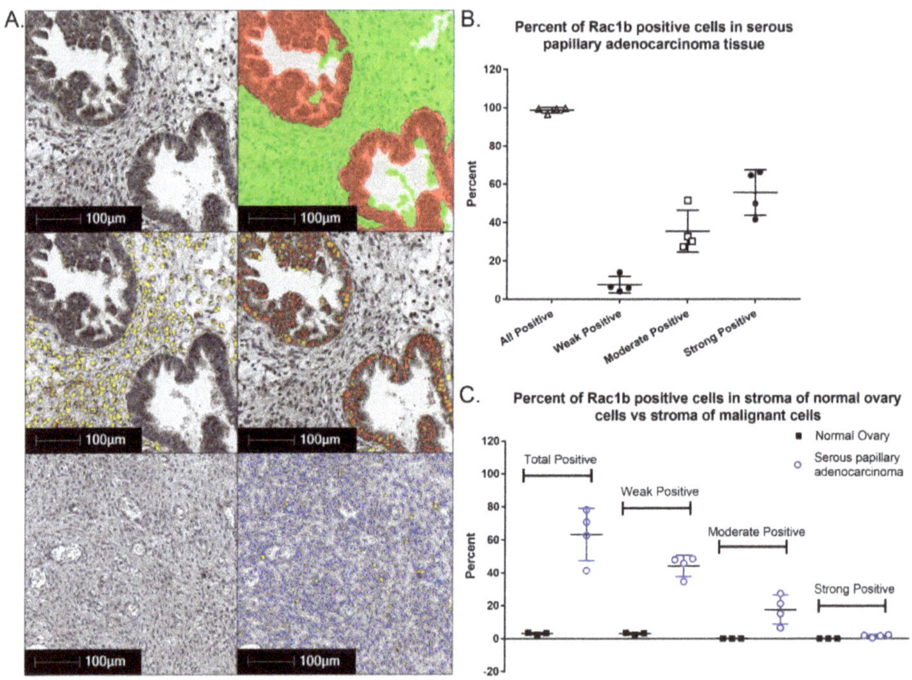

Figure 5. The constitutively active Rac1b splice variant is overexpressed in ovarian cancer. Ovarian cancer tissue microarrays were stained for Rac1b, a constitutively active Rac1 splice variant. Slides were imaged using an Aperio slide scanner and analysis was performed using HALO software. (**A**) Top panel: Malignant tissue stained with DAB for Rac1b. Analysis to identify tumor cells (red) and stromal cells (green). Middle panel: Quantification of the amount of Rac1b expression in tumor cells (right) vs. stromal cells (left) in malignant tissue. Blue-no staining, yellow-weak staining, orange-moderate staining, red-strong staining. Bottom panel: Quantification of Rac1b expression in stromal cells in normal ovary tissue, colors as for middle panel. (**B**) The majority of serous papillary ovarian adenocarcinoma cells were moderately to strongly positive for Rac1b, while stromal cells were weakly positive. (**C**) Quantitative comparisons of normal ovary tissue and serous papillary ovarian adenocarcinoma tissue evidences an elevated expression of Rac1b in the stromal cells adjacent to the malignant tumor cells relative to normal tissue.

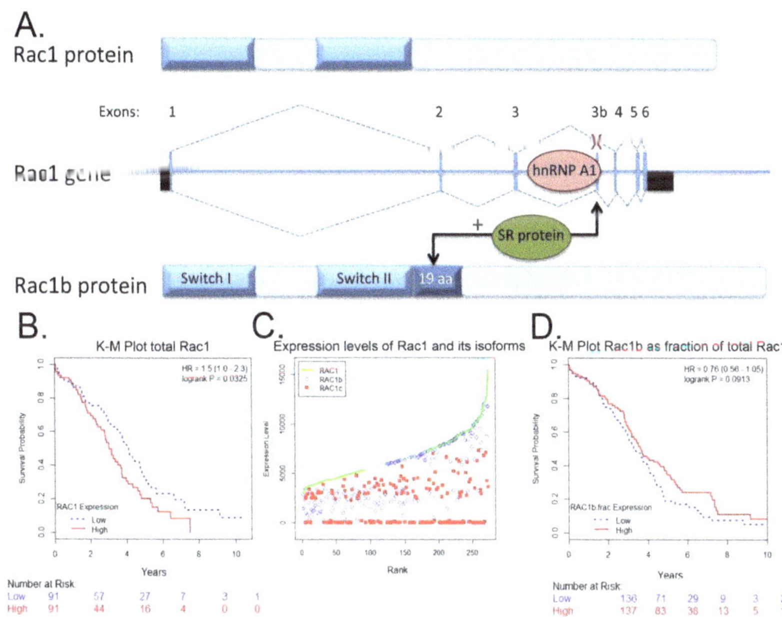

Figure 6. High total RAC1 expression predicts reduced ovarian cancer patient survival. (**A**) Rac1 undergoes regulated splicing in response to growth factor signaling, which is subject to positive and negative regulation by hnRNP A1 and SR protein. The resulting splice variant is called Rac1b and contains a 19 amino acid insert adjacent to the Switch II region. Rac1b is a fast cycling, constitutively active and frequently overexpressed in cancer, including ovarian cancer. (**B**) Kaplan-Meier plot of high vs. low total RAC1 mRNA expression. TCGA datasets for total and RAC1 isoform mRNA expression in ovarian cancer patients from ISOexpresso; uc003spx.3 (canonical RAC1) and uc003spw.3 (RAC1b containing exon 3b/4). Analyses were restricted to 298 patients with Stage III and Stage IV disease. Patients were divided into 3 groups based on total Rac1 expression. Upper tertile values represent high total Rac1 expression and lower tertile values represent low expression, middle values were excluded. Patients with high RAC1 expression have worse survival outcomes than those with low RAC1 expression (HR = 1.5, p = 0.0325); analogous results obtained using data direct from TCGA and CASViewer. No evidence for an association between isoform RAC1b and survival outcomes (HR = 0.96, p = 0.82, not shown). Higher expression of the canonical RAC1 isoform trended to lower survival probability, though was not statistically significant (HR = 1.37, p = 0.121). (**C**) Plot of total RAC1 (green line), canonical RAC1c (blue diamond), and RAC1b (red square) expression in each patient ranked according to expression levels. (**D**) Kaplan-Meier plot of RAC1b as a fraction of total RAC1, with two groups defined based on median expression. High RAC1b expression (HR = 0.76, p = 0.0913).

5. Potential Benefits of Targeting Aberrant Rac1 Activity in Ovarian Cancer

The broad impact of Rac1 on tumor cell behavior has led to consideration of Rac1 as a potential therapeutic target [25,28,95,161–163]. In ovarian cancer cell lines, knock down of Rac1 expression decreased fibronectin production [13], reversed EMT as measured by increased E-cadherin and decreased vimentin expression [46,47], inhibited tumor cell migration and invasion [47] and reduced tumor growth in a xenograft model [47]. An inhibitor of Rac1 (NSC23766) decreased ovarian tumor cell migration, invasion and matrix-metalloproteinase production [48,49,95].

Although a number of small molecule inhibitors have been developed to inhibit Rac1 activity (e.g., NSC23766, EHT 1864, EHop-016 and its derivative MBQ-167), these agents have not been translated to human use [66,164–166]. A high-throughput screen of the Prestwick library of off patent, FDA-approved drugs identified activators and inhibitors of Rho GTPases [95]. The resultant findings coupled with cheminformatics approaches identified the R-enantiomers of a limited number of non-steroidal anti-inflammatory drugs (NSAIDs), R-naproxen and R-ketorolac, as inhibitors of Rac1 [95]. The S-enantiomers are pharmacologic NSAIDs based on cyclooxygenase (COX) inhibition. GTPase inhibition by the R-enantiomers represents a previously unidentified pharmacologic activity. R-naproxen and R-ketorolac inhibit serum and EGF-stimulated Rac1 and Cdc42 activation and downstream signaling through a proposed allosteric mechanism [48,95]. R-ketorolac was tested using ovarian tumor cell lines and primary ovarian tumor cells isolated from patient ascites fluids [48]. R-ketorolac was an effective Rac1 inhibitor and decreased downstream signaling as demonstrated by reduction of PAK1 and PAK2 phosphorylation. R-ketorolac inhibited Rac1-dependent cellular functions in ovarian cancer cell lines and primary cells including inhibition of growth factor-stimulated formation of filopodia, cell adhesion to fibronectin and type I collagen, development of invadopodia and tumor cell migration [48]. The inhibitory effects of R-ketorolac in cells are comparable to those of established Rac1 and Cdc42 selective inhibitors [48,167].

In Phase 0 human studies, ovarian cancer patients received racemic ketorolac for its FDA-approved indication in postoperative analgesia [113] then blood and peritoneal fluids were collected at intervals for 24h. After administration of the racemic drug, R-ketorolac was detected in patient peritoneal fluids. The concentration of R-ketorolac was sufficient to inhibit Rac1 activity in cells retrieved from the peritoneal compartment of these post-surgical ovarian cancer patients. Potential benefit of R-ketorolac is suggested by the results of a medical record review to compare the ovarian cancer–specific survival of ovarian cancer patients who did or did not receive ketorolac [113]. The medical record review revealed increased survival of patients receiving ketorolac and this observation is consistent with other reports of improved clinical outcomes associated with ketorolac usage in breast cancer patients [168–170]. The overall findings suggest that ovarian cancer patients may benefit from inhibition of Rac1 in the clinical setting.

6. Other Ovarian Tumor Microenvironments: Extraperitoneal Dissemination and Bone Niche as a Sanctuary Site and Potential Reservoir for Relapse

While ovarian cancer metastasis is largely confined to the peritoneal cavity and localized to the omentum, there is strong evidence for extra-peritoneal dissemination [171,172]. As illustrated in preceding sections, Rac1 plays a critical role in the key processes that impact tumor dissemination and as such, Rac1 may contribute to ovarian tumor cell escape from the peritonium. Particularly in advanced disease, ovarian carcinoma can spread to distant organs by both hematogenous dissemination and lymphatic invasion [173]. In a well-designed parabiosis study, ovarian tumor cells were found to spread in anastomosed mice within two weeks of ovary injection [18], clearly illustrating hematogenous spread of the disease. Additionally, circulating tumor cells (CTCs) are frequently detected in patients [53,174]. In fact, CTCs were detected in 90% (98/109) of newly diagnosed ovarian cancer patients, where the number of CTCs correlated with disease stage and was altered with treatment [175]. Lymph node involvement of the disease is also common and has been proposed as a potential prognostic factor with site-specific prognostic differences identified between the ovary and lymph node [176]. However, this study was unable to rule out the "safe haven" hypothesis for metastatic ovarian tumor cells in retroperitoneal lymph nodes and suggested that lymph node dissection after complete cytoreduction is warranted pending further prospective data collection [176]. Interestingly, recent work comparing the survival of patients with distant lymph node metastases to patients with pleural metastases or other distant ovarian cancer metastases found increased survival in women having lymph nodes as their only distant metastatic site [177]. A follow up study investigating the relationship between site-specific patterns of distant metastases and overall survival also found that patients with lymph node metastasis had the longest survival when compared to women with other metastatic disease [173]. Collectively, these data suggest that disease dissemination through the lymphatics may have a less aggressive phenotype than disease that spreads hematogenously. Future studies will be necessary to quantitatively compare the aggressive nature of ovarian cancer cells with respect to their route of disease dissemination.

Once outside the peritoneum, other common sites of distant metastatic ovarian cancer include the liver, lung, and bone [6]. While frank bone metastases are rare in ovarian cancer [173,178], prognosis of cases with bone metastasis is poor. A recent publication [179] followed up on previous observations of bone marrow disseminated tumor cells (DTCs) in ovarian cancer patients [51,52,180,181] and affirmed that bone DTCs correlated with reduced progression free and overall survival [182,183]. Bone marrow was isolated from 79 ovarian cancer patients pre- and post-platinum-based chemotherapy. Bone DTCs were detected in 42% and 41% of patients before and after chemotherapy, respectively, illustrating the chemoresistance of cells in the bone niche [179]. Alterations in the bone microenvironment caused by irradiation and cisplatin therapy can further promote and increase metastatic spread that may be ameliorated by non-steroidal anti-inflammatory agents [184]. Additionally, tumor secreted factors such as CCL2 can activate cells in the bone marrow promoting a premetastatic niche and paving the way for successful tumor dissemination at a secondary site [98]. The predominant signaling axis that promotes bone marrow homing is the CXCR4/SDF-1α signaling cascade [185]. The expression and secretion of SDF-1α is abundant in the bone marrow microenvironment (expressed by osteoblasts and endothelial cells) and promotes the homing and maintenance of CXCR4+ cells within the bone marrow. In addition to driving hematopoietic cells as well as breast and prostate cancer cells to the bone [186–189], CXCR4/SDF-1α signaling has also been shown to promote ovarian cancer metastasis and is a predictor of poor prognosis in ovarian cancer [190,191]. Overexpression of CXCR4 is associated with cisplatin resistant ovarian cancer [192] as well as the peritoneal [193], hematogenous [194] and lymph node [195] dissemination of the disease. Moreover, CXCR4 can modulate cancer cell migration through interactions with the downstream effector Rac1 [196]. In fact, blocking or silencing of CXCR4 was found to significantly reduce RhoA and Rac-1/Cdc42 expression levels and decrease ovarian cancer cell migration [197]. Additionally, CXCR4 blockade reduced ovarian tumor growth in animal models [198,199]. Therefore, CXCR4 appears to be a shared signaling mechanism that facilitates homing and engraftment within the peritoneal cavity and the bone marrow microenvironment. How Rac1 specifically influences ovarian tumor cells within these two separate environments remains to be explored.

Ovarian cancer metastasis has long been studied in the context of the peritoneal compartment where the bulk of the tumor grows. However, as we improve our systemic and palliative therapy for ovarian cancer patients, an increasing occurrence of unusual distant metastases is being reported. Despite compelling human findings, the overall significance of the bone niche with respect to ovarian cancer prognosis remains ill-defined and suggests that a shift in research focus to understudied metastatic sites such as the bone will be critical to improving patient outcomes. Moreover, the bone marrow dissemination of ovarian cancer cells has been largely overlooked as a potential mechanism for relapse, where the persistence of tumor cells in the protected bone niche could contribute to disease recurrence. Therefore, future studies should be directed at identifying factors that enable tumor cells to be harbored in specialized niche sites that include the bone. By targeting bone marrow-resident tumor cells, we may uncover mechanistic strategies to eradicate distant tumor cell reservoirs that contribute to ovarian cancer relapse and poor overall patient survival.

7. Conclusions

Ovarian cancer remains a leading cause of death in women resulting from gynecologic malignancy principally due to recurrent, drug resistant disease, and limited options for targeted therapies. Greater understanding of signaling proteins that mediate tumor microenvironmental drivers of disease and resistance may provide new avenues for therapeutic development. Rac1 is at the nexus of numerous signaling pathways stimulated by the ovarian cancer TME and has broad roles in cancer beyond the well-recognized regulation of actin remodeling, tumor cell adhesion and migration. Rac1-dependent functions in EMT, stem cell phenotypes, angiogenesis and chemoresistance all have high relevance to ovarian cancer. Although more research is needed regarding specific contributions of aberrant Rac1 activity in ovarian cancer and disease dissemination with respect to specialized microenvironments, current knowledge suggests benefits of targeting Rac1, alone or in combination, for disease treatment.

Author Contributions: Conceptualization, L.G.H., J.M.G. and A.W.-N.; Methodology, A.W.-N.; Formal Analysis, H.K.; Investigation, M.R.R.; Data Curation, A.W.-N.; Writing—Original Draft Preparation, L.G.H., J.M.G. and A.W.-N.; Writing—Review & Editing, L.G.H., J.M.G., H.K., M.R.R., A.W.-N.; Visualization, H.K., A.W.-N., M.R.R.; Supervision, L.G.H., J.M.G. and A.W.-N.; Project Administration, L.G.H., J.M.G. and A.W.-N.; Funding Acquisition, L.G.H., J.M.G. and A.W.-N.

Funding: This research was funded by the National Institutes of Health; grant numbers are R21 TR001731, P50 GM085273 and P30 CA118100.

Acknowledgments: Rac1b imaging and analyses were conducted using the University of New Mexico Pathology Dept. & Comprehensive Cancer Center Human Tissue Repository & Tissue Analysis Shared Resource. We thank Fred Schultz, M.A. for Aperio scanning of IHC slides and training in the use of Halo quantitative imaging software (Indica Labs).

Conflicts of Interest: The authors declare no conflict of interest.

References

1. Christie, E.L.; Bowtell, D.D.L. Acquired chemotherapy resistance in ovarian cancer. *Ann. Oncol.* **2017**, *28*, viii13–viii15. [CrossRef] [PubMed]
2. DiSilvestro, P.; Alvarez Secord, A. Maintenance treatment of recurrent ovarian cancer: Is it ready for prime time? *Cancer Treat. Rev.* **2018**, *69*, 53–65. [CrossRef] [PubMed]
3. Kroeger, P.T., Jr.; Drapkin, R. Pathogenesis and heterogeneity of ovarian cancer. *Curr. Opin. Obstet. Gynecol.* **2017**, *29*, 26–34. [CrossRef] [PubMed]
4. Previs, R.A.; Sood, A.K.; Mills, G.B.; Westin, S.N. The rise of genomic profiling in ovarian cancer. *Expert Rev. Mol. Diagn.* **2016**, *16*, 1337–1351. [CrossRef] [PubMed]
5. Kim, S.; Kim, B.; Song, Y.S. Ascites modulates cancer cell behavior, contributing to tumor heterogeneity in ovarian cancer. *Cancer Sci.* **2016**, *107*, 1173–1178. [CrossRef] [PubMed]
6. Thibault, B.; Castells, M.; Delord, J.P.; Couderc, B. Ovarian cancer microenvironment: Implications for cancer dissemination and chemoresistance acquisition. *Cancer Metastasis Rev.* **2014**, *33*, 17–39. [CrossRef] [PubMed]
7. Ghoneum, A.; Afify, H.; Salih, Z.; Kelly, M.; Said, N. Role of tumor microenvironment in ovarian cancer pathobiology. *Oncotarget* **2018**, *9*, 22832–22849. [CrossRef] [PubMed]
8. Ghoneum, A.; Afify, H.; Salih, Z.; Kelly, M.; Said, N. Role of tumor microenvironment in the pathobiology of ovarian cancer: Insights and therapeutic opportunities. *Cancer Med.* **2018**. [CrossRef] [PubMed]
9. Weidle, U.H.; Birzele, F.; Kollmorgen, G.; Rueger, R. Mechanisms and Targets Involved in Dissemination of Ovarian Cancer. *Cancer Genom. Proteom.* **2016**, *13*, 407–423. [CrossRef]
10. Worzfeld, T.; Pogge von Strandmann, E.; Huber, M.; Adhikary, T.; Wagner, U.; Reinartz, S.; Muller, R. The Unique Molecular and Cellular Microenvironment of Ovarian Cancer. *Front. Oncol.* **2017**, *7*, 24. [CrossRef] [PubMed]
11. Steinkamp, M.P.; Winner, K.K.; Davies, S.; Muller, C.; Zhang, Y.; Hoffman, R.M.; Shirinifard, A.; Moses, M.; Jiang, Y.; Wilson, B.S. Ovarian tumor attachment, invasion, and vascularization reflect unique microenvironments in the peritoneum: Insights from xenograft and mathematical models. *Front. Oncol* **2013**, *3*, 97. [CrossRef] [PubMed]
12. Giridhar, P.V.; Funk, H.M.; Gallo, C.A.; Porollo, A.; Mercer, C.A.; Plas, D.R.; Drew, A.F. Interleukin-6 receptor enhances early colonization of the murine omentum by upregulation of a mannose family receptor, LY75, in ovarian tumor cells. *Clin. Exp. Metastasis* **2011**, *28*, 887–897. [CrossRef] [PubMed]
13. Kenny, H.A.; Chiang, C.Y.; White, E.A.; Schryver, E.M.; Habis, M.; Romero, I.L.; Ladanyi, A.; Penicka, C.V.; George, J.; Matlin, K.; et al. Mesothelial cells promote early ovarian cancer metastasis through fibronectin secretion. *J. Clin. Investig.* **2014**, *124*, 4614–4628. [CrossRef] [PubMed]
14. Nieman, K.M.; Kenny, H.A.; Penicka, C.V.; Ladanyi, A.; Buell-Gutbrod, R.; Zillhardt, M.R.; Romero, I.L.; Carey, M.S.; Mills, G.B.; Hotamisligil, G.S.; et al. Adipocytes promote ovarian cancer metastasis and provide energy for rapid tumor growth. *Nat. Med.* **2011**, *17*, 1498–1503. [CrossRef] [PubMed]

15. McGrail, D.J.; Kieu, Q.M.; Dawson, M.R. The malignancy of metastatic ovarian cancer cells is increased on soft matrices through a mechanosensitive Rho-ROCK pathway. *J. Cell Sci.* **2014**, *127*, 2621–2626. [CrossRef] [PubMed]
16. Lungchukiet, P.; Sun, Y.; Kasiappan, R.; Quarni, W.; Nicosia, S.V.; Zhang, X.; Bai, W. Suppression of epithelial ovarian cancer invasion into the omentum by 1alpha,25-dihydroxyvitamin D3 and its receptor. *J. Steroid Biochem. Mol. Biol.* **2015**, *148*, 138–147. [CrossRef] [PubMed]
17. Khan, S.M.; Funk, H.M.; Thiolloy, S.; Lotan, T.L.; Hickson, J.; Prins, G.S.; Drew, A.F.; Rinker-Schaeffer, C.W. In vitro metastatic colonization of human ovarian cancer cells to the omentum. *Clin. Exp. Metastasis* **2010**, *27*, 185–196. [CrossRef] [PubMed]
18. Pradeep, S.; Kim, S.W.; Wu, S.Y.; Nishimura, M.; Chaluvally-Raghavan, P.; Miyake, T.; Pecot, C.V.; Kim, S.J.; Choi, H.J.; Bischoff, F.Z.; et al. Hematogenous metastasis of ovarian cancer: Rethinking mode of spread. *Cancer Cell* **2014**, *26*, 77–91. [CrossRef] [PubMed]
19. Cerami, E.; Gao, J.; Dogrusoz, U.; Gross, B.E.; Sumer, S.O.; Aksoy, B.A.; Jacobsen, A.; Byrne, C.J.; Heuer, M.L.; Larsson, E.; et al. The cBio cancer genomics portal: An open platform for exploring multidimensional cancer genomics data. *Cancer Discov.* **2012**, *2*, 401–404. [CrossRef] [PubMed]
20. Cook, D.R.; Rossman, K.L.; Der, C.J. Rho guanine nucleotide exchange factors: Regulators of Rho GTPase activity in development and disease. *Oncogene* **2014**, *33*, 4021–4035. [CrossRef] [PubMed]
21. Haga, R.B.; Ridley, A.J. Rho GTPases: Regulation and roles in cancer cell biology. *Small GTPases* **2016**, *7*, 207–221. [CrossRef] [PubMed]
22. Bid, H.K.; Roberts, R.D.; Manchanda, P.K.; Houghton, P.J. RAC1: An emerging therapeutic option for targeting cancer angiogenesis and metastasis. *Mol. Cancer Ther.* **2013**, *12*, 1925–1934. [CrossRef] [PubMed]
23. Cardama, G.A.; Gonzalez, N.; Maggio, J.; Menna, P.L.; Gomez, D.E. Rho GTPases as therapeutic targets in cancer (Review). *Int. J. Oncol.* **2017**, *51*, 1025–1034. [CrossRef] [PubMed]
24. Hodge, R.G.; Ridley, A.J. Regulating Rho GTPases and their regulators. *Nat. Rev. Mol. Cell Biol.* **2016**, *17*, 496–510. [CrossRef] [PubMed]
25. Jansen, S.; Gosens, R.; Wieland, T.; Schmidt, M. Paving the Rho in cancer metastasis: Rho GTPases and beyond. *Pharmacol. Ther.* **2018**, *183*, 1–21. [CrossRef] [PubMed]
26. Lawson, C.D.; Ridley, A.J. Rho GTPase signaling complexes in cell migration and invasion. *J. Cell Biol.* **2018**, *217*, 447–457. [CrossRef] [PubMed]
27. Porter, A.P.; Papaioannou, A.; Malliri, A. Deregulation of Rho GTPases in cancer. *Small GTPases* **2016**, *7*, 123–138. [CrossRef] [PubMed]
28. Zandvakili, I.; Lin, Y.; Morris, J.C.; Zheng, Y. Rho GTPases: Anti- or pro-neoplastic targets? *Oncogene* **2017**, *36*, 3213–3222. [CrossRef] [PubMed]
29. Troeger, A.; Williams, D.A. Hematopoietic-specific Rho GTPases Rac2 and RhoH and human blood disorders. *Exp. Cell Res.* **2013**, *319*, 2375–2383. [CrossRef] [PubMed]
30. Kazanietz, M.G.; Caloca, M.J. The Rac GTPase in Cancer: From Old Concepts to New Paradigms. *Cancer Res.* **2017**, *77*, 5445–5451. [CrossRef] [PubMed]
31. Maldonado, M.D.M.; Dharmawardhane, S. Targeting Rac and Cdc42 GTPases in Cancer. *Cancer Res.* **2018**, *78*, 3101–3111. [CrossRef] [PubMed]
32. Sadok, A.; Marshall, C.J. Rho GTPases: Masters of cell migration. *Small GTPases* **2014**, *5*, e29710. [CrossRef] [PubMed]
33. Orgaz, J.L.; Herraiz, C.; Sanz-Moreno, V. Rho GTPases modulate malignant transformation of tumor cells. *Small GTPases* **2014**, *5*, e29019. [CrossRef] [PubMed]
34. Akunuru, S.; Palumbo, J.; Zhai, Q.J.; Zheng, Y. Rac1 targeting suppresses human non-small cell lung adenocarcinoma cancer stem cell activity. *PLoS ONE* **2011**, *6*, e16951. [CrossRef] [PubMed]
35. Wang, J.Y.; Yu, P.; Chen, S.; Xing, H.; Chen, Y.; Wang, M.; Tang, K.; Tian, Z.; Rao, Q.; Wang, J. Activation of Rac1 GTPase promotes leukemia cell chemotherapy resistance, quiescence and niche interaction. *Mol. Oncol.* **2013**, *7*, 907–916. [CrossRef] [PubMed]
36. Yoon, C.H.; Hyun, K.H.; Kim, R.K.; Lee, H.; Lim, E.J.; Chung, H.Y.; An, S.; Park, M.J.; Suh, Y.; Kim, M.J.; et al. The small GTPase Rac1 is involved in the maintenance of stemness and malignancies in glioma stem-like cells. *FEBS Lett.* **2011**, *585*, 2331–2338. [CrossRef] [PubMed]

37. Cardama, G.A.; Alonso, D.F.; Gonzalez, N.; Maggio, J.; Gomez, D.E.; Rolfo, C.; Menna, P.L. Relevance of small GTPase Rac1 pathway in drug and radio-resistance mechanisms: Opportunities in cancer therapeutics. *Crit. Rev. Oncol. Hematol.* **2018**, *124*, 29–36. [CrossRef] [PubMed]
38. Lou, S.; Wang, P.; Yang, J.; Ma, J.; Liu, C.; Zhou, M. Prognostic and Clinicopathological Value of Rac1 in Cancer Survival: Evidence from a Meta-Analysis. *J. Cancer* **2018**, *9*, 2571–2579. [CrossRef] [PubMed]
39. Nayak, R.C.; Chang, K.H.; Vaitinadin, N.S.; Cancelas, J.A. Rho GTPases control specific cytoskeleton-dependent functions of hematopoietic stem cells. *Immunol. Rev.* **2013**, *256*, 255–268. [CrossRef] [PubMed]
40. Reimer, D.; Boesch, M.; Wolf, D.; Marth, C.; Sopper, S.; Hatina, J.; Altevogt, P.; Parson, W.; Hackl, H.; Zeimet, A.G. Truncated isoform Vav3.1 is highly expressed in ovarian cancer stem cells and clinically relevant in predicting prognosis and platinum-response. *Int. J. Cancer* **2018**, *142*, 1640–1651. [CrossRef] [PubMed]
41. Klymenko, Y.; Kim, O.; Stack, M.S. Complex Determinants of Epithelial: Mesenchymal Phenotypic Plasticity in Ovarian Cancer. *Cancers* **2017**, *9*, 104. [CrossRef] [PubMed]
42. Deng, J.; Wang, L.; Chen, H.; Hao, J.; Ni, J.; Chang, L.; Duan, W.; Graham, P.; Li, Y. Targeting epithelial-mesenchymal transition and cancer stem cells for chemoresistant ovarian cancer. *Oncotarget* **2016**, *7*, 55771–55788. [CrossRef] [PubMed]
43. Barbolina, M.V.; Moss, N.M.; Westfall, S.D.; Liu, Y.; Burkhalter, R.J.; Marga, F.; Forgacs, G.; Hudson, L.G.; Stack, M.S. Microenvironmental regulation of ovarian cancer metastasis. *Cancer Treat. Res.* **2009**, *149*, 319–334. [PubMed]
44. da Silva-Diz, V.; Lorenzo-Sanz, L.; Bernat-Peguera, A.; Lopez-Cerda, M.; Munoz, P. Cancer cell plasticity: Impact on tumor progression and therapy response. *Semin. Cancer Biol.* **2018**. [CrossRef] [PubMed]
45. Zhang, Y.; Weinberg, R.A. Epithelial-to-mesenchymal transition in cancer: Complexity and opportunities. *Front. Med.* **2018**. [CrossRef] [PubMed]
46. Fang, D.; Chen, H.; Zhu, J.Y.; Wang, W.; Teng, Y.; Ding, H.F.; Jing, Q.; Su, S.B.; Huang, S. Epithelial-mesenchymal transition of ovarian cancer cells is sustained by Rac1 through simultaneous activation of MEK1/2 and Src signaling pathways. *Oncogene* **2017**, *36*, 1546–1558. [CrossRef] [PubMed]
47. Leng, R.; Liao, G.; Wang, H.; Kuang, J.; Tang, L. Rac1 expression in epithelial ovarian cancer: Effect on cell EMT and clinical outcome. *Med. Oncol.* **2015**, *32*, 329. [CrossRef] [PubMed]
48. Guo, Y.; Kenney, S.R.; Muller, C.Y.; Adams, S.; Rutledge, T.; Romero, E.; Murray-Krezan, C.; Prekeris, R.; Sklar, L.A.; Hudson, L.G.; et al. R-Ketorolac Targets Cdc42 and Rac1 and Alters Ovarian Cancer Cell Behaviors Critical for Invasion and Metastasis. *Mol. Cancer Ther.* **2015**, *14*, 2215–2227. [CrossRef] [PubMed]
49. Zhou, G.; Peng, F.; Zhong, Y.; Chen, Y.; Tang, M.; Li, D. Rhein suppresses matrix metalloproteinase production by regulating the Rac1/ROS/MAPK/AP-1 pathway in human ovarian carcinoma cells. *Int. J. Oncol.* **2017**, *50*, 933–941. [CrossRef] [PubMed]
50. Romero-Laorden, N.; Olmos, D.; Fehm, T.; Garcia-Donas, J.; Diaz-Padilla, I. Circulating and disseminated tumor cells in ovarian cancer: A systematic review. *Gynecol. Oncol.* **2014**, *133*, 632–639. [CrossRef] [PubMed]
51. Fehm, T.; Banys, M.; Rack, B.; Janni, W.; Marth, C.; Blassl, C.; Hartkopf, A.; Trope, C.; Kimmig, R.; Krawczyk, N.; et al. Pooled analysis of the prognostic relevance of disseminated tumor cells in the bone marrow of patients with ovarian cancer. *Int. J. Gynecol. Cancer* **2013**, *23*, 839–845. [CrossRef] [PubMed]
52. Fehm, T.; Becker, S.; Bachmann, C.; Beck, V.; Gebauer, G.; Banys, M.; Wallwiener, D.; Solomayer, E.F. Detection of disseminated tumor cells in patients with gynecological cancers. *Gynecol. Oncol.* **2006**, *103*, 942–947. [CrossRef] [PubMed]
53. Gasparri, M.L.; Savone, D.; Besharat, R.A.; Farooqi, A.A.; Bellati, F.; Ruscito, I.; Panici, P.B.; Papadia, A. Circulating tumor cells as trigger to hematogenous spreads and potential biomarkers to predict the prognosis in ovarian cancer. *Tumour Biol. J. Int. Soc. Oncodev. Biol. Med.* **2016**, *37*, 71–75. [CrossRef] [PubMed]
54. Yeung, T.L.; Leung, C.S.; Yip, K.P.; Au Yeung, C.L.; Wong, S.T.; Mok, S.C. Cellular and molecular processes in ovarian cancer metastasis. A Review in the Theme: Cell and Molecular Processes in Cancer Metastasis. *Am. J. Physiol. Cell Physiol.* **2015**, *309*, C444–C456. [CrossRef] [PubMed]
55. Zhou, Y.; Bian, B.; Yuan, X.; Xie, G.; Ma, Y.; Shen, L. Prognostic Value of Circulating Tumor Cells in Ovarian Cancer: A Meta-Analysis. *PLoS ONE* **2015**, *10*, e0130873. [CrossRef] [PubMed]

56. Blassl, C.; Kuhlmann, J.D.; Webers, A.; Wimberger, P.; Fehm, T.; Neubauer, H. Gene expression profiling of single circulating tumor cells in ovarian cancer—Establishment of a multi-marker gene panel. *Mol. Oncol.* **2016**, *10*, 1030–1042. [CrossRef] [PubMed]
57. Nwani, N.G.; Sima, L.E.; Nieves-Neira, W.; Matei, D. Targeting the Microenvironment in High Grade Serous Ovarian Cancer. *Cancers* **2018**, *10*, 266. [CrossRef] [PubMed]
58. Monk, B.J.; Minion, L.E.; Coleman, R.L. Anti-angiogenic agents in ovarian cancer: Past, present, and future. *Ann. Oncol.* **2016**, *27*, i33–i39. [CrossRef] [PubMed]
59. Hansen, J.M.; Coleman, R.L.; Sood, A.K. Targeting the tumour microenvironment in ovarian cancer. *Eur. J. Cancer* **2016**, *56*, 131–143. [CrossRef] [PubMed]
60. Nohata, N.; Uchida, Y.; Stratman, A.N.; Adams, R.H.; Zheng, Y.; Weinstein, B.M.; Mukouyama, Y.S.; Gutkind, J.S. Temporal-specific roles of Rac1 during vascular development and retinal angiogenesis. *Dev. Biol.* **2016**, *411*, 183–194. [CrossRef] [PubMed]
61. Galan Moya, E.M.; Le Guelte, A.; Gavard, J. PAKing up to the endothelium. *Cell. Signal.* **2009**, *21*, 1727–1737. [CrossRef] [PubMed]
62. Abraham, S.; Scarcia, M.; Bagshaw, R.D.; McMahon, K.; Grant, G.; Harvey, T.; Yeo, M.; Esteves, F.O.; Thygesen, H.H.; Jones, P.F.; et al. A Rac/Cdc42 exchange factor complex promotes formation of lateral filopodia and blood vessel lumen morphogenesis. *Nat. Commun.* **2015**, *6*, 7286. [CrossRef] [PubMed]
63. Fryer, B.H.; Field, J. Rho, Rac, Pak and angiogenesis: Old roles and newly identified responsibilities in endothelial cells. *Cancer Lett.* **2005**, *229*, 13–23. [CrossRef] [PubMed]
64. Soga, N.; Connolly, J.O.; Chellaiah, M.; Kawamura, J.; Hruska, K.A. Rac regulates vascular endothelial growth factor stimulated motility. *Cell Commun. Adhes.* **2001**, *8*, 1–13. [CrossRef] [PubMed]
65. Xue, Y.; Bi, F.; Zhang, X.; Pan, Y.; Liu, N.; Zheng, Y.; Fan, D. Inhibition of endothelial cell proliferation by targeting Rac1 GTPase with small interference RNA in tumor cells. *Biochem. Biophys. Res. Commun.* **2004**, *320*, 1309–1315. [CrossRef] [PubMed]
66. Humphries-Bickley, T.; Castillo-Pichardo, L.; Hernandez-O'Farrill, E.; Borrero-Garcia, L.D.; Forestier-Roman, I.; Gerena, Y.; Blanco, M.; Rivera-Robles, M.J.; Rodriguez-Medina, J.R.; Cubano, L.A.; et al. Characterization of a Dual Rac/Cdc42 Inhibitor MBQ-167 in Metastatic Cancer. *Mol. Cancer Ther.* **2017**, *16*, 805–818. [CrossRef] [PubMed]
67. Castillo-Pichardo, L.; Humphries-Bickley, T.; De La Parra, C.; Forestier-Roman, I.; Martinez-Ferrer, M.; Hernandez, E.; Vlaar, C.; Ferrer-Acosta, Y.; Washington, A.V.; Cubano, L.A.; et al. The Rac Inhibitor EHop-016 Inhibits Mammary Tumor Growth and Metastasis in a Nude Mouse Model. *Transl. Oncol.* **2014**, *7*, 546–555. [CrossRef] [PubMed]
68. Nagahashi, M.; Ramachandran, S.; Rashid, O.M.; Takabe, K. Lymphangiogenesis: A new player in cancer progression. *World J. Gastroenterol.* **2010**, *16*, 4003–4012. [CrossRef] [PubMed]
69. Schulz, M.M.; Reisen, F.; Zgraggen, S.; Fischer, S.; Yuen, D.; Kang, G.J.; Chen, L.; Schneider, G.; Detmar, M. Phenotype-based high-content chemical library screening identifies statins as inhibitors of in vivo lymphangiogenesis. *Proc. Natl. Acad. Sci. USA* **2012**, *109*, E2665–E2674. [CrossRef] [PubMed]
70. Sorensen, E.W.; Gerber, S.A.; Sedlacek, A.L.; Rybalko, V.Y.; Chan, W.M.; Lord, E.M. Omental immune aggregates and tumor metastasis within the peritoneal cavity. *Immunol. Res.* **2009**, *45*, 185–194. [CrossRef] [PubMed]
71. Kuerti, S.; Oliveira-Ferrer, L.; Milde-Langosch, K.; Schmalfeldt, B.; Legler, K.; Woelber, L.; Prieske, K.; Mahner, S.; Trillsch, F. VEGF-C expression attributes the risk for lymphatic metastases to ovarian cancer patients. *Oncotarget* **2017**, *8*, 43218–43227. [CrossRef] [PubMed]
72. Hisamatsu, T.; Mabuchi, S.; Sasano, T.; Kuroda, H.; Takahashi, R.; Matsumoto, Y.; Kawano, M.; Kozasa, K.; Takahashi, K.; Sawada, K.; et al. The significance of lymphatic space invasion and its association with vascular endothelial growth factor-C expression in ovarian cancer. *Clin. Exp. Metastasis* **2015**, *32*, 789–798. [CrossRef] [PubMed]
73. Wang, Y.; Nakayama, M.; Pitulescu, M.E.; Schmidt, T.S.; Bochenek, M.L.; Sakakibara, A.; Adams, S.; Davy, A.; Deutsch, U.; Luthi, U.; et al. Ephrin-B2 controls VEGF-induced angiogenesis and lymphangiogenesis. *Nature* **2010**, *465*, 483–486. [CrossRef] [PubMed]
74. Liu, S.; Li, Y.; Qi, W.; Zhao, Y.; Huang, A.; Sheng, W.; Lei, B.; Lin, P.; Zhu, H.; Li, W.; et al. Expression of Tiam1 predicts lymph node metastasis and poor survival of lung adenocarcinoma patients. *Diagn. Pathol.* **2014**, *9*, 69. [CrossRef] [PubMed]

75. Zhong, D.; Li, Y.; Peng, Q.; Zhou, J.; Zhou, Q.; Zhang, R.; Liang, H. Expression of Tiam1 and VEGF-C correlates with lymphangiogenesis in human colorectal carcinoma. *Cancer Biol. Ther.* **2009**, *8*, 689–695. [CrossRef] [PubMed]
76. Lim, J.J.; Yang, K.; Taylor-Harding, B.; Wiedemeyer, W.R.; Buckanovich, R.J. VEGFR3 inhibition chemosensitizes ovarian cancer stemlike cells through down-regulation of BRCA1 and BRCA2. *Neoplasia* **2014**, *16*, 343–353. [CrossRef] [PubMed]
77. Hofbauer, S.W.; Krenn, P.W.; Ganghammer, S.; Asslaber, D.; Pichler, U.; Oberascher, K.; Henschler, R.; Wallner, M.; Kerschbaum, H.; Greil, R.; et al. Tiam1/Rac1 signals contribute to the proliferation and chemoresistance, but not motility, of chronic lymphocytic leukemia cells. *Blood* **2014**, *123*, 2181–2188. [CrossRef] [PubMed]
78. Ikram, M.; Lim, Y.; Baek, S.Y.; Jin, S.; Jeong, Y.H.; Kwak, J.Y.; Yoon, S. Co-targeting of Tiam1/Rac1 and Notch ameliorates chemoresistance against doxorubicin in a biomimetic 3D lymphoma model. *Oncotarget* **2018**, *9*, 2058–2075. [CrossRef] [PubMed]
79. Steg, A.D.; Bevis, K.S.; Katre, A.A.; Ziebarth, A.; Dobbin, Z.C.; Alvarez, R.D.; Zhang, K.; Conner, M.; Landen, C.N. Stem cell pathways contribute to clinical chemoresistance in ovarian cancer. *Clin. Cancer Res.* **2012**, *18*, 869–881. [CrossRef] [PubMed]
80. Huang, R.Y.; Chung, V.Y.; Thiery, J.P. Targeting pathways contributing to epithelial-mesenchymal transition (EMT) in epithelial ovarian cancer. *Curr. Drug Targets* **2012**, *13*, 1649–1653. [CrossRef] [PubMed]
81. Ahmed, N.; Abubaker, K.; Findlay, J.; Quinn, M. Epithelial mesenchymal transition and cancer stem cell-like phenotypes facilitate chemoresistance in recurrent ovarian cancer. *Curr. Cancer Drug Targets* **2010**, *10*, 268–278. [CrossRef] [PubMed]
82. Bustelo, X.R.; Ojeda, V.; Barreira, M.; Sauzeau, V.; Castro-Castro, A. Rac-ing to the plasma membrane: The long and complex work commute of Rac1 during cell signaling. *Small GTPases* **2012**, *3*, 60–66. [CrossRef] [PubMed]
83. Vazquez-Prado, J.; Bracho-Valdes, I.; Cervantes-Villagrana, R.D.; Reyes-Cruz, G. Gbetagamma Pathways in Cell Polarity and Migration Linked to Oncogenic GPCR Signaling: Potential Relevance in Tumor Microenvironment. *Mol. Pharmacol.* **2016**, *90*, 573–586. [CrossRef] [PubMed]
84. Xu, Y. Lysophospholipid Signaling in the Epithelial Ovarian Cancer Tumor Microenvironment. *Cancers* **2018**, *10*, 227. [CrossRef] [PubMed]
85. Devine, K.M.; Smicun, Y.; Hope, J.M.; Fishman, D.A. S1P induced changes in epithelial ovarian cancer proteolysis, invasion, and attachment are mediated by Gi and Rac. *Gynecol. Oncol.* **2008**, *110*, 237–245. [CrossRef] [PubMed]
86. Chen, H.; Wu, X.; Pan, Z.K.; Huang, S. Integrity of SOS1/EPS8/ABI1 tri-complex determines ovarian cancer metastasis. *Cancer Res.* **2010**, *70*, 9979–9990. [CrossRef] [PubMed]
87. Ward, J.D.; Dhanasekaran, D.N. LPA Stimulates the Phosphorylation of p130Cas via Galphai2 in Ovarian Cancer Cells. *Genes Cancer* **2012**, *3*, 578–591. [CrossRef] [PubMed]
88. Ward, J.D.; Ha, J.H.; Jayaraman, M.; Dhanasekaran, D.N. LPA-mediated migration of ovarian cancer cells involves translocalization of Galphai2 to invadopodia and association with Src and beta-pix. *Cancer Lett.* **2015**, *356*, 382–391. [CrossRef] [PubMed]
89. Fritz, R.D.; Pertz, O. The dynamics of spatio-temporal Rho GTPase signaling: Formation of signaling patterns. *F1000Research* **2016**, *5*. [CrossRef] [PubMed]
90. Yang, C.; Liu, Y.; Lemmon, M.A.; Kazanietz, M.G. Essential role for Rac in heregulin beta1 mitogenic signaling: A mechanism that involves epidermal growth factor receptor and is independent of ErbB4. *Mol. Cell. Biol.* **2006**, *26*, 831–842. [CrossRef] [PubMed]
91. Zhu, G.; Fan, Z.; Ding, M.; Zhang, H.; Mu, L.; Ding, Y.; Zhang, Y.; Jia, B.; Chen, L.; Chang, Z.; et al. An EGFR/PI3K/AKT axis promotes accumulation of the Rac1-GEF Tiam1 that is critical in EGFR-driven tumorigenesis. *Oncogene* **2015**, *34*, 5971–5982. [CrossRef] [PubMed]
92. Davies, S.; Holmes, A.; Lomo, L.; Steinkamp, M.P.; Kang, H.; Muller, C.Y.; Wilson, B.S. High incidence of ErbB3, ErbB4, and MET expression in ovarian cancer. *Int. J. Gynecol. Pathol.* **2014**, *33*, 402–410. [CrossRef] [PubMed]
93. Lopez-Haber, C.; Barrio-Real, L.; Casado-Medrano, V.; Kazanietz, M.G. Heregulin/ErbB3 Signaling Enhances CXCR4-Driven Rac1 Activation and Breast Cancer Cell Motility via Hypoxia-Inducible Factor 1alpha. *Mol. Cell. Biol.* **2016**, *36*, 2011–2026. [CrossRef] [PubMed]

94. Li, W.; Xiong, X.; Abdalla, A.; Alejo, S.; Zhu, L.; Lu, F.; Sun, H. HGF-induced formation of the MET-AXL-ELMO2-DOCK180 complex promotes RAC1 activation, receptor clustering, and cancer cell migration and invasion. *J. Biol. Chem.* **2018**. [CrossRef] [PubMed]
95. Oprea, T.I.; Sklar, L.A.; Agola, J.O.; Guo, Y.; Silberberg, M.; Roxby, J.; Vestling, A.; Romero, E.; Surviladze, Z.; Waller, A.; et al. Novel Activities of Select NSAID R-Enantiomers against Rac1 and Cdc42 GTPases. *PLoS ONE* **2015**, *10*, e0142812. [CrossRef] [PubMed]
96. Tan, W.; Palmby, T.R.; Gavard, J.; Amornphimoltham, P.; Zheng, Y.; Gutkind, J.S. An essential role for Rac1 in endothelial cell function and vascular development. *FASEB J.* **2008**, *22*, 1829–1838. [CrossRef] [PubMed]
97. Hoang, M.V.; Nagy, J.A.; Senger, D.R. Active Rac1 improves pathologic VEGF neovessel architecture and reduces vascular leak: Mechanistic similarities with angiopoietin-1. *Blood* **2011**, *117*, 1751–1760. [CrossRef] [PubMed]
98. Peinado, H.; Zhang, H.; Matei, I.R.; Costa-Silva, B.; Hoshino, A.; Rodrigues, G.; Psaila, B.; Kaplan, R.N.; Bromberg, J.F.; Kang, Y.; et al. Pre-metastatic niches: Organ-specific homes for metastases. *Nat. Rev. Cancer* **2017**, *17*, 302–317. [CrossRef] [PubMed]
99. Marei, H.; Carpy, A.; Woroniuk, A.; Vennin, C.; White, G.; Timpson, P.; Macek, B.; Malliri, A. Differential Rac1 signalling by guanine nucleotide exchange factors implicates FLII in regulating Rac1-driven cell migration. *Nat. Commun.* **2016**, *7*, 10664. [CrossRef] [PubMed]
100. Xu, S.; Yang, Y.; Dong, L.; Qiu, W.; Yang, L.; Wang, X.; Liu, L. Construction and characteristics of an E-cadherin-related three-dimensional suspension growth model of ovarian cancer. *Sci. Rep.* **2014**, *4*, 5646. [CrossRef] [PubMed]
101. Shishido, A.; Mori, S.; Yokoyama, Y.; Hamada, Y.; Minami, K.; Qian, Y.; Wang, J.; Hirose, H.; Wu, X.; Kawaguchi, N.; et al. Mesothelial cells facilitate cancer stemlike properties in spheroids of ovarian cancer cells. *Oncol. Rep.* **2018**, *40*, 2105–2114. [PubMed]
102. Ween, M.P.; Oehler, M.K.; Ricciardelli, C. Role of versican, hyaluronan and CD44 in ovarian cancer metastasis. *Int. J. Mol. Sci.* **2011**, *12*, 1009–1029. [CrossRef] [PubMed]
103. Chen, C.; Zhao, S.; Karnad, A.; Freeman, J.W. The biology and role of CD44 in cancer progression: Therapeutic implications. *J. Hematol. Oncol.* **2018**, *11*, 64. [CrossRef] [PubMed]
104. Cai, J.; Tang, H.; Xu, L.; Wang, X.; Yang, C.; Ruan, S.; Guo, J.; Hu, S.; Wang, Z. Fibroblasts in omentum activated by tumor cells promote ovarian cancer growth, adhesion and invasiveness. *Carcinogenesis* **2012**, *33*, 20–29. [CrossRef] [PubMed]
105. Curtis, M.; Kenny, H.A.; Ashcroft, B.; Mukherjee, A.; Johnson, A.; Zhang, Y.; Helou, Y.; Batlle, R.; Liu, X.; Gutierrez, N.; et al. Fibroblasts Mobilize Tumor Cell Glycogen to Promote Proliferation and Metastasis. *Cell MeTable* **2018**. [CrossRef] [PubMed]
106. Raptis, L.; Arulanandam, R.; Geletu, M.; Turkson, J. The R(h)oads to Stat3: Stat3 activation by the Rho GTPases. *Exp. Cell Res.* **2011**, *317*, 1787–1795. [CrossRef] [PubMed]
107. Savant, S.S.; Sriramkumar, S.; O'Hagan, H.M. The Role of Inflammation and Inflammatory Mediators in the Development, Progression, Metastasis, and Chemoresistance of Epithelial Ovarian Cancer. *Cancers* **2018**, *10*, 251. [CrossRef] [PubMed]
108. Lawson, C.D.; Burridge, K. The on-off relationship of Rho and Rac during integrin-mediated adhesion and cell migration. *Small GTPases* **2014**, *5*, e27958. [CrossRef] [PubMed]
109. Pearce, O.M.T.; Delaine-Smith, R.M.; Maniati, E.; Nichols, S.; Wang, J.; Bohm, S.; Rajeeve, V.; Ullah, D.; Chakravarty, P.; Jones, R.R.; et al. Deconstruction of a Metastatic Tumor Microenvironment Reveals a Common Matrix Response in Human Cancers. *Cancer Discov.* **2018**, *8*, 304–319. [CrossRef] [PubMed]
110. Montfort, A.; Pearce, O.; Maniati, E.; Vincent, B.G.; Bixby, L.; Bohm, S.; Dowe, T.; Wilkes, E.H.; Chakravarty, P.; Thompson, R.; et al. A Strong B-cell Response Is Part of the Immune Landscape in Human High-Grade Serous Ovarian Metastases. *Clin. Cancer Res.* **2017**, *23*, 250–262. [CrossRef] [PubMed]
111. Flies, D.B.; Higuchi, T.; Harris, J.C.; Jha, V.; Gimotty, P.A.; Adams, S.F. Immune checkpoint blockade reveals the stimulatory capacity of tumor-associated CD103(+) dendritic cells in late-stage ovarian cancer. *Oncoimmunology* **2016**, *5*, e1185583. [CrossRef] [PubMed]
112. Higuchi, T.; Flies, D.B.; Marjon, N.A.; Mantia-Smaldone, G.; Ronner, L.; Gimotty, P.A.; Adams, S.F. CTLA-4 Blockade Synergizes Therapeutically with PARP Inhibition in BRCA1-Deficient Ovarian Cancer. *Cancer Immunol. Res.* **2015**, *3*, 1257–1268. [CrossRef] [PubMed]

113. Guo, Y.; Kenney, S.R.; Cook, L.; Adams, S.F.; Rutledge, T.; Romero, E.; Oprea, T.I.; Sklar, L.A.; Bedrick, E.; Wiggins, C.L.; et al. A Novel Pharmacologic Activity of Ketorolac for Therapeutic Benefit in Ovarian Cancer Patients. *Clin. Cancer Res.* **2015**, *21*, 5064–5072. [CrossRef] [PubMed]
114. Hodis, E.; Watson, I.R.; Kryukov, G.V.; Arold, S.T.; Imielinski, M.; Theurillat, J.P.; Nickerson, E.; Auclair, D.; Li, L.; Place, C.; et al. A landscape of driver mutations in melanoma. *Cell* **2012**, *150*, 251–263. [CrossRef] [PubMed]
115. Davis, M.J.; Ha, B.H.; Holman, E.C.; Halaban, R.; Schlessinger, J.; Boggon, T.J. RAC1P29S is a spontaneously activating cancer-associated GTPase. *Proc. Natl. Acad. Sci. USA* **2013**, *110*, 912–917. [CrossRef] [PubMed]
116. Kumar, A.; Rajendran, V.; Sethumadhavan, R.; Purohit, R. Molecular dynamic simulation reveals damaging impact of RAC1 F28L mutation in the switch I region. *PLoS ONE* **2013**, *8*, e77453. [CrossRef] [PubMed]
117. Chang, M.T.; Asthana, S.; Gao, S.P.; Lee, B.H.; Chapman, J.S.; Kandoth, C.; Gao, J.; Socci, N.D.; Solit, D.B.; Olshen, A.B.; et al. Identifying recurrent mutations in cancer reveals widespread lineage diversity and mutational specificity. *Nat. Biotechnol.* **2016**, *34*, 155–163. [CrossRef] [PubMed]
118. Gao, J.; Chang, M.T.; Johnsen, H.C.; Gao, S.P.; Sylvester, B.E.; Sumer, S.O.; Zhang, H.; Solit, D.B.; Taylor, B.S.; Schultz, N.; et al. 3D clusters of somatic mutations in cancer reveal numerous rare mutations as functional targets. *Genome Med.* **2017**, *9*, 4. [CrossRef] [PubMed]
119. Marei, H.; Malliri, A. GEFs: Dual regulation of Rac1 signaling. *Small GTPases* **2017**, *8*, 90–99. [CrossRef] [PubMed]
120. Jordan, P.; Brazao, R.; Boavida, M.G.; Gespach, C.; Chastre, E. Cloning of a novel human Rac1b splice variant with increased expression in colorectal tumors. *Oncogene* **1999**, *18*, 6835–6839. [CrossRef] [PubMed]
121. Schnelzer, A.; Prechtel, D.; Knaus, U.; Dehne, K.; Gerhard, M.; Graeff, H.; Harbeck, N.; Schmitt, M.; Lengyel, E. Rac1 in human breast cancer: Overexpression, mutation analysis, and characterization of a new isoform, Rac1b. *Oncogene* **2000**, *19*, 3013–3020. [CrossRef] [PubMed]
122. Singh, A.; Karnoub, A.E.; Palmby, T.R.; Lengyel, E.; Sondek, J.; Der, C.J. Rac1b, a tumor associated, constitutively active Rac1 splice variant, promotes cellular transformation. *Oncogene* **2004**, *23*, 9369–9380. [CrossRef] [PubMed]
123. Matos, P.; Jordan, P. Expression of Rac1b stimulates NF-kappaB-mediated cell survival and G1/S progression. *Exp. Cell Res.* **2005**, *305*, 292–299. [CrossRef] [PubMed]
124. Radisky, D.C.; Levy, D.D.; Littlepage, L.E.; Liu, H.; Nelson, C.M.; Fata, J.E.; Leake, D.; Godden, E.L.; Albertson, D.G.; Nieto, M.A.; et al. Rac1b and reactive oxygen species mediate MMP-3-induced EMT and genomic instability. *Nature* **2005**, *436*, 123–127. [CrossRef] [PubMed]
125. Orlichenko, L.; Geyer, R.; Yanagisawa, M.; Khauv, D.; Radisky, E.S.; Anastasiadis, P.Z.; Radisky, D.C. The 19-amino acid insertion in the tumor-associated splice isoform Rac1b confers specific binding to p120 catenin. *J. Biol. Chem.* **2010**, *285*, 19153–19161. [CrossRef] [PubMed]
126. Silva, A.L.; Carmo, F.; Bugalho, M.J. RAC1b overexpression in papillary thyroid carcinoma: A role to unravel. *Eur. J. Endocrinol. Eur. Fed. Endocr. Soc.* **2013**, *168*, 795–804. [CrossRef] [PubMed]
127. Zhou, C.; Licciulli, S.; Avila, J.L.; Cho, M.; Troutman, S.; Jiang, P.; Kossenkov, A.V.; Showe, L.C.; Liu, Q.; Vachani, A.; et al. The Rac1 splice form Rac1b promotes K-ras-induced lung tumorigenesis. *Oncogene* **2013**, *32*, 903–909. [CrossRef] [PubMed]
128. Li, G.; Ying, L.; Wang, H.; Wei, S.S.; Chen, J.; Chen, Y.H.; Xu, W.P.; Jie, Q.Q.; Zhou, Q.; Li, Y.G.; et al. Rac1b enhances cell survival through activation of the JNK2/c-JUN/Cyclin-D1 and AKT2/MCL1 pathways. *Oncotarget* **2016**, *7*, 17970–17985. [CrossRef] [PubMed]
129. Fu, X.D. Both sides of the same coin: Rac1 splicing regulating by EGF signaling. *Cell Res.* **2017**, *27*, 455–456. [CrossRef] [PubMed]
130. Wang, F.; Fu, X.; Chen, P.; Wu, P.; Fan, X.; Li, N.; Zhu, H.; Jia, T.T.; Ji, H.; Wang, Z.; et al. SPSB1-mediated HnRNP A1 ubiquitylation regulates alternative splicing and cell migration in EGF signaling. *Cell Res.* **2017**, *27*, 540–558. [CrossRef] [PubMed]
131. Gao, J.; Aksoy, B.A.; Dogrusoz, U.; Dresdner, G.; Gross, B.; Sumer, S.O.; Sun, Y.; Jacobsen, A.; Sinha, R.; Larsson, E.; et al. Integrative analysis of complex cancer genomics and clinical profiles using the cBioPortal. *Sci. Signal.* **2013**, *6*, pl1. [CrossRef] [PubMed]
132. Krauthammer, M.; Kong, Y.; Ha, B.H.; Evans, P.; Bacchiocchi, A.; McCusker, J.P.; Cheng, E.; Davis, M.J.; Goh, G.; Choi, M.; et al. Exome sequencing identifies recurrent somatic RAC1 mutations in melanoma. *Nat. Genet.* **2012**, *44*, 1006–1014. [CrossRef] [PubMed]

133. DeLano, W.L. *PyMOL: An Open-Source Molecular Graphics Tool*; Delano Scientific: San Carlos, CA, USA, 2002.
134. Kawazu, M.; Ueno, T.; Kontani, K.; Ogita, Y.; Ando, M.; Fukumura, K.; Yamato, A.; Soda, M.; Takeuchi, K.; Miki, Y.; et al. Transforming mutations of RAC guanosine triphosphatases in human cancers. *Proc. Natl. Acad. Sci. USA* **2013**, *110*, 3029–3034. [CrossRef] [PubMed]
135. Watson, I.R.; Li, L.; Cabeceiras, P.K.; Mahdavi, M.; Gutschner, T.; Genovese, G.; Wang, G.; Fang, Z.; Tepper, J.M.; Stemke-Hale, K.; et al. The RAC1 P29S hotspot mutation in melanoma confers resistance to pharmacological inhibition of RAF. *Cancer Res.* **2014**, *74*, 4845–4852. [CrossRef] [PubMed]
136. Cheung, H.W.; Cowley, G.S.; Weir, B.A.; Boehm, J.S.; Rusin, S.; Scott, J.A.; East, A.; Ali, L.D.; Lizotte, P.H.; Wong, T.C.; et al. Systematic investigation of genetic vulnerabilities across cancer cell lines reveals lineage-specific dependencies in ovarian cancer. *Proc. Natl. Acad. Sci. USA* **2011**, *108*, 12372–12377. [CrossRef] [PubMed]
137. Cowley, G.S.; Weir, B.A.; Vazquez, F.; Tamayo, P.; Scott, J.A.; Rusin, S.; East-Seletsky, A.; Ali, L.D.; Gerath, W.F.; Pantel, S.E.; et al. Parallel genome-scale loss of function screens in 216 cancer cell lines for the identification of context-specific genetic dependencies. *Sci. Data* **2014**, *1*, 140035. [CrossRef] [PubMed]
138. Cancer Genome Atlas Research Network. Integrated genomic analyses of ovarian carcinoma. *Nature* **2011**, *474*, 609–615. [CrossRef] [PubMed]
139. Vogelstein, B.; Papadopoulos, N.; Velculescu, V.E.; Zhou, S.; Diaz, L.A., Jr.; Kinzler, K.W. Cancer genome landscapes. *Science* **2013**, *339*, 1546–1558. [CrossRef] [PubMed]
140. Zoughlami, Y.; van Stalborgh, A.M.; van Hennik, P.B.; Hordijk, P.L. Nucleophosmin1 is a negative regulator of the small GTPase Rac1. *PLoS ONE* **2013**, *8*, e68477. [CrossRef] [PubMed]
141. Payapilly, A.; Malliri, A. Compartmentalisation of RAC1 signalling. *Curr. Opin. Cell Biol.* **2018**, *54*, 50–56. [CrossRef] [PubMed]
142. Mettouchi, A.; Lemichez, E. Ubiquitylation of active Rac1 by the E3 ubiquitin-ligase HACE1. *Small GTPases* **2012**, *3*, 102–106. [CrossRef] [PubMed]
143. Szklarczyk, D.; Franceschini, A.; Wyder, S.; Forslund, K.; Heller, D.; Huerta-Cepas, J.; Simonovic, M.; Roth, A.; Santos, A.; Tsafou, K.P.; et al. STRING v10: Protein-protein interaction networks, integrated over the tree of life. *Nucleic Acids Res.* **2015**, *43*, D447–D452. [CrossRef] [PubMed]
144. Bamford, S.; Dawson, E.; Forbes, S.; Clements, J.; Pettett, R.; Dogan, A.; Flanagan, A.; Teague, J.; Futreal, P.A.; Stratton, M.R.; et al. The COSMIC (Catalogue of Somatic Mutations in Cancer) database and website. *Br. J. Cancer* **2004**, *91*, 355–358. [CrossRef] [PubMed]
145. Chakravarty, D.; Gao, J.; Phillips, S.M.; Kundra, R.; Zhang, H.; Wang, J.; Rudolph, J.E.; Yaeger, R.; Soumerai, T.; Nissan, M.H.; et al. OncoKB: A Precision Oncology Knowledge Base. *JCO Precis. Oncol.* **2017**, *2017*. [CrossRef] [PubMed]
146. Armenia, J.; Wankowicz, S.A.M.; Liu, D.; Gao, J.; Kundra, R.; Reznik, E.; Chatila, W.K.; Chakravarty, D.; Han, G.C.; Coleman, I.; et al. The long tail of oncogenic drivers in prostate cancer. *Nat. Genet.* **2018**, *50*, 645–651. [CrossRef] [PubMed]
147. Futreal, P.A.; Coin, L.; Marshall, M.; Down, T.; Hubbard, T.; Wooster, R.; Rahman, N.; Stratton, M.R. A census of human cancer genes. *Nat. Rev. Cancer* **2004**, *4*, 177–183. [CrossRef] [PubMed]
148. Ryan, M.B.; Finn, A.J.; Pedone, K.H.; Thomas, N.E.; Der, C.J.; Cox, A.D. ERK/MAPK Signaling Drives Overexpression of the Rac-GEF, PREX1, in BRAF- and NRAS-Mutant Melanoma. *Mol. Cancer Res. MCR* **2016**, *14*, 1009–1018. [CrossRef] [PubMed]
149. Misek, S.A.; Chen, J.; Schroeder, L.; Rattanasinchai, C.; Sample, A.; Sarkaria, J.N.; Gallo, K.A. EGFR Signals through a DOCK180-MLK3 Axis to Drive Glioblastoma Cell Invasion. *Mol. Cancer Res. MCR* **2017**, *15*, 1085–1095. [CrossRef] [PubMed]
150. Lissanu Deribe, Y.; Shi, Y.; Rai, K.; Nezi, L.; Amin, S.B.; Wu, C.C.; Akdemir, K.C.; Mahdavi, M.; Peng, Q.; Chang, Q.E.; et al. Truncating PREX2 mutations activate its GEF activity and alter gene expression regulation in NRAS-mutant melanoma. *Proc. Natl. Acad. Sci. USA* **2016**, *113*, E1296–E1305. [CrossRef] [PubMed]
151. Earp, M.; Tyrer, J.P.; Winham, S.J.; Lin, H.Y.; Chornokur, G.; Dennis, J.; Aben, K.K.H.; Anton-Culver, H.; Antonenkova, N.; Bandera, E.V.; et al. Variants in genes encoding small GTPases and association with epithelial ovarian cancer susceptibility. *PLoS ONE* **2018**, *13*, e0197561. [CrossRef] [PubMed]
152. Winkler, S.; Mohl, M.; Wieland, T.; Lutz, S. GrinchGEF—A novel Rho-specific guanine nucleotide exchange factor. *Biochem. Biophys. Res. Commun.* **2005**, *335*, 1280–1286. [CrossRef] [PubMed]

153. Zoughlami, Y.; Voermans, C.; Brussen, K.; van Dort, K.A.; Kootstra, N.A.; Maussang, D.; Smit, M.J.; Hordijk, P.L.; van Hennik, P.B. Regulation of CXCR4 conformation by the small GTPase Rac1: Implications for HIV infection. *Blood* **2012**, *119*, 2024–2032. [CrossRef] [PubMed]
154. Han, S.; Kim, D.; Kim, Y.; Choi, K.; Miller, J.E.; Kim, D.; Lee, Y. CAS-viewer: Web-based tool for splicing-guided integrative analysis of multi-omics cancer data. *BMC Med Genom.* **2018**, *11*, 25. [CrossRef] [PubMed]
155. Yang, I.S.; Son, H.; Kim, S.; Kim, S. ISOexpresso: A web-based platform for isoform-level expression analysis in human cancer. *BMC Genom.* **2016**, *17*, 631. [CrossRef] [PubMed]
156. Mehner, C.; Miller, E.; Khauv, D.; Nassar, A.; Oberg, A.L.; Bamlet, W.R.; Zhang, L.; Waldmann, J.; Radisky, E.S.; Crawford, H.C.; et al. Tumor cell-derived MMP3 orchestrates Rac1b and tissue alterations that promote pancreatic adenocarcinoma. *Mol. Cancer Res. MCR* **2014**, *12*, 1430–1439. [CrossRef] [PubMed]
157. Mehner, C.; Miller, E.; Nassar, A.; Bamlet, W.R.; Radisky, E.S.; Radisky, D.C. Tumor cell expression of MMP3 as a prognostic factor for poor survival in pancreatic, pulmonary, and mammary carcinoma. *Genes Cancer* **2015**, *6*, 480–489. [PubMed]
158. Stallings-Mann, M.L.; Waldmann, J.; Zhang, Y.; Miller, E.; Gauthier, M.L.; Visscher, D.W.; Downey, G.P.; Radisky, E.S.; Fields, A.P.; Radisky, D.C. Matrix metalloproteinase induction of Rac1b, a key effector of lung cancer progression. *Sci. Transl. Med.* **2012**, *4*, 142ra195. [CrossRef] [PubMed]
159. Alonso-Espinaco, V.; Cuatrecasas, M.; Alonso, V.; Escudero, P.; Marmol, M.; Horndler, C.; Ortego, J.; Gallego, R.; Codony-Servat, J.; Garcia-Albeniz, X.; et al. RAC1b overexpression correlates with poor prognosis in KRAS/BRAF WT metastatic colorectal cancer patients treated with first-line FOLFOX/XELOX chemotherapy. *Eur. J. Cancer* **2014**, *50*, 1973–1981. [CrossRef] [PubMed]
160. Huff, L.P.; Decristo, M.J.; Trembath, D.; Kuan, P.F.; Yim, M.; Liu, J.; Cook, D.R.; Miller, C.R.; Der, C.J.; Cox, A.D. The Role of Ect2 Nuclear RhoGEF Activity in Ovarian Cancer Cell Transformation. *Genes Cancer* **2013**, *4*, 460–475. [CrossRef] [PubMed]
161. Lin, Y.; Zheng, Y. Approaches of targeting Rho GTPases in cancer drug discovery. *Expert Opin. Drug Discov.* **2015**, *10*, 991–1010. [CrossRef] [PubMed]
162. Pajic, M.; Herrmann, D.; Vennin, C.; Conway, J.R.; Chin, V.T.; Johnsson, A.K.; Welch, H.C.; Timpson, P. The dynamics of Rho GTPase signaling and implications for targeting cancer and the tumor microenvironment. *Small GTPases* **2015**, *6*, 123–133. [CrossRef] [PubMed]
163. Smithers, C.C.; Overduin, M. Structural Mechanisms and Drug Discovery Prospects of Rho GTPases. *Cells* **2016**, *5*, 26. [CrossRef] [PubMed]
164. Nassar, N.; Cancelas, J.; Zheng, J.; Williams, D.A.; Zheng, Y. Structure-function based design of small molecule inhibitors targeting Rho family GTPases. *Curr. Top. Med. Chem.* **2006**, *6*, 1109–1116. [CrossRef] [PubMed]
165. Dharmawardhane, S.; Hernandez, E.; Vlaar, C. Development of EHop-016: A small molecule inhibitor of Rac. *Enzymes* **2013**, *33 Pt A*, 117–146.
166. Shutes, A.; Onesto, C.; Picard, V.; Leblond, B.; Schweighoffer, F.; Der, C.J. Specificity and mechanism of action of EHT 1864, a novel small molecule inhibitor of Rac family small GTPases. *J. Biol. Chem.* **2007**, *282*, 35666–35678. [CrossRef] [PubMed]
167. Hong, L.; Kenney, S.R.; Phillips, G.K.; Simpson, D.; Schroeder, C.E.; Noth, J.; Romero, E.; Swanson, S.; Waller, A.; Strouse, J.J.; et al. Characterization of a Cdc42 protein inhibitor and its use as a molecular probe. *J. Biol. Chem.* **2013**, *288*, 8531–8543. [CrossRef] [PubMed]
168. Forget, P.; Bentin, C.; Machiels, J.P.; Berliere, M.; Coulie, P.G.; De Kock, M. Intraoperative use of ketorolac or diclofenac is associated with improved disease-free survival and overall survival in conservative breast cancer surgery. *Br. J. Anaesth.* **2014**. [CrossRef] [PubMed]
169. Retsky, M.; Demicheli, R.; Hrushesky, W.J.; Forget, P.; De Kock, M.; Gukas, I.; Rogers, R.A.; Baum, M.; Sukhatme, V.; Vaidya, J.S. Reduction of breast cancer relapses with perioperative non-steroidal anti-inflammatory drugs: New findings and a review. *Curr. Med. Chem.* **2013**, *20*, 4163–4176. [CrossRef] [PubMed]
170. Forget, P.; Vandenhende, J.; Berliere, M.; Machiels, J.P.; Nussbaum, B.; Legrand, C.; De Kock, M. Do intraoperative analgesics influence breast cancer recurrence after mastectomy? A retrospective analysis. *Anesth. Analg.* **2010**, *110*, 1630–1635. [CrossRef] [PubMed]

171. Bowtell, D.D.; Bohm, S.; Ahmed, A.A.; Aspuria, P.J.; Bast, R.C., Jr.; Beral, V.; Berek, J.S.; Birrer, M.J.; Blagden, S.; Bookman, M.A.; et al. Rethinking ovarian cancer II: Reducing mortality from high-grade serous ovarian cancer. *Nat. Rev. Cancer* **2015**, *15*, 668–679. [CrossRef] [PubMed]
172. Lengyel, E. Ovarian cancer development and metastasis. *Am. J. Pathol.* **2010**, *177*, 1053–1064. [CrossRef] [PubMed]
173. Deng, K.; Yang, C.; Tan, Q.; Song, W.; Lu, M.; Zhao, W.; Lou, G.; Li, Z.; Li, K.; Hou, Y. Sites of distant metastases and overall survival in ovarian cancer: A study of 1481 patients. *Gynecol. Oncol.* **2018**, *150*, 460–465. [CrossRef] [PubMed]
174. Obermayr, E.; Bednarz-Knoll, N.; Orsetti, B.; Weier, H.U.; Lambrechts, S.; Castillo-Tong, D.C.; Reinthaller, A.; Braicu, E.I.; Mahner, S.; Sehouli, J.; et al. Circulating tumor cells: Potential markers of minimal residual disease in ovarian cancer? a study of the OVCAD consortium. *Oncotarget* **2017**, *8*, 106415–106428. [CrossRef] [PubMed]
175. Zhang, X.; Li, H.; Yu, X.; Li, S.; Lei, Z.; Li, C.; Zhang, Q.; Han, Q.; Li, Y.; Zhang, K.; et al. Analysis of Circulating Tumor Cells in Ovarian Cancer and Their Clinical Value as a Biomarker. *Cell. Physiol. Biochem. Int. J. Exp. Cell. Physiol. Biochem. Pharmacol.* **2018**, *48*, 1983–1994. [CrossRef] [PubMed]
176. Keyver-Paik, M.D.; Arden, J.M.; Luders, C.; Thiesler, T.; Abramian, A.; Hoeller, T.; Hecking, T.; Ayub, T.H.; Doeser, A.; Kaiser, C.; et al. Impact of Chemotherapy on Retroperitoneal Lymph Nodes in Ovarian Cancer. *Anticancer Res.* **2016**, *36*, 1815–1824. [PubMed]
177. Hjerpe, E.; Staf, C.; Dahm-Kahler, P.; Stalberg, K.; Bjurberg, M.; Holmberg, E.; Borgfeldt, C.; Tholander, B.; Hellman, K.; Kjolhede, P.; et al. Lymph node metastases as only qualifier for stage IV serous ovarian cancer confers longer survival than other sites of distant disease—A Swedish Gynecologic Cancer Group (SweGCG) study. *Acta Oncol.* **2018**, *57*, 331–337. [CrossRef] [PubMed]
178. Sehouli, J.; Olschewski, J.; Schotters, V.; Fotopoulou, C.; Pietzner, K. Prognostic role of early versus late onset of bone metastasis in patients with carcinoma of the ovary, peritoneum and fallopian tube. *Ann. Oncol.* **2013**, *24*, 3024–3028. [CrossRef] [PubMed]
179. Chebouti, I.; Blassl, C.; Wimberger, P.; Neubauer, H.; Fehm, T.; Kimmig, R.; Kasimir-Bauer, S. Analysis of disseminated tumor cells before and after platinum based chemotherapy in primary ovarian cancer. Do stem cell like cells predict prognosis? *Oncotarget* **2016**. [CrossRef] [PubMed]
180. Pantel, K.; Alix-Panabieres, C. Bone marrow as a reservoir for disseminated tumor cells: A special source for liquid biopsy in cancer patients. *BoneKEy Rep.* **2014**, *3*, 584. [CrossRef] [PubMed]
181. Banys, M.; Solomayer, E.F.; Becker, S.; Krawczyk, N.; Gardanis, K.; Staebler, A.; Neubauer, H.; Wallwiener, D.; Fehm, T. Disseminated tumor cells in bone marrow may affect prognosis of patients with gynecologic malignancies. *Int. J. Gynecol. Cancer* **2009**, *19*, 948–952. [CrossRef] [PubMed]
182. Wimberger, P.; Heubner, M.; Otterbach, F.; Fehm, T.; Kimmig, R.; Kasimir-Bauer, S. Influence of platinum-based chemotherapy on disseminated tumor cells in blood and bone marrow of patients with ovarian cancer. *Gynecol. Oncol.* **2007**, *107*, 331–338. [CrossRef] [PubMed]
183. Wimberger, P.; Roth, C.; Pantel, K.; Kasimir-Bauer, S.; Kimmig, R.; Schwarzenbach, H. Impact of platinum-based chemotherapy on circulating nucleic acid levels, protease activities in blood and disseminated tumor cells in bone marrow of ovarian cancer patients. *Int. J. Cancer* **2011**, *128*, 2572–2580. [CrossRef] [PubMed]
184. Gunjal, P.M.; Schneider, G.; Ismail, A.A.; Kakar, S.S.; Kucia, M.; Ratajczak, M.Z. Evidence for induction of a tumor metastasis-receptive microenvironment for ovarian cancer cells in bone marrow and other organs as an unwanted and underestimated side effect of chemotherapy/radiotherapy. *J. Ovarian Res.* **2015**, *8*, 20. [CrossRef] [PubMed]
185. Sharma, M.; Afrin, F.; Satija, N.; Tripathi, R.P.; Gangenahalli, G.U. Stromal-derived factor-1/CXCR4 signaling: Indispensable role in homing and engraftment of hematopoietic stem cells in bone marrow. *Stem Cells Dev.* **2011**, *20*, 933–946. [CrossRef] [PubMed]
186. Gupta, N.; Duda, D.G. Role of stromal cell-derived factor 1alpha pathway in bone metastatic prostate cancer. *J. Biomed. Res.* **2016**, *30*, 181–185. [PubMed]
187. Lapidot, T.; Kollet, O. The essential roles of the chemokine SDF-1 and its receptor CXCR4 in human stem cell homing and repopulation of transplanted immune-deficient NOD/SCID and NOD/SCID/B2m(null) mice. *Leukemia* **2002**, *16*, 1992–2003. [CrossRef] [PubMed]

188. Peled, A.; Petit, I.; Kollet, O.; Magid, M.; Ponomaryov, T.; Byk, T.; Nagler, A.; Ben-Hur, H.; Many, A.; Shultz, L.; et al. Dependence of human stem cell engraftment and repopulation of NOD/SCID mice on CXCR4. *Science* **1999**, *283*, 845–848. [CrossRef] [PubMed]
189. Price, T.T.; Burness, M.L.; Sivan, A.; Warner, M.J.; Cheng, R.; Lee, C.H.; Olivere, L.; Comatas, K.; Magnani, J.; Kim Lyerly, H.; et al. Dormant breast cancer micrometastases reside in specific bone marrow niches that regulate their transit to and from bone. *Sci. Transl. Med.* **2016**, *8*, 340ra373. [CrossRef] [PubMed]
190. Liu, C.F.; Liu, S.Y.; Min, X.Y.; Ji, Y.Y.; Wang, N.; Liu, D.; Ma, N.; Li, Z.F.; Li, K. The prognostic value of CXCR4 in ovarian cancer: A meta-analysis. *PLoS ONE* **2014**, *9*, e92629. [CrossRef] [PubMed]
191. Guo, Q.; Gao, B.L.; Zhang, X.J.; Liu, G.C.; Xu, F.; Fan, Q.Y.; Zhang, S.J.; Yang, B.; Wu, X.H. CXCL12-CXCR4 Axis Promotes Proliferation, Migration, Invasion, and Metastasis of Ovarian Cancer. *Oncol. Res.* **2015**, *22*, 247–258. [CrossRef] [PubMed]
192. Li, J.; Jiang, K.; Qiu, X.; Li, M.; Hao, Q.; Wei, L.; Zhang, W.; Chen, B.; Xin, X. Overexpression of CXCR4 is significantly associated with cisplatin-based chemotherapy resistance and can be a prognostic factor in epithelial ovarian cancer. *BMB Rep.* **2014**, *47*, 33–38. [CrossRef] [PubMed]
193. Kajiyama, H.; Shibata, K.; Terauchi, M.; Ino, K.; Nawa, A.; Kikkawa, F. Involvement of SDF-1alpha/CXCR4 axis in the enhanced peritoneal metastasis of epithelial ovarian carcinoma. *Int. J. Cancer* **2008**, *122*, 91–99. [CrossRef] [PubMed]
194. Figueras, A.; Alsina-Sanchis, E.; Lahiguera, A.; Abreu, M.; Muinelo-Romay, L.; Moreno-Bueno, G.; Casanovas, O.; Graupera, M.; Matias-Guiu, X.; Vidal, A.; et al. A Role for CXCR4 in Peritoneal and Hematogenous Ovarian Cancer Dissemination. *Mol. Cancer Ther.* **2018**, *17*, 532–543. [CrossRef] [PubMed]
195. Guo, L.; Cui, Z.M.; Zhang, J.; Huang, Y. Chemokine axes CXCL12/CXCR4 and CXCL16/CXCR6 correlate with lymph node metastasis in epithelial ovarian carcinoma. *Chin. J. Cancer* **2011**, *30*, 336–343. [CrossRef] [PubMed]
196. Arnaud, M.P.; Vallee, A.; Robert, G.; Bonneau, J.; Leroy, C.; Varin-Blank, N.; Rio, A.G.; Troadec, M.B.; Galibert, M.D.; Gandemer, V. CD9, a key actor in the dissemination of lymphoblastic leukemia, modulating CXCR4-mediated migration via RAC1 signaling. *Blood* **2015**, *126*, 1802–1812. [CrossRef] [PubMed]
197. Mao, T.L.; Fan, K.F.; Liu, C.L. Targeting the CXCR4/CXCL12 axis in treating epithelial ovarian cancer. *Gene Ther.* **2017**, *24*, 621–629. [CrossRef] [PubMed]
198. Ray, P.; Lewin, S.A.; Mihalko, L.A.; Schmidt, B.T.; Luker, K.E.; Luker, G.D. Noninvasive imaging reveals inhibition of ovarian cancer by targeting CXCL12-CXCR4. *Neoplasia* **2011**, *13*, 1152–1161. [CrossRef] [PubMed]
199. Righi, E.; Kashiwagi, S.; Yuan, J.; Santosuosso, M.; Leblanc, P.; Ingraham, R.; Forbes, B.; Edelblute, B.; Collette, B.; Xing, D.; et al. CXCL12/CXCR4 blockade induces multimodal antitumor effects that prolong survival in an immunocompetent mouse model of ovarian cancer. *Cancer Res.* **2011**, *71*, 5522–5534. [CrossRef] [PubMed]

cancers

MDPI

Review

The Tumor Microenvironment of Epithelial Ovarian Cancer and Its Influence on Response to Immunotherapy

Galaxia M. Rodriguez [1,2], Kristianne J. C. Galpin [1,2], Curtis W. McCloskey [1,2] and Barbara C. Vanderhyden [1,2,*]

1 Cancer Therapeutics Program, Ottawa Hospital Research Institute, 501 Smyth Road, Ottawa, ON K1H 8L6, Canada; garodriguez@toh.ca (G.M.R.); kgalpin@ohri.ca (K.J.C.G.); cmccloskey@ohri.ca (C.W.M.)

2 Department of Cellular and Molecular Medicine, University of Ottawa, 451 Smyth Road, Ottawa, ON K1H 8M5, Canada

* Correspondence: bvanderhyden@ohri.ca; Tel.: +1-613-737-7700 (ext. 70330)

Received: 23 June 2018; Accepted: 20 July 2018; Published: 24 July 2018

Abstract: Immunotherapy as a treatment for cancer is a growing field of endeavor but reports of success have been limited for epithelial ovarian cancer. Overcoming the challenges to developing more effective therapeutic approaches lies in a better understanding of the factors in cancer cells and the surrounding tumor microenvironment that limit response to immunotherapies. This article provides an overview of some ovarian cancer cell features such as tumor-associated antigens, ovarian cancer-derived exosomes, tumor mutational burden and overexpression of immunoinhibitory molecules. Moreover, we describe relevant cell types found in epithelial ovarian tumors including immune cells (T and B lymphocytes, Tregs, NK cells, TAMs, MDSCs) and other components found in the tumor microenvironment including fibroblasts and the adipocytes in the omentum. We focus on how those components may influence responses to standard treatments or immunotherapies.

Keywords: epithelial ovarian cancer; tumor microenvironment; tumor infiltrating lymphocytes; tumor-associated antigens; ascites; immunosuppression; prognostic factors; cancer-associated fibroblasts; exosomes; adipocytes

1. Introduction

An increasing body of evidence strongly suggests that the immune system is able to identify, control and eliminate nascent neoplastic cells in a process known as cancer immunosurveillance [1]. Epithelial ovarian cancers (EOCs) are "immunogenic tumors" that produce spontaneous antitumor immune responses detectable in peripheral blood, tumors and ascites of patients [2–4]. The resulting presence of tumor infiltrating lymphocytes (TILs) is associated with improved survival in EOC [5]. Unfortunately, there are a number of factors in the tumor microenvironment (TME) that can impair the presence or activity of TILs, thereby facilitating cancer progression.

Various immunotherapeutic strategies are attempting to address the challenges posed by the highly immunosuppressive EOC TME. Immunotherapies encompass many modalities, including immune checkpoint blockade, antibody-based therapies, cancer vaccines, cytokines, adoptive cell transfer, and chimeric antigen receptor-modified T cells [6]. However, emerging cancer immunotherapies (blocking antibodies for checkpoint inhibitors) have shown low rates of responses in EOC (reviewed in [2]). Improving this response rate is a major goal, which can only be achieved with a better understanding of the elements in the TME that contribute to treatment failure. Immune cells are the main players in the development of antitumor immunity or tumor progression, but there are also other components in the TME that should be taken into consideration when designing new therapeutic strategies. Those components include EOC-derived exosomes, cancer-associated fibroblasts (CAFs) and adipocytes residing in the omentum.

In this review we will describe those elements of the TME, how they influence the burden of the tumor, the responses to therapies, and their relevance in designing cancer immunotherapies for EOC.

2. Cancer Cells and Tumor Antigens

The success of cancer immunotherapy hinges on the ability to generate cancer-specific antitumor T-cell responses, to both recognize tumor-associated antigens (TAAs) and kill tumor cells, and to generate memory responses. TAAs can be classified into different categories: tissue differentiation, cancer testes antigens (CTAs), neoantigens derived from mutations, overexpressed cellular, splice variant, glycolipid, and viral antigens [7,8]. Ideal TAAs for immunotherapy targets are immunogenic and are expressed or overexpressed in tumor tissue, with restricted expression in associated normal tissues, in a significant percentage of patients [9]. Positive responses to immunotherapies such as immune checkpoint inhibitors [blocking programmed cell death protein 1 (PD-l), programmed death-ligand 1 (PD-L1) and cytotoxic T-lymphocyte-associated antigen 4 (CTLA-4)], have been associated with high mutation/neoantigen burden [10,11]. The initial clinical studies of small numbers of EOC patients treated with immune checkpoint inhibitors have resulted in clinical benefits in less than 20% of patients (Table 1). Unfortunately, little is known about the TME at the start of treatment in most studies, making it impossible to discern the factors that may have blocked any response. The failure to respond could be related to the neoantigen burden in EOC, which may be insufficient to generate a significant antitumoral response [12,13]. There are currently intense research efforts to understand other TAAs (Table 2) recognized by TILs to design informed immunotherapy targets (Table 3).

Table 1. Human studies using immune checkpoint inhibitors in epithelial ovarian cancers (EOC) (completed or partially completed studies).

Target	Agent	EOC Characteristic	Antitumoral Responses	Immune Related Parameters	Clinical Study
PD-1	**Nivolumab** (Opdivo, *BMS-936558, MDX1106*) i.v. infusion every 2 weeks (1 or 3 mg/kg)	Advanced or relapsed platinum-resistant ovarian cancer	A quick antitumor response observed by baseline computed tomographic image, decreased CA-125 blood levels. Overall response: 15%, 2 § pts had a durable CR, disease control rate in all 20 pts was 45%. Median PFS 3.5 months.	Expression of PD-L1 in ovarian cancer tissues was not significantly correlated with objective response but 16/20 patients having a high expression of PDL1 on tumors did not respond to treatment (vs. 2/4 responders in the PD-L1-low expression group).	Phase II UMIN000005714 [14]
	Pembrolizumab (Keytruda, *MK-3475*) i.v. infusion every 2 weeks (10 mg/kg) for up to 2 years	PD-L1+ advanced ovarian cancer	1 * pt CR, 2 pts PR; 6 pts stable disease. Duration of response ≥24 weeks. Overall response was 11.5%. 6/26 (23.1%) had evidence of tumor reduction; 3 had a tumor reduction of at least 30%.	N/A	Phase Ib trial NCT02054806 Active, not recruiting [15]
PD-L1	**Avelumab** (Eavencio, *MSB0010718C*) every 2 weeks (10 mg/kg)	Recurrent or refractory ovarian cancer	4/23 (17.4%) pts achieved an unconfirmed best overall response of PR, 11 pts (47.8%) had stable disease, and 2 pts had >30% tumor shrinkage after progression was reported. Median PFS was 11.9 weeks and the PFS rate at 24 weeks was 33.3%.	Exposure to Avelumab significantly increased the ratio of sCD27/sCD40L #. Some antitumor activity of this antibody may be due to ADCC [16].	Phase Ib study NCT01772004 Active, not recruiting [17]
	BMS-936559 (*MDX-1105*) i.v. infusion every 2 weeks (10 mg/kg) in 6-week cycles	Advanced ovarian cancer	1 of 17 pts (6%) had a PR, and 3 (18%) had stable disease lasting at least 24 weeks.	N/A	Multicenter phase 1 trial NCT00729664 completed [18]
CTLA-4	**Ipilimumab** i.v. infusions (10 mg/kg) once every 3 weeks for 4 doses (Induction Phase). Once every 12 weeks (Maintenance Phase), until disease progression or unacceptable toxicity occurs	Recurrent platinum-sensitive ovarian cancer	N/A	N/A	Phase II study (NCT01611558) Active, not recruiting
	Ipilumimab Periodic infusions (3 mg/kg) after vaccination with irradiated, autologous tumor cells engineered to secrete GM-CSF (GVAX)	Stage IV ovarian carcinoma	1/9 pts had reduction in circulating CA-125 levels, regression of metastasis, increased humoral reactions to NY-ESO-1. 3 pts achieved stable disease of >6, 4, and 2 months' duration, as measured by CA-125 levels and radiographic criteria.	The extent of therapy-induced tumor necrosis was linearly related to the natural logarithm of the ratio of intratumoral CD8+ effector T cells to FoxP3+ Tregs in post-treatment biopsies.	[19,20]

§ The tumor was histologically identified as clear cell carcinoma in one of the two patients who experienced a CR. * 1 Patient with CR had a PD-L1 gene rearrangement leading to gain of function of the PD-L1 gene secondary to gene amplification, high PD-L1 expression was observed in cancer epithelial cells, as well as high T lymphocyte infiltration (CD4, CD8), some B cells (CD20) and macrophages (CD68) [21]. # sCD27 is a marker of T-cell activation [22], sCD40L is a measure of immune suppression [23]. Complete response (CR), partial response (PR), patients (pts), progression-free survival (PFS), intravenous (i.v.), antibody-dependent cell-mediated cytotoxicity (ADCC), not available (N/A), Granulocyte-macrophage colony-stimulating factor (GM-CSF), forkhead box P3 (FoxP3), regulatory T cells (Tregs).

Table 2. Type and prevalence of tumor-associated antigens (TAAs) in EOC.

TAA Category	TAA	Prevalence (% Patients)	FIGO Stage	References
CTA	OY-TES-1	69% (All subtypes)	I-IV	[24]
	SCP-1	15% (All subtypes)	I-IV	[25]
	SPAG9 [1]	88% (HGSC)	I-IV	[26]
	AKAP4 [2]	93% (Serous)	I-IV	[27]
	NY-ESO-1	43% (All Subtypes)	I-IV	[28]
	MAGE-A [3]	~7–55% (All subtypes)	I-IV	[29–31]
Oncogene	p53	Mutation (95% HGSC)/Amplification (35% HGSC)	I-IV	[32,33]
	Her2neu [4]	35–45% (All subtypes)	I-IV	[34–37]
	WT1 [5]	71.4% (LGSC) ~55% (HGSC)	III/IV	[38,39]
	Mesothelin	82% (HGSC)	I-IV	[40,41]
	MUC16 [6] (CA-125)	80% (All subtypes)		[42]
Neoantigen	Patient/tumor site specific	Greater number in HR deficient [7] tumors	I-IV	[12,43]

[1] SPAG9: Sperm-associated antigen 9. [2] AKAP4: A-kinase anchoring protein 4. [3] MAGE-A: Melanoma antigen. [4] Her2-neu: human epidermal growth factor receptor 2-neu. [5] WT1: Wilms' tumor 1. [6] MUC16: Mucin-16. [7] *BRCA1/BRCA2*, Fanconi anemia genes (*PALB2*, *FANCA*, *FANCI*, *FANCL*, and *FANCC*), restriction site associated DNA genes (*RAD50*, *RAD51*, *RAD51C*, and *RAD54L*), DNA damage response genes (*ATM*, *ATR*, *CHEK1*, and *CHEK2*). High-grade serous ovarian cancer (HGSC), low-grade serous ovarian cancer (LGSC).

Table 3. TAA targeted immunotherapies in EOC.

TAA Category	TAA	Immunotherapy	References
CTA	NY-ESO-1	Recombinant protein vaccine (Epitope $ESO_{157-170}$) + Incomplete Freund's Adjuvant	[44]
		Overlapping long peptides + Montanide/Poly-ICLC adjuvants	[45]
		NY-ESO-1b + Montanide	[46]
		Recombinant vaccinia prime-NY-ESO-1 (rV-NY-ESO-1) + recombinant fowlpox boost-NY-ESO-1 (rF-NY-ESO-1)	[47]
		NY-ESO-1-specific engineered T Cells	(NCT03159585, NCT03017131, NCT02457650)
		NYESO-1(C259) transduced autologous T cells	(NCT01567891)
	MAGE-A	Autologous genetically modified MAGE-A4^{c1032}T cells	(NCT03132922)
Oncogene	p53	Modified vaccinia Ankara vaccine vs. wild-type human p53 (p53MVA) + gemcitabine	[48]
		Synthetic long peptide (SLP) vaccine	[49]
	Her2neu	Her2-neu peptide vaccine	(NCT00194714)
		Exvivo Her2-neu specific T-cell expansion	(NCT00228358)
	WT1	Autologous WT1 T Cells + Cyclophosphamide + Fludarabine	(NCT00562640)
		WT1 peptide vaccine + Montanide + GM-CSF + Nivolumab (PD-1)	(NCT02737787)
		WT1 mRNA-loaded DCs [2]	[50]
		WT1 peptide vaccine + Montanide	[51]
	Mesothelin	Anti-Mesothelin CAR-T [1] cells	[41]/(NCT02580747)
	MUC16 (CA-125)	Antibody therapy (Oregovomab, ACA125/Abagovomab)	[52–55]
		CAR-T Therapy + IL-12	[56,57] (NCT02498912)
Neoantigen	Patient/tumor site specific	Autologous DCs pulsed with oxidized autologous whole-tumor cell lysate + bevacizumab + cyclophosphamide	[58]
		Autologous neoantigen engineered T-Cells	(NCT03412877)

[1] Chimeric antigen receptor T cell (CAR-T). [2] Dendritic cells (DCs).

2.1. Neoantigens

Ovarian cancer has been shown to harbor an intermediate neoantigen load by whole exome sequencing/next generation sequencing [12,59]. Whole exome sequencing of tumor cells from ascites samples of three high-grade serous ovarian cancer (HGSC) patients revealed a tumor mutation burden (TMB) of approximately 20–40 mutations across all patients, however only 1/79 mutations (1.3%) were recognized by autologous tumor-associated T cells [60]. Comprehensive genomic profiling of ovarian cancer revealed low overall TMB among subtypes: HGSC (3.6), low-grade serous (LGSOC) (2.7), endometrioid (2.7), mucinous (2.7), and clear cell (2.7). Only a small percentage of patients had a significant TMB (20 or more mutations per Mb), meaning only a small percentage of patients would be predicted to show favorable response to immune therapy [12]. Consequently, in clinical trials of checkpoint inhibitors in EOC, CTLA-4 inhibitors (Ipilimumab), PD1 inhibitors (Nivolumab and Pembrolizumab), and PD-L1 inhibitors (MS-936559 and Avelumab) had response rates of 5–20% [14,20,61] (Table 1). A notable exception is the highly aggressive small cell carcinoma of the ovary, hypercalcemic type which, despite being a monogenic cancer, has responsiveness to anti-PD1 immunotherapy [62].

Neoantigen depletion [63], intratumoral heterogeneity, and clonal evolution of primary tumors and metastases may influence immunosurveillance and response to immunotherapy [64,65]. Epithelial T-cell rich tumors show the lowest amount of clonal diversity, neoantigen diversity and greatest loss of human leukocyte antigen (HLA) expression, which suggests immunoediting in the TME. T-cell poor tumors or "cold tumors" have a higher predicted and more diverse neoantigen load (unedited) [63].

2.2. Cancer Testes Antigens

CTAs are encoded by ~140 genes that are normally only expressed in germ cells (testes, placenta, fetal ovary) and not normal somatic adult cells, but often highly expressed in tumors. This along with their immunogenicity makes them significant targets for cancer immunotherapy [9,66,67]. Vaccination with recombinant MAGE-A3 antigen has been used in Phase I/II clinical trials for melanoma [68] and non-small-cell lung cancer (NSCLC) [69] with a good safety profile and observed humoral response, but only slight effects on survival.

Several CTAs have been described in EOC (Table 2) and have been proposed as immunotherapy targets (Table 3) based on their tissue specificity and high expression in a significant number of EOCs of all subtypes. NY-ESO-1 ($ESO_{157-165}$) specific CD8+ T cells were found in TILs of 71% of (10/14) vaccination naïve seropositive patients, and ex vivo proliferation of NY-ESO-1 specific peripheral blood lymphocytes in 65% of patients suggested that an adaptive immune response against this CTA can be achieved [70,71]. Clinical trials have subsequently tested the feasibility of generating NY-ESO-1 specific immune responses (Table 3). These approaches have generated humoral and CD4+ and CD8+ antigen specific T-cell responses, and in some cases, long lasting/complete responses [44–47]. NY-ESO-1 was not expressed in some recurrent tumors, raising the possibility of immune escape [44]. Furthermore, NY-ESO-1 reactive CD8+ T cells often express higher levels of inhibitory molecules lymphocyte-activation gene 3 (LAG3), PD-1 and CTLA-4, suggesting immunosuppression as a reason for lack of complete response during clinical trials [71].

Many characteristics of CTA epitopes and all TAAs such as (i) immunogenicity; (ii) restriction to HLA-I or -II; (iii) natural processing; (iv) expression; and (v) role in tumor progression remain to be elucidated and require validation in larger sample sizes. While the expression of CTAs does not often correlate with improved survival, their tissue specificity makes CTAs attractive targets for immunotherapies (Table 3) such as peptide vaccines [44,70], antigen-loaded dendritic cell (**DC**) vaccines [72], or oncolytic viral platforms, and for combined interventions with immune checkpoint inhibitors [73] or chemotherapy [74], in order to overcome tumor escape mechanisms.

2.3. Other TAAs

Genetic and epigenetic aberrations in cancer cells, resulting from mutations, amplifications or deletions in genes, provide both therapeutic targets and potential TAAs for immunotherapy design (Table 2). However, the greatest hurdles still remain in designing immunotherapeutic targets for a disease in which such aberrations, with the exception of p53 mutation (95% of HGSC [33,75]), are relatively uncommon (<20% frequency in HGSC cases) and lack antigen specificity to the tumor. Immunogenic oncogenes p53, Her2-neu and WT1 are broadly overexpressed in EOC, particularly HGSC, and targeted immunotherapies have been explored in clinical trials (Table 3). Other common but infrequent amplifications, mutations or deletions occur in *CCNE1*, *NF1*, *PTEN*, *KRAS*, *RB*, *CDK2NA*, *PIK3CA* and *AKT1/2* and provide potential therapeutic targets for EOC immunotherapy [33]. The DCs, T-cells, and peptide-based vaccine strategies against proteins described above have largely demonstrated immunological responses including CD4+ and CD8+ T-cell responses in preliminary clinical trials following vaccination, but often in the absence of clinical responses. This is perhaps due to widespread immunosuppression in the TME preventing T-cell activation and proliferation, as well as tumor heterogeneity and immunogenicity that impede proper TAA presentation to the immune cells.

The EOC immunopeptidome was profiled by isolating HLA molecules primarily from HGSC tumors and which were analyzed by mass spectrometry [57]. The analysis identified relevant proteins including CRABP1/2, FOLR1, and KLK10 presented on major histocompatibility complex (MHC) I molecules, and mesothelin, PTPRS and UBB presented on MHC-II molecules [57]. The most abundantly detected protein presented on MHC-I molecules was MUC16 (CA-125), with 113 different peptides expressed in approximately 80% of patients. MUC16-derived peptides were highly immunogenic (85% T-cell responses in vitro), and consequently it was proposed as the top candidate for targeted immunotherapy moving forward [57]. Although CA-125 is immunogenic, the large number of trials with a monoclonal antibody targeting CA-125 (Table 3) have been mostly unsuccessful as a monotherapy [76]. This failure could be explained by the weak magnitude of the immune response generated, the loss of expression or down-regulation of CA-125 on EOC cells to avoid immune recognition, or the overgrowth of CA-125(-) EOC cells as a consequence of cancer immunoediting process.

A single TAA is generally only expressed in a subset of patients, making the design of a universal immunotherapy challenging. The main barrier of targeting a single TAA is cancer immunoediting, which enables the enrichment of neoplastic cells in tumors that do not express the targeted TAA over time. Chimeric antigen receptor T (CAR-T) cells provides the option of combining multiple antigen specificities, and delivering direct cytokine stimulation (GM-CSF, IL-12) to the TME, irrespective of the MHC status of the patient [8].

2.4. Tumor Immunogenicity and Other Immunoinhibitory Molecules

Loss of immunogenicity is an immune hallmark of cancer that is exploited by tumors to evade immune recognition. This can be triggered by down-regulation or loss of expression of MHC-I and -II, and the antigen processing and presentation machinery (APM) [77–80]. Expression of MHC-I genes is altered by 60–90%, depending on the cancer type. These impairments reduce the antigens presented on the cell surface leading to decreased or lack of recognition and elimination by cytotoxic T lymphocytes.

The mechanisms that are related to immune cell infiltration in EOC are dependent on MHC-I and -II status [3,81]. The presence of neoantigen-reactive T cells in patients with EOC can improve survival [82]. However, as mentioned before, since ovarian tumors possess intermediate/low mutation burdens, the incidence of naturally processed and presented neoantigens generating a significant antitumoral response is very low [13]. The expression of APM components and the presence of intratumoral T-cell infiltrates were significantly associated with improved survival [81]. Han. et al. demonstrated that the majority of ovarian carcinomas analyzed had either heterogeneous or positive expression of peptide transporter 1 (TAP1), TAP2, HLA class I heavy chain, and beta-2 microglobulin [81]. Concurrent expression of HLA-DR and CA-125 on cancer cells correlated with higher frequency of CD8+ TILs and increased survival [83]. Similarly, tumor cell expression of HLA-DMB was associated with increased numbers of CD8+ TILs and both were associated with improved survival in advanced-stage serous EOC [84]. The regulation of APM components and MHC molecules in human cancers is a significant area of research but is beyond the scope of this review (reviewed in [85,86]).

The mutational profile of EOC can also predict immunogenicity. Tumors with deficient homologous recombination (HR) machinery occur with a frequency of up to 50% [33]. These include mutations in *BRCA1/BRCA2* (20% frequency) or non-BRCA HR deficiencies (Fanconi anemia genes, restriction site associated DNA genes, and DNA damage response genes) [33]. HR deficient tumors have higher predicted neoantigen load, and infiltrating and peritumoral lymphocytes in these tumors have increased PD-1/PD-L1 expression [43], which may enhance susceptibility to immune checkpoint therapy. *BRCA1/2* mutated HGSC tumors have more CD3+ and CD8+ TILs compared to HR-proficient tumors, a signature associated with higher overall survival [43,87]. p53 mutations are also associated with higher levels of TILs [87,88]. Non-HR deficient tumors therefore have poorer overall survival [43] and may be less immunogenic, making them more difficult to target with immunotherapies. Alternative strategies and TAAs to target this group of EOC tumors need further investigation.

The expression of immunoinhibitory molecules on cancer cells, including PD-L1 and Indoleamine 2,3-dioxygenase (IDO) are associated with patient prognosis. Higher expression of PD-L1 on tumor cells correlates with poorer prognosis, suggesting that the PD-1/PD-L pathway can be a good target for restoring antitumor immunity in EOC [89,90], although others have suggested that high PD-1/PD-L1 expression in primary tumors may be associated with a favorable progression-free survival [91,92]. Increased infiltration of CD8+ T cells is associated with high PD-L1 expression likely as a result of an adaptive response where infiltrating CD8+ T cells secrete interferon gamma (IFNγ) that subsequently induces PD-L1 expression on cancer cells. This in turn inhibits T-cell activation and proliferation, preventing successful targeting and clearance of the tumor. Immune checkpoint inhibitors (anti-PD-L1 and PD-1) have been FDA approved for melanoma and NSCLC, but only a small percentage (10–33%) of ovarian cancers express PD-L1 [61,92,93], thus only a small percentage of patients may respond to anti-PD-L1 immunotherapy (Table 1). The enzyme IDO is often overexpressed by cancer cells, but is also produced by DCs and macrophages [94,95] in the TME. IDO catabolizes tryptophan, which leads to cell cycle arrest or apoptosis in NK and CD4 T cells [96], and skewed differentiation of regulatory T cells (Tregs) induced by plasmacytoid DCs, leading to immunosuppression in the TME [97]. Positive staining for IDO, observed in 24–57% of patient samples, is associated with poor prognosis of HGSC, decreased CD8+ TILs, as well as resistance to chemotherapy [98,99]. Targeting IDO with inhibitors may improve outcome [100,101].

3. Immune Cells

Most solid tumors are infiltrated by myeloid- and lymphoid lineage-derived immune cells that are differentially distributed within the TME with a crucial role in the establishment of antitumoral responses or tumor progression [1]. Growing tumor cells release "danger signals" that enable the recruitment of immune cells into the tumor niche. TILs such as CD4+ and CD8+ T cells, B lymphocytes, Natural Killer (NK)-T cells, as well as innate immune cells such as NK cells, macrophages and DCs, are then recruited in order to eliminate nascent neoplastic cells, acting as an extrinsic tumor suppression mechanism [102]. However, immunosurveillance promotes the selection of poorly immunogenic cancer cells through cancer immunoediting where neoplastic cells that resist the elimination phase can persist in equilibrium with effector CD4+ and CD8+ T cells under a pro-inflammatory milieu. Over time, cancer cells with the most immunoevasive characteristics are selected, enabling them to eventually escape immune attack [102]. Finally, immunoedited tumors become clinically apparent with variants that trigger the establishment of an immunosuppressive TME containing immunosuppressive immune cells such as myeloid-derived suppressor cells (MDSCs), Tregs, and others [2,103].

3.1. Immune Modulators and Adaptive Immune Cells in the Ovarian Cancer TME

3.1.1. TILs

TILs can localize into the tumor islet (intraepithelial) and in the peritumoral space (stromal) [2]. Several studies have shown a positive correlation between the presence of intraepithelial TILs and tumor regression in many solid cancers [4,5,104–107]. T cells can be found in primary tumor tissue and omental metastases [4,104,105,107–111] and their presence has been correlated with positive prognosis. Dadmarz et al. demonstrated that TILs isolated from EOC patients (primary tumor, metastases or ascites) were tumor-specific and could recognize autologous TAAs. Antitumoral responses were mainly characterized by the secretion of tumor necrosis factor-alpha (TNFα) and granulocyte macrophage-colony stimulating factor (GM-CSF) when stimulated with autologous tumor [112]. Later, Zhang and colleagues showed that intraepithelial CD3+ TILs can be found in >50% of advanced-stage EOC with their presence correlating with a five-year overall survival rate of 38% in contrast to 4.5% in patients whose tumors contained no T cells [5]. Even after debulking and platinum-based chemotherapy, the presence of intraepithelial CD3+ TILs increased the five-year overall survival rate (>70%) in comparison to patients whose tumors contained no T cells in islets (11%) [5]. T cell-rich tumors correlated with delayed recurrence or death and were associated with increased expression of Interleukin-2 (IL-2), IFNγ and lymphocyte-attracting chemokines within the tumor such as CXCL9 [113], CCL21, and CCL22 [5]. Conversely, tumors with no T cells in islets were associated with an increased level of vascular endothelial growth factor (VEGF), an angiogenic regulatory factor in the TME associated with early recurrence and short survival [5]. A more recent study showed that intratumoral accumulation of CXCR3 ligands such as CXCL9 and CXCL10, predicts survival in advanced HGSC [113] (Figure 1). This study also identified the cyclooxygenase (**COX**) metabolite Prostaglandin E2 as a negative regulator of chemokine secretion that contributes to tumor progression by impeding TILs recruitment in ovarian cancer [113]. Further investigation showed that expression of both COX-1 and COX-2 were negatively correlated with intraepithelial CD8+ TILs as well as with EOC patient survival [114].

While some studies have reported that the presence of both intraepithelial CD3+ and CD8+ T-cells correlates with improved disease-specific survival for EOC patients [81,87] others have shown that this beneficial characteristic is attributed to intraepithelial CD8+ TILs [4,104,105,107–110,115]. No association was found for CD3+ TILs or other subtypes of intraepithelial or stromal TILs in EOC overall patient survival. Interestingly, the subgroups displaying high versus low intraepithelial CD8+/CD4+ TIL ratios had favorable survival prognosis (median = 58 versus 23 months) [106]. This was due to the unfavorable effect of CD4+ CD25+ forkhead box P3+ (FOXP3) Tregs [88,104,106] that will be discussed later.

In 2012, a meta-analysis of ten studies with 1815 ovarian cancer patients confirmed the prognostic value of intraepithelial CD8+ TILs in EOC specimens regardless of the tumor grade, stage, or histologic subtype studied [111]. Their presence suggests that spontaneously activated antitumoral responses are present in the tumor niche to control tumor outgrowth [111] as observed by the presence of tumor-reactive antibodies and T cells found in the peripheral blood of advanced stage EOC patients [116–118], and oligoclonal tumor-reactive T cells isolated from blood, ascites or tumors [88,119–123]. Conversely, the lack of intraepithelial TILs is significantly associated with poor survival among EOC patients [111]. Thus, immunotherapies aiming to increase the effector functions of pre-existing antitumoral CD8+ TILs and triggering effector T cell-trafficking to the TME are the holy grail of cancer immunotherapy.

CD4+ T cells as well as CD8+ T cells can specifically recognize TAAs from malignant cells. CD4+ T helper (Th) cells provide cytokine support for CD8+ T-cell proliferation and expansion to eliminate cancer cells and trigger antitumoral responses. In an analysis of ovarian tumors, Tsiatas et al. found that a high percentage of CD4+ CD25hi cells and activated CD4+ T cells were significantly associated with improved median overall survival [124]. Two other studies also showed a positive correlation of the high frequency of CD4+ TILs and EOC patient survival [110,125]. Nesbeth et al., using an animal model for EOC, found that tumor-primed CD4+ T cells produce high levels of CCL5 that enables the recruitment and activation of DCs to the TME. Mature DCs were then able to prime tumor-specific CD8+ T cells and confer long-term protection [126]. Hence, immunotherapies stimulating both effector CD4+ and CD8+ T cells could confer synergistic antitumoral responses.

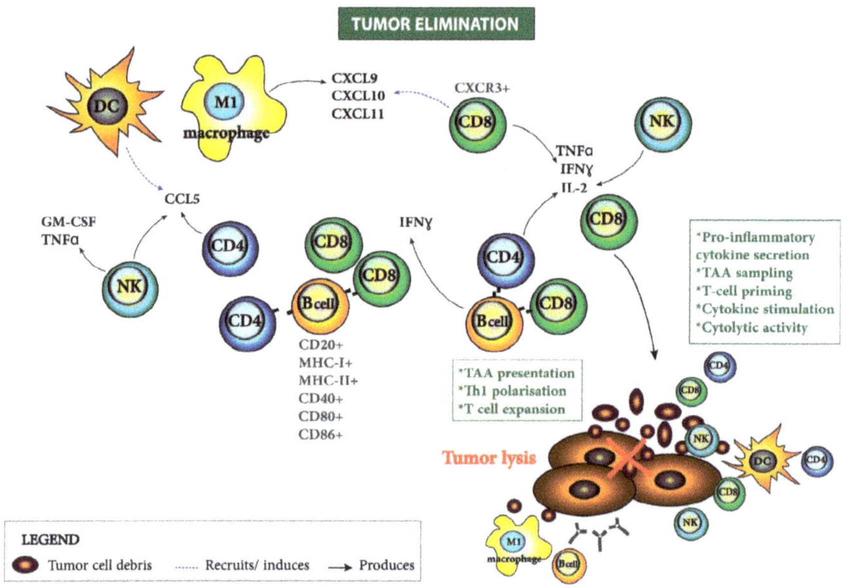

Figure 1. Antitumoral responses in the EOC TME. Immunogenic cell death induces the release of DAMPs mediating the recruitment of innate cells and APCs. Lympho-attracting chemokines produced by APCs such as macrophages enable the recruitment of CD8+ T cells to the tumor niche. DCs are also attracted by the production of CCL5 derived from NK cells and CD4+ T cells. The pro-inflammatory milieu enables TAA sampling and presentation by APCs to T cells to induce their activation and expansion. Pro-inflammatory cytokines released by activated effector T cells, M1 macrophages and DCs allow the amplification of the antitumoral response, enabling the cytolytic death of EOC targeted by CD8+ TILs and NK cells. B cells also participate in antitumor immunity by presenting TAAs to CD8+ T cells, by facilitating Th1 polarization, T-cell expansion and by producing tumor specific antibodies. Danger-associated molecular patterns (DAMPs), Antigen presenting cells (APCs), tumor-associated antigens (TAAs), dendritic cells (DCs), natural killer cells (NKs), CD4+ T helper cell (Th1).

3.1.2. Regulatory T lymphocytes

Tregs negatively regulate antitumoral responses in both a direct and indirect manner, highlighting that Tregs are a fundamental means of tumor immune evasion [127,128]. In healthy tissues, Tregs mediate tolerance by suppressing autoreactive T cells to protect and prevent excessive tissue destruction. Since most TAAs are composed by self-peptides, Tregs are often found in tumors to dampen antitumoral responses. Tregs accumulate and are more frequently present in tumors, with a shift in the median ratio of Tregs to TILs from 3–8% in healthy tissue to 18–25% in all analyzed cancers, including EOC [129]. Curiel et al. analyzed 104 EOC specimens and found that CD4+ CD25+ FOXP3+ Tregs specifically suppress antitumoral T cells in vivo, contributing to tumor growth. In addition, their presence correlates with poor patient outcome [130]. CD4+ Tregs preferentially migrated to tumor and ascites and were rarely found in draining lymph nodes at later cancer stages [131]. Immunotherapies impeding Treg trafficking could release the TME immunosuppression and promote the development of antitumoral responses.

FOXP3+ Tregs express minimal levels of effector cytokines and granzyme B, but are able to induce inhibitory activities through IL-10 and transforming growth factor beta (TGF-β) production [132] and cell–cell interactions [127] (Figure 2). Barnett et al. showed that EOC tumors highly infiltrated by Tregs were associated with poor survival, advanced stage and suboptimal debulking [109]. Investigation of the influence of cytoreduction on the immune system of primary and recurrent EOC found that the ratio of CD4/CD8 is increased in primary but not in recurrent tumors [133]. Primary cytoreduction increased circulating effector CD4+ and CD8+ T cells, but circulating CD4+ Tregs were decreased as well as IL-10 serum levels, but not TGFβ and IL-6 [133]. CD4+ Tregs were also decreased after chemical debulking in patients treated with neoadjuvant chemotherapy. The reduction of the systemic and TME immunosuppression triggered by surgical debulking resulted in an increased capacity of CD8+ T cells to respond to the recall antigens, but not in patients who were previously subjected to chemotherapy or affected by recurrent EOC [133].

Fialová and colleagues studied the dynamics of the tumor-infiltrating immune cells during different stages of EOC [134]. Early stage disease displayed a strong Th17 immune response while stage II patients had responses characterized by the recruitment of Th1 cells. Disseminated disease (stages III and IV) were characterized by high amounts of Tregs, tumor-associated macrophages (TAMs), DCs, and high levels of CCL22, which is secreted by tumor cells, TAMs and DCs to enable further recruitment of Tregs and immunosuppression [134]. Other studies have shown the importance of the TME in facilitating the establishment of tolerance and recruitment of Tregs to sustain tumor growth. Using EOC cell lines in vitro, Facciabene et al. found that tumor hypoxia induces the expression of chemokine ligands such as CCL28, enabling the recruitment of Tregs and triggering angiogenesis [135]. CCL28 overexpression was associated with a poor outcome in patients with EOC [135]. Similarly, CCL22 production by TAMs enabled the recruitment of Tregs [130,134] that induced B7-H4 on antigen-presenting cells including macrophages [136]. CXCR3+ Tregs, able to control type-I T-cell responses, are highly enriched in EOC and represent the majority of Tregs [137]. These Tregs were able to suppress T-cell proliferation and IFNγ secretion [137].

An interesting study analyzed 22 EOC ascites specimens and found significantly elevated levels of IL-6, IL-8, IL-10, IL-15, IP-10, MCP-1, MIP-1β and VEGF and significantly reduced levels of IL-2, IL-5, IL-7, IL-17, PDGF-BB, and CCL5 compared to plasma. Moreover, T cells derived from EOC-associated ascites displayed poor responsiveness when expanded in vitro [138]. The authors claimed that this non-responsiveness could be explained by a high CD4/CD8 ratio that may indicate the presence of Tregs, reduced IL-2 and elevated IL-6 and IL-10 levels triggering a Th2 inhibitory immune response [138]. This high CD4/CD8 ratio was also associated with poor outcome [109,115,136,139], consistent with other studies [124,140,141]. In contrast, a positive correlation between Tregs and patient prognosis has been reported [140]. High tumor grade correlated with higher frequencies of CD3+, CD68+ CD163+ TAMs, and CD25+ FOXP3+ Treg cells, but Treg frequencies were significant predictors of favorable prognosis in patients with familial ovarian cancer (11/73 patients with *BRCA* mutation) [140]. The presence of FOXP3+ TILs may be linked to positive prognostic factors in optimally debulked HGSC patients [141]. Nevertheless, this disease-specific survival was positively associated along with other TIL markers such as CD8, CD3, TIA-1, CD20 (a B cell surface marker), MHC class I and class II [141].

CD8+ Tregs are also found in EOC [142,143]. They regulate the immunosuppressive TME by limiting immunosurveillance mechanisms and contributing to cancer progression [144]. Recently, Zhang and colleagues showed that CD8+ Tregs are found in the stroma and intraepithelial areas of EOC tumors [143]. CD8+ Tregs are characterized by the expression of FOXP3, CTLA-4, and CD25, but decreased expression of CD28 [143]. CD8+ Tregs were able to convert effector CD8+ T cells into suppressor cells [143]. CD8+ Tregs exert their suppressive function through the secretion of TGF-β1 [142].

Overall, Tregs are considered a critical barrier against antitumoral responses along with tolerance-inducing plasmacytoid DCs, B7-H4+ TAMs, MDSCs, IL-10, TGFβ, and VEGF. All these processes act in concert as a tumor evasion mechanism resulting in tumor progression [145]. Barriers to antitumoral responses are summarized in Figure 2.

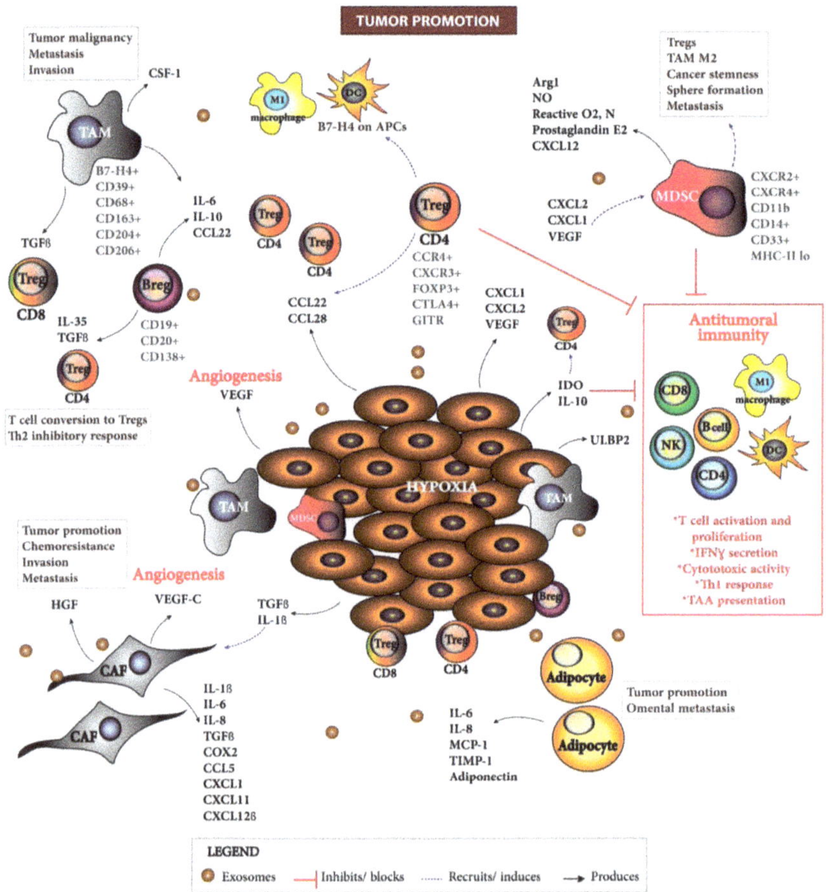

Figure 2. Tumor promoting network in the EOC TME. Outgrowth of EOC provokes hypoxia that induces the expression of chemokines to recruit MDSCs, Tregs, and TAMs. Tregs induce B7-H4 expression on APCs, subsequently blocking cytokine secretion, cytolytic activity, T-cell proliferation and promoting an immunosuppressive TME. EOC cells and MDSCs produce IDO that catabolizes tryptophan, rendering T cells anergic and dysfunctional. MDSCs and TAMs contribute to tumor growth, malignancy, metastasis and stemness. Several tumor promoting cytokines such as IL-6, IL-10 and TGFβ are prominent in the TME. VEGF released by EOC cells and CAFs stimulates angiogenic factors in the TME. CAFs also secrete many factors that mediate tumor cell migration, proliferation, invasion and chemoresistance, and contribute to the immunosuppressive TME. Adipocytes produce FA and cytokines that fuel tumor growth and omental metastasis. Myeloid-derived suppressor cells (MDSCs), tumor-associated macrophages (TAMs), regulatory T cells (Tregs), cancer-associated fibroblasts (CAFs), fatty acids (FA).

3.1.3. B Lymphocytes

B lymphocytes have been reported to have pivotal roles in cancer immunity [146]. Stromal or intraepithelial B lymphocytes have been found in EOC [141]; however their function in tumor development is not yet clear. Their presence is proposed to be associated with a good prognosis depending on the tumor stage and the TME where they are found [4,108,141]. The presence of B cells and CD8+ TILs correlates with increased patient survival compared to CD8+ TILs alone [108]. Nielsen et al. analyzed tumor and serum specimens obtained from patients with HGSC and found that the majority of CD20+ TILs were antigen experienced and suggested to accomplish TAA presentation in the TME since they often co-localized with CD8+ TILs and expressed markers such as MHC-I, MHC-II, CD40, CD80, and CD86 [108]. B cells can achieve antitumor immunity by secreting IFNγ, facilitating CD4+ Th cells to polarize to Th1 responses, and promote T-cell expansion by presenting TAAs [146]. Recently, the positive role of B cells among TILs at metastatic sites from patients with HGSC was reported [147]. B cells were often found in the stroma of metastases and were characterized by a strong memory response against TAAs by production of tumor-specific IgGs (Figure 1). Interestingly, these responses were amplified by chemotherapy [147].

Conversely, a new subset of B cells, regulatory B cells (Bregs), has been recently designated as immunosuppressive cells able to secrete anti-inflammatory mediators such as IL-10, IL-35, and TGF-β, triggering T-cell conversion to Tregs [148] (Figure 2). Indeed, a study that analyzed EOC tumor tissue and omental metastases found that high B cell infiltration negatively correlates with patient survival [149]. High CD20 and CD138 expression correlated with high tumor grade [149]. Analysis of omental specimens from patients with HGSC found that overall survival was 160.6 months in patients with low B-cell expression vs. 47.3 months in those with high B-cell expression, associating increased B-cell infiltration with poorer survival [150]. Similarly, the analysis of post-chemotherapy effusions from ovarian carcinomas revealed that a higher percentage of CD19+ cells (B cell marker) and stage IV disease predicted poor survival for patients [151].

Taken together, it is important to consider that several B-cell subsets with different phenotypes and functions exist, and they may have various roles in modifying the ability of tumors to respond to treatment [146]. Thus, a deep characterization of B-cell subpopulations within the TILs, ascites, and peripheral blood at different stages is crucial in order to provide a better understanding of the capability, importance and therapeutic potential of these cells in EOC.

3.1.4. NK-T Lymphocytes

NK-T cells possess dual-functional capability: as T-cell subsets with a T-cell receptor (TCR)-mediated specific cytotoxicity and as NK cells with acquired killer functions [152,153]. NK-T cells have been found in increased frequencies in EOC tumor ascites compared to blood, but they were decreased at higher tumor grade and in cases of platinum resistance [154]. Moreover, the presence of NK-T cells was inversely correlated with VEGF ascites levels [155]. Since these cells display the most potent cytotoxicity profile, they might be promising agents for adoptive cell immunotherapy [156]. Further studies are needed to better understand the potential antitumoral capacity of these cells and their role in the different EOC TMEs.

3.2. Innate Immune Cells in the Ovarian Cancer TME

3.2.1. NK Cells

Many studies have reported the presence of innate immune cells such as NKs, macrophages and DCs playing important roles in EOC tumorigenesis [103,124,154]. NK cells are crucial effectors in cancer immunosurveillance, recognizing and spontaneously killing virus-infected cells, cancer, and foreign cells hazardous to the host [157]. NK cells mediate antitumoral responses by secreting pro-inflammatory cytokines and chemokines such as IFNγ, TNF, IL-6, GM-CSF and CCL5, which influence antitumor activity and promote innate and adaptive responses in the TME [157–159]

(Figure 1). Tsiatas et al. analyzed 45 fresh specimens from different EOC and found an increased amount of CD56+ NK and NK-T cells along with activated CD4+ and CD8+ CD25+ T cells in serous and endometrioid carcinomas compared with mucinous and clear cell carcinomas [124]. Despite the high concentration of NKs found in ascites compared to peripheral blood, they are functionally impaired [121,160,161]. The influence of infiltrating NK cells on patient outcome is also debated. Analysis of ovarian carcinoma effusions showed that the presence of NK cells at an advanced stage (IV) predicted worse overall survival [151]. However, a positive antitumoral role for NK cells along with effector CD8+ T cells has been reported [162], and NK cell activity of peripheral blood lymphocytes was related to a significant progression-free survival of EOC patients [163]. Importantly, NK cells are activated or not, according to the balance between inhibitory and activating signals through different NK receptors [157]. Like many other cancers, EOC tumors express NK cell receptor ligand ULBP2, which is an indicator of poor prognosis and could promote T-cell dysfunction in the TME [164] (Figure 2). Since NK cells are important players in antitumoral immunity, more studies aiding to characterize their function, phenotypes, incidence and role in the EOC TME are needed to provide new rational for immunotherapies.

3.2.2. Tumor-Associated Macrophages

Both TAMs and MDSCs constitute up to 20% of the EOC TME and are known to maintain and promote an immunosuppressive TME [103] (Figure 2). TAMs are considered the most abundant infiltrating immune cells in EOC tissue and ascites [165,166]. They possess an immunosuppressive M2 phenotype characterized by the expression of CD163, CD204, CD206, and IL-10 [165], and their presence correlates with tumor progression [140,167]. M2 TAMs secrete colony-stimulating factor 1 (CSF-1) that has been found in high levels in malignant EOC [167], and contributes to tumor growth, invasion, and metastasis. Moreover, EOC cells are able to induce an M2 TAM phenotype [168]. TAMs produce the chemokine CCL22 enabling the trafficking of Tregs to the ovarian tumors [130]. EOC cells as well as TAMs are known to express the coinhibitory molecule B7-H4 [169], a member of the B7 family that has a profound inhibitory effect on the growth, cytokine secretion, and development of T-cell cytotoxicity [169]. B7-H4+ TAMs are able to suppress antitumoral responses in EOC [136]. A study of 103 EOC patients showed that enhanced B7-H4 expression in macrophages correlated with Treg cell numbers in the tumor [136]. Tregs and B7-H4+ TAMs were associated with poor patient outcome. Tregs in the TME can induce B7-H4+ TAMs to produce IL-10 and IL-6 [136], further supporting an immunosuppressive milieu. Higher tumor grade correlated with higher frequencies of CD163+ TAMs [140] and worse progression-free survival [170,171]. Importantly, two studies evaluating M1- (HLA-DR, iNOS) and M2-polarization (CD163, VEGF) markers showed that higher M1/M2 TAMs ratio in tumors was associated with a favorable overall survival [172,173], and high serum levels of CD163 predicts poor EOC patient prognosis [174]. In addition, monocyte-derived macrophages in EOC displayed an altered morphology and defective antitumoral functions including defective antibody-dependent cell-mediated cytotoxicity and phagocytosis [175]. Thus, EOC cells and the TME provoke and maintain a strong immunosuppressive M2 phenotype supportive of tumor progression. Immunotherapeutic approaches aiming to switch TAM phenotypes [176] could help the evolution of antitumoral responses and improve patient outcome.

3.2.3. Myeloid-Derived Suppressor Cells

MDSCs are composed of a heterogeneous population of immature myeloid cells that arise in pathologic conditions such as cancer, inflammation and infection, and possess a potent capacity to dampen T-cell responses [177]. MDSCs are considered key inducers of tumor immune evasion and impaired immunity by upregulating arginase-1, nitric oxide, and reactive oxygen species, and by generating reactive nitrogen species [178] (Figure 2). MDSCs also deplete cysteine, induce Tregs, inhibit T-cell activation and proliferation, attenuate the cytolytic ability of NK cells, and trigger a M2 phenotype [103]. Obermajer et al. showed that the frequencies of CD11b+ CD14+ CD33+ CXCR4+

MDSCs in EOC ascites correlated with CXCL12 and prostaglandin E(2) levels [179]. MDSCs derived from EOC patients also increased gene expression of cancer stem cells, sphere formation and metastasis of EOC [180]. Wu et al. characterized typical monocytic CD14+ HLA-DR$^{-/lo}$ MDSCs in peripheral blood and ascites derived from EOC patients and found that MDSCs are enriched in both compartments [181]. Moreover, the density of MDSCs correlated with poor patient prognosis and elevated levels of IL-6 and IL-10 [181,182]. VEGF expression in EOC induced MDSCs recruitment, inhibiting local immunity [182]. A recent study with mouse EOC cells found that *Snai1*, a major transcription factor that induces epithelial-mesenchymal transition (EMT), mediates EOC progression by upregulating CXCR2 ligands, enabling the recruitment of MDSCs [183]. EOC cells also attracts myeloid cells by producing adenosine [184]. Hence, strategies targeting MDSCs could release the brakes against antitumoral responses. Metformin, a drug used to treat type 2 diabetes, may trigger EOC clinical benefit by improving antitumoral T-cell responses that are impeded by MDSCs in the TME, since this drug can block MDSC suppressor functions by decreasing CD39 and CD73 expression [185].

4. Exosomes

Highly proliferating cells such as cancer cells produce large amounts of exosomes which are small (40–100 nm) extracellular vesicles [186]. EOC tumor-derived exosomes carry cell membrane proteins and cargo proteins that could be used for diagnostics (EP-CAM) and immunotherapeutic targeting such as neoantigens and TAAs (Her2-neu, CA-125) [186], proteins (TGF-β1) [187], and miRNAs (miR-21) [188] that are involved in disease progression, metastasis, and chemoresistance [186], as well as immunomodulatory proteins (FAS-L) [189]. Exosomes can be taken up by other cancer cells, CAFs, and immune cells, therefore playing an important role in intercellular communication. Thus far, 2035 exosome cargo molecules have been identified from EOC cells in ExoCarta, a database for exosome cargo [190,191]. Exosomes derived from human patient ascites promotes tumor progression in vivo [189,192], and are proposed to have direct and indirect roles in modulating the immune TME, as exosomes could also be taken up by NKs and B cells [192] (Figure 2). In other disease models, such as melanoma and colorectal cancer, exosomes mediate immunosuppression and immune tolerance by suppressing the activation of T and NK cells, monocytes, modulating T-cell inhibitory molecules expression, and inducing CD8+ T-cell apoptosis [193,194]. FAS-L and TRAIL expression on EOC-derived exosomes inhibit activation of peripheral blood mononuclear cells by DCs through induction of apoptosis [189]. EOC-derived exosomes express ligands (MICA/B and ULBP1-3) for the NK receptor NKG2D, acting as a decoy and interfering with NK-mediated targeting of tumor cells [195]. Greater understanding of the complex network of the intercellular communication between EOC cells, CAFs, and immune cells is needed for the rational design of immunotherapeutic interventions, or leveraged for nanomedicine applications such as TAA loaded-DC-derived exosomes [196] and drug delivery systems [186].

5. Cancer-Associated Fibroblasts

CAFs are activated fibroblasts that express α-smooth muscle actin and fibroblast activation protein. They make up 7–85% [197] of the tumor and are the primary stromal cell type in the TME. Cross-talk between epithelial and stromal compartments creates a positive feedback loop, a supportive hyper-activated storm of cytokines, chemokines, angiogenetic factors, and EMT-promoting factors, to promote tumor progression and chemoresistance (Figure 2). CAFs from ovarian cancer patients secrete high levels of hepatocyte growth factor (HGF) that promotes cancer cell proliferation, chemoresistance, invasion, and migration though constitutive activation of cMet/PI3K/Akt pathways and glucose-regulated protein 78 (GRP78) [198,199]. CAFs produce pro-inflammatory cytokines COX-2 and CXCL1 [200], CCL5 [201], CXCL11 [202], and IL-6 [203], which can promote proliferation and EMT. In addition to their direct actions on cancer cells, CAFs also produce exosomes with high levels of TGF-β1 that subsequently activates normal fibroblasts [187]. Interestingly, Givel et al. identified four CAFs subsets in HGSC, finding an accumulation of one subset, CAF-S1, in the mesenchymal molecular subtype of HGSC. CAF-S1 is associated with an immunosuppressive TME, due to its high levels of expression of CXCL12β, which recruits Tregs to the tumor. The CAF-S1 cells also express CD73, B7-H3 and IL-6, which subsequently promote survival and proliferation of Tregs [204]. Thus, CAFs can make major contributions to the creation of an immunosuppressive TME.

On the other hand, EOC cells can stimulate the activation of CAFs by producing high levels of interleukin-1β (IL-1β) [205] and TGF-β [206], which subsequently induces secretion of IL-8, IL-6, IL-1β, VEGF, and growth regulated oncogene-alpha (GRO-α) by CAFs to promote tumor progression [205]. EOC cells release exosomes not only to activate tumor cells, but also to reprogram normal fibroblasts into CAFs [207]. Furthermore, CAFs act on endothelial cells via the secretion of VEGF-C [208] or by upregulating genes such as lipoma-preferred partner, to promote angiogenesis, which leads to tumor progression and chemoresistance [209]. Cross-talk between CAFs and cancer cells, as well as endothelial cells and immune cells, suggests that targeting signaling mechanisms in this relationship may combat chemoresistance and immune modulation better than singly targeting the epithelial compartment.

6. Adipocytes and the Omentum

The unique TME of the omentum, a large visceral fat pad that covers the bowel and abdomen cavities [210,211], suggests a two-step model of omental metastasis and tumorigenesis where ovarian cancer cells preferentially and rapidly home to "milky spots" [212] in the omentum, prior to spreading throughout non-"milky spot" areas of the omentum and peritoneal cavity [213–217]. "Milky spots" are highly vascularized regions with aggregates of immune cells, capable of innate and adaptive immune functions, and antigen presentation similar to lymph node structures [212]. The involvement of the omentum and adipose tissue suggests the need to develop intraperitoneal immunotherapy similar to the advances seen with intraperitoneal chemotherapy.

Adipocytes in the omentum produce cytokines and chemokines, including highly secreted IL-6, IL-8, MCP-1, tissue inhibitor of metalloproteinases-1 (TIMP-1) and adiponectin, to promote cancer growth and omental metastases. Adipocytes can alter their lipid metabolism via Fatty acid–binding protein 4 (FAB4) to undergo lipolysis providing fatty acids (FA) to cancer cells as a fuel source for rapid tumor growth [216]. Cancer cells themselves can also alter lipid metabolism, often by upregulating FA receptor CD36 [218] and FAB4 in omental metastases at the tumor/adipocyte interface to promote FA and cholesterol uptake from adipocytes [216] to fuel tumor progression.

Many studies have suggested an association between obesity and the incidence of ovarian cancer as well as an association with poor prognosis [219]. Indeed, in a murine model of ovarian cancer, metastasis and tumor growth is supported in obese mice through altered regulation of FA pathway and increased immunosuppression, demonstrated by a decreased ratio of M1/M2 macrophages [220] (Figure 2). Improved understanding of how adipocytes and the omentum support ovarian cancer growth and promote peritoneal metastases will reveal therapeutic targets for both conventional therapy and immunotherapy. It will be important to consider how age and obesity [221–223] may dictate differences in response to immunotherapy and how current models with young, lean mice may fail to accurately model responses to immunotherapy.

7. Conclusions

In summary, in order to develop better immunotherapies for EOC we need to identify and consider all key elements found in the TME of not only primary tumors but also in ascites and metastases with a focus on how these features affect and are affected by different cancer therapies. It is crucial to take into account the quality of the TME (immune-activating vs. immune-suppressing mechanisms), tumor immunogenicity, tumor burden mutations, tumor stage, patient overall condition, and age, as well as treatment effects on the TME (chemotherapy, neoadjuvant chemotherapy, surgery debulking). Each of these factors may influence the outcome of EOC and the responses to cancer immunotherapies. Moreover, to avoid tumor recurrence, EOC characteristics such as TAA presentation, expression of coinhibitory molecules, production of immunosuppressive cytokines and chemokines should all be considered to find therapeutic combinations that could synergize and achieve maximal benefits to eliminate EOC. Other articles in this special issue will address some of these topics, including the exploration of promising immunotherapies for HGSC that are currently under investigation [224].

Author Contributions: All authors contributed to the research, writing, and editing of this review article.

Funding: The authors' research relevant to this article is funded by grants to BCV from BioCanRx and the Canadian Institutes of Health Research, as well as a fellowship from the Fonds de recherche du Québec—Santé (GMR), and a Vanier Canada Graduate Scholarship (CM).

Acknowledgments: The authors are grateful for the donations and support from the local community of ovarian cancer patients, especially the late Margaret Craig, and the Carol Annibale Ovarian Cancer Foundation.

Conflicts of Interest: The authors declare no conflict of interest.

References

1. Vesely, M.D.; Schreiber, R.D. Cancer immunoediting: Antigens, mechanisms, and implications to cancer immunotherapy. *Ann. N. Y. Acad. Sci.* **2013**, *1284*, 1–5. [CrossRef] [PubMed]
2. Santoiemma, P.P.; Powell, D.J. Tumor infiltrating lymphocytes in ovarian cancer. *Cancer Biol. Ther.* **2015**, *16*, 807–820. [CrossRef] [PubMed]
3. Santoiemma, P.P.; Reyes, C.; Wang, L.-P.; McLane, M.W.; Feldman, M.D.; Tanyi, J.L.; Powell, D.J. Systematic evaluation of multiple immune markers reveals prognostic factors in ovarian cancer. *Gynecol. Oncol.* **2016**, *143*, 120–127. [CrossRef] [PubMed]
4. Stumpf, M.; Hasenburg, A.; Riener, M.-O.; Jütting, U.; Wang, C.; Shen, Y.; Orlowska-Volk, M.; Fisch, P.; Wang, Z.; Gitsch, G.; et al. Intraepithelial CD8-positive T lymphocytes predict survival for patients with serous stage III ovarian carcinomas: Relevance of clonal selection of T lymphocytes. *Br. J. Cancer* **2009**, *101*, 1513–1521. [CrossRef] [PubMed]
5. Zhang, L.; Conejo-Garcia, J.R.; Katsaros, D.; Gimotty, P.A.; Massobrio, M.; Regnani, G.; Makrigiannakis, A.; Gray, H.; Schlienger, K.; Liebman, M.N.; et al. Intratumoral T cells, recurrence, and survival in epithelial ovarian cancer. *N. Engl. J. Med.* **2003**, *348*, 203–213. [CrossRef] [PubMed]
6. Coukos, G.; Tanyi, J.; Kandalaft, L.E. Opportunities in immunotherapy of ovarian cancer. *Ann. Oncol. Off. J. Eur. Soc. Med. Oncol.* **2016**, *27* (Suppl. 1), i11–i15. [CrossRef] [PubMed]
7. Odunsi, K. Immunotherapy in ovarian cancer. *Ann. Oncol. Off. J. Eur. Soc. Med. Oncol.* **2017**, *28*, viii1–viii7. [CrossRef] [PubMed]
8. Wang, R.-F.; Wang, H.Y. Immune targets and neoantigens for cancer immunotherapy and precision medicine. *Cell Res.* **2017**, *27*, 11–37. [CrossRef] [PubMed]
9. Ilyas, S.; Yang, J.C. Landscape of Tumor Antigens in T-Cell Immunotherapy. *J. Immunol.* **2015**, *195*, 5117–5122. [CrossRef] [PubMed]
10. Rizvi, N.A.; Hellmann, M.D.; Snyder, A.; Kvistborg, P.; Makarov, V.; Havel, J.J.; Lee, W.; Yuan, J.; Wong, P.; Ho, T.S.; et al. Cancer immunology. Mutational landscape determines sensitivity to PD-1 blockade in non-small cell lung cancer. *Science* **2015**, *348*, 124–128. [CrossRef] [PubMed]
11. Van Allen, E.M.; Miao, D.; Schilling, B.; Shukla, S.A.; Blank, C.; Zimmer, L.; Sucker, A.; Hillen, U.; Foppen, M.H.G.; Goldinger, S.M.; et al. Genomic correlates of response to CTLA-4 blockade in metastatic melanoma. *Science* **2015**, *350*, 207–211. [CrossRef] [PubMed]
12. Chalmers, Z.R.; Connelly, C.F.; Fabrizio, D.; Gay, L.; Ali, S.M.; Ennis, R.; Schrock, A.; Campbell, B.; Shlien, A.; Chmielecki, J.; et al. Analysis of 100,000 human cancer genomes reveals the landscape of tumor mutational burden. *Genome Med.* **2017**, *9*, 34. [CrossRef] [PubMed]
13. Martin, S.D.; Brown, S.D.; Wick, D.A.; Nielsen, J.S.; Kroeger, D.R.; Twumasi-Boateng, K.; Holt, R.A.; Nelson, B.H. Low mutation burden in ovarian cancer may limit the utility of neoantigen-targeted vaccines. *PLoS ONE* **2016**, *11*, e0155189. [CrossRef] [PubMed]
14. Hamanishi, J.; Mandai, M.; Ikeda, T.; Minami, M.; Kawaguchi, A.; Murayama, T.; Kanai, M.; Mori, Y.; Matsumoto, S.; Chikuma, S.; et al. Safety and antitumor activity of anti-PD-1 antibody, Nivolumab, in patients with platinum-resistant Ovarian Cancer. *J. Clin. Oncol. Off. J. Am. Soc. Clin. Oncol.* **2015**, *33*, 4015–4022. [CrossRef] [PubMed]

15. Varga, A.; Piha-Paul, S.A.; Ott, P.A.; Mehnert, J.M.; Berton-Rigaud, D.; Johnson, E.A.; Cheng, J.D.; Yuan, S.; Rubin, E.H.; Matei, D.E. Antitumor activity and safety of pembrolizumab in patients (pts) with PD-L1 positive advanced ovarian cancer: Interim results from a phase Ib study. *J. Clin. Oncol.* **2015**, *33*, 5510. [CrossRef]
16. Donahue, R.N.; Lepone, L.M.; Grenga, I.; Jochems, C.; Fantini, M.; Madan, R.A.; Heery, C.R.; Gulley, J.L.; Schlom, J. Analyses of the peripheral immunome following multiple administrations of avelumab, a human IgG1 anti-PD-L1 monoclonal antibody. *J. Immunother. Cancer* **2017**, *5*, 20. [CrossRef] [PubMed]
17. Disis, M.L.; Patel, M.R.; Pant, S.; Hamilton, E.P.; Lockhart, A.C.; Kelly, K.; Beck, J.T.; Gordon, M.S.; Weiss, G.J.; Taylor, M.H.; et al. Avelumab (MSB0010718C; anti-PD-L1) in patients with recurrent/refractory ovarian cancer from the JAVELIN Solid Tumor phase Ib trial: Safety and clinical activity. *J. Clin. Oncol.* **2016**, *34*, 5533. [CrossRef]
18. Brahmer, J.R.; Tykodi, S.S.; Chow, L.Q.M.; Hwu, W.-J.; Topalian, S.L.; Hwu, P.; Drake, C.G.; Camacho, L.H.; Kauh, J.; Odunsi, K.; et al. Safety and activity of anti–PD-L1 antibody in patients with advanced cancer. *N. Engl. J. Med.* **2012**, *366*, 2455–2465. [CrossRef] [PubMed]
19. Hodi, F.S.; Mihm, M.C.; Soiffer, R.J.; Haluska, F.G.; Butler, M.; Seiden, M.V.; Davis, T.; Henry-Spires, R.; MacRae, S.; Willman, A.; et al. Biologic activity of cytotoxic T lymphocyte-associated antigen 4 antibody blockade in previously vaccinated metastatic melanoma and ovarian carcinoma patients. *Proc. Natl. Acad. Sci. USA* **2003**, *100*, 4712–4717. [CrossRef] [PubMed]
20. Hodi, F.S.; Butler, M.; Oble, D.A.; Seiden, M.V.; Haluska, F.G.; Kruse, A.; MacRae, S.; Nelson, M.; Canning, C.; Lowy, I.; et al. Immunologic and clinical effects of antibody blockade of cytotoxic T lymphocyte-associated antigen 4 in previously vaccinated cancer patients. *Proc. Natl. Acad. Sci. USA* **2008**, *105*, 3005–3010. [CrossRef] [PubMed]
21. Bellone, S.; Buza, N.; Choi, J.; Zammataro, L.; Gay, L.; Elvin, J.; Rimm, D.L.; Liu, Y.; Ratner, E.S.; Schwartz, P.E.; et al. Exceptional response to Pembrolizumab in a metastatic, chemotherapy/radiation-resistant ovarian cancer patient harboring a PD-L1-genetic rearrangement. *Clin. Cancer Res.* **2018**. [CrossRef] [PubMed]
22. Huang, J.; Jochems, C.; Anderson, A.M.; Talaie, T.; Jales, A.; Madan, R.A.; Hodge, J.W.; Tsang, K.Y.; Liewehr, D.J.; Steinberg, S.M.; et al. Soluble CD27-pool in humans may contribute to T cell activation and tumor immunity. *J. Immunol.* **2013**, *190*, 6250–6258. [CrossRef] [PubMed]
23. Huang, J.; Jochems, C.; Talaie, T.; Anderson, A.; Jales, A.; Tsang, K.Y.; Madan, R.A.; Gulley, J.L.; Schlom, J. Elevated serum soluble CD40 ligand in cancer patients may play an immunosuppressive role. *Blood* **2012**, *120*, 3030–3038. [CrossRef] [PubMed]
24. Tammela, J.; Uenaka, A.; Ono, T.; Noguchi, Y.; Jungbluth, A.A.; Mhawech-Fauceglia, P.; Qian, F.; Schneider, S.; Sharma, S.; Driscoll, D.; et al. OY-TES-1 expression and serum immunoreactivity in epithelial ovarian cancer. *Int. J. Oncol.* **2006**, *29*, 903–910. [CrossRef] [PubMed]
25. Tammela, J.; Jungbluth, A.A.; Qian, F.; Santiago, D.; Scanlan, M.J.; Keitz, B.; Driscoll, D.; Rodabaugh, K.; Lele, S.; Old, L.J.; et al. SCP-1 cancer/testis antigen is a prognostic indicator and a candidate target for immunotherapy in epithelial ovarian cancer. *Cancer Immun.* **2004**, *4*, 10. [PubMed]
26. Garg, M.; Chaurasiya, D.; Rana, R.; Jagadish, N.; Kanojia, D.; Dudha, N.; Kamran, N.; Salhan, S.; Bhatnagar, A.; Suri, S.; et al. Sperm-associated antigen 9, a novel cancer testis antigen, is a potential target for immunotherapy in epithelial ovarian cancer. *Clin. Cancer Res.* **2007**, *13*, 1421–1428. [CrossRef] [PubMed]
27. Agarwal, S.; Saini, S.; Parashar, D.; Verma, A.; Sinha, A.; Jagadish, N.; Batra, A.; Suri, S.; Gupta, A.; Ansari, A.S.; et al. The novel cancer-testis antigen A-kinase anchor protein 4 (AKAP4) is a potential target for immunotherapy of ovarian serous carcinoma. *Oncoimmunology* **2013**, *2*. [CrossRef] [PubMed]
28. Odunsi, K.; Jungbluth, A.A.; Stockert, E.; Qian, F.; Gnjatic, S.; Tammela, J.; Intengan, M.; Beck, A.; Keitz, B.; Santiago, D.; et al. NY-ESO-1 and LAGE-1 Cancer-Testis Antigens Are Potential Targets for Immunotherapy in Epithelial Ovarian Cancer. *Cancer Res.* **2003**, *63*, 6076–6083. [PubMed]
29. Daudi, S.; Eng, K.H.; Mhawech-Fauceglia, P.; Morrison, C.; Miliotto, A.; Beck, A.; Matsuzaki, J.; Tsuji, T.; Groman, A.; Gnjatic, S.; et al. Expression and immune responses to MAGE antigens predict survival in epithelial ovarian cancer. *PLoS ONE* **2014**, *9*, e104099. [CrossRef] [PubMed]
30. Gillespie, A.M.; Rodgers, S.; Wilson, A.P.; Tidy, J.; Rees, R.C.; Coleman, R.E.; Murray, A.K. MAGE, BAGE and GAGE: Tumour antigen expression in benign and malignant ovarian tissue. *Br. J. Cancer* **1998**, *78*, 816–821. [CrossRef] [PubMed]
31. Yamada, A.; Kataoka, A.; Shichijo, S.; Kamura, T.; Imai, Y.; Nishida, T.; Itoh, K. Expression of MAGE-1, MAGE-2, MAGE-3/-6 and MAGE-4A/-4B genes in ovarian tumors. *Int. J. Cancer* **1995**, *64*, 388–393. [CrossRef] [PubMed]

32. Hardwick, N.; Chung, V.; Cristea, M.; Ellenhorn, J.D.; Diamond, D.J. Overcoming immunosuppression to enhance a p53MVA vaccine. *Oncoimmunology* **2014**, *3*, e958949. [CrossRef] [PubMed]
33. The Cancer Genome Atlas Research Network. Integrated genomic analyses of ovarian carcinoma. *Nature* **2011**, *474*, 609–615. [CrossRef]
34. Fajac, A.; Benard, J.; Lhomme, C.; Rey, A.; Duvillard, P.; Rochard, F.; Bernaudin, J.F.; Riou, G. c-erbB2 gene amplification and protein expression in ovarian epithelial tumors: Evaluation of their respective prognostic significance by multivariate analysis. *Int. J. Cancer* **1995**, *64*, 146–151. [CrossRef] [PubMed]
35. Ioannides, C.G.; Fisk, B.; Fan, D.; Biddison, W.E.; Wharton, J.T.; O'brian, C.A. Cytotoxic T cells isolated from ovarian malignant ascites recognize a peptide derived from the HER-2/neu proto-oncogene. *Cell. Immunol.* **1993**, *151*, 225–234. [CrossRef] [PubMed]
36. Lanitis, E.; Dangaj, D.; Hagemann, I.S.; Song, D.-G.; Best, A.; Sandaltzopoulos, R.; Coukos, G.; Powell, D.J. Primary human ovarian epithelial cancer cells broadly express HER2 at immunologically-detectable levels. *PLoS ONE* **2012**, *7*, e49829. [CrossRef] [PubMed]
37. Nielsen, J.S.; Jakobsen, E.; HOlund, B.; Bertelsen, K.; Jakobsen, A. Prognostic significance of p53, Her-2, and EGFR overexpression in borderline and epithelial ovarian cancer. *Int. J. Gynecol. Cancer* **2004**, *14*, 1086–1096. [CrossRef] [PubMed]
38. Netinatsunthorn, W.; Hanprasertpong, J.; Dechsukhum, C.; Leetanaporn, R.; Geater, A. WT1 gene expression as a prognostic marker in advanced serous epithelial ovarian carcinoma: An immunohistochemical study. *BMC Cancer* **2006**, *6*, 90. [CrossRef] [PubMed]
39. Sallum, L.F.; Andrade, L.; Ramalho, S.; Ferracini, A.C.; de Andrade Natal, R.; Brito, A.B.C.; Sarian, L.O.; Derchain, S. WT1, p53 and p16 expression in the diagnosis of low- and high-grade serous ovarian carcinomas and their relation to prognosis. *Oncotarget* **2018**, *9*, 15818–15827. [CrossRef] [PubMed]
40. Hassan, R.; Kreitman, R.J.; Pastan, I.; Willingham, M.C. Localization of mesothelin in epithelial ovarian cancer. *Appl. Immunohistochem. Mol. Morphol. AIMM* **2005**, *13*, 243–247. [CrossRef] [PubMed]
41. Tanyi, J.L.; Haas, A.R.; Beatty, G.L.; Stashwick, C.J.; O'Hara, M.H.; Morgan, M.A.; Porter, D.L.; Melenhorst, J.J.; Plesa, G.; Lacey, S.F.; et al. Anti-mesothelin chimeric antigen receptor T cells in patients with epithelial ovarian cancer. *J. Clin. Oncol.* **2016**, *34*, 5511. [CrossRef]
42. Felder, M.; Kapur, A.; Gonzalez-Bosquet, J.; Horibata, S.; Heintz, J.; Albrecht, R.; Fass, L.; Kaur, J.; Hu, K.; Shojaei, H.; et al. MUC16 (CA125): Tumor biomarker to cancer therapy, a work in progress. *Mol. Cancer* **2014**, *13*, 129. [CrossRef] [PubMed]
43. Strickland, K.C.; Howitt, B.E.; Shukla, S.A.; Rodig, S.; Ritterhouse, L.L.; Liu, J.F.; Garber, J.E.; Chowdhury, D.; Wu, C.J.; D'Andrea, A.D.; et al. Association and prognostic significance of BRCA1/2-mutation status with neoantigen load, number of tumor-infiltrating lymphocytes and expression of PD-1/PD-L1 in high grade serous ovarian cancer. *Oncotarget* **2016**, *7*, 13587–13598. [CrossRef] [PubMed]
44. Odunsi, K.; Qian, F.; Matsuzaki, J.; Mhawech-Fauceglia, P.; Andrews, C.; Hoffman, E.W.; Pan, L.; Ritter, G.; Villella, J.; Thomas, B.; et al. Vaccination with an NY-ESO-1 peptide of HLA class I/II specificities induces integrated humoral and T cell responses in ovarian cancer. *Proc. Natl. Acad. Sci. USA* **2007**, *104*, 12837–12842. [CrossRef] [PubMed]
45. Sabbatini, P.; Tsuji, T.; Ferran, L.; Ritter, E.; Sedrak, C.; Tuballes, K.; Jungbluth, A.A.; Ritter, G.; Aghajanian, C.; Bell-McGuinn, K.; et al. Phase I trial of overlapping long peptides from a tumor self-antigen and poly-ICLC shows rapid induction of integrated immune response in ovarian cancer patients. *Clin. Cancer Res.* **2012**, *18*, 6497–6508. [CrossRef] [PubMed]
46. Diefenbach, C.S.M.; Gnjatic, S.; Sabbatini, P.; Aghajanian, C.; Hensley, M.L.; Spriggs, D.R.; Iasonos, A.; Lee, H.; Dupont, B.; Pezzulli, S.; et al. Safety and immunogenicity study of NY-ESO-1b peptide and montanide ISA-51 vaccination of patients with epithelial ovarian cancer in high-risk first remission. *Clin. Cancer Res* **2008**, *14*, 2740–2748. [CrossRef] [PubMed]
47. Odunsi, K.; Matsuzaki, J.; Karbach, J.; Neumann, A.; Mhawech-Fauceglia, P.; Miller, A.; Beck, A.; Morrison, C.D.; Ritter, G.; Godoy, H.; et al. Efficacy of vaccination with recombinant vaccinia and fowlpox vectors expressing NY-ESO-1 antigen in ovarian cancer and melanoma patients. *Proc. Natl. Acad. Sci. USA* **2012**, *109*, 5797–5802. [CrossRef] [PubMed]

48. Hardwick, N.R.; Frankel, P.; Ruel, C.; Kilpatrick, J.; Tsai, W.; Kos, F.; Kaltcheva, T.; Leong, L.; Morgan, R.; Chung, V.; et al. p53-reactive T cells are associated with clinical benefit in patients with platinum-resistant epithelial ovarian cancer after treatment with a p53 vaccine and Gemcitabine chemotherapy. *Clin. Cancer Res.* **2018**. [CrossRef] [PubMed]
49. Leffers, N.; Vermeij, R.; Hoogeboom, B.-N.; Schulze, U.R.; Wolf, R.; Hamming, I.E.; van der Zee, A.G.; Melief, K.J.; van der Burg, S.H.; Daemen, T.; et al. Long-term clinical and immunological effects of p53-SLP® vaccine in patients with ovarian cancer. *Int. J. Cancer* **2012**, *130*, 105–112. [CrossRef] [PubMed]
50. Coosemans, A.; Vanderstraeten, A.; Tuyaerts, S.; Verschuere, T.; Moerman, P.; Berneman, Z.; Vergote, I.; Amant, F.; Gool, S.W.V. Immunological response after WT1 mRNA-loaded dendritic cell immunotherapy in ovarian carcinoma and carcinosarcoma. *Anticancer Res.* **2013**, *33*, 3855–3859. [PubMed]
51. Miyatake, T.; Ueda, Y.; Morimoto, A.; Enomoto, T.; Nishida, S.; Shirakata, T.; Oka, Y.; Tsuboi, A.; Oji, Y.; Hosen, N.; et al. WT1 peptide immunotherapy for gynecologic malignancies resistant to conventional therapies: A phase II trial. *J. Cancer Res. Clin. Oncol.* **2013**, *139*, 457–463. [CrossRef] [PubMed]
52. Sabbatini, P.; Harter, P.; Scambia, G.; Sehouli, J.; Meier, W.; Wimberger, P.; Baumann, K.H.; Kurzeder, C.; Schmalfeldt, B.; Cibula, D.; et al. Abagovomab as maintenance therapy in patients with epithelial ovarian cancer: A phase III trial of the AGO OVAR, COGI, GINECO, and GEICO—The MIMOSA study. *J. Clin. Oncol. Off. J. Am. Soc. Clin. Oncol.* **2013**, *31*, 1554–1561. [CrossRef] [PubMed]
53. Braly, P.; Nicodemus, C.F.; Chu, C.; Collins, Y.; Edwards, R.; Gordon, A.; McGuire, W.; Schoonmaker, C.; Whiteside, T.; Smith, L.M.; et al. The immune adjuvant properties of front-line carboplatin-paclitaxel: A randomized phase 2 study of alternative schedules of intravenous oregovomab chemoimmunotherapy in advanced ovarian cancer. *J. Immunother.* **2009**, *32*, 54–65. [CrossRef] [PubMed]
54. Berek, J.; Taylor, P.; McGuire, W.; Smith, L.M.; Schultes, B.; Nicodemus, C.F. Oregovomab maintenance monoimmunotherapy does not improve outcomes in advanced ovarian cancer. *J. Clin. Oncol. Off. J. Am. Soc. Clin. Oncol.* **2009**, *27*, 418–425. [CrossRef] [PubMed]
55. Reinartz, S.; Köhler, S.; Schlebusch, H.; Krista, K.; Giffels, P.; Renke, K.; Huober, J.; Möbus, V.; Kreienberg, R.; DuBois, A.; et al. Vaccination of patients with advanced ovarian carcinoma with the anti-idiotype ACA125: Immunological response and survival (phase Ib/II). *Clin. Cancer Res.* **2004**, *10*, 1580–1587. [CrossRef] [PubMed]
56. Koneru, M.; O'Cearbhaill, R.; Pendharkar, S.; Spriggs, D.R.; Brentjens, R.J. A phase I clinical trial of adoptive T cell therapy using IL-12 secreting MUC-16(ecto) directed chimeric antigen receptors for recurrent ovarian cancer. *J. Transl. Med.* **2015**, *13*, 102. [CrossRef] [PubMed]
57. Schuster, H.; Peper, J.K.; Bösmüller, H.-C.; Röhle, K.; Backert, L.; Bilich, T.; Ney, B.; Löffler, M.W.; Kowalewski, D.J.; Trautwein, N.; et al. The immunopeptidomic landscape of ovarian carcinomas. *Proc. Natl. Acad. Sci. USA* **2017**, *114*, E9942–E9951. [CrossRef] [PubMed]
58. Tanyi, J.L.; Bobisse, S.; Ophir, E.; Tuyaerts, S.; Roberti, A.; Genolet, R.; Baumgartner, P.; Stevenson, B.J.; Iseli, C.; Dangaj, D.; et al. Personalized cancer vaccine effectively mobilizes antitumor T cell immunity in ovarian cancer. *Sci. Transl. Med.* **2018**, *10*, eaao5931. [CrossRef] [PubMed]
59. Alexandrov, L.B.; Nik-Zainal, S.; Wedge, D.C.; Aparicio, S.A.J.R.; Behjati, S.; Biankin, A.V.; Bignell, G.R.; Bolli, N.; Borg, A.; Børresen-Dale, A.-L.; et al. Signatures of mutational processes in human cancer. *Nature* **2013**, *500*, 415–421. [CrossRef] [PubMed]
60. Wick, D.A.; Webb, J.R.; Nielsen, J.S.; Martin, S.D.; Kroeger, D.R.; Milne, K.; Castellarin, M.; Twumasi-Boateng, K.; Watson, P.H.; Holt, R.A.; et al. Surveillance of the tumor mutanome by T cells during progression from primary to recurrent ovarian cancer. *Clin. Cancer Res.* **2014**, *20*, 1125–1134. [CrossRef] [PubMed]
61. Hamanishi, J.; Mandai, M.; Konishi, I. Immune checkpoint inhibition in ovarian cancer. *Int. Immunol.* **2016**, *28*, 339–348. [CrossRef] [PubMed]
62. Jelinic, P.; Ricca, J.; Van Oudenhove, E.; Olvera, N.; Merghoub, T.; Levine, D.A.; Zamarin, D. Immune-active microenvironment in Small Cell Carcinoma of the Ovary, Hypercalcemic Type: Rationale for immune checkpoint blockade. *JNCI J. Natl. Cancer Inst.* **2018**. [CrossRef] [PubMed]
63. Zhang, A.W.; McPherson, A.; Milne, K.; Kroeger, D.R.; Hamilton, P.T.; Miranda, A.; Funnell, T.; Little, N.; de Souza, C.P.E.; Laan, S.; et al. Interfaces of malignant and immunologic clonal dynamics in ovarian cancer. *Cell* **2018**, *173*, 1755–1769. [CrossRef] [PubMed]
64. McGranahan, N.; Swanton, C. Clonal heterogeneity and tumor evolution: Past, present, and the future. *Cell* **2017**, *168*, 613–628. [CrossRef] [PubMed]

65. Bashashati, A.; Ha, G.; Tone, A.; Ding, J.; Prentice, L.M.; Roth, A.; Rosner, J.; Shumansky, K.; Kalloger, S.; Senz, J.; et al. Distinct evolutionary trajectories of primary high-grade serous ovarian cancers revealed through spatial mutational profiling. *J. Pathol.* **2013**, *231*, 21–34. [CrossRef] [PubMed]
66. Caballero, O.L.; Chen, Y.-T. Cancer/testis (CT) antigens: Potential targets for immunotherapy. *Cancer Sci.* **2009**, *100*, 2014–2021. [CrossRef] [PubMed]
67. Want, M.Y.; Lugade, A.A.; Battaglia, S.; Odunsi, K. Nature of tumor rejection antigens in ovarian cancer. *Immunology* **2018**. [CrossRef] [PubMed]
68. Grob, J.-J.; Mortier, L.; D'Hondt, L.; Grange, F.; Baurain, J.F.; Dréno, B.; Lebbe, C.; Robert, C.; Dompmartin, A.; Neyns, B.; et al. Safety and immunogenicity of MAGE-A3 cancer immunotherapeutic with dacarbazine in patients with MAGE-A3-positive metastatic cutaneous melanoma: An open phase I/II study with a first assessment of a predictive gene signature. *ESMO Open* **2017**, *2*, e000203. [CrossRef] [PubMed]
69. Vansteenkiste, J.; Zielinski, M.; Linder, A.; Dahabreh, J.; Gonzalez, E.E.; Malinowski, W.; Lopez-Brea, M.; Vanakesa, T.; Jassem, J.; Kalofonos, H.; et al. Adjuvant MAGE-A3 immunotherapy in resected non-small-cell lung cancer: Phase II randomized study results. *J. Clin. Oncol. Off. J. Am. Soc. Clin. Oncol.* **2013**, *31*, 2396–2403. [CrossRef] [PubMed]
70. Matsuzaki, J.; Qian, F.; Luescher, I.; Lele, S.; Ritter, G.; Shrikant, P.A.; Gnjatic, S.; Old, L.J.; Odunsi, K. Recognition of naturally processed and ovarian cancer reactive CD8+ T cell epitopes within a promiscuous HLA class II T-helper region of NY-ESO-1. *Cancer Immunol. Immunother. CII* **2008**, *57*, 1185–1195. [CrossRef] [PubMed]
71. Matsuzaki, J.; Gnjatic, S.; Mhawech-Fauceglia, P.; Beck, A.; Miller, A.; Tsuji, T.; Eppolito, C.; Qian, F.; Lele, S.; Shrikant, P.; et al. Tumor-infiltrating NY-ESO-1–specific CD8+ T cells are negatively regulated by LAG-3 and PD-1 in human ovarian cancer. *Proc. Natl. Acad. Sci. USA* **2010**, *107*, 7875–7880. [CrossRef] [PubMed]
72. Hanlon, D.J.; Aldo, P.B.; Devine, L.; Alvero, A.B.; Engberg, A.K.; Edelson, R.; Mor, G. Enhanced stimulation of anti-ovarian cancer CD8(+) T cells by dendritic cells loaded with nanoparticle encapsulated tumor antigen. *Am. J. Reprod. Immunol.* **2011**, *65*, 597–609. [CrossRef] [PubMed]
73. Xue, W.; Metheringham, R.L.; Brentville, V.A.; Gunn, B.; Symonds, P.; Yagita, H.; Ramage, J.M.; Durrant, L.G. SCIB2, an antibody DNA vaccine encoding NY-ESO-1 epitopes, induces potent antitumor immunity which is further enhanced by checkpoint blockade. *Oncoimmunology* **2016**, *5*, e1169353. [CrossRef] [PubMed]
74. Odunsi, K.; Matsuzaki, J.; James, S.R.; Mhawech-Fauceglia, P.; Tsuji, T.; Miller, A.; Zhang, W.; Akers, S.N.; Griffiths, E.A.; Miliotto, A.; et al. Epigenetic potentiation of NY-ESO-1 vaccine therapy in human ovarian cancer. *Cancer Immunol. Res.* **2014**, *2*, 37–49. [CrossRef] [PubMed]
75. Reid, B.M.; Permuth, J.B.; Sellers, T.A. Epidemiology of ovarian cancer: A review. *Cancer Biol. Med.* **2017**, *14*, 9–32. [CrossRef] [PubMed]
76. Leffers, N.; Daemen, T.; Helfrich, W.; Boezen, H.M.; Cohlen, B.J.; Melief, C.J.; Nijman, H.W. Antigen-specific active immunotherapy for ovarian cancer. In *The Cochrane Library*; John Wiley & Sons, Ltd.: Hoboken, NJ, USA, 2014.
77. Marincola, F.M.; Jaffee, E.M.; Hicklin, D.J.; Ferrone, S. Escape of human solid tumors from T-cell recognition: Molecular mechanisms and functional significance. *Adv. Immunol.* **2000**, *74*, 181–273. [PubMed]
78. Campoli, M.; Chang, C.-C.; Ferrone, S. HLA class I antigen loss, tumor immune escape and immune selection. *Vaccine* **2002**, *20* (Suppl. 4), 40–45. [CrossRef]
79. Chang, C.-C.; Campoli, M.; Ferrone, S. Classical and nonclassical HLA class I antigen and NK Cell-activating ligand changes in malignant cells: Current challenges and future directions. *Adv. Cancer Res.* **2005**, *93*, 189–234. [CrossRef] [PubMed]
80. Aptsiauri, N.; Cabrera, T.; Mendez, R.; Garcia-Lora, A.; Ruiz-Cabello, F.; Garrido, F. Role of altered expression of HLA class I molecules in cancer progression. *Adv. Exp. Med. Biol.* **2007**, *601*, 123–131. [PubMed]
81. Han, L.Y.; Fletcher, M.S.; Urbauer, D.L.; Mueller, P.; Landen, C.N.; Kamat, A.A.; Lin, Y.G.; Merritt, W.M.; Spannuth, W.A.; Deavers, M.T.; et al. HLA class I antigen processing machinery component expression and intratumoral T-Cell infiltrate as independent prognostic markers in ovarian carcinoma. *Clin. Cancer Res.* **2008**, *14*, 3372–3379. [CrossRef] [PubMed]
82. Brown, S.D.; Warren, R.L.; Gibb, E.A.; Martin, S.D.; Spinelli, J.J.; Nelson, B.H.; Holt, R.A. Neo-antigens predicted by tumor genome meta-analysis correlate with increased patient survival. *Genome Res.* **2014**, *24*, 743–750. [CrossRef] [PubMed]

83. Matsushita, N.; Ghazizadeh, M.; Konishi, H.; Araki, T. Association of ovarian tumor epithelium coexpressing HLA-DR and CA-125 antigens with tumor infiltrating cytotoxic T lymphocytes. *J. Nippon Med. Sch. Nippon Ika Daigaku Zasshi* **2003**, *70*, 40–44. [CrossRef] [PubMed]
84. Callahan, M.J.; Nagymanyoki, Z.; Bonome, T.; Johnson, M.E.; Litkouhi, B.; Sullivan, E.H.; Hirsch, M.S.; Matulonis, U.A.; Liu, J.; Birrer, M.J.; et al. Increased HLA-DMB expression in the tumor epithelium is associated with increased CTL infiltration and improved prognosis in advanced-stage serous ovarian cancer. *Clin. Cancer Res.* **2008**, *14*, 7667–7673. [CrossRef] [PubMed]
85. Garrido, F.; Perea, F.; Bernal, M.; Sánchez-Palencia, A.; Aptsiauri, N.; Ruiz-Cabello, F. The escape of cancer from T cell-mediated immune surveillance: HLA class I loss and tumor tissue architecture. *Vaccines* **2017**, *5*, 7. [CrossRef] [PubMed]
86. Garrido, F.; Ruiz-Cabello, F.; Aptsiauri, N. Rejection versus escape: The tumor MHC dilemma. *Cancer Immunol. Immunother.* **2017**, *66*, 259–271. [CrossRef] [PubMed]
87. Clarke, B.; Tinker, A.V.; Lee, C.-H.; Subramanian, S.; van de Rijn, M.; Turbin, D.; Kalloger, S.; Han, G.; Ceballos, K.; Cadungog, M.G.; et al. Intraepithelial T cells and prognosis in ovarian carcinoma: Novel associations with stage, tumor type, and BRCA1 loss. *Mod. Pathol.* **2009**, *22*, 393–402. [CrossRef] [PubMed]
88. Shah, C.A.; Allison, K.H.; Garcia, R.L.; Gray, H.J.; Goff, B.A.; Swisher, E.M. Intratumoral T cells, tumor-associated macrophages, and regulatory T cells: Association with p53 mutations, circulating tumor DNA and survival in women with ovarian cancer. *Gynecol. Oncol.* **2008**, *109*, 215–219. [CrossRef] [PubMed]
89. Abiko, K.; Matsumura, N.; Hamanishi, J.; Horikawa, N.; Murakami, R.; Yamaguchi, K.; Yoshioka, Y.; Baba, T.; Konishi, I.; Mandai, M. IFN-γ from lymphocytes induces PD-L1 expression and promotes progression of ovarian cancer. *Br. J. Cancer* **2015**, *112*, 1501–1509. [CrossRef] [PubMed]
90. Hamanishi, J.; Mandai, M.; Iwasaki, M.; Okazaki, T.; Tanaka, Y.; Yamaguchi, K.; Higuchi, T.; Yagi, H.; Takakura, K.; Minato, N.; et al. Programmed cell death 1 ligand 1 and tumor-infiltrating CD8+ T lymphocytes are prognostic factors of human ovarian cancer. *Proc. Natl. Acad. Sci. USA* **2007**, *104*, 3360–3365. [CrossRef] [PubMed]
91. Darb-Esfahani, S.; Kunze, C.A.; Kulbe, H.; Sehouli, J.; Wienert, S.; Lindner, J.; Budczies, J.; Bockmayr, M.; Dietel, M.; Denkert, C.; et al. Prognostic impact of programmed cell death-1 (PD-1) and PD-ligand 1 (PD-L1) expression in cancer cells and tumor-infiltrating lymphocytes in ovarian high grade serous carcinoma. *Oncotarget* **2016**, *7*, 1486–1499. [CrossRef] [PubMed]
92. Webb, J.R.; Milne, K.; Kroeger, D.R.; Nelson, B.H. PD-L1 expression is associated with tumor-infiltrating T cells and favorable prognosis in high-grade serous ovarian cancer. *Gynecol. Oncol.* **2016**, *141*, 293–302. [CrossRef] [PubMed]
93. Drakes, M.L.; Mehrotra, S.; Aldulescu, M.; Potkul, R.K.; Liu, Y.; Grisoli, A.; Joyce, C.; O'Brien, T.E.; Stack, M.S.; Stiff, P.J. Stratification of ovarian tumor pathology by expression of programmed cell death-1 (PD-1) and PD-ligand-1 (PD-L1) in ovarian cancer. *J. Ovarian Res.* **2018**, *11*. [CrossRef] [PubMed]
94. Goyne, H.E.; Stone, P.J.B.; Burnett, A.F.; Cannon, M.J. Ovarian tumor ascites CD14+ cells suppress dendritic cell-activated CD4+ T-cell responses through IL-10 secretion and indoleamine 2,3-dioxygenase. *J. Immunother.* **2014**, *37*, 163–169. [CrossRef] [PubMed]
95. Hennequart, M.; Pilotte, L.; Cane, S.; Hoffmann, D.; Stroobant, V.; Plaen, E.D.; Eynde, B.J.V. den Constitutive IDO1 expression in human tumors is driven by Cyclooxygenase-2 and mediates intrinsic immune resistance. *Cancer Immunol. Res.* **2017**. [CrossRef] [PubMed]
96. Tanizaki, Y.; Kobayashi, A.; Toujima, S.; Shiro, M.; Mizoguchi, M.; Mabuchi, Y.; Yagi, S.; Minami, S.; Takikawa, O.; Ino, K. Indoleamine 2,3-dioxygenase promotes peritoneal metastasis of ovarian cancer by inducing an immunosuppressive environment. *Cancer Sci.* **2014**, *105*, 966–973. [CrossRef] [PubMed]
97. Ino, K. Indoleamine 2,3-dioxygenase and immune tolerance in ovarian cancer. *Curr. Opin. Obstet. Gynecol.* **2011**, *23*, 13–18. [CrossRef] [PubMed]
98. Inaba, T.; Ino, K.; Kajiyama, H.; Yamamoto, E.; Shibata, K.; Nawa, A.; Nagasaka, T.; Akimoto, H.; Takikawa, O.; Kikkawa, F. Role of the immunosuppressive enzyme indoleamine 2,3-dioxygenase in the progression of ovarian carcinoma. *Gynecol. Oncol.* **2009**, *115*, 185–192. [CrossRef] [PubMed]
99. Takao, M.; Okamoto, A.; Nikaido, T.; Urashima, M.; Takakura, S.; Saito, M.; Saito, M.; Okamoto, S.; Takikawa, O.; Sasaki, H.; et al. Increased synthesis of indoleamine-2,3-dioxygenase protein is positively associated with impaired survival in patients with serous-type, but not with other types of, ovarian cancer. *Oncol. Rep.* **2007**, *17*, 1333–1339. [CrossRef] [PubMed]

100. Qian, F.; Villella, J.; Wallace, P.K.; Mhawech-Fauceglia, P.; Tario, J.D.; Andrews, C.; Matsuzaki, J.; Valmori, D.; Ayyoub, M.; Frederick, P.J.; et al. Efficacy of levo-1-methyl tryptophan and dextro-1-methyl tryptophan in reversing indoleamine-2,3-dioxygenase-mediated arrest of T-cell proliferation in human epithelial ovarian cancer. *Cancer Res.* **2009**, *69*, 5498–5504. [CrossRef] [PubMed]
101. Sheridan, C. IDO inhibitors move center stage in immuno-oncology. *Nat. Biotechnol.* **2015**, *33*, 321–322. [CrossRef] [PubMed]
102. Schreiber, R.D.; Old, L.J.; Smyth, M.J. Cancer immunoediting: Integrating immunity's roles in cancer suppression and promotion. *Science* **2011**, *331*, 1565–1570. [CrossRef] [PubMed]
103. Okła, K.; Wertel, I.; Polak, G.; Surówka, J.; Wawruszak, A.; Kotarski, J. Tumor-associated macrophages and myeloid-derived suppressor cells as immunosuppressive mechanism in ovarian cancer patients: Progress and challenges. *Int. Rev. Immunol.* **2016**, *35*, 372–385. [CrossRef] [PubMed]
104. Leffers, N.; Gooden, M.J.M.; de Jong, R.A.; Hoogeboom, B.-N.; ten Hoor, K.A.; Hollema, H.; Boezen, H.M.; van der Zee, A.G.J.; Daemen, T.; Nijman, H.W. Prognostic significance of tumor-infiltrating T-lymphocytes in primary and metastatic lesions of advanced stage ovarian cancer. *Cancer Immunol. Immunother. CII* **2009**, *58*, 449–459. [CrossRef] [PubMed]
105. Raspollini, M.R.; Castiglione, F.; Rossi Degl'innocenti, D.; Amunni, G.; Villanucci, A.; Garbini, F.; Baroni, G.; Taddei, G.L. Tumour-infiltrating gamma/delta T-lymphocytes are correlated with a brief disease-free interval in advanced ovarian serous carcinoma. *Ann. Oncol. Off. J. Eur. Soc. Med. Oncol.* **2005**, *16*, 590–596. [CrossRef] [PubMed]
106. Sato, E.; Olson, S.H.; Ahn, J.; Bundy, B.; Nishikawa, H.; Qian, F.; Jungbluth, A.A.; Frosina, D.; Gnjatic, S.; Ambrosone, C.; et al. Intraepithelial CD8+ tumor-infiltrating lymphocytes and a high CD8+/regulatory T cell ratio are associated with favorable prognosis in ovarian cancer. *Proc. Natl. Acad. Sci. USA* **2005**, *102*, 18538–18543. [CrossRef] [PubMed]
107. Tomsová, M.; Melichar, B.; Sedláková, I.; Steiner, I. Prognostic significance of CD3+ tumor-infiltrating lymphocytes in ovarian carcinoma. *Gynecol. Oncol.* **2008**, *108*, 415–420. [CrossRef] [PubMed]
108. Nielsen, J.S.; Sahota, R.A.; Milne, K.; Kost, S.E.; Nesslinger, N.J.; Watson, P.H.; Nelson, B.H. CD20+ tumor-infiltrating lymphocytes have an atypical CD27- memory phenotype and together with CD8+ T cells promote favorable prognosis in ovarian cancer. *Clin. Cancer Res.* **2012**, *18*, 3281–3292. [CrossRef] [PubMed]
109. Barnett, J.C.; Bean, S.M.; Whitaker, R.S.; Kondoh, E.; Baba, T.; Fujii, S.; Marks, J.R.; Dressman, H.K.; Murphy, S.K.; Berchuck, A. Ovarian cancer tumor infiltrating T-regulatory (T(reg)) cells are associated with a metastatic phenotype. *Gynecol. Oncol.* **2010**, *116*, 556–562. [CrossRef] [PubMed]
110. Hamanishi, J.; Mandai, M.; Abiko, K.; Matsumura, N.; Baba, T.; Yoshioka, Y.; Kosaka, K.; Konishi, I. The comprehensive assessment of local immune status of ovarian cancer by the clustering of multiple immune factors. *Clin. Immunol.* **2011**, *141*, 338–347. [CrossRef] [PubMed]
111. Hwang, W.-T.; Adams, S.F.; Tahirovic, E.; Hagemann, I.S.; Coukos, G. Prognostic significance of tumor-infiltrating T cells in ovarian cancer: A meta-analysis. *Gynecol. Oncol.* **2012**, *124*, 192–198. [CrossRef] [PubMed]
112. Dadmarz, R.D.; Ordoubadi, A.; Mixon, A.; Thompson, C.O.; Barracchini, K.C.; Hijazi, Y.M.; Steller, M.A.; Rosenberg, S.A.; Schwartzentruber, D.J. Tumor-infiltrating lymphocytes from human ovarian cancer patients recognize autologous tumor in an MHC class II-restricted fashion. *Cancer J. Sci. Am.* **1996**, *2*, 263–272. [PubMed]
113. Bronger, H.; Singer, J.; Windmüller, C.; Reuning, U.; Zech, D.; Delbridge, C.; Dorn, J.; Kiechle, M.; Schmalfeldt, B.; Schmitt, M.; et al. CXCL9 and CXCL10 predict survival and are regulated by cyclooxygenase inhibition in advanced serous ovarian cancer. *Br. J. Cancer* **2016**, *115*, 553–563. [CrossRef] [PubMed]
114. Liu, M.; Matsumura, N.; Mandai, M.; Li, K.; Yagi, H.; Baba, T.; Suzuki, A.; Hamanishi, J.; Fukuhara, K.; Konishi, I. Classification using hierarchical clustering of tumor-infiltrating immune cells identifies poor prognostic ovarian cancers with high levels of COX expression. *Mod. Pathol.* **2009**, *22*, 373–384. [CrossRef] [PubMed]
115. Hermans, C.; Anz, D.; Engel, J.; Kirchner, T.; Endres, S.; Mayr, D. Analysis of FoxP3+ T-regulatory cells and CD8+ T-cells in ovarian carcinoma: Location and tumor infiltration patterns are key prognostic markers. *PLoS ONE* **2014**, *9*, e111757. [CrossRef] [PubMed]

116. Shi, J.-X.; Qin, J.-J.; Ye, H.; Wang, P.; Wang, K.-J.; Zhang, J.-Y. Tumor associated antigens or anti-TAA autoantibodies as biomarkers in the diagnosis of ovarian cancer: A systematic review with meta-analysis. *Expert Rev. Mol. Diagn.* **2015**, *15*, 829–852. [CrossRef] [PubMed]
117. Taylor, D.D.; Gercel-Taylor, C.; Parker, L.P. Patient-derived tumor-reactive antibodies as diagnostic markers for ovarian cancer. *Gynecol. Oncol.* **2009**, *115*, 112–120. [CrossRef] [PubMed]
118. Taylor, D.D.; Atay, S.; Metzinger, D.S.; Gercel-Taylor, C. Characterization of humoral responses of ovarian cancer patients: Antibody subclasses and antigenic components. *Gynecol. Oncol.* **2010**, *116*, 213–221. [CrossRef] [PubMed]
119. Jang, M.; Yew, P.-Y.; Hasegawa, K.; Ikeda, Y.; Fujiwara, K.; Fleming, G.F.; Nakamura, Y.; Park, J.-H. Characterization of T cell repertoire of blood, tumor, and ascites in ovarian cancer patients using next generation sequencing. *Oncoimmunology* **2015**, *4*, e1030561. [CrossRef] [PubMed]
120. Landskron, J.; Helland, Ø.; Torgersen, K.M.; Aandahl, E.M.; Gjertsen, B.T.; Bjørge, L.; Taskén, K. Activated regulatory and memory T-cells accumulate in malignant ascites from ovarian carcinoma patients. *Cancer Immunol. Immunother.* **2015**, *64*, 337–347. [CrossRef] [PubMed]
121. Lukesova, S.; Vroblova, V.; Tosner, J.; Kopecky, J.; Sedlakova, I.; Čermáková, E.; Vokurkova, D.; Kopecky, O. Comparative study of various subpopulations of cytotoxic cells in blood and ascites from patients with ovarian carcinoma. *Contemp. Oncol. Poznan Pol.* **2015**, *19*, 290–299. [CrossRef] [PubMed]
122. Martin, S.D.; Wick, D.A.; Nielsen, J.S.; Little, N.; Holt, R.A.; Nelson, B.H. A library-based screening method identifies neoantigen-reactive T cells in peripheral blood prior to relapse of ovarian cancer. *Oncoimmunology* **2017**, *7*, e1371895. [CrossRef] [PubMed]
123. Nelson, B.H. The impact of T-cell immunity on ovarian cancer outcomes. *Immunol. Rev.* **2008**, *222*, 101–116. [CrossRef] [PubMed]
124. Tsiatas, M.L.; Gyftaki, R.; Liacos, C.; Politi, E.; Rodolakis, A.; Dimopoulos, M.-A.; Bamias, A. Study of T lymphocytes infiltrating peritoneal metastases in advanced ovarian cancer: Associations with vascular endothelial growth factor levels and prognosis in patients receiving platinum-based chemotherapy. *Int. J. Gynecol. Cancer Off. J. Int. Gynecol. Cancer Soc.* **2009**, *19*, 1329–1334. [CrossRef] [PubMed]
125. Le Page, C.; Marineau, A.; Bonza, P.K.; Rahimi, K.; Cyr, L.; Labouba, I.; Madore, J.; Delvoye, N.; Mes-Masson, A.-M.; Provencher, D.M.; et al. BTN3A2 expression in epithelial ovarian cancer is associated with higher tumor infiltrating T cells and a better prognosis. *PLoS ONE* **2012**, *7*, e38541. [CrossRef] [PubMed]
126. Nesbeth, Y.C.; Martinez, D.G.; Toraya, S.; Scarlett, U.K.; Cubillos-Ruiz, J.R.; Rutkowski, M.R.; Conejo-Garcia, J.R. CD4+ T cells elicit host immune responses to MHC class II-negative ovarian cancer through CCL5 secretion and CD40-mediated licensing of dendritic cells. *J. Immunol.* **2010**, *184*, 5654–5662. [CrossRef] [PubMed]
127. Hanahan, D.; Coussens, L.M. Accessories to the crime: Functions of cells recruited to the tumor microenvironment. *Cancer Cell* **2012**, *21*, 309–322. [CrossRef] [PubMed]
128. Mittal, D.; Gubin, M.M.; Schreiber, R.D.; Smyth, M.J. New insights into cancer immunoediting and its three component phases–elimination, equilibrium and escape. *Curr. Opin. Immunol.* **2014**, *27*, 16–25. [CrossRef] [PubMed]
129. Sehouli, J.; Loddenkemper, C.; Cornu, T.; Schwachula, T.; Hoffmüller, U.; Grützkau, A.; Lohneis, P.; Dickhaus, T.; Gröne, J.; Kruschewski, M.; et al. Epigenetic quantification of tumor-infiltrating T-lymphocytes. *Epigenetics* **2011**, *6*, 236–246. [CrossRef] [PubMed]
130. Curiel, T.J.; Coukos, G.; Zou, L.; Alvarez, X.; Cheng, P.; Mottram, P.; Evdemon-Hogan, M.; Conejo-Garcia, J.R.; Zhang, L.; Burow, M.; et al. Specific recruitment of regulatory T cells in ovarian carcinoma fosters immune privilege and predicts reduced survival. *Nat. Med.* **2004**, *10*, 942–949. [CrossRef] [PubMed]
131. Woo, E.Y.; Chu, C.S.; Goletz, T.J.; Schlienger, K.; Yeh, H.; Coukos, G.; Rubin, S.C.; Kaiser, L.R.; June, C.H. Regulatory CD4(+)CD25(+) T cells in tumors from patients with early-stage non-small cell lung cancer and late-stage ovarian cancer. *Cancer Res.* **2001**, *61*, 4766–4772. [PubMed]
132. Kryczek, I.; Liu, R.; Wang, G.; Wu, K.; Shu, X.; Szeliga, W.; Vatan, L.; Finlayson, E.; Huang, E.; Simeone, D.; et al. FOXP3 defines regulatory T cells in human tumor and autoimmune disease. *Cancer Res.* **2009**, *69*, 3995–4000. [CrossRef] [PubMed]
133. Napoletano, C.; Bellati, F.; Landi, R.; Pauselli, S.; Marchetti, C.; Visconti, V.; Sale, P.; Liberati, M.; Rughetti, A.; Frati, L.; et al. Ovarian cancer cytoreduction induces changes in T cell population subsets reducing immunosuppression. *J. Cell. Mol. Med.* **2010**, *14*, 2748–2759. [CrossRef] [PubMed]

134. Fialová, A.; Partlová, S.; Sojka, L.; Hromádková, H.; Brtnický, T.; Fučíková, J.; Kocián, P.; Rob, L.; Bartůňková, J.; Spíšek, R. Dynamics of T-cell infiltration during the course of ovarian cancer: The gradual shift from a Th17 effector cell response to a predominant infiltration by regulatory T-cells. *Int. J. Cancer* **2013**, *132*, 1070–1079. [CrossRef] [PubMed]
135. Facciabene, A.; Peng, X.; Hagemann, I.S.; Balint, K.; Barchetti, A.; Wang, L.-P.; Gimotty, P.A.; Gilks, C.B.; Lal, P.; Zhang, L.; et al. Tumour hypoxia promotes tolerance and angiogenesis via CCL28 and T(reg) cells. *Nature* **2011**, *475*, 226–230. [CrossRef] [PubMed]
136. Kryczek, I.; Wei, S.; Zhu, G.; Myers, L.; Mottram, P.; Cheng, P.; Chen, L.; Coukos, G.; Zou, W. Relationship between B7-H4, regulatory T cells, and patient outcome in human ovarian carcinoma. *Cancer Res.* **2007**, *67*, 8900–8905. [CrossRef] [PubMed]
137. Redjimi, N.; Raffin, C.; Raimbaud, I.; Pignon, P.; Matsuzaki, J.; Odunsi, K.; Valmori, D.; Ayyoub, M. CXCR3+ T regulatory cells selectively accumulate in human ovarian carcinomas to limit type I immunity. *Cancer Res.* **2012**, *72*, 4351–4360. [CrossRef] [PubMed]
138. Giuntoli, R.L.; Webb, T.J.; Zoso, A.; Rogers, O.; Diaz-Montes, T.P.; Bristow, R.E.; Oelke, M. Ovarian Cancer-associated Ascites Demonstrates Altered Immune Environment: Implications for Antitumor Immunity. *Anticancer Res.* **2009**, *29*, 2875–2884. [PubMed]
139. Wolf, D.; Wolf, A.M.; Rumpold, H.; Fiegl, H.; Zeimet, A.G.; Muller-Holzner, E.; Deibl, M.; Gastl, G.; Gunsilius, E.; Marth, C. The expression of the regulatory T cell-specific forkhead box transcription factor FoxP3 is associated with poor prognosis in ovarian cancer. *Clin. Cancer Res.* **2005**, *11*, 8326–8331. [CrossRef] [PubMed]
140. Mhawech-Fauceglia, P.; Wang, D.; Ali, L.; Lele, S.; Huba, M.A.; Liu, S.; Odunsi, K. Intraepithelial T cells and tumor-associated macrophages in ovarian cancer patients. *Cancer Immun.* **2013**, *13*, 1. [PubMed]
141. Milne, K.; Köbel, M.; Kalloger, S.E.; Barnes, R.O.; Gao, D.; Gilks, C.B.; Watson, P.H.; Nelson, B.H. Systematic analysis of immune infiltrates in high-grade serous ovarian cancer reveals CD20, FoxP3 and TIA-1 as positive prognostic factors. *PLoS ONE* **2009**, *4*, e6412. [CrossRef] [PubMed]
142. Wu, M.; Chen, X.; Lou, J.; Zhang, S.; Zhang, X.; Huang, L.; Sun, R.; Huang, P.; Wang, F.; Pan, S. TGF-β1 contributes to CD8+ Treg induction through p38 MAPK signaling in ovarian cancer microenvironment. *Oncotarget* **2016**, *7*, 44534–44544. [CrossRef] [PubMed]
143. Zhang, S.; Ke, X.; Zeng, S.; Wu, M.; Lou, J.; Wu, L.; Huang, P.; Huang, L.; Wang, F.; Pan, S. Analysis of CD8+ Treg cells in patients with ovarian cancer: A possible mechanism for immune impairment. *Cell. Mol. Immunol.* **2015**, *12*, 580–591. [CrossRef] [PubMed]
144. Wang, R.-F. CD8+ regulatory T cells, their suppressive mechanisms, and regulation in cancer. *Hum. Immunol.* **2008**, *69*, 811–814. [CrossRef] [PubMed]
145. Yigit, R.; Massuger, L.F.A.G.; Figdor, C.G.; Torensma, R. Ovarian cancer creates a suppressive microenvironment to escape immune elimination. *Gynecol. Oncol.* **2010**, *117*, 366–372. [CrossRef] [PubMed]
146. Sarvaria, A.; Madrigal, J.A.; Saudemont, A. B cell regulation in cancer and anti-tumor immunity. *Cell. Mol. Immunol.* **2017**, *14*, 662–674. [CrossRef] [PubMed]
147. Montfort, A.; Pearce, O.; Maniati, E.; Vincent, B.G.; Bixby, L.; Böhm, S.; Dowe, T.; Wilkes, E.H.; Chakravarty, P.; Thompson, R.; et al. A strong B-cell response is part of the immune landscape in human high-grade serous ovarian metastases. *Clin. Cancer Res.* **2017**, *23*, 250–262. [CrossRef] [PubMed]
148. Rosser, E.C.; Mauri, C. Regulatory B cells: Origin, phenotype, and function. *Immunity* **2015**, *42*, 607–612. [CrossRef] [PubMed]
149. Lundgren, S.; Berntsson, J.; Nodin, B.; Micke, P.; Jirström, K. Prognostic impact of tumour-associated B cells and plasma cells in epithelial ovarian cancer. *J. Ovarian Res.* **2016**, *9*, 21. [CrossRef] [PubMed]
150. Yang, C.; Lee, H.; Jove, V.; Deng, J.; Zhang, W.; Liu, X.; Forman, S.; Dellinger, T.H.; Wakabayashi, M.; Yu, H.; et al. Prognostic significance of B-cells and pSTAT3 in patients with ovarian cancer. *PLoS ONE* **2013**, *8*, e54029. [CrossRef] [PubMed]
151. Dong, H.P.; Elstrand, M.B.; Holth, A.; Silins, I.; Berner, A.; Trope, C.G.; Davidson, B.; Risberg, B. NK- and B-cell infiltration correlates with worse outcome in metastatic ovarian carcinoma. *Am. J. Clin. Pathol.* **2006**, *125*, 451–458. [CrossRef] [PubMed]
152. Lu, P.H.; Negrin, R.S. A novel population of expanded human CD3+CD56+ cells derived from T cells with potent in vivo antitumor activity in mice with severe combined immunodeficiency. *J. Immunol.* **1994**, *153*, 1687–1696. [PubMed]

153. Schmidt-Wolf, I.G.; Lefterova, P.; Mehta, B.A.; Fernandez, L.P.; Huhn, D.; Blume, K.G.; Weissman, I.L.; Negrin, R.S. Phenotypic characterization and identification of effector cells involved in tumor cell recognition of cytokine-induced killer cells. *Exp. Hematol.* **1993**, *21*, 1673–1679. [PubMed]
154. Bamias, A.; Tsiatas, M.L.; Kafantari, E.; Liakou, C.; Rodolakis, A.; Voulgaris, Z.; Vlahos, G.; Papageorgiou, T.; Tsitsilonis, O.; Bamia, C.; et al. Significant differences of lymphocytes isolated from ascites of patients with ovarian cancer compared to blood and tumor lymphocytes. Association of CD3+CD56+ cells with platinum resistance. *Gynecol. Oncol.* **2007**, *106*, 75–81. [CrossRef] [PubMed]
155. Bamias, A.; Koutsoukou, V.; Terpos, E.; Tsiatas, M.L.; Liakos, C.; Tsitsilonis, O.; Rodolakis, A.; Voulgaris, Z.; Vlahos, G.; Papageorgiou, T.; et al. Correlation of NK T-like CD3+CD56+ cells and CD4+CD25+(hi) regulatory T cells with VEGF and TNFalpha in ascites from advanced ovarian cancer: Association with platinum resistance and prognosis in patients receiving first-line, platinum-based chemotherapy. *Gynecol. Oncol.* **2008**, *108*, 421–427. [CrossRef] [PubMed]
156. Wolf, B.J.; Choi, J.E.; Exley, M.A. Novel approaches to exploiting invariant NKT cells in cancer immunotherapy. *Front. Immunol.* **2018**, *9*, 384. [CrossRef] [PubMed]
157. Guillerey, C.; Huntington, N.D.; Smyth, M.J. Targeting natural killer cells in cancer immunotherapy. *Nat. Immunol.* **2016**, *17*, 1025–1036. [CrossRef] [PubMed]
158. Sungur, C.M.; Murphy, W.J. Positive and negative regulation by NK cells in cancer. *Crit. Rev. Oncog.* **2014**, *19*, 57–66. [CrossRef] [PubMed]
159. Voskoboinik, I.; Smyth, M.J.; Trapani, J.A. Perforin-mediated target-cell death and immune homeostasis. *Nat. Rev. Immunol.* **2006**, *6*, 940–952. [CrossRef] [PubMed]
160. Lai, P.; Rabinowich, H.; Crowley-Nowick, P.A.; Bell, M.C.; Mantovani, G.; Whiteside, T.L. Alterations in expression and function of signal-transducing proteins in tumor-associated T and natural killer cells in patients with ovarian carcinoma. *Clin. Cancer Res.* **1996**, *2*, 161–173. [PubMed]
161. Yunusova, N.V.; Stakheyeva, M.N.; Molchanov, S.V.; Afanas'ev, S.G.; Tsydenova, A.A.; Kolomiets, L.A.; Cherdyntseva, N.V. Functional activity of natural killer cells in biological fluids in patients with colorectal and ovarian cancers. *Cent.-Eur. J. Immunol.* **2018**, *43*, 26–32. [CrossRef] [PubMed]
162. Webb, J.R.; Milne, K.; Watson, P.; Deleeuw, R.J.; Nelson, B.H. Tumor-infiltrating lymphocytes expressing the tissue resident memory marker CD103 are associated with increased survival in high-grade serous ovarian cancer. *Clin. Cancer Res.* **2014**, *20*, 434–444. [CrossRef] [PubMed]
163. Garzetti, G.G.; Cignitti, M.; Ciavattini, A.; Fabris, N.; Romanini, C. Natural killer cell activity and progression-free survival in ovarian cancer. *Gynecol. Obstet. Invest.* **1993**, *35*, 118–120. [CrossRef] [PubMed]
164. Li, K.; Mandai, M.; Hamanishi, J.; Matsumura, N.; Suzuki, A.; Yagi, H.; Yamaguchi, K.; Baba, T.; Fujii, S.; Konishi, I. Clinical significance of the NKG2D ligands, MICA/B and ULBP2 in ovarian cancer: High expression of ULBP2 is an indicator of poor prognosis. *Cancer Immunol. Immunother. CII* **2009**, *58*, 641–652. [CrossRef] [PubMed]
165. Colvin, E.K. Tumor-associated macrophages contribute to tumor progression in ovarian cancer. *Front. Oncol.* **2014**, *4*, 137. [CrossRef] [PubMed]
166. Takaishi, K.; Komohara, Y.; Tashiro, H.; Ohtake, H.; Nakagawa, T.; Katabuchi, H.; Takeya, M. Involvement of M2-polarized macrophages in the ascites from advanced epithelial ovarian carcinoma in tumor progression via Stat3 activation. *Cancer Sci.* **2010**, *101*, 2128–2136. [CrossRef] [PubMed]
167. Kawamura, K.; Komohara, Y.; Takaishi, K.; Katabuchi, H.; Takeya, M. Detection of M2 macrophages and colony-stimulating factor 1 expression in serous and mucinous ovarian epithelial tumors. *Pathol. Int.* **2009**, *59*, 300–305. [CrossRef] [PubMed]
168. Hagemann, T.; Wilson, J.; Burke, F.; Kulbe, H.; Li, N.F.; Plüddemann, A.; Charles, K.; Gordon, S.; Balkwill, F.R. Ovarian cancer cells polarize macrophages toward a tumor-associated phenotype. *J. Immunol.* **2006**, *176*, 5023–5032. [CrossRef] [PubMed]
169. Sica, G.L.; Choi, I.H.; Zhu, G.; Tamada, K.; Wang, S.D.; Tamura, H.; Chapoval, A.I.; Flies, D.B.; Bajorath, J.; Chen, L. B7-H4, a molecule of the B7 family, negatively regulates T cell immunity. *Immunity* **2003**, *18*, 849–861. [CrossRef]
170. Yuan, X.; Zhang, J.; Li, D.; Mao, Y.; Mo, F.; Du, W.; Ma, X. Prognostic significance of tumor-associated macrophages in ovarian cancer: A meta-analysis. *Gynecol. Oncol.* **2017**, *147*, 181–187. [CrossRef] [PubMed]

171. Lan, C.; Huang, X.; Lin, S.; Huang, H.; Cai, Q.; Wan, T.; Lu, J.; Liu, J. Expression of M2-polarized macrophages is associated with poor prognosis for advanced epithelial ovarian cancer. *Technol. Cancer Res. Treat.* **2013**, *12*, 259–267. [CrossRef] [PubMed]
172. He, Y.; Zhang, M.; Wu, X.; Sun, X.; Xu, T.; He, Q.; Di, W. High MUC2 expression in ovarian cancer is inversely associated with the M1/M2 ratio of tumor-associated macrophages and patient survival time. *PLoS ONE* **2013**, *8*, e79769. [CrossRef] [PubMed]
173. Zhang, M.; He, Y.; Sun, X.; Li, Q.; Wang, W.; Zhao, A.; Di, W. A high M1/M2 ratio of tumor-associated macrophages is associated with extended survival in ovarian cancer patients. *J. Ovarian Res.* **2014**, *7*, 19. [CrossRef] [PubMed]
174. No, J.H.; Moon, J.M.; Kim, K.; Kim, Y.-B. Prognostic significance of serum soluble CD163 level in patients with epithelial ovarian cancer. *Gynecol. Obstet. Investig.* **2013**, *75*, 263–267. [CrossRef] [PubMed]
175. Gordon, I.O.; Freedman, R.S. Defective antitumor function of monocyte-derived macrophages from epithelial ovarian cancer patients. *Clin. Cancer Res.* **2006**, *12*, 1515–1524. [CrossRef] [PubMed]
176. Genard, G.; Lucas, S.; Michiels, C. Reprogramming of tumor-associated macrophages with anticancer therapies: Radiotherapy versus chemo- and immunotherapies. *Front. Immunol.* **2017**, *8*, 828. [CrossRef] [PubMed]
177. Youn, J.-I.; Gabrilovich, D.I. The biology of myeloid-derived suppressor cells: The blessing and the curse of morphological and functional heterogeneity. *Eur. J. Immunol.* **2010**, *40*, 2969–2975. [CrossRef] [PubMed]
178. Gabrilovich, D.I.; Nagaraj, S. Myeloid-derived suppressor cells as regulators of the immune system. *Nat. Rev. Immunol.* **2009**, *9*, 162–174. [CrossRef] [PubMed]
179. Obermajer, N.; Muthuswamy, R.; Odunsi, K.; Edwards, R.P.; Kalinski, P. PGE(2)-induced CXCL12 production and CXCR4 expression controls the accumulation of human MDSCs in ovarian cancer environment. *Cancer Res.* **2011**, *71*, 7463–7470. [CrossRef] [PubMed]
180. Cui, T.X.; Kryczek, I.; Zhao, L.; Zhao, E.; Kuick, R.; Roh, M.H.; Vatan, L.; Szeliga, W.; Mao, Y.; Thomas, D.G.; et al. Myeloid-derived suppressor cells enhance stemness of cancer cells by inducing microRNA101 and suppressing the corepressor CtBP2. *Immunity* **2013**, *39*, 611–621. [CrossRef] [PubMed]
181. Wu, L.; Deng, Z.; Peng, Y.; Han, L.; Liu, J.; Wang, L.; Li, B.; Zhao, J.; Jiao, S.; Wei, H. Ascites-derived IL-6 and IL-10 synergistically expand CD14+HLA-DR-/low myeloid-derived suppressor cells in ovarian cancer patients. *Oncotarget* **2017**, *8*, 76843–76856. [CrossRef] [PubMed]
182. Horikawa, N.; Abiko, K.; Matsumura, N.; Hamanishi, J.; Baba, T.; Yamaguchi, K.; Yoshioka, Y.; Koshiyama, M.; Konishi, I. Expression of Vascular Endothelial Growth Factor in Ovarian Cancer Inhibits Tumor Immunity through the Accumulation of Myeloid-Derived Suppressor Cells. *Clin. Cancer Res.* **2017**, *23*, 587–599. [CrossRef] [PubMed]
183. Taki, M.; Abiko, K.; Baba, T.; Hamanishi, J.; Yamaguchi, K.; Murakami, R.; Yamanoi, K.; Horikawa, N.; Hosoe, Y.; Nakamura, E.; et al. Snail promotes ovarian cancer progression by recruiting myeloid-derived suppressor cells via CXCR2 ligand upregulation. *Nat. Commun.* **2018**, *9*, 1685. [CrossRef] [PubMed]
184. Montalbán Del Barrio, I.; Penski, C.; Schlahsa, L.; Stein, R.G.; Diessner, J.; Wöckel, A.; Dietl, J.; Lutz, M.B.; Mittelbronn, M.; Wischhusen, J.; et al. Adenosine-generating ovarian cancer cells attract myeloid cells which differentiate into adenosine-generating tumor associated macrophages—A self-amplifying, CD39- and CD73-dependent mechanism for tumor immune escape. *J. Immunother. Cancer* **2016**, *4*, 49. [CrossRef] [PubMed]
185. Li, L.; Wang, L.; Li, J.; Fan, Z.; Yang, L.; Zhang, Z.; Zhang, C.; Yue, D.; Qin, G.; Zhang, T.; et al. Metformin-induced reduction of CD39 and CD73 blocks myeloid-derived suppressor cell activity in patients with ovarian cancer. *Cancer Res.* **2018**, *78*, 1779–1791. [CrossRef] [PubMed]
186. Cheng, L.; Wu, S.; Zhang, K.; Qing, Y.; Xu, T. A comprehensive overview of exosomes in ovarian cancer: Emerging biomarkers and therapeutic strategies. *J. Ovarian Res.* **2017**, *10*, 73. [CrossRef] [PubMed]
187. Li, W.; Zhang, X.; Wang, J.; Li, M.; Cao, C.; Tan, J.; Ma, D.; Gao, Q. TGFβ1 in fibroblasts-derived exosomes promotes epithelial-mesenchymal transition of ovarian cancer cells. *Oncotarget* **2017**, *8*, 96035–96047. [CrossRef] [PubMed]
188. Au Yeung, C.L.; Co, N.-N.; Tsuruga, T.; Yeung, T.-L.; Kwan, S.-Y.; Leung, C.S.; Li, Y.; Lu, E.S.; Kwan, K.; Wong, K.-K.; et al. Exosomal transfer of stroma-derived miR21 confers paclitaxel resistance in ovarian cancer cells through targeting APAF1. *Nat. Commun.* **2016**, *7*, 11150. [CrossRef] [PubMed]

189. Peng, P.; Yan, Y.; Keng, S. Exosomes in the ascites of ovarian cancer patients: Origin and effects on anti-tumor immunity. *Oncol. Rep.* **2011**, *25*, 749–762. [CrossRef] [PubMed]
190. Keerthikumar, S.; Chisanga, D.; Ariyaratne, D.; Al Saffar, H.; Anand, S.; Zhao, K.; Samuel, M.; Pathan, M.; Jois, M.; Chilamkurti, N.; et al. ExoCarta: A web-based compendium of exosomal cargo. *J. Mol. Biol.* **2016**, *428*, 688–692. [CrossRef] [PubMed]
191. Mathivanan, S.; Simpson, R.J. ExoCarta: A compendium of exosomal proteins and RNA. *Proteomics* **2009**, *9*, 4997–5000. [CrossRef] [PubMed]
192. Keller, S.; König, A.-K.; Marmé, F.; Runz, S.; Wolterink, S.; Koensgen, D.; Mustea, A.; Sehouli, J.; Altevogt, P. Systemic presence and tumor-growth promoting effect of ovarian carcinoma released exosomes. *Cancer Lett.* **2009**, *278*, 73–81. [CrossRef] [PubMed]
193. Valenti, R.; Huber, V.; Iero, M.; Filipazzi, P.; Parmiani, G.; Rivoltini, L. Tumor-released microvesicles as vehicles of immunosuppression. *Cancer Res.* **2007**, *67*, 2912–2915. [CrossRef] [PubMed]
194. Robbins, P.D.; Morelli, A.E. Regulation of immune responses by extracellular vesicles. *Nat. Rev. Immunol.* **2014**, *14*, 195–208. [CrossRef] [PubMed]
195. Cai, X.; Caballero-Benitez, A.; Gewe, M.M.; Jenkins, I.C.; Drescher, C.W.; Strong, R.K.; Spies, T.; Groh, V. Control of tumor initiation by NKG2D naturally expressed on ovarian cancer cells. *Neoplasia* **2017**, *19*, 471–482. [CrossRef] [PubMed]
196. Syn, N.L.; Wang, L.; Chow, E.K.-H.; Lim, C.T.; Goh, B.-C. Exosomes in cancer nanomedicine and immunotherapy: Prospects and challenges. *Trends Biotechnol.* **2017**, *35*, 665–676. [CrossRef] [PubMed]
197. Labiche, A.; Heutte, N.; Herlin, P.; Chasle, J.; Gauduchon, P.; Elie, N. Stromal compartment as a survival prognostic factor in advanced ovarian carcinoma. *Int. J. Gynecol. Cancer Off. J. Int. Gynecol. Cancer Soc.* **2010**, *20*, 28–33. [CrossRef] [PubMed]
198. Deying, W.; Feng, G.; Shumei, L.; Hui, Z.; Ming, L.; Hongqing, W. CAF-derived HGF promotes cell proliferation and drug resistance by up-regulating the c-Met/PI3K/Akt and GRP78 signalling in ovarian cancer cells. *Biosci. Rep.* **2017**, *37*. [CrossRef] [PubMed]
199. Kwon, Y.; Smith, B.D.; Zhou, Y.; Kaufman, M.D.; Godwin, A.K. Effective inhibition of c-MET-mediated signaling, growth and migration of ovarian cancer cells is influenced by the ovarian tissue microenvironment. *Oncogene* **2015**, *34*, 144–153. [CrossRef] [PubMed]
200. Erez, N.; Glanz, S.; Raz, Y.; Avivi, C.; Barshack, I. Cancer associated fibroblasts express pro-inflammatory factors in human breast and ovarian tumors. *Biochem. Biophys. Res. Commun.* **2013**, *437*, 397–402. [CrossRef] [PubMed]
201. Zhou, B.; Sun, C.; Li, N.; Shan, W.; Lu, H.; Guo, L.; Guo, E.; Xia, M.; Weng, D.; Meng, L.; et al. Cisplatin-induced CCL5 secretion from CAFs promotes cisplatin-resistance in ovarian cancer via regulation of the STAT3 and PI3K/Akt signaling pathways. *Int. J. Oncol.* **2016**, *48*, 2087–2097. [CrossRef] [PubMed]
202. Lau, T.-S.; Chung, T.K.-H.; Cheung, T.-H.; Chan, L.K.-Y.; Cheung, L.W.-H.; Yim, S.-F.; Siu, N.S.-S.; Lo, K.-W.; Yu, M.M.-Y.; Kulbe, H.; et al. Cancer cell-derived lymphotoxin mediates reciprocal tumour-stromal interactions in human ovarian cancer by inducing CXCL11 in fibroblasts. *J. Pathol.* **2014**, *232*, 43–56. [CrossRef] [PubMed]
203. Wang, L.; Zhang, F.; Cui, J.-Y.; Chen, L.; Chen, Y.-T.; Liu, B.-W. CAFs enhance paclitaxel resistance by inducing EMT through the IL-6/JAK2/STAT3 pathway. *Oncol. Rep.* **2018**, *39*, 2081–2090. [CrossRef] [PubMed]
204. Givel, A.-M.; Kieffer, Y.; Scholer-Dahirel, A.; Sirven, P.; Cardon, M.; Pelon, F.; Magagna, I.; Gentric, G.; Costa, A.; Bonneau, C.; et al. miR200-regulated CXCL12β promotes fibroblast heterogeneity and immunosuppression in ovarian cancers. *Nat. Commun.* **2018**, *9*, 1056. [CrossRef] [PubMed]
205. Schauer, I.G.; Zhang, J.; Xing, Z.; Guo, X.; Mercado-Uribe, I.; Sood, A.K.; Huang, P.; Liu, J. Interleukin-1β promotes ovarian tumorigenesis through a p53/NF-κB-mediated inflammatory response in stromal fibroblasts. *Neoplasia* **2013**, *15*, 409–420. [CrossRef] [PubMed]
206. Yeung, T.-L.; Leung, C.S.; Wong, K.-K.; Samimi, G.; Thompson, M.S.; Liu, J.; Zaid, T.M.; Ghosh, S.; Birrer, M.J.; Mok, S.C. TGF-β modulates ovarian cancer invasion by upregulating CAF-derived versican in the tumor microenvironment. *Cancer Res.* **2013**, *73*, 5016–5028. [CrossRef] [PubMed]
207. Giusti, I.; Francesco, M.D.; Ascenzo, S.D.; Palmerini, M.G.; Macchiarelli, G.; Carta, G.; Dolo, V. Ovarian cancer-derived extracellular vesicles affect normal human fibroblast behavior. *Cancer Biol. Ther.* **2018**, 1–44. [CrossRef] [PubMed]

208. Wei, R.; Lv, M.; Li, F.; Cheng, T.; Zhang, Z.; Jiang, G.; Zhou, Y.; Gao, R.; Wei, X.; Lou, J.; et al. Human CAFs promote lymphangiogenesis in ovarian cancer via the Hh-VEGF-C signaling axis. *Oncotarget* **2017**, *8*, 67315–67328. [CrossRef] [PubMed]
209. Leung, C.S.; Yeung, T.-L.; Yip, K.-P.; Wong, K.-K.; Ho, S.Y.; Mangala, L.S.; Sood, A.K.; Lopez-Berestein, G.; Sheng, J.; Wong, S.T.; et al. Cancer-associated fibroblasts regulate endothelial adhesion protein LPP to promote ovarian cancer chemoresistance. *J. Clin. Investig.* **2018**, *128*, 589–606. [CrossRef] [PubMed]
210. Lengyel, E. Ovarian Cancer Development and Metastasis. *Am. J. Pathol.* **2010**, *177*, 1053–1064. [CrossRef] [PubMed]
211. Lengyel, E.; Makowski, L.; DiGiovanni, J.; Kolonin, M.G. Cancer as a matter of fat: The crosstalk between adipose tissue and tumors. *Trends Cancer* **2018**, *4*, 374–384. [CrossRef] [PubMed]
212. Meza-Perez, S.; Randall, T.D. Immunological functions of the omentum. *Trends Immunol.* **2017**, *38*, 526–536. [CrossRef] [PubMed]
213. Clark, R.; Krishnan, V.; Schoof, M.; Rodriguez, I.; Theriault, B.; Chekmareva, M.; Rinker-Schaeffer, C. Milky spots promote ovarian cancer metastatic colonization of peritoneal adipose in experimental models. *Am. J. Pathol.* **2013**, *183*, 576–591. [CrossRef] [PubMed]
214. Gerber, S.A.; Rybalko, V.Y.; Bigelow, C.E.; Lugade, A.A.; Foster, T.H.; Frelinger, J.G.; Lord, E.M. Preferential attachment of peritoneal tumor metastases to omental immune aggregates and possible role of a unique vascular microenvironment in metastatic survival and growth. *Am. J. Pathol.* **2006**, *169*, 1739–1752. [CrossRef] [PubMed]
215. Krishnan, V.; Clark, R.; Chekmareva, M.; Johnson, A.; George, S.; Shaw, P.; Seewaldt, V.; Rinker-Schaeffer, C. In vivo and Ex vivo approaches to study ovarian cancer metastatic colonization of milky spot structures in peritoneal adipose. *J. Vis. Exp. JoVE* **2015**. [CrossRef] [PubMed]
216. Nieman, K.M.; Kenny, H.A.; Penicka, C.V.; Ladanyi, A.; Buell-Gutbrod, R.; Zillhardt, M.R.; Romero, I.L.; Carey, M.S.; Mills, G.B.; Hotamisligil, G.S.; et al. Adipocytes promote ovarian cancer metastasis and provide energy for rapid tumor growth. *Nat. Med.* **2011**, *17*, 1498–1503. [CrossRef] [PubMed]
217. Oosterling, S.J.; van der Bij, G.J.; Bögels, M.; van der Sijp, J.R.M.; Beelen, R.H.J.; Meijer, S.; van Egmond, M. Insufficient ability of omental milky spots to prevent peritoneal tumor outgrowth supports omentectomy in minimal residual disease. *Cancer Immunol. Immunother.* **2006**, *55*, 1043–1051. [CrossRef] [PubMed]
218. Ladanyi, A.; Mukherjee, A.; Kenny, H.A.; Johnson, A.; Mitra, A.K.; Sundaresan, S.; Nieman, K.M.; Pascual, G.; Benitah, S.A.; Montag, A.; et al. Adipocyte-induced CD36 expression drives ovarian cancer progression and metastasis. *Oncogene* **2018**, *37*, 2285–2301. [CrossRef] [PubMed]
219. Protani, M.M.; Nagle, C.M.; Webb, P.M. Obesity and ovarian cancer survival: A systematic review and meta-analysis. *Cancer Prev. Res.* **2012**, *5*, 901–910. [CrossRef] [PubMed]
220. Liu, Y.; Metzinger, M.N.; Lewellen, K.A.; Cripps, S.N.; Carey, K.D.; Harper, E.I.; Shi, Z.; Tarwater, L.; Grisoli, A.; Lee, E.; et al. Obesity contributes to ovarian cancer metastatic success through increased lipogenesis, enhanced vascularity, and decreased infiltration of M1 macrophages. *Cancer Res.* **2015**, *75*, 5046–5057. [CrossRef] [PubMed]
221. James, B.R.; Tomanek-Chalkley, A.; Askeland, E.J.; Kucaba, T.; Griffith, T.S.; Norian, L.A. Diet-induced obesity alters dendritic cell function in the presence and absence of tumor growth. *J. Immunol.* **2012**, *189*, 1311–1321. [CrossRef] [PubMed]
222. Macia, L.; Delacre, M.; Abboud, G.; Ouk, T.-S.; Delanoye, A.; Verwaerde, C.; Saule, P.; Wolowczuk, I. Impairment of dendritic cell functionality and steady-state number in obese mice. *J. Immunol.* **2006**, *177*, 5997–6006. [CrossRef] [PubMed]
223. Shirakawa, K.; Yan, X.; Shinmura, K.; Endo, J.; Kataoka, M.; Katsumata, Y.; Yamamoto, T.; Anzai, A.; Isobe, S.; Yoshida, N.; et al. Obesity accelerates T cell senescence in murine visceral adipose tissue. *J. Clin. Investig.* **2016**, *126*, 4626–4639. [CrossRef] [PubMed]
224. McCloskey, C.W.; Rodriguez, G.M.; Galpin, K.J.C.; Vanderhyden, B.C. Ovarian cancer immunotherapy: Preclinical models and emerging therapeutics. *Cancers* **2018**, in press.

cancers

MDPI

Review

Ovarian Cancer Immunotherapy: Preclinical Models and Emerging Therapeutics

Curtis W. McCloskey [1,2,†], Galaxia M. Rodriguez [1,2,†], Kristianne J. C. Galpin [1,2] and Barbara C. Vanderhyden [1,2,*]

1 Cancer Therapeutics Program, Ottawa Hospital Research Institute, 501 Smyth Road, Ottawa, ON K1H 8L6, Canada; cmccloskey@ohri.ca (C.W.M.); garodriguez@toh.ca (G.M.R.); kgalpin@ohri.ca (K.J.C.G.)

2 Department of Cellular and Molecular Medicine, University of Ottawa, 451 Smyth Road, Ottawa, ON K1H 8M5, Canada

* Correspondence: bvanderhyden@ohri.ca; Tel.: +1-613-737-7700 (ext. 70330)

† These authors contributed equally to this work.

Received: 29 June 2018; Accepted: 23 July 2018; Published: 26 July 2018

Abstract: Immunotherapy has emerged as one of the most promising approaches for ovarian cancer treatment. The tumor microenvironment (TME) is a key factor to consider when stimulating antitumoral responses as it consists largely of tumor promoting immunosuppressive cell types that attenuate antitumor immunity. As our understanding of the determinants of the TME composition grows, we have begun to appreciate the need to address both inter- and intra-tumor heterogeneity, mutation/neoantigen burden, immune landscape, and stromal cell contributions. The majority of immunotherapy studies in ovarian cancer have been performed using the well-characterized murine ID8 ovarian carcinoma model. Numerous other animal models of ovarian cancer exist, but have been underutilized because of their narrow initial characterizations in this context. Here, we describe animal models that may be untapped resources for the immunotherapy field because of their shared genomic alterations and histopathology with human ovarian cancer. We also shed light on the strengths and limitations of these models, and the knowledge gaps that need to be addressed to enhance the utility of preclinical models for testing novel immunotherapeutic approaches.

Keywords: ovarian cancer; tumor microenvironment; immune infiltrating cells; chemotherapy; immunotherapy; syngeneic; transgenic models; hot vs. cold tumors; immunosuppression

1. Introduction

At present, there are no approved immune therapies for epithelial ovarian cancer (EOC) patients. As EOC is often detected at a late stage, research has mainly focused on the discovery of new treatments. Current first-line treatment is debulking surgery and adjuvant or neoadjuvant chemotherapy. Even though >80% of patients show a positive response to this initial therapy, most patients will relapse with chemotherapy-resistant disease [1]. As the presence of tumor infiltrating lymphocytes (TILs) correlates with increased EOC patient survival [2–9], immunotherapies hold great potential for improving EOC outcomes, as they have for several other types of cancers. The U.S. Food and Drug Administration has approved the use of several immune checkpoint inhibitors for non-small-cell lung cancer (NSCLC), melanoma, bladder cancer, renal cell carcinomas, and Hodgkin lymphoma and recently approved the first chimeric antigen receptor (CAR)-T-cell therapy to treat children with B-cell acute lymphoblastic leukemia [10,11].

Antitumor immunity in EOC patients is robustly attenuated because of the immunosuppressive cells within the tumor microenvironment (TME), as reviewed in the literature [12,13]. Several cell types are found in the tumor niche, including immune cells [effector T and B lymphocytes, regulatory T and B cells, natural killer cells (NKs), tumor-associated macrophages (TAMs), and myeloid-derived suppressor cells (MDSCs), among many others (Table 1)], as well as other components found in the TME, including fibroblasts and the adipocytes in the omentum [12]. MDSCs, TAMs, and regulatory T cells (Tregs) play a critical role in maintaining a highly immunosuppressive TME by producing immunomodulatory molecules [transforming growth factor beta (TGFβ), interleukin (IL)-10, IL-6, etc.] and inducing and recruiting immunoinhibitory cells, which dampens antitumoral immunity and supports tumor promotion [12,14]. Therefore, EOC immunotherapy must combine approaches that aim to reduce the highly immunosuppressive TME, as well as stimulate immune-activating antitumoral responses. This review describes encouraging results from both preclinical and clinical trials and highlights the immunotherapies that offer innovative and combinatorial approaches to circumvent the antitumoral barriers within the TME.

Table 1. Main subsets of immune infiltrating cells in epithelial ovarian cancer (EOC) tumor microenvironment (TME).

Immune Cell Type	Antitumoral Function	Tumor-Promoting Function
CD4+ Th1 cells	Help to CTLs in tumor rejection and production of TNFα, IFNγ, and IL-2	Production of cytokines
CD4+ Th2 cells		Education of macrophages, production of cytokines, B cell activation
CD4+ Treg Cells	Suppression of inflammation (cytokines and other suppressive mechanisms)	Immunosuppression: causes IL-2 and other cytokine deprivation, production of TGFβ, IL-10, impaired activation of CTLs
CD8+ T Cells	Direct lysis of cancer cells and production of pro-inflammatory cytokines TNFα, IFNγ, and IL-2	FOXP3+ CTLA-4+ CD25+, convert effector CD8+ T cells into suppressor cells, suppressive function through TGF-β1
B Cells	Production of tumor specific antibodies, IFNγ, TAA presentation, Th1 polarization, promotes T cell expansion	Production of IL-6, IL-10, IL-35, TGFβ, CCL22, immunosuppression, T cell conversion to Tregs, promotes Th2 inhibitory responses
Macrophages, DCs	TAA sampling and presentation; T-cell priming; and production of IL-12 and type I IFN, lympho-attracting chemokines CXCL9, CXCL10, CXCL11	Promotes metastasis and invasion. Produces CSF-1, arginase, IL-6, IL-10, and CCL22. B7-H4+ TAMs suppress antitumoral responses.
MDSCs		Immunosuppression, induces Tregs differentiation, M2 TAM, cancer stemness, sphere formation, and metastasis. Defective TAA presentation. Production of arginase-1, nitric oxide, reactive oxygen and nitrogen species, prostaglandin E2, CXCL12. Deplete cysteine, induce Tregs, inhibit T-cell activation and proliferation, and attenuate the cytolytic ability of NK cells.
NK Cells	Direct cytotoxicity toward cancer cells and production of pro-inflammatory cytokines GM-CSF, TNFα, IFNγ, IL-2 and chemokine CCL5	

CD4+ helper T cells (Th), cytotoxic T lymphocytes (CTLs), interferon (IFN), interleukin (IL), transforming growth factor beta (TGFβ), forkhead box P3 (FoxP3), cytotoxic T-lymphocyte associated protein 4 (CTLA-4), tumor-associated antigens (TAAs), tumor-associated macrophages (TAMs), dendritic cells (DCs), colony stimulating factor 1 (CSF1), granulocyte-macrophage colony-stimulating factor (GM-CSF), myeloid-derived suppressor cells (MDSCs), natural killer cells (NKs), regulatory T cells (Tregs). See the literature [12] for details.

2. Adoptive Cell Therapy

Adoptive cell therapy (ACT) aims to boost the antitumoral activity of autologous (patient) or allogeneic (from healthy donors) lymphocytes [15]. ACT consists of the isolation of T cells from a patient's tumor or peripheral blood to expand or manipulate those cells ex vivo, away from the influence of the immunosuppressive TME. These ex vivo expanded T cells are reintroduced into the patient along with recombinant IL-2 (rIL-2) after a lympho-depleting chemotherapy regimen [15]. ACT has resulted in complete and durable regressions in patients with melanoma [16,17]. In 1995, Fujita and colleagues used ACT of TILs in 13 patients with advanced-stage EOC who did not show any detectable lesions after primary surgery and cisplatin-containing chemotherapy [18]. All patients who received TILs following post-surgery chemotherapy survived three years, compared with only 67.5% of patients receiving chemotherapy alone. Interestingly, TILs promoted tumor regression even in patients with advanced disease or recurrent platinum-resistant EOC [18]. Among TILs, CD8+ T cells have been shown to migrate and infiltrate tumors and mediate antitumoral responses [19]. An early trial in EOC patients that used ACT with TILs following a single injection of cyclophosphamide showed tumor regression in primary tumors and metastases (ovary, liver, lung, and lymph node), which was stable for up to five months with one out of seven patients showing complete response and four out of seven with >50% reduction in tumor burden [20]. When this study was expanded with 10 additional patients, seven cases showed complete regression without recurrence for up to 15 months [20]. This highlighted the prospect of combined therapy using TILs and cisplatin without rIL-2 administration that has unfavorable toxicity in many cases [20].

In another study, 11 patients with advanced platinum-resistant EOC received intraperitoneal (IP) TILs and low doses of rIL-2 IP [21]. Grade 3 clinical toxicity (peritonitis) and anemia were observed without a significant clinical response in any patient. However, 50% of treated patients had regression of ascites (two patients), tumor and carcinoma antigen (CA)-125 (one patient), and surgically confirmed stable tumor and CA-125 values (one patient) [21]. Thus, ACT efficacy has shown conflicting results depending on the study (reviewed in the literature [22]). Importantly, the most encouraging clinical responses were observed when patients were stratified according to the presence or absence of TILs, with the best clinical response in patients who had TILs or 'hot' tumors, a property known to improve survival [22]. As discussed by Santoiemma and Powell, these early studies were executed prior to our advanced understanding of TIL quality, persistence and specificity, patient disease burden, and pre-conditioning regimens, which have since been significantly improved [22].

As ACT has resulted in response rates up to 72% in metastatic melanoma occurring at all sites and has been durable beyond three years in many patients [17], there remains considerable promise in identifying and optimizing conditions whereby ACT can be consistently successful in EOC.

Chimeric Antigen Receptor T Cells (CAR-T)

New approaches, such as genetically engineered T cells, build on the promising early ACT trials that are constrained by the need to isolate and expand functional tumor-reactive T cells [23]. Genetic engineering of T cells has become a powerful approach to increase tumor immunity [24]. The T-cell receptor (TCR) from lymphocytes and chimeric antigen receptors (CARs) can be adapted to specifically target patient tumor cells. CAR-T cells allow for the recognition of tumor cells in a major histocompatibility complex (MHC)-unrestricted manner, combining antigen-specificity and T-cell activating properties in a single fusion molecule [25]. The first generation of CARs was tested in several cancers, including EOC [26], renal cancer, lymphoma, and neuroblastoma, inducing modest responses [25]. In the first study of CAR-T cells in EOC, autologous T cells specific to the EOC tumor associated antigen (TAA) α-folate receptor (FRα) were generated with a chimeric gene composed of an anti-FRα single-chain antibody linked to the signaling domain of the Fc receptor gamma chain [26]. From this study, no reduction in tumor burden was seen in any patient. Although large numbers of CAR-T cells were well tolerated, they did not persist long-term [26]. Some promising TAAs related to EOC for the generation of CARs are FRα [27], human epidermal growth factor receptor 2 (HER-2) [28], CA-125 (MUC16) [29,30], and mesothelin [31]. Moreover, CAR-T cells can be redirected against NKG2D ligands, which are widely expressed on EOC [32], as well as the epithelial cell adhesion molecule (EpCAM) [33] and 5T4 [34].

CAR translation to solid tumors is actively being investigated at present; however, few facilities have the capacity to produce CARs and many trials have achieved less than expected efficacy. This could be explained by the highly immunosuppressive TME [35]. Koneru and colleagues generated T cells engineered to specifically recognize the MUC-16ecto TAA that is expressed in the majority of ovarian tumors and derived from the cleavage of CA-125 [29,36]. As a strategy to overcome the TME, they developed a construct that co-expressed both MUC16ecto CAR and IL-12, a pro-inflammatory cytokine that has potential roles in anticancer therapy [37]. Some immune cells such as NK cells, dendritic cells (DCs), and macrophages normally produce IL-12 to induce T-cell proliferation and inhibit Tregs. The IL-12 secreting CAR-T cells displayed enhanced antitumor efficacy as determined by increased survival, prolonged persistence of T cells, and higher systemic interferon gamma (IFNγ) in mice with human EOC xenografts [36]. These observations suggest that IP delivery of CAR-T cells may be most beneficial for EOC treatment [38]. The peritoneal cavity is the main locus for EOC metastases, and local treatment seems to be a safer option for patients, because adverse reactions induced by 'on-target off-tumor' toxicities, such as cytokine release syndrome, were reported in a study that used ACT of autologous mesothelin-redirected CAR-T cells in a patient with BRCA1+ advanced recurrent serous EOC [39]. By local administration of CARs, antigens expressed by both EOC and healthy tissue, such as EpCAM, can be targeted in a safer way [40]. Indeed, there is a long-term survival advantage associated with IP chemotherapy in advanced EOC disease [41]. Recent studies have shown that CAR-T cells could be administrated along with cytokines such as IL-2, IL-7, IL-15, and IL-21 to increase their efficacy against hematologic and solid tumors [42].

There are still many barriers to overcome for CAR-T therapy effectiveness, such as T-cell trafficking into the tumor niche, patient selection, cancer-specific TAAs, and the highly immunosuppressive TME, as well as the dose and route of administration [35,43,44]. Combinatorial strategies to circumvent these barriers, such as incorporating immune checkpoint inhibitors into CAR-T cells [45], may be the new frontier in enhancing the tumor elimination efficacy of CAR-T cells for EOC patients.

3. Strategies Targeting Immunosuppression in the TME

Regulatory T cells (Tregs) participate in the establishment of the immunosuppressive TME, attenuating antitumor immunity. Neoplastic cells and TAMs produce CCL22 that mediates Treg tumor infiltration, a mechanism that could be blocked to decrease immunosuppression and enhance antitumoral immunity [46,47]. T cells transduced to express chemokine receptors matching the TME chemokines can improve tumor homing after ACT, as shown by improved migration of tumor ascites lymphocytes to the EOC microenvironment by T-cell CXCR2 transduction [48].

Highly immunosuppressive MDSCs are attractive targets to enhance the efficacy of cancer immunotherapy. Immunizing mice with microparticles containing TLR9 and NOD-2 ligands (MIS416), followed by anti-CD11b administration, was shown to abrogate the immunosuppressive capacity of MDSCs in the ID8 murine model of EOC [49]. This treatment significantly prolonged survival, highlighting the need for more immunotherapies targeting innate immunity within the TME [49]. However, there are still many unknowns concerning the different MDSC phenotypes and levels in tumor tissue, peripheral blood, and/or ascites fluid, and how their presence influences the TME [50].

TGF-β plays a key role in EOC TME by preventing antitumoral T cell responses. Recent studies have identified stromal TGF-β signaling as a determinant of immune exclusion [51,52]. In a model of colorectal cancer, Tauriello and colleagues recently showed that TGF-β inhibition prevents tumor metastasis by increasing cytotoxic T-cell responses [53]. Thus, blocking TGF-β production along with immunotherapy could be a promising pro-immunogenic approach in EOC by promoting strong T cell infiltration and antitumoral immunity [51,54].

A recent study found that the accumulation of effector memory CD8+ T cells (T_{EM}) in EOC ascites was mediated by TAM-derived CXCL9. This accumulation of CD8+ T cells correlated with increased patient survival. However, ascites-derived factors can suppress T_{EM} effector functions through the production of IFNγ and TNFα, and CD107a expression, shortening relapse-free survival of patients. Inducing TAMs to produce CXCL9, CXCL10, and CXCL-11 chemokines may potentially be therapeutic [55]. COX inhibitors also enhance chemokine release, suggesting their use in combinatorial strategies to increase TILs within the EOC TME [56].

Increased expression of immune checkpoint molecules on cancer cells, antigen-presenting cells (APCs), and T cells in the EOC TME leads to immunosuppression upon binding of their corresponding receptor/ligand, effectively putting the brakes on CD8+ effector T and NK cells. Blockage of co-inhibitory molecules has now been exploited in many cancer types to increase pre-existing patient antitumoral responses (reviewed in the literature [57]). These co-inhibitory molecules include cytotoxic T-lymphocyte associated protein 4 (CTLA-4; on T cells to control T cell activation, and binds to CD80 or CD86 on APCs or tumor cells), lymphocyte-activation gene 3 (LAG-3; on T and NK cells, and binds to MHC-II and LSECtin on APCs or tumor cells [58]), programmed cell death protein 1 (PD-1; on activated T, B, and NK cells, and binds to PD-L1 or PD-L2 on APCs or tumor cells), and PD-L1 (on APC or tumor cells, and binds to PD-1 on T cells or CD80 on APC or tumor cells).

Anti-PD-1 and anti-LAG-3 synergized in the ID8-OT-I murine model to prolong survival, reduce tumor burden, and reduce Tregs, while increasing CD4+ and CD8+ TILs [59]. Therefore, blockade of immune inhibitory molecules, such as LAG-3, T cell immunoglobulin and mucin domain 3 (TIM-3); or T cell immunoreceptor with Ig and ITIM domains (TIGIT); as well as fibrinogen-like protein 2 (FGL-2) [60–62], may synergize with PD-1/PD-L1 to target multiple cell types and more potently relieve immunosuppression [58,63]. Interestingly, Lin and colleagues found that the efficacy of anti-PD-L1 therapy was unaltered by tumor cell expression of PD-L1 and instead, DC and macrophage PD-L1 expression was likely to underlie response [64]. Thus, more studies need to explore how these immunotherapies abrogate immunosuppression and the cell types underlying responses.

Beyond the immunosuppressive cytokine milieu present in the TME, antigen persistence also contributes to long-lasting T cell exhaustion (reviewed in the literature [65]), in part through increasing PD-1 expression and through de novo methylation of effector function genes [66]. Thus, targeting aberrant methylation using epigenetic modifiers such as the DNA methyltransferase inhibitor, azacytidine, has offered promising results in rejuvenating T cell responses [66]. Further, azacytidine induces the expression of antigen processing machinery and activates the interferon response in cancer cells by inducing viral mimicry through the MDA5/MAVS/IRF7 pathway [67,68]. However, azacytidine can also upregulate PD-L1 on cancer cells [69], and thus shift the balance from antitumor immunity toward immunosuppression. Combinatorial approaches with epigenetic therapies and immune checkpoint blockade (anti-PD-1) have been employed and shown to enhance CD8+ T and NK cell recruitment and function, and overall survival in murine models [66,70].

4. Increasing Tumor Immunogenicity

4.1. Chemotherapy

Chemotherapeutic agents used in standard EOC treatment can induce an immunogenic cell death of cancer cells by the release of danger signals known as damage-associated molecular patterns (DAMPs) [71,72]. Memory T cells derived from peripheral blood mononuclear cells from EOC patients after cytoreductive surgery and platinum and taxane chemotherapy recognized antigens associated with apoptotic EOC cells, and their presence was correlated with prolonged survival [73]. Responders displayed significant IFNγ or IL-17 functions by both CD4+ and CD8+ T cells in response to apoptotic EOC antigens [73]. Exposure to platinum-based drugs can disrupt the STAT6-mediated immunosuppression in the TME by decreasing the expression of PD-L2 on both human DCs and tumor cells, enhancing antigen-specific proliferation and Th1 cytokine secretion, as well as increasing tumor T cell recognition [74,75]. CD8+ T-cell function is not permanently suppressed in advanced EOC and successful carboplatin/paclitaxel chemotherapy is associated with improved antigen-specific T cell reactivity [76].

Paclitaxel has been previously reported to be a ligand to Toll-like receptor 4 (TLR4), normally found on normal and neoplastic cells, which, under ligation, significantly increase the secretion of IL-6 and IL-8 by human EOC cell lines (SKOV3, OVCAR3), abrogating paclitaxel effects on cell proliferation, and promoting tumor survival and chemoresistance [77–79]. MyD88 expression is more restricted to EOC cells, independent of tumor grade, and is associated with reduced progression-free and overall survival [80]. Strategies aiming to target the TLR4 signaling pathway on TLR4/MyD88(+) EOC patients may hold promise for the treatment of paclitaxel-resistant EOC. For example, atractylenolide-I (AO-I), a naturally occurring sesquiterpene lactone and TLR4-antagonizing agent, inhibits TLR4 signaling by interfering with the binding of paclitaxel to membrane TLR4, thus sensitizing the response of MyD88(+) EOC cells to paclitaxel [81]. AO-I indirectly downregulates MyD88/NF-κB signaling; reduces activation of NF-κB, Akt, and indoleamine 2,3-dioxygenase (IDO)-1; and attenuates the secretion of IL-6, TGF-β1, VEGF, and IL-17A by SKOV3 cells [82]. In another study, Peng and colleagues found that chemotherapy induces local immune suppression by increasing PD-L1 expression in ovarian tumor cells [83]. Paclitaxel treatment is able to increase CD8+ T-cell infiltration of ovarian tumors in a mouse model of ovarian cancer by upregulating MHC-I expression, as well as PD-L1 expression, in an NF-κB-dependent manner [83]. Thus, the authors showed that by combining paclitaxel treatment with anti-PD-L1 or anti-PD1 antibodies, the immunosuppressive TME is attenuated, enabling the achievement of maximal antitumoral responses and increasing survival of ovarian tumor-bearing mice [83].

Agents such as cyclophosphamide can also be used in combination with immune checkpoint inhibitors, such as anti-PD-1, to gain therapeutic synergy by decreasing Treg infiltration and stimulating the generation of CD8+ TILs [84]. Gemcitabine chemotherapy combined with CTLA-4 blockade results in a potent antitumor immune response that is CD4+ and CD8+ T-cell dependent [85]. Oxaliplatin treatment can enhance susceptibility of human EOC cells to NK cell-mediated cytolysis by inducing the production of type I IFN and chemokines, and enhance MHC class I-related chains A/B, UL16-binding protein (ULBP)-3, CD155, and TNF-related apoptosis-inducing ligand (TRAIL)-R1/R2 expression [86].

Chemotherapy possesses immunomodulatory properties by augmenting pre-existing TIL responses in high-grade serous ovarian cancer (HGSC) patients; however, this increase fails to confer significant prognostic benefit [87]. In contrast, platinum-resistant EOC has been shown to generate poor immunologic responses [88]. Several studies have assessed the clinical and cost-effectiveness of different combined chemotherapeutic approaches for advanced recurrent or refractory EOC [89,90]. By combining standard chemotherapeutic agents with immunotherapies, chemotherapy-induced apoptosis can be exploited as an adjuvant to synergistically enhance antitumoral immunity.

EOC-derived ascites offers accessible and plentiful tumor tissue to identify and target molecules that shape the TME. Both the cellular and fluid compartments allow for the investigation of prognostic and predictive biomarkers, pharmacodynamic markers, and molecular profiling analysis [91]. Moreover, ascites may be a useful tool to reveal the immune status of the TME within the peritoneal cavity. For example, IL-6 is enriched in the malignant EOC ascites, enhancing the invasive properties of EOC cells [92]. IL-6 levels are elevated in recurrent compared with primary advanced EOC [93], and because the IL-6R/STAT3/miR-204 feedback loop contributes to cisplatin resistance of EOC [94], targeting IL-6 or IL6R with neutralizing antibodies could increase EOC sensitivity to cisplatin. IL-6 levels can also be modulated using NF-κB inhibitors, like dehydroxymethylepoxyquinomicin (DHMEQ), as showed by Nishio and colleagues, who evaluated the effects of DHMEQ in vitro on human EOC cells and macrophages [95]. DHMEQ was able to inhibit the production of IL-6 and IL-8 by EOC cell lines and enable the release of immunosuppression of human DCs and macrophages incubated with culture supernatant of EOC pretreated cells. In vivo studies in nude mice implanted with human EOC cells showed a reduction of arginase expression and tumor accumulation of MDSCs, demonstrating the important role of NF-κB in maintaining EOC TME immunosuppression [95]. Given NF-κB is a pleiotropic transcription factor and also possesses anti-inflammatory properties [96], further studies are needed to evaluate the therapeutic potential of DHMEQ in EOC in immunocompetent hosts to better evaluate its impact on all the components of the EOC TME.

4.2. Oncolytic Viruses

Oncolytic viruses (OV) can be engineered for transgene expression to enhance their tumor specificity, safety, drug susceptibility, immunogenicity, and oncolytic potency [97]. They can be administered locally or intravenously and spread to the tumor and metastases [98,99]. OVs trigger at least two modes of cancer cell killing: direct oncolysis of infected cells or indirect cell death elicited by the host immune system [97,100]. These phenomena allow the release of viral antigens, DAMPs, and TAAs, which, under a proper inflammatory milieu, enable their recognition and phagocytosis by immune cells, such as macrophages and DCs, to eventually activate T cells in the draining lymph nodes [97,100,101]. OV have tumor specificity because, unlike malignant cells, healthy cells can respond to infection by inducing antiviral IFNs, though tumor heterogeneity in IFN expression has been identified as means of OV resistance [102].

OVs are a new class of immunotherapy that have shown promising results in clinical studies, leading to the approval of the first OV, talimogene laherparepvec (Imlygic®, Amgen, Thousand Oaks, CA, USA) T-VEC, for treatment of metastatic melanoma [103–106]. Notably, OVs have shown synergy when combined with other immunotherapies, such as checkpoint inhibitor antibodies [103,104,106]. IL-12 expressing oncolytic herpes simplex virus was shown to promote eradication of both murine and human ovarian cancer cell lines and promote TAA-specific CD8+ T-cell responses in the peritoneal cavity and omentum, leading to reduced peritoneal metastasis and improved survival in the mouse model tgMISIIRTAg [107]. Thus, the use of OV immunotherapy alone or combined with approaches to increase immunostimulatory or immunogenic responses offers promising strategies for investigation as novel treatments for EOC patients.

In addition to cancer cells, OVs can infect and lyse CAFs [108] and endothelial cells in the TME [109,110], leading to the destruction of these cells and immune infiltration [111]. At present, several viruses are under active investigation in preclinical [112–117] and clinical trials (NCT02028117, NCT00408590, NCT02759588, NCT02068794, NCT03225989, NCT01199263, NCT02285816) as potential therapies for various cancers, including EOC, as well as in combinatorial strategies with other immunotherapies such as checkpoint inhibitors.

A promising new strategy for cancer immunotherapy is to exploit autologous tumor cells as carriers of viruses to the tumor niche [118]. Such oncolytic vaccine platforms consisting of tumor cells infected with OV have been shown to be a favorable strategy in murine models of melanoma and other solid cancers [101,103]. An infected cell vaccine (ICV) platform was developed using irradiated autologous tumor cells infected with oncolytic Maraba (MG1) virus that is engineered to express the immune stimulatory cytokine IL-12 [119]. When delivered directly into the peritoneal cavity in a model of peritoneal carcinomatosis, the vaccine promoted the migration of IFNγ-secreting NK cells, decreased tumor burden, and improved survival. Importantly, the enhanced NK-cell cytotoxicity and migratory capacity driven by ICV-MG1-IL12 was also observed in human lymphocytes exposed to human tumor cell lines infected with MG1-IL12, highlighting the benefit of this approach in patients with abdominal cancers [119]. MG1 is currently being evaluated in phase I/II clinical trials as a stand-alone therapy and in a vaccination strategy for the treatment of late-stage disseminated disease (NCT02285816, NCT02879760). Thus, by using tumor cells as virus carriers, the TME can be remodeled, making a "cold" tumor into an inflamed or "hot" tumor that could support and sustain the generation of significant antitumoral responses. This approach could be advantageous to EOC patients whose tumors have suboptimal immune infiltration and do not respond to standard therapies. Additional studies are needed to assess the conditions (TME quality, tumor grade, etc.) under which this approach could benefit those EOC patients, as well as to determine if this rationale could be exploited with tumor cells derived from EOC ascites. Figure 1 summarizes some emerging EOC immunotherapies.

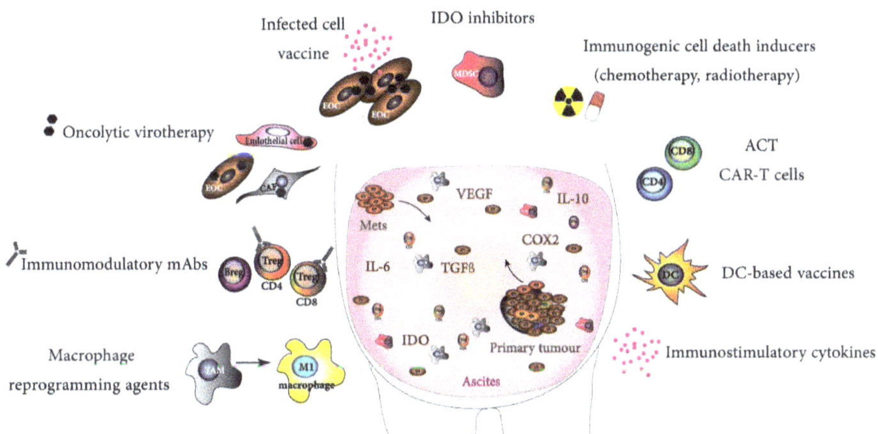

Figure 1. Emerging ovarian cancer immunotherapies. T cell infiltrated EOCs (>50% EOCs [120] can be targeted with therapies such as ACT, CARs cells, co-stimulatory mAbs [121] (like anti-CD137), oncolytic virotherapies, and DC-based vaccines [122], aiming to increase the effector functions of the pre-existing antitumoral immunity. Conversely, strategies aiming to decrease the highly immunosuppressive TME [checkpoint blockers mAbs (anti-PD-L1), IDO inhibitors [123], etc.] can be exploited for 'cold' tumors and/or for advanced stages EOC, to decrease the immunosuppressive functions of MDSCs, TAMs, and Tregs. Radiotherapy and chemotherapy are immunogenic cell death inducers increasing the release of TAAs in the TME, thus augmenting the NK-cell mediated killing, the incidence of TAA presentation by APCs, and eventually T cell priming. Also, chemotherapy can target MDSCs (gemcitabine [124], 5-fluorouracil [125]). EOC cells and TAMs can be targeted with trabectedin, which inhibits CCL2 production and decreases monocyte recruitment in tumors [126]. OVs can infect tumor cells, as well as CAFs and endothelial cells, thereby helping to decrease their immunosuppressive action in the TME. Many approaches can be combined, such as administration of costimulatory cytokines (IL-2, IL-7, IL-15 and IL-21) along with approaches such as CARs, OVs, and ACT. Tumor cells derived from ascites can be exploited for the production of infected cell vaccines with OVs delivering IL-12. Simultaneous targeting of CD137 and PD-1 [127] or TIM-3 [128] with mAbs along with cisplatin treatment [129] can achieve significant antitumoral responses. Adoptive cell transfer (ACT), cancer-associated fibroblasts (CAFs), chimeric antigen receptor (CAR), monoclonal antibodies (mAbs), natural killer (NK), myeloid-derived suppressor cell (MDSC), tumor associated macrophage (TAM), regulatory T cells (Tregs), tumor associated antigen (TAA), antigen presenting cell (APC), oncolytic virus (OV), programmed cell death 1 (PD-1), T-cell immunoglobulin and mucin domain 3 (TIM-3), metastasis (Mets).

5. Preclinical Models for Ovarian Cancer Immunotherapy

The ovarian cancer field has accelerated rapidly since the discovery that ovarian cancer is not one disease, but exists as numerous subtypes that behave differently. For the most common EOC subtype, HGSC, we have only recently begun to appreciate the role of the fallopian tube secretory epithelium (murine oviductal epithelium) as one of the origins of HGSC. These discoveries led researchers to focus on modeling ovarian cancer after specific subtypes or origin(s) of disease, leading to a narrow characterization of many of these preclinical models in relation to their origin and common genomic alterations with HGSC. Commonly studied features include growth rate, tumorigenic potential, immunohistochemical markers, DNA mutations, BRCAness, RNA expression, and copy number variation, with the results correlated to The Cancer Genome Atlas (TCGA) or other large datasets on HGSC. Since the emergence of immunotherapies as promising agents for ovarian cancer treatment, we have a rich reservoir of human and murine-derived ovarian cancer models that have limited characterization for the features that we have come to appreciate as important indicators of immunotherapeutic efficacy. Such features include MHC status and PD-L1/2 expression on cancer cells, total mutational and neoantigen burden, ascites composition, hot versus cold immune landscape, and the contribution of tumor stroma (cancer-associated fibroblasts) to tumor immunosuppression. Given that the current clinical efficacy of immune checkpoint blockade for HGSC ranges from 9.7 to 15% [35], preclinical models will be of great importance as we seek better indicators of response, as well as new immunotherapy development. In this section, we summarize the current knowledge of ovarian cancer models and highlight the models that may be untapped resources for the immunotherapy field.

5.1. *Syngeneic Murine Models*

A syngeneic model is defined by its immunological compatibility such that the host does not reject either the outgrowth or transplant of cancer cells in immunocompetent animals. We have divided syngeneic models into spontaneously occurring and genetically engineered models, as described below.

5.1.1. Spontaneously Transformed Syngeneic Models

The only two non-human animals that are known to spontaneously develop EOC are the egg-laying hen and the jaguar (Table 2). Up to 35% of egg-laying hens develop EOC with similar risk factors to humans such as age and ovulation number, reviewed in the literature [130]. The hen model has yet to be used for immunotherapy development, but has features that make it an exciting candidate for future use. The first evidence that spontaneous hen tumors are immunogenic was the observation of mesothelin auto-antibodies and mRNA in 44% of hens harboring tumors [131]. Serous histology hen tumors contain the most TILs characterized by T and B cell infiltration [132], and highly express immunosuppressive ILT3, which functions to limit T cell proliferation and differentiation, suggesting perturbed antitumor adaptive immunity within the TME [133]. Barua and colleagues showed elevated expression of DR6, a known inhibitor of DC function, with increasing stage of disease, suggesting perturbed innate immunity, as well as adaptive immunity [134]. This group then showed that increased immune infiltration in late-stage disease is restricted to the tumor stroma, while intratumoral immune infiltration largely decreases with the stage, indicating that late-stage tumors acquire mechanisms to limit immune trafficking [135]. These studies support the presence of immunosuppression within the hen TME and highlight the hen model's promise for immunotherapy development. Profiling immune checkpoint expression, total mutation, and neoantigen burden in hen tumors would be an exciting addition to the hen model dataset.

Interestingly, 40% of captive jaguars develop ovarian carcinoma with non-synonymous mutations in *BRCA1* [136,137]. The endangered nature of this species prevents its use as an ovarian cancer model, though immunotherapies that enhance survival of patients with *BRCA1*-associated cancers could later play a role in the conservation of this species.

The ID8 and STOSE models (in the C57BL/6 and FVB/N strains, respectively) are two incidences of spontaneous transformation of primary murine ovarian surface epithelial cell cultures. Both of these models share similar epithelial markers [cytokeratin(+), WT1(+), inhibin(−)], growth rates, expression profiles similar to human HGSC, and tumorigenicity in syngeneic xenografts [138,139], and they both form malignant ascites and disseminated disease following orthotopic intrabursal injection [138,140]. The ID8 model, established by Roby and colleagues in 2000, has been the most commonly used model for immunotherapy development based on its established characterization and reliability in forming syngeneic tumors. Peritoneal tumors generated by IP injection of ID8 cells develop a complex microenvironment with SMA+ fibroblasts, CD3+ T cells, CD68+ macrophages, and neo-vasculature [140]. ID8 cells have been employed in the development of epigenetic modifiers, immune checkpoint inhibitor and oncolytic virus studies, DC and microparticle vaccines, and numerous emerging immunotherapies [112,129,141–144].

Antibody monotherapy has shown little promise in the ID8 model, as neither immune checkpoint inhibitors (anti-PD-1, anti-CTLA-4) or activating antibodies (anti-OX40, anti-CD137) used as monotherapies had any impact on survival [129,142]. Two studies using combination antibody immunotherapy, anti-PD-1 and -OX40 [142] or anti-PD-1 and -CTLA-4 and -CD137 [129] have shown prolonged survival of ID8 tumor-bearing mice and a shift from a CD4+ T helper 2 (Th2) cell milieu to an antitumor Th1 response characterized by an increased ratio of CD8+ and CD4+ T cells over immunosuppressive CD4+FOXP3+ Tregs, and a reduction in CD11b+Gr-1+ MDSCs. In both studies, antibodies were administered within 15 days of ID8 injection, representing early-stage disease before robust tumor formation and ascites develops. It remains to be determined if these therapies promote regression of late-stage disease in the presence of ascites. Turner and colleagues reported enhanced MHC class II expression in subcutaneous ID8 tumors and restricted tumor growth with a combined epigenetic therapy using a histone deacetylase inhibitor (entinostat) and DNA methyltransferase inhibitor (azacytidine). Furthermore, azacytidine enhances recruitment of CD8 T and NK cells in the ID8 model, and shows synergy with anti-PD-1 checkpoint blockade, offering an alternative approach to modify the immune landscape of the TME [70,144].

Numerous OV platforms have been tested in the ID8 model including reovirus, vaccinia, myxoma, vesicular stomatitis, and herpes simplex viruses [109,112,141,143,145]. Many of these OV platforms have been shown to prolong survival, enhance CD8+ T-cell infiltration, and reduce immunosuppression. Oncolytic vaccinia virus encoding a CXCR4 antagonist helped prevent peritoneal spread and reduced Treg recruitment in ID8 tumors [112]. Although promising, this monotherapy failed to cure the ID8 model, highlighting the need for combinatorial therapies [112]. Synergy was observed when myxoma virus was administered prior to cisplatin treatment in an IP ID8 model, generating a T-cell response that could recognize TAAs from ID8 cell lysates [143]. A recent study by Liu and colleagues reported synergy between anti-PD-1 antibody therapy and oncolytic vaccinia virus in the ID8 model [146], opening the door to combining oncolytic platforms and immune checkpoint inhibition or novel antibody therapies.

Among the studies using ID8 cells, none have identified a full curative therapy that generates a robust memory response that can protect against ID8 cell re-challenge, a sought-after goal of immunotherapies. This could be because of the poor immunogenicity of ID8 cells; out of their mutational burden of 92 somatic mutations, only 17 are predicted to generate transcribed neoantigens. Upon vaccination with synthetic peptides carrying these 17 mutations, none induced a neoantigen-specific T-cell response, indicating that they likely do not yield MHC presented epitopes [147]. The use of modified ID8 cell lines may better phenocopy human HGSC, given that parental ID8 cells have a relatively low mutational burden compared with human HGSC [147].

One of the first modifications to ID8 cells was the stable expression of beta-defensin, *Defb29*, and *Vegf-A* yielding ID8-Defb29/Veg-A cells that had increased pro-tumor DC recruitment, neovasculature, and a more aggressive phenotype with reduced survival compared with parental ID8 cells [148]. ID8-Defb29/Veg-A derived tumors are good models for DC dysfunction and recently, Cubillos-Ruiz and colleagues identified the role of the endoplasmic reticulum stress sensor XBP1 in mediating DC dysfunction in this model [149]. A second modification was the addition of the ovalbumin (OVA) peptide, ID8-OVA, a useful tool to assess antitumoral responses mediated by OT-I CD8+ T cells or OT-II CD4+ T cells derived from the transgenic mouse models OT-I and OT-II, where the TCRs were designed to specifically recognize OVA peptides in the context of H2Kb and I-A b, respectively [141,150]. Using a reovirus platform, Chiang and colleagues showed prolonged survival, enhanced expression of MHC class I antigen presentation machinery (beta-2-microglobulin, TAP-1, and TAP-2), reduction of MDSCs and Tregs, and enhanced DC-activation of OVA-specific CD8+ T cells in the ID8-OVA model [141].

One of the notable weaknesses of the ID8 model is that it does not contain a *Trp53* mutation, which is characteristic of 94% of human HGSC [151]. Walton and colleagues generated ID8 cells with both a *Trp53* and *Brca2* mutation using CRISPR-Cas9 [152,153]. ID8-*Trp53*−/− tumors had increased MDSCs recruitment, possibly through increased CCL2 expression. With the addition of *Brca2* deletion, the tumors gained intraepithelial lymphoid aggregates, characteristic of hereditary human HGSC (~9% of cases), making this model relevant to the study of *BRCA*-associated HGSC [154]. Given the observed increase in mutational burden in HGSC of BRCA mutation carriers [155], it would be interesting to profile neoantigen burden and TAA-specific T-cell responses in ID8-Trp53-Brca2 cells, as well as ID8-Defb29/Vegf-A, to further enhance their relevance to HGSC. These modified ID8 cell lines may offer more relevant models for ovarian cancer immunotherapy as they better phenocopy the TME found in human HGSC. Roberts and colleagues have also published a spontaneously transformed murine ovarian surface epithelial cell line, MOSE-L cells, which were highly proliferative, expressed epithelial markers, and are tumorigenic in syngeneic C57BL/6 mice, though uptake of this model has been sparse, likely because of the well-established characterization of the ID8 model [156].

In 2014, our group published the second spontaneously transformed syngeneic model of HGSC-like cancer, the STOSE model, which reliably generates tumors in syngeneic FVB/N mice. Our initial characterization profiled the growth rate, genomic instability, and immunohistochemical markers relevant to human HGSC [138]. Recently, we have expanded the characterization of the STOSE model by profiling the immune landscape of orthotopic intrabursal-derived STOSE tumors and have found STOSE tumors have increased T-cell infiltration and a larger CD4+ Treg population than ID8 tumors (data not shown), suggesting STOSE tumors contain a T-cell-rich or 'hot' TME. In contrast, orthotopic ID8 tumors generate a more myeloid-rich or 'cold' TME. Given their susceptibility to oncolytic virus infection in vitro [107], it will be important to compare the efficacy of immunotherapeutic approaches in these two models, which generate tumors in different murine strains and have contrasting immune landscapes within the TME, and may better reflect the heterogeneity of HGSC seen in the clinic. Characterization of the STOSE model for copy number and mutational and neoantigen burden must be done to assess the utility of the STOSE model.

Both ID8 and STOSE models are derived from the ovarian surface epithelium. Given the contribution of the fallopian tube epithelium to human HGSC, spontaneous and transplantable syngeneic models derived from murine oviductal epithelium are much needed. Endsley and colleagues described spontaneously transformed murine oviductal epithelial cells derived from CD1 mice that exhibited features of transformation, but only generated subcutaneous tumors in athymic nude mice, limiting their use as a syngeneic model for cancer immunotherapy studies [157]. The addition of PTEN loss in these cells resulted in the first and, currently, the only syngeneic model of fallopian tube-derived EOC [158].

5.1.2. Genetically Engineered Mouse Models (GEMM)

The majority of GEMM models were generated to improve our understanding of the origin(s) of ovarian cancer. Consequently, most of the characterization of these models has placed primary emphasis on identifying the location of early lesions and an immunohistochemical panel assessing positivity for PAX8, P53, WT1, and cytokeratins with a lack of inhibin and calretinin staining. With emerging immunotherapies targeting the TME, characterization of the majority of GEMM models has been too narrow to assess their use for testing immunotherapies, although many models may be ideal because of their shared genomic alterations and TME complexity that phenocopy HGSC. The various GEMM models of ovarian cancer have been comprehensively reviewed [159,160]. Here, we discuss some important considerations in using GEMM models for immunotherapy development.

Many models have been made using oncogenic simian-virus 40 T-antigen (SV40TAg) driven from the ovarian or oviductal epithelium (Table 3). These models include the TgMISIIRTAg model, which drives SV40TAg expression from the *MISIIR* gene, leading to ovarian tumor development in 50% of mice at 6–13 weeks of age [161]. The use of this model has revealed a synergistic effect of the viral sensitizer colchicine and vaccinia virotherapy [109]. Epigenetic combination therapy, entinostat and azacytidine, was shown to enhance MHC class II expression in TgMISIIRTAg tumors [144]. The TgCAG-LS-TAg model that drives SV40TAg from the chicken β-actin promoter was used to show that estrogen can accelerate EOC development, though an immune basis for this acceleration was not explored [162]. Another model used the oviduct-specific gene, *Ovgp1*, to drive

SV40TAg expression generating tumors derived from oviductal epithelial cells [163]. Although these results showed promise, the use of SV40TAg for immunotherapy studies should proceed with caution as SV40TAg is both an oncogenic driver and a dominant immunogen. Schietinger and colleagues designed a sophisticated experiment in which SV40TAg-specific T cells ($TCR_{SV40\text{-}1}$) and OT-I T cells were co-injected into liver tumor-bearing ASTxAlb–Cre mice that constitutively express SV40TAg. Only $TCR_{SV40\text{-}1}$ cells became dysfunctional in the presence of cognate SV40TAg with enhanced expression of PD-1, TIM-3, LAG-3, and 2B4, while OT-I T cells maintained their functional expression of IFNγ and TNFα [164]. This showed that SV40TAg expression strongly inhibited effective T cell responses with little contribution from the immunosuppressive TME, because OT-I T cells maintained their functional phenotype within the TME. Thus, a strong clonal response to a persistent dominant antigen was enough to reduce the influence of the TME, which does not reflect the normal contribution of the TME in suppressing antitumor immunity in HGSC (reviewed in the literature [165]). Further, McGranahan and colleagues recently showed that effective cancer immunotherapy goes beyond total mutational burden and requires a clonal neoantigen T-cell response [166]. Patients who exhibit increased intratumoral neoantigen heterogeneity (subclonality) have reduced clinical benefit to immune checkpoint inhibition. They further showed that chemotherapy can induce neoantigen heterogeneity [166]. Thus, using models with a dominant antigen, such as SV40TAg, which can abrogate the effects of the TME, could confound the interpretation of an immunotherapy's efficacy. Models that have neoantigen heterogeneity or lack authentic neoantigens (like the ID8 cells) may better phenocopy HGSC, as only 12% of HGSC are likely to express $\geq$ 1 neoantigen [147].

Given that 94% of HGSCs possess *TP53* mutations, with 35% of tumors expressing high levels of TP53 and 62% expressing little to no detectable TP53 [151], immunotherapies should be tested in models that represent both high and low p53 expression. Multiple GEMMs have either Tr*p53* knockout or *Trp53* mutation, driven from both the ovarian and oviductal epithelium [167–170]. HGSCs with mutant *TP53* have higher PD-L1 expression than tumors with wild-type *TP53*, indicating a role for mutant TP53 in modulating the TME, though the exact *TP53* mutations were not specified [171]. The effect of *TP53* loss was further corroborated by Son and colleagues who showed *p53* loss enhanced pro-inflammatory cytokine (CXCL1, CXCL2, CXCL3, and CXCL8; and TNFα) expression in HGSC [172]. More studies are needed to elucidate the effects of specific *TP53* mutations on the ovarian cancer TME.

Interestingly, hereditary *BRCA1/2* mutations lead to higher mutational burden and neoantigen burden that correlates with improved survival, increased TILs, and increased PD-1/PD-L1 expression, indicating that these tumors may especially benefit from immune checkpoint inhibition [155,171]. Perets and colleagues generated GEMMs with doxycycline-inducible Cre-recombinase mediated deletion of *Brca1* or *2* and *Pten*, and Tr*p53* loss or mutation, driven from the oviductal epithelium-specific *Pax8* promoter. All combinations yielded HGSC-like tumors with high mutational burden and genomic alterations similar to the TCGA dataset on ovarian carcinoma such as c-myc amplification [169]. Similarly, Zhai and colleagues characterized a model of tamoxifen-inducible deletion of *Brca1*, *Pten*, *Rb1*, and *Nf1* driven from the *Ovgp1* promoter, which generated serous-tubal intraepithelial carcinomas that progressed to HGSC. They also characterized a similar model with deletion of *Brca1*, *Pten*, and p53 that also developed precursor lesions and HGSC, but with a mixed tumor phenotype with mucinous metaplasia [173]. These models have numerous features relevant to human disease and profiling the immune landscape and mutational and neoantigen burden would be exciting additions to extend the use of these models into the realm of cancer immunotherapy.

One GEMM study that profiled the TME was done by Budiu and colleagues, using mice that express human MUC1 from the endogenous promoter that were then crossed with mice containing conditional alleles for *Pten* deletion and activation of KrasG12D [174]. MUC1 is overexpressed in 75–90% of human EOC and, interestingly, tumor-associated MUC1 is more immunogenic because of the loss of glycosylation, revealing epitopes that can be targeted by antibodies and specific CD8+ T cell responses [175–177]. The MUC1KrasPTEN model generates ovarian tumors with high serum levels of human MUC1, robust CD4+FOXP3+ TILs, and dysfunctional DCs [174]. Mice that expressed human

MUC1 had a larger splenic Treg population than KrasPTEN mice alone. Using a MUC1 vaccination strategy along with a type 1 DC polarizing cocktail, they were able to reduce Tregs and enhance survival in MUC1KrasPTEN tumor-bearing mice [174]. This model allows for a MUC1-directed antitumor response that could be modulated by a vaccination strategy targeting the TME. In this aspect, this model improves upon SV40TAg models where the immunogen is too dominant to see any effect of the TME modulation [164].

Although GEMM models enable us to better model the origins of disease and genomic alterations characteristic of HGSC, they have two weaknesses that limit their use for immunotherapy. Firstly, most GEMMs have been generated on a mixed strain background, preventing the generation of transplantable syngeneic cell lines. Secondly, although GEMMs may reproducibly generate tumors, they tend to arise over a wide course of time. The difficulty in controlling tumor onset and size in GEMMs introduces a logistical challenge for immunotherapy studies that rely on flow cytometric analysis of immune populations that are generally performed simultaneously at one point in time. In contrast, this limitation is easily overcome with transplantable syngeneic models, such as the ID8 and STOSE models, where tumor onset is uniform and controlled.

Table 2. The utility of spontaneous and syngeneic models of ovarian cancer.

Model	Genetic Engineering	Key Features of Tumor Immune Landscape	Mutation/ Neoantigen Burden	Advantages	Limitations	References
Laying Hen	None	- T and B cell infiltration - Immunosuppressive DR6 and ILT3 expression	Unknown	- Shared risk factors with human disease - Tumors classified from Stage I–IV similar to HGSC - Ascites develops in later stages II–IV	- Time > 2 years for tumor development - Lack of reagents for species	[131–134,178]
Jaguar	None	- Unknown	Familial BRCA mutations	- Shared risk factors and familial BRCA mutations similar to high-risk women	- Endangered species - Lack of reagents for species	[136,137]
ID8-(original)	None	- Fully profiled - Predominant innate cell infiltration	Low	- Reliable and fast tumorigenesis - Well characterized - Develops ascites	- Lacking mutations common to human HGSC	[139,147]
ID8-Defb29/ Vegf-A	Stable Defb29 and Vegf-A expression	- Robust DC infiltration	Unknown	- Dysfunctional DCs characteristic of human HGSC - Aggressive - Forms neovasculature	- Lacking mutations common to HGSC	[148,149,179,180]
ID8-OVA	Stable ovalbumin expression	- Not profiled	OVA	- Immunogenic with DC vaccination strategy - Can track T cell responses against OVA - Allow antitumoral T cell studies with transgenic mice	- OVA dominance may not reflect the nature of TAAs in HGSC	[141]
ID8-Trp53−/− Brca2−/−	*Trp53* and *Brca2* deletion	- Increased MDSCs recruitment - Develops intraepithelial lymphoid aggregates	Unknown	- Shared genomic alterations with human HGSC - Complex immune landscape similar to human HGSC	- *Trp53* deletion may not reflect biology of *TP53* mutations seen in human HGSC	[152,153]

Table 2. *Cont.*

Model	Genetic Engineering	Key Features of Tumor Immune Landscape	Mutation/ Neoantigen Burden	Advantages	Limitations	References
ID8-NGL	NF-kappaB-dependent GFP/luciferase expression	M2 macrophages dominant immune cell type in ascites	Unknown	- Track tumor cells in vivo - Assess role of NF-kappaB on immune function	- Ascites fluid interferes with luciferase signal - Lacking mutations common to human HGSC - Luciferase can act as a neoantigen	[181,182]
STOSE	None	- Not profiled - Predominant Treg infiltration	Unknown	- Reliable and fast tumorigenesis - Different mouse strain than ID8 model - Give rise to T cell inflamed tumors - Develops ascites	- Lacking mutations common to human HGSC	[138]

High-grade serous ovarian cancer (HGSC), ovalbumin (OVA).

Table 3. The utility of genetically engineered mouse models (GEMM) of ovarian cancers.

Model	Genetic Engineering	Key Features of Tumor Immune Landscape	Mutation/ Neoantigen Burden	Advantages	Disadvantages	References
TgMISIIRTAg	SV40TAg driven from reproductive tract-specific MISIIR (*Amhr2*) promoter during development	Epigenetic modifiers enhance MHCII expression on cancer cells	Unknown	- Forms ascites	- SV40TAg - Slow tumor development (6–13 weeks) - Non-inducible tumorigenesis	[144,161]
TgCAG-LS-TAg	SV40TAg with lox-stop cassette driven from ubiquitous CAG promoter *	Unknown	Unknown	- Ascites develops in a subset of mice - Inducible SV40TAg	- SV40TAg - Slow tumor development—>22 weeks - Surgical administration of Ad-Cre	[162]
mogp-TAg	SV40TAg driven from oviduct-specific *Ovgp1* promoter	Unknown	Unknown	- Oviduct tumor origin	- SV40TAg - Non-inducible tumorigenesis - Slow tumor development (>6 weeks) - Fails to develop ascites	[163,183]
TgK18-GT121-Brca-Trp53	Inducible SV40TAg and either *Trp53*−/− or *Trp53*mut and *Brca1* or 2 deletions driven from epithelial specific cytokeratin 18 expression *	Unknown	Unknown	- R172H *Trp53* mutation that phenocopies human R175H *TP53* mutation - Inducible SV40TAg	- SV40TAg - Surgical administration of Ad-Cre	[170]
Trp53loxP/loxP-Rb1loxP/loxP	Inducible deletion of *Trp53* and *Rb1* *	Unknown	Unknown	- Inducible gene deletions - Genomic alterations similar to human HGSC	- *Trp53* deletion may not reflect biology of all *TP53* mutations seen in HGSC - Slow tumor development (median survival 227 days)	[167,168]

Table 3. *Cont.*

Model	Genetic Engineering	Key Features of Tumor Immune Landscape	Mutation/ Neoantigen Burden	Advantages	Disadvantages	References
Pax8-Cre-Brca1(2) −/−; Trp53mut(−/−);Pten −/− *	Doxycyline inducible Cre-mediated deletion of *Brca*, *Pten*, and *Trp53*, driven from oviduct-specific *Pax8* promoter.	Unknown	Copy number alterations similar to HGSC, Neoantigen and mutation burden unknown	- Inducible gene deletions from oviduct origin - Genomic alterations similar to human HGSC - Models with both *Trp53* deletion and mutation	- Fails to develop ascites - *Pten* deletion induces endometrial lesions	[169]
Ovgp1-iCre-ERT2 + tumor suppressor genes	Conditional deletion of *Brca1*, *Pten*, *Rb1*, and *Nf1* (BPRN mice) or *Brca1*, *Pten*, and *p53* (BPP mice), driven from the oviduct-specific *Ovgp1* promoter	Unknown	Unknown	- Inducible gene deletions from oviduct origin - Genomic alterations similar to human HGSC - Models with both Trp53 deletion and mutation	- Ascites only in 12% of mice - BPP mice develop a mixed tumor phenotype with mucinous metaplasia	[173]
MUC1KrasPTEN	Constitutive expression of human MUC1 and inducible oncogenic $KRAS^{G12D}$ and *Pten* deletion. *	Robust Tregs among TILs and dysfunctional DCs	unknown	- Expression of human TAA MUC1 - Inducible activation of $KRAS^{G12D}$ and deletion of *Pten* - Tumor development from both ovary and fallopian tube - Shared genomic alterations with endometroid ovarian cancer	- Surgical administration of Ad-Cre - Lacking common genetic alterations with human HGSC	[174,184]

* Model tissue-specificity governed by the site of administration of adenovirus expressing Cre recombinase (Ad-Cre). Tumor infiltrating lymphocytes (TIL).

6. Human-Derived and Autologous Cultures

Numerous ovarian cancer cell lines have been used historically with inconsistencies in their relevance to human HGSC, particularly A2780 and SKOV3 cells, which are unlikely to represent HGSC (reviewed in the literature [185]). Recently, HGSC primary cultures have been established that have genomic alterations, TAA expression, and gene expression profiles consistent with TCGA datasets [186–188]. These primary cultures offer resources for TAA discovery, infectivity with oncolytic viruses, and developing methods to increase immunogenicity. The major weakness of using primary cultures is that tumorigenesis can only be studied in xenografts using immunodeficient mice that fail to develop a complex TME with the immune subsets seen in patient tumors.

Ovarian cancers are rich resources for easily accessible cancer and immune cells from ascites fluid. Ascites fluid is remarkably immunosuppressive, containing high levels of Tregs and MDSCs [47,143]. Ascites have been a source for NK cells, where ex vivo expansion restored their cytotoxicity against autologous CD45-EpCAM+ cells [189]. Nounamo and colleagues showed that myxoma virus can prevent the secretion of IL-10 from ascites-derived CD14+ myeloid cells, thereby providing in vitro evidence that myxoma virus may remodel the ascites microenvironment to facilitate stronger antitumor immunity [143]. An interesting approach was developed by Freedman and colleagues using an oncolytic adenovirus expressing a bispecific T cell engager (BiTE) that targets autologous CD8+ T cells to EpCAM+ ascites cells and pleural effusions. Remarkably, even in the presence of ascitic fluid, EnAdEpCAMBiTE stimulated T cell proliferation and cytotoxicity against EpCAM+ ascites cells [190]. A co-culture system has also been developed to assess the efficacy of CAR-T cell therapy by culturing dissociated primary ovarian tumors with autologous derived anti-5T4 CAR-T cells [34]. Even though the use of human samples has limitations for studying the ovarian cancer TME, they offer invaluable resources to determine the specificity of both innate and adaptive targeted immunotherapies.

7. Summary

Further studies about EOC TME evolution during disease and treatment are needed. Importantly, heterogeneity among metastatic and primary tumors within a single patient can coexist [191], and this heterogeneity can influence the immune cell landscape, thereby affecting prognosis and therapeutic responses. Similarly, the ascites TME can respond differently to therapy, because EOC ascites contain another complex immunosuppressive network. Therefore, the challenges of tumor heterogeneity must be considered when designing therapeutic strategies for EOC patients.

In this review, we described some of the current emerging immunotherapies that have shown promising results in animal models and other cancer types and that could be exploited in EOC. The major barrier in EOC immunotherapy is the highly immunosuppressive TME. Thus, therapies aiming to decrease immunosuppression as a first line therapy combined with immunostimulating strategies could succeed in the fight against EOC.

We have also highlighted spontaneous and GEMM syngeneic models of ovarian cancer that offer promising characteristics for use in immunotherapy research. In moving forward, it will be important to characterize many of these models for immune infiltration, neoantigen burden, TAA, and immune checkpoint expression, as well as stromal features, in order to generate meaningful data for the immunotherapy field that goes beyond survival. Models using dominant immunogens such as SV40TAg should proceed with caution and require validation in separate models. The use of ovalbumin may offer a superior model that allows for the modulation of the ovarian cancer TME. Thus far, the ID8 model has been the most widely used, but given recent findings on the weak immunogenicity of ID8 cells [147], novel therapeutics should be tested in both spontaneous and GEMM models that cover a wide range of the tumor heterogeneity seen in the clinic.

New sequencing technologies have enhanced our ability to look deeper into tumors, stroma, and TIL compartments. These studies have revealed the impact of genetic heterogeneity and epigenetic plasticity in cancer evolution during treatment (drug resistance) and clinical outcome [192,193]. At present, we know that a tumor is not a single entity determined solely by genetic alterations, but a whole complex network that affects surrounding healthy cells provoking tumorigenesis advantage. Thus, in order to achieve significant responses to eradicate neoplastic cells, TME screening (TIL composition and quality) must be considered to better assign a therapeutic strategy to a patient, especially in advanced EOC stages. The detection of key biomarkers allowing the prediction of responsiveness to an immunotherapeutic approach is also necessary to select the best strategies and combined therapies that have the maximum potential to fully eradicate cancers [1].

Author Contributions: All authors contributed to the research, writing, and editing of this review article.

Funding: The authors' research relevant to this article is funded by grants to BCV from BioCanRx and the Canadian Institutes of Health Research, as well as a fellowship from the Fonds de recherche du Québec—Santé (GR), and a Vanier Canada Graduate Scholarship (CM).

Acknowledgments: The authors are grateful for the donations and support from the local community of ovarian cancer patients, especially the late Margaret Craig and the Carol Annibale Ovarian Cancer Foundation.

Conflicts of Interest: The authors declare no conflict of interest.

References

1. Odunsi, K. Immunotherapy in ovarian cancer. *Ann. Oncol. Off. J. Eur. Soc. Med. Oncol.* **2017**, *28*, viii1–viii7. [CrossRef] [PubMed]
2. Barnett, J.C.; Bean, S.M.; Whitaker, R.S.; Kondoh, E.; Baba, T.; Fujii, S.; Marks, J.R.; Dressman, H.K.; Murphy, S.K.; Berchuck, A. Ovarian cancer tumor infiltrating T-regulatory (T(reg)) cells are associated with a metastatic phenotype. *Gynecol. Oncol.* **2010**, *116*, 556–562. [CrossRef] [PubMed]
3. Hamanishi, J.; Mandai, M.; Abiko, K.; Matsumura, N.; Baba, T.; Yoshioka, Y.; Kosaka, K.; Konishi, I. The comprehensive assessment of local immune status of ovarian cancer by the clustering of multiple immune factors. *Clin. Immunol.* **2011**, *141*, 338–347. [CrossRef] [PubMed]
4. Hermans, C.; Anz, D.; Engel, J.; Kirchner, T.; Endres, S.; Mayr, D. Analysis of FoxP3+ T-regulatory cells and CD8+ T-cells in ovarian carcinoma: Location and tumor infiltration patterns are key prognostic markers. *PLoS ONE* **2014**, *9*, e111757. [CrossRef] [PubMed]
5. Leffers, N.; Gooden, M.J.M.; de Jong, R.A.; Hoogeboom, B.-N.; ten Hoor, K.A.; Hollema, H.; Boezen, H.M.; van der Zee, A.G.J.; Daemen, T.; Nijman, H.W. Prognostic significance of tumor-infiltrating T-lymphocytes in primary and metastatic lesions of advanced stage ovarian cancer. *Cancer Immunol. Immunother.* **2009**, *58*, 449–459. [CrossRef] [PubMed]
6. Nielsen, J.S.; Sahota, R.A.; Milne, K.; Kost, S.E.; Nesslinger, N.J.; Watson, P.H.; Nelson, B.H. $CD20^+$ tumor-infiltrating lymphocytes have an atypical $CD27^-$ memory phenotype and together with $CD8^+$ T cells promote favorable prognosis in ovarian cancer. *Clin. Cancer Res.* **2012**, *18*, 3281–3292. [CrossRef] [PubMed]
7. Raspollini, M.R.; Castiglione, F.; Rossi Degl'innocenti, D.; Amunni, G.; Villanucci, A.; Garbini, F.; Baroni, G.; Taddei, G.L. Tumour-infiltrating gamma/delta T-lymphocytes are correlated with a brief disease-free interval in advanced ovarian serous carcinoma. *Ann. Oncol.* **2005**, *16*, 590–596. [CrossRef] [PubMed]
8. Stumpf, M.; Hasenburg, A.; Riener, M.-O.; Jütting, U.; Wang, C.; Shen, Y.; Orlowska-Volk, M.; Fisch, P.; Wang, Z.; Gitsch, G.; et al. Intraepithelial CD8-positive T lymphocytes predict survival for patients with serous stage III ovarian carcinomas: Relevance of clonal selection of T lymphocytes. *Br. J. Cancer* **2009**, *101*, 1513–1521. [CrossRef] [PubMed]
9. Tomsová, M.; Melichar, B.; Sedláková, I.; Steiner, I. Prognostic significance of CD3+ tumor-infiltrating lymphocytes in ovarian carcinoma. *Gynecol. Oncol.* **2008**, *108*, 415–420. [CrossRef] [PubMed]
10. Liu, Y.; Chen, X.; Han, W.; Zhang, Y. Tisagenlecleucel, an approved anti-CD19 chimeric antigen receptor T-cell therapy for the treatment of leukemia. *Drugs Today* **2017**, *53*, 597–608. [CrossRef] [PubMed]
11. Maude, S.L.; Laetsch, T.W.; Buechner, J.; Rives, S.; Boyer, M.; Bittencourt, H.; Bader, P.; Verneris, M.R.; Stefanski, H.E.; Myers, G.D.; et al. Tisagenlecleucel in children and young adults with B-Cell lymphoblastic leukemia. *N. Engl. J. Med.* **2018**, *378*, 439–448. [CrossRef] [PubMed]
12. Rodriguez, G.M.; Galpin, K.J.C.; McCloskey, C.W.; Vanderhyden, B.C. The tumour microenvironment of epithelial ovarian cancer and its influence on response to immunotherapy. *Cancers* **2018**, *10*, 230. [CrossRef]
13. Yigit, R.; Massuger, L.F.A.G.; Figdor, C.G.; Torensma, R. Ovarian cancer creates a suppressive microenvironment to escape immune elimination. *Gynecol. Oncol.* **2010**, *117*, 366–372. [CrossRef] [PubMed]

14. Hanahan, D.; Coussens, L.M. Accessories to the crime: Functions of cells recruited to the tumor microenvironment. *Cancer Cell* **2012**, *21*, 309–322. [CrossRef] [PubMed]
15. Rosenberg, S.A.; Restifo, N.P.; Yang, J.C.; Morgan, R.A.; Dudley, M.E. Adoptive cell transfer: A clinical path to effective cancer immunotherapy. *Nat. Rev. Cancer* **2008**, *8*, 299–308. [CrossRef] [PubMed]
16. Besser, M.J.; Shapira-Frommer, R.; Treves, A.J.; Zippel, D.; Itzhaki, O.; Hershkovitz, L.; Levy, D.; Kubi, A.; Hovav, E.; Chermoshniuk, N.; et al. Clinical responses in a phase II study using adoptive transfer of short-term cultured tumor infiltration lymphocytes in metastatic melanoma patients. *Clin. Cancer Res.* **2010**, *16*, 2646–2655. [CrossRef] [PubMed]
17. Rosenberg, S.A.; Dudley, M.E. Adoptive cell therapy for the treatment of patients with metastatic melanoma. *Curr. Opin. Immunol.* **2009**, *21*, 233–240. [CrossRef] [PubMed]
18. Fujita, K.; Ikarashi, H.; Takakuwa, K.; Kodama, S.; Tokunaga, A.; Takahashi, T.; Tanaka, K. Prolonged disease-free period in patients with advanced epithelial ovarian cancer after adoptive transfer of tumor-infiltrating lymphocytes. *Clin. Cancer Res.* **1995**, *1*, 501–507.
19. Dudley, M.E.; Wunderlich, J.R.; Robbins, P.F.; Yang, J.C.; Hwu, P.; Schwartzentruber, D.J.; Topalian, S.L.; Sherry, R.; Restifo, N.P.; Hubicki, A.M.; et al. Cancer regression and autoimmunity in patients after clonal repopulation with antitumor lymphocytes. *Science* **2002**, *298*, 850–854. [CrossRef] [PubMed]
20. Aoki, Y.; Takakuwa, K.; Kodama, S.; Tanaka, K.; Takahashi, M.; Tokunaga, A.; Takahashi, T. Use of adoptive transfer of tumor-infiltrating lymphocytes alone or in combination with cisplatin-containing chemotherapy in patients with epithelial ovarian cancer. *Cancer Res.* **1991**, *51*, 1934–1939. [PubMed]
21. Freedman, R.S.; Edwards, C.L.; Kavanagh, J.J.; Kudelka, A.P.; Katz, R.L.; Carrasco, C.H.; Atkinson, E.N.; Scott, W.; Tomasovic, B.; Templin, S. Intraperitoneal adoptive immunotherapy of ovarian carcinoma with tumor-infiltrating lymphocytes and low-dose recombinant interleukin-2: A pilot trial. *J. Immunother. Emphas. Tumor Immunol.* **1994**, *16*, 198–210. [CrossRef]
22. Santoiemma, P.P.; Powell, D.J. Tumor infiltrating lymphocytes in ovarian cancer. *Cancer Biol. Ther.* **2015**, *16*, 807–820. [CrossRef] [PubMed]
23. Yang, J.C.; Rosenberg, S.A. Adoptive T-Cell therapy for cancer. *Adv. Immunol.* **2016**, *130*, 279–294. [CrossRef] [PubMed]
24. Hinrichs, C.S.; Rosenberg, S.A. Exploiting the curative potential of adoptive T-cell therapy for cancer. *Immunol. Rev.* **2014**, *257*, 56–71. [CrossRef] [PubMed]
25. Sadelain, M.; Brentjens, R.; Rivière, I. The promise and potential pitfalls of chimeric antigen receptors. *Curr. Opin. Immunol.* **2009**, *21*, 215–223. [CrossRef] [PubMed]
26. Kershaw, M.H.; Westwood, J.A.; Parker, L.L.; Wang, G.; Eshhar, Z.; Mavroukakis, S.A.; White, D.E.; Wunderlich, J.R.; Canevari, S.; Rogers-Freezer, L.; et al. A phase I study on adoptive immunotherapy using gene-modified T cells for ovarian cancer. *Clin. Cancer Res.* **2006**, *12*, 6106–6115. [CrossRef] [PubMed]
27. Kandalaft, L.E.; Powell, D.J.; Coukos, G. A phase I clinical trial of adoptive transfer of folate receptor-alpha redirected autologous T cells for recurrent ovarian cancer. *J. Transl. Med.* **2012**, *10*, 157. [CrossRef] [PubMed]
28. Lanitis, E.; Dangaj, D.; Hagemann, I.S.; Song, D.-G.; Best, A.; Sandaltzopoulos, R.; Coukos, G.; Powell, D.J. Primary human ovarian epithelial cancer cells broadly express HER2 at immunologically-detectable levels. *PLoS ONE* **2012**, *7*, e49829. [CrossRef] [PubMed]
29. Chekmasova, A.A.; Rao, T.D.; Nikhamin, Y.; Park, K.J.; Levine, D.A.; Spriggs, D.R.; Brentjens, R.J. Successful eradication of established peritoneal ovarian tumors in SCID-Beige mice following adoptive transfer of T cells genetically targeted to the MUC16 antigen. *Clin. Cancer Res.* **2010**, *16*, 3594–3606. [CrossRef] [PubMed]
30. Felder, M.; Kapur, A.; Gonzalez-Bosquet, J.; Horibata, S.; Heintz, J.; Albrecht, R.; Fass, L.; Kaur, J.; Hu, K.; Shojaei, H.; et al. MUC16 (CA125): Tumor biomarker to cancer therapy, a work in progress. *Mol. Cancer* **2014**, *13*, 129. [CrossRef] [PubMed]
31. Chang, K.; Pastan, I. Molecular cloning of mesothelin, a differentiation antigen present on mesothelium, mesotheliomas, and ovarian cancers. *Proc. Natl. Acad. Sci. USA* **1996**, *93*, 136–140. [CrossRef] [PubMed]
32. Song, D.-G.; Ye, Q.; Santoro, S.; Fang, C.; Best, A.; Powell, D.J. Chimeric NKG2D CAR-expressing T cell-mediated attack of human ovarian cancer is enhanced by histone deacetylase inhibition. *Hum. Gene Ther.* **2013**, *24*, 295–305. [CrossRef] [PubMed]
33. Patriarca, C.; Macchi, R.M.; Marschner, A.K.; Mellstedt, H. Epithelial cell adhesion molecule expression (CD326) in cancer: A short review. *Cancer Treat. Rev.* **2012**, *38*, 68–75. [CrossRef] [PubMed]

34. Owens, G.L.; Sheard, V.E.; Kalaitsidou, M.; Blount, D.; Lad, Y.; Cheadle, E.J.; Edmondson, R.J.; Kooner, G.; Gilham, D.E.; Harrop, R. Preclinical assessment of Car T-cell therapy targeting the tumor antigen 5t4 in ovarian cancer. *J. Immunother.* **2018**, *41*, 130–140. [CrossRef] [PubMed]
35. Genta, S.; Ghisoni, E.; Giannone, G.; Mittica, G.; Valabrega, G. Reprogramming T-cells for adoptive immunotherapy of ovarian cancer. *Expert Opin. Biol. Ther.* **2018**, *18*, 359–367. [CrossRef] [PubMed]
36. Koneru, M.; Purdon, T.J.; Spriggs, D.; Koneru, S.; Brentjens, R.J. IL-12 secreting tumor-targeted chimeric antigen receptor T cells eradicate ovarian tumors in vivo. *Oncoimmunology* **2015**, *4*, e994446. [CrossRef] [PubMed]
37. Lu, X. Impact of IL-12 in Cancer. *Curr. Cancer Drug Targets* **2017**, *17*, 682–697. [CrossRef] [PubMed]
38. Koneru, M.; O'Cearbhaill, R.; Pendharkar, S.; Spriggs, D.R.; Brentjens, R.J. A phase I clinical trial of adoptive T cell therapy using IL-12 secreting MUC-16(ecto) directed chimeric antigen receptors for recurrent ovarian cancer. *J. Transl. Med.* **2015**, *13*, 102. [CrossRef] [PubMed]
39. Tanyi, J.L.; Stashwick, C.; Plesa, G.; Morgan, M.A.; Porter, D.; Maus, M.V.; June, C.H. Possible compartmental cytokine release syndrome in a patient with recurrent ovarian cancer after Treatment with Mesothelin-targeted CAR-T Cells. *J. Immunother.* **2017**, *40*, 104–107. [CrossRef] [PubMed]
40. Ang, W.X.; Li, Z.; Chi, Z.; Du, S.-H.; Chen, C.; Tay, J.C.K.; Toh, H.C.; Connolly, J.E.; Xu, X.H.; Wang, S. Intraperitoneal immunotherapy with T cells stably and transiently expressing anti-EpCAM CAR in xenograft models of peritoneal carcinomatosis. *Oncotarget* **2017**, *8*, 13545–13559. [CrossRef] [PubMed]
41. Tewari, D.; Java, J.J.; Salani, R.; Armstrong, D.K.; Markman, M.; Herzog, T.; Monk, B.J.; Chan, J.K. Long-term survival advantage and prognostic factors associated with intraperitoneal chemotherapy treatment in advanced ovarian cancer: A gynecologic oncology group study. *J. Clin. Oncol.* **2015**, *33*, 1460–1466. [CrossRef] [PubMed]
42. Xu, X.-J.; Song, D.-G.; Poussin, M.; Ye, Q.; Sharma, P.; Rodríguez-García, A.; Tang, Y.-M.; Powell, D.J. Multiparameter comparative analysis reveals differential impacts of various cytokines on CART cell phenotype and function ex vivo and in vivo. *Oncotarget* **2016**, *7*, 82354–82368. [CrossRef] [PubMed]
43. Jindal, V.; Arora, E.; Gupta, S.; Lal, A.; Masab, M.; Potdar, R. Prospects of chimeric antigen receptor T cell therapy in ovarian cancer. *Med. Oncol.* **2018**, *35*, 70. [CrossRef] [PubMed]
44. Xu, X.; Qiu, J.; Sun, Y. The basics of CAR T design and challenges in immunotherapy of solid tumors—Ovarian cancer as a model. *Hum. Vaccines Immunother.* **2017**, *13*, 1548–1555. [CrossRef] [PubMed]
45. Yoon, D.H.; Osborn, M.J.; Tolar, J.; Kim, C.J. Incorporation of immune checkpoint blockade into chimeric antigen receptor T cells (CAR-Ts): Combination or built-in CAR-T. *Int. J. Mol. Sci.* **2018**, *19*, 340. [CrossRef] [PubMed]
46. Chang, D.-K.; Peterson, E.; Sun, J.; Goudie, C.; Drapkin, R.I.; Liu, J.F.; Matulonis, U.; Zhu, Q.; Marasco, W.A. Anti-CCR4 monoclonal antibody enhances antitumor immunity by modulating tumor-infiltrating Tregs in an ovarian cancer xenograft humanized mouse model. *Oncoimmunology* **2015**, *5*, e1090075. [CrossRef] [PubMed]
47. Curiel, T.J.; Coukos, G.; Zou, L.; Alvarez, X.; Cheng, P.; Mottram, P.; Evdemon-Hogan, M.; Conejo-Garcia, J.R.; Zhang, L.; Burow, M.; et al. Specific recruitment of regulatory T cells in ovarian carcinoma fosters immune privilege and predicts reduced survival. *Nat. Med.* **2004**, *10*, 942–949. [CrossRef] [PubMed]
48. Idorn, M.; Olsen, M.; Halldórsdóttir, H.R.; Skadborg, S.K.; Pedersen, M.; Høgdall, C.; Høgdall, E.; Met, Ö.; Thor Straten, P. Improved migration of tumor ascites lymphocytes to ovarian cancer microenvironment by CXCR2 transduction. *Oncoimmunology* **2018**, *7*, e1412029. [CrossRef] [PubMed]
49. Khan, A.N.H.; Kolomeyevskaya, N.; Singel, K.L.; Grimm, M.J.; Moysich, K.B.; Daudi, S.; Grzankowski, K.S.; Lele, S.; Ylagan, L.; Webster, G.A.; et al. Targeting myeloid cells in the tumor microenvironment enhances vaccine efficacy in murine epithelial ovarian cancer. *Oncotarget* **2015**, *6*, 11310–11326. [CrossRef] [PubMed]
50. Okła, K.; Wertel, I.; Polak, G.; Surówka, J.; Wawruszak, A.; Kotarski, J. Tumor-associated macrophages and myeloid-derived suppressor cells as immunosuppressive mechanism in ovarian cancer patients: Progress and challenges. *Int. Rev. Immunol.* **2016**, *35*, 372–385. [CrossRef] [PubMed]
51. Ganesh, K.; Massagué, J. TGF-β inhibition and immunotherapy: Checkmate. *Immunity* **2018**, *48*, 626–628. [CrossRef] [PubMed]
52. Mariathasan, S.; Turley, S.J.; Nickles, D.; Castiglioni, A.; Yuen, K.; Wang, Y.; Kadel, E.E.; Koeppen, H.; Astarita, J.L.; Cubas, R.; et al. TGFβ attenuates tumour response to PD-L1 blockade by contributing to exclusion of T cells. *Nature* **2018**, *554*, 544–548. [CrossRef] [PubMed]

53. Tauriello, D.V.F.; Palomo-Ponce, S.; Stork, D.; Berenguer-Llergo, A.; Badia-Ramentol, J.; Iglesias, M.; Sevillano, M.; Ibiza, S.; Cañellas, A.; Hernando-Momblona, X.; et al. TGFβ drives immune evasion in genetically reconstituted colon cancer metastasis. *Nature* **2018**, *554*, 538–543. [CrossRef] [PubMed]

54. Ros, X.R.; Vermeulen, L. Turning cold rumors hot by blocking TGF-β. *Trends Cancer* **2018**, *4*, 335–337. [CrossRef] [PubMed]

55. Lieber, S.; Reinartz, S.; Raifer, H.; Finkernagel, F.; Dreyer, T.; Bronger, H.; Jansen, J.M.; Wagner, U.; Worzfeld, T.; Müller, R.; et al. Prognosis of ovarian cancer is associated with effector memory CD8+ T cell accumulation in ascites, CXCL9 levels and activation-triggered signal transduction in T cells. *Oncoimmunology* **2018**, *7*, e1424672. [CrossRef] [PubMed]

56. Bronger, H.; Singer, J.; Windmüller, C.; Reuning, U.; Zech, D.; Delbridge, C.; Dorn, J.; Kiechle, M.; Schmalfeldt, B.; Schmitt, M.; et al. CXCL9 and CXCL10 predict survival and are regulated by cyclooxygenase inhibition in advanced serous ovarian cancer. *Br. J. Cancer* **2016**, *115*, 553–563. [CrossRef] [PubMed]

57. Topalian, S.L.; Drake, C.G.; Pardoll, D.M. Immune checkpoint blockade: A common denominator approach to cancer therapy. *Cancer Cell* **2015**, *27*, 450–461. [CrossRef] [PubMed]

58. Anderson, A.C.; Joller, N.; Kuchroo, V.K. Lag-3, Tim-3, and TIGIT co-inhibitory receptors with specialized functions in immune regulation. *Immunity* **2016**, *44*, 989–1004. [CrossRef] [PubMed]

59. Huang, R.-Y.; Eppolito, C.; Lele, S.; Shrikant, P.; Matsuzaki, J.; Odunsi, K. LAG3 and PD1 co-inhibitory molecules collaborate to limit CD8+ T cell signaling and dampen antitumor immunity in a murine ovarian cancer model. *Oncotarget* **2015**, *6*, 27359–27377. [CrossRef] [PubMed]

60. Latha, K.; Yan, J.; Yang, Y.; Gressot, L.V.; Kong, L.-Y.; Manyam, G.; Ezhilarasan, R.; Wang, Q.; Sulman, E.P.; Eric Davis, R.; et al. The Role of Fibrinogen-Like Protein 2 on Immunosuppression and Malignant Progression in Glioma. *J. Natl. Cancer Inst.* **2018**. [CrossRef] [PubMed]

61. Yan, J.; Kong, L.-Y.; Hu, J.; Gabrusiewicz, K.; Dibra, D.; Xia, X.; Heimberger, A.B.; Li, S. FGL2 as a multimodality regulator of tumor-mediated immune suppression and therapeutic target in gliomas. *J. Natl. Cancer Inst.* **2015**, *107*. [CrossRef] [PubMed]

62. Zhu, Y.; Zhang, L.; Zha, H.; Yang, F.; Hu, C.; Chen, L.; Guo, B.; Zhu, B. Stroma-derived fibrinogen-like protein 2 activates cancer-associated fibroblasts to promote tumor growth in lung cancer. *Int. J. Biol. Sci.* **2017**, *13*, 804–814. [CrossRef] [PubMed]

63. Nguyen, L.T.; Ohashi, P.S. Clinical blockade of PD1 and LAG3—Potential mechanisms of action. *Nat. Rev. Immunol.* **2015**, *15*, 45–56. [CrossRef] [PubMed]

64. Lin, H.; Wei, S.; Hurt, E.M.; Green, M.D.; Zhao, L.; Vatan, L.; Szeliga, W.; Herbst, R.; Harms, P.W.; Fecher, L.A.; et al. Host expression of PD-L1 determines efficacy of PD-L1 pathway blockade-mediated tumor regression. *J. Clin. Investig.* **2018**, *128*, 1708. [CrossRef] [PubMed]

65. Davoodzadeh Gholami, M.; Kardar, G.A.; Saeedi, Y.; Heydari, S.; Garssen, J.; Falak, R. Exhaustion of T lymphocytes in the tumor microenvironment: Significance and effective mechanisms. *Cell. Immunol.* **2017**, *322*, 1–14. [CrossRef] [PubMed]

66. Ghoneim, H.E.; Fan, Y.; Moustaki, A.; Abdelsamed, H.A.; Dash, P.; Dogra, P.; Carter, R.; Awad, W.; Neale, G.; Thomas, P.G.; et al. De novo epigenetic programs inhibit PD-1 blockade-Mediated T cell rejuvenation. *Cell* **2017**, *170*, 142–157. [CrossRef] [PubMed]

67. Li, H.; Chiappinelli, K.B.; Guzzetta, A.A.; Easwaran, H.; Yen, R.-W.C.; Vatapalli, R.; Topper, M.J.; Luo, J.; Connolly, R.M.; Azad, N.S.; et al. Immune regulation by low doses of the DNA methyltransferase inhibitor 5-azacitidine in common human epithelial cancers. *Oncotarget* **2014**, *5*, 587–598. [CrossRef] [PubMed]

68. Roulois, D.; Loo Yau, H.; Singhania, R.; Wang, Y.; Danesh, A.; Shen, S.Y.; Han, H.; Liang, G.; Jones, P.A.; Pugh, T.J.; et al. DNA-demethylating agents target colorectal cancer cells by inducing viral mimicry by endogenous transcripts. *Cell* **2015**, *162*, 961–973. [CrossRef] [PubMed]

69. Wrangle, J.; Wang, W.; Koch, A.; Easwaran, H.; Mohammad, H.P.; Vendetti, F.; Vancriekinge, W.; Demeyer, T.; Du, Z.; Parsana, P.; et al. Alterations of immune response of Non-Small Cell Lung Cancer with Azacytidine. *Oncotarget* **2013**, *4*, 2067–2079. [CrossRef] [PubMed]

70. Stone, M.L.; Chiappinelli, K.B.; Li, H.; Murphy, L.M.; Travers, M.E.; Topper, M.J.; Mathios, D.; Lim, M.; Shih, I.-M.; Wang, T.-L.; et al. Epigenetic therapy activates type I interferon signaling in murine ovarian cancer to reduce immunosuppression and tumor burden. *Proc. Natl. Acad. Sci. USA* **2017**, *114*, E10981–E10990. [CrossRef] [PubMed]

71. Bezu, L.; Gomes-da-Silva, L.C.; Dewitte, H.; Breckpot, K.; Fucikova, J.; Spisek, R.; Galluzzi, L.; Kepp, O.; Kroemer, G. Combinatorial strategies for the induction of immunogenic cell death. *Front. Immunol.* **2015**, *6*, 187. [CrossRef] [PubMed]
72. Gebremeskel, S.; Johnston, B. Concepts and mechanisms underlying chemotherapy induced immunogenic cell death: Impact on clinical studies and considerations for combined therapies. *Oncotarget* **2015**, *6*, 41600–41619. [CrossRef] [PubMed]
73. Paroli, M.; Bellati, F.; Videtta, M.; Focaccetti, C.; Mancone, C.; Donato, T.; Antonilli, M.; Perniola, G.; Accapezzato, D.; Napoletano, C.; et al. Discovery of chemotherapy-associated ovarian cancer antigens by interrogating memory T cells. *Int. J. Cancer* **2014**, *134*, 1823–1834. [CrossRef] [PubMed]
74. Hato, S.V.; Khong, A.; de Vries, I.J.M.; Lesterhuis, W.J. Molecular pathways: The immunogenic effects of platinum-based chemotherapeutics. *Clin. Cancer Res.* **2014**, *20*, 2831–2837. [CrossRef] [PubMed]
75. Lesterhuis, W.J.; Punt, C.J.A.; Hato, S.V.; Eleveld-Trancikova, D.; Jansen, B.J.H.; Nierkens, S.; Schreibelt, G.; de Boer, A.; Van Herpen, C.M.L.; Kaanders, J.H.; et al. Platinum-based drugs disrupt STAT6-mediated suppression of immune responses against cancer in humans and mice. *J. Clin. Investig.* **2011**, *121*, 3100–3108. [CrossRef] [PubMed]
76. Coleman, S.; Clayton, A.; Mason, M.D.; Jasani, B.; Adams, M.; Tabi, Z. Recovery of CD8+ T-cell function during systemic chemotherapy in advanced ovarian cancer. *Cancer Res.* **2005**, *65*, 7000–7006. [CrossRef] [PubMed]
77. Szajnik, M.; Szczepanski, M.J.; Czystowska, M.; Elishaev, E.; Mandapathil, M.; Nowak-Markwitz, E.; Spaczynski, M.; Whiteside, T.L. TLR4 signaling induced by lipopolysaccharide or paclitaxel regulates tumor survival and chemoresistance in ovarian cancer. *Oncogene* **2009**, *28*, 4353–4363. [CrossRef] [PubMed]
78. Wang, A.C.; Su, Q.B.; Wu, F.X.; Zhang, X.L.; Liu, P.S. Role of TLR4 for paclitaxel chemotherapy in human epithelial ovarian cancer cells. *Eur. J. Clin. Investig.* **2009**, *39*, 157–164. [CrossRef] [PubMed]
79. Wang, A.-C.; Ma, Y.-B.; Wu, F.-X.; Ma, Z.-F.; Liu, N.-F.; Gao, R.; Gao, Y.-S.; Sheng, X.-G. TLR4 induces tumor growth and inhibits paclitaxel activity in MyD88-positive human ovarian carcinoma in vitro. *Oncol. Lett.* **2014**, *7*, 871–877. [CrossRef] [PubMed]
80. D'Adhemar, C.J.; Spillane, C.D.; Gallagher, M.F.; Bates, M.; Costello, K.M.; Barry-O'Crowley, J.; Haley, K.; Kernan, N.; Murphy, C.; Smyth, P.C.; et al. The MyD88+ phenotype is an adverse prognostic factor in epithelial ovarian cancer. *PLoS ONE* **2014**, *9*, e100816. [CrossRef] [PubMed]
81. Huang, J.-M.; Zhang, G.-N.; Shi, Y.; Zha, X.; Zhu, Y.; Wang, M.-M.; Lin, Q.; Wang, W.; Lu, H.-Y.; Ma, S.-Q.; et al. Atractylenolide-I sensitizes human ovarian cancer cells to paclitaxel by blocking activation of TLR4/MyD88-dependent pathway. *Sci. Rep.* **2014**, *4*, 3840. [CrossRef] [PubMed]
82. Liu, H.; Zhang, G.; Huang, J.; Ma, S.; Mi, K.; Cheng, J.; Zhu, Y.; Zha, X.; Huang, W. Atractylenolide I modulates ovarian cancer cell-mediated immunosuppression by blocking MD-2/TLR4 complex-mediated MyD88/NF-κB signaling in vitro. *J. Transl. Med.* **2016**, *14*, 104. [CrossRef] [PubMed]
83. Peng, J.; Hamanishi, J.; Matsumura, N.; Abiko, K.; Murat, K.; Baba, T.; Yamaguchi, K.; Horikawa, N.; Hosoe, Y.; Murphy, S.K.; et al. Chemotherapy induces Programmed Cell Death-Ligand 1 overexpression via the Nuclear Factor-κB to foster an immunosuppressive tumor microenvironment in ovarian cancer. *Cancer Res.* **2015**, *75*, 5034–5045. [CrossRef] [PubMed]
84. Mkrtichyan, M.; Najjar, Y.G.; Raulfs, E.C.; Abdalla, M.Y.; Samara, R.; Rotem-Yehudar, R.; Cook, L.; Khleif, S.N. Anti-PD-1 synergizes with cyclophosphamide to induce potent anti-tumor vaccine effects through novel mechanisms. *Eur. J. Immunol.* **2011**, *41*, 2977–2986. [CrossRef] [PubMed]
85. Lesterhuis, W.J.; Salmons, J.; Nowak, A.K.; Rozali, E.N.; Khong, A.; Dick, I.M.; Harken, J.A.; Robinson, B.W.; Lake, R.A. Synergistic effect of CTLA-4 blockade and cancer chemotherapy in the induction of anti-tumor immunity. *PLoS ONE* **2013**, *8*, e61895. [CrossRef] [PubMed]
86. Siew, Y.-Y.; Neo, S.-Y.; Yew, H.-C.; Lim, S.-W.; Ng, Y.-C.; Lew, S.-M.; Seetoh, W.-G.; Seow, S.-V.; Koh, H.-L. Oxaliplatin regulates expression of stress ligands in ovarian cancer cells and modulates their susceptibility to natural killer cell-mediated cytotoxicity. *Int. Immunol.* **2015**, *27*, 621–632. [CrossRef] [PubMed]
87. Lo, C.S.; Sanii, S.; Kroeger, D.R.; Milne, K.; Talhouk, A.; Chiu, D.S.; Rahimi, K.; Shaw, P.A.; Clarke, B.A.; Nelson, B.H. Neoadjuvant chemotherapy of ovarian cancer results in three patterns of tumor-Infiltrating lymphocyte response with distinct implications for immunotherapy. *Clin. Cancer Res.* **2017**, *23*, 925–934. [CrossRef] [PubMed]

88. Mariya, T.; Hirohashi, Y.; Torigoe, T.; Asano, T.; Kuroda, T.; Yasuda, K.; Mizuuchi, M.; Sonoda, T.; Saito, T.; Sato, N. Prognostic impact of human leukocyte antigen class I expression and association of platinum resistance with immunologic profiles in epithelial ovarian cancer. *Cancer Immunol. Res.* **2014**, *2*, 1220–1229. [CrossRef] [PubMed]
89. Edwards, S.J.; Barton, S.; Thurgar, E.; Trevor, N. Topotecan, pegylated liposomal doxorubicin hydrochloride, paclitaxel, trabectedin and gemcitabine for advanced recurrent or refractory ovarian cancer: A systematic review and economic evaluation. *Health Technol. Assess.* **2015**, *19*, 1–480. [CrossRef] [PubMed]
90. Salman, L.; Ben-Haroush, A.; Raban, O.; Yeoshoua, E.; Sabah, G.; Jakobson-Setton, A.; Tsoref, D.; Eitan, R. Neoadjuvant chemotherapy treatment modifications in ovarian carcinoma: The impact on surgical outcome and progression-free survival. *Am. J. Clin. Oncol.* **2018**. [CrossRef] [PubMed]
91. Kipps, E.; Tan, D.S.P.; Kaye, S.B. Meeting the challenge of ascites in ovarian cancer: New avenues for therapy and research. *Nat. Rev. Cancer* **2013**, *13*, 273–282. [CrossRef] [PubMed]
92. Kim, S.; Gwak, H.; Kim, H.S.; Kim, B.; Dhanasekaran, D.N.; Song, Y.S. Malignant ascites enhances migratory and invasive properties of ovarian cancer cells with membrane bound IL-6R in vitro. *Oncotarget* **2016**, *7*, 83148–83159. [CrossRef] [PubMed]
93. Yigit, R.; Figdor, C.G.; Zusterzeel, P.L.M.; Pots, J.M.; Torensma, R.; Massuger, L.F.A.G. Cytokine analysis as a tool to understand tumour-host interaction in ovarian cancer. *Eur. J. Cancer* **2011**, *47*, 1883–1889. [CrossRef] [PubMed]
94. Zhu, X.; Shen, H.; Yin, X.; Long, L.; Chen, X.; Feng, F.; Liu, Y.; Zhao, P.; Xu, Y.; Li, M.; et al. IL-6R/STAT3/miR-204 feedback loop contributes to cisplatin resistance of epithelial ovarian cancer cells. *Oncotarget* **2017**, *8*, 39154–39166. [CrossRef] [PubMed]
95. Nishio, H.; Yaguchi, T.; Sugiyama, J.; Sumimoto, H.; Umezawa, K.; Iwata, T.; Susumu, N.; Fujii, T.; Kawamura, N.; Kobayashi, A.; et al. Immunosuppression through constitutively activated NF-κB signalling in human ovarian cancer and its reversal by an NF-κB inhibitor. *Br. J. Cancer* **2014**, *110*, 2965–2974. [CrossRef] [PubMed]
96. Sidthipong, K.; Ma, J.; Yu, W.L.; Wang, Y.F.; Kobayashi, S.; Kishino, S.; Koide, N.; Yokochi, T.; Kato, K.; Okada, S.; et al. Rational design, synthesis and in vitro evaluation of novel exo-methylene butyrolactone salicyloylamide as NF-κB inhibitor. *Bioorg. Med. Chem. Lett.* **2017**, *27*, 562–566. [CrossRef] [PubMed]
97. Russell, S.J.; Peng, K.-W. Oncolytic virotherapy: A contest between apples and oranges. *Mol. Ther. J. Am. Soc. Gene Ther.* **2017**, *25*, 1107–1116. [CrossRef] [PubMed]
98. Russell, S.J.; Peng, K.-W.; Bell, J.C. Oncolytic virotherapy. *Nat. Biotechnol.* **2012**, *30*, 658–670. [CrossRef] [PubMed]
99. Seymour, L.W.; Fisher, K.D. Oncolytic viruses: Finally delivering. *Br. J. Cancer* **2016**, *114*, 357–361. [CrossRef] [PubMed]
100. Galluzzi, L.; Buqué, A.; Kepp, O.; Zitvogel, L.; Kroemer, G. Immunogenic cell death in cancer and infectious disease. *Nat. Rev. Immunol.* **2017**, *17*, 97–111. [CrossRef] [PubMed]
101. Lemay, C.G.; Rintoul, J.L.; Kus, A.; Paterson, J.M.; Garcia, V.; Falls, T.J.; Ferreira, L.; Bridle, B.W.; Conrad, D.P.; Tang, V.A.; et al. Harnessing oncolytic virus-mediated antitumor immunity in an infected cell vaccine. *Mol. Ther. J. Am. Soc. Gene Ther.* **2012**, *20*, 1791–1799. [CrossRef] [PubMed]
102. Forbes, N.E.; Abdelbary, H.; Lupien, M.; Bell, J.C.; Diallo, J.-S. Exploiting tumor epigenetics to improve oncolytic virotherapy. *Front. Genet.* **2013**, *4*, 184. [CrossRef] [PubMed]
103. Dummer, R.; Hoeller, C.; Gruter, I.P.; Michielin, O. Combining talimogene laherparepvec with immunotherapies in melanoma and other solid tumors. *Cancer Immunol. Immunother.* **2017**, *66*, 683–695. [CrossRef] [PubMed]
104. Puzanov, I.; Milhem, M.M.; Minor, D.; Hamid, O.; Li, A.; Chen, L.; Chastain, M.; Gorski, K.S.; Anderson, A.; Chou, J.; et al. Talimogene Laherparepvec in Combination With Ipilimumab in Previously Untreated, Unresectable Stage IIIB-IV Melanoma. *J. Clin. Oncol.* **2016**, *34*, 2619–2626. [CrossRef] [PubMed]
105. Rehman, H.; Silk, A.W.; Kane, M.P.; Kaufman, H.L. Into the clinic: Talimogene laherparepvec (T-VEC), a first-in-class intratumoral oncolytic viral therapy. *J. Immunother. Cancer* **2016**, *4*, 53. [CrossRef] [PubMed]
106. Ribas, A.; Dummer, R.; Puzanov, I.; VanderWalde, A.; Andtbacka, R.H.I.; Michielin, O.; Olszanski, A.J.; Malvehy, J.; Cebon, J.; Fernandez, E.; et al. Oncolytic virotherapy promotes intratumoral T cell infiltration and improves anti-PD-1 immunotherapy. *Cell* **2017**, *170*, 1109–1119. [CrossRef] [PubMed]

107. Thomas, E.D.; Meza-Perez, S.; Bevis, K.S.; Randall, T.D.; Gillespie, G.Y.; Langford, C.; Alvarez, R.D. IL-12 Expressing oncolytic herpes simplex virus promotes anti-tumor activity and immunologic control of metastatic ovarian cancer in mice. *J. Ovarian Res.* **2016**, *9*, 70. [CrossRef] [PubMed]
108. Ilkow, C.S.; Marguerie, M.; Batenchuk, C.; Mayer, J.; Neriah, D.B.; Cousineau, S.; Falls, T.; Jennings, V.A.; Boileau, M.; Bellamy, D.; et al. Reciprocal cellular cross-talk within the tumor microenvironment promotes oncolytic virus activity. *Nat. Med.* **2015**, *21*, 530–536. [CrossRef] [PubMed]
109. Arulanandam, R.; Batenchuk, C.; Angarita, F.A.; Ottolino-Perry, K.; Cousineau, S.; Mottashed, A.; Burgess, E.; Falls, T.J.; De Silva, N.; Tsang, J.; et al. VEGF-mediated induction of PRD1-BF1/Blimp1 expression sensitizes tumor vasculature to oncolytic virus infection. *Cancer Cell* **2015**, *28*, 210–224. [CrossRef] [PubMed]
110. Breitbach, C.J.; Arulanandam, R.; De Silva, N.; Thorne, S.H.; Patt, R.; Daneshmand, M.; Moon, A.; Ilkow, C.; Burke, J.; Hwang, T.-H.; et al. Oncolytic vaccinia virus disrupts tumor-associated vasculature in humans. *Cancer Res.* **2013**, *73*, 1265–1275. [CrossRef] [PubMed]
111. Achard, C.; Surendran, A.; Wedge, M.-E.; Ungerechts, G.; Bell, J.; Ilkow, C.S. Lighting a Fire in the Tumor Microenvironment Using Oncolytic Immunotherapy. *EBioMedicine* **2018**, *31*, 17–24. [CrossRef] [PubMed]
112. Gil, M.; Komorowski, M.P.; Seshadri, M.; Rokita, H.; McGray, A.J.R.; Opyrchal, M.; Odunsi, K.O.; Kozbor, D. CXCL12/CXCR4 blockade by oncolytic virotherapy inhibits ovarian cancer growth by decreasing immunosuppression and targeting cancer initiating cells. *J. Immunol.* **2014**, *193*, 5327–5337. [CrossRef] [PubMed]
113. Hartkopf, A.D.; Fehm, T.; Wallwiener, M.; Lauer, U. Oncolytic viruses to treat ovarian cancer patients—A review of results from clinical trials. *Geburtshilfe Frauenheilkd.* **2012**, *72*, 132–136. [CrossRef] [PubMed]
114. Jennings, V.A.; Ilett, E.J.; Scott, K.J.; West, E.J.; Vile, R.; Pandha, H.; Harrington, K.; Young, A.; Hall, G.D.; Coffey, M.; et al. Lymphokine-activated killer and dendritic cell carriage enhances oncolytic reovirus therapy for ovarian cancer by overcoming antibody neutralization in ascites. *Int. J. Cancer* **2014**, *134*, 1091–1101. [CrossRef] [PubMed]
115. Kuhn, I.; Bauzon, M.; Green, N.; Seymour, L.; Fisher, K.; Hermiston, T. OvAd1, a novel, Potent, and selective chimeric oncolytic virus developed for ovarian cancer by 3D-directed evolution. *Mol. Ther. Oncolytics* **2017**, *4*, 55–66. [CrossRef] [PubMed]
116. Orzechowska, B.U.; Jędryka, M.; Zwolińska, K.; Matkowski, R. VSV based virotherapy in ovarian cancer: The past, the present and … future? *J. Cancer* **2017**, *8*, 2369–2383. [CrossRef] [PubMed]
117. Uusi-Kerttula, H.; Davies, J.A.; Thompson, J.; Wongthida, P.; Evgin, L.; Shim, K.G.; Bradshaw, A.; Baker, A.T.; Rizkallah, P.J.; Jones, R.; et al. Ad5NULL-A20—A tropism-modified, αvβ6 integrin-selective oncolytic adenovirus for epithelial ovarian cancer therapies. *Clin. Cancer Res.* **2018**. [CrossRef] [PubMed]
118. Power, A.T.; Wang, J.; Falls, T.J.; Paterson, J.M.; Parato, K.A.; Lichty, B.D.; Stojdl, D.F.; Forsyth, P.A.J.; Atkins, H.; Bell, J.C. Carrier cell-based delivery of an oncolytic virus circumvents antiviral immunity. *Mol. Ther. J. Am. Soc. Gene Ther.* **2007**, *15*, 123–130. [CrossRef] [PubMed]
119. Alkayyal, A.A.; Tai, L.-H.; Kennedy, M.A.; de Souza, C.T.; Zhang, J.; Lefebvre, C.; Sahi, S.; Ananth, A.A.; Mahmoud, A.B.; Makrigiannis, A.P.; et al. NK-Cell recruitment is necessary for eradication of peritoneal carcinomatosis with an IL12-expressing Maraba virus cellular vaccine. *Cancer Immunol. Res.* **2017**, *5*, 211–221. [CrossRef] [PubMed]
120. Zhang, L.; Conejo-Garcia, J.R.; Katsaros, D.; Gimotty, P.A.; Massobrio, M.; Regnani, G.; Makrigiannakis, A.; Gray, H.; Schlienger, K.; Liebman, M.N.; et al. Intratumoral T cells, recurrence, and survival in epithelial ovarian cancer. *N. Engl. J. Med.* **2003**, *348*, 203–213. [CrossRef] [PubMed]
121. Sanmamed, M.F.; Pastor, F.; Rodriguez, A.; Perez-Gracia, J.L.; Rodriguez-Ruiz, M.E.; Jure-Kunkel, M.; Melero, I. Agonists of co-stimulation in cancer immunotherapy directed against CD137, OX40, GITR, CD27, CD28, and ICOS. *Semin. Oncol.* **2015**, *42*, 640–655. [CrossRef] [PubMed]
122. Tanyi, J.L.; Bobisse, S.; Ophir, E.; Tuyaerts, S.; Roberti, A.; Genolet, R.; Baumgartner, P.; Stevenson, B.J.; Iseli, C.; Dangaj, D.; et al. Personalized cancer vaccine effectively mobilizes antitumor T cell immunity in ovarian cancer. *Sci. Transl. Med.* **2018**, *10*, eaao5931. [CrossRef] [PubMed]
123. Sheridan, C. IDO inhibitors move center stage in immuno-oncology. *Nat. Biotechnol.* **2015**, *33*, 321–322. [CrossRef] [PubMed]
124. Suzuki, E.; Kapoor, V.; Jassar, A.S.; Kaiser, L.R.; Albelda, S.M. Gemcitabine selectively eliminates splenic Gr-1+/CD11b+ myeloid suppressor cells in tumor-bearing animals and enhances antitumor immune activity. *Clin. Cancer Res.* **2005**, *11*, 6713–6721. [CrossRef] [PubMed]

125. Vincent, J.; Mignot, G.; Chalmin, F.; Ladoire, S.; Bruchard, M.; Chevriaux, A.; Martin, F.; Apetoh, L.; Rébé, C.; Ghiringhelli, F. 5-Fluorouracil selectively kills tumor-associated myeloid-derived suppressor cells resulting in enhanced T cell-dependent antitumor immunity. *Cancer Res.* **2010**, *70*, 3052–3061. [CrossRef] [PubMed]
126. Allavena, P.; Germano, G.; Belgiovine, C.; D'Incalci, M.; Mantovani, A. Trabectedin: A drug from the sea that strikes tumor-associated macrophages. *Oncoimmunology* **2013**, *2*, e24614. [CrossRef] [PubMed]
127. Wei, H.; Zhao, L.; Hellstrom, I.; Hellstrom, K.E.; Guo, Y. Dual targeting of CD137 co-stimulatory and PD-1 co-inhibitory molecules for ovarian cancer immunotherapy. *Oncoimmunology* **2014**, *3*, e28248. [CrossRef] [PubMed]
128. Guo, Z.; Cheng, D.; Xia, Z.; Luan, M.; Wu, L.; Wang, G.; Zhang, S. Combined TIM-3 blockade and CD137 activation affords the long-term protection in a murine model of ovarian cancer. *J. Transl. Med.* **2013**, *11*, 215. [CrossRef] [PubMed]
129. Wei, H.; Zhao, L.; Li, W.; Fan, K.; Qian, W.; Hou, S.; Wang, H.; Dai, M.; Hellstrom, I.; Hellstrom, K.E.; et al. Combinatorial PD-1 Blockade and CD137 activation has therapeutic efficacy in murine cancer models and synergizes with cisplatin. *PLoS ONE* **2013**, *8*, e84927. [CrossRef] [PubMed]
130. Hawkridge, A.M. The chicken model of spontaneous ovarian cancer. *Proteom. Clin. Appl.* **2014**, *8*, 689–699. [CrossRef] [PubMed]
131. Yu, Y.; Edassery, S.L.; Barua, A.; Abramowicz, J.S.; Bahr, J.M.; Hellstrom, I.; Luborsky, J.L. The hen model of human ovarian cancer develops anti-mesothelin autoantibodies in response to mesothelin expressing tumors. *J. Ovarian Res.* **2011**, *4*, 12. [CrossRef] [PubMed]
132. Bradaric, M.J.; Penumatsa, K.; Barua, A.; Edassery, S.L.; Yu, Y.; Abramowicz, J.S.; Bahr, J.M.; Luborsky, J.L. Immune cells in the normal ovary and spontaneous ovarian tumors in the laying hen (*Gallus domesticus*) model of human ovarian cancer. *PLoS ONE* **2013**, *8*, e74147. [CrossRef] [PubMed]
133. Khan, M.F.; Bahr, J.M.; Yellapa, A.; Bitterman, P.; Abramowicz, J.S.; Edassery, S.L.; Basu, S.; Rotmensch, J.; Barua, A. Expression of leukocyte inhibitory immunoglobulinlike transcript 3 receptors by ovarian tumors in laying hen model of spontaneous ovarian cancer. *Transl. Oncol.* **2012**, *5*, 85–91. [CrossRef] [PubMed]
134. Barua, A.; Yellapa, A.; Bahr, J.M.; Abramowicz, J.S.; Edassery, S.L.; Basu, S.; Rotmensch, J.; Bitterman, P. Expression of death receptor 6 by ovarian rumors in laying hens, a preclinical model of spontaneous ovarian cancer. *Transl. Oncol.* **2012**, *5*, 260–268. [CrossRef] [PubMed]
135. McNeal, S.; Bitterman, P.; Bahr, J.M.; Edassery, S.L.; Abramowicz, J.S.; Basu, S.; Barua, A. Association of Immunosuppression with DR6 Expression during the Development and Progression of Spontaneous Ovarian Cancer in Laying Hen Model. Available online: https://www.hindawi.com/journals/jir/2016/6729379/ (accessed on 29 May 2018).
136. Corner, S.; Parys, M.; Agnew, D.; Moresco, A.; Yuzbasiyan-Gurkan, V. Ovarian Adenocarcinoma in Captive North American Jaguars (Panthera Onca): Tumor Characterization and Investigation of Brca1 Mutations. In Proceedings of the 47th Annual Conference of the American Association of Zoo Veterinarians, Portland, OR, USA, 26 September–2 October 2016.
137. McLean, K.; Mehta, G. Tumor microenvironment and models of ovarian cancer: The 11th Biennial Rivkin Center Ovarian Cancer Research Symposium. *Int. J. Gynecol. Cancer* **2017**, *27*, S2–S9. [CrossRef] [PubMed]
138. McCloskey, C.W.; Goldberg, R.L.; Carter, L.E.; Gamwell, L.F.; Al-Hujaily, E.M.; Collins, O.; Macdonald, E.A.; Garson, K.; Daneshmand, M.; Carmona, E.; et al. A new spontaneously transformed syngeneic model of high-grade serous ovarian cancer with a tumor-initiating cell population. *Front. Oncol.* **2014**, *4*, 53. [CrossRef] [PubMed]
139. Roby, K.F.; Taylor, C.C.; Sweetwood, J.P.; Cheng, Y.; Pace, J.L.; Tawfik, O.; Persons, D.L.; Smith, P.G.; Terranova, P.F. Development of a syngeneic mouse model for events related to ovarian cancer. *Carcinogenesis* **2000**, *21*, 585–591. [CrossRef] [PubMed]
140. Leinster, D.A.; Kulbe, H.; Everitt, G.; Thompson, R.; Perretti, M.; Gavins, F.N.E.; Cooper, D.; Gould, D.; Ennis, D.P.; Lockley, M.; et al. The peritoneal tumour microenvironment of high-grade serous ovarian cancer. *J. Pathol.* **2012**, *227*, 136–145. [CrossRef] [PubMed]
141. Chiang, C.L.-L.; Kandalaft, L.E.; Tanyi, J.; Hagemann, A.R.; Motz, G.T.; Svoronos, N.; Montone, K.; Mantia-Smaldone, G.M.; Smith, L.; Nisenbaum, H.L.; et al. A dendritic cell vaccine pulsed with autologous hypochlorous acid-oxidized ovarian cancer lysate primes effective broad antitumor immunity: From bench to bedside. *Clin. Cancer Res.* **2013**, *19*, 4801–4815. [CrossRef] [PubMed]

142. Guo, Z.; Wang, X.; Cheng, D.; Xia, Z.; Luan, M.; Zhang, S. PD-1 Blockade and OX40 Triggering synergistically protects against tumor growth in a murine model of ovarian cancer. *PLoS ONE* **2014**, *9*, e89350. [CrossRef] [PubMed]
143. Nounamo, B.; Liem, J.; Cannon, M.; Liu, J. Myxoma Virus Optimizes Cisplatin for the treatment of ovarian cancer in vitro and in a syngeneic murine dissemination model. *Mol. Ther. Oncolytics* **2017**, *6*, 90–99. [CrossRef] [PubMed]
144. Turner, T.B.; Meza-Perez, S.; Londoño, A.; Katre, A.; Peabody, J.E.; Smith, H.J.; Forero, A.; Norian, L.A.; Straughn, J.M.; Buchsbaum, D.J.; et al. Epigenetic modifiers upregulate MHC II and impede ovarian cancer tumor growth. *Oncotarget* **2017**, *8*, 44159–44170. [CrossRef] [PubMed]
145. Benencia, F.; Courrèges, M.C.; Fraser, N.W.; Coukos, G. Herpes virus oncolytic therapy reverses tumor immune dysfunction and facilitates tumor antigen presentation. *Cancer Biol. Ther.* **2008**, *7*, 1194–1205. [CrossRef] [PubMed]
146. Liu, Z.; Ravindranathan, R.; Kalinski, P.; Guo, Z.S.; Bartlett, D.L. Rational combination of oncolytic vaccinia virus and PD-L1 blockade works synergistically to enhance therapeutic efficacy. *Nat. Commun.* **2017**, *8*, 14754. [CrossRef] [PubMed]
147. Martin, S.D.; Brown, S.D.; Wick, D.A.; Nielsen, J.S.; Kroeger, D.R.; Twumasi-Boateng, K.; Holt, R.A.; Nelson, B.H. Low mutation burden in ovarian cancer may limit the utility of neoantigen-targeted vaccines. *PLoS ONE* **2016**, *11*, e0155189. [CrossRef] [PubMed]
148. Conejo-Garcia, J.R.; Benencia, F.; Courreges, M.-C.; Kang, E.; Mohamed-Hadley, A.; Buckanovich, R.J.; Holtz, D.O.; Jenkins, A.; Na, H.; Zhang, L.; et al. Tumor-infiltrating dendritic cell precursors recruited by a β-defensin contribute to vasculogenesis under the influence of Vegf-A. *Nat. Med.* **2004**, *10*, 950. [CrossRef] [PubMed]
149. Cubillos-Ruiz, J.R.; Silberman, P.C.; Rutkowski, M.R.; Chopra, S.; Perales-Puchalt, A.; Song, M.; Zhang, S.; Bettigole, S.E.; Gupta, D.; Holcomb, K.; et al. ER stress sensor XBP1 controls anti-tumor immunity by disrupting dendritic cell homeostasis. *Cell* **2015**, *161*, 1527–1538. [CrossRef] [PubMed]
150. Clarke, S.R.; Barnden, M.; Kurts, C.; Carbone, F.R.; Miller, J.F.; Heath, W.R. Characterization of the ovalbumin-specific TCR transgenic line OT-I: MHC elements for positive and negative selection. *Immunol. Cell Biol.* **2000**, *78*, 110–117. [CrossRef] [PubMed]
151. Cole, A.J.; Dwight, T.; Gill, A.J.; Dickson, K.-A.; Zhu, Y.; Clarkson, A.; Gard, G.B.; Maidens, J.; Valmadre, S.; Clifton-Bligh, R.; et al. Assessing mutant p53 in primary high-grade serous ovarian cancer using immunohistochemistry and massively parallel sequencing. *Sci. Rep.* **2016**, *6*, 26191. [CrossRef] [PubMed]
152. Walton, J.; Blagih, J.; Ennis, D.; Leung, E.; Dowson, S.; Farquharson, M.; Tookman, L.A.; Orange, C.; Athineos, D.; Mason, S.; et al. CRISPR/Cas9-mediated Trp53 and Brca2 knockout to generate improved murine models of ovarian high-grade serous carcinoma. *Cancer Res.* **2016**, *76*, 6118–6129. [CrossRef] [PubMed]
153. Walton, J.B.; Farquharson, M.; Mason, S.; Port, J.; Kruspig, B.; Dowson, S.; Stevenson, D.; Murphy, D.; Matzuk, M.; Kim, J.; et al. CRISPR/Cas9-derived models of ovarian high grade serous carcinoma targeting Brca1, Pten and Nf1, and correlation with platinum sensitivity. *Sci. Rep.* **2017**, *7*, 16827. [CrossRef] [PubMed]
154. Kroeger, D.R.; Milne, K.; Nelson, B.H. Tumor-infiltrating plasma cells are associated with tertiary lymphoid structures, cytolytic T-Cell responses, and superior prognosis in ovarian cancer. *Clin. Cancer Res.* **2016**, *22*, 3005–3015. [CrossRef] [PubMed]
155. Strickland, K.C.; Howitt, B.E.; Shukla, S.A.; Rodig, S.; Ritterhouse, L.L.; Liu, J.F.; Garber, J.E.; Chowdhury, D.; Wu, C.J.; D'Andrea, A.D.; et al. Association and prognostic significance of BRCA1/2-mutation status with neoantigen load, number of tumor-infiltrating lymphocytes and expression of PD-1/PD-L1 in high grade serous ovarian cancer. *Oncotarget* **2016**, *7*, 13587–13598. [CrossRef] [PubMed]
156. Roberts, P.C.; Mottillo, E.P.; Baxa, A.C.; Heng, H.H.Q.; Doyon-Reale, N.; Gregoire, L.; Lancaster, W.D.; Rabah, R.; Schmelz, E.M. Sequential molecular and cellular events during neoplastic progression: A mouse syngeneic ovarian cancer model. *Neoplasia* **2005**, *7*, 944–956. [CrossRef] [PubMed]
157. Endsley, M.P.; Moyle-Heyrman, G.; Karthikeyan, S.; Lantvit, D.D.; Davis, D.A.; Wei, J.-J.; Burdette, J.E. Spontaneous transformation of murine oviductal epithelial cells: A model system to investigate the onset of fallopian-derived tumors. *Front. Oncol.* **2015**, *5*, 154. [CrossRef] [PubMed]

158. Russo, A.; Czarnecki, A.A.; Dean, M.; Modi, D.A.; Lantvit, D.D.; Hardy, L.; Baligod, S.; Davis, D.A.; Wei, J.-J.; Burdette, J.E. PTEN loss in the fallopian tube induces hyperplasia and ovarian tumor formation. *Oncogene* **2018**, *37*, 1976–1990. [CrossRef] [PubMed]
159. Garson, K.; Gamwell, L.F.; Pitre, E.M.; Vanderhyden, B.C. Technical challenges and limitations of current mouse models of ovarian cancer. *J. Ovarian Res.* **2012**, *5*, 39. [CrossRef] [PubMed]
160. Morin, P.J.; Weeraratna, A.T. Genetically defined ovarian cancer mouse models. *J. Pathol.* **2016**, *238*, 180–191. [CrossRef] [PubMed]
161. Connolly, D.C.; Bao, R.; Nikitin, A.Y.; Stephens, K.C.; Poole, T.W.; Hua, X.; Harris, S.S.; Vanderhyden, B.C.; Hamilton, T.C. Female mice chimeric for expression of the simian virus 40 TAg under control of the MISIIR promoter develop epithelial ovarian cancer. *Cancer Res.* **2003**, *63*, 1389–1397. [PubMed]
162. Laviolette, L.A.; Garson, K.; Macdonald, E.A.; Senterman, M.K.; Courville, K.; Crane, C.A.; Vanderhyden, B.C. 17β-Estradiol accelerates tumor onset and decreases survival in a transgenic mouse model of ovarian cancer. *Endocrinology* **2010**, *151*, 929–938. [CrossRef] [PubMed]
163. Sherman-Baust, C.A.; Kuhn, E.; Valle, B.L.; Shih, I.-M.; Kurman, R.J.; Wang, T.-L.; Amano, T.; Ko, M.S.H.; Miyoshi, I.; Araki, Y.; et al. A genetically engineered ovarian cancer mouse model based on fallopian tube transformation mimics human high-grade serous carcinoma development. *J. Pathol.* **2014**, *233*, 228–237. [CrossRef] [PubMed]
164. Schietinger, A.; Philip, M.; Krisnawan, V.E.; Chiu, E.Y.; Delrow, J.J.; Basom, R.S.; Lauer, P.; Brockstedt, D.G.; Knoblaugh, S.E.; Hämmerling, G.J.; et al. Tumor-Specific T Cell Dysfunction Is a Dynamic Antigen-Driven Differentiation Program Initiated Early during Tumorigenesis. *Immunity* **2016**, *45*, 389–401. [CrossRef] [PubMed]
165. Charbonneau, B.; Goode, E.L.; Kalli, K.R.; Knutson, K.L.; DeRycke, M.S. The immune system in the pathogenesis of ovarian cancer. *Crit. Rev. Immunol.* **2013**, *33*, 137–164. [CrossRef] [PubMed]
166. McGranahan, N.; Furness, A.J.S.; Rosenthal, R.; Ramskov, S.; Lyngaa, R.; Saini, S.K.; Jamal-Hanjani, M.; Wilson, G.A.; Birkbak, N.J.; Hiley, C.T.; et al. Clonal neoantigens elicit T cell immunoreactivity and sensitivity to immune checkpoint blockade. *Science* **2016**, *351*, 1463–1469. [CrossRef] [PubMed]
167. Flesken-Nikitin, A.; Choi, K.-C.; Eng, J.P.; Shmidt, E.N.; Nikitin, A.Y. Induction of Carcinogenesis by Concurrent Inactivation of p53 and Rb1 in the Mouse Ovarian Surface Epithelium. *Cancer Res.* **2003**, *63*, 3459–3463. [PubMed]
168. Flesken-Nikitin, A.; Hwang, C.-I.; Cheng, C.-Y.; Michurina, T.V.; Enikolopov, G.; Nikitin, A.Y. Ovarian surface epithelium at the junction area contains cancer-prone stem cell niche. *Nature* **2013**, *495*, 241–245. [CrossRef] [PubMed]
169. Perets, R.; Wyant, G.A.; Muto, K.W.; Bijron, J.G.; Poole, B.B.; Chin, K.T.; Chen, J.Y.H.; Ohman, A.W.; Stepule, C.D.; Kwak, S.; et al. Transformation of the fallopian tube secretory epithelium leads to high-grade serous ovarian cancer in *Brca;Tp53;Pten* Models. *Cancer Cell* **2013**, *24*, 751–765. [CrossRef] [PubMed]
170. Szabova, L.; Yin, C.; Bupp, S.; Guerin, T.M.; Schlomer, J.J.; Householder, D.B.; Baran, M.L.; Yi, M.; Song, Y.; Sun, W.; et al. Perturbation of Rb, p53, and Brca1 or Brca2 cooperate in inducing metastatic serous epithelial ovarian cancer. *Cancer Res.* **2012**, *72*, 4141–4153. [CrossRef] [PubMed]
171. Wieser, V.; Gaugg, I.; Fleischer, M.; Shivalingaiah, G.; Wenzel, S.; Sprung, S.; Lax, S.F.; Zeimet, A.G.; Fiegl, H.; Marth, C. BRCA1/2 and TP53 mutation status associates with PD-1 and PD-L1 expression in ovarian cancer. *Oncotarget* **2018**, *9*, 17501–17511. [CrossRef] [PubMed]
172. Son, D.-S.; Kabir, S.M.; Dong, Y.-L.; Lee, E.; Adunyah, S.E. Inhibitory effect of tumor suppressor p53 on proinflammatory chemokine expression in ovarian cancer cells by Reducing Proteasomal Degradation of IκB. *PLoS ONE* **2012**, *7*, e51116. [CrossRef] [PubMed]
173. Zhai, Y.; Wu, R.; Kuick, R.; Sessine, M.S.; Schulman, S.; Green, M.; Fearon, E.R.; Cho, K.R. High-grade serous carcinomas arise in the mouse oviduct via defects linked to the human disease. *J. Pathol.* **2017**, *243*, 16–25. [CrossRef] [PubMed]
174. Budiu, R.A.; Elishaev, E.; Brozick, J.; Lee, M.; Edwards, R.P.; Kalinski, P.; Vlad, A.M. Immunobiology of human mucin 1 in a preclinical ovarian tumor model. *Oncogene* **2013**, *32*, 3664. [CrossRef] [PubMed]
175. Deng, J.; Wang, L.; Chen, H.; Li, L.; Ma, Y.; Ni, J.; Li, Y. The role of tumour-associated MUC1 in epithelial ovarian cancer metastasis and progression. *Cancer Metastasis Rev.* **2013**, *32*, 535–551. [CrossRef] [PubMed]

176. Snijdewint, F.G.; von Mensdorff-Pouilly, S.; Karuntu-Wanamarta, A.H.; Verstraeten, A.A.; van Zanten-Przybysz, I.; Hummel, P.; Nijman, H.W.; Kenemans, P.; Hilgers, J. Cellular and humoral immune responses to MUC1 mucin and tandem-repeat peptides in ovarian cancer patients and controls. *Cancer Immunol. Immunother.* **1999**, *48*, 47–55. [CrossRef] [PubMed]
177. Vlad, A.M.; Diaconu, I.; Gantt, K.R. MUC1 in endometriosis and ovarian cancer. *Immunol. Res.* **2006**, *36*, 229–236. [CrossRef]
178. Barua, A.; Bitterman, P.; Abramowicz, J.S.; Dirks, A.L.; Bahr, J.M.; Hales, D.B.; Bradaric, M.J.; Edassery, S.L.; Rotmensch, J.; Luborsky, J.L. Histopathology of ovarian tumors in laying hens, a preclinical model of human ovarian cancer. *Int. J. Gynecol. Cancer Off. J. Int. Gynecol. Cancer Soc.* **2009**, *19*, 531–539. [CrossRef] [PubMed]
179. Cubillos-Ruiz, J.R.; Engle, X.; Scarlett, U.K.; Martinez, D.; Barber, A.; Elgueta, R.; Wang, L.; Nesbeth, Y.; Durant, Y.; Gewirtz, A.T.; et al. Polyethylenimine-based siRNA nanocomplexes reprogram tumor-associated dendritic cells via TLR5 to elicit therapeutic antitumor immunity. *J. Clin. Investig.* **2009**, *119*, 2231–2244. [CrossRef] [PubMed]
180. Cubillos-Ruiz, J.R.; Baird, J.R.; Tesone, A.J.; Rutkowski, M.R.; Scarlett, U.K.; Camposeco-Jacobs, A.L.; Anadon-Arnillas, J.; Harwood, N.M.; Korc, M.; Fiering, S.N.; et al. Reprogramming tumor-associated dendritic cells in vivo using microrna mimetics triggers protective immunity against ovarian cancer. *Cancer Res.* **2012**, *72*, 1683–1693. [CrossRef] [PubMed]
181. Baert, T.; Verschuere, T.; Van Hoylandt, A.; Gijsbers, R.; Vergote, I.; Coosemans, A. The dark side of ID8-Luc2: Pitfalls for luciferase tagged murine models for ovarian cancer. *J. Immunother. Cancer* **2015**, *3*, 57. [CrossRef] [PubMed]
182. Wilson, A.J.; Barham, W.; Saskowski, J.; Tikhomirov, O.; Chen, L.; Lee, H.-J.; Yull, F.; Khabele, D. Tracking NF-κB activity in tumor cells during ovarian cancer progression in a syngeneic mouse model. *J. Ovarian Res.* **2013**, *6*, 63. [CrossRef] [PubMed]
183. Miyoshi, I.; Takahashi, K.; Kon, Y.; Okamura, T.; Mototani, Y.; Araki, Y.; Kasai, N. Mouse transgenic for murine oviduct-specific glycoprotein promoter-driven simian virus 40 large T-antigen: Tumor formation and its hormonal regulation. *Mol. Reprod. Dev.* **2002**, *63*, 168–176. [CrossRef] [PubMed]
184. Tirodkar, T.S.; Budiu, R.A.; Elishaev, E.; Zhang, L.; Mony, J.T.; Brozick, J.; Edwards, R.P.; Vlad, A.M. MUC1 positive, Kras and Pten driven mouse gynecologic tumors replicate human tumors and vary in survival and nuclear grade based on anatomical location. *PLoS ONE* **2014**, *9*, e102409. [CrossRef] [PubMed]
185. Domcke, S.; Sinha, R.; Levine, D.A.; Sander, C.; Schultz, N. Evaluating cell lines as tumour models by comparison of genomic profiles. *Nat. Commun.* **2013**, *4*, 2126. [CrossRef] [PubMed]
186. Fleury, H.; Communal, L.; Carmona, E.; Portelance, L.; Arcand, S.L.; Rahimi, K.; Tonin, P.N.; Provencher, D.; Mes-Masson, A.-M. Novel high-grade serous epithelial ovarian cancer cell lines that reflect the molecular diversity of both the sporadic and hereditary disease. *Genes Cancer* **2015**, *6*, 378–398. [PubMed]
187. Ince, T.A.; Sousa, A.D.; Jones, M.A.; Harrell, J.C.; Agoston, E.S.; Krohn, M.; Selfors, L.M.; Liu, W.; Chen, K.; Yong, M.; et al. Characterization of twenty-five ovarian tumour cell lines that phenocopy primary tumours. *Nat. Commun.* **2015**, *6*, 7419. [CrossRef] [PubMed]
188. Kloudová, K.; Hromádková, H.; Partlová, S.; Brtnický, T.; Rob, L.; Bartůňková, J.; Hensler, M.; Halaška, M.J.; Špíšek, R.; Fialová, A. Expression of tumor antigens on primary ovarian cancer cells compared to established ovarian cancer cell lines. *Oncotarget* **2016**, *7*, 46120–46126. [CrossRef] [PubMed]
189. Nham, T.; Poznanski, S.M.; Fan, I.Y.; Shenouda, M.M.; Chew, M.V.; Lee, A.J.; Vahedi, F.; Karimi, Y.; Butcher, M.; Lee, D.A.; et al. Ex vivo-expanded NK cells from blood and ascites of ovarian cancer patients are cytotoxic against autologous primary ovarian cancer cells. *Cancer Immunol. Immunother.* **2018**, *67*, 575–587. [CrossRef] [PubMed]
190. Freedman, J.D.; Hagel, J.; Scott, E.M.; Psallidas, I.; Gupta, A.; Spiers, L.; Miller, P.; Kanellakis, N.; Ashfield, R.; Fisher, K.D.; et al. Oncolytic adenovirus expressing bispecific antibody targets T-cell cytotoxicity in cancer biopsies. *EMBO Mol. Med.* **2017**, *9*, 1067–1087. [CrossRef] [PubMed]
191. Jiménez-Sánchez, A.; Memon, D.; Pourpe, S.; Veeraraghavan, H.; Li, Y.; Vargas, H.A.; Gill, M.B.; Park, K.J.; Zivanovic, O.; Konner, J.; et al. Heterogeneous tumor-immune microenvironments among differentially growing metastases in an ovarian cancer patient. *Cell* **2017**, *170*, 927–938. [CrossRef] [PubMed]

192. Greaves, M. Evolutionary determinants of cancer. *Cancer Discov.* **2015**, *5*, 806–820. [CrossRef] [PubMed]
193. McGranahan, N.; Swanton, C. Cancer evolution constrained by the immune microenvironment. *Cell* **2017**, *170*, 825–827. [CrossRef] [PubMed]

cancers

MDPI

Review

Regulation of Ovarian Cancer Prognosis by Immune Cells in the Tumor Microenvironment

Maureen L. Drakes * and Patrick J. Stiff

Cardinal Bernardin Cancer Center, Department of Medicine, Loyola University Chicago, Building 112, 2160 South First Avenue, Maywood, IL 60153, USA; pstiff@lumc.edu

* Correspondence: mdrakes@luc.edu; Tel.: +1-708-327-3125

Received: 8 August 2018; Accepted: 29 August 2018; Published: 1 September 2018

Abstract: It is estimated that in the United States in 2018 there will be 22,240 new cases of ovarian cancer and 14,070 deaths due to this malignancy. The most common subgroup of this disease is high-grade serous ovarian cancer (HGSOC), which is known for its aggressiveness, high recurrence rate, metastasis to other sites, and the development of resistance to conventional therapy. It is important to understand the ovarian cancer tumor microenvironment (TME) from the viewpoint of the function of pre-existing immune cells, as immunocompetent cells are crucial to mounting robust antitumor responses to prevent visible tumor lesions, disease progression, or recurrence. Networks consisting of innate and adaptive immune cells, metabolic pathways, intracellular signaling molecules, and a vast array of soluble factors, shape the pathogenic nature of the TME and are useful prognostic indicators of responses to conventional therapy and immunotherapy, and subsequent survival rates. This review highlights key immune cells and soluble molecules in the TME of ovarian cancer, which are important in the development of effective antitumor immunity, as well as those that impair effector T cell activity. A more insightful knowledge of the HGSOC TME will reveal potential immune biomarkers to aid in the early detection of this disease, as well as biomarkers that may be targeted to advance the design of novel therapies that induce potent antitumor immunity and survival benefit.

Keywords: tumor microenvironment; immune inhibition; tumor-infiltrating lymphocytes; tumor-associated macrophages; dendritic cells; antitumor immunity; immunotherapy

1. Introduction

Ovarian cancer may be divided into six subgroups, namely, serous, mucinous, endometroid, transitional-cell, clear-cell, and squamous carcinoma [1]. The most common group is high-grade serous ovarian carcinoma (HGSOC), a disease that escapes detection and diagnosis until after it is disseminated to areas of the abdomen and beyond. At this advanced stage, survival is dismal, with only about twenty percent of patients diagnosed at International Federation of Gynecology and Obstetrics (FIGO) stage III or IV disease fortunate enough to reach a five-year survival time point, since most of these individuals become resistant to conventional therapy and succumb to disease. This disease grows aggressively, often recurs at the primary or metastatic sites, and is the most deadly of gynecologic cancers [2,3].

HGSOC is believed to arise from the ovarian-surface epithelium and/or the fallopian epithelium [4]. Most patients (96%) with this disease have TP53 mutations, with BRCA1/2 (22% patients) mutations also common [5,6]. An accompanying feature of HGSOC is an accumulation of ascites fluid in the peritoneal cavity, which allows the adhesion of cancer cells to the omentum (connective and fatty tissue covering the ventral surface of the intestines) and serous membranes lining the peritoneal organs [7], thereby increasing the potential of cancer lesions at these sites soon after the primary disease is established [8,9].

The tumor microenvironment (TME) in HGSOC is comprised of an intricate system of immune cells, including subsets of T cells, dendritic cells, macrophages, and NK cells, as well as soluble factors elaborated by myriads of existing cell types, both spontaneously and as a result of their networking interactions [10,11]. Studies on the ovarian TME in HGSOC have been prompted by the need to understand the disease biology, with the goal of targeting cancer-promoting immune mechanisms, and providing effective therapies for the management and ultimately a cure for HGSOC. The full significance of the ovarian TME in determining disease progression, recurrence, or regression is yet to be revealed. This review focuses on the dynamic and diverse immune components in the ovarian TME and how they mediate the balance between protumor and antitumor immunity, and patient survival.

2. Immune Regulation by T Cells in the TME

Tumor-associated/infiltrated lymphocytes (TILs) are found in the tumor stroma or in the tumor islets (intraepithelial TILs). CD3+, CD4+, and CD8+ TILs are usually associated with a positive outcome [12–16]. Notably, in a study of 186 samples of advanced-stage ovarian cancer, it was found that 55% patients with CD3+ TILs had a five-year survival of 38%, whereas only 4.5% patients without detectable TILs reached his survival mark. Moreover, taking into consideration patients who had surgical debulking and platinum-based chemotherapy, 73.9% of patients with pre-existing TILs had complete response (CR), while only 11.9% patients without TILs exhibited CR [15].

Some groups have demonstrated that in ovarian-cancer tumor-tissue sections, intraepithelial CD8+ TILs correlate with good outcome, and others have shown that a high ratio of CD8+/FoxP3+ T regulatory cells (Tregs) is beneficial to survival [17]. In a meta-analysis of 10 studies and 1815 patients, both CD3 and CD8 TILS were found to be associated with survival, but CD8+ TILS were the more significant of these two subsets. Interestingly, in these studies the prognostic value of TILS was more significant in some geographic regions studied compared with others, raising the possibility that genetic factors or different levels of access to healthcare may also be relevant factors to consider when measuring survival in such studies [13].

In detailed investigations with over 5500 patients, including 3196 with HGSOC, it was found that among the five invasive histotypes studied, HGSOC showed the most infiltration of CD8+ T cells. Patients were followed over 24,650 person-years. Analysis of CD8+ TILS in the tumor epithelium on a scale of negative, low, moderate, and high revealed distinct survival differences in HGSOC patients based on the density of CD8+ TILS in the epithelial components of tumor islets. The median survival for patients with no CD8+ TILS was 2.8 years, whereas with low, moderate, or high TILS, survival was 3.0 years, 3.8 years, and 5.1 years, respectively. The presence of CD8+ TILS was favorable to outcome regardless of extent of residual disease, standard therapy or BRCA1 mutation [18].

Others report that the CD8+CD103+ T cell subset are found in abundance in the ovarian-cancer epithelium, and are associated with a better outcome [19,20]. Together, these studies of CD8+ T cells in the ovarian TME further emphasize the relevance of these TILS as a prognostic indicator in HGSOC.

Another subset of TILs populating the HGSOC TME are FoxP3+ T regulatory cells. These cells were initially regarded as a potent immunosuppressive mechanism, limiting the potency of antitumor immune responses. CD4+CD25+FoxP3+ T regulatory cells may act by elaborating protumor cytokines such as IL-10 or TGF-β, or by cell–cell contact mechanisms [21]. Despite several early reports associating this subset of T regulatory cells with a poor outcome [22,23], a meta-analysis of 869 patients over several studies did not conclude that FoxP3 Tregs in the tumors of ovarian-cancer patients are a significant prognostic indicator of survival [24]. Yet there are other T cell subsets in the ovarian TME that may negatively impact survival. These include T cells expressing cytotoxic T lymphocyte-associated antigen 4 (CTLA-4), glucocorticoid-induced TNF receptor family-related protein (GITR), or CCR4, CD8+CD28- T regs [25–28], as well as exhausted CD8+T cells expressing immune checkpoint inhibitory molecules programmed cell death-1 (PD-1) or lymphocyte activation gene-3 (LAG-3; CD223) [29–31]. These cells may all confer immunosuppression in the TME or limit antitumor responses (Figure 1). There is still debate in the literature concerning the role of the

Th17 CD4+ T cell subset in ovarian cancer, but some have reported that these cells have an inverse relationship with Tregs, and correlate with survival [32,33].

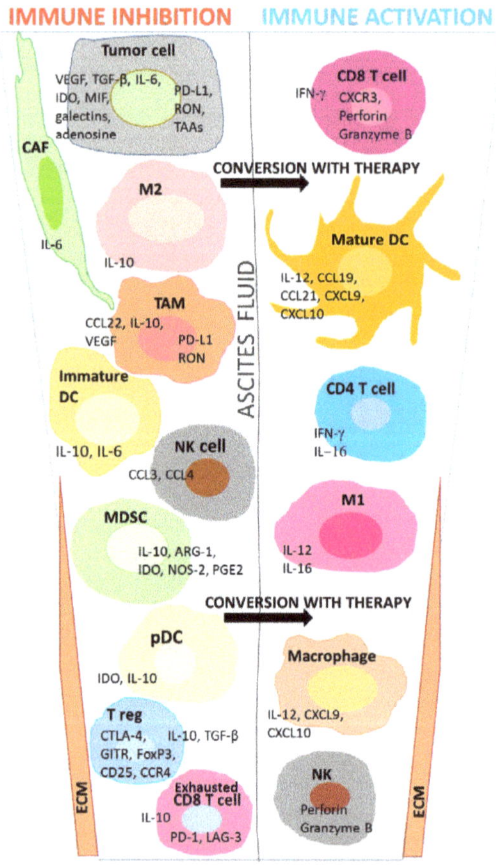

Figure 1. Schematic representation of the primary immune components in the tumor microenvironment (TME). Several cell types in the TME of high-grade serous ovarian carcinoma (HGSOC) elaborate factors that can lead to immune dysregulation and inhibition of antitumor responses. The ascites of these patients is rich in TGF-β, IL-6, IL-8, IL-10, vascular endothelial growth factor (VEGF), and CCL22 and other factors released by contributing cell types as shown in the graphic. CCL22 (the ligand for CCR4) preferentially recruits Tregs into tumors. Exhausted CD8 T cells in tumors express PD-1 and LAG-3 and secrete low quantities of IFN-γ. Several Treg subsets exist in the TME, each bearing some of the phenotypic markers, CD4, CD8, CCR4, FoxP3, CD25, GITR, or CTLA-4, and primarily release TGF-β and IL-10. Molecules such as recepteur d'origine nantais (RON) on tumor cells are associated with invasiveness, and tumor associated antigens (TAAs) such as New York Esophageal antigen-1 (NY-ESO-1), human epidermal growth factor receptor 2 (HER-2), and Wilm's tumor-1 (WT-1) are immunogenic targets. Immune-suppressive mechanisms in the TME that foster tumor initiation, progression, and recurrence may be reversed with combinations of conventional and novel therapies, designed to potentiate antitumor immune responses. Parameters consistent with disease improvement include CD8+ T cells secreting IFN-γ, perforin, and granzyme B, which facilitate the killing of tumor cells. Additionally, DC-secreted chemokines, such as CXCL9 and CXCL10, can recruit CD4+ and CD8+ immunocompetent T cells, and IL-16-a-cytokine secreted by T cells, macrophages, and dendritic cells, is a primary chemoattractant for CD4+T cells in ovarian cancer.

In addition to their prime role in immune surveillance limiting the initiation of ovarian cancer and other cancers [34–37], immunocompetent TILs can recognize cancer antigens or overexpressed self-antigens that have been processed by antigen-presenting cells, and mount potent antitumor immune responses. CD4+ TILs can recruit dendritic cells that can prime T cells to exert their cytotoxic effects by secreting perforin, granzyme B, or Fas ligand (cell death receptor ligand; FasL; CD95L), which may directly kill cancer cells. Both cytotoxic CD8+ and CD4+ TILs cells secrete cytokines such as IFN-γ and IL-2 [38] that can induce other cells in the TME to mount antitumor immunity, and promote longer survival. In the ovarian tumor, IL-16, primarily a Th1 cytokine, has been reported to be a critical chemoattractant for the recruitment of CD4+ T cells into the tumor [39].

However, there is an ongoing interplay between TILs and the TME, and by suppressing the function and limiting the infiltration of CD3+, CD4+, or CD8+ TILs, tumors can circumvent antitumor immunity, especially in TME where TILs were already low in numbers at the time of diagnosis. Exclusion or inhibitory mechanisms imposed on TILs in the ovarian TME are as follows. Increased angiogenesis in ovarian-cancer cells presents a great barrier to the infiltration of tumor-specific T cells, thereby reducing the numbers of TILs in patients' tumors. TGF-β increases angiogenesis directly as well as suppresses the proliferation and activation of TILs [40,41]. Overexpressed vascular endothelial growth factor (VEGF) can enhance the proliferation, migration, and invasion of endothelial cells and is associated with poor outcomes in ovarian cancer [42]. VEGF-A decreases the adhesive interaction between lymphocytes and tumor vascular endothelial cells, and reduces TIL penetration through deregulation of intercellular adhesion molecule-1 (ICAM-1) and vascular cell adhesion-molecule-1 (VCAM-1) [43]. Furthermore, VEGF-A in concert with IL-10 and PGE2 induces FasL in endothelial cells. Increased FasL in endothelial cells favors the selective trafficking of Tregs above CD8+ immunocompetent T cells [44]. Several other molecules in the tumor endothelium, such as programmed cell death-1 ligand (PD-L1), B7-H3, arginase-1 (ARG-1), indoleamine 2,3, dioxygenase (IDO), IL-10, and PGE2, released by endothelial cells, downregulate TIL function or kill CD8+ effector TILs [45–49].

The density of TILs in tumors has recently been used to categorize tumors. Tumors are termed "hot" or immunogenic if they consist of high numbers of TILs, whereas "cold tumors" have much fewer TILs [50,51], and patients in this latter group are likely to have a poor response to therapy. An understanding of T cell subsets in relation to inhibitory mechanisms in the ovarian TME at baseline diagnosis is crucial to effectively designing novel therapies for HGSOC, and to predict outcome after treatment regimens, as T cells (especially CD8+ T cells) may be the critical antitumor effector mechanism in the disease. The success of these therapies may depend largely on the ability of T cells to reverse immune dysregulation at the site of the disease.

3. Multifaceted Nature of Macrophages in the TME

In contrast to immunocompetent T cells, the majority of myeloid lineage cells in the ovarian TME are generally of a protumor propensity [52,53]. Subpopulations of myeloid lineage cells in the TME of patients consist of a variety of phenotypes and nomenclature [54,55] (Table 1).

Tumor-associated macrophages (TAMS) are the major subpopulation of this lineage cells in the ovarian TME. TAMS can readily change phenotype and function in the presence of soluble molecules in the surrounding milieu [56,57]. These cells can be recruited from blood monocytes, or arise from resident peritoneal macrophages [54,58–60]. TAMS from both of these origins have some phenotypes in common such as the expression of molecules CD163 and CD206, as well as similar levels of genes for phagocytosis and antigen presentation. However, a distinctive feature of TAMS in the TME is an upregulation of genes linked to extracellular-matrix (ECM) remodeling [61]. In ovarian cancer, TAMS are mostly immunosuppressive, and associated with tumor cell invasion, angiogenesis, metastasis and early relapse [11,62,63].

Table 1. Phenotypic characterization of myeloid lineage cells in the ovarian TME.

Myeloid Group [a]	Cell Classification	Phenotype
TAMS (monocytes/ macrophages)	Inflammatory monocyte	CD14+, HLA-DR high, CD11c+, CD64+
	M1 macrophage	HLA-DR+, CD68+, CD80+, CD86+
	M2 macrophage	HLA-DR+, CD68+, CD163+, CD206+, CD200R
	M-MDSC	CD11b+, CD33+, CD14+, HLA-DR low
	G-MDSC	CD11b+, CD33+, CD15+, CD66b+, HLA-DR low
Dendritic cells	Immature DC	CD80 low, CD86 low, CD40 low, CD14+, CXCR3+
	Mature DC	C80 high, CD86 high, CD40 high, CD83+, HLA DR high, CCR7+, CD103+

[a] The primary identification markers of TAM subsets and of myeloid DC in the TME are shown.

In tumors, the benign-to-malignant state is associated with angiogenesis (increase in vascularization). VEGF, TGF-β, matrix metalloproteinases (MMPs), hypoxia-inducible factor (HIF), and adrenomedullin (ADM) secreted by TAMS enhance the process of vascularization [64–66]. TAMS are also critical in mediating epithelial–mesenchymal transition (EMT), which is essential to tumor progression. In this process, polarized epithelial cells change their phenotype to motile mesenchymal cells. The downregulation of epithelial markers, such as E-cadherin, is replaced by the upregulation of mesenchymal markers such as vimentin, Slug, Snail, fibronectin, zinc-finger E-box binding homeobox 1 (ZEB1), ZEB2, and α-smooth muscle actin, allowing cells to migrate and invade [61,66,67]. These changes correlate with metastasis, recurrence, chemoresistant tumors, and poor outcome. EMT is mediated by several TAM products, such as TGF-β, hepatocyte growth factor (HGF), and epidermal growth factor (EGF) [66,68–71].

Studies in ovarian-cancer tissue showed that there is a significant elevation in the numbers of CD68+ and CD206+ TAMS, and of MMPs expression, in comparison with benign ovarian tissues [72]. This difference was due to higher levels of these parameters in patients with stage III/IV in comparison with those at stage I/II disease. Furthermore, in patients with positive lymphatic invasion, the numbers of CD68+, CD206+, and MMP-positive cells was significantly higher than in patients without lymphatic invasion [72]. Additional studies with SKOV3 ovarian-tumor cells demonstrated that TAMS promoted upregulation of TLRs 1, 2, 4 and 6, MMP-2, MMP-9, and MMP-10 expression. Ovarian-cancer cell invasion was enhanced via TLRs signaling pathway and activation of downstream nuclear transcription factor (NF)-KB p65 and microtubule-associated proteins (MAPs) kinases pathway in SKOV3 cells [72].

In other studies, TAMS enhanced spheroid formation and tumor growth and early ovarian-cancer metastasis by secreting EGF [62]. TAMS can also be high secretors of CCL18, a chemokine that promotes tumor migration and metastasis in ovarian cancer [73]. IFN-γ treatment reduces CCL18 secretion and can switch TAMS to an immunostimulatory phenotype [74]. TAMS in the ovarian-cancer TME are very low IL-12 secreting, a cytokine that is positively associated with outcome in this disease [61,75,76]. A brief summary of the primary protumor processes regulated by TAMS in the TME is outlined in Table 2. A more detailed account of TAM activities in the TME is reported elsewhere [66,70].

Table 2. Immune dysregulation by TAMS in the TME.

Mediators [a]	Cell Targets	Major Actions
IL-10	CTL	Inhibits activation
TGF-β	Treg	Induces differentiation
TGF-β, HGF, collagen, cathepsin and serine proteases, EGF, CSF-1	Tumor	Increases adhesion, invasion, and EMT
IL-6, TNF-α, WNT, JAG	Tumor	Promotes survival, growth, stemness
ADM, VEGF, COX-2, MMPs, HIF-1α, TGF-β	Endothelial	Angiogenesis

[a] TAMS elaborate a range of immune molecules and soluble mediators that are involved in the initiation and progression of cancer.

Macrophages and monocytes in the ovarian tumor may exhibit polarization to an M2, protumor, and immunosuppressive state, under the influence of colony-stimulating factor (CSF), IL-4, IL-13, IL-10, TGF-β and other soluble molecules. M2 (alternatively activated) macrophages secrete IL-10 and TGF-β and play an active role in tissue remodeling and tumor progression [77,78].

The presence of CD4+ or CD8+ T cells secreting IFN-γ in the tumor can promote the presence of M1 immunocompetent classically activated macrophages, above an M2 phenotype. M1 macrophages are stimulated by Toll-like receptors (TLR) ligands and by IFN-γ to give a Th1 response secreting IL-12, IL-23, and TNF-α. M1 macrophages are highly potent against micro-organisms and tumors, and are associated with survival in HGSOC [79].

Another subgroup of TAMS that also merits special mention is the myeloid-derived suppressor cell (MDSC, M-MDSC). This is a group of immature myeloid cells in the TME that correlate well with heightened disease, increased tumor burden, and resistance to immune therapy in HGSOC [80,81]. These cells have a role in enhancing stemness and promoting metastasis of ovarian-cancer cells by inducing miRNA101 expression, subsequently repressing the corepressor gene C-terminal binding protein-2 (CtBP2) [82].

MDSC are also recruited to the ovarian TME under the influence of chemokine receptor CXCR4. PGE2 is required for the production of chemokine CXCL12, and for the expression of its binding receptor CXCR4 in these cells [83]. The CXCR4–CXCL12 axis and PGE2 are critical to the progression of HGSOC [84–86], and negatively impact the function of several immune cells in the TME, as we will discuss in the subsequent text.

Cyclooxygenase-2 (COX-2, an enzyme required for PGE2 synthesis) and PGE2 drive the differentiation of CD1a+ DC to CD14+CD33+CD34+ MDSC, and induce the expression of immunosuppressive molecules IDO, arginase-1 (ARG-1), IL-10, nitric oxide synthase-2 (NOS-2) and COX-2 by MDSC, molecules that limit CD8+ cytotoxic T cell responses [49,87,88]. Blocking COX-2/PGE2 suppression in MDSC prevents the accumulation of MDSC and enhances antitumor immunity [85]. Additional stimuli that may recruit MDSC to the ovarian TME are soluble factors such as VEGF, which are secreted by tumor cells in the microenvironment [81]. Granulocytic-MDSC (G-MDSC) are also of the myeloid lineage but do not appear to be a critical factor in the progression of most cancers.

Taken together, TAMS play critical roles in the establishment of cancers, including HGSOC. Targeting of these cells with anti-CCL2 antibody, anti-CSF-IR inhibitors, anti-CD52 antibody, and anti-CD11b antibody for therapy of ovarian cancer has been investigated in preclinical models of ovarian cancer [89–91]. There has also been a limited number of Phase I/II clinical trials blocking TAM activity in patients [92,93], but to date there is no such approved therapy.

4. Dendritic Cell Function in the Ovarian TME

Dendritic cells capture antigen, process and present antigenic peptide to cells in the immune system [94]. DC present exogenously captured peptides to CD4+ T cells via MHC class-II, and endogenous peptide antigens via major histocompatibility complex class-I (MHC-I) to CD8+ T cells. DC can also present exogenously captured antigens as MHC class I associated peptides (cross presentation), consequently facilitating more efficient CD4+ and CD8+ T cell activity [95,96]. Potent activation of T cells requires a cognate antigen (signal 1), costimulatory molecules (such as CD80, CD86, CD40) on DC or other antigen-presenting cells (signal 2), and proinflammatory cytokines (signal 3). If this process is sequential and efficient the outcome is Th1 (antitumor) immunity by CD4+ and CD8+ T cells. Lack of any of these signals can result in Th2 immunity or immune suppression mediated by Tregs [97–101]. Tumors can disrupt these signals by strategies such as loss of tumor antigens, and by the abundance of immunosuppressive soluble factors in the TME that can induce DC dysfunction [102,103].

Immature myeloid DC are derived from hematopoietic bone-marrow (BM) progenitor cells. These cells leave the BM enter the bloodstream and reside in lymph nodes or other tissue. They express costimulatory molecules at low levels, release low levels of cytokines, and are capable of mounting only limited immune responses. These cells express chemokine receptors CXCR3, CXCR4, CCR1, 2, 5, and 6. On stimulation by antigen, immature DC migrate to lymph nodes from tissues and present the specific antigen to other immune cells [104,105].

DC exposed to antigen undergo a process of maturation, characterized by an increase in costimulatory molecules, downregulation of existing chemokine receptors, and the acquisition of CCR7, the latter of which recruits DC to LN, attracted by CCL19 (MIP-3β) and CCL21, chemokines secreted by DC. Mature DC can activate naïve CD8+ T cells, crosslink with CD40 ligand on other cells, and secrete IL-12 [94,104–107].

Myeloid DC in tumors are found in low numbers and exhibit many features of immature DC. The immune-suppressive environment in the ovarian tumor, rich in TGF-β, IL-10, VEGF, ARG-1, along with inhibitory molecules such as IDO, PD-1, and PD-L1, drives the differentiation of CD14+CD1a- immature myeloid cells, anergic T cells and Tregs, induces tolerance, and promotes tumor growth [49,108,109]. In one study, depletion of DC in mice at advanced stages of ovarian cancer delayed tumor growth [110]. The benefit of mature myeloid DC function in inducing antitumor immune responses has been exploited in DC vaccine therapy clinical trials in ovarian and other cancers (NCT00703105) [111–113].

Recent evidence indicates that in melanoma, tumor-residing CD103+ DC were necessary for CD8+ effector T cell recruitment in the TME. These CD103+ DC have high expression of CXCR3 and of the transcription molecule Batf3, which possibly controls the development and maintenance of the DC1 lineage [114–116]. The presence of Baft3-lineage CD103+ DC correlated with the presence of CXCR3-binding chemokines CXCL9, 10 and 11, which increase the trafficking of effector T cells into tumors, and are associated with survival in cancers such as HGSOC [117,118]. The lack of conventional DC as in Batf3 −/− mice abolishes the rejection of immunogenic tumors, the response to adoptive T cell therapy, and to immune checkpoint blockade [114,115,119,120]. It is plausible that in the ovarian TME a similar mechanism of recruitment of effector T cells by this DC lineage would be an immune-enhancing mechanism to counteract the underlying immunosuppressive myeloid networks that favor disease progression, recurrence, and death.

In addition to immature myeloid-derived DC, plasmacytoid DC also contribute to the immunosuppressive network in HGSOC. CXCR4 expressing plasmacytoid DC (pDC) precursor cells are recruited into the ovarian TME by CXCL12 and IL-10 in the tumor [121]. Plasmacytoid DC (CD4+CD123+BDCA2+) in tumors such as HGSOC are often tolerogenic, and are noted for the release of IDO, an enzyme that catalyzes tryptophan degradation [47,48]. IDO promotes tumor angiogenesis and metastasis, and downregulates the proliferation and other functions of TILS [122].

In ovarian cancer, pDC induced IL-10 secreting CD4+ and CD8+ Tregs and enhanced angiogenesis, mediated by the secretion of TNF-α and IL-8. Tumor pDC produced low quantities of IL-6, TNF-α, IFN-α, macrophage inflammatory protein-1β (MIP-1β), and RANTES (CCL5) in response to TLR stimulation, in contrast to pDC from ascites or peripheral blood. In a cohort of 44 ovarian-cancer patients, pDC were the most abundant DC subset in tumor and malignant ascites, but they were almost depleted in peripheral blood. The presence of pDC in the tumor only (but not in ascites) was associated with early relapse [123].

5. Tumor-Associated Neutrophils

Polymorphonuclear neutrophils (PMNS, neutrophils) are of the myeloid lineage of cells and exhibit some of the phenotypes of G-MDSC (CD33+CD66b+). However, transcriptome analysis shows these cell types to be two distinct populations [124]. Neutrophils are a heterogenous group of cells that may be classified into two main functional groups, antitumor (N1) and protumor (N2) [125]. Neutrophils move into tissues from blood under the influence of CXCL1 and CXCL2 and other mediators [126,127].

The role of tumor-associated neutrophils (TANS) in the ovarian TME is not yet fully elucidated. Recent investigations showed that coincubation of ovarian-cancer SKOV3 cells with either PMNS or PMN lysate changed the polygonal epithelial phenotype of the cells to a spindle shape, causing a cribform cell growth. This PMN-induced alteration was due to elastase, a prominent protease of PMN. PMN elastase induced changes in cells were consistent with an EMT process of the cancer cells, and a more migratory phenotype. These authors also studied 213 HGSOC patient samples and showed that PMN are a significant portion of TILS in many patients. Some biopsies showed a definite clustering of PMN and ZEB1 (EMT transcription factor)-positive cancer cells, especially in areas of low E-cadherin [128]. Transition from an epithelial to mesenchymal profile is characteristic of a more aggressive nature in cancers.

In the TME, TGF-β appears to be the predominant soluble molecule responsible for tumor associated neutrophil (TAN) polarization, and inhibition of this molecule favors the accumulation of N1 TANS [125,129]. Neutrophils have a prime role in the initiation of tumors as they act to alter the ECM and the TME. MMP-9 secreted by neutrophils is a key upregulator of carcinogenesis [130]. Additional destructive roles of neutrophils in the TME, such as contributing to angiogenesis, extravasation, and metastases, and suppression of the adaptive immune response are well-reported, as summarized elsewhere [131–133].

N1 TANS exhibit protection against tumor development through several mechanisms. They may directly kill tumor cells, or they can promote CD8+ T cell recruitment and activation by elaborating T cell-attracting chemokines such as CXCL9 and CXCL10, and Th1 cytokines such as IL-12 [134–137]. Several other mechanisms whereby N1 TANS potentiate antitumor immunity have also been reported [133,138].

The prognostic value of neutrophils in ovarian cancer is further underscored by the findings of a recent meta-analysis study, which showed that a high neutrophil-to-lymphocyte ratio (NLR) was associated with worse overall survival (O/S) in some groups of patients (Asians, but not in Caucasians) [139].

6. Natural Killer Cells

Natural killer (NK) cells are an integral part of the innate immune system. These cells do not rely on HLA-mediated recognition of tumor targets, rather, the CD16 receptor, the NKG2D receptor and the NKp30 cytotoxicity receptor on NK cells mediate the death of tumor cells. CD56 high CD16- NK cells have low cytotoxic potential, whereas CD56 low CD16+ NK cells are more efficient at killing tumor cells. In ovarian cancer, there may be defects in NK-cell function such as aberrant receptor or ligand expression, fewer NK cells, or inability of these cells to effectively secrete cytotoxicity molecules or cytokines, which are all possible mechanisms of immune escape [58]. For example, cancer cells from ovarian cancer ascites fluid release macrophage migration inhibitory factor (MIF), a chemokine that stimulates tumor-cell proliferation, migration, and metastasis. MIF transcriptionally downregulates NKG2D in NK cells and lowers the ability of these cells to kill tumor cells [140]. Additionally, high expression of soluble B7-H6 (a ligand for the NKp30 receptor) was associated with lowered NKp30 expression on NK cells and reduced NK-cell activity [141]. It has also been reported that lower B7-H6 expression correlates with reduced metastasis and disease progression, and better overall survival in ovarian cancer [142].

In the presence of IL-18, NK cells can release chemokines CCL3 and CCL4, which attract immature DC. Efficient NK–DC interaction in the tumor can lead to increase of CXCR3 and CCR5 on DC, which can recruit CD8+ effector T cells to tumors, in the presence of chemokines CXCL9, CXCL10, and CCL5 [143]. Gene-analysis expression from the Immunological Genome Project showed that NK cells can secrete CCL5, CCL3, XCL1, CXCL1, CCL4, and CCL27A [144,145]. In tumors, NK cells were strong inducers of conventional DC chemoattractants XCL1 and CCL5. Tumor production of PGE2 could disrupt this process and the ability of DC to secrete chemokines [145]. Taken together, NK cells can directly regulate tumor-cell numbers through cytotoxic mechanisms, or NK cells can potentiate the efficacy of antitumor T cell responses through adaptive immune mechanisms.

Investigations have been conducted using an IL-15 superagonist complex, IL-15N72D/IL-15Rα-Fc (ALT-803; Altor Bioscience Corporation, FL, USA), which inhibits complement activation, and includes the addition of a domain to mediate IL-15/IL-15Rα transpresentation to NK cells. In this study, NOD/SCID/γc−/− (NSG, which do not contain NK/NKT/γδ T/B cells) mice were xenografted with firefly luciferase-expressing MA148 tumor cells, and sublethally irradiated. Mice were then administered overnight activated human NK cells, followed by ALT-803, and analyzed for tumor cells at different time points. When mice were euthanized, a peritoneal lavage was performed and NK-cell function evaluated [146].

Mice treated with ALT-803 resulted in an NK-dependent significant decrease in tumor. ALT-803 also enhanced the cytotoxic function (as measured by increases in CD107a, IFN-γ, and TNF-α) of NK cells from PBMC or ascites, when coincubated with ovarian-cancer cell lines [146]. Targeting of NK cells in a clinical setting may be a promising therapy strategy in HGSOC.

7. Other Components of the Ovarian TME

7.1. TME Architecture

An underlying factor in metastasis involves the attachment of ovarian-cancer cells in ascites to areas of the abdomen. The mesothelium, the squamous epithelium that covers organs of the peritoneal cavity, consists of a single layer of mesothelial cells, below which is a basement membrane of collagen, fibronectin, and laminin, components of the ECM. Some studies showed that cancer cells from ascites preferentially attach to the basement membrane rather than to mesothelial cells [147], suggesting that this mesothelial layer may be a limited frontline defence against ovarian-cancer progression. However, it is also known that ovarian-cancer cells also directly attach to mesothelial cells via β1 integrin and CD44 [71,148–150]. During this process, ovarian-cancer cells upregulate mesenchymal genes such as *TWIST1* and *ZEB1* [149], and decrease the expression of genes such as *CDH1*, an epithelial gene for E-cadherin [71]. There are several other processes whereby ovarian-cancer cells may invade the mesothelial cell layer, such as by actively killing mesothelial cells. In colon-cancer cells for example, a Fas (expressed on mesothelial cell)- Fas ligand (expressed on cancer cells) mediated mechanism of killing mesothelial cells has been described [150].

As earlier addressed, TAMS also play a central role in altering the ECM, thereby contributing to the adhesion, invasion, and proliferation of ovarian-cancer cells. Additionally, adipocytes of the omentum contribute to a protumor TME by secreting IL-6, IL-8, CCL2, and adiponectin, which support ovarian-cancer cell metastasis [151].

Cancer-associated fibroblasts (CAFs) contribute to excessive deposition and alteration of the ECM, creating a barrier that blocks efficient delivery of anticancer drugs and enhancing chemoresistance [152]. CAFs also secrete a range of protumor molecules that create an immunosuppressive milieu in the ovarian TME, and support the proliferation, invasion, and migration of cancer cells [153–157]. In an epithelial ovarian-cancer (EOC) xenograft model, human bone-marrow mesenchymal stem cells were shown to give rise to CAFs that produced IL-6 to enhance tumor growth [158].

7.2. Exosomal Vesicles (EVs)

These vesicles are released by tumor cells and most other cells types of the TME [159,160]. They mediate the transfer of proteins, lipids, and nucleic acids such as DNAs, mRNAs, and miRNAs between tumor and stroma [161]. EVs range from 30 to 150 nm, whereas microvesicular bodies (MVBs) are 100 nm to 1 μm [162]. EVs carry molecules such as CD24, and epithelial cell adhesion molecule (EPCAM1), which directly regulate cancer-cell migration, proteases (MMP2, MMP9), which promote ECM degradation and cancer invasiveness [160,163,164], or EV-associated mRNAs, such as miR21, which may induce resistance to paclitaxel [163,165,166].

8. Interactive Communication in the TME

Characteristics of HGSOC are aggressive growth and recurrence of tumors within the peritoneal cavity as well as metastasis to other sites. Novel therapy to manage ovarian cancer is tailored to overcome immune suppressive mechanisms in the TME that contribute to reduced immune surveillance and immune evasion by tumor cells. Since the TME in each HGSOC patient is both heterogenous and unique [167], there is the need for a better understanding of the contribution of the TME to disease outcome, and more adequate tools to evaluate patients in this present era of personalized therapy.

Blank and colleagues [168] proposed an immunogram model, consisting of seven parameters, which describes interactions between cancers and the immune system that may occur in individual patients. In this framework, the assumption is that T cell activity is the ultimate effector mechanism in therapy response, and that even though other cells, or other factors such as modulation of the microbiome, may contribute to outcome, the contribution to disease improvement will ultimately be mediated by enhanced T cell activity. In some patients, overcoming T cell inhibition may be the only factor that needs to be addressed for disease improvement. The parameters addressed in this immunogram model, as briefly outlined below, are also helpful for understanding the interactions between other solid cancers and the immune system.

Tumor foreignness: for example, it is reported that the outcome to anti-CTLA-4 blockade therapy correlates with increased tumor mutational burden (a measure of neoantigen load) [169].

General Immune status: this may include a study of changes in immune cells in peripheral blood [170].

Immune cell infiltration: chemokines CXCL9 and CXCL10 that recruit CD8+ effector T cells are part of a gene signature associated with improved outcome to PD-1 blockade [18,171,172].

Checkpoint molecules: molecules such as PD-1 and PD-L1 on tumor cells or immune cells present potent immunosupression in TMEs [173–175].

Soluble inhibitors: IDO, a soluble molecule produced by TAMS or pDC, interferes with anti-CTLA-4 antibody efficacy in mice [176].

Absence of inhibitory tumor metabolism: high serum lactate dehydrogenase concentrations correlate with poor outcome to anti-CTLA-4 and anti-PD-1 antibody immunotherapy [177].

Tumor sensitivity to immune effectors: Tumor cells have developed several immune evasion mechanisms, such as inactivation of antigen-presentation machinery [102]. Additionally, by epigenetic post-translational mechanisms, the TME can select for cancer cells that can downregulate the expression of some tumor antigens, which would normally be recognized by T cells [178].

Other factors in the TME that regulate communication between cancers and the immune system include many of the parameters outlined in the preceding text, such as the maturation level and function of DC, the density and immunosuppressive nature of TAMS, NK-cell activity, and the TME architecture (Figure 1).

9. Conclusions and Perspectives

A better understanding of the TME in HGSOC will reveal useful diagnostic and prognostic biomarkers, and advance the development of suitable bioassays for routine clinical use for the detection and diagnosis of this malignancy. With such a heterogenous disease and multiple immune and biochemical networks, success in diagnosing this disease and predicting outcome will require multiple biomarkers, and more sensitive and precise methods of imaging to detect early lesions.

Current tools used to study the TME involve the use of genomics to investigate gene-expression signatures in the tumors of HGSOC. Verkaak and colleagues described four different gene classifications in a study of ovarian tumors as differentiated, immunoreactive, mesenchymal, and proliferative [179]. By IHC, the immunoreactive group had increased T lymphocytes, whereas desmoplasia associated with infiltrating stromal cells was in the mesenchymal group. Patients in the immunoreactive group had the best survival outcome. Some tumors also exhibited more than one of the 4 gene clusters.

Findings were validated on an independent dataset of 879 HGSOC-expression profiles. Additional information to survival outcome and platinum resistance rates was obtained by using survival-outcome prediction models for association with BRCA1/2 mutation status, residual disease after surgery and stage of disease [179]. Similar gene-classification models may be useful for the selection of patients for targeted or immunotherapy, or to predict patient outcome. It is likely that patients exhibiting mesenchymal signatures may respond better to treatments such as angiogenesis inhibitors.

Additional methods to study the HGSOC TME include combinations of proteomic and other genomic data output [180,181] and a study approach addressing multiple parameters (such as gene expression, matrix proteomics, cytokine and chemokine expression, ECM parameters, and biomechanical properties) on a single biopsy sample for a better understanding of the events occurring in tumor tissue [182]. Other novel tools to study the ovarian TME include the use of artificial microenvironments to monitor ovarian-cancer progressiveness [183].

The HGSOC TME is a complex and dynamic interactive entity, which may vary between the primary disease and at the time of recurrence, and in the quest for more effective therapy design one needs to take into account pre-existing immunosuppression, as well as emerging resistance mechanisms with therapy [184]. Attempts to manage ovarian cancer with immunotherapy has not been as successful as for some other cancers [174,185]. We are hopeful that combining immunotherapy, such as PD-1 blockade, with other checkpoint inhibitory molecules (such as anti-CTLA-4, anti-TIM-3, anti-LAG-3), PARP inhibitors, kinase inhibitors, chemotherapeutics [186], dendritic-cell vaccines, CAR T cell therapy [187,188], or other treatments, will prove to be successful measures to overcome the multiple immunosuppressive mechanisms in the TME. As a cautionary measure, combination therapy will require optimizing doses and schedules of regimens, while limiting adverse effects. However, we anticipate that a combined therapy approach will be the way forward, towards providing effective therapy for improved survival, and ultimately a cure for HGSOC.

Author Contributions: M.L.D. wrote the original draft and reviewed the manuscript. P.J.S. reviewed the manuscript.

Funding: This research received no external funding.

Acknowledgments: This work was supported by the Cronk Foundation funds to M.L.D. and the Perritt Charitable Foundation funds to P.J.S. We are also grateful for additional support from the Leo and Anne Albert Charitable Trust, all of which made this work possible.

Conflicts of Interest: The authors declare no conflict of interest.

References

1. Colombo, N.; Peiretti, M.; Parma, G.; Lapresa, M.; Mancari, R.; Carinelli, S.; Sessa, C.; Castiglione, M. Newly Diagnosed and Relapsed Epithelial Ovarian Carcinoma: ESMO Clinical Practice Guidelines for Diagnosis, Treatment and Follow-Up. *Ann. Oncol.* **2010**, *21*, v23–v30. [CrossRef] [PubMed]
2. Pogge von Strandmann, E.; Reinartz, S.; Wager, U.; Müller, R. Tumor–Host Cell Interactions in Ovarian Cancer: Pathways to Therapy Failure. *Trends Cancer* **2017**, *3*, 137–148. [CrossRef] [PubMed]
3. Siegel, R.L.; Miller, K.D.; Jemal, A. Cancer Statistics, 2018. *CA Cancer J. Clin.* **2018**, *68*, 7–30. [CrossRef] [PubMed]
4. Perets, R.; Wyant, G.; Muto, K.; Bijron, J.; Poole, B.; Chin, K.; Chen, J.; Ohman, A.; Stepule, C.; Kwak, S.; et al. Transformation of the Fallopian Tube Secretory Epithelium Leads to High-Grade Serous Ovarian Cancer in Brca;Tp53;Pten Models. *Cancer Cell* **2013**, *24*, 751–765. [CrossRef] [PubMed]
5. Cole, A.J.; Dwight, T.; Gill, A.J.; Dickson, K.A.; Zhu, Y.; Clarkson, A.; Gard, G.B.; Maidens, J.; Valmadre, S.; Clifton-Bligh, R.; et al. Assessing Mutant p53 in Primary High-Grade Serous Ovarian Cancer using Immunohistochemistry and Massively Parallel Sequencing. *Sci. Rep.* **2016**, *6*, 26191. [CrossRef] [PubMed]
6. Bell, D.; Berchuck, A.; Birrer, M.; Chien, J.; Cramer, D.W.; Dao, F.; Dhir, R.; Disaia, P.; Gabra, H.; Glenn, P.; et al. Integrated Genomic Analyses of Ovarian Carcinoma. *Nature* **2011**, *474*, 609–615. [CrossRef] [PubMed]
7. Meza-Perez, S.; Randall, T.D. Immunological Functions of the Omentum. *Trends Immunol.* **2017**, *38*, 526–536. [CrossRef] [PubMed]

8. Lengyel, E. Ovarian Cancer Development and Metastasis. *Am. J. Pathol.* **2010**, *177*, 1053–1064. [CrossRef] [PubMed]
9. Steinkamp, M.P.; Winner, K.K.; Davies, S.; Muller, C.; Zhang, Y.; Hoffman, R.M.; Shirinifard, A.; Moses, M.; Jiang, Y.; Wilson, B.S. Ovarian Tumor Attachment, Invasion, and Vascularization Reflect Unique Microenvironments in the Peritoneum: Insights from Xenograft and Mathematical Models. *Front. Oncol.* **2013**, *3*, 97. [CrossRef] [PubMed]
10. Kulbe, H.; Chakravarty, P.; Leinster, D.A.; Charles, K.A.; Kwong, J.; Thompson, R.G.; Coward, J.I.; Schioppa, T.; Robinson, S.C.; Gallagher, W.M.; et al. A Dynamic Inflammatory Cytokine Network in the Human Ovarian Cancer Microenvironment. *Cancer Res.* **2012**, *72*, 66–75. [CrossRef] [PubMed]
11. Reinartz, S.; Schumann, T.; Finkernagel, F.; Wortmann, A.; Jansen, J.M.; Meissner, W.; Krause, M.; Schwörer, A.M.; Wagner, U.; Müller-Brüsselbach, S.; et al. Mixed-Polarization Phenotype of Ascites-Associated Macrophages in Human Ovarian Carcinoma: Correlation of CD163 Expression, Cytokine Levels and Early Relapse. *Int. J. Cancer* **2014**, *134*, 32–42. [CrossRef] [PubMed]
12. Fridman, W.H.; Pagès, F.; Sautès-Fridman, C.; Galon, J. The Immune Contexture in Human Tumours: Impact on Clinical Outcome. *Nat. Rev. Cancer* **2012**, *12*, 298–306. [CrossRef] [PubMed]
13. Hwang, W.T.; Adams, S.F.; Tahirovic, E.; Hagemann, I.S.; Coukos, G. Prognostic Significance of Tumor-Infiltrating T Cells in Ovarian Cancer: A Meta-Analysis. *Gynecol. Oncol.* **2012**, *124*, 192–198. [CrossRef] [PubMed]
14. Santoiemma, P.P.; Powell, D.J. Tumor Infiltrating Lymphocytes in Ovarian Cancer. *Cancer Biol. Ther.* **2015**, *16*, 807–820. [CrossRef] [PubMed]
15. Zhang, L.; Conejo-Garcia, J.; Katsaros, D.; Gimotty, P.A.; Massobrio, M.; Regnani, G.; Makrigiannakis, A.; Gray, H.; Schlienger, K.; Liebman, M.N.; et al. Intratumoral T Cells, Recurrence, and Survival in Epithelial Ovarian Cancer. *N. Engl. J. Med.* **2003**, *348*, 203–213. [CrossRef] [PubMed]
16. Hamanishi, J.; Mandai, M.; Abiko, K.; Matsumura, N.; Baba, T.; Yoshioka, Y.; Kosaka, K.; Konishi, I. The Comprehensive Assessment of Local Immune Status of Ovarian Cancer by the Clustering of Multiple Immune Factors. *Clin. Immunol.* **2011**, *141*, 338–347. [CrossRef] [PubMed]
17. Sato, E.; Olson, S.H.; Ahn, J.; Bundy, B.; Nishikawa, H.; Qian, F.; Jungbluth, A.A.; Frosina, D.; Gnjatic, S.; Ambrosone, C.; et al. Intraepithelial CD8+ Tumor-Infiltrating Lymphocytes and a High CD8+/Regulatory T Cell Ratio are Associated with Favorable Prognosis in Ovarian Cancer. *Proc. Natl. Acad. Sci. USA* **2005**, *102*, 18538–18543. [CrossRef] [PubMed]
18. Ovarian Tumor Tissue Analysis (OTTA) Consortium; Goode, E.L.; Block, M.S.; Kalli, K.R.; Vierkant, R.A.; Chen, W.; Fogarty, Z.C.; Gentry-Maharaj, A.; Toloczko, A.; Hein, A.; et al. Dose-Response Association of CD8+ Tumor-Infiltrating Lymphocytes and Survival Time in High-Grade Serous Ovarian Cancer. *JAMA Oncol.* **2017**, *3*, e173290.
19. Komdeur, F.L.; Wouters, M.C.A.; Workel, H.H.; Tijans, A.M.; Terwindt, A.L.J.; Brunekreeft, K.L.; Plat, A.; Klip, H.G.; Eggink, F.A.; Leffers, N.; et al. CD103+ Intraepithelial T Cells in High-Grade Serous Ovarian Cancer are Phenotypically Diverse TCRaß+ CD8aß+ T Cells that can be Targeted for Cancer Immunotherapy. *Oncotarget* **2016**, *7*, 75130–75144. [CrossRef] [PubMed]
20. Webb, J.R.; Milne, K.; Watson, P.; Deleeuw, R.J.; Nelson, B.H. Tumor-Infiltrating Lymphocytes Expressing the Tissue Resident Memory Marker CD103 are Associated with Increased Survival in High-Grade Serous Ovarian Cancer. *Clin. Cancer Res.* **2014**, *20*, 434–444. [CrossRef] [PubMed]
21. Manrique, S.Z.; Correa, M.A.; Hoelzinger, D.B.; Dominguez, A.L.; Mirza, N.; Lin, H.H.; Stein-Streilein, J.; Gordon, S.; Lustgarten, J. Foxp3-Positive Macrophages Display Immunosuppressive Properties and Promote Tumor Growth. *J. Exp. Med.* **2011**, *208*, 1485–1499. [CrossRef] [PubMed]
22. Curiel, T.J.; Coukos, G.; Zou, L.; Alvarez, X.; Cheng, P.; Mottram, P.; Evdemon-Hogan, M.; Conejo-Garcia, J.; Zhang, L.; Burow, M.; et al. Specific Recruitment of Regulatory T Cells in Ovarian Carcinoma Fosters Immune Privilege and Predicts Reduced Survival. *Nat. Med.* **2004**, *10*, 942–949. [CrossRef] [PubMed]
23. Wolf, D.; Wolf, A.M.; Rumpold, H.; Fiegl, H.; Zeimet, A.G.; Muller-Holzner, E.; Deibl, M.; Gastl, G.; Gunsilius, E.; Marth, C. The Expression of the Regulatory T Cell-Specific Forkhead Box Transcription Factor FoxP3 is Associated with Poor Prognosis in Ovarian Cancer. *Clin. Cancer Res.* **2005**, *11*, 8326–8331. [CrossRef] [PubMed]
24. Shang, B.; Liu, Y.; Jiang, S.J.; Liu, Y. PMC4604472; Prognostic Value of Tumor-Infiltrating FoxP3+ Regulatory T Cells in Cancers: A Systematic Review and Meta-Analysis. *Sci. Rep.* **2015**, *5*, 1–9. [CrossRef] [PubMed]

25. Filaci, G.; Fenoglio, D.; Fravega, M.; Ansaldo, G.; Borgonovo, G.; Traverso, P.; Villaggio, B.; Ferrera, A.; Kunkl, A.; Rizzi, M.; et al. CD8+ CD28- T Regulatory Lymphocytes Inhibiting T Cell Proliferative and Cytotoxic Functions Infiltrate Human Cancers. *J. Immunol.* **2007**, *179*, 4323–4334. [CrossRef]
26. Tanaka, A.; Sakaguchi, S. Regulatory T Cells in Cancer Immunotherapy. *Cell Res.* **2017**, *27*, 109–118. [CrossRef] [PubMed]
27. Wu, M.; Chen, X.; Lou, J.; Zhang, S.; Zhang, X.; Huang, L.; Sun, R.; Huang, P.; Wang, F.; Pan, S. TGF-Beta1 Contributes to CD8+ Treg Induction through p38 MAPK Signaling in Ovarian Cancer Microenvironment. *Oncotarget* **2016**, *7*, 44534–44544. [PubMed]
28. Wang, R.F. CD8+ Regulatory T Cells, their Suppressive Mechanisms, and Regulation in Cancer. *Hum. Immunol.* **2008**, *69*, 811–814. [CrossRef] [PubMed]
29. Alsaab, H.O.; Sau, S.; Alzhrani, R.; Tatiparti, K.; Bhise, K.; Kashaw, S.K.; Iyer, A.K. PD-1 and PD-L1 Checkpoint Signaling Inhibition for Cancer Immunotherapy: Mechanism, Combinations, and Clinical Outcome. *Front. Pharmacol.* **2017**, *8*, 561. [CrossRef] [PubMed]
30. Fujio, K.; Yamamoto, K.; Okamura, T. Overview of LAG-3-Expressing, IL-10-Producing Regulatory T Cells. *Curr. Top. Microbiol. Immunol.* **2017**, *410*, 29–45. [PubMed]
31. Huang, R.Y.; Eppolito, C.; Lele, S.; Shrikant, P.; Matsuzaki, J.; Odunsi, K. LAG3 and PD1 Co-Inhibitory Molecules Collaborate to Limit CD8+ T Cell Signaling and Dampen Antitumor Immunity in a Murine Ovarian Cancer Model. *Oncotarget* **2015**, *6*, 27359–27377. [CrossRef] [PubMed]
32. Zeng, Y.; Zhang, Q.; Wang, H.; Lu, M.; Kong, H.; Zhang, Y.; Shi, H. Prognostic Significance of Interleukin-17 in Solid Tumors: A Meta-Analysis. *Int. J. Clin. Exp. Med.* **2015**, *8*, 10515–10536. [PubMed]
33. Fialova, A.; Partlova, S.; Sojka, L.; Hromadkova, H.; Brtnicky, T.; Fucikova, J.; Kocian, P.; Rob, L.; Bartunkova, J.; Spisek, R. Dynamics of T-Cell Infiltration during the Course of Ovarian Cancer: The Gradual Shift from a Th17 Effector Cell Response to a Predominant Infiltration by Regulatory T-Cells. *Int. J. Cancer* **2013**, *132*, 1070–1079. [CrossRef] [PubMed]
34. Burnet, F.M. The Concept of Immunological Surveillance. *Prog. Exp. Tumor Res.* **1969**, *13*, 1–27.
35. Dunn, G.P.; Old, L.J.; Schreiber, R.D. The Immunobiology of Cancer Immunosurveillance and Immunoediting. *Immunity* **2004**, *21*, 137–148. [CrossRef] [PubMed]
36. Ehrlich, P. Ueber Den Jetzigen. *Stand der Karzinomforschung Ned Tijdschr Geneeskd* **1909**, *5*, 273–290.
37. Hamai, A.; Benlalam, H.; Meslin, F.; Hasmim, M.; Carre, T.; Akalay, I.; Janji, B.; Berchem, G.; Noman, M.Z.; Chouaib, S. Immune Surveillance of Human Cancer: If the Cytotoxic T-lymphocytes Play the Music, does the Tumoral System Call the Tune? *Tissue Antigens* **2010**, *75*, 1–8. [CrossRef] [PubMed]
38. Kaech, S.M.; Cui, W. Transcriptional Control of Effector and Memory CD8+ T Cell Differentiation. *Nat. Rev. Immunol.* **2012**, *12*, 749–761. [CrossRef] [PubMed]
39. Lynch, E.A.; Heijens, C.A.W.; Horst, N.F.; Center, D.M.; Cruikshank, W.W. Cutting Edge: IL-16/CD4 Preferentially Induces Th1 Cell Migration: Requirement of CCR5. *J. Immunol.* **2003**, *171*, 4965–4968. [CrossRef] [PubMed]
40. Mariathasan, S.; Turley, S.J.; Nickles, D.; Castiglioni, A.; Yuen, K.; Wang, Y.; Kadel, E.E., III; Koeppen, H.; Astarita, J.L.; Cubas, R.; et al. TGFbeta Attenuates Tumour Response to PD-L1 Blockade by Contributing to Exclusion of T Cells. *Nature* **2018**, *554*, 544–548. [CrossRef] [PubMed]
41. Pepper, M.S. Transforming Growth Factor-Beta: Vasculogenesis, Angiogenesis, and Vessel Wall Integrity. *Cytokine Growth Factor Rev.* **1997**, *8*, 21–43. [CrossRef]
42. Kassim, S.K.; El-Salahy, E.M.; Fayed, S.T.; Helal, S.A.; Helal, T.; Azzam, E.E.; Khalifa, A. Vascular Endothelial Growth Factor and Interleukin-8 are Associated with Poor Prognosis in Epithelial Ovarian Cancer Patients. *Clin. Biochem.* **2004**, *37*, 363–369. [CrossRef] [PubMed]
43. Bouzin, C.; Brouet, A.; De Vriese, J.; Dewever, J.; Feron, O. Effects of Vascular Endothelial Growth Factor on the Lymphocyte-Endothelium Interactions: Identification of Caveolin-1 and Nitric Oxide as Control Points of Endothelial Cell Anergy. *J. Immunol.* **2007**, *178*, 1505–1511. [CrossRef] [PubMed]
44. Motz, G.T.; Santoro, S.P.; Wang, L.P.; Garrabrant, T.; Lastra, R.R.; Hagemann, I.S.; Lal, P.; Feldman, M.D.; Benencia, F.; Coukos, G. Tumor Endothelium FasL Establishes a Selective Immune Barrier Promoting Tolerance in Tumors. *Nat. Med.* **2014**, *20*, 607–615. [CrossRef] [PubMed]

45. Hamanishi, J.; Mandai, M.; Iwasaki, M.; Okazaki, T.; Tanaka, Y.; Yamaguchi, K.; Higuchi, T.; Yagi, H.; Takakura, K.; Minato, N.; et al. Programmed Cell Death 1 Ligand 1 and Tumor-Infiltrating CD8+ T Lymphocytes are Prognostic Factors of Human Ovarian Cancer. *Proc. Natl. Acad. Sci. USA* **2007**, *104*, 3360–3365. [CrossRef] [PubMed]
46. Gordon, S.R.; Maute, R.L.; Dulken, B.W.; Hutter, G.; George, B.M.; McCracken, M.N.; Gupta, R.; Tsai, J.M.; Sinha, R.; Corey, D.; et al. PD-1 Expression by Tumour-Associated Macrophages Inhibits Phagocytosis and Tumour Immunity. *Nature* **2017**, *545*, 495–499. [CrossRef] [PubMed]
47. Inaba, T.; Ino, K.; Kajiyama, H.; Yamamoto, E.; Shibata, K.; Nawa, A.; Nagasaka, T.; Akimoto, H.; Takikawa, O.; Kikkawa, F. Role of the Immunosuppressive Enzyme Indoleamine 2,3-Dioxygenase in the Progression of Ovarian Carcinoma. *Gynecol. Oncol.* **2009**, *115*, 185–192. [CrossRef] [PubMed]
48. Ino, K. Indoleamine 2,3-Dioxygenase and Immune Tolerance in Ovarian Cancer. *Curr. Opin. Obstet. Gynecol.* **2011**, *23*, 13–18. [CrossRef] [PubMed]
49. Cai, D.L.; Jin, L. Immune Cell Population in Ovarian Tumor Microenvironment. *J. Cancer* **2017**, *8*, 2915–2923. [CrossRef] [PubMed]
50. Ros, X.R.; Vermeulen, L. Turning Cold Tumors Hot by Blocking TGF-Beta. *Trends Cancer* **2018**, *4*, 335–337. [CrossRef] [PubMed]
51. Haanen John, B.A.G. Converting Cold into Hot Tumors by Combining Immunotherapies. *Cell* **2017**, *170*, 1055–1056. [CrossRef] [PubMed]
52. Colvin, E.K. Tumor-Associated Macrophages Contribute to Tumor Progression in Ovarian Cancer. *Front. Oncol.* **2014**, *4*, 137. [CrossRef] [PubMed]
53. Yuan, X.; Zhang, J.; Li, D.; Mao, Y.; Mo, F.; Du, W.; Ma, X. Prognostic Significance of Tumor-Associated Macrophages in Ovarian Cancer: A Meta-Analysis. *Gynecol. Oncol.* **2017**, *147*, 181–187. [CrossRef] [PubMed]
54. Elliott, L.A.; Doherty, G.A.; Sheahan, K.; Ryan, E.J. Human Tumor-Infiltrating Myeloid Cells: Phenotypic and Functional Diversity. *Front. Immunol.* **2017**, *8*, 86. [CrossRef] [PubMed]
55. Barros, M.H.M.; Hauck, F.; Dreyer, J.H.; Kempkes, B.; Niedobitek, G. Macrophage Polarisation: An Immunohistochemical Approach for Identifying M1 and M2 Macrophages. *PLoS ONE* **2013**, *8*, e80908. [CrossRef] [PubMed]
56. Okla, K.; Wertel, I.; Polak, G.; Surowka, J.; Wawruszak, A.; Kotarski, J. Tumor-Associated Macrophages and Myeloid-Derived Suppressor Cells as Immunosuppressive Mechanism in Ovarian Cancer Patients: Progress and Challenges. *Int. Rev. Immunol.* **2016**, *35*, 372–385. [CrossRef] [PubMed]
57. Gabrilovich, D.I.; Nagaraj, S. Myeloid-Derived Suppressor Cells as Regulators of the Immune System. *Nat. Rev. Immunol.* **2009**, *9*, 162–174. [CrossRef] [PubMed]
58. Worzfeld, T.; von Strandmann, E.P.; Huber, M.; Adhikary, T.; Wagner, U.; Reinartz, S.; Müller, R. The Unique Molecular and Cellular Microenvironment of Ovarian Cancer. *Front. Oncol.* **2017**, *7*, 24. [CrossRef] [PubMed]
59. Liu, Y.; Cao, X. The Origin and Function of Tumor-Associated Macrophages. *Cell. Mol. Immunol.* **2015**, *12*, 1–4. [CrossRef] [PubMed]
60. Franklin, R.A.; Liao, W.; Sarkar, A.; Kim, M.V.; Bivona, M.R.; Liu, K.; Pamer, E.G.; Li, M.O. The Cellular and Molecular Origin of Tumor-Associated Macrophages. *Science* **2014**, *344*, 921–925. [CrossRef] [PubMed]
61. Finkernagel, F.; Reinartz, S.; Lieber, S.; Adhikary, T.; Wortmann, A.; Hoffmann, N.; Bieringer, T.; Nist, A.; Stiewe, T.; Jansen, J.M.; et al. The Transcriptional Signature of Human Ovarian Carcinoma Macrophages is Associated with Extracellular Matrix Reorganization. *Oncotarget* **2016**, *7*, 75339–75352. [CrossRef] [PubMed]
62. Yin, M.; Li, X.; Tan, S.; Zhou, H.J.; Ji, W.; Bellone, S.; Xu, X.; Zhang, H.; Santin, A.D.; Lou, G.; et al. Tumor-Associated Macrophages Drive Spheroid Formation during Early Transcoelomic Metastasis of Ovarian Cancer. *J. Clin. Investig.* **2016**, *126*, 4157–4173. [CrossRef] [PubMed]
63. Pollard, J.W. Tumour-Educated Macrophages Promote Tumour Progression and Metastasis. *Nat. Rev. Cancer* **2004**, *4*, 71–78. [CrossRef] [PubMed]
64. Henze, A.T.; Mazzone, M. The Impact of Hypoxia on Tumor-Associated Macrophages. *J. Clin. Investig.* **2016**, *126*, 3672–3679. [CrossRef] [PubMed]
65. Wang, Y.; Ma, J.; Shen, H.; Wang, C.; Sun, Y.; Howell, S.B.; Lin, X. Reactive Oxygen Species Promote Ovarian Cancer Progression Via the HIF-1a/LOX/E-Cadherin Pathway. *Oncol. Rep.* **2014**, *32*, 2150–2158. [CrossRef] [PubMed]
66. Aras, S.; Zaidi, M.R. TAMeless Traitors: Macrophages in Cancer Progression and Metastasis. *Br. J. Cancer* **2017**, *117*, 1583–1591. [CrossRef] [PubMed]

67. Ghoneum, A.; Afify, H.; Salih, Z.; Kelly, M.; Said, N. Role of Tumor Microenvironment in Ovarian Cancer Pathobiology. *Oncotarget* **2018**, *9*, 22832–22849. [CrossRef] [PubMed]
68. Zhou, J.; Jain, S.; Azad, A.K.; Xu, X.; Yu, H.C.; Xu, Z.; Godbout, R.; Fu, Y. Notch and TGFß Form a Positive Regulatory Loop and Regulate EMT in Epithelial Ovarian Cancer Cells. *Cell. Signal.* **2016**, *28*, 838–849. [CrossRef] [PubMed]
69. Straussman, R.; Morikawa, T.; Shee, K.; Barzily-Rokni, M.; Qian, Z.R.; Du, J.; Davis, A.; Mongare, M.M.; Gould, J.; Frederick, D.T.; et al. Tumour Micro-Environment Elicits Innate Resistance to RAF Inhibitors through HGF Secretion. *Nature* **2012**, *487*, 500–504. [CrossRef] [PubMed]
70. Zhang, J.; Yao, H.; Song, G.; Liao, X.; Xian, Y.; Li, W. Regulation of Epithelial-Mesenchymal Transition by Tumor-Associated Macrophages in Cancer. *Am. J. Transl. Res.* **2015**, *7*, 1699–1711. [PubMed]
71. Sawada, K.; Mitra, A.K.; Radjabi, A.R.; Bhaskar, V.; Kistner, E.O.; Tretiakova, M.; Jagadeeswaran, S.; Montag, A.; Becker, A.; Kenny, H.A.; et al. Loss of E-Cadherin Promotes Ovarian Cancer Metastasis Via A5-Integrin, which is a Therapeutic Target. *Cancer Res.* **2008**, *68*, 2329–2339. [CrossRef] [PubMed]
72. Ke, X.; Zhang, S.; Wu, M.; Lou, J.; Zhang, J.; Xu, T.; Huang, L.; Huang, P.; Wang, F.; Pan, S. Tumor-Associated Macrophages Promote Invasion Via Toll-Like Receptors Signaling in Patients with Ovarian Cancer. *Int. Immunopharmacol.* **2016**, *40*, 184–195. [CrossRef] [PubMed]
73. Lane, D.; Matte, I.; Laplante, C.; Garde-Granger, P.; Carignan, A.; Bessette, P.; Rancourt, C.; Piche, A. CCL18 from Ascites Promotes Ovarian Cancer Cell Migration through Proline-Rich Tyrosine Kinase 2 Signaling. *Mol. Cancer* **2016**, *15*, 58. [CrossRef] [PubMed]
74. Duluc, D.; Corvaisier, M.; Blanchard, S.; Catala, L.; Descamps, P.; Gamelin, E.; Ponsoda, S.; Delneste, Y.; Hebbar, M.; Jeannin, P. Interferon-Γ Reverses the Immunosuppressive and Protumoral Properties and Prevents the Generation of Human Tumor-Associated Macrophages. *Int. J. Cancer* **2009**, *125*, 367–373. [CrossRef] [PubMed]
75. Cohen, C.A.; Shea, A.A.; Heffron, C.L.; Schmelz, E.M.; Roberts, P.C. Interleukin-12 Immunomodulation Delays the Onset of Lethal Peritoneal Disease of Ovarian Cancer. *J. Interferon Cytokine Res.* **2016**, *36*, 62–73. [CrossRef] [PubMed]
76. Koneru, M.; Purdon, T.J.; Spriggs, D.; Koneru, S.; Brentjens, R.J. IL-12 Secreting Tumor-Targeted Chimeric Antigen Receptor T Cells Eradicate Ovarian Tumors in Vivo. *OncoImmunology* **2015**, *4*, 1–11. [CrossRef] [PubMed]
77. Lan, C.; Huang, X.; Lin, S.; Huang, H.; Cai, Q.; Wan, T.; Lu, J.; Liu, J. Expression of M2-Polarized Macrophages is Associated with Poor Prognosis for Advanced Epithelial Ovarian Cancer. *Technol. Cancer Res. Treat.* **2013**, *12*, 259–267. [CrossRef] [PubMed]
78. Carroll, M.J.; Kapur, A.; Felder, M.; Patankar, M.S.; Kreeger, P.K. M2 Macrophages Induce Ovarian Cancer Cell Proliferation Via a Heparin Binding Epidermal Growth Factor/Matrix Metalloproteinase 9 Intercellular Feedback Loop. *Oncotarget* **2016**, *7*, 86608–86620. [CrossRef] [PubMed]
79. Zhang, M.; He, Y.; Sun, X.; Li, Q.; Wang, W.; Zhao, A.; Di, W. A High M1/M2 Ratio of Tumor-Associated Macrophages is Associated with Extended Survival in Ovarian Cancer Patients. *J. Ovarian Res.* **2014**, *7*, 19. [CrossRef] [PubMed]
80. Wu, L.; Deng, Z.; Peng, Y.; Han, L.; Liu, J.; Wang, L.; Li, B.; Zhao, J.; Jiao, S.; Wei, H. Ascites-Derived IL-6 and IL-10 Synergistically Expand CD14(+)HLA-DR(-/Low) Myeloid-Derived Suppressor Cells in Ovarian Cancer Patients. *Oncotarget* **2017**, *8*, 76843–76856. [CrossRef] [PubMed]
81. Horikawa, N.; Abiko, K.; Matsumura, N.; Hamanishi, J.; Baba, T.; Yamaguchi, K.; Yoshioka, Y.; Koshiyama, M.; Konishi, I. Expression of Vascular Endothelial Growth Factor in Ovarian Cancer Inhibits Tumor Immunity through the Accumulation of Myeloid-Derived Suppressor Cells. *Clin. Cancer Res.* **2017**, *23*, 587–599. [CrossRef] [PubMed]
82. Cui, T.X.; Kryczek, I.; Zhao, L.; Zhao, E.; Kuick, R.; Roh, M.H.; Vatan, L.; Szeliga, W.; Mao, Y.; Thomas, D.G.; et al. Myeloid-Derived Suppressor Cells Enhance Stemness of Cancer Cells by Inducing microRNA101 and Suppressing the Corepressor CTBP2. *Immunity* **2013**, *39*, 611–621. [CrossRef] [PubMed]
83. Obermajer, N.; Muthuswamy, R.; Odunsi, K.; Edwards, R.P.; Kalinski, P. PGE 2-Induced CXCL 12 Production and CXCR4 Expression Controls the Accumulation of Human MDSCs in Ovarian Cancer Environment. *Cancer Res.* **2011**, *71*, 7463–7470. [CrossRef] [PubMed]

84. Kajiyama, H.; Shibata, K.; Terauchi, M.; Ino, K.; Nawa, A.; Kikkawa, F. Involvement of SDF-1a/CXCR4 Axis in the Enhanced Peritoneal Metastasis of Epithelial Ovarian Carcinoma. *Int. J. Cancer* **2008**, *122*, 91–99. [CrossRef] [PubMed]
85. Liu, M.; Matsumura, N.; Mandai, M.; Li, K.; Yagi, H.; Baba, T.; Suzuki, A.; Hamanishi, J.; Fukuhara, K.; Konishi, I. Classification using Hierarchical Clustering of Tumor-Infiltrating Immune Cells Identifies Poor Prognostic Ovarian Cancers with High Levels of COX Expression. *Mod. Pathol.* **2009**, *22*, 873–880. [CrossRef] [PubMed]
86. Righi, E.; Kashiwagi, S.; Yuan, J.; Santosuosso, M.; Leblanc, P.; Ingraham, R.; Forbes, B.; Edelblute, B.; Collette, B.; Xing, D.; et al. CXCL12/CXCR4 Blockade Induces Multimodal Antitumor Effects that Prolong Survival in an Immunocompetent Mouse Model of Ovarian Cancer. *Cancer Res.* **2011**, *71*, 5522–5534. [CrossRef] [PubMed]
87. Obermajer, N.; Muthuswamy, R.; Lesnock, J.; Edwards, R.P.; Kalinski, P. Positive Feedback between PGE2 and COX2 Redirects the Differentiation of Human Dendritic Cells Toward Stable Myeloid-Derived Suppressor Cells. *Blood* **2011**, *118*, 5498–5505. [CrossRef] [PubMed]
88. Vasquez-Dunddel, D.; Pan, F.; Zeng, Q.; Gorbounov, M.; Albesiano, E.; Fu, J.; Blosser, R.L.; Tam, A.J.; Bruno, T.; Zhang, H.; et al. STAT3 Regulates Arginase-I in Myeloid-Derived Suppressor Cells from Cancer Patients. *J. Clin. Investig.* **2013**, *123*, 1580–1589. [CrossRef] [PubMed]
89. Khan, A.N.H.; Kolomeyevskaya, N.; Singel, K.L.; Grimm, M.J.; Moysich, K.B.; Daudi, S.; Grzankowski, K.S.; Lele, S.; Ylagan, L.; Webster, G.A.; et al. Targeting Myeloid Cells in the Tumor Microenvironment Enhances Vaccine Efficacy in Murine Epithelial Ovarian Cancer. *Oncotarget* **2015**, *6*, 11310–11326. [CrossRef] [PubMed]
90. Pulaski, H.L.; Spahlinger, G.; Silva, I.A.; McLean, K.; Kueck, A.S.; Reynolds, R.K.; Coukos, G.; Conejo-Garcia, J.R.; Buckanovich, R.J. Identifying Alemtuzumab as an Anti-Myeloid Cell Antiangiogenic Therapy for the Treatment of Ovarian Cancer. *J. Transl. Med.* **2009**, *7*, 49. [CrossRef] [PubMed]
91. Moughon, D.L.; He, H.; Schokrpur, S.; Jiang, Z.K.; Yaqoob, M.; David, J.; Lin, C.; Iruela-Arispe, M.L.; Dorigo, O.; Wu, L. Macrophage Blockade using CSF1R Inhibitors Reverses the Vascular Leakage Underlying Malignant Ascites in Late-Stage Epithelial Ovarian Cancer. *Cancer Res.* **2015**, *75*, 4742–4752. [CrossRef] [PubMed]
92. Brana, I.; Calles, A.; LoRusso, P.M.; Yee, L.K.; Puchalski, T.A.; Seetharam, S.; Zhong, B.; de Boer, C.J.; Tabernero, J.; Calvo, E. Carlumab, an Anti-C-C Chemokine Ligand 2 Monoclonal Antibody, in Combination with Four Chemotherapy Regimens for the Treatment of Patients with Solid Tumors: An Open-Label, Multicenter Phase 1b Study. *Target. Oncol.* **2015**, *10*, 111–123. [CrossRef] [PubMed]
93. Sawa-Wejksza, K.; Kandefer-Szerszen, M. Tumor-Associated Macrophages as Target for Antitumor Therapy. *Arch. Immunol. Ther. Exp.* **2018**, *66*, 97–111. [CrossRef] [PubMed]
94. Banchereau, J.; Briere, F.; Caux, C.; Davoust, J.; Lebecque, S.; Liu, Y.J.; Pulendran, B.; Palucka, K. Immunobiology of Dendritic Cells. *Annu. Rev. Immunol.* **2000**, *18*, 767–811. [CrossRef] [PubMed]
95. Dudek, A.M.; Martin, S.; Garg, A.D.; Agostinis, P. Immature, Semi-Mature, and Fully Mature Dendritic Cells: Toward a DC-Cancer Cells Interface that Augments Anticancer Immunity. *Front. Immunol.* **2013**, *4*. [CrossRef] [PubMed]
96. Sabado, R.L.; Balan, S.; Bhardwaj, N. Dendritic Cell-Based Immunotherapy. *Cell Res.* **2017**, *27*, 74–95. [CrossRef] [PubMed]
97. Bol, K.F.; Schreibelt, G.; Gerritsen, W.R.; De Vries, I.J.M.; Figdor, C.G. Dendritic Cell-Based Immunotherapy: State of the Art and Beyond. Clin. *Cancer Res.* **2016**, *22*, 1897–1906. [CrossRef] [PubMed]
98. Henrickson, S.E.; Perro, M.; Loughhead, S.M.; Senman, B.; Stutte, S.; Quigley, M.; Alexe, G.; Iannacone, M.; Flynn, M.P.; Omid, S.; et al. Antigen Availability Determines CD8+ T Cell-Dendritic Cell Interaction Kinetics and Memory Fate Decisions. *Immunity* **2013**, *39*, 496–507. [CrossRef] [PubMed]
99. Lutz, M.B.; Schuler, G. Immature, Semi-Mature and Fully Mature Dendritic Cells: Which Signals Induce Tolerance or Immunity? *Trends Immunol.* **2002**, *23*, 445–449. [CrossRef]
100. Mempel, T.R.; Henrickson, S.E.; Von Andrian, U.H. T-Cell Priming by Dendritic Cells in Lymph Nodes Occurs in Three Distinct Phases. *Nature* **2004**, *427*, 154–159. [CrossRef] [PubMed]
101. Garg, A.D.; Coulie, P.G.; Van den Eynde, B.J.; Agostinis, P. Integrating Next-Generation Dendritic Cell Vaccines into the Current Cancer Immunotherapy Landscape. *Trends Immunol.* **2017**, *38*, 577–593. [CrossRef] [PubMed]

102. Garrido, F.; Perea, F.; Bernal, M.; Sanchez-Palencia, A.; Aptsiauri, N.; Ruiz-Cabello, F. The Escape of Cancer from T Cell-Mediated Immune Surveillance: HLA Class I Loss and Tumor Tissue Architecture. *Vaccines (Basel)* **2017**, *5*, 7. [CrossRef] [PubMed]
103. Yang, D.H.; Park, J.S.; Jin, C.J.; Kang, H.K.; Nam, J.H.; Rhee, J.H.; Kim, Y.K.; Chung, S.Y.; Choi, S.J.; Kim, H.J.; et al. The Dysfunction and Abnormal Signaling Pathway of Dendritic Cells Loaded by Tumor Antigen can be Overcome by Neutralizing VEGF in Multiple Myeloma. *Leuk. Res.* **2009**, *33*, 665–670. [CrossRef] [PubMed]
104. Volovitz, I.; Melzer, S.; Amar, S.; Bocsi, J.; Bloch, M.; Efroni, S.; Ram, Z.; Tárnok, A. Dendritic Cells in the Context of Human Tumors: Biology and Experimental Tools. *Int. Rev. Immunol.* **2016**, *35*, 116–135. [CrossRef] [PubMed]
105. Strioga, M.; Schijns, V.; Powell, D.J.; Pasukoniene, V.; Dobrovolskiene, N.; Michalek, J. Dendritic Cells and their Role in Tumor Immunosurveillance. *Innate Immun.* **2013**, *19*, 98–111. [CrossRef] [PubMed]
106. Braun, A.; Worbs, T.; Moschovakis, G.L.; Halle, S.; Hoffmann, K.; Bolter, J.; Munk, A.; Forster, R. Afferent Lymph-Derived T Cells and DCs use Different Chemokine Receptor CCR7-Dependent Routes for Entry into the Lymph Node and Intranodal Migration. *Nat. Immunol.* **2011**, *12*, 879–887. [CrossRef] [PubMed]
107. Comerford, I.; Harata-Lee, Y.; Bunting, M.D.; Gregor, C.; Kara, E.E.; McColl, S.R. A Myriad of Functions and Complex Regulation of the CCR7/CCL19/CCL21 Chemokine Axis in the Adaptive Immune System. *Cytokine Growth Factor Rev.* **2013**, *24*, 269–283. [CrossRef] [PubMed]
108. Chen, F.; Hou, M.; Ye, F.; Lv, W.; Xie, X. Ovarian Cancer Cells Induce Peripheral Mature Dendritic Cells to Differentiate into Macrophagelike Cells in Vitro. *Int. J. Gynecol. Cancer* **2009**, *19*, 1487–1493. [CrossRef] [PubMed]
109. Lin, H.; Wei, S.; Hurt, E.M.; Green, M.D.; Zhao, L.; Vatan, L.; Szeliga, W.; Herbst, R.; Harms, P.W.; Fecher, L.A.; et al. Host Expression of PD-L1 Determines Efficacy of PD-L1 Pathway Blockade-Mediated Tumor Regression. *J. Clin. Investig.* **2018**, *128*, 1708. [CrossRef] [PubMed]
110. Scarlett, U.K.; Rutkowski, M.R.; Rauwerdink, A.M.; Fields, J.; Escovar-Fadul, X.; Baird, J.; Cubillos-Ruiz, J.R.; Jacobs, A.C.; Gonzalez, J.L.; Weaver, J.; et al. Ovarian Cancer Progression is Controlled by Phenotypic Changes in Dendritic Cells. *J. Exp. Med.* **2012**, *209*, 495–506. [CrossRef] [PubMed]
111. Drakes, M.L.; Stiff, P.J. Understanding Dendritic Cell Immunotherapy in Ovarian Cancer. *Expert Rev. Anticancer Ther.* **2016**, *16*, 643–652. [CrossRef] [PubMed]
112. Palucka, K.; Banchereau, J. Cancer Immunotherapy Via Dendritic Cells. *Nat. Rev. Cancer* **2012**, *12*, 265–277. [CrossRef] [PubMed]
113. Tanyi, J.L.; Bobisse, S.; Ophir, E.; Tuyaerts, S.; Roberti, A.; Genolet, R.; Baumgartner, P.; Stevenson, B.J.; Iseli, C.; Dangaj, D.; et al. Personalized Cancer Vaccine Effectively Mobilizes Antitumor T Cell Immunity in Ovarian Cancer. *Sci. Transl. Med.* **2018**, *10*, eaao5931. [CrossRef] [PubMed]
114. Hildner, K.; Edelson, B.T.; Purtha, W.E.; Diamond, M.; Matsushita, H.; Kohyama, M.; Calderon, B.; Schraml, B.U.; Unanue, E.R.; Diamond, M.S.; et al. Batf3 Deficiency Reveals a Critical Role for CD8a+ Dendritic Cells in Cytotoxic T Cell Immunity. *Science* **2008**, *322*, 1097–1100. [CrossRef] [PubMed]
115. Spranger, S.; Dai, D.; Horton, B.; Gajewski, T.F. Tumor-Residing Batf3 Dendritic Cells are Required for Effector T Cell Trafficking and Adoptive T Cell Therapy. *Cancer Cell* **2017**, *31*, 711–723. [CrossRef] [PubMed]
116. Shortman, K. How does Batf3 Determine Dendritic Cell Development? *Immunol. Cell Biol.* **2015**, *93*, 681–682. [CrossRef] [PubMed]
117. Bronger, H.; Singer, J.; Windmuller, C.; Reuning, U.; Zech, D.; Delbridge, C.; Dorn, J.; Kiechle, M.; Schmalfeldt, B.; Schmitt, M.; et al. CXCL9 and CXCL10 Predict Survival and are Regulated by Cyclooxygenase Inhibition in Advanced Serous Ovarian Cancer. *Br. J. Cancer* **2016**, *115*, 553–563. [CrossRef] [PubMed]
118. Lieber, S.; Reinartz, S.; Raifer, H.; Finkernagel, F.; Dreyer, T.; Bronger, H.; Jansen, J.M.; Wagner, U.; Worzfeld, T.; Muller, R.; et al. Prognosis of Ovarian Cancer is Associated with Effector Memory CD8(+) T Cell Accumulation in Ascites, CXCL9 Levels and Activation-Triggered Signal Transduction in T Cells. *Oncoimmunology* **2018**, *7*, e1424672. [CrossRef] [PubMed]
119. Broz, M.L.; Binnewies, M.; Boldajipour, B.; Nelson, A.E.; Pollack, J.L.; Erle, D.J.; Barczak, A.; Rosenblum, M.D.; Daud, A.; Barber, D.L.; et al. Dissecting the Tumor Myeloid Compartment Reveals Rare Activating Antigen-Presenting Cells Critical for T Cell Immunity. *Cancer Cell* **2014**, *26*, 638–652. [CrossRef] [PubMed]

120. Salmon, H.; Idoyaga, J.; Rahman, A.; Leboeuf, M.; Remark, R.; Jordan, S.; Casanova-Acebes, M.; Khudoynazarova, M.; Agudo, J.; Tung, N.; et al. Expansion and Activation of CD103+ Dendritic Cell Progenitors at the Tumor Site Enhances Tumor Responses to Therapeutic PD-L1 and BRAF Inhibition. *Immunity* **2016**, *44*, 924–938. [CrossRef] [PubMed]
121. Zoul, W.; Machelon, V.; Coulomb-L'Hermin, A.; Borvakz, J.; Nome, F.; Isaeva, T.; Wei, S.; Krzysieks, R.; Durand-Gasselin, I.; Gordon, A.; et al. Stromal-Derived Factor-1 in Human Tumors Recruits and Alters the Function of Plasmacytoid Precursor Dendritic Cells. *Nat. Med.* **2001**, *7*, 1339–1346.
122. Tanizaki, Y.; Kobayashi, A.; Toujima, S.; Shiro, M.; Mizoguchi, M.; Mabuchi, Y.; Yagi, S.; Minami, S.; Takikawa, O.; Ino, K. Indoleamine 2,3-Dioxygenase Promotes Peritoneal Metastasis of Ovarian Cancer by Inducing an Immunosuppressive Environment. *Cancer Sci.* **2014**, *105*, 966–973. [CrossRef] [PubMed]
123. Labidi-Galy, S.I.; Sisirak, V.; Meeus, P.; Gobert, M.; Treilleux, I.; Bajard, A.; Combes, J.D.; Faget, J.; Mithieux, F.; Cassignol, A.; et al. Quantitative and Functional Alterations of Plasmacytoid Dendritic Cells Contribute to Immune Tolerance in Ovarian Cancer. *Cancer Res.* **2011**, *71*, 5423–5434. [CrossRef] [PubMed]
124. Fridlender, Z.G.; Sun, J.; Mishalian, I.; Singhal, S.; Cheng, G.; Kapoor, V.; Horng, W.; Fridlender, G.; Bayuh, R.; Worthen, G.S.; et al. Transcriptomic Analysis Comparing Tumor-Associated Neutrophils with Granulocytic Myeloid-Derived Suppressor Cells and Normal Neutrophils. *PLoS ONE* **2012**, *7*, e31524. [CrossRef] [PubMed]
125. Fridlender, Z.G.; Sun, J.; Kim, S.; Kapoor, V.; Cheng, G.; Ling, L.; Worthen, G.S.; Albelda, S.M. Polarization of Tumor-Associated Neutrophil Phenotype by TGF-Beta: "N1" Versus "N2" TAN. *Cancer Cell* **2009**, *16*, 183–194. [CrossRef] [PubMed]
126. Kobayashi, Y. The Role of Chemokines in Neutrophil Biology. *Front. Biosci.* **2008**, *13*, 2400–2407. [CrossRef] [PubMed]
127. Wengner, A.M.; Pitchford, S.C.; Furze, R.C.; Rankin, S.M. The Coordinated Action of G-CSF and ELR + CXC Chemokines in Neutrophil Mobilization during Acute Inflammation. *Blood* **2008**, *111*, 42–49. [CrossRef] [PubMed]
128. Mayer, C.; Darb-Esfahani, S.; Meyer, A.S.; Hubner, K.; Rom, J.; Sohn, C.; Braicu, I.; Sehouli, J.; Hansch, G.M.; Gaida, M.M. Neutrophil Granulocytes in Ovarian Cancer-Induction of Epithelial-to-Mesenchymal-Transition and Tumor Cell Migration. *J. Cancer* **2016**, *7*, 546–554. [CrossRef] [PubMed]
129. Shaul, M.E.; Levy, L.; Sun, J.; Mishalian, I.; Singhal, S.; Kapoor, V.; Horng, W.; Fridlender, G.; Albelda, S.M.; Fridlender, Z.G. Tumor-Associated Neutrophils Display a Distinct N1 Profile Following TGFbeta Modulation: A Transcriptomics Analysis of Pro- Vs. Antitumor TANs. *Oncoimmunology* **2016**, *5*, e1232221. [CrossRef] [PubMed]
130. Acuff, H.B.; Carter, K.J.; Fingleton, B.; Gorden, D.L.; Matrisian, L.M. Matrix Metalloproteinase-9 from Bone Marrow-Derived Cells Contributes to Survival but Not Growth of Tumor Cells in the Lung Microenvironment. *Cancer Res.* **2006**, *66*, 259–266. [CrossRef] [PubMed]
131. Loukinova, E.; Dong, G.; Enamorado-Ayalya, I.; Thomas, G.R.; Chen, Z.; Schreiber, H.; Van Waes, C. Growth Regulated Oncogene-Alpha Expression by Murine Squamous Cell Carcinoma Promotes Tumor Growth, Metastasis, Leukocyte Infiltration and Angiogenesis by a Host CXC Receptor-2 Dependent Mechanism. *Oncogene* **2000**, *19*, 3477–3486. [CrossRef] [PubMed]
132. Houghton, A.M.; Rzymkiewicz, D.M.; Ji, H.; Gregory, A.D.; Egea, E.E.; Metz, H.E.; Stolz, D.B.; Land, S.R.; Marconcini, L.A.; Kliment, C.R.; et al. Neutrophil Elastase-Mediated Degradation of IRS-1 Accelerates Lung Tumor Growth. *Nat. Med.* **2010**, *16*, 219–223. [CrossRef] [PubMed]
133. Fridlender, Z.G.; Albelda, S.M. Tumor-Associated Neutrophils: Friend or Foe? *Carcinogenesis* **2012**, *33*, 949–955. [CrossRef] [PubMed]
134. Katano, M.; Torisu, M. Neutrophil-Mediated Tumor Cell Destruction in Cancer Ascites. *Cancer* **1982**, *50*, 62–68. [CrossRef]
135. Lichtenstein, A.; Seelig, M.; Berek, J.; Zighelboim, J. Human Neutrophil-Mediated Lysis of Ovarian Cancer Cells. *Blood* **1989**, *74*, 805–809. [PubMed]
136. Granot, Z.; Henke, E.; Comen, E.A.; King, T.A.; Norton, L.; Benezra, R. Tumor Entrained Neutrophils Inhibit Seeding in the Premetastatic Lung. *Cancer Cell* **2011**, *20*, 300–314. [CrossRef] [PubMed]
137. Scapini, P.; Lapinet-Vera, J.A.; Gasperini, S.; Calzetti, F.; Bazzoni, F.; Cassatella, M.A. The Neutrophil as a Cellular Source of Chemokines. *Immunol. Rev.* **2000**, *177*, 195–203. [CrossRef] [PubMed]
138. Beauvillain, C.; Delneste, Y.; Scotet, M.; Peres, A.; Gascan, H.; Guermonprez, P.; Barnaba, V.; Jeannin, P. Neutrophils Efficiently Cross-Prime Naive T Cells in Vivo. *Blood* **2007**, *110*, 2965–2973. [CrossRef] [PubMed]

139. Chen, S.; Zhang, L.; Yan, G.; Cheng, S.; Fathy, A.H.; Yan, N.; Zhao, Y. Neutrophil-to-Lymphocyte Ratio is a Potential Prognostic Biomarker in Patients with Ovarian Cancer: A Meta-Analysis. *Biomed. Res. Int.* **2017**, *2017*, 7943467. [CrossRef] [PubMed]
140. Krockenberger, M.; Dombrowski, Y.; Weidler, C.; Ossadnik, M.; Hönig, A.; Häusler, S.; Voigt, H.; Becker, J.C.; Leng, L.; Steinle, A.; et al. Macrophage Migration Inhibitory Factor Contributes to the Immune Escape of Ovarian Cancer by Down-Regulating NKG2D. *J. Immunol.* **2008**, *180*, 7338–7348. [CrossRef] [PubMed]
141. Pesce, S.; Tabellini, G.; Cantoni, C.; Patrizi, O.; Coltrini, D.; Rampinelli, F.; Matta, J.; Vivier, E.; Moretta, A.; Parolini, S.; et al. B7-H6-Mediated Downregulation of NKp30 in NK Cells Contributes to Ovarian Carcinoma Immune Escape. *OncoImmunology* **2015**, *4*, e1001224. [CrossRef] [PubMed]
142. Zhou, Y.; Xu, Y.; Chen, L.; Xu, B.; Wu, C.; Jiang, J. B7-H6 Expression Correlates with Cancer Progression and Patient's Survival in Human Ovarian Cancer. *Int. J. Clin. Exp. Pathol.* **2015**, *8*, 9428–9433. [PubMed]
143. Wong, J.L.; Berk, E.; Edwards, R.P.; Kalinski, P. IL-18-Primed Helper NK Cells Collaborate with Dendritic Cells to Promote Recruitment of Effector CD8+ T Cells to the Tumor Microenvironment. *Cancer Res.* **2013**, *73*, 4653–4662. [CrossRef] [PubMed]
144. Heng, T.S.; Painter, M.W.; Immunological Genome Project Consortium. The Immunological Genome Project: Networks of Gene Expression in Immune Cells. *Nat. Immunol.* **2008**, *9*, 1091–1094. [CrossRef] [PubMed]
145. Böttcher, J.P.; Bonavita, E.; Chakravarty, P.; Blees, H.; Cabeza-Cabrerizo, M.; Sammicheli, S.; Rogers, N.C.; Sahai, E.; Zelenay, S.; Reis e Sousa, C. NK Cells Stimulate Recruitment of cDC1 into the Tumor Microenvironment Promoting Cancer Immune Control. *Cell* **2018**, *172*, 1022–1037. [CrossRef] [PubMed]
146. Felices, M.; Chu, S.; Kodal, B.; Bendzick, L.; Ryan, C.; Lenvik, A.J.; Boylan, K.L.M.; Wong, H.C.; Skubitz, A.P.N.; Miller, J.S.; et al. IL-15 Super-Agonist (ALT-803) Enhances Natural Killer (NK) Cell Function Against Ovarian Cancer. *Gynecol. Oncol.* **2017**, *145*, 453–461. [CrossRef] [PubMed]
147. Niedbala, M.J.; Crickard, K.; Bernacki, R.J. Interactions of Human Ovarian Tumor Cells with Human Mesothelial Cells Grown on Extracellular Matrix. an in Vitro Model System for Studying Tumor Cell Adhesion and Invasion. *Exp. Cell Res.* **1985**, *160*, 499–513. [CrossRef]
148. Lessan, K.; Aguiar, D.J.; Oegema, T.; Siebenson, L.; Skubitz, A.P. CD44 and Beta1 Integrin Mediate Ovarian Carcinoma Cell Adhesion to Peritoneal Mesothelial Cells. *Am. J. Pathol.* **1999**, *154*, 1525–1537. [CrossRef]
149. Davidowitz, R.A.; Selfors, L.M.; Iwanicki, M.P.; Elias, K.M.; Karst, A.; Piao, H.; Ince, T.A.; Drage, M.G.; Dering, J.; Konecny, G.E.; et al. Mesenchymal Gene Program-Expressing Ovarian Cancer Spheroids Exhibit Enhanced Mesothelial Clearance. *J. Clin. Investig.* **2014**, *124*, 2611–2625. [CrossRef] [PubMed]
150. Heath, R.M.; Jayne, D.G.; O'Leary, R.; Morrison, E.E.; Guillou, P.J. Tumour-Induced Apoptosis in Human Mesothelial Cells: A Mechanism of Peritoneal Invasion by Fas Ligand/Fas Interaction. *Br. J. Cancer* **2004**, *90*, 1437–1442. [CrossRef] [PubMed]
151. Nieman, K.M.; Kenny, H.A.; Penicka, C.V.; Ladanyi, A.; Buell-Gutbrod, R.; Zillhardt, M.R.; Romero, I.L.; Carey, M.S.; Mills, G.B.; Hotamisligil, G.S.; et al. Adipocytes Promote Ovarian Cancer Metastasis and Provide Energy for Rapid Tumor Growth. *Nat. Med.* **2011**, *17*, 1498–1503. [CrossRef] [PubMed]
152. Wei, D.; Geng, F.; Liang, S.; Zhao, H.; Liu, M.; Wang, H. CAF-Derived HGF Promotes Cell Proliferation and Drug Resistance by Up-Regulating the C-Met/PI3K/Akt and GRP78 Signalling in Ovarian Cancer Cells. *Biosci. Rep.* **2017**, *37*. [CrossRef]
153. Givel, A.M.; Kieffer, Y.; Scholer-Dahirel, A.; Sirven, P.; Cardon, M.; Pelon, F.; Magagna, I.; Gentric, G.; Costa, A.; Bonneau, C.; et al. miR200-Regulated CXCL12beta Promotes Fibroblast Heterogeneity and Immunosuppression in Ovarian Cancers. *Nat. Commun.* **2018**, *9*, 1056. [CrossRef] [PubMed]
154. Kraman, M.; Bambrough, P.J.; Arnold, J.N.; Roberts, E.W.; Magiera, L.; Jones, J.O.; Gopinathan, A.; Tuveson, D.A.; Fearon, D.T. Suppression of Antitumor Immunity by Stromal Cells Expressing Fibroblast Activation Protein-A. *Science* **2010**, *330*, 827–830. [CrossRef] [PubMed]
155. Zhang, B.; Chen, F.; Xu, Q.; Han, L.; Xu, J.; Gao, L.; Sun, X.; Li, Y.; Li, Y.; Qian, M.; et al. Revisiting Ovarian Cancer Microenvironment: A Friend or a Foe? *Protein Cell.* **2018**, *9*, 674–692. [CrossRef] [PubMed]
156. Rodriguez, G.M.; Galpin, K.J.C.; McCloskey, C.W.; Vanderhyden, B.C. The Tumor Microenvironment of Epithelial Ovarian Cancer and its Influence on Response to Immunotherapy. *Cancers (Basel)* **2018**, *10*, 242. [CrossRef] [PubMed]
157. Leung, C.S.; Yeung, T.L.; Yip, K.P.; Wong, K.K.; Ho, S.Y.; Mangala, L.S.; Sood, A.K.; Lopez-Berestein, G.; Sheng, J.; Wong, S.T.; et al. Cancer-Associated Fibroblasts Regulate Endothelial Adhesion Protein LPP to Promote Ovarian Cancer Chemoresistance. *J. Clin. Investig.* **2018**, *128*, 589–606. [CrossRef] [PubMed]

158. Spaeth, E.L.; Dembinski, J.L.; Sasser, A.K.; Watson, K.; Klopp, A.; Hall, B.; Andreeff, M.; Marini, F. Mesenchymal Stem Cell Transition to Tumor-Associated Fibroblasts Contributes to Fibrovascular Network Expansion and Tumor Progression. *PLoS ONE* **2009**, *4*, e4992. [CrossRef] [PubMed]
159. Peng, P.; Yan, Y.; Keng, S. Exosomes in the Ascites of Ovarian Cancer Patients: Origin and Effects on Anti-Tumor Immunity. *Oncol. Rep.* **2011**, *25*, 749–762. [PubMed]
160. Cheng, L.; Wu, S.; Zhang, K.; Qing, Y.; Xu, T. A Comprehensive Overview of Exosomes in Ovarian Cancer: Emerging Biomarkers and Therapeutic Strategies. *J. Ovarian Res.* **2017**, *10*, 6. [CrossRef] [PubMed]
161. Han, L.; Xu, J.; Xu, Q.; Zhang, B.; Lam, E.W.; Sun, Y. Extracellular Vesicles in the Tumor Microenvironment: Therapeutic Resistance, Clinical Biomarkers, and Targeting Strategies. *Med. Res. Rev.* **2017**, *37*, 1318–1349. [CrossRef] [PubMed]
162. Budnik, V.; Ruiz-Cañada, C.; Wendler, F. Extracellular Vesicles Round Off Communication in the Nervous System. *Nat. Rev. Neurosci.* **2016**, *17*, 160–172. [CrossRef] [PubMed]
163. Yokoi, A.; Yoshioka, Y.; Yamamoto, Y.; Ishikawa, M.; Ikeda, S.-I.; Kato, T.; Kiyono, T.; Takeshita, F.; Kajiyama, H.; Kikkawa, F.; et al. Malignant Extracellular Vesicles Carrying MMP1 mRNA Facilitate Peritoneal Dissemination in Ovarian Cancer. *Nat. Commun.* **2017**, *8*, 14470. [CrossRef] [PubMed]
164. Runz, S.; Keller, S.; Rupp, C.; Stoeck, A.; Issa, Y.; Koensgen, D.; Mustea, A.; Sehouli, J.; Kristiansen, G.; Altevogt, P. Malignant Ascites-Derived Exosomes of Ovarian Carcinoma Patients Contain CD24 and EpCAM. *Gynecol. Oncol.* **2007**, *107*, 563–571. [CrossRef] [PubMed]
165. Strandmann, E.P.V.; Müller, R. Shipping Drug Resistance: Extracellular Vesicles in Ovarian Cancer. *Trends Mol. Med.* **2016**, *22*, 741–743. [CrossRef] [PubMed]
166. Au Yeung, C.L.; Co, N.N.; Tsuruga, T.; Yeung, T.L.; Kwan, S.Y.; Leung, C.S.; Li, Y.; Lu, E.S.; Kwan, K.; Wong, K.K.; et al. Exosomal Transfer of Stroma-Derived miR21 Confers Paclitaxel Resistance in Ovarian Cancer Cells through Targeting APAF1. *Nat. Commun.* **2016**, *7*, 11150. [CrossRef] [PubMed]
167. Jiménez-Sánchez, A.; Memon, D.; Pourpe, S.; Veeraraghavan, H.; Li, Y.; Vargas, H.A.; Gill, M.B.; Park, K.J.; Zivanovic, O.; Konner, J.; et al. Heterogeneous Tumor-Immune Microenvironments among Differentially Growing Metastases in an Ovarian Cancer Patient. *Cell* **2017**, *170*, 938.e20. [CrossRef] [PubMed]
168. Blank, C.U.; Haanen, J.B.; Ribas, A.; Schumacher, T.N. The "Cancer Immunogram". *Science* **2016**, *352*, 658–660. [CrossRef] [PubMed]
169. Van Allen, E.M.; Miao, D.; Schilling, B.; Shukla, S.A.; Blank, C.; Zimmer, L.; Sucker, A.; Hillen, U.; Foppen, M.H.G.; Goldinger, S.M.; et al. Genomic Correlates of Response to CTLA-4 Blockade in Metastatic Melanoma. *Science* **2015**, *350*, 207–211. [CrossRef] [PubMed]
170. Delyon, J.; Mateus, C.; Lefeuvre, D.; Lanoy, E.; Zitvogel, L.; Chaput, N.; Roy, S.; Eggermont, A.M.M.; Routier, E.; Robert, C. Experience in Daily Practice with Ipilimumab for Thetreatment of Patients with Metastatic Melanoma: Anearly Increase in Lymphocyte and Eosinophil Countsis Associated with Improved Survival. *Ann. Oncol.* **2013**, *24*, 1697–1703. [CrossRef] [PubMed]
171. Peske, J.D.; Woods, A.B.; Engelhard, V.H. Control of CD8 T-Cell Infiltration into Tumors by Vasculature and Microenvironment. *Adv. Cancer Res.* **2015**, *128*, 263–307. [PubMed]
172. Chen, D.; Mellman, I. Oncology Meets Immunology: The Cancer-Immunity Cycle. *Immunity* **2013**, *39*, 1–10. [CrossRef] [PubMed]
173. Larkin, J.; Chiarion-Sileni, V.; Gonzalez, R.; Grob, J.J.; Cowey, C.L.; Lao, C.D.; Schadendorf, D.; Dummer, R.; Smylie, M.; Rutkowski, P.; et al. Combined Nivolumab and Ipilimumab Or Monotherapy in Untreated Melanoma. *N. Engl. J. Med.* **2015**, *373*, 23–34. [CrossRef] [PubMed]
174. Hamanishi, J.; Mandai, M.; Ikeda, T.; Minami, M.; Kawaguchi, A.; Murayama, T.; Kanai, M.; Mori, Y.; Matsumoto, S.; Chikuma, S.; et al. Safety and Antitumor Activity of Anti-PD-1 Antibody, Nivolumab, in Patients with Platinum-Resistant Ovarian Cancer. *J. Clin. Oncol.* **2015**, *33*, 4015–4022. [CrossRef] [PubMed]
175. Hamanishi, J.; Mandai, M.; Konishi, I. Immune Checkpoint Inhibition in Ovarian Cancer. *Int. Immunol.* **2016**, *28*, 339–348. [CrossRef] [PubMed]
176. Holmgaard, R.B.; Zamarin, D.; Munn, D.H.; Wolchok, J.D.; Allison, J.P. Indoleamine 2,3-Dioxygenase is a Critical Resistance Mechanism in Antitumor T Cell Immunotherapy Targeting CTLA-4. *J. Exp. Med.* **2013**, *210*, 1389–1402. [CrossRef] [PubMed]

177. Kelderman, S.; Heemskerk, B.; van Tinteren, H.; van den Brom, R.R.; Hospers, G.A.; van den Eertwegh, A.J.; Kapiteijn, E.W.; de Groot, J.W.; Soetekouw, P.; Jansen, R.L.; et al. Lactate Dehydrogenase as a Selection Criterion for Ipilimumab Treatment in Metastatic Melanoma. *Cancer Immunol. Immunother.* **2014**, *63*, 449–458. [CrossRef] [PubMed]
178. Seliger, B. The Link between MHC Class I Abnormalities of Tumors, Oncogenes, Tumor Suppressor Genes, and Transcription Factors. *J. Immunotoxicol.* **2014**, *11*, 308–310. [CrossRef] [PubMed]
179. Verhaak, R.G.; Tamayo, P.; Yang, J.Y.; Hubbard, D.; Zhang, H.; Creighton, C.J.; Fereday, S.; Lawrence, M.; Carter, S.L.; Mermel, C.H.; et al. Prognostically Relevant Gene Signatures of High-Grade Serous Ovarian Carcinoma. *J. Clin. Investig.* **2013**, *123*, 517–525. [CrossRef] [PubMed]
180. Reinartz, S.; Finkernagel, F.; Adhikary, T.; Rohnalter, V.; Schumann, T.; Schober, Y.; Nockher, W.A.; Nist, A.; Stiewe, T.; Jansen, J.M.; et al. A Transcriptome-Based Global Map of Signaling Pathways in the Ovarian Cancer Microenvironment Associated with Clinical Outcome. *Genome Biol.* **2016**, *17*, 108. [CrossRef] [PubMed]
181. Zhang, H.; Liu, T.; Zhang, Z.; Payne, S.H.; Zhang, B.; McDermott, J.E.; Zhou, J.Y.; Petyuk, V.A.; Chen, L.; Ray, D.; et al. Integrated Proteogenomic Characterization of Human High-Grade Serous Ovarian Cancer. *Cell* **2016**, *166*, 755–765. [CrossRef] [PubMed]
182. Pearce, O.M.T.; Delaine-Smith, R.M.; Maniati, E.; Nichols, S.; Wang, J.; Böhm, S.; Rajeeve, V.; Ullah, D.; Chakravarty, P.; Jones, R.R.; et al. Deconstruction of a Metastatic Tumor Microenvironment Reveals a Common Matrix Response in Human Cancers. *Cancer Discov.* **2018**, *8*, 304–319. [CrossRef] [PubMed]
183. Dhawan, U.; Wang, S.M.; Chu, Y.H.; Huang, G.S.; Lin, Y.R.; Hung, Y.C.; Chen, W.L. Nanochips of Tantalum Oxide Nanodots as Artificial-Microenvironments for Monitoring Ovarian Cancer Progressiveness. *Sci. Rep.* **2016**, *6*, 31998. [CrossRef] [PubMed]
184. Bu, X.; Mahoney, K.M.; Freeman, G.J. Learning from PD-1 Resistance: New Combination Strategies. *Trends Mol. Med.* **2016**, *22*, 448–451. [CrossRef] [PubMed]
185. Chester, C.; Dorigo, O.; Berek, J.S.; Kohrt, H. PMC4372273; Immunotherapeutic Approaches to Ovarian Cancer Treatment. *J. Immunother. Cancer* **2015**, *3*, 1–10. [CrossRef] [PubMed]
186. Tewari, D.; Java, J.J.; Salani, R.; Armstrong, D.K.; Markman, M.; Herzog, T.; Monk, B.J.; Chan, J.K. Long-Term Survival Advantage and Prognostic Factors Associated with Intraperitoneal Chemotherapy Treatment in Advanced Ovarian Cancer: A Gynecologic Oncology Group Study. *J. Clin. Oncol.* **2015**, *33*, 1460–1466. [CrossRef] [PubMed]
187. Jindal, V.; Arora, E.; Gupta, S.; Lal, A.; Masab, M.; Potdar, R. Prospects of Chimeric Antigen Receptor T Cell Therapy in Ovarian Cancer. *Med. Oncol.* **2018**, *35*, 6. [CrossRef] [PubMed]
188. Owens, G.L.; Sheard, V.E.; Kalaitsidou, M.; Blount, D.; Lad, Y.; Cheadle, E.J.; Edmondson, R.J.; Kooner, G.; Gilham, D.E.; Harrop, R. Preclinical Assessment of CAR T-Cell Therapy Targeting the Tumor Antigen 5T4 in Ovarian Cancer. *J. Immunother.* **2018**, *41*, 130–140. [CrossRef] [PubMed]

cancers

MDPI

Review

Heated Intraperitoneal Chemotherapy in the Management of Advanced Ovarian Cancer

Andrea Jewell *, Megan McMahon and Dineo Khabele

Department of Obstetrics and Gynecology, Division of Gynecologic Oncology, University of Kansas Medical Center, Kansas, KS 66160, USA; mmcmahon3@kumc.edu (M.M.); dkhabele@kumc.edu (D.K.)

* Correspondence: AJEWELL@kumc.edu

Received: 5 June 2018; Accepted: 16 August 2018; Published: 1 September 2018

Abstract: Heated intraperitoneal chemotherapy (HIPEC) has several potential benefits. Higher doses of chemotherapy can be used with HIPEC because the plasma-peritoneal barrier results in little absorption into the blood stream. HIPEC offers higher peritoneal penetration in comparison to an intravenous (IV) regimen and does not have the traditional normothermic intraperitoneal (IP) regimen limitation of post-operative adhesions. Hyperthermia itself has cytotoxic effects and can potentiate antineoplastic effects of chemotherapy in part by increasing the depth of tumor penetration by up to 3 mm. For the treatment of ovarian cancer, HIPEC has been evaluated in the recurrent setting with secondary cytoreduction. Recent studies, including a prospective trial, have evaluated its role in primary management of ovarian cancer. This review summarizes previous and ongoing studies regarding the use of HIPEC in the management of ovarian cancer.

Keywords: ovarian cancer; heated intraperitoneal chemotherapy (HIPEC); intraperitoneal chemotherapy (IP); cytoreductive surgery; secondary cytoreduction; interval cytoreduction

1. Introduction

Epithelial ovarian cancer (EOC) is the deadliest gynecologic malignancy [1]. The majority of women are diagnosed at advanced stage with widely metastatic peritoneal disease. Standard of care involves a combination of surgery and chemotherapy. The ability to surgically resect tumors with optimal cytoreduction surgery (CRS), ideally to no gross residual disease (R0), is an important positive prognostic factor [2]. Despite the improvements seen in median survival time with the current standard of radical tumor CRS and IV carboplatin and paclitaxel, long term survival rates for patients with advanced epithelial ovarian carcinoma remain disappointing and efforts continue to develop more effective primary therapy.

For most patients with EOC, the majority of disease burden is in the peritoneal cavity and can be quantified by the peritoneal cancer index (PCI) [3]. The PCI is a measure of the extent of disease burden in the peritoneal cavity. Due to this location, normothermic IP chemotherapy has been studied in prospective clinical trials in the post-operative treatment of primary EOC, and NCCN has noted the combined IV/IP regimen as preferred regimen for optimally cytoreduced Stage III EOC. In the setting of recurrence, treatment guidelines are determined by the time to recurrence and location of metastatic disease. HIPEC during CRS for EOC has been gaining more attention in the treatment of metastatic peritoneal disease. Specifically, HIPEC has more frequently been utilized in the recurrent setting with secondary CRS, but recent studies have evaluated its role in primary management of ovarian cancer. The aim of this article is to review previous and ongoing studies regarding the use of HIPEC in context of the overall use of IP chemotherapy for the treatment of EOC.

2. Normothermic Intraperitoneal Chemotherapy

In normothermic IP chemotherapy, cisplatin and paclitaxel are injected into the patient's peritoneal cavity through an intraabdominal port. IP chemotherapy is administered in the post-operative period over a course of up to six cycles. Three large prospective randomized studies support the use of IP chemotherapy in the primary treatment of EOC. In the Gynecologic Oncology Group (GOG) 104 study, patients were randomized to two arms: the control arm of cisplatin and cyclophosphamide IV and the experimental arm of cisplatin IP and cyclophosphamide IV. While there was a statistically significant overall survival (OS) benefit to the IP regimen of 49 months in comparison to 41 months for the IV regimen, consensus was the benefits of IP chemo are not greater than the benefits of new agent paclitaxel [4]. In GOG 114, patients in the control arm received six cycles of cisplatin and paclitaxel IV with an OS of 52.5 months and the experimental arm received two cycles of carboplatin IV, followed by six cycles of cisplatin and paclitaxel IP with an OS of 63.2 months. Progression free survival (PFS) and OS were statistically significant, but were partially attributed to the addition of two extra cycles of chemotherapy in the IP arm [5].

GOG 172 influenced practice patterns in the United States. The IV/IP regimen of IP cisplatin and paclitaxel, plus IV paclitaxel demonstrated the longest median OS compared to IV carboplatin and paclitaxel in patients with optimally cytoreduced stage III ovarian cancer. The median PFS for the IV alone and IV/IP regimens was 18.3 and 23.8 months, respectively. The median OS for the IV and the IP regimens was 49.7 and 66.9 months, respectively. Due to chemotherapy-associated toxicities, only 42% of women on the IP regimen actually received six cycles of therapy, and 49% received three or fewer IP cycles [6]. Because the OS benefit outweighed the toxicity of the regimen, the NCI Clinical Announcement recognized the superiority of IP chemotherapy in the optimal disease setting [7].

In a follow up analysis of the mature data of GOG 114 and GOG 172 combined, an OS benefit remains significant for IP regimens after 10 years of follow up. This benefit in OS was most pronounced in patients who underwent optimal CRS to R0 treated with the IP regimen. Specifically, in GOG 172, the OS was 127 months in this subset of patients [8]. There was also a correlation noted between survival and the number of IP cycles completed in a separate follow up analysis [9].

Despite the favorable OS for IP chemotherapy, the IP cisplatin-based chemotherapy regimen has not been universally accepted as a standard treatment for EOC secondary to regimen toxicity and IP catheter access device problems. A more recent large prospective trial, GOG 252, compared weekly IV chemotherapy regimens to varying dose reduced IP regimens. All arms of the trial had bevacizumab added during treatment and as maintenance. No significant differences in PFS were observed between the three arms. In comparison to GOG 172, more patients were able to complete the IP regimens, but all arms had excessive toxicity. One concern in interpreting the data from GOG 252 is the addition of bevacizumab to all arms could have influenced the results and analysis [10]. With the inability to replicate the results from GOG 172 and the limitation to access IP chemotherapy outside of the tertiary setting, there has been increased interest in HIPEC as a treatment alternative in the primary and recurrent ovarian cancer setting.

3. HIPEC

In HIPEC, heated intraabdominal chemotherapy is administered at the time of CRS. HIPEC has several potential benefits. High-dose chemotherapy can be used because the plasma-peritoneal barrier results in little absorption into the blood stream [11,12]. In addition, there is higher peritoneal penetration in comparison to IV regimen, and HIPEC does not have the limitation of traditional IP regimen of post-operative adhesions [13,14]. Hyperthermia itself has cytotoxic effects and can increase the depth of tumor penetration by the chemotherapeutic agent up to 3 mm and moreover can potentiate its antineoplastic effects [15–18].

A major historic limitation to HIPEC is the previously reported morbidity and mortality and thus its use was often discouraged [19]. To proceed with HIPEC, CRS to R0, CC0 (non-visible disease remaining) or CC1 (less than 2.5 mm visible disease remaining) is required and involves radical and complex surgeries that are associated with higher complication rates. Currently, particularly in high-volume centers with HIPEC specialists, morbidity and mortality has drastically improved [20,21]. One large retrospective review of 694 patients, treated between 2005 and 2011, utilizing the American College of Surgeons National Surgical Quality Improvement Program (ACS NSQUIP) database, demonstrated a complication rate of 33% and 30-day mortality of 2.3%, both rates consistent with outcomes for other major complex abdominal operations [21].

In EOC, HIPEC has been evaluated in the primary and recurrent setting. The majority of published data regarding this treatment modality is retrospective, but recently some prospective data has been published. Here we will review study outcomes with HIPEC in the management of primary and recurrent ovarian cancer as well as review ongoing trials.

4. HIPEC in the Primary Treatment of Ovarian Cancer

The largest prospective randomized clinical trial demonstrated a survival advantage for patients who received HIPEC, compared to standard IV chemotherapy, for the treatment of primary EOC. (Table 1) All patients received neoadjuvant chemotherapy after determining they were not eligible for primary CRS and had to have at least stable disease after receiving up-front IV chemotherapy. The control arm received standard IV chemotherapy before and after CRS (PFS = 10.7 months, OS = 33.9 months). The experimental arm received the same standard IV chemotherapy but also received HIPEC with cisplatin during CRS (PFS = 14.2 months ($p = 0.01$), OS = 45.7 months ($p = 0.02$)). Over 90% of patients completed full six cycles of IV chemotherapy in both arms [22]. While the PFS and OS in this trial are shorter than the previous mentioned normothermic IP trials, it should be noted that this is a different patient population. The PFS and OS survival in the control arm of this trial are similar to established data in patients receiving NACT and interval CRS [23]. Similarly, a large retrospective study from Italy showed improved outcomes in patients who underwent HIPEC after having a complete or partial response to neoadjuvant IV chemotherapy in comparison to HIPEC at primary CRS. [24,25] In addition to standardizing the HIPEC procedure, the time of administration of HIPEC is another important factor.

Table 1. HIPEC primary trials.

Author	Study type	N [1]	Chemotherapy	PFS	OS
Van Driel, et al. [23]	Prospective	245	Cisplatin	14.2 months	45.7 months
Bakrin, et al. [26]	Retrospective cohort	92	Cisplatin (80%) [2]	n/a	CC0: 41.5 months
Gonzalez Bayon, et al. [27]	Prospective	15	Cisplatin and Doxorubicin	n/a	77.8 months
Cascales-Campos, et al. [28]	Retrospective Series	52	Paclitaxel	1 year: 81% 3 years: 63%	n/a
Bae, et al. [29]	Retrospective Case Control	67	Carboplatin or Paclitaxel	3 years: 56.3%	3 years: 66.1%

[1]: Number of HIPEC patients in trial. [2]: Chemotherapy included in analysis: included cisplatin, doxorubicin, oxaliplatin, mitomycin, cisplatin and mitomycin, and cisplatin and doxorubicin

A retrospective cohort study from France looked at 92 patients receiving HIPEC for primary EOC treatment. The majority (60.8%) received consolidation HIPEC treatment after receiving 6–9 cycles of IV carboplatin and paclitaxel. The rest received HIPEC at primary CRS (13%) and at interval CRS (26.1%.) The majority of patients received cisplatin HIPEC (80.4%,) but 35.9% did receive a second agent with HIPEC, either doxorubicin (19.6%) or mitomycin (18.5%). Significant to survival were timing of HIPEC, peritoneal cancer index (PCI), and R0 CRS. Longest median OS was seen in the primary CRS group at 52.7 months, followed by interval CRS at 36.5 months and then consolidation HIPEC at 33.4 months ($p = 0.03$.) Of all primary HIPEC patients, those able to be optimally cytoreduced to less than 2.5 millimeters (mm) had a median survival of 41.5 months compared to 21.2 months in those with residual disease greater than 2.5 mm ($p < 0.01$) [26]. Again, this is a different patient population than was evaluated in previous normothermic IP trials; therefore we cannot make direct comparisons.

A trial from Spain prospectively evaluated a smaller series of primary, first recurrence and second recurrence EOC patients. Fifteen patients received HIPEC in the primary setting, and all received neoadjuvant chemotherapy. All patients received a combination of cisplatin and doxorubicin chemotherapy during HIPEC. The majority (73%) of patients were optimally cytoreduced to no gross residual cancer, and the median OS was remarkably 77.8 months in this patient population. This survival is similar to previously published normothermic IP chemotherapy data, but, again, we cannot compare such a small series of patients with different parameters [27].

Another larger trial from Spain was a case control series evaluating HIPEC in both the primary and interval CRS setting. Twenty three patients underwent primary cytoreduction with HIPEC and 29 patients underwent neoadjuvant chemotherapy and then interval CRS with HIPEC. All patients had CC0 CRS to no visible residual disease. Interestingly, the PCI was significantly higher in the HIPEC arm meaning that these patients had a larger tumor burden at the beginning of surgery. Also, a higher rate of bowel anastomosis and peritoneal stripping was observed in the HIPEC arm, but these cases were performed after data was published showing that aggressive CRS is associated with improved survival. In contrast, most of the control arm cases were performed before this time period. While the ovarian histology was not categorized, they did identify tumor grade. Up to 30% of tumors in the HIPEC arm were low grade which is a higher than typical ratio. No information was provided of how many cycles of IV chemotherapy was completed. While unable to complete analysis of OS, the disease free survival (DFS) was followed at 1, 2 and 3 years. In the control arm, respectively, the DFS was 66%, 33%, 18%; and in the HIPEC arm, the DFS was 81%, 67%, 63% ($p < 0.01$). It was noted that the survival benefit of HIPEC was not significant in undifferentiated tumors [28].

A retrospective review from South Korea evaluated the role of HIPEC as consolidation treatment at the end of primary IV chemotherapy. All patients underwent primary CRS (included both CC0 and suboptimal patients in analysis) then received adjuvant IV chemotherapy. Patients then underwent a planned secondary CRS. There were 29 patients in the control arm and 67 in the HIPEC arm. HIPEC patients received either single agent carboplatin or paclitaxel at time of CRS. Early stage EOC did not show a survival advantage with HIPEC treatment. However, for stage III control and HIPEC patients, PFS at 3 years was, respectively, 16.7 % and 56.3% ($p < 0.01$) and OS 32.8% and 66.1% ($p < 0.01$.) There was no survival difference between the carboplatin HIPEC and paclitaxel HIPEC subgroups. A higher hematologic toxicity was seen in the carboplatin HIPEC arm, however [29].

5. HIPEC in the Treatment of Recurrent Ovarian Cancer

Substantially more studies have been published regarding the use of HIPEC in the management of recurrent ovarian cancer. Although, a significant amount are retrospective, evaluating a small series of patients or inconsistent with patient parameters and HIPEC dosing. Platinum agents are one of the most commonly used during HIPEC for ovarian cancer, but the dose varies in trials. A phase I trial was published regarding the maximum tolerated dose of (MTD) of cisplatin for HIPEC at time of first recurrence (Table 2). The MTD established was 100 mg/m^2 with 25% of patients experiencing Gr 3–4 toxicity. Notably no severe hematologic toxicity at this dose, and over 90% of patients completed all 6 cycles of adjuvant IV chemotherapy. The median PFS of 13.6 months was comparable to previously published PFS in recurrent ovarian patients treated with IV chemotherapy alone. Peritoneal platinum concentration was significantly elevated in comparison to plasma levels, and platinum DNA adducts were found in tumor biopsies after HIPEC confirming cytotoxic activity immediately after a single dose of cisplatin. A Phase II trial is currently open to further evaluate the efficacy of this dose and regimen [30].

The retrospective cohort study from France also looked at the role of HIPEC in recurrent ovarian cancer. The paper included 247 chemo-sensitive (defined as a recurrence interval of greater than six months after completing IV chemotherapy) and 223 chemo-resistant (defined as a recurrence interval of less than six months) EOC patients. Similarly, the majority of patients received cisplatin HIPEC (75.3%,) but 36.4% did receive a second agent with HIPEC, either doxorubicin (28.1%) or mitomycin (9.1%). Significant to survival were lower PCI and CC0 CRS. Longest median OS was in patients with PCI score of 0–8, 59.3 months, followed by patients achieving CC0 CRS, 51.5 months. Interestingly, there was not a significant difference in survival between the chemo-sensitive (42.2 months) and chemo-resistant (48.0 months) subgroups. This could signify the benefit of hyperthermia in chemo-resistant tumors [26]. Other studies, however, have shown no benefit to HIPEC in chemo-resistant patients, and this needs to be further evaluated [25].

The previous trial from Spain prospectively evaluated a smaller series of primary, as well as 19 first recurrence and eight second recurrence EOC patients. All patients received a combination of cisplatin and doxorubicin chemotherapy during HIPEC. The majority (74%, 75% respectively) of patients were optimally CRS to CC0. The median OS was 62.8 months in the first recurrence group and 35.7 months in the second recurrence group. There was no difference in survival between patients reduced to no gross residual disease (CC0) and those with less than 2.5 mm of disease (CC1) [27]. The survival in this study is similar to previously published data of patients being treated with secondary surgery for recurrent ovarian cancer [31,32].

In a second trial from Spain, a case control review was performed on chemo-sensitive disease at first recurrence. Chemo-sensitive defined as recurrence greater than 12 months from completion of treatment. Twenty two patients underwent CRS solely and 39 patients underwent CRS with HIPEC. All patients included underwent CC0 CRS to no residual disease. Median PFS was 22 months in the CRS alone group and 21 months in the CRS with HIPEC group. While both groups were optimally cytoreduced, the HIPEC had a significantly higher PCI score. This could indicate a more aggressive group of tumors and explain the similar PFS even with the addition of HIPEC. Also, paclitaxel rather than a platinum agent was used in the trial, and, due to the cell cycle dependent mechanism of action, it was theorized that it may not be the most effective agent for use during HIPEC. Reassuringly, both groups had similar post-operative toxicity [33].

Table 2. HIPEC recurrent trials in ovarian cancer.

Author	Study type	N [1]	Chemotherapy	PFS	OS
Zivanovic et al. [30]	Phase I prospective	12	Cisplatin	13.6 months	n/a
Bakrin et al. [26]	Retrospective Cohort	470	Cisplatin (76%) [2]	n/a	CC0: 51.5 months
Gonzalez Bayon et al. [27]	Prospective	27	Cisplatin and Doxorubicin	n/a	1st recurrence: 62.8 months 2nd recurrence: 35.7 months
Cascales-Campos et al. [28]	Case control	39	Paclitaxel	21 months	n/a
Fagotti et al. [34]	Case Control	30	Oxaliplatin	26 months	5 years: 42.7%
Spiliotis et al. [35]	Prospective	60	Multiagent [3]	n/a	26.7 months

[1]: Number of HIPEC patients in trial. [2]: Chemotherapy included in analysis: included cisplatin, doxorubicin, oxaliplatin, mitomycin, cisplatin and mitomycin, and cisplatin and doxorubicin. [3]: Chemo-sensitive—Cisplatin and paclitaxel; Chemo-resistant—Doxorubicin with paclitaxel or mitomycin

A similar patient population was studied in Italy. A case control study with 37 patient controls receiving either CRS and IV chemotherapy (13 patients) or IV chemotherapy alone (24 patients) versus 30 patients undergoing CRS and HIPEC. All patients were experiencing a first recurrence, and the initial PFS was similar in both the control and case arms. The only significant difference between the arms was pattern of recurrence. The control arm had significantly more patients with single nodule or localized recurrence. All control patients achieved CC0 CRS, and 96.7% of HIPEC patients achieved CC0 CRS. PFS was 15 months in the control arm and 26 months in the HIPEC arm. Interestingly, over half of the HIPEC patients had a longer secondary PFS after HIPEC than the primary PFS after initial treatment. The HIPEC patients had significantly longer OS, secondary PFS, and deaths than the control group [34].

A prospective trial from Greece evaluated the role of HIPEC at first recurrence. Sixty patients were randomized to each arm; CRS followed by IV chemotherapy versus CRS with HIPEC followed by IV chemotherapy. The trial included both chemo-sensitive and chemo-resistant patients. The HIPEC chemo-sensitive patients were treated with cisplatin and paclitaxel during CRS and the chemo-resistant were treated with doxorubicin and paclitaxel or mitomycin. Mean OS was 26.7 months in the HIPEC group versus 13.4 months in the control group ($p < 0.01$.) The OS was similar in both the HIPEC chemo-sensitive (26.8 months) and chemo-resistant (26.6 months) subgroups. In comparison, the OS was significantly different in the control arm chemo-sensitive (15.2 months) and chemo-resistant (10.2 months) subgroups ($p < 0.01$.) Both arms achieved similar rates of CC0 CRS. However, the overall survival in the HIPEC CC0 group was significantly higher (30.9 months) than the control CC0 group (16.9 months) [35].

6. Discussion

Ovarian cancer is the deadliest gynecologic malignancy in the United States. Normothermic IP chemotherapy for primary EOC has a known benefit in the optimal CRS setting. Unfortunately, widespread use has not occurred due to concern for toxicity and patient access to tertiary care centers. Due to these concerns, there is interest in HIPEC therapy for the management of primary and recurrent EOC.

The largest HIPEC study published to date was in the setting of primary EOC. A survival benefit in patients undergoing interval CRS was found with the addition of HIPEC, and there was no difference in toxicity between the control and HIPEC arms [22]. A critique of the study is that it did not have an IP chemotherapy arm for comparison. The role of normothermic IP chemotherapy is unclear in the interval CRS patient population. A phase II randomized trial, OV21/PETROC, was completed and the IP regimen was found to be well tolerated with reasonable toxicity and no reduction in QOL. There was a noted decrease in progression of disease at nine months in the IP group, however, as the study was underpowered, there was no difference found in PFS and OS between the IV and IP arms [36].

More studies have been published in the recurrent setting, however, most are small and retrospective. A primary critique of HIPEC therapy in EOC is that there is not a standardized regimen. Platinum agents, specifically cisplatin, are frequently used but at varying doses. The phase I trial published defining cisplatin 100 mg/m^2 as the maximum tolerated dose (MTD) will be important to consider when moving forward with designing HIPEC trials in EOC. This was the same dose utilized in the above mentioned primary EOC prospective trial.

Along with varying doses in the recurrent setting, there were varying responses to HIPEC therapy. Prolonged disease free intervals have been shown in both the first and second recurrence settings. Interestingly, some trials have shown similar response in both chemo-sensitive and chemo-resistant recurrences [26]. In one study, the HIPEC arm of patients had a significantly higher PCI at time of CRS yet similar survival to the control arm [33]. A higher PCI is concerning for a more aggressive tumor biology, and could mean that the HIPEC played a role in the similar survival. Overall, there has been a positive significant survival response to HIPEC in the recurrent setting, but almost all published data is from small, retrospective studies.

A significant concern of HIPEC is the toxicity associated with the regimen. Prospective data published shows HIPEC to have similar toxicity to CRS followed by IV therapy [22,33]. Again, these are a limited number of studies, and further evaluation of morbidity and mortality needs to be performed. Another concern of HIPEC therapy is the increased cost associated with frequent ICU admissions and length of hospital stay. The inpatient IP regimen was found not cost effective in the short term in comparison to the traditional IV regimen, but when long term survival analysis was considered it became more cost effective due to the improved survival [37]. There has been no cost analysis performed for HIPEC in EOC. The addition of targeted or immunotherapies to IV regimens is another popular treatment option being considered. The addition of bevacizumab has been found not cost effective when considering all advanced stage EOC receiving IV therapy [38]. However, the cost effectiveness of bevacizumab was improved when looking at a subgroup of patients [39]. This illustrates the significance of identifying appropriate patient populations for specific treatment modalities. It will be important in future trials to perform comparative cost analysis, especially if survival outcomes are similar.

7. Conclusions

In conclusion, there is now high quality prospective data suggesting a survival benefit to HIPEC therapy for patients undergoing primary treatment of EOC after receipt of neoadjuvant chemotherapy and optimal cytoreduction. Poorer quality data exists supporting its use in other clinical contexts such as recurrent disease. This treatment has not been studied in multiple clinical contexts, the regimen and toxicity management has not been standardized and HIPEC has not yet been compared to other standard treatments such as normothermic IP chemotherapy. Therefore, the treatment of EOC with HIPEC outside of clinical trial would not be recommended. Further trials are undergoing (Table 3) and are needed to assess the appropriate patient population and mechanisms of action for HIPEC therapy.

Table 3. Ongoing randomized HIPEC trials in ovarian cancer.

Country	PI	Phase	Time Point	Sample Size	Chemotherapy	Clinicaltrials.gov Identifier
South Korea	Chang	N/A	Primary	204	Paclitaxel	NCT03448354
United States	Momeni	1	Recurrent	20	Carboplatin	NCT02672098
Italy	Not provided	N/A	Recurrent	158	Cisplatin	NCT01538785
Spain	Villarejo Campos	3	Primary or recurrent	94	Paclitaxel	NCT02681432
China	Cui	3	Primary or recurrent	214	Paclitaxeland cisplatin	NCT03373058
United States	Jewell	2	Primary	20	Cisplatin	NCT03321188
Italy, Germany	Ansaloni	3	Primary	94	Cisplatin and paclitaxel	NCT01628380
Mexico	Salcedo-Hernandez	2	Primary	100	Cisplatin and doxorubicin	NCT03275194
Spain	Villarejo Campos	3	Primary or recurrent	32	Cisplatin	NCT02328716
Belgium, France, Spain	Classe	3	Recurrent	444	Cisplatin	NCT01376752
France	not provided	3	Recurrent	220	Cisplatin	NCT03220932
United States	Zivanovic	2	Recurrent	98	Carboplatin	NCT01767675
India	Solanki	N/A	Primary or recurrent	150	Not provided	NCT02754115
United States	Sardi	2	Primary	48	Carboplatin	NCT02124421
United States	Kelly	2	Primary or recurrent	40	Carboplatin	NCT03188432
United States	Dellinger	1	Primary or recurrent	5	Cisplatin	NCT01970722
Norway	Flatmark	Observational	Primary or recurrent	200	Not provided	NCT02073500
Belgium	Ceelen	2	Primary or recurrent	48	Cisplatin	NCT02567253
United States	Lilja	2	Recurrent	200	Cisplatin	NCT02349958
France	Bereder	N/A	Primary or recurrent	44	Not provided	NCT02803515

Author Contributions: Conceptualization, A.J. and D.K.; Writing-Original Draft Preparation, M.M. and A.J.; Writing-Review & Editing, M.M., A.J. and D.K.; Supervision, A.J. and D.K.

Funding: This research received no external funding.

Conflicts of Interest: The authors declare no conflict of interest.

References

1. Siegel, R.L.; Miller, K.D.; Jemal, A. Cancer statistics, 2018. *CA Cancer J. Clin.* **2018**, *68*, 7–30. [CrossRef] [PubMed]
2. Chang, S.J.; Hodeib, M.; Chang, J.; Bristow, R.E. Survival impact of complete cytoreduction to no gross residual disease for advanced-stage ovarian cancer: A meta-analysis. *Gynecol. Oncol.* **2013**, *130*, 493–498. [CrossRef] [PubMed]
3. Jayson, G.C.; Kohn, E.C.; Kitchener, H.C.; Ledermann, J.A. Ovarian cancer. *Lancet* **2014**, *384*, 1376–1388. [CrossRef]
4. Alberts, D.S.; Liu, P.Y.; Hannigan, E.V.; O'Toole, R.; Williams, S.D.; Young, J.A.; Franklin, E.W. Intraperitoneal cisplatin plus intravenous cyclophosphamide versus intravenous cisplatin plus intravenous cyclophosphamide for stage III ovarian cancer. *N. Engl. J. Med.* **1996**, *335*, 1950–1955. [CrossRef] [PubMed]
5. Markman, M.; Bundy, B.N.; Alberts, D.S.; Fowler, J.M.; Clark-Pearson, D.L.; Carson, L.F.; Wadler, S.; Sickel, J. Phase III trial of standard-dose intravenous cisplatin plus paclitaxel versus moderately high-dose carboplatin followed by intravenous paclitaxel and intraperitoneal cisplatin in small-volume stage III ovarian carcinoma: An intergroup study of the Gynecologic Oncology Group, Southwestern Oncology Group, and Eastern Cooperative Oncology Group. *J. Clin. Oncol.* **2001**, *19*, 1001–1007. [PubMed]
6. Armstrong, D.K.; Bundy, B.; Wenzel, L.; Huang, H.Q.; Baergen, R.; Lele, S.; Copeland, L.J. Intraperitoneal cisplatin and paclitaxel in ovarian cancer. *N. Engl. J. Med.* **2006**, *354*, 34–43. [CrossRef] [PubMed]
7. NCI Clinical Announcement on Intraperitoneal Chemotherapy in Ovarian Cancer. 2006. Available online: https://ctep.cancer.gov/highlights/20060105_ovarian.htm (accesssed on 20 August 2018).
8. Landrum, L.M.; Java, J.; Mathews, C.A.; Lanneau, G.S., Jr.; Copeland, L.J.; Armstrong, D.K.; Walker, J.L. Prognostic factors for stage III epithelial ovarian cancer treated with intraperitoneal chemotherapy: A Gynecologic Oncology Group study. *Gynecol. Oncol.* **2013**, *130*, 12–18. [CrossRef] [PubMed]
9. Tewari, D.; Java, J.J.; Salani, R.; Armstrong, D.K.; Markman, M.; Herzog, T.; Monk, B.J.; Chan, J.K. Long-term survival advantage and prognostic factors associated with intraperitoneal chemotherapy treatment in advanced ovarian cancer: A gynecologic oncology group study. *J. Clin. Oncol.* **2015**, *33*, 1460–1466. [CrossRef] [PubMed]
10. Walker, J.; Brady, M.F.; DiSilvestro, P.A.; Fujiwara, K.; Alberts, D.; Zheng, W.; Tewari, K.; Cohn, D.E.; Powell, M.; Van Le, L.; et al. A phase III clinical trial of bevacizumab with IV versus IP chemotherapy in ovarian, fallopian tube, and primary peritoneal carcinoma. NCI-supplied agent: Bevacizumab. NCT01167712, a GOG/NRG trial (GOG 252), in Society of Gynecologic Oncologists. In Proceedings of the 2016 Annual Meeting on Women's Cancer, San Diego, CA, USA, 21 March 2016.
11. Sugarbaker, P.H.; Graves, T.; DeBruijn, E.A.; Cunliffe, W.J.; Mullins, R.E.; Hull, W.E.; Oliff, L.; Schlag, P. Early postoperative intraperitoneal chemotherapy as an adjuvant therapy to surgery for peritoneal carcinomatosis from gastrointestinal cancer: Pharmacological studies. *Cancer Res.* **1990**, *50*, 5790–5794. [PubMed]
12. Katz, M.H.; Barone, R.M. The rationale of perioperative intraperitoneal chemotherapy in the treatment of peritoneal surface malignancies. *Surg. Oncol. Clin. N. Am.* **2003**, *12*, 673–688. [CrossRef]
13. Sun, X.; Li, X.F.; Russell, J.; Xing, L.; Urano, M.; Li, G.C.; Humm, J.L.; Ling, C.C. Changes in tumor hypoxia induced by mild temperature hyperthermia as assessed by dual-tracer immunohistochemistry. *Radiother. Oncol.* **2008**, *88*, 269–276. [CrossRef] [PubMed]
14. Petrillo, M.; De Iaco, P.; Cianci, S.; Perrone, M.; Costantini, B.; Ronsini, C.; Scambia, G.; Fagotti, A. Long-Term Survival for Platinum-Sensitive Recurrent Ovarian Cancer Patients Treated with Secondary Cytoreductive Surgery Plus Hyperthermic Intraperitoneal Chemotherapy (HIPEC). *Ann. Surg. Oncol.* **2016**, *23*, 1660–1665. [CrossRef] [PubMed]
15. Teicher, B.A.; Kowal, C.D.; Kennedy, K.A.; Sartorelli, A.C. Enhancement by hyperthermia of the in vitro cytotoxicity of mitomycin C toward hypoxic tumor cells. *Cancer Res.* **1981**, *41*, 1096–1099. [PubMed]
16. El-Kareh, A.W.; Secomb, T.W. A theoretical model for intraperitoneal delivery of cisplatin and the effect of hyperthermia on drug penetration distance. *Neoplasia* **2004**, *6*, 117–1127. [CrossRef] [PubMed]
17. VanderWaal, R.; Thampy, G.; Wright, W.D.; Roti Roti, J.L. Heat-induced modifications in the association of specific proteins with the nuclear matrix. *Radiat. Res.* **1996**, *145*, 746–753. [CrossRef] [PubMed]
18. Roti Roti, J.L.; Kampinga, H.H.; Malyapa, R.S.; Wright, W.D.; vanderWaal, R.P.; Xu, M. Nuclear matrix as a target for hyperthermic killing of cancer cells. *Cell Stress Chaperones* **1998**, *3*, 245–255. [CrossRef]

19. Chua, T.C.; Yan, T.D.; Saxena, A.; Morris, D.L. Should the treatment of peritoneal carcinomatosis by cytoreductive surgery and hyperthermic intraperitoneal chemotherapy still be regarded as a highly morbid procedure? A systematic review of morbidity and mortality. *Ann. Surg.* **2009**, *249*, 900–907. [CrossRef] [PubMed]
20. Voron, T.; Eveno, C.; Jouvin, I.; Beaugerie, A.; Lo Dico, R.; Dagois, S.; Soyer, P.; Pocard, M. Cytoreductive surgery with a hyperthermic intraperitoneal chemotherapy program: Safe after 40 cases, but only controlled after 140 cases. *Eur. J. Surg. Oncol.* **2015**, *41*, 1071–1077. [CrossRef] [PubMed]
21. Jafari, M.D.; Halabi, W.J.; Stamos, M.J.; Nguyen, V.Q.; Carmichael, J.C.; Mills, S.D.; Pigazzi, A. Surgical outcomes of hyperthermic intraperitoneal chemotherapy: Analysis of the american college of surgeons national surgical quality improvement program. *JAMA Surg.* **2014**, *149*, 170–175. [CrossRef] [PubMed]
22. Van Driel, W.J.; Koole, S.N.; Sonke, G.S. Hyperthermic Intraperitoneal Chemotherapy in Ovarian Cancer. *N. Engl. J. Med.* **2018**, *378*, 1363–1364. [CrossRef] [PubMed]
23. Vergote, I.; Trope, C.; Amant, F.; Kristensen, G.; Ehlen, T.; Johnson, N.; Verheijen, R.; van der Burg, M.; Lacave, A.; Benedettie Panici, P.; et al. Neoadjuvant chemotherapy or primary surgery in stage IIIC or IV ovarian cancer. *N. Engl. J. Med.* **2010**, *363*, 943–953. [CrossRef] [PubMed]
24. Kireeva, G.S.; Gafton, G.I.; Guseynov, K.D.; Senchik, K.Y.; Belyaeva, O.A.; Bespalov, V.G.; Panchenko, A.V.; Maydin, M.A.; Belyaev, A.M. HIPEC in patients with primary advanced ovarian cancer: Is there a role? A systematic review of short- and long-term outcomes. *Surg. Oncol.* **2018**, *27*, 251–258. [CrossRef] [PubMed]
25. Di Giorgio, A.; De Iaco, P.; De Simone, M.; Garofalo, A.; Scambia, G.; Pinna, A.D.; Verdecchia, G.M. Cytoreduction (Peritonectomy Procedures) Combined with Hyperthermic Intraperitoneal Chemotherapy (HIPEC) in Advanced Ovarian Cancer: Retrospective Italian Multicenter Observational Study of 511 Cases. *Ann. Surg. Oncol.* **2017**, *24*, 914–922. [CrossRef] [PubMed]
26. Bakrin, N.; Bereder, J.M.; Decullier, E.; Classe, J.M.; Msika, S.; Lorimier, G.; Abboud, K. Peritoneal carcinomatosis treated with cytoreductive surgery and Hyperthermic Intraperitoneal Chemotherapy (HIPEC) for advanced ovarian carcinoma: A French multicentre retrospective cohort study of 566 patients. *Eur. J. Surg. Oncol.* **2013**, *39*, 1435–1443. [CrossRef] [PubMed]
27. Gonzalez Bayon, L.; Steiner, M.A.; Vasquez Jimenez, W.; Asencio, J.M.; Alvarez de Sierra, P.; Atahualpa Arenas, F.; Rodriguez del Campo, J.; Garcia Sabrido, J.L. Cytoreductive surgery and hyperthermic intraperitoneal chemotherapy for the treatment of advanced epithelial ovarian carcinoma: Upfront therapy, at first recurrence, or later? *Eur. J. Surg. Oncol.* **2013**, *39*, 1109–1115. [CrossRef] [PubMed]
28. Cascales-Campos, P.A.; Gil, J.; Gil, E.; Feliciangeli, E.; Gonzalez-Gil, A.; Parrilla, J.J.; Parrilla, P. Treatment of microscopic disease with hyperthermic intraoperative intraperitoneal chemotherapy after complete cytoreduction improves disease-free survival in patients with stage IIIC/IV ovarian cancer. *Ann. Surg. Oncol.* **2014**, *21*, 2383–2389. [CrossRef] [PubMed]
29. Bae, J.H.; Lee, J.M.; Ryu, K.S.; Lee, Y.S.; Park, Y.G.; Hur, S.Y.; Ahn, W.S.; Namkoong, S.E. Treatment of ovarian cancer with paclitaxel- or carboplatin-based intraperitoneal hyperthermic chemotherapy during secondary surgery. *Gynecol. Oncol.* **2007**, *106*, 193–200. [CrossRef] [PubMed]
30. Zivanovic, O.; Abramian, A.; Kullmann, M.; Fuhrmann, C.; Coch, C.; Hoeller, T.; Ruehs, H. HIPEC ROC I: A phase I study of cisplatin administered as hyperthermic intraoperative intraperitoneal chemoperfusion followed by postoperative intravenous platinum-based chemotherapy in patients with platinum-sensitive recurrent epithelial ovarian cancer. *Int. J. Cancer* **2015**, *136*, 699–708. [CrossRef] [PubMed]
31. Benedetti Panici, P.; De Vivo, A.; Bellati, F.; Manci, N.; Perniola, G.; Basile, S.; Muzii, L.; Angioli, R. Secondary cytoreductive surgery in patients with platinum-sensitive recurrent ovarian cancer. *Ann. Surg. Oncol.* **2007**, *14*, 1136–1142. [CrossRef] [PubMed]
32. Tian, W.J.; Jiang, R.; Cheng, X.; Tang, J.; Xing, Y.; Zang, R.Y. Surgery in recurrent epithelial ovarian cancer: Benefits on Survival for patients with residual disease of 0.1–1 cm after secondary cytoreduction. *J. Surg. Oncol.* **2010**, *101*, 244–250. [CrossRef] [PubMed]
33. Cascales-Campos, P.A.; Gil, J.; Feliciangeli, E.; Gil, E.; Gonzalez-Gil, A.; Lopez, V.; Ruiz-Pardo, J.; Nieto, A.; Parrilla, J.J.; Parrilla, P. The role of hyperthermic intraperitoneal chemotherapy using paclitaxel in platinum-sensitive recurrent epithelial ovarian cancer patients with microscopic residual disease after cytoreduction. *Ann. Surg. Oncol.* **2015**, *22*, 987–993. [CrossRef] [PubMed]

34. Fagotti, A.; Costantini, B.; Petrillo, M.; Vizzielli, G.; Fanfani, F.; Margariti, P.A.; Turco, L.C.; Piovano, E.; Scambia, G. Cytoreductive surgery plus HIPEC in platinum-sensitive recurrent ovarian cancer patients: A case-control study on survival in patients with two year follow-up. *Gynecol. Oncol.* **2012**, *127*, 502–505. [CrossRef] [PubMed]
35. Spiliotis, J.; Halkia, E.; Lianos, E.; Kalantzi, N.; Grivas, A.; Efstathiou, E.; Giassas, S. Cytoreductive surgery and HIPEC in recurrent epithelial ovarian cancer: A prospective randomized phase III study. *Ann. Surg. Oncol.* **2015**, *22*, 1570–1575. [CrossRef] [PubMed]
36. Provencher, D.M.; Gallagher, C.J.; Parulekar, W.R.; Ledermann, J.A.; Armstrong, D.K.; Brundage, M.; Gourley, C. OV21/PETROC: A randomized Gynecologic Cancer Intergroup phase II study of intraperitoneal versus intravenous chemotherapy following neoadjuvant chemotherapy and optimal debulking surgery in epithelial ovarian cancer. *Ann. Oncol.* **2018**, *29*, 431–438. [CrossRef] [PubMed]
37. Havrilesky, L.J.; Secord, A.A.; Darcy, K.M.; Armstrong, D.K.; Kulasingam, S.; Group Gynecologic Oncology. Cost effectiveness of intraperitoneal compared with intravenous chemotherapy for women with optimally resected stage III ovarian cancer: A Gynecologic Oncology Group study. *J. Clin. Oncol.* **2008**, *26*, 4144–4150. [CrossRef] [PubMed]
38. Cohn, D.E.; Kim, K.H.; Resnick, K.E.; O'Malley, D.M.; Straughn, J.M., Jr. At what cost does a potential survival advantage of bevacizumab make sense for the primary treatment of ovarian cancer? A cost-effectiveness analysis. *J. Clin. Oncol.* **2011**, *29*, 1247–1251. [CrossRef] [PubMed]
39. Chan, J.K.; Herzog, T.J.; Hu, L.; Monk, B.J.; Kiet, T.; Blansit, K.; Kapp, D.S.; Yu, X. Bevacizumab in treatment of high-risk ovarian cancer—A cost-effectiveness analysis. *Oncologist* **2014**, *19*, 523–527. [CrossRef] [PubMed]

cancers

MDPI

Review

Bipolar Tumor-Associated Macrophages in Ovarian Cancer as Targets for Therapy

Vijayalaxmi Gupta [1], Fiona Yull [2,3] and Dineo Khabele [1,4,*]

1 Division of Gynecologic Oncology, Department of Obstetrics and Gynecology, University of Kansas Medical Center, Kansas City, KS 66160, USA; vgupta@kumc.edu

2 Department of Pharmacology, Vanderbilt University, Nashville, TN 37232, USA; fiona.yull@vanderbilt.edu

3 Vanderbilt-Ingram Cancer Center, Nashville, TN 37232, USA

4 The University of Kansas Cancer Center, Kansas City, KS 66160, USA

* Correspondence: dkhabele@kumc.edu; Tel.: +1-913-588-6225

Received: 22 August 2018; Accepted: 23 September 2018; Published: 29 September 2018

Abstract: Ovarian cancer, a rare but fatal disease, has been a challenging area in the field of gynecological cancer. Ovarian cancer is characterized by peritoneal metastasis, which is facilitated by a cross-talk between tumor cells and other cells in the tumor microenvironment (TME). In epithelial ovarian cancer, tumor-associated macrophages (TAMs) constitute over 50% of cells in the peritoneal TME and malignant ascites, and are potential targets for therapy. Here, we review the bipolar nature of TAMs and the evolving strategies to target TAMs in ovarian cancer.

Keywords: ovarian cancer; tumor-associated macrophages; peritoneal metastasis; tumor microenvironment

---

1. Introduction

Ovarian cancer is a rare but often fatal disease. Despite accounting for only 2.5% of all female cancers, ovarian cancer represents 5% of cancer deaths, and is the leading cause of gynecologic cancer death, in the United States [1,2]. The primary cause of death and the most common presence, in a high-grade serous epithelial ovarian cancer, is a peritoneal metastasis. Metastasis in epithelial ovarian cancer is characterized by ascites and tumor implants, that typically disseminate throughout the peritoneal cavity, along the lining of the peritoneum, the omentum, and the serosal surfaces of the viscera.

Peritoneal metastasis is regulated by cross-talk between tumor cells and the tumor microenvironment (TME). The TME is a dynamic cellular environment within an extracellular matrix surrounding the tumors, which contain a heterogeneous group of cells, including macrophages, lymphocytes, mesenchymal stem cells, fibroblasts, blood vessels, pericytes, and adipocytes [3,4]. Macrophages are converted into tumor-associated macrophages (TAMs), primarily through the release of cytokines, chemokines, and growth factors, secreted from tumor cells and other cells in the TME.

In epithelial ovarian cancer, TAMs constitute over 50% of cells in the peritoneal tumor implants and the ascites. TAMs are plastic and heterogeneous. Depending on the TME and the extracellular stimuli, macrophages exhibit two main phenotypes along a spectrum, the anti-tumorigenic (M1-like) and the pro-tumorigenic (M2-like). M2-like macrophages contribute to an immune suppressive TME and promote cross-talk between tumor cells and other cells leading to an enhanced tumor-cell growth, invasion, and metastasis [3,4]. This bipolar and plastic nature of the TAMs has the potential to be harnessed for therapeutic purposes. Indeed, the concept of re-educating M2-like macrophages to convert them into M1-like tumoricidal phenotypes was introduced as a therapeutic strategy, almost two decades ago [5]. Here, we review the bipolar nature of the TAMs and the evolving strategies to target TAMs in ovarian cancer.

2. Macrophages in Epithelial Ovarian Cancer

Macrophages are of myeloid lineage. They contribute to physiological homeostasis and constitute critical components of the innate immune response. Macrophages are involved in antigen presentation, phagocytosis, and other immuno-modulatory processes. Epithelial ovarian cancer TAMs originate from two main sources: (1) resident macrophages that arise from the embryonic yolk sac during development and (2) infiltrating macrophages that arise from the bone marrow monocytes (Figure 1) [6–8]. Both resident and infiltrating macrophages are heavily influenced by their cellular niche and transform into specific phenotypes, based on the signals they receive from the TME.

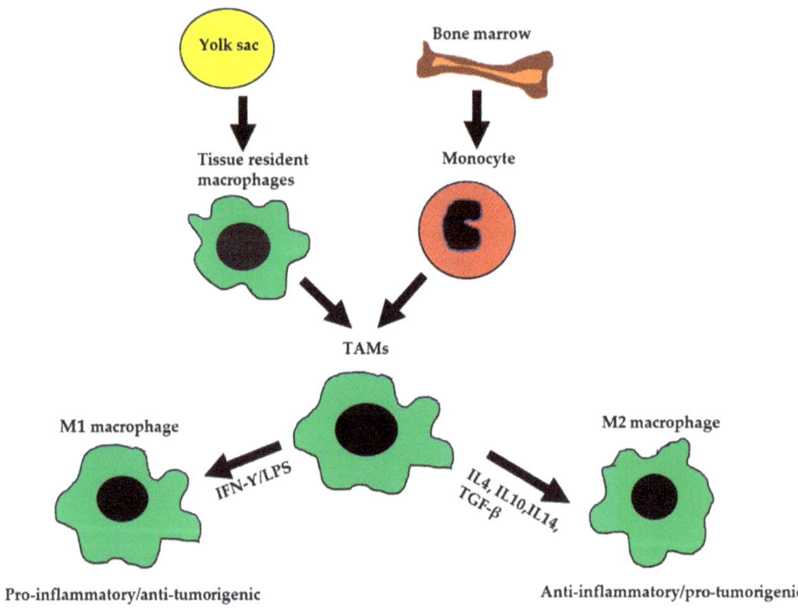

Figure 1. The ontogeny and polarization of M1 and M2 macrophages. Tissue-resident macrophages are mainly derived from yolk sac during development. Tumor-associated macrophages (TAMs) are derived from tissue-resident macrophages, or by differentiation of monocytes from the bone marrow. TAMs are polarized into M1-like or M2-like phenotypes based on signals received from the microenvironment (TME).

Resident macrophages are long-lived and maintained by local signals and the primary colony stimulating factor 1 (CSF-1), via the CSF-1 receptor (CSF-1R). Resident macrophages regulate immune responses and metabolic functions in a tissue-specific manner [9]. In ovarian cancer in mouse models, resident macrophages in the peritoneum are associated with GATA-6 [10]. In the omentum, one of the favored sites of the ovarian cancer peritoneal metastasis, resident macrophages are found in leukocyte-rich "milky spots" [11] and contribute to ovarian cancer cell invasion, both in the omentum and the rest of the peritoneal cavity [12,13]. In a mouse model of ovarian cancer, omental macrophages serve as a source of retinoic acid and other inducers to transport resident macrophages from the omentum to the peritoneum [10].

Infiltrating macrophages are short-lived and recruited from bone marrow monocytes. Infiltrating macrophages arrive in local tissue microenvironment and differentiate further into tissue-specific macrophages, which under homeostatic conditions, abide by the signals they receive from the surrounding microenvironment. In cancer, both the resident and the infiltrating macrophages in the TME, typically differentiate into pro-tumorigenic M2-like phenotypes.

Ascites is a hallmark of epithelial ovarian cancer, and its presence and volume are negatively related to prognosis [14]. TAMs, representing both the resident macrophages and the infiltrating macrophages, constitute a major fraction of the cells in epithelial ovarian cancer ascites [3,15,16]. TAMs in ovarian cancer ascites are primarily M2-like and pro-tumorigenic, with features similar to resident peritoneal macrophages, expressing genes involved in extracellular matrix remodeling, such as PCOLCE2 [17]. A sub-set of TAMS, found in ascites, are more similar to infiltrating macrophages. These are M1-like, expressing high levels of IFN-Ɣ, which induces an IL-12-mediated cytotoxic response against tumor cells [18].

3. Bipolar Macrophages

A mixed population of TAMs exists in the TME of epithelial ovarian cancer [19]. Classically activated M1 and alternatively activated M2 are the two main phenotypes that represent a spectrum of functions (Figure 1 and Table 1) [20–22]. TAMs in the peritoneal cavity and ascites are primarily M2-like and are pro-tumorigenic. This polarization of TAMs towards M2 and M2/M1 ratios, has the potential use as a predictive and prognostic marker. For example, the ratio of M1/M2 is associated with an improved ovarian cancer prognosis [23]. In contrast, the ratio of CD163+ M2-like macrophages to the total CD68 macrophages (CD163/CD68) is a poor prognostic factor [24]. In addition, higher levels of CD163+ M2-like macrophages correlates with elevated IL6, and IL-10, and a shorter relapse-free survival [25].

Table 1. Comparison of characteristics of M1 and M2 macrophages. Adapted from Krishnan, 2018 [19], Mantovani, 2002 [21], and Mantovani, 2004 [22].

Characteristics	M1 Macrophage	M2 Macrophage
Activation pathway	Th1 (Classical)	Th2 (Alternative)
Tumor relation	Anti-tumorigenic	Pro-tumorigenic
Inducers	LPS, IFN-gamma, IL-12	IL4, IL10, IL13, TGF-β, CCL2, CXCL4
Chemokines	CXCL9, CXCL10, CCL4, CCL10, CCL11	CCL17, CCL22, CCL24
Markers	CD86, CD80, iNOS, TLR2, TLR4, IL-1R, MHC-II	CD163, CD206, CCl18, IL-1RII, TGM2
Antigen processing/presentation	Yes–Increased MHCII, STAT1, NO production	No-Decreased MHCII, STAT-1, NO production
Function	Pro-inflammatory/Tissue damage/Pathogenic clearance/Anti-angiogenic	Anti-inflammatory/Tissue repair and remodeling/Fibrosis/Pro-angiogenic

Abbreviations: Major histocompatibility complex (MHC); signal transducer and activator or transcription 1 (STAT1); Nitric oxide (NO).

3.1. M1 Macrophages

M1-like polarized macrophages are classically activated via the Th1 immune pathway. Th1 cells are mainly type 1 immune cells, which secrete cytokines, such as IFN-Ɣ, IL-12, and TNF. They activate macrophages to induce inflammatory signaling pathways that exert tumoricidal effects [26]. Thus, the role of M1 macrophages is primarily pro-inflammatory, and they aid in killing pathogens and cancer cells.

M1 macrophages are critical for recruiting tumor-infiltrating lymphocytes that exhibit tumoricidal properties [27]. M1 macrophages secrete chemokines and cytokines to recruit T cells. Tumors with T cells have 14 times higher levels of macrophage-secreted chemokine mRNA levels, compared to tumors without T cells. Macrophage-derived chemokines delay the recurrence of ovarian cancer from 6 months to later than 40 months. Patients diagnosed with metastatic ovarian cancer, and tumors that contain tumor-infiltrating T cells, have a significantly improved clinical response to treatment, and a 38% overall five-year survival rate, as compared to a 4.5% survival rate in patients whose tumors had no T cells [27,28]. Tumor-infiltrating T cells recruited by M1-like macrophages induce tumors to express high levels of IFN-γ, IL-2, and other anti-tumorigenic cytokines. Tumors devoid of T cells have high levels of vascular endothelial growth factor (VEGF) expression, which contributes to the angiogenesis and a pro-tumorigenic TME. Glypican-3 (GPC3) enhances M1 macrophage recruitment and increases the secretion of IL-12 and TNF-alpha in ascites of GPC3 expressing mouse models of ovarian cancer [29]. Further, GPC3 is associated with an increased CD8+ T cell infiltration into the TME, induction of apoptosis of tumor cells, decreased ascites formation, and improved survival. Thus, M1 macrophage-derived chemokines, play a key role in recruiting cytotoxic T cells into the tumor microenvironment.

3.2. M2 Macrophages

M2-like macrophages are alternatively activated via the type 2 (Th2) immune pathway. Th2 immune cells secrete cytokines such as IL-4 and IL-5, that induce antibody formation. In general, M2 macrophages are anti-inflammatory and are involved in wound healing via tissue remodeling and the secretion of the extracellular matrix. In the setting of the TME in ovarian cancer, TAMs are primarily M2-like and are pro-tumorigenic. M2 TAMs support angiogenesis, tumor cell growth, migration, invasion, and metastasis [30,31]. This observation was supported by another group that showed that advanced ovarian cancers, with infiltration of M2 macrophages, are associated with poor survival [32].

M2 macrophages enhance cell proliferation in epithelial ovarian cancer cells via the MMP9/HB-EGF axis [33]. Sphere-forming ability is one of the hallmarks of cancer cells that are capable of metastasis. TAMs aid sphere formation and tumor growth, by secreting the epidermal growth factor (EGF) [34]. The EGF leads to integrin (αMβ2) upregulation, on TAMs, and an increased EGFR and ICAM-1 expression, on cancer cells. The elevated EGFR, in tumors, further activates the VEGF/VEGFR pathway in neighboring tumor cells, and thus supports cell proliferation and metastasis. M2-like macrophages facilitate the cell adhesion of ovarian cancer cells to mesothelial cells by causing the mesothelial cells to over-express P-selectin [35]. This mechanism likely supports the epithelial ovarian cancer spread, along the mesothelial-lined peritoneal cavity.

3.3. Molecular Mechanisms of Macrophage Polarization

The precise mechanism that regulates TAM polarization is an area of ongoing investigation. Although interferon regulatory factor IRF5 is the main transcription factor for M1 macrophages [36], an advanced transcriptome analysis unveiled additional transcription factors including, IRF3, Signal transducer, and activator of transcription (STAT) STAT1, STAT5. Hypoxia-inducible factor (HIF-1), nuclear factor kappa B (NF-κB) heterodimer, containing p65–p50, are major regulators of inflammatory chemokines and cytokines that polarize macrophages M1 phenotypes [37].

The main transcription factor for M2 polarization is IRF4 [38]. Proteomic analysis, comparing the proteins and transcripts of the resting and the M2 macrophages, revealed other transcription factors, such as STAT3 and STAT6, the NF-κB homodimer p50–p50, HIF-2, PI3K, AKT, and transglutaminase 2 (TGM2), which were associated with M2 polarization [39]. TGM2 is an enzyme with multiple functions, including cross-linking proteins, cell proliferation, and apoptosis [40]. Together, these transcription factors produce anti-inflammatory cytokines and chemokines typical of Th2 type immune cells.

MicroRNAs are also involved in macrophage polarization [41]. miR-216a is associated with M1 macrophage polarization, through telomerase activation, via the Smad3/NF-κB pathway [42]. Interestingly, miR-216a enhances p53 and p16 expression, which are suppressed in ovarian cancer. This suggests that increasing miR-216a through indirect means could be exploited therapeutically, in ovarian cancer. Reactive oxygen species (ROS) polarize macrophages to M1-like phenotypes [43,44]. HOXA9 polarizes peritoneal macrophages to M2-like phenotypes [45]. Thus, these molecular pathways offer additional means for therapeutically exploiting the bipolar nature of macrophages.

4. Inflammation and TAMs in Ovarian Cancer

Inflammation is one of the classic characteristics of cancer and is integral to cancer initiation, progression, and metastasis. Macrophages facilitate ovarian cancer peritoneal metastasis, via inflammatory pathways, mediated by cytokines and chemokines [46]. The NF-κB pathway provides an important link between inflammation and many types of cancer, including ovarian cancer [47]

Ascites derived from a syngeneic mouse model of ovarian cancer, contains macrophages as dominant cell populations [48]. Macrophage cell density increases proportionately to the volume of ascites and tumor progression. In this model, tumor cells at advanced stages have enhanced NF-κB activation. The peritoneal spread of cancer cells during tumor progression is associated with an increase in the number of M2 macrophages, but had a marginal effect on the number of M1 macrophages. Further, M2 macrophage levels are reduced by inhibiting NF-κB, in the tumors. The p50 component of NF-κB regulates M2-dependent inflammation and a lack of p50, leads to the elevated M1-associated inflammatory processes [49]. This provides an encouraging evidence that the ratio of M1/M2 macrophages can be shifted by targeting NF-κB.

Other factors linking inflammation and epithelial ovarian cancer, include serum amyloid A (SAA1/2) and macrophage migration inhibitory factors (MIF). Accumulation of serum amyloid A (SAA1/2) is associated with inflammation in epithelial ovarian cancer, via the TNF-alpha mediated activation of NF-κB [50]. Normal human ovarian tissues express little or no SAA1/2, whereas, ovarian cancers express high levels of SAA1/2 [43]. Elevated levels of MIF are found in ascites and in the circulation of ovarian cancer patients [51,52]. MIF levels correlate with the histological grade of the cancer tissue, disease prognosis, and platinum sensitivity [53]. MIF reduces natural-killer group 2, member D (NKG2D) expression, and prevents the natural killer (NK) cells from exerting their tumoricidal effects. NKG2D, under normal circumstances, activate the tumoricidal properties of NK and T cells. TME releases ligands for NKG2D and depletes NK cells, which in turn, increases the ratio of anti-tumorigenic CD163+ CD206+ M2-like macrophages in the TME. Soluble NKG2D ligands in ovarian cancer ascites indicated poor prognosis and decreased memory effector T cells [54].

5. TAMs as Therapeutic Targets

TAMs play a critical role in epithelial ovarian cancer tumorigenesis and, therefore, are promising targets for therapy. Evolving therapeutic approaches fall into three broad categories, that include strategies to (1) Block migration of monocytes to the TME; (2) re-polarize macrophages to increase the ratio of M1 to M2-like macrophages; and (3) inhibit immune-signaling pathways in macrophages.

5.1. Block Migration of Monocytes to the TME

Tumor cells and other cells in the TME release cytokines, chemokines, and growth factors that attract monocytes to the TME. This has been demonstrated, in vitro, using ovarian cancer cell lines, as well as, in vivo, using mouse models and some clinical settings.

5.1.1. CSF-1 and CSF-1R

In clinical studies, CSF-1 and CSF-1R expression upregulation in epithelial ovarian cancer have been associated with poor prognosis [55]. The survival, proliferation, and differentiation of monocytes and macrophages are dependent on the CSF1R pathway [56]. In the syngeneic mouse model of ovarian cancer, GW2580, a selective CSF1R kinase inhibitor significantly reduces ascites fluid buildup and the infiltration of M2 TAMs [57]. Further, inhibiting the CSF-1R, partly overcomes anti-VEGF resistance [58], and the CSF-1R disruption results in macrophage depletion, which supports a direct role of the CSF-1R in macrophage recruitment [59]. Currently, active clinical trials targeting CSF1R on M2 macrophages, involve PLX3397, in combination with anti-PD-1 pembrolizumab (Clinical trial # NCT02452424), and Cabiralizumab (antibody against CSF1R), in combination with anti-PD-1 monoclonal antibody Nivolumab (NCT02526017). A clinical trial using LY3022855, a CSF1R inhibitor in combination with anti-PDl1 monoclonal antibody Durvalumab, or anti-cytotoxic T-lymphocyte associated protein 4 monoclonal antibody Tremelimumab, is currently recruiting (NCT02718911).

5.1.2. CCL2

CCL2 is also known as MCP-1 (CC motif ligand 2 or macrophage chemoattractant protein-1), is a chemokine that plays a key role in monocyte recruitment to the TME. Epithelial ovarian cancer cells release CCL2/MCP-1 to attract monocytes and convert them to TAMs, within the TME [60]. A plant-derived product, 9-hydroxycanthin-6-one reduces the MCP-1 expression in ovarian cancer cells and inhibits macrophage recruitment [61]. Interestingly, using a mouse model, it was seen that CCL2/MCP-1 is crucial for Th2 immune responses. MCP-1$^{-/-}$ mice do not induce the Th2 response and express low levels of IL-4, IL-5, and IL-10 [62]. Monocytes and macrophages express CCR2, which is a receptor for CCL2. Thus, the CCL2/CCR2 axis represents an attractive target for ovarian cancer therapy. A CCR2 antagonist RS504303 that is under development, significantly reduces bone-marrow derived monocyte cell migration, in mouse [63]. A clinical trial using an anti-CCl2 antibody, known as CNTO 888, in combination with gemcitabine or paclitaxel, and carboplatin or docetaxel, has been completed (NCT01204996).

5.1.3. Drugs

Bisphosphonates deplete monocytes/macrophages in ovarian cancer. In a syngeneic mouse model of ovarian cancer, clodronate reduces TAMs by inhibiting cytokine secretion, which decreases angiogenesis [64]. In patients with epithelial ovarian cancer, transient depletion of peritoneal macrophages using liposomal alendronic acid potentiates an adoptive immunotherapy [65].

Trabectedin, a marine-derived anti-tumor compound, depletes macrophages in mouse models [66]. A phase 2 clinical trial of trabectedin in ovarian cancer patients, showed a significant depletion of blood monocytes, as well as a reduction in CCL2 levels, in TAMs and ovarian tumor cells [67]. However, trabectedin as a single agent has limited efficacy. An alternate strategy to deplete TAMs is to exploit elevated expression levels of folate receptor-2 (FOLR2) that has been found in human and murine ovarian cancer TAMs, and use G-5 methotrexate nanoparticles to target these TAMs [68].

5.2. Re-Polarize Macrophages to Increase the Ratio of M1 to M2-Like Macrophages

Notch signaling plays a crucial role in M1 polarization in a mouse model, where macrophages with an active Notch display anti-tumor properties. Most of the following studies, unless otherwise stated, were carried out using a mouse model. When Notch signaling is blocked, M2 macrophages are polarized and resist M1 activators [69]. CCL2, apart from recruiting monocytes, enhances M2 polarization as well [70]. Activation of the peroxisome proliferator-activated receptor γ (PPARγ)/NF-κB axis, in ovarian cancer stem cells, induces M2 polarization [71].

An unexpected observation made by our group revealed that inhibition of NF-κB in a syngeneic mouse model of ovarian cancer, increased pro-tumorigenic M2 macrophages, which promoted ascites, an increased expression of pro-tumorigenic soluble factors (such as VEGF in ascites fluid), and an infiltration of more M2 macrophages into the TME [48,72]. These results suggest that the activation of NF-κB in TAMs, not tumor cells, could be a viable therapeutic strategy. Indeed, NF-κB transfected TAMs, display anti-tumorigenic properties in mice harboring solid tumors, which on treatment showed elevated M1 phenotype favoring Th1 cytokines and reduced Th2 cytokines [73].

Antibiotics and natural products modulate macrophages. Doxycycline is a common antibiotic that reduces pro-angiogenic properties of M2 macrophages, in neovascular age-related macular degeneration models [74]. Among natural products, deoxyschizandrin, a phytochemical extracted from berries, significantly reduces the pro-tumorigenic activity of TAMs by inhibiting M2 macrophages [75]. In addition to blocking macrophage recruitment to tumor sites, 9-hydroxycanthin-6-one, inhibits M2 polarization in ovarian cancer [61]. Neferine, another plant-derived product, was found to inhibit M2-macrophages in an OVHM xenograft mouse model [76].

5.3. Inhibit Immune Signaling Pathways in Macrophages

The tumor-associated PD-L1 expression has been investigated by several investigators [77,78]. In addition, macrophages associated with primary and metastatic high-grade serous, ovarian cancer express PD-L1 [77]. A comparison of the TME of primary and recurrent epithelial ovarian cancer showed interesting trends, regarding the effect of T cells and macrophages, on survival. Recurrent tumor TME, with higher immune cell recruitment and higher TAMs, have better survival [79]. In clinical studies, expression of PD-L1, by both immune cells and tumor cells in recurrent tumors, leads to an active immune response and imparts better survival in recurrent cancer, as compared to primary cancer, where only the immune cells express PD-L1.

They further explained that the phenotype of regulatory T cells (Tregs), in primary and recurrent cancer, is different, with recurrent cancer expressing more CD25+ Tregs, which are indicators of better prognosis. Thus, they concluded that the dynamics between TAM PD-L1 expression and cytotoxic vs. Tregs create an imbalance that favors survival. PD-L1 expression was significantly higher in ovarian cancer than in other cancers and coincided with poor prognosis [78]. Although PD-L2 expression was associated with poor prognosis, there was no significant difference in PD-L2 between primary and recurrent ovarian cancer. Their most interesting finding was that the tumor cell PD-L1 expression was inversely proportional to the CD8 expression of intraepithelial tumor-infiltrating lymphocytes (TILs). Further, their findings supported CD8+ TILs as a positive predictor of overall survival and progression-free survival, in ovarian cancer. B7-H4 protein and mRNA is highly expressed in ovarian cancers and is involved in epithelial cell transformation [80]. B7-H4 inhibits T cell activation, thereby, halting host anti-tumor response, leading to a tumor escape from immune surveillance. Earlier reports from this group showed that B7-H4 is also expressed in ovarian tumor-associated macrophages, and similar to tumor B7-H4, these macrophages also suppress tumor immunity [81]. When normal blood monocytes were incubated with tumor ascites, elevated levels of B7-H4 was observed, whereas, the serum-free medium showed no such effect, thereby suggesting that B7-H4 expression is regulated by the tumor microenvironment, specifically IL-6 and IL-7.

Current clinical trials that target PDL1/2 and PD1/2 axis include patients diagnosed with ovarian cancer. A clinical trial for platinum-resistant ovarian cancers involves a combination of the anti-PD-L1 antibody Atezoliuzub with Bevacizumab (NCT02659384). Another clinical trial, for advanced ovarian tumors and recurrent ovarian cancer is investigating a combination of an anti-PDL1 antibody MEDI4736 with Olaparib and/or Cedinarib (NCT024844004). Designing strategies to alleviate immune suppression, by reducing monocyte recruitment, decreasing the M2/M1 ratio, and targeting TAMs in combination with with immune checkpoint inhibitors, could represent attractive targets that switch the innate immunity balance in favor of tumor cell death (Figure 2).

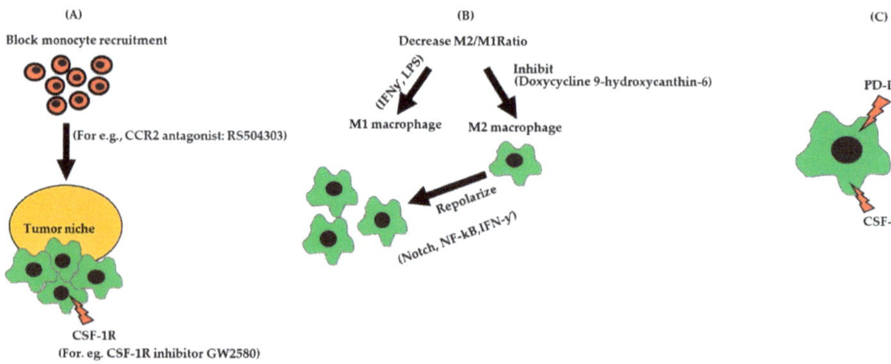

Figure 2. Strategies for targeting TAMs in ovarian cancer. (**A**) Block monocyte recruitment to the tumor niche. (**B**) Chemical intervention to increase M1/M2 ratio by inhibiting M2 polarization, increasing M1 polarization by using Interferon gamma (IFN-γ, Lipopolysaccharide (LPS)) or by repolarizing M2 to M1 by adding IFN-γ or regulating the Notch, NF-κB. (**C**) Inhibit immune signaling pathways on macrophages, for e.g., CSF-1, VEGFR, which promotes angiogenesis, and PD-L1, which inhibits T cell activity.

6. Conclusions

In epithelial ovarian cancer, TAMs mediate progressive ovarian cancer and thus present an attractive target to develop anti-cancer regimens, as they are involved in all stages of ovarian cancer development. M1 macrophages, on the other hand, represent anti-tumorigenic TAMs. An advantage of using TAMs as anti-cancer targets is their genomic stability, which could provide a means of alleviating drug resistance. A deeper understanding of mechanisms behind macrophage polarization, will aid in developing strategies to enhance M1 macrophage polarization or shift the balance between M1 and M2 towards anti-tumorigenic M1 macrophage population. TAMs represent a plastic immune cell population amenable to manipulation and re-education and repolarizing M2 to M1 tumoricidal phenotypes. The affinity of TAMs to the peritoneal TME and the ascites in epithelial ovarian cancer, offers a future potential for targeted intraperitoneal treatment, in combination with chemotherapy drugs.

Despite their promise, clinical implementation of macrophage-based therapies has been limited. The main challenges in targeting TAMs, are their complexity and heterogeneity in the context of the TME and the likely need to combine macrophage-based therapies with other anti-tumor agents. Cross-talk between tumor cells and other cells in the TME is complex. Ongoing research to 'deconvolute' elements of the TME will lead to a better understanding of how to strategically target dominant cell populations, such as macrophages in different TME niches [4]. Classic definitions of M1 and M2 do not fully encompass the full spectrum of macrophage function. Next generation single cell sequencing and flow methods will be required to better understand the most important functions of the sub-types of TAMs, in ovarian cancer.

Author Contributions: Conceptualization, F.Y. and D.K.; Writing-Original Draft Preparation, V.G.; Writing-Review & Editing, V.G., F.Y. and D.K.; Supervision, F.Y. and D.K.; Funding Acquisition, F.Y. and D.K.

Funding: This research was funded by 5R01CA214043-02, F.Y., D.K., 2P30CA168524-06, D.K. and 5P30CA068485-21, F.Y.

Conflicts of Interest: The authors declare no conflicts of interest.

References

1. Siegel, R.L.; Miller, K.D.; Jemal, A. Cancer statistics, 2018. *CA Cancer J. Clin.* **2018**, *68*, 7–30. [CrossRef] [PubMed]
2. Torre, L.A.; Trabert, B.; DeSantis, C.E.; Miller, K.D.; Samimi, G.; Runowicz, C.D.; Gaudet, M.M.; Jemal, A.; Siegel, R.L. Ovarian cancer statistics, 2018. *CA Cancer J. Clin.* **2018**, *68*, 284–296. [CrossRef] [PubMed]
3. Balkwill, F.R.; Capasso, M.; Hagemann, T. The tumor microenvironment at a glance. *J. Cell Sci.* **2012**, *125*, 5591–5596. [CrossRef] [PubMed]
4. Pearce, O.M.T.; Delaine-Smith, R.M.; Maniati, E.; Nichols, S.; Wang, J.; Bohm, S.; Rajeeve, V.; Ullah, D.; Chakravarty, P.; Jones, R.R.; et al. Deconstruction of a metastatic tumor microenvironment reveals a common matrix response in human cancers. *Cancer Discov.* **2018**, *8*, 304–319. [CrossRef] [PubMed]
5. Mantovani, A.; Bottazzi, B.; Colotta, F.; Sozzani, S.; Ruco, L. The origin of tumor-associated macrophages. *Immunol. Today* **1992**, *13*, 265–270. [CrossRef]
6. Ghosn, E.E.; Cassado, A.A.; Govoni, G.R.; Fukuhara, T.; Yang, Y.; Monack, D.M.; Bortoluci, K.R.; Almeida, S.R.; Herzenberg, L.A.; Herzenberg, L.A. Two physically, functionally, and developmentally distinct peritoneal macrophage subsets. *Proc. Natl. Acad. Sci. USA* **2010**, *107*, 2568–2573. [CrossRef] [PubMed]
7. Davies, L.C.; Jenkins, S.J.; Allen, J.E.; Taylor, P.R. Tissue-resident macrophages. *Nat. Immunol.* **2013**, *14*, 986–995. [CrossRef] [PubMed]
8. Rosas, M.; Davies, L.C.; Giles, P.J.; Liao, C.T.; Kharfan, B.; Stone, T.C.; O'Donnell, V.B.; Fraser, D.J.; Jones, S.A.; Taylor, P.R. The transcription factor gata6 links tissue macrophage phenotype and proliferative renewal. *Science* **2014**, *344*, 645–648. [CrossRef] [PubMed]
9. Gordon, S.; Pluddemann, A. Tissue macrophages: Heterogeneity and functions. *BMC Biol.* **2017**, *15*, 53. [CrossRef] [PubMed]
10. Okabe, Y.; Medzhitov, R. Tissue-specific signals control reversible program of localization and functional polarization of macrophages. *Cell* **2014**, *157*, 832–844. [CrossRef] [PubMed]
11. Meza-Perez, S.; Randall, T.D. Immunological functions of the omentum. *Trends Immunol.* **2017**, *38*, 526–536. [CrossRef] [PubMed]
12. Cohen, C.A.; Shea, A.A.; Heffron, C.L.; Schmelz, E.M.; Roberts, P.C. The parity-associated microenvironmental niche in the omental fat band is refractory to ovarian cancer metastasis. *Cancer Prev. Res.* **2013**, *6*, 1182–1193. [CrossRef] [PubMed]
13. Clark, R.; Krishnan, V.; Schoof, M.; Rodriguez, I.; Theriault, B.; Chekmareva, M.; Rinker-Schaeffer, C. Milky spots promote ovarian cancer metastatic colonization of peritoneal adipose in experimental models. *Am. J. Pathol.* **2013**, *183*, 576–591. [CrossRef] [PubMed]
14. Lane, D.; Matte, I.; Laplante, C.; Garde-Granger, P.; Carignan, A.; Bessette, P.; Rancourt, C.; Piche, A. Ccl18 from ascites promotes ovarian cancer cell migration through proline-rich tyrosine kinase 2 signaling. *Mol. Cancer* **2016**, *15*, 58. [CrossRef] [PubMed]
15. Ortega, R.A.; Barham, W.; Sharman, K.; Tikhomirov, O.; Giorgio, T.D.; Yull, F.E. Manipulating the nf-kappab pathway in macrophages using mannosylated, sirna-delivering nanoparticles can induce immunostimulatory and tumor cytotoxic functions. *Int. J. Nanomed.* **2016**, *11*, 2163–2177. [CrossRef] [PubMed]
16. Wang, X.P.; Deavers, M.; Patenia, R.; Bassett, R.L.; Mueller, P.; Ma, Q.; Wang, E.; Freedman, R.S. Monocyte/macrophage and t-cell infiltrates in peritoneum of patients with ovarian cancer or benign pelvic disease. *J. Transl. Med.* **2006**, *4*, 30. [CrossRef] [PubMed]
17. Finkernagel, F.; Reinartz, S.; Lieber, S.; Adhikary, T.; Wortmann, A.; Hoffmann, N.; Bieringer, T.; Nist, A.; Stiewe, T.; Jansen, J.M.; et al. The transcriptional signature of human ovarian carcinoma macrophages is associated with extracellular matrix reorganization. *Oncotarget* **2016**, *7*, 75339–75352. [CrossRef] [PubMed]
18. Adhikary, T.; Wortmann, A.; Finkernagel, F.; Lieber, S.; Nist, A.; Stiewe, T.; Wagner, U.; Muller-Brusselbach, S.; Reinartz, S.; Muller, R. Interferon signaling in ascites-associated macrophages is linked to a favorable clinical outcome in a subgroup of ovarian carcinoma patients. *BMC Genom.* **2017**, *18*, 243. [CrossRef] [PubMed]
19. Krishnan, V.; Schaar, B.; Tallapragada, S.; Dorigo, O. Tumor associated macrophages in gynecologic cancers. *Gynecol. Oncol.* **2018**, *149*, 205–213. [CrossRef] [PubMed]
20. Biswas, S.K.; Mantovani, A. Macrophage plasticity and interaction with lymphocyte subsets: Cancer as a paradigm. *Nat. Immunol.* **2010**, *11*, 889–896. [CrossRef] [PubMed]

21. Mantovani, A.; Sica, A.; Sozzani, S.; Allavena, P.; Vecchi, A.; Locati, M. The chemokine system in diverse forms of macrophage activation and polarization. *Trends Immunol.* **2004**, *25*, 677–686. [CrossRef] [PubMed]
22. Mantovani, A.; Sozzani, S.; Locati, M.; Allavena, P.; Sica, A. Macrophage polarization: Tumor-associated macrophages as a paradigm for polarized m2 mononuclear phagocytes. *Trends Immunol.* **2002**, *23*, 549–555. [CrossRef]
23. Zhang, M.Y.; He, Y.F.; Sun, X.J.; Li, Q.; Wang, W.J.; Zhao, A.M.; Di, W. A high m1/m2 ratio of tumor-associated macrophages is associated with extended survival in ovarian cancer patients. *J. Ovarian Res.* **2014**, *7*, 19. [CrossRef] [PubMed]
24. Yafei, Z.; Jun, G.; Guolan, G. Correlation between macrophage infiltration and prognosis of ovarian cancer—A preliminary study. *Biomed. Res.* **2016**, *27*, 305.
25. Reinartz, S.; Schumann, T.; Finkernagel, F.; Wortmann, A.; Jansen, J.M.; Meissner, W.; Krause, M.; Schworer, A.M.; Wagner, U.; Muller-Brusselbach, S.; et al. Mixed-polarization phenotype of ascites-associated macrophages in human ovarian carcinoma: Correlation of cd163 expression, cytokine levels and early relapse. *Int. J. Cancer* **2014**, *134*, 32–42. [CrossRef] [PubMed]
26. Kidd, P. Th1/th2 balance: The hypothesis, its limitations, and implications for health and disease. *Altern. Med. Rev.* **2003**, *8*, 223–246. [PubMed]
27. Zhang, L.; Conejo-Garcia, J.R.; Katsaros, D.; Gimotty, P.A.; Massobrio, M.; Regnani, G.; Makrigiannakis, A.; Gray, H.; Schlienger, K.; Liebman, M.N.; et al. Intratumoral t cells, recurrence, and survival in epithelial ovarian cancer. *N. Engl. J. Med.* **2003**, *348*, 203–213. [CrossRef] [PubMed]
28. Adams, S.F.; Levine, D.A.; Cadungog, M.G.; Hammond, R.; Facciabene, A.; Olvera, N.; Rubin, S.C.; Boyd, J.; Gimotty, P.A.; Coukos, G. Intraepithelial t cells and tumor proliferation: Impact on the benefit from surgical cytoreduction in advanced serous ovarian cancer. *Cancer* **2009**, *115*, 2891–2902. [CrossRef] [PubMed]
29. Luo, C.; Shibata, K.; Suzuki, S.; Kajiyama, H.; Senga, T.; Koya, Y.; Daimon, M.; Yamashita, M.; Kikkawa, F. Gpc3 expression in mouse ovarian cancer induces gpc3specific t cell-mediated immune response through m1 macrophages and suppresses tumor growth. *Oncol. Rep.* **2014**, *32*, 913–921. [CrossRef] [PubMed]
30. Bingle, L.; Brown, N.J.; Lewis, C.E. The role of tumour-associated macrophages in tumour progression: Implications for new anticancer therapies. *J. Pathol.* **2002**, *196*, 254–265. [CrossRef] [PubMed]
31. Qian, B.Z.; Pollard, J.W. Macrophage diversity enhances tumor progression and metastasis. *Cell* **2010**, *141*, 39–51. [CrossRef] [PubMed]
32. Lan, C.Y.; Huang, X.; Lin, S.X.; Huang, H.Q.; Cai, Q.C.; Wan, T.; Lu, J.B.; Liu, J.H. Expression of m2-polarized macrophages is associated with poor prognosis for advanced epithelial ovarian cancer. *Technol. Cancer Res. Treat.* **2013**, *12*, 259–267. [CrossRef] [PubMed]
33. Carroll, M.J.; Kapur, A.; Felder, M.; Patankar, M.S.; Kreeger, P.K. M2 macrophages induce ovarian cancer cell proliferation via a heparin binding epidermal growth factor/matrix metalloproteinase 9 intercellular feedback loop. *Oncotarget* **2016**, *7*, 86608–86620. [CrossRef] [PubMed]
34. Yin, M.Z.; Li, X.; Tan, S.; Zhou, H.J.; Ji, W.D.; Bellone, S.; Xu, X.C.; Zhang, H.F.; Santin, A.D.; Lou, G.; et al. Tumor-associated macrophages drive spheroid formation during early transcoelomic metastasis of ovarian cancer. *J. Clin. Investig.* **2016**, *126*, 4157–4173. [CrossRef] [PubMed]
35. Carroll, M.J.; Fogg, K.C.; Patel, H.A.; Krause, H.B.; Mancha, A.S.; Patankar, M.S.; Weisman, P.S.; Barroilhet, L.; Kreeger, P.K. Alternatively-activated macrophages upregulate mesothelial expression of p-selectin to enhance adhesion of ovarian cancer cells. *Cancer Res.* **2018**, *78*, 3560–3573. [CrossRef] [PubMed]
36. Krausgruber, T.; Blazek, K.; Smallie, T.; Alzabin, S.; Lockstone, H.; Sahgal, N.; Hussell, T.; Feldmann, M.; Udalova, I.A. Irf5 promotes inflammatory macrophage polarization and th1-th17 responses. *Nat. Immunol.* **2011**, *12*, 231–238. [CrossRef] [PubMed]
37. Xue, J.; Schmidt, S.V.; Sander, J.; Draffehn, A.; Krebs, W.; Quester, I.; De Nardo, D.; Gohel, T.D.; Emde, M.; Schmidleithner, L.; et al. Transcriptome-based network analysis reveals a spectrum model of human macrophage activation. *Immunity* **2014**, *40*, 274–288. [CrossRef] [PubMed]
38. Satoh, T.; Takeuchi, O.; Vandenbon, A.; Yasuda, K.; Tanaka, Y.; Kumagai, Y.; Miyake, T.; Matsushita, K.; Okazaki, T.; Saitoh, T.; et al. The jmjd3-irf4 axis regulates m2 macrophage polarization and host responses against helminth infection. *Nat. Immunol.* **2010**, *11*, 936–944. [CrossRef] [PubMed]
39. Martinez, F.O.; Helming, L.; Milde, R.; Varin, A.; Melgert, B.N.; Draijer, C.; Thomas, B.; Fabbri, M.; Crawshaw, A.; Ho, L.P.; et al. Genetic programs expressed in resting and il-4 alternatively activated mouse and human macrophages: Similarities and differences. *Blood* **2013**, *121*, e57–e69. [CrossRef] [PubMed]

40. Griffin, M.; Casadio, R.; Bergamini, C.M. Transglutaminases: Nature's biological glues. *Biochem. J.* **2002**, *368*, 377–396. [CrossRef] [PubMed]
41. Li, H.; Jiang, T.; Li, M.Q.; Zheng, X.L.; Zhao, G.J. Transcriptional regulation of macrophages polarization by micrornas. *Front. Immunol.* **2018**, *9*, 1175. [CrossRef] [PubMed]
42. Yang, S.; Li, J.; Chen, Y.; Zhang, S.; Feng, C.; Hou, Z.; Cai, J.; Wang, Y.; Hui, R.; Lv, B.; et al. Microrna-216a promotes m1 macrophages polarization and atherosclerosis progression by activating telomerase via the smad3/nf-kappab pathway. *Biochim. Biophys. Acta* **2018**. [CrossRef] [PubMed]
43. Zhou, Y.; Que, K.T.; Zhang, Z.; Yi, Z.J.; Zhao, P.X.; You, Y.; Gong, J.P.; Liu, Z.J. Iron overloaded polarizes macrophage to proinflammation phenotype through ros/acetyl-p53 pathway. *Cancer Med.* **2018**, *7*, 4012–4022. [CrossRef] [PubMed]
44. Tan, H.Y.; Wang, N.; Li, S.; Hong, M.; Wang, X.; Feng, Y. The reactive oxygen species in macrophage polarization: Reflecting its dual role in progression and treatment of human diseases. *Oxid. Med. Cell. Longev.* **2016**, *2016*, 2795090. [CrossRef] [PubMed]
45. Ko, S.Y.; Barengo, N.; Ladanyi, A.; Lee, J.S.; Marini, F.; Lengyel, E.; Naora, H. Hoxa9 promotes ovarian cancer growth by stimulating cancer-associated fibroblasts. *J. Clin. Investig.* **2012**, *122*, 3603–3617. [CrossRef] [PubMed]
46. Robinson-Smith, T.M.; Isaacsohn, I.; Mercer, C.A.; Zhou, M.; Van Rooijen, N.; Husseinzadeh, N.; McFarland-Mancini, M.M.; Drew, A.F. Macrophages mediate inflammation-enhanced metastasis of ovarian tumors in mice. *Cancer Res.* **2007**, *67*, 5708–5716. [CrossRef] [PubMed]
47. Escárcega, R.O.; Fuentes-Alexandro, S.; García-Carrasco, M.; Gatica, A.; Zamora, A. The transcription factor nuclear factor-kappa b and cancer. *Clin. Oncol.* **2007**, *19*, 154–161. [CrossRef]
48. Wilson, A.J.; Barham, W.; Saskowski, J.; Tikhomirov, O.; Chen, L.; Lee, H.J.; Yull, F.; Khabele, D. Tracking nf-kappab activity in tumor cells during ovarian cancer progression in a syngeneic mouse model. *J. Ovarian Res.* **2013**, *6*, 63. [CrossRef] [PubMed]
49. Porta, C.; Rimoldi, M.; Raes, G.; Brys, L.; Ghezzi, P.; Di Liberto, D.; Dieli, F.; Ghisletti, S.; Natoli, G.; De Baetselier, P.; et al. Tolerance and m2 (alternative) macrophage polarization are related processes orchestrated by p50 nuclear factor kappab. *Proc. Natl. Acad. Sci. USA* **2009**, *106*, 14978–14983. [CrossRef] [PubMed]
50. Choi, H.; Ignacio, R.M.C.; Lee, E.S.; Wilson, A.J.; Khabele, D.; Son, D.S. Augmented serum amyloid a1/2 mediated by tnf-induced nf-kappab in human serous ovarian epithelial tumors. *Immun. Netw.* **2017**, *17*, 121–127. [CrossRef] [PubMed]
51. Krockenberger, M.; Dombrowski, Y.; Weidler, C.; Ossadnik, M.; Honig, A.; Hausler, S.; Voigt, H.; Becker, J.C.; Leng, L.; Steinle, A.; et al. Macrophage migration inhibitory factor contributes to the immune escape of ovarian cancer by down-regulating nkg2d. *J. Immunol.* **2008**, *180*, 7338–7348. [CrossRef] [PubMed]
52. Krockenberger, M.; Kranke, P.; Hausler, S.; Engel, J.B.; Horn, E.; Nurnberger, K.; Wischhusen, J.; Dietl, J.; Honig, A. Macrophage migration-inhibitory factor levels in serum of patients with ovarian cancer correlates with poor prognosis. *Anticancer Res.* **2012**, *32*, 5233–5238. [PubMed]
53. Krockenberger, M.; Engel, J.; Hausler, S.; Wischhusen, J.; Heuer, S.; Dietl, J.; Honig, A. Macrophage migration inhibitory factor levels in serum of ovarian cancer patients correlate with poor prognostic parameters in ovarian cancer. *Anticancer Res.* **2011**, *31*, 1987–1988.
54. Vyas, M.; Reinartz, S.; Hoffmann, N.; Reiners, K.S.; Lieber, S.; Jansen, J.M.; Wagner, U.; Muller, R.; von Strandmann, E.P. Soluble nkg2d ligands in the ovarian cancer microenvironment are associated with an adverse clinical outcome and decreased memory effector t cells independent of nkg2d downregulation. *Oncoimmunology* **2017**, *6*, e1339854. [CrossRef] [PubMed]
55. Chambers, S.K.; Kacinski, B.M.; Ivins, C.M.; Carcangiu, M.L. Overexpression of epithelial macrophage colony-stimulating factor (csf-1) and csf-1 receptor: A poor prognostic factor in epithelial ovarian cancer, contrasted with a protective effect of stromal csf-1. *Clin. Cancer Res.* **1997**, *3*, 999–1007. [PubMed]
56. Stanley, E.R.; Chitu, V. Csf-1 receptor signaling in myeloid cells. *Cold Spring Harb. Perspect. Biol.* **2014**, *6*, a021857. [CrossRef] [PubMed]
57. Moughon, D.L.; He, H.; Schokrpur, S.; Jiang, Z.K.; Yaqoob, M.; David, J.; Lin, C.; Iruela-Arispe, M.L.; Dorigo, O.; Wu, L. Macrophage blockade using csf1r inhibitors reverses the vascular leakage underlying malignant ascites in late-stage epithelial ovarian cancer. *Cancer Res.* **2015**, *75*, 4742–4752. [CrossRef] [PubMed]
58. Lyons, Y.A.; Pradeep, S.; Wu, S.Y.; Haemmerle, M.; Hansen, J.M.; Wagner, M.J.; Villar-Prados, A.; Nagaraja, A.S.; Dood, R.L.; Previs, R.A.; et al. Macrophage depletion through colony stimulating factor

1 receptor pathway blockade overcomes adaptive resistance to anti-vegf therapy. *Oncotarget* **2017**, *8*, 96496–96505. [CrossRef] [PubMed]

59. Dai, X.M.; Ryan, G.R.; Hapel, A.J.; Dominguez, M.G.; Russell, R.G.; Kapp, S.; Sylvestre, V.; Stanley, E.R. Targeted disruption of the mouse colony-stimulating factor 1 receptor gene results in osteopetrosis, mononuclear phagocyte deficiency, increased primitive progenitor cell frequencies, and reproductive defects. *Blood* **2002**, *99*, 111–120. [CrossRef] [PubMed]
60. Negus, R.P.; Stamp, G.W.; Relf, M.G.; Burke, F.; Malik, S.T.; Bernasconi, S.; Allavena, P.; Sozzani, S.; Mantovani, A.; Balkwill, F.R. The detection and localization of monocyte chemoattractant protein-1 (mcp-1) in human ovarian cancer. *J. Clin. Investig.* **1995**, *95*, 2391–2396. [CrossRef] [PubMed]
61. Jeong, M.; Kim, H.M.; Ahn, J.H.; Lee, K.T.; Jang, D.S.; Choi, J.H. 9-hydroxycanthin-6-one isolated from stem bark of ailanthus altissima induces ovarian cancer cell apoptosis and inhibits the activation of tumor-associated macrophages. *Chem. Biol. Interact.* **2018**, *280*, 99–108. [CrossRef] [PubMed]
62. Gu, L.; Tseng, S.; Horner, R.M.; Tam, C.; Loda, M.; Rollins, B.J. Control of th2 polarization by the chemokine monocyte chemoattractant protein-1. *Nature* **2000**, *404*, 407–411. [CrossRef] [PubMed]
63. Song, S.; Kong, X.; Acosta, S.; Sava, V.; Borlongan, C.V.; Sanchez-Ramos, J. Effects of an inhibitor of monocyte recruitment on recovery from traumatic brain injury in mice treated with granulocyte colony-stimulating factor. *Int. J. Mol. Sci.* **2017**, *18*, 1418. [CrossRef] [PubMed]
64. Reusser, N.M.; Dalton, H.J.; Pradeep, S.; Gonzalez-Villasana, V.; Jennings, N.B.; Vasquez, H.G.; Wen, Y.; Rupaimoole, R.; Nagaraja, A.S.; Gharpure, K.; et al. Clodronate inhibits tumor angiogenesis in mouse models of ovarian cancer. *Cancer Biol. Ther.* **2014**, *15*, 1061–1067. [CrossRef] [PubMed]
65. Parente-Pereira, A.C.; Shmeeda, H.; Whilding, L.M.; Zambirinis, C.P.; Foster, J.; van der Stegen, S.J.; Beatson, R.; Zabinski, T.; Brewig, N.; Sosabowski, J.K.; et al. Adoptive immunotherapy of epithelial ovarian cancer with vgamma9vdelta2 t cells, potentiated by liposomal alendronic acid. *J. Immunol.* **2014**, *193*, 5557–5566. [CrossRef] [PubMed]
66. Germano, G.; Frapolli, R.; Belgiovine, C.; Anselmo, A.; Pesce, S.; Liguori, M.; Erba, E.; Uboldi, S.; Zucchetti, M.; Pasqualini, F.; et al. Role of macrophage targeting in the antitumor activity of trabectedin. *Cancer Cell* **2013**, *23*, 249–262. [CrossRef] [PubMed]
67. Allavena, P.; Signorelli, M.; Chieppa, M.; Erba, E.; Bianchi, G.; Marchesi, F.; Olimpio, C.O.; Bonardi, C.; Garbi, A.; Lissoni, A.; et al. Anti-inflammatory properties of the novel antitumor agent yondelis (trabectedin): Inhibition of macrophage differentiation and cytokine production. *Cancer Res.* **2005**, *65*, 2964–2971. [CrossRef] [PubMed]
68. Penn, C.A.; Yang, K.; Zong, H.; Lim, J.Y.; Cole, A.; Yang, D.; Baker, J.; Goonewardena, S.N.; Buckanovich, R.J. Therapeutic impact of nanoparticle therapy targeting tumor-associated macrophages. *Mol. Cancer Ther.* **2018**, *17*, 96–106. [CrossRef] [PubMed]
69. Wang, Y.C.; He, F.; Feng, F.; Liu, X.W.; Dong, G.Y.; Qin, H.Y.; Hu, X.B.; Zheng, M.H.; Liang, L.; Feng, L.; et al. Notch signaling determines the m1 versus m2 polarization of macrophages in antitumor immune responses. *Cancer Res.* **2010**, *70*, 4840–4849. [CrossRef] [PubMed]
70. Sierra-Filardi, E.; Nieto, C.; Dominguez-Soto, A.; Barroso, R.; Sanchez-Mateos, P.; Puig-Kroger, A.; Lopez-Bravo, M.; Joven, J.; Ardavin, C.; Rodriguez-Fernandez, J.L.; et al. Ccl2 shapes macrophage polarization by gm-csf and m-csf: Identification of ccl2/ccr2-dependent gene expression profile. *J. Immunol.* **2014**, *192*, 3858–3867. [CrossRef] [PubMed]
71. Deng, X.C.; Zhang, P.; Liang, T.T.; Deng, S.Y.; Chen, X.J.; Zhu, L. Ovarian cancer stem cells induce the m2 polarization of macrophages through the ppar gamma and nf-kappa b pathways. *Int. J. Mol. Med.* **2015**, *36*, 449–454. [CrossRef] [PubMed]
72. Wilson, A.J.; Saskowski, J.; Barham, W.; Khabele, D.; Yull, F. Microenvironmental effects limit efficacy of thymoquinone treatment in a mouse model of ovarian cancer. *Mol. Cancer* **2015**, *14*, 192. [CrossRef] [PubMed]
73. Kono, Y.; Kawakami, S.; Higuchi, Y.; Maruyama, K.; Yamashita, F.; Hashida, M. Tumour-associated macrophages targeted transfection with nf-kappab decoy/mannose-modified bubble lipoplexes inhibits tumour growth in tumour-bearing mice. *J. Drug Target.* **2014**, *22*, 439–449. [CrossRef] [PubMed]
74. He, L.; Marneros, A.G. Doxycycline inhibits polarization of macrophages to the proangiogenic m2-type and subsequent neovascularization. *J. Biol. Chem.* **2014**, *289*, 8019–8028. [CrossRef] [PubMed]

75. Lee, K.; Ahn, J.H.; Lee, K.T.; Jang, D.S.; Choi, J.H. Deoxyschizandrin, isolated from schisandra berries, induces cell cycle arrest in ovarian cancer cells and inhibits the protumoural activation of tumour-associated macrophages. *Nutrients* **2018**, *10*, 91. [CrossRef] [PubMed]
76. Zhang, Q.; Li, Y.; Miao, C.; Wang, Y.; Xu, Y.; Dong, R.; Zhang, Z.; Griffin, B.B.; Yuan, C.; Yan, S.; et al. Anti-angiogenesis effect of neferine via regulating autophagy and polarization of tumor-associated macrophages in high grade serous ovarian carcinoma. *Cancer Lett.* **2018**, *432*, 144–155. [CrossRef] [PubMed]
77. Gottlieb, C.E.; Mills, A.M.; Cross, J.V.; Ring, K.L. Tumor-associated macrophage expression of pd-l1 in implants of high grade serous ovarian carcinoma: A comparison of matched primary and metastatic tumors. *Gynecol. Oncol.* **2017**, *144*, 607–612. [CrossRef] [PubMed]
78. Hamanishi, J.; Mandai, M.; Iwasaki, M.; Okazaki, T.; Tanaka, Y.; Yamaguchi, K.; Higuchi, T.; Yagi, H.; Takakura, K.; Minato, N.; et al. Programmed cell death 1 ligand 1 and tumor-infiltrating cd8+ t lymphocytes are prognostic factors of human ovarian cancer. *Proc. Natl. Acad. Sci. USA* **2007**, *104*, 3360–3365. [CrossRef] [PubMed]
79. Ojalvo, L.S.; Thompson, E.D.; Wang, T.L.; Meeker, A.K.; Shih, I.M.; Fader, A.N.; Cimino-Mathews, A.; Emens, L.A. Tumor-associated macrophages and the tumor immune microenvironment of primary and recurrent epithelial ovarian cancer. *Hum. Pathol.* **2018**, *74*, 135–147. [CrossRef] [PubMed]
80. Kryczek, I.; Wei, S.; Zhu, G.; Myers, L.; Mottram, P.; Cheng, P.; Chen, L.; Coukos, G.; Zou, W. Relationship between b7-h4, regulatory t cells, and patient outcome in human ovarian carcinoma. *Cancer Res.* **2007**, *67*, 8900–8905. [CrossRef] [PubMed]
81. Kryczek, I.; Zou, L.; Rodriguez, P.; Zhu, G.; Wei, S.; Mottram, P.; Brumlik, M.; Cheng, P.; Curiel, T.; Myers, L.; et al. B7-h4 expression identifies a novel suppressive macrophage population in human ovarian carcinoma. *J. Exp. Med.* **2006**, *203*, 871–881. [CrossRef] [PubMed]

cancers

MDPI

Review

Cell Origins of High-Grade Serous Ovarian Cancer

Jaeyeon Kim [1,3,*], Eun Young Park [1], Olga Kim [1], Jeanne M. Schilder [2,3], Donna M. Coffey [4], Chi-Heum Cho [5] and Robert C. Bast Jr. [6]

1 Department of Biochemistry and Molecular Biology, Indiana University School of Medicine, Indianapolis, IN 46202, USA; ep11@iu.edu (E.Y.P.); olgakim@iu.edu (O.K.)
2 Department of Obstetrics and Gynecology, Indiana University School of Medicine, Indianapolis, IN 46202, USA; jschilde@iupui.edu
3 Indiana University Melvin & Bren Simon Cancer Center, Indianapolis, IN 46202, USA
4 Department of Pathology and Genomic Medicine, Houston Methodist and Weill Cornell Medical College, Houston, TX 77030, USA; dcoffey@houstonmethodist.org
5 Department of Obstetrics and Gynecology, School of Medicine, Keimyung University, Daegu 41931, Korea; c0035@dsmc.or.kr
6 Department of Experimental Therapeutics, The University of Texas MD Anderson Cancer Center, Houston, TX 77030, USA; rbast@mdanderson.org
* Correspondence: jaeyeonk@iu.edu; Tel.: +1-317-278-9740

Received: 22 October 2018; Accepted: 7 November 2018; Published: 12 November 2018

Abstract: High-grade serous ovarian cancer, also known as high-grade serous carcinoma (HGSC), is the most common and deadliest type of ovarian cancer. HGSC appears to arise from the ovary, fallopian tube, or peritoneum. As most HGSC cases present with widespread peritoneal metastases, it is often not clear where HGSC truly originates. Traditionally, the ovarian surface epithelium (OSE) was long believed to be the origin of HGSC. Since the late 1990s, the fallopian tube epithelium has emerged as a potential primary origin of HGSC. Particularly, serous tubal intraepithelial carcinoma (STIC), a noninvasive tumor lesion formed preferentially in the distal fallopian tube epithelium, was proposed as a precursor for HGSC. It was hypothesized that STIC lesions would progress, over time, to malignant and metastatic HGSC, arising from the fallopian tube or after implanting on the ovary or peritoneum. Many clinical studies and several mouse models support the fallopian tube STIC origin of HGSC. Current evidence indicates that STIC may serve as a precursor for HGSC in high-risk women carrying germline *BRCA1* or *2* mutations. Yet not all STIC lesions appear to progress to clinical HGSCs, nor would all HGSCs arise from STIC lesions, even in high-risk women. Moreover, the clinical importance of STIC remains less clear in women in the general population, in which 85–90% of all HGSCs arise. Recently, increasing attention has been brought to the possibility that many potential precursor or premalignant lesions, though composed of microscopically—and genetically—cancerous cells, do not advance to malignant tumors or lethal malignancies. Hence, rigorous causal evidence would be crucial to establish that STIC is a bona fide premalignant lesion for metastatic HGSC. While not all STICs may transform into malignant tumors, these lesions are clearly associated with increased risk for HGSC. Identification of the molecular characteristics of STICs that predict their malignant potential and clinical behavior would bolster the clinical importance of STIC. Also, as STIC lesions alone cannot account for all HGSCs, other potential cellular origins of HGSC need to be investigated. The fallopian tube stroma in mice, for instance, has been shown to be capable of giving rise to metastatic HGSC, which faithfully recapitulates the clinical behavior and molecular aspect of human HGSC. Elucidating the precise cell(s) of origin of HGSC will be critical for improving the early detection and prevention of ovarian cancer, ultimately reducing ovarian cancer mortality.

Keywords: ovarian cancer; epithelial ovarian cancer; high-grade serous ovarian cancer (HGSOC); high-grade serous carcinoma (HGSC); ovarian cancer origin; fallopian tube; ovarian surface epithelium (OSE); serous tubal intraepithelial carcinoma (STIC)

1. Ovarian Cancer

"Ovarian cancer" is an umbrella term that refers to a heterogeneous group of malignancies arising from or involving the ovary [1–3]. Morphologically, ovarian cancer is classified into two broad categories: (i) non-epithelial ovarian cancer (NEOC) and (ii) epithelial ovarian cancer (EOC). There are two types of NEOC: germ-cell tumors (GCT) and sex cord-stromal tumors (SCST) [4–6]. While 10–15% of ovarian cancer cases are NEOC [4,5], the vast majority (85–90%) belong to EOC [2,3]. According to morphology, molecular alterations, and clinical behavior, EOC is further divided into two groups: type I and type II [7]. Type I tumors are low-grade, slow-growing ovarian carcinomas. Type II tumors are high-grade, aggressive malignancies. The most common type II malignancy is high-grade serous ovarian cancer, also known as high-grade serous carcinoma (HGSC). Hence, an alternative, clinically useful way to categorize EOCs would be to simply split them into two groups: high-grade serous ovarian cancer (HGSC) and non-high-grade-serous ovarian cancer (non-HGSC). Though both are epithelial ovarian cancers, these two groups are biologically-distinct malignancies [8]. Non-HGSCs are mostly indolent tumors confined to the ovary at the time of diagnosis [9]. In contrast, HGSC is an inherently aggressive malignancy, which commonly presents as advanced-stage disease and accounts for the majority of ovarian cancer deaths [9–14]. Thus, "ovarian cancer" is also largely synonymous with "high-grade serous ovarian cancer (HGSC)." While significant tumor burden often involves the ovaries at the time of diagnosis, also typical is widespread metastatic disease involving the fallopian tubes, peritoneal surfaces, and omentum, obscuring the tissue and cell of origin of ovarian cancer [7,15,16].

1.1. Non-Epithelial Ovarian Cancer (NEOC)

Implicit in the name, "ovarian cancer" was thought to be tumors originating in the ovary [17–19]. Certainly, the ovary can be a site of tumor origin. The ovary is composed of follicles, each containing an egg, embedded in interstitial (stromal) tissue and encircled by a single layer of the ovarian surface epithelium (OSE) [20,21]. During each menstrual cycle, in response to the pituitary gonadotropins follicle-stimulating hormone (FSH) and luteinizing hormone (LH), a cohort of preantral follicles in the ovary grow to become antral follicles with multiple layers of granulosa cells and theca cells [21–27]. In antral follicles, the granulosa cells, theca cells, and stromal cells together make up somatic cells of the ovary [21]. These ovarian somatic cells, particularly granulosa cells (90% of SCSTs) as well as theca cells, can transform into sex cord-stromal tumors (SCST), which represent 5–8% of all primary ovarian tumors [4,28–30]. Ovarian tumors can also arise from the egg, which leads to germ-cell tumors (GCT) [6,31–33]. Germ-cell tumors and sex cord-stromal tumors are major types of non-epithelial ovarian malignancies, which account for 10–15% of all ovarian tumors [4,5]. Non-epithelial ovarian cancer (NEOC) is generally diagnosed at an early stage, where tumors are confined to the ovary without distant metastasis (GCT: 60–70%; SCST: 60–95% of cases) [4,5]. Early-stage diagnosis and chemosensitivity present a favorable prognosis to patients with NEOC (five-year survival rates: 90–100%) [4,5,28,34].

1.2. Epithelial Ovarian Cancer (EOC)

Besides non-epithelial tumors, epithelial tumors can form in the ovary. The vast majority of ovarian tumors are epithelial ovarian cancer (EOC), which accounts for 85–90% of ovarian malignancies [2,3]. According to morphologic resemblance to normal epithelial cells lining the reproductive tract, EOC can be subdivided into four major types: serous (~70%), endometrioid (10%), mucinous (3–10%), and clear-cell carcinomas (10%) [2,3,10,11,14,35]. Serous carcinoma cells resemble fallopian tube epithelium; endometrioid carcinoma is likened to endometrial epithelium; mucinous carcinoma resembles the epithelium in the endocervix; and clear-cell carcinoma is similar to clumps of normal glycogen-rich epithelial cells found in the vagina [2,3]. Also, serous carcinomas are of two types: high grade (90–96%) and low grade (4–10%) [7,14].

Recent advances in molecular and genetic analyses on ovarian carcinomas, in conjunction with clinical behavior and histopathology, have EOCs classified into type I and type II [7,36–38]. Type I tumors are low grade, indolent (slow growing), genetically stable, and devoid of p53 (*TP53*) mutations, mostly presenting at early stage. They include low-grade serous carcinoma, low-grade endometrioid carcinoma, clear-cell carcinoma, mucinous carcinoma, and malignant Brenner tumors [7,36]. Type I tumors are associated with wild-type p53 (*TP53*), but often contain mutations in genes such as *KRAS*, *BRAF*, *PTEN*, and β-catenin [7,36,39]. In contrast, Type II tumors are high grade, inherently aggressive, genetically unstable, typically harboring p53 (*TP53*) mutations, and presenting at advanced stage [40,41]. Included in this group are high-grade serous carcinoma (HGSC), high-grade endometrioid carcinoma, undifferentiated carcinomas, and malignant mixed-mesodermal tumors (MMMT; carcinosarcoma) [7,36]. Low-grade cancers resemble normal cells cytologically, whereas high-grade cancers show variation in cellular size and shape, large and irregular nuclei, more frequent mitoses, and loss of polarity [42].

Unlike non-epithelial ovarian tumors, EOCs are diagnosed predominantly at advanced stage (stage III or IV: 60–80%) with widespread metastases throughout the peritoneal cavity, which is associated with high mortality [11,14,43]. In contrast, a smaller fraction of EOCs (20–40% of EOCs) are diagnosed at early stage (stage I or II) [8,11,14]. A recent, large comprehensive histotype analysis of 28,118 cases of EOCs—diagnosed in 2004–2014, drawn from the U.S. Surveillance, Epidemiology, and End Results (SEER) cancer registry data—indicates that 39.2% (11,009/28,118) of EOCs are diagnosed in early stages (stage I and II) and 60.8% (17,109/28,118) of EOCs in advanced stages (stage III and IV) [14]. These early-stage tumors are dominated by type I (low-grade) tumors (61.1%: 6728/11,009) [14] (~85%) [11]: low-grade serous, endometrioid, mucinous, and clear-cell carcinomas. Two-thirds (75.6%: 6728/8900) of the type I, low-grade carcinomas are diagnosed in early stages [14,44]. Generally, the type I tumors are clinically indolent tumors, and thus a relatively minor contributor to ovarian cancer deaths [9,45]. Still, type I tumors account for 18.6% (2235/12,045) of EOC deaths [14]. In contrast, while constituting a minor fraction (38.9%: 4281/11,009) of early-stage EOCs, type II tumors (HGSC and carcinosarcoma) account for the vast majority (87.3%: 14,937/17,109) of advanced-stage EOCs and for most EOC deaths (81.4%: 9810/12,045) [14]. Among the type II tumors, HGSC is the predominant type. HGSC alone accounts for 81.1% (13,898/17,109) of all advanced-stage EOCs and is responsible for nearly three-quarters (73.9%: 8900/12,045) of all EOC deaths [14].

Overall, high-grade serous carcinoma (HGSC), also known as high-grade serous ovarian cancer, is estimated to be 50–60% of all ovarian malignancies [10,11]. Moreover, HGSC accounts for a large majority (63.4%: 7837/28,118) of all ovarian carcinomas, and advanced-stage HGSC represents nearly a half (49.4%: 13,898/28,118) of all EOCs [14]. Hence, when the term "ovarian cancer" or "epithelial ovarian cancer" is used without specific subtype elaboration, most often, it refers to "high-grade serous ovarian cancer (HGSC)".

High-Grade Serous Ovarian Cancer (High-Grade Serous Carcinoma: HGSC)

High-grade serous ovarian cancer refers to the HGSC arising from the ovary, fallopian tube, or peritoneum [9,46] (Figure 1). Yet HGSC (high-grade serous ovarian cancer) should be distinguished from (high-grade) endometrial serous carcinoma. Serous carcinoma arising from the endometrium is also classified as high-grade serous carcinoma, but of uterine origin; hence, it is an endometrial cancer, not an ovarian cancer [47]. This uterine cancer is also commonly called uterine papillary serous carcinoma (UPSC). To distinguish from (high-grade) endometrial serous carcinoma (UPSC), the HGSC of primary ovarian, tubal, or peritoneal malignancy is also called high-grade pelvic (nonuterine) serous carcinoma [48]. HGSC accounts for more than 60% of epithelial ovarian cancers and over 70% of all ovarian cancer deaths [10,14,41]. Thus, HGSC is not only the most common, but also deadliest ovarian cancer [41].

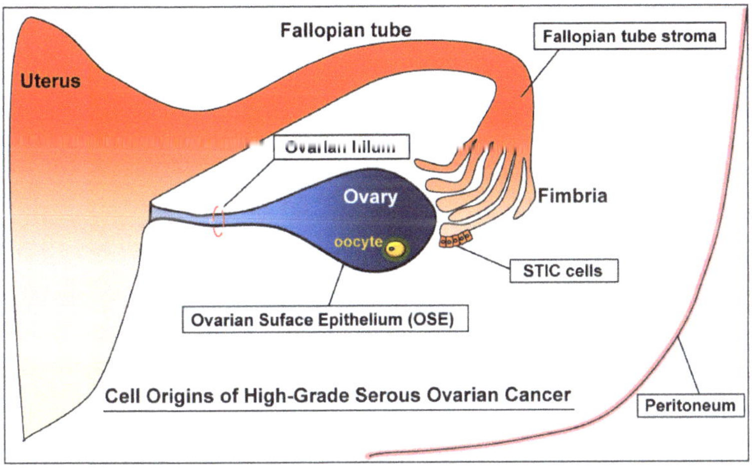

Figure 1. Cells of origin for high-grade serous ovarian cancer (HGSC).

The primary reason for the high mortality associated with HGSC is its diagnosis predominantly at advanced stage [11,14,45,49]. Generally, despite overall high mortality, ovarian cancer has a much better prognosis when diagnosed in the early stages [50]. When ovarian cancer is diagnosed in stage I, and when tumors are still localized to the ovary (15–20% of all cases [2]), the five-year survival rate is 92.3% after treatment with conventional surgery and platinum-based chemotherapy [50]. The five-year survival is still 74.5%, even when the disease has spread to the pelvis in stage II. For advanced-stage ovarian cancer, however, the five-year survival falls to 29.2% [50]. These observation suggest that early detection of ovarian cancer would improve treatment outcomes and survival [51]. Challenging this seemingly obvious notion, however, is the fact that most of early-stage diagnoses are indolent tumors [8,11,44]. The vast majority (>80%) of ovarian cancers detected in stage I are non-epithelial tumors and low-grade epithelial tumors, which are mostly indolent, portending a favorable prognosis [8,11,44]. In contrast, HGSC, which accounts for the majority of ovarian cancer deaths, is not frequently detected in early stages (<25%), with most cases of HGSC being diagnosed in stage III or IV (>75%) [11,14,45,49]. As HGSC account for more than 80% of advanced-stage (III–IV) ovarian cancers and over 70% of ovarian cancer deaths, effective early detection would require detection of a greater fraction of HGSC in early stages prior to distant metastasis [14,45].

HGSC can be detected in early stages. In the Normal Risk Ovarian Cancer Screening Study (NROSS) in the US with more than 5000 women [52], 21 operations were prompted by the screening strategy to detect 13 ovarian cancers with 9 in early stage (I/II). Two of the 9 were borderline, and 7 were invasive. Of the 7 invasive early-stage ovarian cancers detected, 6 were high-grade serous or endometrioid. While the numbers are small, these updated results from the NROSS suggest that HGSC can be detected in the early stage, although there is clearly room to improve the sensitivity of currently-available serum biomarkers and imaging techniques [52].

In addition, early-stage diagnosis of HGSC is not rare. In the large study of the U.S SEER data, HGSC accounts for more than a third (35.8%: 3939/11,009) of early-stage (I and II) EOCs: 19.1% (88/4621) in stage IA/IB (localized) and 47.9% (3057/6388) in stage IC/II (regional) [14]. This study also shows that patients with advanced-stage HGSC have a poor prognosis: 32.1% for 5-year survival and 15% for 10-year survival. In contrast, in patients diagnosed with early-stage HGSC, survival rates improve to 71.4% (5-year) and 53% (10-year), respectively. These improvements in survival among patients with early-stage HGSCs are fairly comparable to survival rates in patients with early-stage type I (indolent) tumors (80.4% for 5-year and 68.0% for 10-year survival). This suggests that effective early-stage detections of HGSC could improve overall patient survival in ovarian cancer.

By far, the largest screening trial of ovarian cancer has been a randomized clinical trial from the United Kingdom Collaborative Trial of Ovarian Cancer Screening (UKCTOCS) [53], in which a total of 202,638 postmenopausal women ages 50 to 74 had been followed for a median of 11.1 years and evaluated for ovarian cancer mortality, after a randomization into a control (no screening) group and two screening groups: (i) serum CA125 and (ii) transvaginal ultrasound (TVUS). The primary analysis shows no significant reduction in ovarian cancer mortality, though ovarian mortality rates are reduced in screening groups by 15% (CA125) and 11% (TVUS), compared with no screening group. There is no difference in all-cause mortality between screening and control groups. Curiously, a secondary analysis, after exclusion of deaths in years 0–7, produces a mortality reduction of 21% in the CA125 group ($p = 0.021$; 95% confidence interval: −2 to 40%). Overall, the results are in line with those from previous ovarian cancer screening trials, which have not shown a significant decrease in ovarian cancer mortality [44,51,54–56].

As HGSC is responsible for more than 70% of ovarian cancer deaths and over 80% of advanced-stage ovarian cancer diagnoses, key to effective ovarian cancer screening is an effective detection of early-stage HGSC. This appears to be challenging. When women with symptoms were promptly diagnosed by CA125 blood test and TVUS, seven of nine (78%) HGSC cases were already in advanced stages (III–IV) with two HGSC cases (22%) in early stages [49], illustrating the challenge of diagnosing HGSC at an early stage with the currently available detection methods.

In principle, the goal of cancer screening is straightforward: detection of cancer at an early, curable stage to reduce cancer mortality and improve patient survival. In practice, however, it is profoundly challenging to detect eventual life-threatening malignancies in their early stages among asymptomatic individuals, who may be at average risk or genetically high risk. Considering the relatively low incidence of ovarian cancer (lifetime risk: 1.3% for ovarian cancer vs. e.g., 12.4% for breast cancer), an effective screening test needs to be equipped with high sensitivity as well as robust specificity. Presently, less than a quarter (22.1%: 3939/17,837) of HGSCs are diagnosed in early stages, while more than three-quarters (77.9%: 13,898/17,837) of HGSCs are not diagnosed until advanced stage [14]. The future success of ovarian cancer screening will therefore depend on how effectively these advanced-stage diagnoses of HGSC could be shifted to early-stage diagnoses.

Achieving effective early-stage detection of HGSC will also require a better understanding of the mechanism of HGSC, particularly early mechanisms, including the cell(s) of origin, cancer initiation and development, early progression, and metastatic transition. As ovarian cancer, particularly HGSC, presents mostly as an advanced-stage disease with widespread peritoneal metastases, it would often not be clear, at the time of diagnosis, where the tumors actually originate in a given patient. Nevertheless, ovarian cancer, including HGSC, was presumed, by convention, to originate in the ovary [18,19,57].

2. Origins of High-Grade Serous Ovarian Cancer (HGSC)

2.1. Ovary

Ovarian Surface Epithelium (OSE)

Epithelial ovarian cancer has been thought to arise from epithelium in the ovary—the ovarian surface epithelium (OSE). Most ovarian cancer patients present with advanced-stage disease, in which tumors are found in the ovary and other peritoneal tissues, including the fallopian tube, mesentery, omentum, and diaphragm. In some patients, however, tumors are confined to the ovary. Though observed in different patients, viewing these tumors as different phases of the same malignancy, ovarian cancer with advanced-stage disease was assumed to have originated in the ovary [18,19,57]. The vast majority of advanced-stage ovarian cancers are high-grade serous carcinoma (HGSC). Hence, HGSC is also thought to arise from the ovary [18,58]. In the ovary, epithelial cells reside in the OSE.

During ovulation, as the mature ovarian follicle ruptures and releases the egg, it also causes a local breakage of the OSE at the ruptured site. As the ruptured follicle differentiates into a corpus luteum, the damaged OSE may invaginate during the repair process. The inward movement of the damaged OSE toward the ovarian cortex (stroma) could result in the formation of a cyst with an epithelial lining inside (termed an "inclusion cyst") [57,59,60]. Alternatively, it was suggested that inclusion cysts could form, without ovulation, as a result of an interaction between the OSE and the ovarian stroma [57]. It has been theorized that OSE and cortical inclusion cysts derived from the OSE may be the origin of all epithelial ovarian cancers [18,19,57,58].

OSE: Developmental View

The OSE is a single layer of squamous (flattened) epithelial cells derived from the coelomic epithelium, which lines the body cavity (coelom) of a developing embryo [59]. In the peritoneal cavity, the coelomic epithelium differentiates into mesothelium, the epithelial cells lining the peritoneum [45]. The part of the coelomic epithelium overlying the gonadal ridge, in which the ovary forms, differentiates into the OSE [59,61]. The OSE is thus a type of mesothelium covering the ovary; hence, the OSE is also known as ovarian mesothelium (OM) [59]. Interestingly, though both tissues are histologically mesothelium, the peritoneal mesothelium and the ovarian mesothelium (OSE) give rise to distinct malignancies. When the mesothelium lining the peritoneum undergoes a malignant transformation, the resulting tumor is called (peritoneal) mesothelioma [45]. On the other hand, transformation of the OSE (ovarian mesothelium) leads to ovarian carcinoma.

The coelomic epithelium also invaginates at the upper lateral part of the gonadal ridge, forming the Müllerian ducts [19,59]. The Müllerian ducts give rise to the epithelia of the fallopian tube, uterine endometrium, and endocervix (the upper part of the cervix) [19,59]. Thus, the peritoneal mesothelium, the ovarian surface epithelium (ovarian mesothelium), and the epithelium derived from the Müllerian ducts share the same embryonic origin: the coelomic epithelium [19,59]. Despite this, the peritoneal and ovarian mesothelium and the Müllerian duct-derived epithelium are phenotypically different [19]. The normal epithelial cells in the fallopian tube, uterus, and endocervix are columnar (tall) epithelium, whereas the peritoneal and ovarian mesothelium are flat-to-cuboidal epithelial cells [59]. Unlike the peritoneal mesothelium, the OSE (ovarian mesothelium) is prone to metaplasia (replacement of one mature cell type with another mature cell type) [19,57]. The OSE tends to spontaneously undergo metaplasia to resemble the normal epithelium of the fallopian tube, uterus, and endocervix [19,57]. This metaplastic capability of OSE to Müllerian epithelium, when combined with gene mutations, could prompt the formation of ovarian carcinomas bearing a morphological similarity to the normal epithelium of the fallopian tube, uterus, and endocervix [62]. Hence, though not naturally resembling the cellular morphology (Müllerian epithelium) of ovarian carcinomas, the OSE appears to have the capacity to transform into ovarian carcinomas bearing the morphology of Müllerian epithelium.

OSE: Mouse Models

Mouse studies have provided abundant evidence that the OSE can transform into ovarian carcinomas resembling the morphology of Müllerian epithelium [62–68] (Table 1). In an allograft study, individual expression of *Hoxa9*, *10*, and *11* in the transformed mouse OSE results in ovarian carcinomas resembling serous, endometrioid, and mucinous ovarian carcinoma, respectively, when these *Hoxa*-expressing OSE cells are injected into mice [62]. The OSE has also been genetically modified by targeted mutations in genetically engineered mouse models. When *Pten* is inactivated and simultaneously *Kras* ^{G12D} mutant is expressed specifically in the OSE by a delivery of adenovirus-cre to the ovarian bursa, these mice develop endometrioid ovarian carcinoma with peritoneal metastases [66]. Adenovirus-cre-mediated inactivation of *Pten* and *Apc* in the murine OSE leads to endometrioid ovarian carcinoma [68]. When *Pten* deletion and *Kras*^{G12D} expression are induced by *Amhr2* $^{cre/+}$, the OSE transforms to produce low-grade serous carcinoma [67]. With an addition of p53 (*Trp53*) $^{R172H/+}$

mutant expression to *Pten* deletion and *Kras*^{G12D} expression, these mice form mucinous carcinomas from the OSE, coexisting with cells of serous features [69].

Table 1. Mouse models of ovarian cancer: ovarian origins.

Targeted Genes	Promoter	Ovarian Tumor	Metastasis	Ascites	Ref.
p53, Myc, Kras ^{G12D}; p53, Kras ^{G12D}, Akt; p53, Akt, Myc	RCAS viral vector	Poorly differentiated or serous carcinoma (in nude mice)	Peritoneal lining, Omentum, Diaphragm, Liver, Pancreas, Intestines, Kidneys	Yes	[64]
p53, Rb1	Adenovirus cre	Serous carcinoma: 97% (33/34 mice)	Peritoneal: 27% (9/33) Lungs: 18% (6/33) Liver: 6% (1/33)	24% (8/33)	[65]
p53, Rb1; p53; Brca1, Rb1; p53, Rb1, Brca1	Adenovirus cre	Leiomyosarcoma: 100% (44/44)	No	27% (12/44)	[70]
p53, Brca1	Adenovirus cre	Leiomyosarcoma or high-grade sarcoma: 56% (23/41)	No	No	[71]
SV40 TAg	Amhr2 (MISIIR)	Serous carcinoma: 50% (18/36)	Peritoneal metastasis including omentum: ?%	Yes (?%)	[63]
Hox9; Hox10; Hox11	pCMV-Tag	HGSC (Hox9) Endometrioid (Hox10) Mucinous (Hox11) (in nude mice)	No description (ND)	ND	[62]
Pten, Apc	Adenovirus cre	Endometrioid carcinoma: 100% (29/29)	Peritoneal: 21% (6/29)	76% (22/29)	[68]
Pten, Kras ^{G12D}	Adenovirus cre	Endometrioid carcinoma: ?% (?/9)	Peritoneal: ?% Lungs: 43%	Yes (?%)	[66]
Pten, Kras ^{G12D}	Amhr2 $^{cre/+}$	Low-grade serous carcinoma:100% (8/8)	Omentum: 100% (8/8)	No	[67,69]
Rb1, p53, Brca1; Rb1, p53 ^{R172H}, Brca1/2	Adenovirus cre	HGSC: Stage I and II: 29% (46/158; 21–32%)	Peritoneal: 16% (25/158; 0–26%) Liver or lung or pleural: 17% (28/158; 0–25%)	Yes (?%)	[72]
p53, Rb1 in the OSE hilum	Adenovirus cre	HGSC: 88% (7/8) (in NOD scid mice)	Lungs: 71% (5/7)	No	[73]
Pten, Kras ^{G12D}, p53 ^{R172H}	Amhr2 $^{cre/+}$	Mucinous carcinoma: 80% (8/10) Mucinous & serous: 100% (10/10)	Omentum: 100% (36/36)	No	[69]
Lkb1, Pten	Amhr2 $^{cre/+}$	HGSC: 100% (12/12)	No description	25% (3/12)	[74]
p53 ^{R172H}, Pten	Amhr2 $^{cre/+}$	HGSC: 30% (15/50) mixed with granulosa cell tumor	Peritoneal HGSC: 100% (15/15); omentum, diaphragm, mesentery, peritoneal lining	80% (12/15)	[75]
p53 ^{R172H}, Pten	Amhr2 $^{cre/+}$	Granulosa cell tumor: 70% (35/50)	Lungs: 53.3% (19/35)	No	[75]

?: information not described in the cited reference.

Ovarian carcinomas histopathologically resembling human HGSC can also arise from the OSE [65,72,74] (Table 1). Mutations in the p53 gene (*TP53* in humans; *Trp53* in mice) is the most common genetic event observed in human HGSC [41]. Inactivation of *p53* (*Trp53*) and *Rb1* in the OSE, via an intrabursal injection of recombinant adenovirus cre, leads to metastatic serous ovarian carcinoma [65]. In this model, nearly all mice (97%: 33/34) develop ovarian tumors which may be histologically classified as serous or undifferentiated carcinoma. Though the majority of these ovarian tumors appear to remain at stage I, 27% (9/33) of the mice exhibit peritoneal metastases with accompanying ascites. Lung metastases are noted in 18% (6/33) of the mice, and 6% (one mouse) develop liver metastases. This reported phenotype, however, was not reproducible in another independent study [70,71]. In this second study, the same experimental approach produced leiomyosarcoma in the ovary—a smooth muscle tumor—instead of the reported metastatic (high-grade) serous ovarian carcinoma [70]. In another mouse study, adenovirus cre was delivered inside the bursa to delete or express in the OSE the following genes: (i) inactivation of *Rb1*, *p53*, and *Brca1*; (ii) expression of a *p53* mutant ($p53^{R172H}$) with *Rb1* and *Brca1/2* deletion [72]. These mice developed ovarian tumors with high-grade serous morphology, ranging from stage I to IV [72]. Examined closely, some of these ovarian tumors remained at stage I or II (29%: 21–32%), lacking metastatic capability. Peritoneal metastases were not common: 16% (0–26%) of the mice were at stage III. Without accompanying peritoneal metastasis, liver, lung, or plural metastases were observed in 17% (0–25%) of the mice.

In another model, loss of *Lkb1* and *Pten* in the OSE leads to ovarian HGSC (100%: 12/12), but with seemingly weak metastatic potential: ascites noted in 25% (3/12) of the mice [74]. In addition, in a transgenic mouse model expressing SV40 large T antigen (TAg) in the OSE, ~50% of the mice produce metastatic ovarian carcinoma of serous morphology [63]. Also, ovarian carcinoma with peritoneal metastasis is formed in an allograft model, in which mice are injected with the OSE harboring p53-null mutation and any two oncogenic mutations of *Myc*, $Kras^{G12D}$, and *Akt* [64].

Evidently, the mouse OSE, incited by gene mutations, is capable of transforming into an ovarian carcinoma histopathologically resembling human HGSC. Yet, the OSE-derived HGSCs from mouse models seem to exhibit weak metastatic potential. Conversely, human HGSC appears to be inherently aggressive, and to be capable of metastasis from a relatively early stage of development [8,9]. It is possible that additional mutations could enable aggressiveness of OSE-derived murine HGSCs. However, a number of mouse models, exhibiting generally weak or deficient metastatic ability of OSE-derived HGSCs, also suggest another possibility: though likely serving as the origin for some HGSCs, the OSE may not be the only primary site from which human HGSC arises [9,73,75–79].

2.2. Fallopian Tube

2.2.1. Serous Tubal Intraepithelial Carcinoma (STIC)

Beginning in the late 1990s, the fallopian tube has emerged as another likely site of origin for high-grade serous carcinoma (HGSC) [9,76,80–82]. This insight came from studies of women carrying germline *BRCA1* or *2* mutations, which make these women more prone to malignancies, particularly ovarian cancer as well as breast cancer [83–85]. In these *BRCA* carriers, the lifetime risk of ovarian cancer ranges from 40–60% for *BRCA1* and 10–30% for *BRCA2*, respectively [2,76,84,86]. The most common type of ovarian cancer noted in these *BRCA1* or *2* carriers is HGSC [87,88]. When *BRCA* mutations were first recognized as conferring high risk for ovarian cancer, ovarian cancer was believed to arise solely from the ovary. Further evaluation, however, demonstrated that *BRCA* carriers were also susceptible to peritoneal and fallopian tube malignancies, as well as ovarian cancer [89–94]. Hence, the standard risk-reducing prophylactic procedure has become surgical removal of the ovaries and fallopian tubes (bilateral salpingo-oophorectomy), preferably by age 40 [84,95].

An early observation implicating the fallopian tube as a site of origin for HGSC came from a study of the fallopian tubes, prophylactically removed by surgery, from high-risk women including *BRCA1* carriers [80]. In this study, half of the high-risk women (6/12) exhibited dysplasia (preneoplastic change) in the epithelium of their fallopian tubes [80]. In contrast, little abnormality was found in the prophylactically-removed ovaries from *BRCA1* carriers [96,97]. These observations prompted more extensive histopathological examinations of the ovaries and fallopian tubes prophylactically removed from germline *BRCA*-mutation-carrying women [96]. These studies had led to the discovery of a potential premalignant lesion in the epithelium of the distal fallopian tube (fimbria), termed "serous tubal intraepithelial carcinoma (STIC)" [98,99].

STIC: Clinical and Molecular Observations

STIC is a noninvasive premalignant lesion with malignant cellular features [45,48], including enlarged nuclei, dark staining of the nucleus (hyperchromasia), coarse chromatin aggregates, and prominent nucleoli, which are also characteristic histopathological features of HGSC [9]. STIC is closely associated with HGSC. STIC lesions are found in prophylactically removed fallopian tubes from asymptomatic germline *BRCA*-mutation carriers (0–11.5%) [100–103]. In addition, STICs are also identified in the fallopian tubes of patients with sporadic (nonhereditary) HGSC (21–59%) [45,48,104–106], as well as germline-*BRCA*-mutation-positive women with hereditary HGSC (3–31%) [107,108]. Most STICs exhibit robust immunostaining of p53 [109,110] and harbor p53 mutations [111] (collectively termed the "p53 signature"). The p53 signature refers to benign-appearing secretory cells in the distal fallopian tube that exhibit intense nuclear p53 (TP53) staining, positive γ-H2AX staining (indicative of DNA damage), and lack of Ki-67 (MIB-1) staining (indicative of minimal proliferative activity) [98,109,112,113]. As p53 signatures lack histological features of STIC, p53 signatures are considered earlier lesions preceding STICs [101,114]. Besides the histopathological resemblance and the association of its occurrence to HGSC, STIC lesions also exhibit genomic instability, a characteristic genomic feature of HGSC, indicating genetic similarity to HGSC [114,115]. In addition, genomic analysis of STIC lesions, fallopian tube tumors, ovarian tumors, and peritoneal metastases from the same patients reveals an evolutionary relationship, suggesting that STICs and p53 signatures are likely early events in the progression of HGSC [114,116]. Furthermore, the evolutionary analysis of genetic changes observed in these various tumor tissues has identified alterations in *BRCA1*, *BRCA2*, *TP53*, and *PTEN* as critical early events in the initiation of STICs and subsequent development of HGSC [114]. Also, gene-expression profile of HGSC exhibits a greater similarity to that of the fallopian tube epithelium than to the ovarian surface epithelium, suggesting a fallopian tube origin of HGSC [117]. Together, these observations have led to the hypothesis that the fallopian tube STIC is a precursor of HGSC arising from the ovary, fallopian tube, or peritoneum [9,76,118,119].

According to this hypothesis, STICs could develop into invasive tumors (HGSC) in the fallopian tube, and the fallopian tube HGSCs then spread to the ovary and peritoneal cavity. Alternatively, STICs could shed and implant on the surface of the ovary or peritoneum where they could progress to HGSC [9,120].

STIC: Mouse Models

Crucial to this hypothesis is the demonstration of a cause-effect relationship between STIC and HGSC: i.e., whether STIC can transform into malignant and metastatic HGSC. Partial evidence has come from studies with genetically engineered mice, in which the fallopian tube epithelium is preferentially, or in combination with other tissues, targeted with gene mutations [78,79,121–125] (Table 2). In one mouse study, expression of mutant p53 (*Trp53*) and inactivation of *Pten* and *Brca1/2*, together, generated STICs (83.9%: 26/31 mice) as well as HGSCs in the ovary (73.1%: 19/26) and peritoneum (73.1%: 19/26) [78]. In another study, STIC formed in 35% (28/80) of mice harboring the inactivation of four genes (*Brca1*, *p53*, *Rb1*, and *Nf1*: 48 mice) or three genes (*Brca1*, *p53*, and *Rb1*: 29 mice; *Brca1*, *p53*, and *Nf1*: 3 mice) in the fallopian tube epithelium [79]. Also, in this study, more than two-thirds of the mice (68.8%: 55/80) formed fallopian tube tumors (STIC, early-stage HGSC, and HGSC); some of these mice also developed ovarian tumors (38.2%: 21/55) or peritoneal tumors (12.7%: 7/55) or ascites (12.7%: 7/55), or a combination of the three [79]. In the same study [79], inactivation of *Brca1*, *p53*, and *Pten* in the fallopian tube epithelium also produced STIC or early-stage HGSC or both or HGSC in the fallopian tube in 90% of mice: STIC (40%: 4/10 mice); early-stage HGSC (60%: 6/10); and HGSC (20%: 2/10). In these mouse models [78,79], as the gene mutations occur in the fallopian tube epithelium—but not in the ovary or peritoneum—ovarian and peritoneal HGSCs are likely tumors resulting from STICs formed in the fallopian tube.

Table 2. Mouse models of ovarian cancer: fallopian tube origins.

Targeted Genes	Promoter	STIC	Fallopian Tube HGSC	Ovarian HGSC Metastasis	Peritoneal HGSC Metastasis	Ascites	Ref.
SV40 TAg	Ovgp1	-	Oviductal tumors (?%)	No ovarian tumor	No; Uterine tumor: 100% (26/26) Vaginal tumor: 62% (16/26)	No	[122]
- Monitoring of tumor development: 6–13 weeks of age							
Brca1, p53 ^{R172H}, Pten	Pax8	100% (4/4)	No	25% (1/4)	25% (1/4): peritoneal mass	No	[78]
- Monitoring of tumor development: 5–7 weeks of age							
Brca2, p53 ^{R172H}, Pten	Pax8	75% (9/12)	No	75% (9/12)	67% (8/12): peritoneal mass	No	[78]
- Monitoring of tumor development: 7–15 weeks of age							
p53 ^{R172H}, Pten	Pax8	67% (4/6)	No	0% (0/6)	0% (0/6)	No	[78]
- Monitoring of tumor development: 19–38 weeks of age							
SV40 TAg	Ovgp1	Yes (?%)	No	Adeno-carcinoma (56%)	No	No	[123]
- Monitoring of tumor development: 8–10 weeks of age							
Brca1, p53, Rb1, Nf1	Ovgp1-iCreER	37.5% (18/48)	HGSC: 60% (29/48) MMMT: 25% (12/48)	HGSC or MMMT: 40% (19/48)	HGSC or MMMT: 13% (6/48)	13% (6/48)	[125]
- Monitoring of tumor development: 3.5–26 months of age							
Brca1, p53, Rb1	Ovgp1-iCreER	34.5% (10/29)	HGSC: 17% (5/29) MMMT: 7% (2/29)	0%	0%	0%	[125]
- Monitoring of tumor development: 3.5–26 months of age							
Brca1, p53, Nf1	Ovgp1-iCreER	0% (0/3)	HGSC: 67% (2/3) MMMT: 67% (2/2)	HGSC or MMMT: 100% (3/3)	HGSC or MMMT: 33% (1/3)	0%	[125]
- Monitoring of tumor development: 3.5–26 months of age							
Brca1, p53, Pten	Ovgp1-iCreER	40% (4/10)	HGSC: 80% (8/10) MMMT: 10% (1/10)	MMMT: 10% (1/10)	0%	10% (1/10)	[125]
- Monitoring of tumor development: 3–8 months of age							
Dicer1, Pten	Amhr2 $^{cre/+}$	No	100% (24/24)	100% (24/24)	100% (24/24): omentum, diaphragm, mesentery, peritoneal lining	100% (24/24)	[77]
- Survival range: 6.2–13 months of age (mean survival = 9.4 months; *n* = 24)							

MMMT: malignant mixed mesodermal tumor (carcinosarcoma); HGSC: high-grade serous carcinoma or high-grade serous ovarian cancer; p53: *Trp53*; ?: information not described in the cited reference.

Overall, these mouse models, targeting fallopian tube epithelium with gene mutations, develop STICs and HGSCs which closely recapitulate many of clinical and molecular features of STIC and HGSC in humans. Also, notably, *Brca1* or *2* inactivation appears to be necessary for STIC to advance to HGSC in mice [78]. These mouse models therefore support a notion that STIC can be a precursor lesion for HGSC in genetically high-risk women carrying germline *BRCA1/2* mutations.

It remains unclear, however, whether STIC could also serve as a precursor for sporadic (nonhereditary) HGSCs, which account for 85–90% of all HGSC cases [2,76]. In mouse models without *Brca1* or *2* mutation, *p53* mutation and *Pten* deletion together can produce STIC lesions in the fallopian tube epithelium (67%: 4/6 mice) (Table 2) [78]. Yet these STICs do not progress to invasive or metastatic HGSCs in the ovary or peritoneal cavity [78]. In another study, expression of the SV40 large T antigen (TAg), driven by the oviductal glycoprotein 1 (*Ovgp1*) promoter, also results in STIC lesions in the fallopian tube epithelium along with adenocarcinoma in the ovary in some mice [123]. As *Ovgp1* is highly expressed in the fallopian tube and TAg expression is not detected in the ovarian surface epithelium, this ovarian carcinoma is presumed to have resulted from the spread and transformation of fallopian tube STICs in the ovary [123]. Nevertheless, these STIC lesions do not advance to peritoneal HGSC [123]. Thus, these mouse studies suggest that many STIC lesions may not progress to invasive, and more critically, metastatic malignancies [78,79,123]. Studies of human HGSCs also note that most STIC lesions likely do not advance to metastatic HGSC, and may thus be classified as low grade [98,126]. Hence, it remains to be elucidated whether STIC could be a bona fide precursor lesion for HGSC in women in the general population who are at average risk, and yet who account for most cases of HGSC.

STIC: Clinical Significance

Overall, though existence of STIC and its association to HGSC are extensively described in human and mouse studies, the clinical significance of STIC remains uncertain [127]. In the early studies, in which the fallopian tube was first proposed as a potential primary origin of HGSC, tubal dysplastic lesions, later termed STICs, were reported in 37% and 50% of high-risk women whose fallopian tubes were prophylactically removed [80,118] (Table 3). In most clinical studies, however, including ones with larger sample sizes, the prevalence of STICs (or occult tubal carcinomas) in high-risk women generally varies from 0–11.5% [100–103,128–139] (Table 3). It is unknown what proportion of the STICs would progress to malignant and metastatic HGSC. It is possible that HGSC could also develop from precursors distinct from STICs. Human and murine studies have indicated that metastatic HGSC can arise from the fallopian tube without evidence of STIC [77,79,120].

Table 3. Incidence of STIC in high-risk women and in the general population.

Sample Tissue	Population	Incidence of STIC or Occult Tubal Carcinoma	Number of Cases	Reference
Fallopian tubes from prophylactic salpingo-oophorectomy	High risk	50% (6)	12	Piek et al., 2001 [80]
		37% (16?)	44	Piek et al., 2003 [118]
		6.7% (4)	60	Colgan et al., 2001 [133]
		10% (3)	30	Leeper et al., 2002 [93]
		6% (4)	67	Powell et al., 2005 [137]
		8% (4)	50	Carcangiu et al., 2006 [131]
		3.8% (6)	159	Finch et al., 2006 [100]
		5.7% (7)	122	Callahan et al., 2007 [130]
		8.5% (15)	176	Shaw et al., 2009 [101]
		8.9% (4)	45	Hirst et al., 2009 [134]
		8.1% (9)	111	Powell et al., 2011 [128]
		8.5% (10)	117	Manchanda et al., 2011 [135]
		7.1% (16)	226	Mingels et al., 2012 [136]
		1.7% (5)	303	Reitsma et al., 2013 [138]
		4.2% (17)	405	Powell et al., 2013 [129]
		2.0% (12)	593	Wethington et al., 2013 [102]
		11.5% (9)	78	Cass et al., 2014 [132]
		2.6% (25)	966	Sherman et al., 2014 [139]
		0% (0)	111	Seidman et al., 2016 [140]
		5.6% (2)	36	Lee et al., 2017 [103]
Fallopian tubes from HGSC cases	High risk	30.8% (8)	26	Howitt et al., 2015 [108]
		3.3% (2)	60	Malmberg et al., 2016 [107]

Table 3. *Cont.*

Sample Tissue	Population	Incidence of STIC or Occult Tubal Carcinoma	Number of Cases	Reference
Fallopian tubes from HGSC cases	General	47.6% (20)	42	Kindelberger et al., 2007 [104]
		58.5% (24)	41	Przybycin et al., 2010 [48]
		37.3% (19)	51	Seidman et al., 2011 [45]
		20.5% (8)	39	Tang et al., 2012 [106]
		38.3% (23)	60	Mingels et al., 2014 [141]
		38.2% (13)	34	Koc et al., 2014 [105]
		33.3% (6)	18	Malmberg et al., 2016 [107]
Fallopian tubes from non-ovarian-cancer or benign cases	General	3.1% (2)	64	Shaw et al., 2009 [101]
		0.8% (4)	522	Rabban et al., 2014 [142]
		1.1% (3)	277	Seidman et al., 2016 [140]
Fallopian tubes from endometrial serous carcinoma cases	General	22.7% (5)	22	Jarboe et al., 2009 [47]
		21.8% (12)	55	Stewart et al., 2010 [143]
		14.3% (4)	28	Tang et al., 2012 [106]
		7.9% (3)	38	Tolcher et al., 2015 [144]
Fallopian tubes from endometrial carcinoma or hyperplasia cases	General	1.7% (3)	175	Seidman et al., 2016 [140]

?: information not described in the cited reference.

Also, STIC lesions may not be unique to HGSC. Though STICs are associated chiefly with HGSC, they are not exclusive to HGSCs of the ovary, fallopian tube, and peritoneum. STICs are also found in the fallopian tubes of patients with (high-grade) endometrial serous carcinoma (8–23%) [47,106,143,144], as well as ones from patients with nonserous endometrial carcinoma or endometrial hyperplasia (1.1%) [140] (Table 3). STICs are also found in the fallopian tubes of 3% (2/64) of women undergoing surgery not related to ovarian, tubal, or peritoneal malignancy [101]. In addition, STIC lesions are not found in a large fraction of clinical HGSC cases. In advanced human HGSC, the occurrence of STIC ranges from 21–59% in sporadic HGSC [45,48,104–106,141] and 3–31% in (germline-*BRCA*-mutation-positive) hereditary HGSC [107,108]. These findings suggest that a significant number of HGSCs may derive from precursors independent of STICs [36,108]. Also, p53 (*TP53*) signatures, another potential precursor lesions in the fallopian tube closely associated with STICs, is not unique to high-risk women, but is also commonly seen in women who are at low risk of ovarian cancer [101,109]. The p53 (*TP53*) signatures were present in the fallopian tubes of 19% (12/64) and 33% (19/58) of women who are at average risk of ovarian cancer, compared with 11% (19/176) and 24% (10/41) of high-risk women undergoing risk-reducing salpingo-oophorectomy (RRSO) [101,109]. The fact that STIC and p53 (*TP53*) signatures are not unique to HGSC and HGSC can arise without evidence of STIC suggests that there may exist additional, yet undiscovered, precursor lesions for HGSC [36]. Given the highly aggressive nature of HGSC, the existence of novel precursor lesions would not be inconceivable.

STIC is evidently a risk factor for HGSC. Yet, more causal evidence would be needed to affirm that STIC is a bona fide precursor lesion for hereditary (10–15%) and sporadic (85–90%) HGSCs in women. The key issue is whether STIC, a lesion of noninvasive neoplastic cells, could evolve into an invasive tumor and advance to aggressive metastatic HGSC. In the aforementioned mouse models [78,79], though some fallopian tube STICs lead to ovarian and peritoneal HGSCs, the extent and spectrum of metastases do not appear to fully match the clinical metastases observed in human HGSC.

Regarding the fallopian tube as an origin of HGSC, currently available data indicate that STIC likely serves as a premalignant lesion that could develop into metastatic HGSC in high-risk women carrying *germline BRCA1/2* mutations [145]. The overall fraction of ovarian cancers that originate in the fallopian tube is not known. In one study examining specimens from risk-reducing salpingo-oophorectomy (RRSO), microscopic HGSC was identified in the fallopian tube in six (4.5%) of 133 *BRCA1/2* carriers [113]. Four of the HGSCs were confined to the fallopian tube, one to the ovary only, and one was identified from peritoneal washings only. In contrast, when early-stage (I and II) HGSCs (14/131) from *BRCA1/2* carriers were examined, the majority of early-stage HGSCs (78.6%: 11/14) were diagnosed as ovarian primaries, while three cases (21.4%: 3/14) as fallopian tube primaries [113]. These results suggest that the fallopian tube may be the primary site of origin of HGSC, but the ovary is a preferred site of tumor growth and progression in high-risk women. Temptingly, this notion might be extended to HGSC in the general population, but there is not yet sufficient causal evidence [127].

It seems logical to predict that a precursor lesion such as STIC, which exhibits malignant morphological and genetic features, would gradually progress and eventually manifest as a full-blown metastatic malignancy [146]. However, it is also increasingly recognized that many of microscopically cancerous precursor or early-tumor lesions may not proceed to clinical, lethal malignancies [81,147–150]. Thus, rigorous evaluations of the causal relationship between STIC and metastatic HGSC would be critical for establishing STIC as an origin of metastatic HGSC.

Insights from Ductal Carcinoma In Situ (DCIS) in Breast Cancer

Ductal carcinoma in situ (DCIS), a microscopic malignancy confined to the breast ductal epithelium without invasion of the duct wall, is widely believed to be a precursor or premalignant lesion for all breast cancer malignancies [151]. In 2018, DCIS was estimated to be 24% (63,960) of newly diagnosed breast tumors (266,120) [152]. Since the 1970s, when screening mammography was introduced, detection of DCIS had risen by 700% from 1976 to 2008 (7 DCIS/100,000 women in 1976 and 56 in 2008) [153]. An additional 573,000 cases of DCIS were estimated to be diagnosed during the three decades. If DCIS is a true premalignant lesion for invasive and lethal breast cancer, more cases of DCIS should lead to a proportional decline in the incidence of advanced-stage breast cancer. However, late-stage breast cancer cases had fallen by only 8% during the same period (102/100,000 women in 1976 and 94 in 2008; diagnosis of additional 573,000 cases of DCIS vs. an estimated decrease of 67,000 cases of late-stage breast cancer during this period) [153]. These clinical observations suggest that though some DCIS can advance to invasive breast cancer, the vast majority of DCIS lesions are unlikely to progress to metastatic, life-threatening malignancies [154].

In an autopsy study in which women died from causes other than breast cancer, DCIS was identified in about 30% of women between the ages of 40 and 54 years [155]. As one in eight women (12.4%) develops breast cancer during her lifetime in the U.S. [156], this common occurrence of DCIS in the general population also bolsters the notion that many DCIS lesions do not lead to malignant breast cancers [155]. Some DCIS lesions do proceed to invasive, metastatic breast cancer [157]. It is unknown, however, which DCIS would advance to invasive, metastatic cancer and which would not; the estimates vary widely from 0–50% [154,157,158]. For this reason, though noninvasive, DCIS is still regarded as breast cancer (stage 0) and treated as aggressively as invasive breast cancer with surgery and hormonal therapy [154]. There is a growing perception that the rise of DCIS diagnoses has led to overdiagnosis and overtreatment [147,159].

Reflecting an increasing awareness of the overdiagnosis and overtreatment, there has also been a growing level of recognition that many of premalignant lesions, albeit histopathologically classified as cancer, often do not progress to invasive or metastatic tumors [81,147,148]. Hence, it has been proposed that premalignant lesions, such as ductal carcinoma in situ (DCIS), not be labeled as cancer or carcinoma [147]. Instead, they may be reclassified using a more appropriate term reflecting their indolent clinical behavior, such as "indolent lesions of epithelial origin (IDLE)" [147]. Also, appreciating the generally low malignant potential of most STICs, a new term, such as "low-grade serous tubal intraepithelial neoplasia," was suggested in place of "STIC" [81].

Like DCIS, STIC is also a noninvasive tumor of epithelial origin [45,48]. Though not invasive, these STIC lesions possess malignant cellular features [98] as well as widespread genomic alterations, and are therefore considered preinvasive or premalignant lesions [114–116]. It is intuitive to assume that these lesions would eventually become invasive in the local tissue and ultimately spread to other parts of the body. Yet many microscopic cancers may not, though some do, proceed to clinical malignancy [149,150,160,161]. Current evidence suggests that some STIC lesions would be capable of progressing to invasive tumors leading to peritoneal metastases in the context of germline *BRCA*-mutation carriers [78,79]. Other STIC lesions may not undergo malignant transformation, remaining noninvasive [78,79,123]. Particularly critical, yet unknown, is the clinical significance of STIC in the general population where 85–90% of HGSC cases occur. The information on the prevalence of STICs in the general population is limited, but several studies suggest that it could vary from 0.8–3.1% [101,140,142], while the lifetime risk of ovarian cancer is 1.3% (one in 78 women) [50,162]. Like DCIS, most likely, some STICs would possess malignant potential, while more STICs would not.

Advances in technology would increasingly facilitate detection of these noninvasive yet potentially cancerous lesions, such as DCIS and STIC, in the general population as well as high-risk individuals. The challenge is to be able to predict how these potential precancerous lesions would behave in the course of tumor progression: would these precancerous lesions cause little harm or turn deadly if left untreated? As STIC has emerged as a potential precursor for human HGSC, it is crucial to elucidate the natural progression of STIC in the context of HGSC development at the molecular as well as biological levels. A deeper understanding of the natural history of STIC would help develop ways to clinically assess the malignant potential of STIC lesions.

2.2.2. Fallopian Tube Stroma

Fallopian tube origin of HGSC is also supported by a serendipitous phenotype in a mouse model, in which *Dicer1* and *Pten* are inactivated in the fallopian tube [77] (Table 2). These *Dicer1-Pten* double-knockout (DKO) mice faithfully and reliably reproduce the clinical behavior of human HGSC with 100% penetrance [77]. In these mice, HGSC forms in the fallopian tube, and then spreads to envelop the ovaries, and also aggressively metastasizes throughout the peritoneal cavity. Peritoneal metastases occur preferentially to the omentum and diaphragm with widespread tumors in the mesentery and peritoneal membrane, invariably accompanied by ascites. All of the mice die from peritoneal metastases of fallopian tube HGSC (n = 24/24; 6.2–13 months of age; mean survival: 9.4 months) [77]. Though metastases are generally confined to the peritoneal cavity in these mice, occasionally, HGSC metastasizes to the lungs (stage IV), as well as the peritoneal cavity (stage III). Besides phenotypic and histopathologic resemblance, there are also significant correlations in gene expression between mouse and human HGSCs [77]. Moreover, these mouse HGSCs also exhibit widespread genomic instability resembling human HGSC (unpublished).

However, while modeling the clinical behavior of human HGSC, histopathologically classified as HGSC, exhibiting marked genomic disarray, and unmistakably stemming from the fallopian tube, the HGSC in this model does not appear to originate in the fallopian tube epithelium, but rather, in the fallopian tube stroma [77]. In this model, *Amhr2-cre* (*Amhr2*$^{cre/+}$), in which the insertion of *cre* recombinase gene is targeted to an endogenous *Amhr2* locus, would direct the deletion of *Dicer1* and *Pten* specific to the fallopian tube stroma, not in the epithelium [163]. Accordingly, no histopathologic abnormality or STIC lesions were present in fallopian tube epithelium [77]. In this mouse model, in the absence of *Dicer1* and *Pten*, stromal stem cells residing in the fallopian tube [163] may transform into HGSC in the fallopian tube, leading to widespread peritoneal metastases as well as the ovarian metastasis. Thus, despite fallopian tube origin, the stromal origin of HGSC in this mouse model is at odds with the STIC hypothesis, which predicts that the fallopian epithelium, particularly distal tubal epithelium, is the primary cell of origin of metastatic HGSC in humans. There is yet no clinical evidence that human HGSC could originate in the stroma of the fallopian tube.

Also, the genetic relevance of *DICER1* and *PTEN* in human HGSC is not clear. Low expression of DICER1 is associated with advanced stages and reduced survival in human ovarian cancer (HGSC) [164], suggesting that DICER1 may function as a tumor suppressor. Yet the role of *DICER1* loss in ovarian cancer remains to be clarified. In human HGSC, homozygous deletion of *DICER1* (*DICER1*$^{-/-}$) is extremely rare (0.3%: 1/316 patients) [41,165,166]. Relatively common, however, is a single-copy loss of *DICER1* (44%). Similarly, PTEN deletion is found in 38.9% of the cases for combined homozygous (6.6%) and heterozygous losses (32.3%) [41]. Though PI3K signaling is frequently activated in human HGSC, it is not clear how this partial loss of *PTEN* impacts the development of HGSC.

Thus, most cases of human HGSC are unlikely to occur as a direct consequence of loss of *DICER1* and *PTEN*. Rather, this mouse model reveals critical pathways—activated by loss of *Dicer1* and *Pten*—what are essential to the development of metastatic HGSC. Accordingly, this mouse HGSC reveals significant alterations of known critical pathways for HGSC [41], including PI3K signaling [77], FOXM1 signaling (unpublished), and homologous recombination (unpublished).

Despite many features in common between this mouse model and human HGSC—most notably, the striking clinical resemblance to human HGSC—the major limitation of this model is the precise cell origin of this murine HGSC: fallopian tube stroma. According to current understanding, it is difficult to envision that human HGSC, an epithelial malignancy, could originate, not in epithelium, but rather in stroma. In the mouse uterus, a fraction of stromal cells of *Amhr2* lineage, likely stromal stem cells, are capable of differentiating into epithelial cells during the endometrial regeneration after parturition [163]. Plausibly, stem cells residing in the fallopian tube stroma, in the absence of *Dicer1* and *Pten*, transform into HGSC. Whether this potential stromal-to-epithelial transition could occur during the development of human HGSC remains to be clarified. Nevertheless, it should also be acknowledged that the natural course of human HGSC—initiation, development, early progression, and ultimate peritoneal metastasis—remains poorly understood.

Rarely does an animal model manifest a full spectrum of clinical disease of a human disorder. If an animal model, however, develops a cancer that behaves like the human cancer with a nearly identical metastatic pattern, one could reason that the mouse and human malignancies likely share similar mechanisms of development and tumor progression. This mouse model could be a valuable tool in understanding the mechanisms underlying the development, early progression, and metastatic progression of human HGSC—and also serve as a useful preclinical model for evaluating new therapies.

2.2.3. Ovarian Cancer Prevention: Salpingo-Oophorectomy vs. Salpingectomy

Over the last decade, clinical and mouse studies have bolstered the idea that the fallopian tube is the predominant origin of HGSC. This novel concept has spawned new thinking in ovarian cancer prevention in women at high genetic risk and also in the general population.

Salpingo-oophorectomy—surgical removal of the ovaries and fallopian tubes—is the standard preventive surgery recommended for germline *BRCA1/2* carriers, who are at high risk of ovarian cancer [9,85,167,168]. This prophylactic surgery has proven to be highly effective, as it reduces the risk of ovarian cancer by 72–96% [86,89,169], ovarian-cancer-specific mortality by 79–95% [86,170], and overall mortality by 60–66% [86,170]. Besides ovarian cancer protection, oophorectomy also significantly decreases the risk of breast cancer (50% risk reduction) in *BRCA1/2* carriers [84,86,89,170–173] (though some studies indicate no difference in breast cancer risk, or a selective risk reduction only among *BRCA2* carriers [174,175]).

Despite the proven benefits of risk and mortality reductions in ovarian and breast cancer, salpingo-oophorectomy has a major drawback of premature menopause, which hinders wider acceptance. Prophylactic surgery involves removal of the ovaries and fallopian tubes, and is recommended between the ages of 35 and 40 years, when women are premenopausal [84,95,168,176]. Consequently, these women undergo premature surgical menopause with an increased risk of experiencing postmenopausal symptoms, including hot flashes, sleep disturbance, mood changes, vaginal dryness, sexual dysfunction, cognitive decline, osteoporosis, and cardiovascular disease [177–179]. To avoid these adverse health outcomes, approximately 30–50% of *BRCA1* and *2* carriers choose not to undergo prophylactic salpingo-oophorectomy even after completion of child bearing [95,180–185].

With the emerging concept of the fallopian tube as the principal site of origin for HGSC, an alternative preventive approach has been proposed to mitigate the adverse consequences of ovary removal in women at high genetic risk. A two-step surgery has been proposed, in which salpingectomy (removal of the fallopian tubes) is performed early upon completion of childbearing, followed by subsequent oophorectomy (removal of the ovaries) at ages 40–45 for *BRCA1* and ages 45–50 for *BRCA2* carriers [180,186,187]. Although salpingectomy alone with ovary conservation could theoretically reduce the risk of HGSC, this approach warrants careful consideration.

While the majority of HGSCs arise in the fallopian tube, it may not be that all originate there [113,139]. Thus, retaining the ovaries with salpingectomy alone would continue to pose ovarian cancer risk to some high-risk women until the completion of oophorectomy [180,188]. Another concern is a delay in oophorectomy. Breast cancer protection appears to occur when oophorectomy is performed before menopause in high-risk women [86,89,170,172,173]. Delaying ovary removal may diminish the benefit of breast cancer protection [176,188]. Ovarian cancer risk appears to be higher in salpingectomy alone than salpingo-oophorectomy in the general population [189,190]. Clinical trials are underway to evaluate the effectiveness of salpingectomy alone or salpingectomy followed by delayed oophorectomy in high-risk women [176,186]. Presently, bilateral salpingo-oophorectomy is still the recommended prophylactic surgery for high-risk women carrying germline *BRCA1* or *2* mutations [176].

This new paradigm of fallopian tube origin also presents the option for an opportunistic salpingectomy as an alternative preventive measure in the general population (who are at average risk) undergoing hysterectomy for benign disease or pelvic surgery [95,191–193]. In the United States, approximately 600,000 women undergo a hysterectomy for benign (uterine) disease [194]. About half (300,000) the women undergoing hysterectomy also opt for a prophylactic salpingo-oophorectomy for a variety of health reasons, including risk reduction of ovarian cancer; the other half choose to retain their ovaries and fallopian tubes to avert the adverse health consequences associated with removal of the ovaries [191,195]. As expected, the risk of developing ovarian and breast cancer is reduced (by 96% and 25%, respectively) in average-risk women undergoing hysterectomy and salpingo-oophorectomy, compared with women undergoing hysterectomy but conserving the ovaries [196]. However, adverse effects of ovary removal are also pronounced in these women. Beyond premature menopausal symptoms, ovary removal has also been associated, in some studies, with increased mortality in women with hysterectomy (28% increase in the risk of death for coronary heart disease, and 12% increase in overall mortality), compared with women whose ovaries are retained at the time of hysterectomy [195,196]. Observational studies suggest that ovary removal may do more harm than good in average-risk women in the general population [195].

Germline *BRCA1/2* carriers have high risk of both ovarian and breast cancer (lifetime risk: 10–60% for ovarian cancer [2,86]; 66–82% for breast cancer [84,85]). In these high-risk women, removal of the ovaries offers dual benefits: not only does it prevent ovarian cancer, it also significantly lowers breast cancer risk (by 47–64%) [84,86,171] and mortality (by 56–90%) [84,86,170]. The benefits of removing the ovaries (as well as the fallopian tubes), therefore, appear to be evident in these women at high genetic risk. In the general population, however, women are at average risk for ovarian and breast cancer (lifetime risk: 1.3% for ovarian cancer [50,162]; 12.4% for breast cancer [50,152]). In these average-risk women, though ovary removal appears to reduce the risk of breast cancer, it seems to have little impact on breast cancer mortality [195,196]. In contrast to a clear risk reduction in high-risk women, ovary removal seems to bring to average-risk women a relatively modest benefit in relation to breast cancer.

These average-risk women may benefit from fallopian tube removal alone and retaining the ovaries. Salpingectomy alone is shown to reduce the risk of ovarian cancer by 35–64% in the general population [189,190]. Conservation of the ovaries would improve quality of life by averting premature menopausal symptoms, and also extend overall survival by reducing ovary-loss-associated mortality. Hence, salpingectomy with ovary retention, in lieu of salpingo-oophorectomy, can be an option to these average-risk women undergoing hysterectomy or pelvic surgery [95,193].

In summary, in high-risk women, until further evidence is available, risk-reducing salpingo-oophorectomy (RRSO) is recommended by age 40 for *BRCA1* mutation carriers and by age 45 for *BRCA2* mutation carriers [95,176]. On the other hand, for women at average risk of ovarian and breast cancer, salpingectomy alone—and keeping the ovaries—may be a prophylactic option to consider when undergoing hysterectomy or pelvic surgery [193].

2.3. Other Potential Origins of HGSC

2.3.1. Secondary Müllerian System

Epithelial ovarian cancer may also arise in the secondary Müllerian system [197], which refers to the presence of Müllerian epithelium (i.e., epithelium of the fallopian tube, uterus, and endocervix) outside the indigenous locations (i.e., the fallopian tube, uterus, and endocervix). The secondary Müllerian system includes endometriosis (endometrium-like tissue present outside the uterus), endosalpingiosis (fallopian tube-epithelium-like epithelium on or beneath the peritoneal surface), and the rete ovarii (Müllerian epithelium-resembling tubular structures near the ovarian hilum, a junctional area between the ovary and the fallopian tube). As HGSC resembles normal fallopian tube epithelium, the secondary Müllerian system could serve as an origin of HGSC [198]. After prophylactic removal of the ovaries, some of these ovary-deficient high-risk women develop HGSC in the peritoneum [89,199]. This primary peritoneal HGSC is believed to arise from endosalpingiosis [45] or Müllerian metaplasia of peritoneal mesothelium [76].

2.3.2. Ovarian Hilum

In addition, the epithelium (particularly stem cells) lining the hilum, a junctional area between the ovary and the fallopian tube, has been suggested to be a cell of origin for HGSC [61,73]. In a mouse allograft study, epithelial cells in the mouse ovarian hilum that are also positive for stem-cell markers are isolated and cultured followed by inactivation of p53 (*Trp53*) and *Rb1* [73]. When injected into the peritoneal cavity of mice, these hilum-derived potential stem cells lacking p53 and *Rb1* are able to form HGSC in the ovary ($n = 7/8$ mice) [73]. However, the metastatic behavior of this HGSC does not seem to align with the typical metastatic pattern of human HGSC, which is characterized by widespread peritoneal metastases. Though five (71%) of the seven mice with ovarian HGSC also developed lung metastasis, there seem no other peritoneal tumors besides the tumors formed in the ovary [73]. While it is intriguing to see that hilum-derived, stem-cell-like epithelial cells lacking p53 and *Rb1* have potential to form serous carcinoma in the ovary, this serous tumor appears to lack peritoneal metastatic potential, a characteristic clinical feature of human HGSC [43].

3. Conclusions

Once it seemed obvious to think that all ovarian cancers, including high-grade serous ovarian cancer (HGSC), originated in the ovarian surface epithelium (OSE) [18,19,57]. Presently, it seems equally compelling to think that HGSC arises from serous tubal intraepithelial carcinoma (STIC) formed in the distal fallopian tube epithelium [9,46,200]. Both may be the cell of origin in different fractions of cases. Abundant clinical observations, genetic evidence, and increasing consensus in the field all point to STIC as the precursor lesion, particularly in women at increased genetic risk [9,76,114,116]. Yet, it is still imperative to rigorously establish a cause-effect relationship between STIC and clinical HGSC: i.e., whether STICs progress to malignant and metastatic HGSCs. Particularly, the clinical significance of STIC remains uncertain in women in the general population, who account for 85–90% of HGSC cases. Undoubtedly, STICs pose high risk for HGSC, and certain STIC lesions could transform into clinical HGSC. However, logical and intuitive as it may seem, many of precursor or premalignant lesions, despite consisting of microscopically and genetically cancerous cells, would not progress to lethal malignancies [147,149,150]. Hence, it is crucial to understand the natural progression of STIC lesions at the molecular and biological levels. In particular, the focus should be directed toward assessing the clinical consequences of STIC lesions: i.e., those likely to remain benign vs. those with the ability to undergo full malignant transformation, progressing to metastatic HGSC. A deeper and more comprehensive molecular characterization of STICs in different stages of HGSC may help predict or determine the malignant and metastatic potential of STICs. Also, as STICs do not account for all HGSCs, it is also important to search for other potential cell origins for HGSC. The fallopian tube stroma in mice, for instance, has the capability to develop metastatic HGSC which faithfully reproduces the clinical metastasis of human HGSC.

Determining the precise cell(s) of origin of HGSC is crucial for improving early-detection and prevention rates of ovarian cancer, and could offer insight into devising effective treatment against advanced ovarian cancer.

Author Contributions: Conceptualization: J.K.; Writing-Original Draft Preparation: J.K.; Figure illustration: E.Y.P.; Writing-Review & Editing: E.Y.P., O.K., J.M.S., D.M.C., C.-H.C., and R.C.B.Jr.

Funding: This work is funded by R00CA179137 (J.K. and E.Y.P.) from the U.S. National Cancer Institute (NCI) and the Vera Bradley Scholarship (O.K.) from the Vera Bradley Foundation for Breast Cancer Research. It is also supported by the Young Investigator Award (J.K.) from Showalter Trust as well as startup funds from Indiana University Melvin & Bren Simon Cancer Center. This work was also supported by funds (R.C.B.Jr) from the Early Detection Research Network (5 U01 CA200462-02) and the MD Anderson Ovarian SPOREs (P50 CA83639 and P50 CA217685) from the NCI; the Cancer Prevention Research Institute of Texas (RP160145); Golfer's Against Cancer, the Mossy Foundation, the Roberson Endowment, National Foundation for Cancer Research; The University of Texas MD Anderson Women's Moon Shot; and generous donations from Stuart and Gaye Lynn Zarrow.

Acknowledgments: We are grateful to Karen H. Lu, Chair of the Department of Gynecologic Oncology and Reproductive Medicine at the University of Texas MD Anderson Cancer Center, for kindly reviewing the manuscript and providing valuable comments and clinical insights. We appreciate support from the Leo and Anne Albert Charitable Trust Fund for Ovarian Cancer Research and the Vera Bradley Foundation for Breast Cancer Research.

Conflicts of Interest: Bast receives royalties for discovery of CA125 from Fujirebio Diagnostics Inc.

References

1. Karnezis, A.N.; Cho, K.R.; Gilks, C.B.; Pearce, C.L.; Huntsman, D.G. The disparate origins of ovarian cancers: Pathogenesis and prevention strategies. *Nat. Rev. Cancer* **2017**, *17*, 65–74. [CrossRef] [PubMed]
2. Bast, R.C., Jr.; Hennessy, B.; Mills, G.B. The biology of ovarian cancer: New opportunities for translation. *Nat. Rev. Cancer* **2009**, *9*, 415–428. [CrossRef] [PubMed]
3. Cho, K.R.; Shih Ie, M. Ovarian cancer. *Annu. Rev. Pathol.* **2009**, *4*, 287–313. [CrossRef] [PubMed]
4. Boussios, S.; Zarkavelis, G.; Seraj, E.; Zerdes, I.; Tatsi, K.; Pentheroudakis, G. Non-epithelial Ovarian Cancer: Elucidating Uncommon Gynaecological Malignancies. *AntiCancer Res.* **2016**, *36*, 5031–5042. [CrossRef] [PubMed]
5. Colombo, N.; Peiretti, M.; Garbi, A.; Carinelli, S.; Marini, C.; Sessa, C. Non-epithelial ovarian cancer: ESMO Clinical Practice Guidelines for diagnosis, treatment and follow-up. *Ann. Oncol.* **2012**, *23*, VII20–VII26. [CrossRef] [PubMed]
6. Koulouris, C.R.; Penson, R.T. Ovarian stromal and germ cell tumors. *Semin. Oncol.* **2009**, *36*, 126–136. [CrossRef] [PubMed]
7. Kurman, R.J.; Shih Ie, M. Pathogenesis of ovarian cancer: Lessons from morphology and molecular biology and their clinical implications. *Int. J. Gynecol. Pathol.* **2008**, *27*, 151–160. [CrossRef] [PubMed]
8. Yemelyanova, A.V.; Cosin, J.A.; Bidus, M.A.; Boice, C.R.; Seidman, J.D. Pathology of stage I versus stage III ovarian carcinoma with implications for pathogenesis and screening. *Int. J. Gynecol. Cancer* **2008**, *18*, 465–469. [CrossRef] [PubMed]
9. Kurman, R.J.; Shih Ie, M. The origin and pathogenesis of epithelial ovarian cancer: A proposed unifying theory. *Am. J. Surg. Pathol.* **2010**, *34*, 433–443. [CrossRef] [PubMed]
10. Gurung, A.; Hung, T.; Morin, J.; Gilks, C.B. Molecular abnormalities in ovarian carcinoma: Clinical, morphological and therapeutic correlates. *Histopathology* **2013**, *62*, 59–70. [CrossRef] [PubMed]
11. Seidman, J.D.; Horkayne-Szakaly, I.; Haiba, M.; Boice, C.R.; Kurman, R.J.; Ronnett, B.M. The histologic type and stage distribution of ovarian carcinomas of surface epithelial origin. *Int. J. Gynecol. Pathol.* **2004**, *23*, 41–44. [CrossRef] [PubMed]
12. Bowtell, D.D.; Bohm, S.; Ahmed, A.A.; Aspuria, P.J.; Bast, R.C., Jr.; Beral, V.; Berek, J.S.; Birrer, M.J.; Blagden, S.; Bookman, M.A.; et al. Rethinking ovarian cancer II: Reducing mortality from high-grade serous ovarian cancer. *Nat. Rev. Cancer* **2015**, *15*, 668–679. [CrossRef] [PubMed]

13. Vaughan, S.; Coward, J.I.; Bast, R.C., Jr.; Berchuck, A.; Berek, J.S.; Brenton, J.D.; Coukos, G.; Crum, C.C.; Drapkin, R.; Etemadmoghadam, D.; et al. Rethinking ovarian cancer: Recommendations for improving outcomes. *Nat. Rev. Cancer* **2011**, *11*, 719–725. [CrossRef] [PubMed]
14. Peres, L.C.; Cushing-Haugen, K.L.; Kobel, M.; Harris, H.R.; Berchuck, A.; Rossing, M.A.; Schildkraut, J.M.; Doherty, J.A. Invasive Epithelial Ovarian Cancer Survival by Histotype and Disease Stage. *J. Natl. Cancer Inst.* **2018**. [CrossRef] [PubMed]
15. Boyd, J. Whence epithelial ovarian carcinoma? *Gynecol. Oncol.* **2008**, *109*, 161–163. [CrossRef] [PubMed]
16. Bowtell, D.D. The genesis and evolution of high-grade serous ovarian cancer. *Nat. Rev. Cancer* **2010**, *10*, 803–808. [CrossRef] [PubMed]
17. Cannistra, S.A. Cancer of the ovary. *N. Engl. J. Med.* **2004**, *351*, 2519–2529. [CrossRef] [PubMed]
18. Bell, D.A. Origins and molecular pathology of ovarian cancer. *Mod. Pathol.* **2005**, *18*, S19–S32. [CrossRef] [PubMed]
19. Feeley, K.M.; Wells, M. Precursor lesions of ovarian epithelial malignancy. *Histopathology* **2001**, *38*, 87–95. [CrossRef] [PubMed]
20. Oktem, O.; Oktay, K. The ovary: Anatomy and function throughout human life. *Ann. N. Y. Acad. Sci.* **2008**, *1127*, 1–9. [CrossRef] [PubMed]
21. Richards, J.S.; Pangas, S.A. The ovary: Basic biology and clinical implications. *J. Clin. Investig.* **2010**, *120*, 963–972. [CrossRef] [PubMed]
22. Hsueh, A.J.; Kawamura, K.; Cheng, Y.; Fauser, B.C. Intraovarian control of early folliculogenesis. *Endocr. Rev.* **2015**, *36*, 1–24. [CrossRef] [PubMed]
23. Gougeon, A. Human ovarian follicular development: From activation of resting follicles to preovulatory maturation. *Ann. Endocrinol.* **2010**, *71*, 132–143. [CrossRef] [PubMed]
24. Baerwald, A.R.; Adams, G.P.; Pierson, R.A. Ovarian antral folliculogenesis during the human menstrual cycle: A review. *Hum. Reprod. Update* **2012**, *18*, 73–91. [CrossRef] [PubMed]
25. Edson, M.A.; Nagaraja, A.K.; Matzuk, M.M. The mammalian ovary from genesis to revelation. *Endocr. Rev.* **2009**, *30*, 624–712. [CrossRef] [PubMed]
26. Matzuk, M.M.; Lamb, D.J. The biology of infertility: Research advances and clinical challenges. *Nat. Med.* **2008**, *14*, 1197–1213. [CrossRef] [PubMed]
27. Hawkins, S.M.; Matzuk, M.M. The menstrual cycle: Basic biology. *Ann. N. Y. Acad. Sci.* **2008**, *1135*, 10–18. [CrossRef] [PubMed]
28. Horta, M.; Cunha, T.M. Sex cord-stromal tumors of the ovary: A comprehensive review and update for radiologists. *Diagn. Interv. Radiol.* **2015**, *21*, 277–286. [CrossRef] [PubMed]
29. Fuller, P.J.; Leung, D.; Chu, S. Genetics and genomics of ovarian sex cord-stromal tumors. *Clin. Genet.* **2017**, *91*, 285–291. [CrossRef] [PubMed]
30. Lim, D.; Oliva, E. Ovarian sex cord-stromal tumours: An update in recent molecular advances. *Pathology* **2018**, *50*, 178–189. [CrossRef] [PubMed]
31. Brown, J.; Friedlander, M.; Backes, F.J.; Harter, P.; O'Connor, D.M.; de la Motte Rouge, T.; Lorusso, D.; Maenpaa, J.; Kim, J.W.; Tenney, M.E.; et al. Gynecologic Cancer Intergroup (GCIG) consensus review for ovarian germ cell tumors. *Int. J. Gynecol. Cancer* **2014**, *24*, S48–S54. [CrossRef] [PubMed]
32. Low, J.J.; Ilancheran, A.; Ng, J.S. Malignant ovarian germ-cell tumours. *Best Pract. Res. Clin. Obstet. Gynaecol.* **2012**, *26*, 347–355. [CrossRef] [PubMed]
33. Pectasides, D.; Pectasides, E.; Kassanos, D. Germ cell tumors of the ovary. *Cancer Treat. Rev.* **2008**, *34*, 427–441. [CrossRef] [PubMed]
34. Parkinson, C.A.; Hatcher, H.M.; Ajithkumar, T.V. Management of malignant ovarian germ cell tumors. *Obstet. Gynecol. Surv.* **2011**, *66*, 507–514. [CrossRef] [PubMed]
35. Wendel, J.R.H.; Wang, X.; Hawkins, S.M. The Endometriotic Tumor Microenvironment in Ovarian Cancer. *Cancers* **2018**, *10*, 261. [CrossRef] [PubMed]
36. Lim, D.; Oliva, E. Precursors and pathogenesis of ovarian carcinoma. *Pathology* **2013**, *45*, 229–242. [CrossRef] [PubMed]
37. Nezhat, F.R.; Apostol, R.; Nezhat, C.; Pejovic, T. New insights in the pathophysiology of ovarian cancer and implications for screening and prevention. *Am. J. Obstet. Gynecol.* **2015**, *213*, 262–267. [CrossRef] [PubMed]

38. Rojas, V.; Hirshfield, K.M.; Ganesan, S.; Rodriguez-Rodriguez, L. Molecular Characterization of Epithelial Ovarian Cancer: Implications for Diagnosis and Treatment. *Int. J. Mol. Sci.* **2016**, *17*, 2113. [CrossRef] [PubMed]
39. Malpica, A.; Wong, K.K. The molecular pathology of ovarian serous borderline tumors. *Ann. Oncol.* **2016**, *27*, i16–i19. [CrossRef] [PubMed]
40. Ahmed, A.A.; Etemadmoghadam, D.; Temple, J.; Lynch, A.G.; Riad, M.; Sharma, R.; Stewart, C.; Fereday, S.; Caldas, C.; Defazio, A.; et al. Driver mutations in TP53 are ubiquitous in high grade serous carcinoma of the ovary. *J. Pathol.* **2010**, *221*, 49–56. [CrossRef] [PubMed]
41. TCGA. Integrated genomic analyses of ovarian carcinoma. *Nature* **2011**, *474*, 609–615. [CrossRef] [PubMed]
42. Kumar, V.; Abbas, A.K.; Fausto, N.; Aster, J.C. *Robbins and Cotran Pathologic Basics of Disease*, 9th ed.; Elsevier Saunders: Philadelphia, PA, USA, 2014; p. 270.
43. Lengyel, E. Ovarian cancer development and metastasis. *Am. J. Pathol.* **2010**, *177*, 1053–1064. [CrossRef] [PubMed]
44. Hogg, R.; Friedlander, M. Biology of epithelial ovarian cancer: Implications for screening women at high genetic risk. *J. Clin. Oncol.* **2004**, *22*, 1315–1327. [CrossRef] [PubMed]
45. Seidman, J.D.; Zhao, P.; Yemelyanova, A. "Primary peritoneal" high-grade serous carcinoma is very likely metastatic from serous tubal intraepithelial carcinoma: Assessing the new paradigm of ovarian and pelvic serous carcinogenesis and its implications for screening for ovarian cancer. *Gynecol. Oncol.* **2011**, *120*, 470–473. [CrossRef] [PubMed]
46. Crum, C.P.; Drapkin, R.; Miron, A.; Ince, T.A.; Muto, M.; Kindelberger, D.W.; Lee, Y. The distal fallopian tube: A new model for pelvic serous carcinogenesis. *Curr. Opin. Obstet. Gynecol.* **2007**, *19*, 3–9. [CrossRef] [PubMed]
47. Jarboe, E.A.; Miron, A.; Carlson, J.W.; Hirsch, M.S.; Kindelberger, D.; Mutter, G.L.; Crum, C.P.; Nucci, M.R. Coexisting intraepithelial serous carcinomas of the endometrium and fallopian tube: Frequency and potential significance. *Int. J. Gynecol. Pathol.* **2009**, *28*, 308–315. [CrossRef] [PubMed]
48. Przybycin, C.G.; Kurman, R.J.; Ronnett, B.M.; Shih Ie, M.; Vang, R. Are all pelvic (nonuterine) serous carcinomas of tubal origin? *Am. J. Surg. Pathol.* **2010**, *34*, 1407–1416. [CrossRef] [PubMed]
49. Gilbert, L.; Basso, O.; Sampalis, J.; Karp, I.; Martins, C.; Feng, J.; Piedimonte, S.; Quintal, L.; Ramanakumar, A.V.; Takefman, J.; et al. Assessment of symptomatic women for early diagnosis of ovarian cancer: Results from the prospective DOvE pilot project. *Lancet Oncol.* **2012**, *13*, 285–291. [CrossRef]
50. Noone, A.M.; Howlader, N.; Krapcho, M.; Miller, D.; Brest, A.; Yu, M.; Ruhl, J.; Tatalovich, Z.; Mariotto, A.; Lewis, D.R.; et al. (Eds.) *SEER Cancer Statistics Review, 1975–2015*; Based on November 2017 SEER Data Submission, Posted to the SEER Web Site; National Cancer Institute: Bethesda, MD, USA, 2018.
51. Chan, A.; Gilks, B.; Kwon, J.; Tinker, A.V. New insights into the pathogenesis of ovarian carcinoma: Time to rethink ovarian cancer screening. *Obstet. Gynecol.* **2012**, *120*, 935–940. [CrossRef] [PubMed]
52. Lu, K.H.; Skates, S.; Hernandez, M.A.; Bedi, D.; Bevers, T.; Leeds, L.; Moore, R.; Granai, C.; Harris, S.; Newland, W.; et al. A 2-stage ovarian cancer screening strategy using the Risk of Ovarian Cancer Algorithm (ROCA) identifies early-stage incident cancers and demonstrates high positive predictive value. *Cancer* **2013**, *119*, 3454–3461. [CrossRef] [PubMed]
53. Jacobs, I.J.; Menon, U.; Ryan, A.; Gentry-Maharaj, A.; Burnell, M.; Kalsi, J.K.; Amso, N.N.; Apostolidou, S.; Benjamin, E.; Cruickshank, D.; et al. Ovarian cancer screening and mortality in the UK Collaborative Trial of Ovarian Cancer Screening (UKCTOCS): A randomised controlled trial. *Lancet* **2016**, *387*, 945–956. [CrossRef]
54. Buys, S.S.; Partridge, E.; Black, A.; Johnson, C.C.; Lamerato, L.; Isaacs, C.; Reding, D.J.; Greenlee, R.T.; Yokochi, L.A.; Kessel, B.; et al. Effect of screening on ovarian cancer mortality: The Prostate, Lung, Colorectal and Ovarian (PLCO) Cancer Screening Randomized Controlled Trial. *JAMA* **2011**, *305*, 2295–2303. [CrossRef] [PubMed]
55. Henderson, J.T.; Webber, E.M.; Sawaya, G.F. Screening for Ovarian Cancer: Updated Evidence Report and Systematic Review for the US Preventive Services Task Force. *JAMA* **2018**, *319*, 595–606. [CrossRef] [PubMed]
56. Kobayashi, H.; Yamada, Y.; Sado, T.; Sakata, M.; Yoshida, S.; Kawaguchi, R.; Kanayama, S.; Shigetomi, H.; Haruta, S.; Tsuji, Y.; et al. A randomized study of screening for ovarian cancer: A multicenter study in Japan. *Int. J. Gynecol. Cancer* **2008**, *18*, 414–420. [CrossRef] [PubMed]
57. Scully, R.E. Pathology of ovarian cancer precursors. *J. Cell. Biochem. Suppl.* **1995**, *23*, 208–218. [CrossRef] [PubMed]

58. Scott, M.; McCluggage, W.G. Current concepts in ovarian epithelial tumorigenesis: Correlation between morphological and molecular data. *Histol. Histopathol.* **2006**, *21*, 81–92. [PubMed]
59. Auersperg, N.; Wong, A.S.; Choi, K.C.; Kang, S.K.; Leung, P.C. Ovarian surface epithelium: Biology, endocrinology, and pathology. *Endocr. Rev.* **2001**, *22*, 255–288. [CrossRef] [PubMed]
60. Burdette, J.E.; Oliver, R.M.; Ulyanov, V.; Kilen, S.M.; Mayo, K.E.; Woodruff, T.K. Ovarian epithelial inclusion cysts in chronically superovulated CD1 and Smad2 dominant-negative mice. *Endocrinology* **2007**, *148*, 3595–3604. [CrossRef] [PubMed]
61. Auersperg, N. The origin of ovarian carcinomas: A unifying hypothesis. *Int. J. Gynecol. Pathol.* **2011**, *30*, 12–21. [CrossRef] [PubMed]
62. Cheng, W.; Liu, J.; Yoshida, H.; Rosen, D.; Naora, H. Lineage infidelity of epithelial ovarian cancers is controlled by HOX genes that specify regional identity in the reproductive tract. *Nat. Med.* **2005**, *11*, 531–537. [CrossRef] [PubMed]
63. Connolly, D.C.; Bao, R.; Nikitin, A.Y.; Stephens, K.C.; Poole, T.W.; Hua, X.; Harris, S.S.; Vanderhyden, B.C.; Hamilton, T.C. Female mice chimeric for expression of the simian virus 40 TAg under control of the MISIIR promoter develop epithelial ovarian cancer. *Cancer Res.* **2003**, *63*, 1389–1397. [PubMed]
64. Orsulic, S.; Li, Y.; Soslow, R.A.; Vitale-Cross, L.A.; Gutkind, J.S.; Varmus, H.E. Induction of ovarian cancer by defined multiple genetic changes in a mouse model system. *Cancer Cell* **2002**, *1*, 53–62. [CrossRef]
65. Flesken-Nikitin, A.; Choi, K.C.; Eng, J.P.; Shmidt, E.N.; Nikitin, A.Y. Induction of carcinogenesis by concurrent inactivation of p53 and Rb1 in the mouse ovarian surface epithelium. *Cancer Res.* **2003**, *63*, 3459–3463. [PubMed]
66. Dinulescu, D.M.; Ince, T.A.; Quade, B.J.; Shafer, S.A.; Crowley, D.; Jacks, T. Role of K-ras and Pten in the development of mouse models of endometriosis and endometrioid ovarian cancer. *Nat. Med.* **2005**, *11*, 63–70. [CrossRef] [PubMed]
67. Fan, H.Y.; Liu, Z.; Paquet, M.; Wang, J.; Lydon, J.P.; DeMayo, F.J.; Richards, J.S. Cell type-specific targeted mutations of Kras and Pten document proliferation arrest in granulosa cells versus oncogenic insult to ovarian surface epithelial cells. *Cancer Res.* **2009**, *69*, 6463–6472. [CrossRef] [PubMed]
68. Wu, R.; Hendrix-Lucas, N.; Kuick, R.; Zhai, Y.; Schwartz, D.R.; Akyol, A.; Hanash, S.; Misek, D.E.; Katabuchi, H.; Williams, B.O.; et al. Mouse model of human ovarian endometrioid adenocarcinoma based on somatic defects in the Wnt/beta-catenin and PI3K/Pten signaling pathways. *Cancer Cell* **2007**, *11*, 321–333. [CrossRef] [PubMed]
69. Ren, Y.A.; Mullany, L.K.; Liu, Z.; Herron, A.J.; Wong, K.K.; Richards, J.S. Mutant p53 Promotes Epithelial Ovarian Cancer by Regulating Tumor Differentiation, Metastasis, and Responsiveness to Steroid Hormones. *Cancer Res.* **2016**, *76*, 2206–2218. [CrossRef] [PubMed]
70. Clark-Knowles, K.V.; Senterman, M.K.; Collins, O.; Vanderhyden, B.C. Conditional inactivation of Brca1, p53 and Rb in mouse ovaries results in the development of leiomyosarcomas. *PLoS ONE* **2009**, *4*, e8534. [CrossRef] [PubMed]
71. Quinn, B.A.; Brake, T.; Hua, X.; Baxter-Jones, K.; Litwin, S.; Ellenson, L.H.; Connolly, D.C. Induction of ovarian leiomyosarcomas in mice by conditional inactivation of Brca1 and p53. *PLoS ONE* **2009**, *4*, e8404. [CrossRef] [PubMed]
72. Szabova, L.; Yin, C.; Bupp, S.; Guerin, T.M.; Schlomer, J.J.; Householder, D.B.; Baran, M.L.; Yi, M.; Song, Y.; Sun, W.; et al. Perturbation of Rb, p53, and BRCA1 or BRCA2 cooperate in inducing metastatic serous epithelial ovarian cancer. *Cancer Res.* **2012**, *72*, 4141–4153. [CrossRef] [PubMed]
73. Flesken-Nikitin, A.; Hwang, C.I.; Cheng, C.Y.; Michurina, T.V.; Enikolopov, G.; Nikitin, A.Y. Ovarian surface epithelium at the junction area contains a cancer-prone stem cell niche. *Nature* **2013**, *495*, 241–245. [CrossRef] [PubMed]
74. Tanwar, P.S.; Mohapatra, G.; Chiang, S.; Engler, D.A.; Zhang, L.; Kaneko-Tarui, T.; Ohguchi, Y.; Birrer, M.J.; Teixeira, J.M. Loss of LKB1 and PTEN tumor suppressor genes in the ovarian surface epithelium induces papillary serous ovarian cancer. *Carcinogenesis* **2014**, *35*, 546–553. [CrossRef] [PubMed]
75. Kim, J.; Coffey, D.M.; Ma, L.; Matzuk, M.M. The ovary is an alternative site of origin for high-grade serous ovarian cancer in mice. *Endocrinology* **2015**, *156*, 1975–1981. [CrossRef] [PubMed]
76. Crum, C.P.; Drapkin, R.; Kindelberger, D.; Medeiros, F.; Miron, A.; Lee, Y. Lessons from BRCA: The tubal fimbria emerges as an origin for pelvic serous cancer. *Clin. Med. Res.* **2007**, *5*, 35–44. [CrossRef] [PubMed]

77. Kim, J.; Coffey, D.M.; Creighton, C.J.; Yu, Z.; Hawkins, S.M.; Matzuk, M.M. High-grade serous ovarian cancer arises from fallopian tube in a mouse model. *Proc. Natl. Acad. Sci. USA* **2012**, *109*, 3921–3926. [CrossRef] [PubMed]
78. Perets, R.; Wyant, G.A.; Muto, K.W.; Bijron, J.G.; Poole, B.B.; Chin, K.T.; Chen, J.Y.; Ohman, A.W.; Stepule, C.D.; Kwak, S.; et al. Transformation of the fallopian tube secretory epithelium leads to high-grade serous ovarian cancer in Brca;Tp53;Pten models. *Cancer Cell* **2013**, *24*, 751–765. [CrossRef] [PubMed]
79. Zhai, Y.; Wu, R.; Kuick, R.; Sessine, M.S.; Schulman, S.; Green, M.; Fearon, E.R.; Cho, K.R. High-grade serous carcinomas arise in the mouse oviduct via defects linked to the human disease. *J. Pathol.* **2017**, *243*, 16–25. [CrossRef] [PubMed]
80. Piek, J.M.; van Diest, P.J.; Zweemer, R.P.; Jansen, J.W.; Poort-Keesom, R.J.; Menko, F.H.; Gille, J.J.; Jongsma, A.P.; Pals, G.; Kenemans, P.; et al. Dysplastic changes in prophylactically removed Fallopian tubes of women predisposed to developing ovarian cancer. *J. Pathol.* **2001**, *195*, 451–456. [CrossRef] [PubMed]
81. Crum, C.P.; Herfs, M.; Ning, G.; Bijron, J.G.; Howitt, B.E.; Jimenez, C.A.; Hanamornroongruang, S.; McKeon, F.D.; Xian, W. Through the glass darkly: Intraepithelial neoplasia, top-down differentiation, and the road to ovarian cancer. *J. Pathol.* **2013**, *231*, 402–412. [CrossRef] [PubMed]
82. George, S.H.; Shaw, P. BRCA and Early Events in the Development of Serous Ovarian Cancer. *Front. Oncol.* **2014**, *4*, 5. [CrossRef] [PubMed]
83. Roy, R.; Chun, J.; Powell, S.N. BRCA1 and BRCA2: Different roles in a common pathway of genome protection. *Nat. Rev. Cancer* **2012**, *12*, 68–78. [CrossRef] [PubMed]
84. Hartmann, L.C.; Lindor, N.M. The Role of Risk-Reducing Surgery in Hereditary Breast and Ovarian Cancer. *N. Engl. J. Med.* **2016**, *374*, 454–468. [CrossRef] [PubMed]
85. King, M.C.; Marks, J.H.; Mandell, J.B. Breast and ovarian cancer risks due to inherited mutations in BRCA1 and BRCA2. *Science* **2003**, *302*, 643–646. [CrossRef] [PubMed]
86. Domchek, S.M.; Friebel, T.M.; Singer, C.F.; Evans, D.G.; Lynch, H.T.; Isaacs, C.; Garber, J.E.; Neuhausen, S.L.; Matloff, E.; Eeles, R.; et al. Association of risk-reducing surgery in BRCA1 or BRCA2 mutation carriers with cancer risk and mortality. *JAMA* **2010**, *304*, 967–975. [CrossRef] [PubMed]
87. Mavaddat, N.; Barrowdale, D.; Andrulis, I.L.; Domchek, S.M.; Eccles, D.; Nevanlinna, H.; Ramus, S.J.; Spurdle, A.; Robson, M.; Sherman, M.; et al. Pathology of breast and ovarian cancers among BRCA1 and BRCA2 mutation carriers: Results from the Consortium of Investigators of Modifiers of BRCA1/2 (CIMBA). *Cancer Epidemiol. Biomark. Prev.* **2012**, *21*, 134–147. [CrossRef] [PubMed]
88. Press, J.Z.; De Luca, A.; Boyd, N.; Young, S.; Troussard, A.; Ridge, Y.; Kaurah, P.; Kalloger, S.E.; Blood, K.A.; Smith, M.; et al. Ovarian carcinomas with genetic and epigenetic BRCA1 loss have distinct molecular abnormalities. *BMC Cancer* **2008**, *8*, 17. [CrossRef] [PubMed]
89. Rebbeck, T.R.; Lynch, H.T.; Neuhausen, S.L.; Narod, S.A.; Van't Veer, L.; Garber, J.E.; Evans, G.; Isaacs, C.; Daly, M.B.; Matloff, E.; et al. Prophylactic oophorectomy in carriers of BRCA1 or BRCA2 mutations. *N. Engl. J. Med.* **2002**, *346*, 1616–1622. [CrossRef] [PubMed]
90. Finch, A.P.; Lubinski, J.; Moller, P.; Singer, C.F.; Karlan, B.; Senter, L.; Rosen, B.; Maehle, L.; Ghadirian, P.; Cybulski, C.; et al. Impact of oophorectomy on cancer incidence and mortality in women with a BRCA1 or BRCA2 mutation. *J. Clin. Oncol.* **2014**, *32*, 1547–1553. [CrossRef] [PubMed]
91. Zweemer, R.P.; van Diest, P.J.; Verheijen, R.H.; Ryan, A.; Gille, J.J.; Sijmons, R.H.; Jacobs, I.J.; Menko, F.H.; Kenemans, P. Molecular evidence linking primary cancer of the fallopian tube to BRCA1 germline mutations. *Gynecol. Oncol.* **2000**, *76*, 45–50. [CrossRef] [PubMed]
92. Colgan, T.J.; Boerner, S.L.; Murphy, J.; Cole, D.E.; Narod, S.; Rosen, B. Peritoneal lavage cytology: An assessment of its value during prophylactic oophorectomy. *Gynecol. Oncol.* **2002**, *85*, 397–403. [CrossRef] [PubMed]
93. Leeper, K.; Garcia, R.; Swisher, E.; Goff, B.; Greer, B.; Paley, P. Pathologic findings in prophylactic oophorectomy specimens in high-risk women. *Gynecol. Oncol.* **2002**, *87*, 52–56. [CrossRef] [PubMed]
94. Paley, P.J.; Swisher, E.M.; Garcia, R.L.; Agoff, S.N.; Greer, B.E.; Peters, K.L.; Goff, B.A. Occult cancer of the fallopian tube in BRCA-1 germline mutation carriers at prophylactic oophorectomy: A case for recommending hysterectomy at surgical prophylaxis. *Gynecol. Oncol.* **2001**, *80*, 176–180. [CrossRef] [PubMed]

95. Walker, J.L.; Powell, C.B.; Chen, L.M.; Carter, J.; Bae Jump, V.L.; Parker, L.P.; Borowsky, M.E.; Gibb, R.K. Society of Gynecologic Oncology recommendations for the prevention of ovarian cancer. *Cancer* **2015**, *121*, 2108–2120. [CrossRef] [PubMed]
96. Medeiros, F.; Muto, M.G.; Lee, Y.; Elvin, J.A.; Callahan, M.J.; Feltmate, C.; Garber, J.E.; Cramer, D.W.; Crum, C.P. The tubal fimbria is a preferred site for early adenocarcinoma in women with familial ovarian cancer syndrome. *Am. J. Surg. Pathol.* **2006**, *30*, 230–236. [CrossRef] [PubMed]
97. Barakat, R.R.; Federici, M.G.; Saigo, P.E.; Robson, M.E.; Offit, K.; Boyd, J. Absence of premalignant histologic, molecular, or cell biologic alterations in prophylactic oophorectomy specimens from BRCA1 heterozygotes. *Cancer* **2000**, *89*, 383–390. [CrossRef]
98. Jarboe, E.; Folkins, A.; Nucci, M.R.; Kindelberger, D.; Drapkin, R.; Miron, A.; Lee, Y.; Crum, C.P. Serous carcinogenesis in the fallopian tube: A descriptive classification. *Int. J. Gynecol. Pathol.* **2008**, *27*, 1–9. [CrossRef] [PubMed]
99. Visvanathan, K.; Vang, R.; Shaw, P.; Gross, A.; Soslow, R.; Parkash, V.; Shih Ie, M.; Kurman, R.J. Diagnosis of serous tubal intraepithelial carcinoma based on morphologic and immunohistochemical features: A reproducibility study. *Am. J. Surg. Pathol.* **2011**, *35*, 1766–1775. [CrossRef] [PubMed]
100. Finch, A.; Shaw, P.; Rosen, B.; Murphy, J.; Narod, S.A.; Colgan, T.J. Clinical and pathologic findings of prophylactic salpingo-oophorectomies in 159 BRCA1 and BRCA2 carriers. *Gynecol. Oncol.* **2006**, *100*, 58–64. [CrossRef] [PubMed]
101. Shaw, P.A.; Rouzbahman, M.; Pizer, E.S.; Pintilie, M.; Begley, H. Candidate serous cancer precursors in fallopian tube epithelium of BRCA1/2 mutation carriers. *Mod. Pathol.* **2009**, *22*, 1133–1138. [CrossRef] [PubMed]
102. Wethington, S.L.; Park, K.J.; Soslow, R.A.; Kauff, N.D.; Brown, C.L.; Dao, F.; Otegbeye, E.; Sonoda, Y.; Abu-Rustum, N.R.; Barakat, R.R.; et al. Clinical outcome of isolated serous tubal intraepithelial carcinomas (STIC). *Int. J. Gynecol. Cancer* **2013**, *23*, 1603–1611. [CrossRef] [PubMed]
103. Lee, Y.J.; Lee, S.W.; Kim, K.R.; Jung, K.H.; Lee, J.W.; Kim, Y.M. Pathologic findings at risk-reducing salpingo-oophorectomy (RRSO) in germline BRCA mutation carriers with breast cancer: Significance of bilateral RRSO at the optimal age in germline BRCA mutation carriers. *J. Gynecol. Oncol.* **2017**, *28*, e3. [CrossRef] [PubMed]
104. Kindelberger, D.W.; Lee, Y.; Miron, A.; Hirsch, M.S.; Feltmate, C.; Medeiros, F.; Callahan, M.J.; Garner, E.O.; Gordon, R.W.; Birch, C.; et al. Intraepithelial carcinoma of the fimbria and pelvic serous carcinoma: Evidence for a causal relationship. *Am. J. Surg. Pathol.* **2007**, *31*, 161–169. [CrossRef] [PubMed]
105. Koc, N.; Ayas, S.; Uygur, L. The association of serous tubal intraepithelial carcinoma with gynecologic pathologies and its role in pelvic serous cancer. *Gynecol. Oncol.* **2014**, *134*, 486–491. [CrossRef] [PubMed]
106. Tang, S.; Onuma, K.; Deb, P.; Wang, E.; Lytwyn, A.; Sur, M.; Daya, D. Frequency of serous tubal intraepithelial carcinoma in various gynecologic malignancies: A study of 300 consecutive cases. *Int. J. Gynecol. Pathol.* **2012**, *31*, 103–110. [CrossRef] [PubMed]
107. Malmberg, K.; Klynning, C.; Floter-Radestad, A.; Carlson, J.W. Serous tubal intraepithelial carcinoma, chronic fallopian tube injury, and serous carcinoma development. *Virchows Arch. Int. J. Pathol.* **2016**, *468*, 707–713. [CrossRef] [PubMed]
108. Howitt, B.E.; Hanamornroongruang, S.; Lin, D.I.; Conner, J.E.; Schulte, S.; Horowitz, N.; Crum, C.P.; Meserve, E.E. Evidence for a dualistic model of high-grade serous carcinoma: BRCA mutation status, histology, and tubal intraepithelial carcinoma. *Am. J. Surg. Pathol.* **2015**, *39*, 287–293. [CrossRef] [PubMed]
109. Lee, Y.; Miron, A.; Drapkin, R.; Nucci, M.R.; Medeiros, F.; Saleemuddin, A.; Garber, J.; Birch, C.; Mou, H.; Gordon, R.W.; et al. A candidate precursor to serous carcinoma that originates in the distal fallopian tube. *J. Pathol.* **2007**, *211*, 26–35. [CrossRef] [PubMed]
110. Folkins, A.K.; Jarboe, E.A.; Saleemuddin, A.; Lee, Y.; Callahan, M.J.; Drapkin, R.; Garber, J.E.; Muto, M.G.; Tworoger, S.; Crum, C.P. A candidate precursor to pelvic serous cancer (p53 signature) and its prevalence in ovaries and fallopian tubes from women with BRCA mutations. *Gynecol. Oncol.* **2008**, *109*, 168–173. [CrossRef] [PubMed]
111. Kuhn, E.; Kurman, R.J.; Vang, R.; Sehdev, A.S.; Han, G.; Soslow, R.; Wang, T.L.; Shih Ie, M. TP53 mutations in serous tubal intraepithelial carcinoma and concurrent pelvic high-grade serous carcinoma–evidence supporting the clonal relationship of the two lesions. *J. Pathol.* **2012**, *226*, 421–426. [CrossRef] [PubMed]

112. Jarboe, E.A.; Folkins, A.K.; Drapkin, R.; Ince, T.A.; Agoston, E.S.; Crum, C.P. Tubal and ovarian pathways to pelvic epithelial cancer: A pathological perspective. *Histopathology* **2008**, *53*, 127–138. [CrossRef] [PubMed]
113. Yates, M.S.; Meyer, L.A.; Deavers, M.T.; Daniels, M.S.; Keeler, E.R.; Mok, S.C.; Gershenson, D.M.; Lu, K.H. Microscopic and early-stage ovarian cancers in BRCA1/2 mutation carriers: Building a model for early BRCA-associated tumorigenesis. *Cancer Prev. Res.* **2011**, *4*, 463–470. [CrossRef] [PubMed]
114. Labidi-Galy, S.I.; Papp, E.; Hallberg, D.; Niknafs, N.; Adleff, V.; Noe, M.; Bhattacharya, R.; Novak, M.; Jones, S.; Phallen, J.; et al. High grade serous ovarian carcinomas originate in the fallopian tube. *Nat. Commun.* **2017**, *8*, 1093. [CrossRef] [PubMed]
115. Salvador, S.; Rempel, A.; Soslow, R.A.; Gilks, B.; Huntsman, D.; Miller, D. Chromosomal instability in fallopian tube precursor lesions of serous carcinoma and frequent monoclonality of synchronous ovarian and fallopian tube mucosal serous carcinoma. *Gynecol. Oncol.* **2008**, *110*, 408–417. [CrossRef] [PubMed]
116. Eckert, M.A.; Pan, S.; Hernandez, K.M.; Loth, R.M.; Andrade, J.; Volchenboum, S.L.; Faber, P.; Montag, A.; Lastra, R.; Peter, M.E.; et al. Genomics of Ovarian Cancer Progression Reveals Diverse Metastatic Trajectories Including Intraepithelial Metastasis to the Fallopian Tube. *Cancer Discov.* **2016**, *6*, 1342–1351. [CrossRef] [PubMed]
117. Xiang, L.; Rong, G.; Zhao, J.; Wang, Z.; Shi, F. Identification of candidate genes associated with tubal origin of high-grade serous ovarian cancer. *Oncol. Lett.* **2018**, *15*, 7769–7775. [CrossRef] [PubMed]
118. Piek, J.M.; Verheijen, R.H.; Kenemans, P.; Massuger, L.F.; Bulten, H.; van Diest, P.J. BRCA1/2-related ovarian cancers are of tubal origin: A hypothesis. *Gynecol. Oncol.* **2003**, *90*, 491. [CrossRef]
119. Kroeger, P.T., Jr.; Drapkin, R. Pathogenesis and heterogeneity of ovarian cancer. *Curr. Opin. Obstet. Gynecol.* **2017**, *29*, 26–34. [CrossRef] [PubMed]
120. Bijron, J.G.; Seldenrijk, C.A.; Zweemer, R.P.; Lange, J.G.; Verheijen, R.H.; van Diest, P.J. Fallopian tube intraluminal tumor spread from noninvasive precursor lesions: A novel metastatic route in early pelvic carcinogenesis. *Am. J. Surg. Pathol.* **2013**, *37*, 1123–1130. [CrossRef] [PubMed]
121. Stuckelberger, S.; Drapkin, R. Precious GEMMs: Emergence of faithful models for ovarian Cancer Research. *J. Pathol.* **2018**, *245*, 129–131. [CrossRef] [PubMed]
122. Miyoshi, I.; Takahashi, K.; Kon, Y.; Okamura, T.; Mototani, Y.; Araki, Y.; Kasai, N. Mouse transgenic for murine oviduct-specific glycoprotein promoter-driven simian virus 40 large T-antigen: Tumor formation and its hormonal regulation. *Mol. Reprod. Dev.* **2002**, *63*, 168–176. [CrossRef] [PubMed]
123. Sherman-Baust, C.A.; Kuhn, E.; Valle, B.L.; Shih, I.M.; Kurman, R.J.; Wang, T.L.; Amano, T.; Ko, M.S.; Miyoshi, I.; Araki, Y.; et al. A genetically engineered ovarian cancer mouse model based on fallopian tube transformation mimics human high-grade serous carcinoma development. *J. Pathol.* **2014**, *2014*, 4353. [CrossRef] [PubMed]
124. van der Horst, P.H.; van der Zee, M.; Heijmans-Antonissen, C.; Jia, Y.; DeMayo, F.J.; Lydon, J.P.; van Deurzen, C.H.; Ewing, P.C.; Burger, C.W.; Blok, L.J. A mouse model for endometrioid ovarian cancer arising from the distal oviduct. *Int. J. Cancer* **2014**, *135*, 1028–1037. [CrossRef] [PubMed]
125. Wu, R.; Zhai, Y.; Kuick, R.; Karnezis, A.N.; Garcia, P.; Naseem, A.; Hu, T.C.; Fearon, E.R.; Cho, K.R. Impact of oviductal versus ovarian epithelial cell of origin on ovarian endometrioid carcinoma phenotype in the mouse. *J. Pathol.* **2016**, *240*, 341–351. [CrossRef] [PubMed]
126. Carlson, J.W.; Miron, A.; Jarboe, E.A.; Parast, M.M.; Hirsch, M.S.; Lee, Y.; Muto, M.G.; Kindelberger, D.; Crum, C.P. Serous tubal intraepithelial carcinoma: Its potential role in primary peritoneal serous carcinoma and serous cancer prevention. *J. Clin. Oncol.* **2008**, *26*, 4160–4165. [CrossRef] [PubMed]
127. Chen, F.; Gaitskell, K.; Garcia, M.J.; Albukhari, A.; Tsaltas, J.; Ahmed, A.A. Serous tubal intraepithelial carcinomas associated with high-grade serous ovarian carcinomas: A systematic review. *BJOG* **2017**, *124*, 872–878. [CrossRef] [PubMed]
128. Powell, C.B.; Chen, L.M.; McLennan, J.; Crawford, B.; Zaloudek, C.; Rabban, J.T.; Moore, D.H.; Ziegler, J. Risk-reducing salpingo-oophorectomy (RRSO) in BRCA mutation carriers: Experience with a consecutive series of 111 patients using a standardized surgical-pathological protocol. *Int. J. Gynecol. Cancer* **2011**, *21*, 846–851. [CrossRef] [PubMed]
129. Powell, C.B.; Swisher, E.M.; Cass, I.; McLennan, J.; Norquist, B.; Garcia, R.L.; Lester, J.; Karlan, B.Y.; Chen, L. Long term follow up of BRCA1 and BRCA2 mutation carriers with unsuspected neoplasia identified at risk reducing salpingo-oophorectomy. *Gynecol. Oncol.* **2013**, *129*, 364–371. [CrossRef] [PubMed]

130. Callahan, M.J.; Crum, C.P.; Medeiros, F.; Kindelberger, D.W.; Elvin, J.A.; Garber, J.E.; Feltmate, C.M.; Berkowitz, R.S.; Muto, M.G. Primary fallopian tube malignancies in BRCA-positive women undergoing surgery for ovarian cancer risk reduction. *J. Clin. Oncol.* **2007**, *25*, 3985–3990. [CrossRef] [PubMed]
131. Carcangiu, M.L.; Peissel, B.; Pasini, B.; Spatti, G.; Radice, P.; Manoukian, S. Incidental carcinomas in prophylactic specimens in BRCA1 and BRCA2 germ-line mutation carriers, with emphasis on fallopian tube lesions: Report of 6 cases and review of the literature. *Am. J. Surg. Pathol.* **2006**, *30*, 1222–1230. [CrossRef] [PubMed]
132. Cass, I.; Walts, A.E.; Barbuto, D.; Lester, J.; Karlan, B. A cautious view of putative precursors of serous carcinomas in the fallopian tubes of BRCA mutation carriers. *Gynecol. Oncol.* **2014**, *134*, 492–497. [CrossRef] [PubMed]
133. Colgan, T.J.; Murphy, J.; Cole, D.E.; Narod, S.; Rosen, B. Occult carcinoma in prophylactic oophorectomy specimens: Prevalence and association with BRCA germline mutation status. *Am. J. Surg. Pathol.* **2001**, *25*, 1283–1289. [CrossRef] [PubMed]
134. Hirst, J.E.; Gard, G.B.; McIllroy, K.; Nevell, D.; Field, M. High rates of occult fallopian tube cancer diagnosed at prophylactic bilateral salpingo-oophorectomy. *Int. J. Gynecol. Cancer* **2009**, *19*, 826–829. [CrossRef] [PubMed]
135. Manchanda, R.; Abdelraheim, A.; Johnson, M.; Rosenthal, A.N.; Benjamin, E.; Brunell, C.; Burnell, M.; Side, L.; Gessler, S.; Saridogan, E.; et al. Outcome of risk-reducing salpingo-oophorectomy in BRCA carriers and women of unknown mutation status. *BJOG* **2011**, *118*, 814–824. [CrossRef] [PubMed]
136. Mingels, M.J.; Roelofsen, T.; van der Laak, J.A.; de Hullu, J.A.; van Ham, M.A.; Massuger, L.F.; Bulten, J.; Bol, M. Tubal epithelial lesions in salpingo-oophorectomy specimens of BRCA-mutation carriers and controls. *Gynecol. Oncol.* **2012**, *127*, 88–93. [CrossRef] [PubMed]
137. Powell, C.B.; Kenley, E.; Chen, L.M.; Crawford, B.; McLennan, J.; Zaloudek, C.; Komaromy, M.; Beattie, M.; Ziegler, J. Risk-reducing salpingo-oophorectomy in BRCA mutation carriers: Role of serial sectioning in the detection of occult malignancy. *J. Clin. Oncol.* **2005**, *23*, 127–132. [CrossRef] [PubMed]
138. Reitsma, W.; de Bock, G.H.; Oosterwijk, J.C.; Bart, J.; Hollema, H.; Mourits, M.J. Support of the 'fallopian tube hypothesis' in a prospective series of risk-reducing salpingo-oophorectomy specimens. *Eur. J. Cancer* **2013**, *49*, 132–141. [CrossRef] [PubMed]
139. Sherman, M.E.; Piedmonte, M.; Mai, P.L.; Ioffe, O.B.; Ronnett, B.M.; Van Le, L.; Ivanov, I.; Bell, M.C.; Blank, S.V.; DiSilvestro, P.; et al. Pathologic findings at risk-reducing salpingo-oophorectomy: Primary results from Gynecologic Oncology Group Trial GOG-0199. *J. Clin. Oncol.* **2014**, *32*, 3275–3283. [CrossRef] [PubMed]
140. Seidman, J.D.; Krishnan, J.; Yemelyanova, A.; Vang, R. Incidental Serous Tubal Intraepithelial Carcinoma and Non-Neoplastic Conditions of the Fallopian Tubes in Grossly Normal Adnexa: A Clinicopathologic Study of 388 Completely Embedded Cases. *Int. J. Gynecol. Pathol.* **2016**, *35*, 423–429. [CrossRef] [PubMed]
141. Mingels, M.J.; van Ham, M.A.; de Kievit, I.M.; Snijders, M.P.; van Tilborg, A.A.; Bulten, J.; Massuger, L.F. Mullerian precursor lesions in serous ovarian cancer patients: Using the SEE-Fim and SEE-End protocol. *Mod. Pathol.* **2014**, *27*, 1002–1013. [CrossRef] [PubMed]
142. Rabban, J.T.; Garg, K.; Crawford, B.; Chen, L.M.; Zaloudek, C.J. Early detection of high-grade tubal serous carcinoma in women at low risk for hereditary breast and ovarian cancer syndrome by systematic examination of fallopian tubes incidentally removed during benign surgery. *Am. J. Surg. Pathol.* **2014**, *38*, 729–742. [CrossRef] [PubMed]
143. Stewart, C.J.; Armstrong, M.; Brennan, B.A.; Hammond, I.G.; Havlat, M.; Rene Kee, A.; Koay, E.; Leung, Y.; Netreba, A.N.; Ruba, S. Coexisting serous carcinoma of the endometrium and the fallopian tube. *Int. J. Gynecol. Pathol.* **2010**, *29*, 278–281. [CrossRef] [PubMed]
144. Tolcher, M.C.; Swisher, E.M.; Medeiros, F.; Lima, J.F.; Hilderbrand, J.L.; Donovan, J.L.; Garcia, R.L.; Cliby, W.A.; Dowdy, S.C. Characterization of precursor lesions in the endometrium and fallopian tube epithelium of early-stage uterine serous carcinoma. *Int. J. Gynecol. Pathol.* **2015**, *34*, 57–64. [CrossRef] [PubMed]
145. Klotz, D.M.; Wimberger, P. Cells of origin of ovarian cancer: Ovarian surface epithelium or fallopian tube? *Arch. Gynecol. Obstet.* **2017**, *296*, 1055–1062. [CrossRef] [PubMed]

146. McCluggage, W.G.; Hirschowitz, L.; Gilks, C.B.; Wilkinson, N.; Singh, N. The Fallopian Tube Origin and Primary Site Assignment in Extrauterine High-grade Serous Carcinoma: Findings of a Survey of Pathologists and Clinicians. *Int. J. Gynecol. Pathol.* **2017**, *36*, 230–239. [CrossRef] [PubMed]
147. Esserman, L.J.; Thompson, I.M., Jr.; Reid, B. Overdiagnosis and overtreatment in cancer: An opportunity for improvement. *JAMA* **2013**, *310*, 797–798. [CrossRef] [PubMed]
148. Esserman, L.J.; Thompson, I.M.; Reid, B.; Nelson, P.; Ransohoff, D.F.; Welch, H.G.; Hwang, S.; Berry, D.A.; Kinzler, K.W.; Black, W.C.; et al. Addressing overdiagnosis and overtreatment in cancer: A prescription for change. *Lancet Oncol.* **2014**, *15*, e234–242. [CrossRef]
149. Brawley, O.W. Accepting the Existence of Breast Cancer Overdiagnosis. *Ann. Intern. Med.* **2017**, *166*, 364–365. [CrossRef] [PubMed]
150. Nikiforov, Y.E.; Seethala, R.R.; Tallini, G.; Baloch, Z.W.; Basolo, F.; Thompson, L.D.; Barletta, J.A.; Wenig, B.M.; Al Ghuzlan, A.; Kakudo, K.; et al. Nomenclature Revision for Encapsulated Follicular Variant of Papillary Thyroid Carcinoma: A Paradigm Shift to Reduce Overtreatment of Indolent Tumors. *JAMA Oncol.* **2016**, *2*, 1023–1029. [CrossRef] [PubMed]
151. Burstein, H.J.; Polyak, K.; Wong, J.S.; Lester, S.C.; Kaelin, C.M. Ductal carcinoma in situ of the breast. *N. Engl. J. Med.* **2004**, *350*, 1430–1441. [CrossRef] [PubMed]
152. Siegel, R.L.; Miller, K.D.; Jemal, A. Cancer statistics, 2018. *CA Cancer J. Clin.* **2018**, *68*, 7–30. [CrossRef] [PubMed]
153. Bleyer, A.; Welch, H.G. Effect of three decades of screening mammography on breast-cancer incidence. *N. Engl. J. Med.* **2012**, *367*, 1998–2005. [CrossRef] [PubMed]
154. Shieh, Y.; Eklund, M.; Sawaya, G.F.; Black, W.C.; Kramer, B.S.; Esserman, L.J. Population-based screening for cancer: Hope and hype. *Nat. Rev. Clin. Oncol.* **2016**, *13*, 550–565. [CrossRef] [PubMed]
155. Nielsen, M.; Thomsen, J.L.; Primdahl, S.; Dyreborg, U.; Andersen, J.A. Breast cancer and atypia among young and middle-aged women: A study of 110 medicolegal autopsies. *Br. J. Cancer* **1987**, *56*, 814–819. [CrossRef] [PubMed]
156. *SEER Cancer Statistics Review, 1975–2013*; Howlader, N., Noone, A.M., Krapcho, M., Miller, D., Bishop, K., Altekruse, S.F., Kosary, C.L., Yu, M., Ruhl, J., Tatalovich, Z., et al. (Eds.) National Cancer Institute: Bethesda, MD, USA, 2016.
157. Duffy, S.W.; Dibden, A.; Michalopoulos, D.; Offman, J.; Parmar, D.; Jenkins, J.; Collins, B.; Robson, T.; Scorfield, S.; Green, K.; et al. Screen detection of ductal carcinoma in situ and subsequent incidence of invasive interval breast cancers: A retrospective population-based study. *Lancet Oncol.* **2016**, *17*, 109–114. [CrossRef]
158. Sopik, V.; Nofech-Mozes, S.; Sun, P.; Narod, S.A. The relationship between local recurrence and death in early-stage breast cancer. *Breast Cancer Res. Treat.* **2016**, *155*, 175–185. [CrossRef] [PubMed]
159. Welch, H.G.; Prorok, P.C.; O'Malley, A.J.; Kramer, B.S. Breast-Cancer Tumor Size, Overdiagnosis, and Mammography Screening Effectiveness. *N. Engl. J. Med.* **2016**, *375*, 1438–1447. [CrossRef] [PubMed]
160. Autier, P.; Boniol, M.; Koechlin, A.; Pizot, C.; Boniol, M. Effectiveness of and overdiagnosis from mammography screening in the Netherlands: Population based study. *BMJ* **2017**, *359*, j5224. [CrossRef] [PubMed]
161. Jorgensen, K.J.; Gotzsche, P.C.; Kalager, M.; Zahl, P.H. Breast Cancer Screening in Denmark: A Cohort Study of Tumor Size and Overdiagnosis. *Ann. Intern. Med.* **2017**, *166*, 313–323. [CrossRef] [PubMed]
162. Torre, L.A.; Trabert, B.; DeSantis, C.E.; Miller, K.D.; Samimi, G.; Runowicz, C.D.; Gaudet, M.M.; Jemal, A.; Siegel, R.L. Ovarian cancer statistics, 2018. *CA Cancer J. Clin.* **2018**. [CrossRef] [PubMed]
163. Huang, C.C.; Orvis, G.D.; Wang, Y.; Behringer, R.R. Stromal-to-Epithelial Transition during Postpartum Endometrial Regeneration. *PLoS ONE* **2012**, *7*, e44285. [CrossRef] [PubMed]
164. Merritt, W.M.; Lin, Y.G.; Han, L.Y.; Kamat, A.A.; Spannuth, W.A.; Schmandt, R.; Urbauer, D.; Pennacchio, L.A.; Cheng, J.F.; Nick, A.M.; et al. Dicer, Drosha, and outcomes in patients with ovarian cancer. *N. Engl. J. Med.* **2008**, *359*, 2641–2650. [CrossRef] [PubMed]
165. Cerami, E.; Gao, J.; Dogrusoz, U.; Gross, B.E.; Sumer, S.O.; Aksoy, B.A.; Jacobsen, A.; Byrne, C.J.; Heuer, M.L.; Larsson, E.; et al. The cBio cancer genomics portal: An open platform for exploring multidimensional cancer genomics data. *Cancer Discov.* **2012**, *2*, 401–404. [CrossRef] [PubMed]

166. Gao, J.; Aksoy, B.A.; Dogrusoz, U.; Dresdner, G.; Gross, B.; Sumer, S.O.; Sun, Y.; Jacobsen, A.; Sinha, R.; Larsson, E.; et al. Integrative analysis of complex cancer genomics and clinical profiles using the cBioPortal. *Sci. Signal.* **2013**, *6*, pl1. [CrossRef] [PubMed]
167. NIH Consensus Conference. Ovarian cancer. Screening, treatment, and follow-up. NIH Consensus Development Panel on Ovarian Cancer. *JAMA* **1995**, *273*, 491–497. [CrossRef]
168. Lancaster, J.M.; Powell, C.B.; Chen, L.M.; Richardson, D.L. Society of Gynecologic Oncology statement on risk assessment for inherited gynecologic cancer predispositions. *Gynecol. Oncol.* **2015**, *136*, 3–7. [CrossRef] [PubMed]
169. Kauff, N.D.; Domchek, S.M.; Friebel, T.M.; Robson, M.E.; Lee, J.; Garber, J.E.; Isaacs, C.; Evans, D.G.; Lynch, H.; Eeles, R.A.; et al. Risk-reducing salpingo-oophorectomy for the prevention of BRCA1- and BRCA2-associated breast and gynecologic cancer: A multicenter, prospective study. *J. Clin. Oncol.* **2008**, *26*, 1331–1337. [CrossRef] [PubMed]
170. Domchek, S.M.; Friebel, T.M.; Neuhausen, S.L.; Wagner, T.; Evans, G.; Isaacs, C.; Garber, J.E.; Daly, M.B.; Eeles, R.; Matloff, E.; et al. Mortality after bilateral salpingo-oophorectomy in BRCA1 and BRCA2 mutation carriers: A prospective cohort study. *Lancet Oncol.* **2006**, *7*, 223–229. [CrossRef]
171. Eisen, A.; Lubinski, J.; Klijn, J.; Moller, P.; Lynch, H.T.; Offit, K.; Weber, B.; Rebbeck, T.; Neuhausen, S.L.; Ghadirian, P.; et al. Breast cancer risk following bilateral oophorectomy in BRCA1 and BRCA2 mutation carriers: An international case-control study. *J. Clin. Oncol.* **2005**, *23*, 7491–7496. [CrossRef] [PubMed]
172. Rebbeck, T.R.; Kauff, N.D.; Domchek, S.M. Meta-analysis of risk reduction estimates associated with risk-reducing salpingo-oophorectomy in BRCA1 or BRCA2 mutation carriers. *J. Natl. Cancer Inst.* **2009**, *101*, 80–87. [CrossRef] [PubMed]
173. Kramer, J.L.; Velazquez, I.A.; Chen, B.E.; Rosenberg, P.S.; Struewing, J.P.; Greene, M.H. Prophylactic oophorectomy reduces breast cancer penetrance during prospective, long-term follow-up of BRCA1 mutation carriers. *J. Clin. Oncol.* **2005**, *23*, 8629–8635. [CrossRef] [PubMed]
174. Heemskerk-Gerritsen, B.A.; Seynaeve, C.; van Asperen, C.J.; Ausems, M.G.; Collee, J.M.; van Doorn, H.C.; Gomez Garcia, E.B.; Kets, C.M.; van Leeuwen, F.E.; Meijers-Heijboer, H.E.; et al. Breast cancer risk after salpingo-oophorectomy in healthy BRCA1/2 mutation carriers: Revisiting the evidence for risk reduction. *J. Natl. Cancer Inst.* **2015**, *107*, djv033. [CrossRef] [PubMed]
175. Kotsopoulos, J.; Huzarski, T.; Gronwald, J.; Singer, C.F.; Moller, P.; Lynch, H.T.; Armel, S.; Karlan, B.; Foulkes, W.D.; Neuhausen, S.L.; et al. Bilateral Oophorectomy and Breast Cancer Risk in BRCA1 and BRCA2 Mutation Carriers. *J. Natl. Cancer Inst.* **2017**, *109*, djw177. [CrossRef] [PubMed]
176. Practice Bulletin No 182: Hereditary Breast and Ovarian Cancer Syndrome. *Obstet. Gynecol.* **2017**, *130*, e110–e126. [CrossRef] [PubMed]
177. Russo, A.; Calo, V.; Bruno, L.; Rizzo, S.; Bazan, V.; Di Fede, G. Hereditary ovarian cancer. *Crit. Rev. Oncol. Hematol.* **2009**, *69*, 28–44. [CrossRef] [PubMed]
178. Madalinska, J.B.; van Beurden, M.; Bleiker, E.M.; Valdimarsdottir, H.B.; Hollenstein, J.; Massuger, L.F.; Gaarenstroom, K.N.; Mourits, M.J.; Verheijen, R.H.; van Dorst, E.B.; et al. The impact of hormone replacement therapy on menopausal symptoms in younger high-risk women after prophylactic salpingo-oophorectomy. *J. Clin. Oncol.* **2006**, *24*, 3576–3582. [CrossRef] [PubMed]
179. Stan, D.L.; Shuster, L.T.; Wick, M.J.; Swanson, C.L.; Pruthi, S.; Bakkum-Gamez, J.N. Challenging and complex decisions in the management of the BRCA mutation carrier. *J. Women's Health* **2013**, *22*, 825–834. [CrossRef] [PubMed]
180. Greene, M.H.; Mai, P.L.; Schwartz, P.E. Does bilateral salpingectomy with ovarian retention warrant consideration as a temporary bridge to risk-reducing bilateral oophorectomy in BRCA1/2 mutation carriers? *Am. J. Obstet. Gynecol.* **2011**, *204*, 19.e1–19.e6. [CrossRef] [PubMed]
181. Beattie, M.S.; Crawford, B.; Lin, F.; Vittinghoff, E.; Ziegler, J. Uptake, time course, and predictors of risk-reducing surgeries in BRCA carriers. *Genet. Test. Mol. Biomark.* **2009**, *13*, 51–56. [CrossRef] [PubMed]
182. Greene, M.H.; Piedmonte, M.; Alberts, D.; Gail, M.; Hensley, M.; Miner, Z.; Mai, P.L.; Loud, J.; Rodriguez, G.; Basil, J.; et al. A prospective study of risk-reducing salpingo-oophorectomy and longitudinal CA-125 screening among women at increased genetic risk of ovarian cancer: Design and baseline characteristics: A Gynecologic Oncology Group study. *Cancer Epidemiol. Biomark. Prev.* **2008**, *17*, 594–604. [CrossRef] [PubMed]

183. Kram, V.; Peretz, T.; Sagi, M. Acceptance of preventive surgeries by Israeli women who had undergone BRCA testing. *Fam. Cancer* **2006**, *5*, 327–335. [CrossRef] [PubMed]
184. Metcalfe, K.A.; Birenbaum-Carmeli, D.; Lubinski, J.; Gronwald, J.; Lynch, H.; Moller, P.; Ghadirian, P.; Foulkes, W.D.; Klijn, J.; Friedman, E.; et al. International variation in rates of uptake of preventive options in BRCA1 and BRCA2 mutation carriers. *Int. J. Cancer* **2008**, *122*, 2017–2022. [CrossRef] [PubMed]
185. Schmeler, K.M.; Sun, C.C.; Bodurka, D.C.; White, K.G.; Soliman, P.T.; Uyei, A.R.; Erlichman, J.L.; Arun, B.K.; Daniels, M.S.; Rimes, S.A.; et al. Prophylactic bilateral salpingo-oophorectomy compared with surveillance in women with BRCA mutations. *Obstet. Gynecol.* **2006**, *108*, 515–520. [CrossRef] [PubMed]
186. Harmsen, M.G.; Arts-de Jong, M.; Hoogerbrugge, N.; Maas, A.H.; Prins, J.B.; Bulten, J.; Teerenstra, S.; Adang, E.M.; Piek, J.M.; van Doorn, H.C.; et al. Early salpingectomy (TUbectomy) with delayed oophorectomy to improve quality of life as alternative for risk-reducing salpingo-oophorectomy in BRCA1/2 mutation carriers (TUBA study): A prospective non-randomised multicentre study. *BMC Cancer* **2015**, *15*, 593. [CrossRef] [PubMed]
187. Kwon, J.S.; Tinker, A.; Pansegrau, G.; McAlpine, J.; Housty, M.; McCullum, M.; Gilks, C.B. Prophylactic salpingectomy and delayed oophorectomy as an alternative for BRCA mutation carriers. *Obstet. Gynecol.* **2013**, *121*, 14–24. [CrossRef] [PubMed]
188. Narod, S.A. Salpingectomy to prevent ovarian cancer: A Countercurrents Series. *Curr. Oncol.* **2013**, *20*, 145–147. [CrossRef] [PubMed]
189. Falconer, H.; Yin, L.; Gronberg, H.; Altman, D. Ovarian cancer risk after salpingectomy: A nationwide population-based study. *J. Natl. Cancer Inst.* **2015**, *107*, dju410. [CrossRef] [PubMed]
190. Lessard-Anderson, C.R.; Handlogten, K.S.; Molitor, R.J.; Dowdy, S.C.; Cliby, W.A.; Weaver, A.L.; Sauver, J.S.; Bakkum-Gamez, J.N. Effect of tubal sterilization technique on risk of serous epithelial ovarian and primary peritoneal carcinoma. *Gynecol. Oncol.* **2014**, *135*, 423–427. [CrossRef] [PubMed]
191. Dietl, J.; Wischhusen, J.; Hausler, S.F. The post-reproductive Fallopian tube: Better removed? *Hum. Reprod.* **2011**, *26*, 2918–2924. [CrossRef] [PubMed]
192. Miller, D.M.; McAlpine, J.N.; Gilks, C.B.; Huntsman, D.G. Opportunistic salpingectomy: The way forward-response to Steven Narod. *Curr. Oncol.* **2013**, *20*, 143–144. [CrossRef] [PubMed]
193. American College of Obstetricians and Gynecologists. *Salpingectomy for Ovarian Cancer Prevention*; Number 620; American Congress of Obstetricians and Gynecologists: Washington, DC, USA, 2015.
194. Whiteman, M.K.; Hillis, S.D.; Jamieson, D.J.; Morrow, B.; Podgornik, M.N.; Brett, K.M.; Marchbanks, P.A. Inpatient hysterectomy surveillance in the United States, 2000–2004. *Am. J. Obstet. Gynecol.* **2008**, *198*, 34.e1–34.e7.
195. Parker, W.H. Bilateral oophorectomy versus ovarian conservation: Effects on long-term women's health. *J. Minim. Invasive Gynecol.* **2010**, *17*, 161–166. [CrossRef] [PubMed]
196. Parker, W.H.; Broder, M.S.; Chang, E.; Feskanich, D.; Farquhar, C.; Liu, Z.; Shoupe, D.; Berek, J.S.; Hankinson, S.; Manson, J.E. Ovarian conservation at the time of hysterectomy and long-term health outcomes in the nurses' health study. *Obstet. Gynecol.* **2009**, *113*, 1027–1037. [CrossRef] [PubMed]
197. Dubeau, L. The cell of origin of ovarian epithelial tumors and the ovarian surface epithelium dogma: Does the emperor have no clothes? *Gynecol. Oncol.* **1999**, *72*, 437–442. [CrossRef] [PubMed]
198. Dubeau, L. The cell of origin of ovarian epithelial tumours. *Lancet Oncol.* **2008**, *9*, 1191–1197. [CrossRef]
199. Kauff, N.D.; Satagopan, J.M.; Robson, M.E.; Scheuer, L.; Hensley, M.; Hudis, C.A.; Ellis, N.A.; Boyd, J.; Borgen, P.I.; Barakat, R.R.; et al. Risk-reducing salpingo-oophorectomy in women with a BRCA1 or BRCA2 mutation. *N. Engl. J. Med.* **2002**, *346*, 1609–1615. [CrossRef] [PubMed]
200. Tone, A.A.; Salvador, S.; Finlayson, S.J.; Tinker, A.V.; Kwon, J.S.; Lee, C.H.; Cohen, T.; Ehlen, T.; Lee, M.; Carey, M.S.; et al. The role of the fallopian tube in ovarian cancer. *Clin. Adv. Hematol. Oncol.* **2012**, *10*, 296–306. [PubMed]

cancers

MDPI

Review

Molecular Mechanisms Regulating Organ-Specific Metastases in Epithelial Ovarian Carcinoma

Maria V. Barbolina

Department of Biopharmaceutical Sciences, University of Illinois at Chicago, Chicago, IL 60607, USA; mvb@uic.edu; Tel.: +1-312-355-0670

Received: 13 September 2018; Accepted: 13 November 2018; Published: 15 November 2018

Abstract: Epithelial ovarian carcinoma is the most predominant type of ovarian carcinoma, the deadliest gynecologic malignancy. It is typically diagnosed late when the cancer has already metastasized. Transcoelomic metastasis is the most predominant mechanism of dissemination from epithelial ovarian carcinoma, although both hematogenously and lymphogenously spread metastases also occur. In this review, we describe molecular mechanisms known to regulate organ-specific metastasis from epithelial ovarian carcinoma. We begin by discussing the sites colonized by metastatic ovarian carcinoma and rank them in the order of prevalence. Next, we review the mechanisms regulating the transcoelomic metastasis. Within this chapter, we specifically focus on the mechanisms that were demonstrated to regulate peritoneal adhesion—one of the first steps in the transcoelomic metastatic cascade. Furthermore, we describe mechanisms of the transcoelomic metastasis known to regulate colonization of specific sites within the peritoneal cavity, including the omentum. Mechanisms underlying hematogenous and lymphogenous metastatic spread are less comprehensively studied in ovarian cancer, and we summarize mechanisms that were identified to date. Lastly, we discuss the outcomes of the clinical trials that attempted to target some of the mechanisms described in this review.

Keywords: ovarian cancer; organ-specific metastases; peritoneal adhesion; mesothelium; omentum; peritoneal wall; lymph node; liver; lung; gene expression

1. Introduction

Ovarian carcinoma is the fifth leading cause of death from female cancers [1] and comprises several malignancies of epithelial and non-epithelial origins. Epithelial ovarian carcinoma (EOC) is the most predominant type, which, in turn, encompasses several distinct histotypes that are thought to originate in epithelial cells of the female reproductive tract [2–4]. High-grade serous ovarian carcinoma (HGSOC) is the most predominant histotype of EOC, and is thought to originate in epithelial cells of the ovaries and fallopian tubes.

HGSOC metastases spread via several distinct routes, including transcoelomic, hematogenous, and lymphogenous, with the former being the most predominant [5,6]. Transcoelomic spread refers to a route of tumor metastasis across a body cavity, such as the peritoneal cavity, as in the case of ovarian cancer. Transcoelomic metastases from ovarian cancer primarily seed the viscera of organs and tissues of the peritoneal cavity; metastasizing cells first attach to the mesothelial monolayer of intraperitoneal tissues and subsequently invade the extracellular matrix of the underlying stroma. The majority of patients with EOC are first diagnosed when peritoneal metastases have already formed, because the disease at early stages (when the tumor is confined to the ovary) is nearly asymptomatic.

These patients are typically treated by surgery and neoadjuvant chemotherapy (NACT) consisting of a combination of a platinum drug and a taxane. The Gynecologic Oncology Group defined optimal debulking as residual implants less than 1 cm [7]. Optimal debulking is often not possible due to the vast spread of metastatic lesions across the viscera of the peritoneal cavity [8]. Analysis of 3126 patients demonstrated that one-third each had complete resection, a small residual tumor burden of 1–10 mm, or macroscopic residual disease >1 cm in diameter [9], indicating that optimal resection could be achieved in approximately two-thirds of patients. In cases when optimal debulking is not feasible, the preferred treatment route includes interval debulking surgery after NACT [10,11]. Although most cases are initially responsive to chemotherapy, most become less and less sensitive with every recurrence, and eventually develop chemoresistance [12,13]. Moreover, metastatic ovarian cancer is thought to contain a subpopulation of cells with self-renewing capacity, or cancer stem cells, which are not affected by cytotoxic chemotherapy [14,15]. Patients who relapse more than six months after completion of platinum/taxane initial therapy are considered platinum-sensitive. Patients who respond to primary treatment and relapse within six months are characterized as platinum-resistant. Patients who relapse within 3 months of treatment are regarded as platinum-refractory [16]. The five-year survival for patients with advanced EOC is below 30%. Recurrent chemotherapy-resistant EOC is incurable. Moreover, peritoneal metastases are known to cause malnutrition and cachexia, which is associated with metabolic changes and bowel obstruction in patients with ovarian carcinoma [17–19]. Cachexia is strongly associated with worse prognosis [20–22].

For these reasons, the mechanisms regulating peritoneal metastasis from EOC are studied most extensively with the aim of identifying ways of preventing re-colonization of mesothelial linings and blocking or retarding the growth of intraperitoneal lesions using novel targeted molecular therapies that could be applied either alone or in conjunction with conventional chemotherapies. The hematogenous route of metastatic colonization also contributes to formation of intraperitoneal metastases [23,24]. The lymphatic system is often involved by the disseminating EOC cells as well [25–28]. Formation of thoracic metastases from EOC is thought to occur, in part, via direct extension of the peritoneal metastases or via lymphatics, although the relative contribution of these mechanisms is yet to be experimentally established.

In this review, we focus on the molecular mechanisms currently known to underlie the formation and development of organ-specific metastases from EOC.

2. Sites of EOC Metastases

Several studies assessed the patterns of metastatic spread from EOC at relapse, as well as at autopsy. These studies demonstrated that, although EOC typically colonized a wide variety of organs and tissues, there was not one single metastatic site that was always involved by metastasis in all studied cases. The main site of metastasis was the peritoneum, including the parietal and visceral peritoneum and omentum, which was involved in 77% of cases on average among several reports (ranging between 53% and 99%, Table 1) [25–27,29–31]. Other commonly colonized sites identified by autopsy studies included lymph nodes (38–77% of cases), large and small intestine (44–86% of cases), liver parenchyma (45–59% of cases), and lung (33–39% of cases). Pancreas, spleen, stomach, and ureter were involved in 3–24% of cases, while organs, such as the thyroid, bone, brain, skin, heart, breast, and kidney were typically colonized in 1–12% of the cases [25–27,32]. Notably, although distant metastases are rarely the main cause of death from metastatic EOC, their presence usually indicates worse prognosis [33,34]. For example, the presence of parenchymal splenic metastasis was an independent predictor of decreased overall survival [35]. Brain metastasis also correlated with worse prognosis among older patients [36,37].

Table 1. Prevalence of peritoneal metastasis in patients with epithelial ovarian carcinoma (EOC). USA—United States of America; UK—United Kingdom.

Patient Population, Institution	Number of Cases	Method of Study	Number of Cases with Peritoneal Metastasis (Percentage of Total)	Time of Assessment	Reference
USA, Roswell Park Institute	381*	Autopsy	316 (83%)	At autopsy	[27]
USA, University of Rochester Medical School	100	Autopsy	73 (73%)	At autopsy	[25]
USA, National Cancer Institute	73	Autopsy	39 (53%)	At autopsy	[26]
Switzerland, University of Basel	166 **	Autopsy	164 (99%)	At autopsy	[32]
UK, St. Bartholomew's Hospital	67 ***	Computed tomography	59 (88%)	At relapse	[30]
Japan, The Jikei University School of Medicine	70 ****	Imaging, cytometry, CA125 level	49 (70%)	At relapse	[31]

* The number of cases with *epithelial* ovarian carcinoma only. ** The number of patients who were analyzed. *** The number of cases for which complete imaging data are available. **** The number of cases with recurrent ovarian cancer.

Interestingly, the mutational status of tumor protein P53 (TP53) is linked with the propensity to seed either mainly peritoneal or peritoneal and distant metastases. Ninety-six percent of all cases belonging to HGSOC carry mutations in the *tp53* gene, which could occur at multiple locations within the gene sequence [34,38,39]. Null mutations of p53 were predictive of distant metastasis to liver, thorax, spleen, brain, and lymph nodes at initial diagnosis, and they occurred eightfold more frequently compared with cases containing either missense mutations of *tp53* or those expressing wild-type TP53 [34], although the detailed mechanisms are not known. Furthermore, cases with serous histology displayed slightly higher preponderance to having distant metastases (22 out of 66) in comparison to cases with other histologies, in which seven out of 35 cases had distant metastases [34].

In summary, the studies showed that metastases from ovarian carcinoma predominantly formed locally within the peritoneal cavity (peritoneum, omentum, and mesothelium); however, a large number of cases also presented with distant metastases, most commonly at lymph nodes, liver, and lung.

3. Mechanisms Regulating Transcoelomic Metastasis from EOC

Transcoelomic dissemination is thought to be the major route via which EOC metastasis spreads [6]. Peritoneal metastases can reach very large sizes and are often accompanied by the presence of the malignant ascites. These metastases are seeded by individual cells and multicellular aggregates or spheroids, which first adhere to mesothelial cells outlining the peritoneal cavity and then invade the submesothelial extracellular matrix (Figure 1). Studies of cell cultures of ovarian cancer cell lines demonstrated that cells could spontaneously detach from monolayers and remain as individual cells or form spheroids [40,41]. Recent in vivo studies demonstrated that spheroids predominantly formed by collective detachment from the primary tumor [40]. Mechanistically, it was suggested that a membrane type-1 matrix metalloproteinase plays a pivotal role in the spontaneous release of cell–cell adherent sheets, which later form spheroids [42]. Another study showed that individual cells could also aggregate together prior to mesothelial adhesion [43].

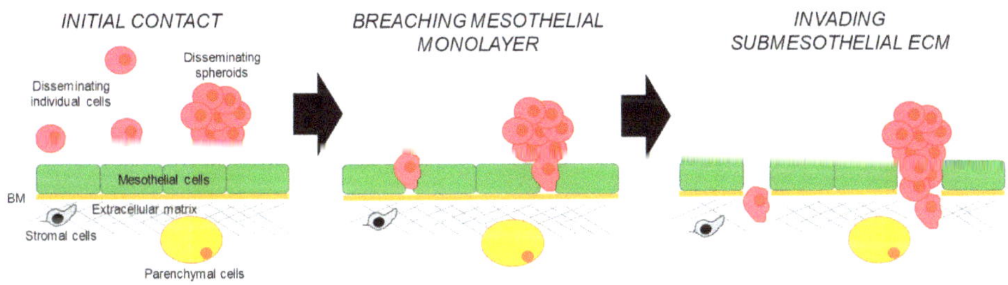

Figure 1. A scheme of the peritoneal metastasis through the transcoelomic route. Disseminating epithelial ovarian carcinoma (EOC) cells and spheroids are shown in mauve, mesothelial cells are shown in green, the basement membrane is shown in orange, stromal cells are shown in grey, parenchymal cells are shown in yellow, and the extracellular matrix is shown in grey.

Progression of ovarian carcinoma is often accompanied by the presence of the malignant ascites, a fluid that accumulates within the peritoneal cavity [44,45]. It was estimated that more than one-third of ovarian cancer patients present with ascites at diagnosis, and nearly all contain ascites at the time of recurrence [46]. It is thought that ascites forms as a result of impaired drainage of the peritoneal cavity due to the obstruction of the lymphatic system by metastasizing tumor cells or an increased filtration rate to the peritoneal cavity [44]. The presence of the malignant ascites predicts poor prognosis in different ovarian cancer patient cohorts irrespective of the tumor's histological type [47–50]. Ascites contains many soluble molecules, such as survival factors, including cytokines, chemokines, growth factors, and fragments of the extracellular matrix (ECM). In addition, ascites is an abundant source of the cells of the immune system, stromal and mesothelial cells, and cancer stem cells [46]. It is thought that the microenvironment within the ascites contributes to survival of the metastasizing ovarian cancer cells, provides support for tumor growth, and contributes to tumor heterogeneity [51].

3.1. Mechanisms Regulating Peritoneal Adhesion

The peritoneal mesothelium is a monolayer of mesothelial cells that lines the abdominal cavity [52]. Disseminating individual cells and spheroids adhere to the mesothelial lining of the intraperitoneal cavity to establish metastatic lesions. Studies described below demonstrated that both ovarian cancer and mesothelial cells play active roles in this process.

Mesothelial cells produce hyaluronan, which is thought to serve as a protective layer preventing attachment of the malignant EOC cells [53]. Inflammation associated with cancer may result in production of low-molecular-weight hyaluronan fragments and destruction of the protective hyaluronan coat consisting of the high-molecular-weight hyaluronan [54]. Different molecular pathways associated with both mesothelial and disseminating tumor cells could participate in promoting peritoneal adhesion.

The majority of EOC cases affect the elderly, as the median age at diagnosis is 63 [55]. Thus, several studies focused on characterization of senescent stromal cells, including peritoneal mesothelial cells in the metastatic process. Senescent mesothelial cells play a critical role in the development of peritoneal carcinomatosis in several cancer models, including ovarian cancer [56]. In studies of syngeneic ovarian cancer models, aged mice were more prone to formation of metastases than their younger counterparts [57]. Increased metastatic burden in aged mice also corresponded to significant changes in the cellular composition of the native immune system within the peritoneal adipose tissue such that the presence of tumor-infiltrating lymphocytes was higher and B-cell-related pathways were altered in comparison to younger mice [57].

EOC cells can co-opt mesothelial cells in order to facilitate peritoneal adhesion. It was demonstrated that EOC cells can secrete exosomes enriched for CD44 molecule (CD44). These exosomes could be internalized by mesothelial cells, resulting in the secretion of matrix metallopeptidase 9 (MMP-9), which helps EOC with cell invasion [58]. Mesothelial cells also secrete lysophosphatidic acid (LPA), which aids in mesothelial adhesion of EOC cells expressing receptors for LPA [59].

Numerous in vitro and in vivo studies used different experimental approaches to show that disseminating ovarian cancer cells themselves express a number of membranous receptors and adhesion molecules that facilitate their adhesion to mesothelial cells expressing ligands for these receptors (Figure 2).

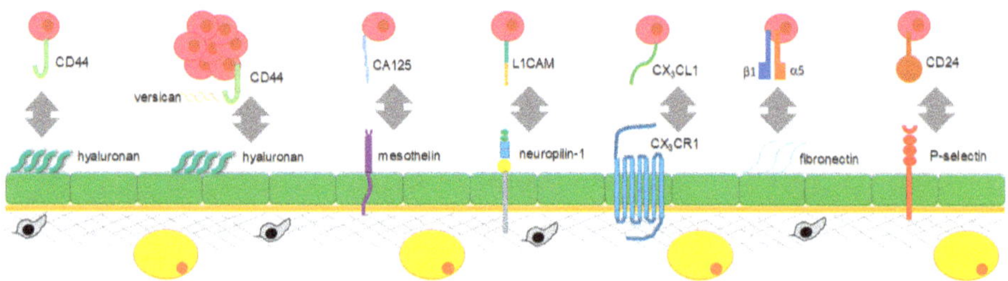

Figure 2. Molecular mechanisms regulating peritoneal adhesion. Disseminating epithelial ovarian cancer cells and spheroids are shown in mauve, mesothelial cells are shown in green, the basement membrane is shown in orange, stromal cells are shown in grey, parenchymal cells are shown in yellow, and the extracellular matrix is shown in grey. Only one interaction between a cancer cell and a mesothelial cell is shown for simplicity and a clearer presentation of the known mechanisms. CD44: CD44 molecule; CA125: mucin 16, cell surface associated or ovarian carcinoma antigen CA125; L1CAM: L1 cell adhesion molecule; CX_3CL1: C-X3-C motif chemokine ligand 1; CX_3CR1: C-X3-C motif chemokine receptor 1; CD24: CD24 molecule.

The first interaction reported to facilitate peritoneal adhesion was between hyaluronan (HA) expressed by mesothelial cells and CD44 expressed by EOC cells [60,61]. It was also reported that extracellular tissue transglutaminase (TG2) expressed by ovarian cancer cells results in upregulation of CD44, which further promotes peritoneal adhesion [62]. Reduction of expression of a secreted glycoprotein, versican, which is thought to facilitate the CD44–HA interaction, in EOC cells, resulted in reduction of tumor formation by individual cells and abrogation of metastatic ability of spheroids [63,64]. These studies revealed an important role of CD44 and molecular interactions supporting its function, notably, TG2 and versican, in mesothelial adhesion. Further studies indicated that treatment with neutralizing anti-CD44 antibodies reduced tumor burden on the peritoneal wall and diaphragm in a xenograft mouse model [65], suggesting that targeting CD44 is a promising approach for the reduction of peritoneal tumor.

Other molecular interactions supporting peritoneal adhesion are fostered by a glycoprotein on the surface of EOC cells, ovarian carcinoma antigen CA125 (MUC16), and mesothelin expressed on the surface of mesothelial cells, both of which were demonstrated by in vitro studies using EOC cell lines [66,67].

A series of in vitro studies that utilized EOC cell lines and mesothelial cells demonstrated that neuropilin-1 expressed in mesothelial cells can interact with a glycoprotein expressed in EOC cells, the L1 adhesion molecule (L1CAM) [68].

Both in vitro studies of EOC and mesothelial cell co-cultures and in vivo studies of short-term adhesion and survival xenograft studies showed that a chemokine receptor fractalkine (CX_3CR1) expressed in EOC cells can interact with its ligand CX_3CL1 expressed (in its transmembrane form) by peritoneal mesothelial cells [69,70].

Studies using co-cultures of EOC cells and mesothelial cells pre-treated with β1-integrin-specific neutralizing antibodies demonstrated that β1-integrins expressed by EOC cells could interact with fibronectin expressed by mesothelial cells [71]. The role of α5β1-integrin-mediated adhesion of

EOC cells to fibronectin-expressing mesothelial cells was further confirmed with a series of in vitro experiments utilizing primary mesothelial cells as well as in vivo using xenograft models of the disease [72]. The latter study highlighted the role of EOC cells in inducing fibronectin expression in mesothelial cells via transforming growth factor beta 1 (TGFβ1)-mediated signaling.

Another interactive loop facilitating peritoneal adhesion is initiated by alternatively activated macrophages (AAMs) occurring in the peritoneal microenvironment of EOC, which involved stimulation of expression of a calcium-dependent receptor, P-selectin, on mesothelial cells by a C–C motif chemokine ligand 4 (CCL4 or MIP-1β) secreted by the AAMs; EOC cells expressing CD24 interacted with P-selectin-expressing mesothelial cells, as demonstrated by ex vivo and in vivo studies of a syngeneic ovarian cancer model [73].

Once metastasizing EOC cells adhered to the mesothelial monolayer, to anchor metastatic lesions, they need to disrupt the mesothelial lining and invade submesothelial parenchymal tissues consisting of the ECM, stromal cells, and cells of the organ parenchyma. EOC cells express various molecules that assist their invasion into the parenchyma of the organs and tissues to which they have adhered. Overexpression of alcohol dehydrogenase 1B (ADH1B) was demonstrated to regulate EOC cell attachment and clearance of the mesothelial lining, as well as subsequent matrix invasion [74]. Adherent spheroids could utilize integrin- and talin-dependent activation of myosin and traction force to clear the mesothelial monolayer [75]. Expression of a transmembrane glycoprotein prominin-1 (PROM1) correlated with the ability of EOC cells to adhere to and clear the mesothelial monolayer as well [76]. It was shown that N-cadherin, but not E-cadherin, is essential for the lateral dispersal of spheroids onto extracellular matrix and invasion; individual cells also depended on N-cadherin for their dispersal and penetration into the collagen gels [77].

After breaching the mesothelial monolayer, ovarian cancer cells quickly adhere to the submesothelial matrix, which is predominantly composed of collagens type I and III, using both α2β1- and α3β1-integrins [78,79]. MT1-MMP is a major interstitial collagenase enabling invasion and anchorage of metastatic ovarian cancer cells in the submesothelial matrix [80].Three-dimensional collagen I is instrumental in upregulating the transmembrane collagenase membrane type 1 matrix metalloproteinase (MT1-MMP) via several mechanisms, including integrin-dependent activation of an Src proto-oncogene, non-receptor tyrosine kinase (Src)-dependent pathway, and subsequent induction of a transcription factor early growth response 1 (EGR1), as well as matrix rigidity-dependent activation of wingless (Wnt) signaling through downregulation of dickkopf-1 expression [81,82]. Epidermal growth factor receptor (EGFR)-dependent modulation of MT1-MMP surface dynamics was also found to contribute to transition to a more invasive phenotype of ovarian cancer cells [83].

In summary, several molecular interactions between cancer and mesothelial cells establish successful cell–cell adhesion during mesothelial adhesion. Disseminating cancer cells take advantage of secreted molecules produced by mesothelial cells and can reprogram their gene expression to aid peritoneal adhesion. Likewise, aging can amplify the process of peritoneal carcinomatosis by providing more permissive conditions for cancer cell adhesion. Importantly, EOC cells themselves express proteins that enable their attachment and tissue invasion.

3.2. Mechanisms Regulating the Transcoelomic Omental Metastasis

The omentum is a peritoneal fold that connects the stomach with abdominal organs [84]. The omentum functions to protect and support abdominal organs and to limit intraperitoneal infection. In addition to a mesothelial monolayer covering this tissue, omentum mainly consists of adipocytes. Other prominent structures within the omentum are milky spots that are the areas of lymphoid tissue containing macrophages, lymphocytes, and mast cells [85]. The omentum also contains other stromal cells, such as fibroblasts, and it is supplied by the gastroepiploic arteries [86]. Studies showed that invading EOC cells successfully establish metastatic lesions in the omentum by taking advantage of the unique microenvironment within this tissue [87,88].

In metastasis from EOC, omentum plays a central role as one of the major tissues hosting peritoneal metastatic lesions [5,25,27,32]. Omentum is also a major site of recurrent metastasis in patients whose omentum was not completely resected. According to the current standard of care, omentum may be partially or completely resected in medically fit patients in the process of the debulking surgery depending on the degree of its involvement with the metastasis [89]. Due to the importance of omentum as a major secondary site, several studies addressed the mechanisms supporting survival and proliferation of metastatic EOC cells within the omental tissues, and uncovered the role of various omental stromal cells in supporting this process.

Recent studies suggested that, as cells detach from the primary tumor and become suspended in the ascites, they undergo a metabolic shift from glycolysis to lipid metabolism, which later affords and facilitates their survival within the omental tissue [90]. Metastasizing EOC cells are also attracted to the omentum by adipokines expressed by the adipocytes, such as adiponectin, interleukin-6 (IL-6), interleukin-8 (IL-8), C–X–C motif chemokine ligand 1 (CXCL1, GRO-α), and others [87,91].

Milky spots are mainly composed of macrophages and lymphocytes, and represent initial lymphatics of the omentum that drain into lymph collectors [92]. Preclinical studies that used ex vivo and in vivo syngeneic and xenograft mouse models demonstrated that disseminating EOC cells can lodge onto milky spots and further spread through the adipose-rich tissue [88,93]. In vivo studies with both syngeneic (ID8 mouse-derived ovarian cancer cell line in C57BL/6 mice) and xenograft (Caov-3, HEYA8, and SKOV3i.p.1 human-derived ovarian cancer cell lines in athymic nude mice) models of ovarian carcinoma suggested that disseminating cells preferentially lodge onto milky spot-containing adipose tissue as opposed to peritoneal fat, while the number and size of the milky spots did not depend on the mouse genetic background [93]. The study also showed that conditioned media collected from milky spot-containing adipose tissue significantly increased cell migration in comparison to the conditioned media from milky spot-deficient adipose tissue [93].

Once EOC cells lodge onto the omentum, proximity to adipocytes results in upregulation of fatty-acid-binding protein 4 (FABP4) and a fatty-acid receptor CD36, followed by transfer of lipids from adipocytes to EOC cells, and induction of lipolysis in adipocytes and β-oxidation in cancer cells [87,94]. Interaction of EOC cells with mesothelial cells reduced expression of microRNA-193 (miR-193) in the former, resulting in increased ability to colonize the omentum [95]. Consistent with cancer cell utilization of lipids stored in adipocytes as an energy source, a study that described the role of milky spots in metastatic colonization of the omentum also reported reduction of the adipose tissue as the tumors grew over time [93].

Fibroblasts in omentum also play a prominent role in regulating this organ-specific metastasis. A study uncovered interaction between tumor necrosis factor alpha (TNFα), transforming growth factor alpha (TGFα), and epidermal growth factor receptor (EGFR); this TNFα–TGFα–EGFR interacting loop is thought to form between EOC cells and fibroblasts that reside in omentum, and it is suggested that it functions to promote peritoneal metastasis [96].

Interactions between chemokine receptors expressed by cancer cells and their corresponding chemokines at the metastatic sites was suggested to regulate homing of metastasizing cells to their niches. Among these interactions, association between the C–X–C motif chemokine receptor 4 (CXCR4) and its ligand, stromal-derived factor 1, was demonstrated to regulate pro-metastatic functions of cells from several cancer types, including ovarian [97–100]. A specific inhibitor of CXCR4, AMD3100, nearly completely blocked EOC cell dissemination to the omentum in a rodent syngeneic model, supporting the importance of this chemokine axis in development of the omental metastasis [101]. Another study demonstrated that omentum-secreted IL-8 and GRO-α can activate C-X-C motif chemokine receptor 2 (CXCR2) in ovarian carcinoma cells and facilitate EOC cell spreading in the peritoneal cavity [91].

Thus, studies of the mechanisms of omental metastasis to date demonstrated important roles of the omentum itself and metastasizing EOC cells in facilitating formation and development of this major type of metastatic lesions.

3.3. Mechanisms Regulating Transcoelomic Metastasis to Other Intraperitoneal Organs and Tissues

The mechanisms regulating organ-specific intraperitoneal dissemination to other organs, including peritoneal wall, viscera of the bowels, viscera of the liver, etc., are less well understood as compared to the mechanisms regulating formation of the omental metastasis. Several studies uncovered the pivotal role of the chemokine–receptor interactions in regulating these organ-specific metastases. Inhibition of CXCR4 with AMD3100 significantly reduced colonization of the colon, peritoneal wall, diaphragm, and liver [101]. Downregulation of another chemokine receptor, X-C motif chemokine receptor 1, or lymphotactin (XCR1), almost completely abrogated colonization of diaphragm and peritoneal wall [102]. Further, it was demonstrated that yet another chemokine axis, between fractalkine (CX_3CL1) and its receptor (CX_3CR1), regulates dissemination of the CX_3CR1-positive EOC cells to the surfaces of the CX_3CL1-positive tissues, including peritoneal wall, diaphragm, liver, mesentery, and retroperitoneal kidneys [69,70].

4. Mechanisms Regulating Hematogenous Metastasis from EOC

Peritoneovenous shunting is a procedure in which a shunt could be used to return the peritoneal fluid from the peritoneal cavity into veins, such as the superior vena cava or the internal jugular vein, by means of a one-way valved anastomosis [103–105]. This method was attempted on a cohort of patients with ovarian cancer and other malignancies who had intractable ascites for the purpose of palliative care [106,107]. A study that described the autopsy findings of the patients that underwent this procedure concluded that most patients either did not develop distant metastases or grew very small isolated lesions as a result of this procedure [108]. In the ovarian cancer field, this was interpreted as suggesting that the hematogenous route has little relevance as a mechanism via which the metastasis forms. However, a close examination of the presented data suggests that this conclusion was overgeneralized. Eight out of nine ovarian cancer patients did not survive longer than about four months on average (survival ranged from one to seven months) after the initiation of this procedure; moreover, even over this short period of survival, evidence of distant metastases at lung, liver, spleen, brain, and other distant sites was found in three of the eight patients. Only one out of nine ovarian cancer patients survived for 27 months after the procedure without developing distant metastasis [108,109]. Importantly, although distant metastases were not the cause of death in this study, they did occur, even though all but one patient survived between one and seven months after insertion of the shunts.

In another patient-based study that focused on investigating the outcomes of inferior vena cava filter placement in patients with epithelial ovarian, fallopian, and primary peritoneal cancer, the authors reported that patients who underwent this procedure had significantly lower survival and significantly higher incidence of deep vein thrombosis and distant metastasis [110], supporting the role of a hematogenous route in seeding distant metastases from EOC. Additionally, seeding of distant organs, including brain [36], is likely to occur via this mechanism.

Experimentally, evidence of development of the hematogenous metastasis within the omentum was recently presented [23]. A novel parabiosis mouse model was used to demonstrate that the molecular interaction between the Erb-B2 receptor tyrosine kinase 3 (ERBB3) expressed by ovarian cancer cells and its ligand neuregulin-1 expressed by the omentum is the main driving force of the hematogenously spread omental metastasis. Parabiosis is a surgical union of two organisms that allows sharing of the blood circulation [111]. In the study on ovarian carcinoma, the parabiosis model was created by excising the skin of female mice from the shoulder to the hip joint followed by surgical anastomosis to make new connections between blood vessels of pairs of mice [23].

In another study, three approaches were employed to investigate the role of the hematogenous route of EOC metastases, including an intravascular tail-vein injection of ovarian cancer cells, as well as subcutaneous engraftment of murine and human tumors. Primary ovarian cancer cells were co-injected with mesenchymal stem cells subcutaneously. To promote formation of blood vessels in the tumor, human infantile hemangioma stem cells were co-injected as well. This protocol resulted in

100% engraftment rate and macroscopic ovarian metastases by the time of sacrifice [24]. The authors observed development of tumors not only within the ovary, but also at other distant sites, including the lung [24], further supporting existence of mechanisms driving hematogenous dissemination to different organ sites. A study on the role of CXCR4 in EOC metastases showed that downregulation of CXCR4 by short hairpin RNA (shRNA) resulted in a robust reduction of the circulating tumor cells, suggesting a possible role of the stromal cell-derived factor 1 (SDF1)/CXCR4 axis in the hematogenous route of dissemination [112].

In summary, many patient studies reported occurrence of the distant metastasis, which could have arrived at these sites, notably the brain, likely via the hematogenous route. Although these distant hematogenously spread metastases are not considered to be the cause of death from EOC by themselves, their presence is significantly correlated with worse survival. Presently, the peritoneal metastases from EOC is still an unsolved problem in clinical management of this disease. However, it is very likely that continuous progress in the treatment of the peritoneal metastasis and increased survival could allow for more time for development of the distant metastasis, which could become clinically relevant in long-term survivors of metastatic EOC.

5. Mechanisms Regulating Lymphatic Metastasis from EOC

The International Federation of Gynecology and Obstetrics (FIGO) ovarian cancer staging states that metastatic involvement of the retroperitoneal lymph nodes indicates FIGO Stage IIIC of the disease, and colonization of the inguinal lymph nodes and lymph nodes outside of the abdominal cavity by metastases signifies Stage IVB of the disease [113]. Although EOC metastases frequently involve lymph nodes, autopsy studies reported that the frequency of colonization differed by their anatomic location with the abdominal lymph nodes being most frequently colonized among others (Table 2).

Table 2. Most frequently colonized lymph nodes identified in EOC patients by autopsy studies.

Study	Abdominal Lymph Nodes	Pelvic Lymph Node	Thoracic Lymph Node
[27]	58%	48%	28%
[25]	47%	17%	29%
[32]	74.1%	27.7	34.9
Average number of patients with indicated metastasis	60	31	31

Vascular endothelial growth factor receptor 3 (VEGFR3) is the major receptor involved in lymphangiogenesis and maintenance of the lymphatic endothelium [114]. The ligands activating this receptor are vascular endothelial growth factors C and D (VEGFC, VEGFD). Immunohistochemical analysis of expression of VEGFA, VEGFC and VEGFD in ovarian carcinoma patients, most of which (92/100) were diagnosed with FIGO Stage III disease with retroperitoneal metastases or those with predominantly intraperitoneal metastasis, demonstrated that high expression of VEGFC corresponded to the presence of the retroperitoneal metastasis, while low VEGFC correlated with mostly intraperitoneal metastatic spread, supporting the role of VEGFC–VEGFR3 interaction in EOC cell tropism to the lymph node. High VEGFC also correlated with shorter overall survival [115]. Another study was performed to characterize the patterns of expression of the ubiquitin-specific protease 7 (USP7); it was found that high expression of USP7 significantly correlated with lymph node metastases [116]. Upregulation of focal adhesion kinase (FAK) in EOC cells strongly correlated with incidence of lymph node metastases as well [117].

In summary, lymphatic involvement is correlated with worse outcomes. Patient studies demonstrated preferential colonization of the abdominal over other lymph nodes in the human body.

6. Targeted Therapies in Ovarian Cancer

Development of targeted therapies against ovarian carcinoma, although still mainly at the stage of characterization of new potential targets, is an actively growing field. Several proteins that were demonstrated to play a role in progression and metastasis of ovarian cancer were or currently are investigated as novel drug targets in clinical trials. The targeting agents used in these studies vary widely from small-molecule inhibitors to monoclonal antibodies, antibody–drug conjugates, and immunotherapy.

For example, a CD44-targeting compound SPL-108 is being investigated in conjunction with paclitaxel in a phase I trial against epithelial ovarian carcinoma (ClinicalTrials.gov identifier: NCT03078400).

Several clinical trials are attempting to target mesothelin. Among those, one clinical trial (ClinicalTrials.gov identifier: NCT03692637) that is currently in phase I is aiming to use anti-mesothelin chimeric antigen receptor (CAR) natural killer (NK) cells in epithelial ovarian carcinoma. This study is taking advantage of the new targeting technology based on the use of CAR-NK therapy consisting of chimeric antigen receptor (CAR)-expressing natural killer cells of the immune system [118]. Another mesothelin-targeting approach is being investigated in a phase I clinical trial involving patients with recurrent mesothelin-expressing platinum-resistant ovarian cancer, and this study will test the efficacy of anetumab ravtansine in combination with polyethylene glycol (PEG) PEGylated liposomal doxorubicin (ClinicalTrials.gov identifier: NCT02751918). Anetumab ravtansine is an antibody–drug conjugate, in which an anti-mesothelin antibody (anetumab) is attached to a tubulin inhibitor (ravtansine) [119].

As β1-integrins play a pivotal role in different mechanisms underlying progression of ovarian carcinoma, they are targeted in clinical trials in ovarian carcinoma. Volociximab, a chimeric monoclonal antibody that binds to and inhibits the functional activity of α5β1-integrins, was studied in a phase II trial as a monotherapy in patients with platinum-resistant advanced epithelial ovarian or primary peritoneal cancer. Although the agent alone did not provide sufficient clinical activity, it was well tolerated, prompting the development of improved strategies to target α5β1-integrins [120]. A phase II clinical trial (ClinicalTrials.gov identifier: NCT00635193) was conducted in patients with advanced ovarian cancer or those who relapsed after platinum/taxane therapy that tested volociximab in a combination with liposomal doxorubicin. Preliminary data from this trial suggest that the combination was well tolerated [121], while no data on the efficacy against the relapsed disease was reported.

Epidermal growth factor receptor (EGFR) is overexpressed in epithelial ovarian carcinoma and its high expression is associated with poor prognosis [122]. EGFR is important in progression of many cancer types [123]; thus, it became one of the major targets in cancer following the development of several targeting agents [124,125]. Therefore, many clinical trials addressed targeting EGFR using tyrosine kinase inhibitors or antibodies as monotherapy in ovarian cancer, although the results of these trials demonstrated no difference in survival [126]. While new EGFR-targeting agents, such as a monoclonal antibody matuzumab (EMD 72000), are still being investigated as monotherapy (ClinicalTrials.gov identifier: NCT00073541), combinations of EGFR-targeting therapy with the standard chemotherapy are being tested in other clinical trials. In one such combination trial, gefitinib (Iressa), a small-molecule inhibitor of EGFR, is being investigated in combination with topotecan, a topoisomerase inhibitor, in patients with relapsed ovarian, peritoneal, and fallopian tube cancers (ClinicalTrials.gov identifier: NCT00317772).

A small-molecule inhibitor of CXCR4, plerixafor (Mozobil), will be investigated in patients with advanced cancers, including ovarian (ClinicalTrials.gov identifier: NCT02179970). VEGFRs were studied as targets of angiogenesis in ovarian carcinoma, resulting in approval of a monoclonal antibody against VEGFA, bevacizumab (Avastin), by The Food and Drug Administration in 2014 [127]; however, these agents were not directed at targeting the lymphogenous spread.

In summary, although many targeted therapies directed at mechanisms described in this review are yet to show any advantage in the treatment of relapsed ovarian carcinoma when given as a monotherapy, optimism still remains that their combinations with other chemotherapeutic agents will benefit the patients at terminal stages of the disease. The successes behind targeted therapies for breast cancer gene (*BRCA*)-deficient ovarian carcinoma with the inhibitors of poly (ADP)-ribose polymerase [128] provide further confidence in the approach.

7. Conclusions

Many studies addressed mechanisms regulating the formation and development of intraperitoneal metastases. Foremost, the findings demonstrate the importance of the receptor–ligand interactions between metastatic EOC cells with other cells and molecules in the microenvironment, such as mesothelial cells, omental adipocytes, fibroblasts, ECM, and others, in the development of metastatic lesions. Disseminating EOC cells are endowed with expression of several types of receptors, including those for chemokines, tyrosine kinases, integrins, and glycoproteins. Expression of these receptors proved to be essential for successful colonization of the various tissues and organs. These findings pointed research efforts toward the development of targeted therapies against disseminating EOC cells. However, further studies into EOC metastasis-related mechanisms are crucial for the development of personalized therapies against this highly heterogeneous and deadly disease. Several studies highlighted the role of the microenvironment and stromal cells (both naïve and tumor-associated) in the intraperitoneal milieu, as well as the role of the molecular changes during aging that support peritoneal metastases. Thus, these causal aspects of the microenvironment should also be viewed as potential molecular targets for reduction of the metastatic spread and prevention of recurrences. Both hematogenous and lymphatic routes of dissemination are relatively less studied due to their perceived limited impact on the outcomes, as most EOC patients succumb to the intraperitoneal metastasis. Nonetheless, it is likely that their clinical relevance, however unfortunate for the patients, will increase as treatment of the intraperitoneal metastasis improves in the future. Therefore, understanding of the mechanisms regulating the distant metastasis is essential for ultimately blocking this deadly metastatic disease. Overall, a more comprehensive characterization of the mechanisms regulating metastatic ovarian carcinoma and therapy response is required for the development of new targeted therapies and improvement of currently used treatment regimens.

Author Contributions: M.V.B. conceived the topic of the review and wrote the manuscript.

Funding: M.V.B. gratefully acknowledges funding support from Tell Every Amazing Lady about Ovarian Cancer, Louisa M. McGregor Ovarian Cancer Foundation (T.E.A.L.).

Conflicts of Interest: The authors declare no conflict of interest.

References

1. Siegel, R.L.; Miller, K.D.; Jemal, A. Cancer statistics, 2018. *CA Cancer J. Clin.* **2018**, *68*, 7–30. [CrossRef] [PubMed]
2. Karnezis, A.N.; Cho, K.R. Of mice and women—Non-ovarian origins of "ovarian" cancer. *Gynecol. Oncol.* **2017**, *144*, 5–7. [CrossRef] [PubMed]
3. Karnezis, A.N.; Cho, K.R.; Gilks, C.B.; Pearce, C.L.; Huntsman, D.G. The disparate origins of ovarian cancers: Pathogenesis and prevention strategies. *Nat. Rev. Cancer* **2017**, *17*, 65–74. [CrossRef] [PubMed]
4. Kurman, R.J.; Ie, M.S. The Dualistic Model of Ovarian Carcinogenesis: Revisited, Revised, and Expanded. *Am. J. Pathol.* **2016**, *186*, 733–747. [CrossRef] [PubMed]
5. Lengyel, E. Ovarian cancer development and metastasis. *Am. J. Pathol.* **2010**, *177*, 1053–1064. [CrossRef] [PubMed]
6. Tan, D.S.; Agarwal, R.; Kaye, S.B. Mechanisms of transcoelomic metastasis in ovarian cancer. *Lancet Oncol.* **2006**, *7*, 925–934. [CrossRef]
7. Whitney, C.W.; Spirtos, N. *Gynecologic Oncology Group Surgical Procedures Manual*; Gynecologic Oncology Group: Philadelphia, PA, USA, 2010.
8. Shih, K.K.; Chi, D.S. Maximal cytoreductive effort in epithelial ovarian cancer surgery. *J. Gynecol. Oncol.* **2010**, *21*, 75–80. [CrossRef] [PubMed]

9. du Bois, A.; Reuss, A.; Pujade-Lauraine, E.; Harter, P.; Ray-Coquard, I.; Pfisterer, J. Role of surgical outcome as prognostic factor in advanced epithelial ovarian cancer: A combined exploratory analysis of 3 prospectively randomized phase 3 multicenter trials: By the Arbeitsgemeinschaft Gynaekologische Onkologie Studiengruppe Ovarialkarzinom (AGO-OVAR) and the Groupe d'Investigateurs Nationaux Pour les Etudes des Cancers de l'Ovaire (GINECO). *Cancer* **2009**, *115*, 1234–1244. [PubMed]
10. Wright, A.A.; Bohlke, K.; Armstrong, D.K.; Bookman, M.A.; Cliby, W.A.; Coleman, R.L.; Dizon, D.S.; Kash, J.J.; Meyer, L.A.; Moore, K.N.; et al. Neoadjuvant Chemotherapy for Newly Diagnosed, Advanced Ovarian Cancer: Society of Gynecologic Oncology and American Society of Clinical Oncology Clinical Practice Guideline. *J. Clin. Oncol.* **2016**, *34*, 3460–3473. [CrossRef] [PubMed]
11. Vergote, I.; Trope, C.G.; Amant, F.; Kristensen, G.B.; Ehlen, T.; Johnson, N.; Verheijen, R.H.; van der Burg, M.E.; Lacave, A.J.; Panici, P.B.; et al. Neoadjuvant chemotherapy or primary surgery in stage IIIC or IV ovarian cancer. *N. Engl. J. Med.* **2010**, *363*, 943–953. [CrossRef] [PubMed]
12. Matulonis, U.A.; Sood, A.K.; Fallowfield, L.; Howitt, B.E.; Sehouli, J.; Karlan, B.Y. Ovarian cancer. *Nat. Rev. Dis Primers* **2016**, *2*, 16061. [CrossRef] [PubMed]
13. Goldie, J.H.; Coldman, A.J. A mathematic model for relating the drug sensitivity of tumors to their spontaneous mutation rate. *Cancer Treat. Rep.* **1979**, *63*, 1727–1733. [PubMed]
14. Curley, M.D.; Garrett, L.A.; Schorge, J.O.; Foster, R.; Rueda, B.R. Evidence for cancer stem cells contributing to the pathogenesis of ovarian cancer. *Front. Biosci.* **2011**, *16*, 368–392. [CrossRef]
15. Zhang, S.; Balch, C.; Chan, M.W.; Lai, H.C.; Matei, D.; Schilder, J.M.; Yan, P.S.; Huang, T.H.; Nephew, K.P. Identification and characterization of ovarian cancer-initiating cells from primary human tumors. *Cancer Res.* **2008**, *68*, 4311–4320. [CrossRef] [PubMed]
16. Armstrong, D.K. Relapsed ovarian cancer: Challenges and management strategies for a chronic disease. *Oncologist* **2002**, *7* (Suppl. 5), 20–28. [CrossRef] [PubMed]
17. Yim, G.W.; Eoh, K.J.; Kim, S.W.; Nam, E.J.; Kim, Y.T. Malnutrition Identified by the Nutritional Risk Index and Poor Prognosis in Advanced Epithelial Ovarian Carcinoma. *Nutr. Cancer* **2016**, *68*, 772–779. [CrossRef] [PubMed]
18. Balogun, N.; Forbes, A.; Widschwendter, M.; Lanceley, A. Noninvasive nutritional management of ovarian cancer patients: Beyond intestinal obstruction. *Int. J. Gynecol. Cancer* **2012**, *22*, 1089–1095. [CrossRef] [PubMed]
19. Gadducci, A.; Cosio, S.; Fanucchi, A.; Genazzani, A.R. Malnutrition and cachexia in ovarian cancer patients: Pathophysiology and management. *Anticancer Res.* **2001**, *21*, 2941–2947. [PubMed]
20. Daniele, A.; Ferrero, A.; Fuso, L.; Mineccia, M.; Porcellana, V.; Vassallo, D.; Biglia, N.; Menato, G. Palliative care in patients with ovarian cancer and bowel obstruction. *Support. Care Cancer* **2015**, *23*, 3157–3163. [CrossRef] [PubMed]
21. Aust, S.; Knogler, T.; Pils, D.; Obermayr, E.; Reinthaller, A.; Zahn, L.; Radlgruber, I.; Mayerhoefer, M.E.; Grimm, C.; Polterauer, S. Skeletal Muscle Depletion and Markers for Cancer Cachexia Are Strong Prognostic Factors in Epithelial Ovarian Cancer. *PLoS ONE* **2015**, *10*, e0140403. [CrossRef] [PubMed]
22. Dell, D.D. Cachexia in patients with advanced cancer. *Clin. J. Oncol. Nurs.* **2002**, *6*, 235–238. [PubMed]
23. Pradeep, S.; Kim, S.W.; Wu, S.Y.; Nishimura, M.; Chaluvally-Raghavan, P.; Miyake, T.; Pecot, C.V.; Kim, S.J.; Choi, H.J.; Bischoff, F.Z.; et al. Hematogenous metastasis of ovarian cancer: Rethinking mode of spread. *Cancer Cell* **2014**, *26*, 77–91. [CrossRef] [PubMed]
24. Coffman, L.G.; Burgos-Ojeda, D.; Wu, R.; Cho, K.; Bai, S.; Buckanovich, R.J. New models of hematogenous ovarian cancer metastasis demonstrate preferential spread to the ovary and a requirement for the ovary for abdominal dissemination. *Transl. Res.* **2016**, *175*, 92–102. [CrossRef] [PubMed]
25. Dvoretsky, P.M.; Richards, K.A.; Angel, C.; Rabinowitz, L.; Stoler, M.H.; Beecham, J.B.; Bonfiglio, T.A. Distribution of disease at autopsy in 100 women with ovarian cancer. *Hum. Pathol.* **1988**, *19*, 57–63. [CrossRef]
26. Reed, E.; Zerbe, C.S.; Brawley, O.W.; Bicher, A.; Steinberg, S.M. Analysis of autopsy evaluations of ovarian cancer patients treated at the National Cancer Institute 1972–1988. *Am. J. Clin. Oncol.* **2000**, *23*, 107–116. [CrossRef] [PubMed]
27. Rose, P.G.; Piver, M.S.; Tsukada, Y.; Lau, T.S. Metastatic patterns in histologic variants of ovarian cancer. An autopsy study. *Cancer* **1989**, *64*, 1508–1513. [CrossRef]

28. Bachmann, C.; Bachmann, R.; Fend, F.; Wallwiener, D. Incidence and Impact of Lymph Node Metastases in Advanced Ovarian Cancer: Implications for Surgical Treatment. *J. Cancer* **2016**, *7*, 2241–2246. [CrossRef] [PubMed]
29. Guth, U.; Arndt, V.; Stadlmann, S.; Huang, D.J.; Singer, G. Epidemiology in ovarian carcinoma: Lessons from autopsy. *Gynecol. Oncol.* **2015**, *138*, 417–420. [CrossRef] [PubMed]
30. Sahdev, A.; Hughes, J.H.; Barwick, T.; Rockall, A.G.; Gallagher, C.J.; Reznek, R.H. Computed tomography features of recurrent ovarian carcinoma according to time to relapse. *Acta Radiol.* **2007**, *48*, 1038–1044. [CrossRef] [PubMed]
31. Hirose, S.; Tanabe, H.; Nagayoshi, Y.; Hirata, Y.; Narui, C.; Ochiai, K.; Isonishi, S.; Takano, H.; Okamoto, A. Retrospective analysis of sites of recurrence in stage I epithelial ovarian cancer. *J. Gynecol. Oncol.* **2018**, *29*, e37. [CrossRef] [PubMed]
32. Guth, U.; Huang, D.; Bauer, G.; Stieger, M.; Wight, E.; Singer, G. Metastatic patterns at autopsy in patients with ovarian carcinoma. *Cancer* **2007**, *110*, 1272–1280. [CrossRef] [PubMed]
33. Cormio, G.; Rossi, C.; Cazzolla, A.; Resta, L.; Loverro, G.; Greco, P.; Selvaggi, L. Distant metastases in ovarian carcinoma. *Int. J. Gynecol. Cancer* **2003**, *13*, 125–129. [CrossRef] [PubMed]
34. Sood, A.K.; Sorosky, J.I.; Dolan, M.; Anderson, B.; Buller, R.E. Distant metastases in ovarian cancer: Association with p53 mutations. *Clin. Cancer Res.* **1999**, *5*, 2485–2490. [PubMed]
35. Tanner, E.J.; Long, K.C.; Feffer, J.B.; Leitao, M.M., Jr.; Abu-Rustum, N.R.; Barakat, R.R.; Chi, D.S.; Gardner, G.J. Parenchymal splenic metastasis is an independent negative predictor of overall survival in advanced ovarian, fallopian tube, and primary peritoneal cancer. *Gynecol. Oncol.* **2013**, *128*, 28–33. [CrossRef] [PubMed]
36. Pakneshan, S.; Safarpour, D.; Tavassoli, F.; Jabbari, B. Brain metastasis from ovarian cancer: A systematic review. *J. Neurooncol.* **2014**, *119*, 1–6. [CrossRef] [PubMed]
37. Pectasides, D.; Pectasides, M.; Economopoulos, T. Brain metastases from epithelial ovarian cancer: A review of the literature. *Oncologist* **2006**, *11*, 252–260. [CrossRef] [PubMed]
38. Cerami, E.; Gao, J.; Dogrusoz, U.; Gross, B.E.; Sumer, S.O.; Aksoy, B.A.; Jacobsen, A.; Byrne, C.J.; Heuer, M.L.; Larsson, E.; et al. The cBio Cancer Genomics Portal: An Open Platform for Exploring Multidimensional Cancer Genomics Data. *Cancer Discov.* **2012**, *2*, 401–404. [CrossRef] [PubMed]
39. Gao, J.; Aksoy, B.A.; Dogrusoz, U.; Dresdner, G.; Gross, B.; Sumer, S.O.; Sun, Y.; Jacobsen, A.; Sinha, R.; Larsson, E.; et al. Integrative analysis of complex cancer genomics and clinical profiles using the cBioPortal. *Sci. Signal.* **2013**, *6*, pl1. [CrossRef] [PubMed]
40. Al Habyan, S.; Kalos, C.; Szymborski, J.; McCaffrey, L. Multicellular detachment generates metastatic spheroids during intra-abdominal dissemination in epithelial ovarian cancer. *Oncogene* **2018**, *37*, 5127–5135. [CrossRef] [PubMed]
41. Pease, J.C.; Brewer, M.; Tirnauer, J.S. Spontaneous spheroid budding from monolayers: A potential contribution to ovarian cancer dissemination. *Biol. Open* **2012**, *1*, 622–628. [CrossRef] [PubMed]
42. Moss, N.M.; Barbolina, M.V.; Liu, Y.; Sun, L.; Munshi, H.G.; Stack, M.S. Ovarian cancer cell detachment and multicellular aggregate formation are regulated by membrane type 1 matrix metalloproteinase: A potential role in I.p. metastatic dissemination. *Cancer Res.* **2009**, *69*, 7121–7129. [CrossRef] [PubMed]
43. Steinkamp, M.P.; Winner, K.K.; Davies, S.; Muller, C.; Zhang, Y.; Hoffman, R.M.; Shirinifard, A.; Moses, M.; Jiang, Y.; Wilson, B.S. Ovarian tumor attachment, invasion, and vascularization reflect unique microenvironments in the peritoneum: Insights from xenograft and mathematical models. *Front. Oncol.* **2013**, *3*, 97. [CrossRef] [PubMed]
44. Cohen, M.; Petignat, P. The bright side of ascites in ovarian cancer. *Cell. Cycle* **2014**, *13*, 2319. [CrossRef] [PubMed]
45. Kipps, E.; Tan, D.S.; Kaye, S.B. Meeting the challenge of ascites in ovarian cancer: New avenues for therapy and research. *Nat. Rev. Cancer* **2013**, *13*, 273–282. [CrossRef] [PubMed]
46. Ahmed, N.; Stenvers, K.L. Getting to know ovarian cancer ascites: Opportunities for targeted therapy-based translational research. *Front. Oncol.* **2013**, *3*, 256. [CrossRef] [PubMed]
47. Sorbe, B.; Frankendal, B. Prognostic importance of ascites in ovarian carcinoma. *Acta Obstet. Gynecol. Scand.* **1983**, *62*, 415–418. [CrossRef] [PubMed]
48. Sigurdsson, K.; Alm, P.; Gullberg, B. Prognostic factors in malignant epithelial ovarian tumors. *Gynecol. Oncol.* **1983**, *15*, 370–380. [CrossRef]

49. Ansell, S.M.; Rapoport, B.L.; Falkson, G.; Raats, J.I.; Moeken, C.M. Survival determinants in patients with advanced ovarian cancer. *Gynecol. Oncol.* **1993**, *50*, 215–220. [CrossRef] [PubMed]
50. Roszkowski, P.; Wronkowski, Z.; Szamborski, J.; Romejko, M. Evaluation of selected prognostic factors in ovarian cancer. *Eur. J. Gynaecol. Oncol.* **1993**, *14*, 140–145. [PubMed]
51. Kim, S.; Kim, B.; Song, Y.S. Ascites modulates cancer cell behavior, contributing to tumor heterogeneity in ovarian cancer. *Cancer Sci.* **2016**, *107*, 1173–1178. [CrossRef] [PubMed]
52. Yung, S.; Li, F.K.; Chan, T.M. Peritoneal mesothelial cell culture and biology. *Perit. Dial. Int.* **2006**, *26*, 162–173. [PubMed]
53. Jones, L.M.; Gardner, M.J.; Catterall, J.B.; Turner, G.A. Hyaluronic acid secreted by mesothelial cells: A natural barrier to ovarian cancer cell adhesion. *Clin. Exp. Metastasis* **1995**, *13*, 373–380. [CrossRef] [PubMed]
54. Yung, S.; Chan, T.M. Pathophysiology of the peritoneal membrane during peritoneal dialysis: The role of hyaluronan. *J. Biomed. Biotechnol.* **2011**, *2011*, 180594. [CrossRef] [PubMed]
55. Howlader, N.; Krapcho, A.M.; Miller, D.; Bishop, K.; Kosary, C.L.; Yu, M.; Ruhl, J.; Tatalovich, Z.; Mariotto, A.; Lewis, D.R.; et al. (Eds.) *SEER Cancer Statistics Review, 1975–2014*; National Cancer Institute: Bethesda, MD, USA, 2017.
56. Mikula-Pietrasik, J.; Stryczynski, L.; Uruski, P.; Tykarski, A.; Ksiazek, K. Procancerogenic activity of senescent cells: A case of the peritoneal mesothelium. *Ageing Res. Rev.* **2018**, *43*, 1–9. [CrossRef] [PubMed]
57. Loughran, E.A.; Leonard, A.K.; Hilliard, T.S.; Phan, R.C.; Yemc, M.G.; Harper, E.; Sheedy, E.; Klymenko, Y.; Asem, M.; Liu, Y.; et al. Aging Increases Susceptibility to Ovarian Cancer Metastasis in Murine Allograft Models and Alters Immune Composition of Peritoneal Adipose Tissue. *Neoplasia* **2018**, *20*, 621–631. [CrossRef] [PubMed]
58. Nakamura, K.; Sawada, K.; Kinose, Y.; Yoshimura, A.; Toda, A.; Nakatsuka, E.; Hashimoto, K.; Mabuchi, S.; Morishige, K.I.; Kurachi, H.; et al. Exosomes Promote Ovarian Cancer Cell Invasion through Transfer of CD44 to Peritoneal Mesothelial Cells. *Mol. Cancer Res.* **2017**, *15*, 78–92. [CrossRef] [PubMed]
59. Ren, J.; Xiao, Y.J.; Singh, L.S.; Zhao, X.; Zhao, Z.; Feng, L.; Rose, T.M.; Prestwich, G.D.; Xu, Y. Lysophosphatidic acid is constitutively produced by human peritoneal mesothelial cells and enhances adhesion, migration, and invasion of ovarian cancer cells. *Cancer Res.* **2006**, *66*, 3006–3014. [CrossRef] [PubMed]
60. Cannistra, S.A.; Kansas, G.S.; Niloff, J.; DeFranzo, B.; Kim, Y.; Ottensmeier, C. Binding of ovarian cancer cells to peritoneal mesothelium in vitro is partly mediated by CD44H. *Cancer Res.* **1993**, *53*, 3830–3838. [PubMed]
61. Gardner, M.J.; Catterall, J.B.; Jones, L.M.; Turner, G.A. Human ovarian tumour cells can bind hyaluronic acid via membrane CD44: A possible step in peritoneal metastasis. *Clin. Exp. Metastasis* **1996**, *14*, 325–334. [CrossRef] [PubMed]
62. Yakubov, B.; Chelladurai, B.; Schmitt, J.; Emerson, R.; Turchi, J.J.; Matei, D. Extracellular tissue transglutaminase activates noncanonical NF-kappaB signaling and promotes metastasis in ovarian cancer. *Neoplasia* **2013**, *15*, 609–619. [CrossRef] [PubMed]
63. Desjardins, M.; Xie, J.; Gurler, H.; Muralidhar, G.G.; Sacks, J.D.; Burdette, J.E.; Barbolina, M.V. Versican regulates metastasis of epithelial ovarian carcinoma cells and spheroids. *J. Ovarian Res.* **2014**, *7*, 70. [CrossRef] [PubMed]
64. Ween, M.P.; Hummitzsch, K.; Rodgers, R.J.; Oehler, M.K.; Ricciardelli, C. Versican induces a pro-metastatic ovarian cancer cell behavior which can be inhibited by small hyaluronan oligosaccharides. *Clin. Exp. Metastasis* **2011**, *28*, 113–125. [CrossRef] [PubMed]
65. Strobel, T.; Swanson, L.; Cannistra, S.A. In vivo inhibition of CD44 limits intra-abdominal spread of a human ovarian cancer xenograft in nude mice: A novel role for CD44 in the process of peritoneal implantation. *Cancer Res.* **1997**, *57*, 1228–1232. [PubMed]
66. Rump, A.; Morikawa, Y.; Tanaka, M.; Minami, S.; Umesaki, N.; Takeuchi, M.; Miyajima, A. Binding of ovarian cancer antigen CA125/MUC16 to mesothelin mediates cell adhesion. *J. Biol. Chem.* **2004**, *279*, 9190–9198. [CrossRef] [PubMed]
67. Gubbels, J.A.; Belisle, J.; Onda, M.; Rancourt, C.; Migneault, M.; Ho, M.; Bera, T.K.; Connor, J.; Sathyanarayana, B.K.; Lee, B.; et al. Mesothelin-MUC16 binding is a high affinity, N-glycan dependent interaction that facilitates peritoneal metastasis of ovarian tumors. *Mol. Cancer* **2006**, *5*, 50. [CrossRef] [PubMed]

68. Stoeck, A.; Schlich, S.; Issa, Y.; Gschwend, V.; Wenger, T.; Herr, I.; Marme, A.; Bourbie, S.; Altevogt, P.; Gutwein, P. L1 on ovarian carcinoma cells is a binding partner for Neuropilin-1 on mesothelial cells. *Cancer Lett.* **2006**, *239*, 212–226. [CrossRef] [PubMed]
69. Gurler Main, H.; Xie, J.; Muralidhar, G.G.; Elfituri, O.; Xu, H.; Kajdacsy-Balla, A.A.; Barbolina, M.V. Emergent role of the fractalkine axis in dissemination of peritoneal metastasis from epithelial ovarian carcinoma. *Oncogene* **2017**, *36*, 3025–3036. [CrossRef] [PubMed]
70. Kim, M.; Rooper, L.; Xie, J.; Kajdacsy-Balla, A.A.; Barbolina, M.V. Fractalkine receptor CX(3)CR1 is expressed in epithelial ovarian carcinoma cells and required for motility and adhesion to peritoneal mesothelial cells. *Mol. Cancer Res.* **2012**, *10*, 11–24. [CrossRef] [PubMed]
71. Strobel, T.; Cannistra, S.A. Beta1-integrins partly mediate binding of ovarian cancer cells to peritoneal mesothelium in vitro. *Gynecol. Oncol.* **1999**, *73*, 362–367. [CrossRef] [PubMed]
72. Kenny, H.A.; Chiang, C.Y.; White, E.A.; Schryver, E.M.; Habis, M.; Romero, I.L.; Ladanyi, A.; Penicka, C.V.; George, J.; Matlin, K.; et al. Mesothelial cells promote early ovarian cancer metastasis through fibronectin secretion. *J. Clin. Investig.* **2014**, *124*, 4614–4628. [CrossRef] [PubMed]
73. Carroll, M.J.; Fogg, K.C.; Patel, H.A.; Krause, H.B.; Mancha, A.S.; Patankar, M.S.; Weisman, P.S.; Barroilhet, L.; Kreeger, P.K. Alternatively-Activated Macrophages Upregulate Mesothelial Expression of P-Selectin to Enhance Adhesion of Ovarian Cancer Cells. *Cancer Res.* **2018**, *78*, 3560–3573. [CrossRef] [PubMed]
74. Gharpure, K.M.; Lara, O.D.; Wen, Y.; Pradeep, S.; LaFargue, C.; Ivan, C.; Rupaimoole, R.; Hu, W.; Mangala, L.S.; Wu, S.Y.; et al. ADH1B promotes mesothelial clearance and ovarian cancer infiltration. *Oncotarget* **2018**, *9*, 25115–25126. [CrossRef] [PubMed]
75. Iwanicki, M.P.; Davidowitz, R.A.; Ng, M.R.; Besser, A.; Muranen, T.; Merritt, M.; Danuser, G.; Ince, T.A.; Brugge, J.S. Ovarian cancer spheroids use myosin-generated force to clear the mesothelium. *Cancer Discov.* **2011**, *1*, 144–157. [CrossRef] [PubMed]
76. Roy, L.; Bobbs, A.; Sattler, R.; Kurkewich, J.L.; Dausinas, P.B.; Nallathamby, P.; Cowden Dahl, K.D. CD133 Promotes Adhesion to the Ovarian Cancer Metastatic Niche. *Cancer Growth Metastasis* **2018**, *11*. [CrossRef] [PubMed]
77. Klymenko, Y.; Kim, O.; Loughran, E.; Yang, J.; Lombard, R.; Alber, M.; Stack, M.S. Cadherin composition and multicellular aggregate invasion in organotypic models of epithelial ovarian cancer intraperitoneal metastasis. *Oncogene* **2017**, *36*, 5840–5851. [CrossRef] [PubMed]
78. Ellerbroek, S.M.; Wu, Y.I.; Overall, C.M.; Stack, M.S. Functional interplay between type I collagen and cell surface matrix metalloproteinase activity. *J. Biol. Chem.* **2001**, *276*, 24833–24842. [CrossRef] [PubMed]
79. Moser, T.L.; Pizzo, S.V.; Bafetti, L.M.; Fishman, D.A.; Stack, M.S. Evidence for preferential adhesion of ovarian epithelial carcinoma cells to type I collagen mediated by the alpha2beta1 integrin. *Int. J. Cancer* **1996**, *67*, 695–701. [CrossRef]
80. Sodek, K.L.; Ringuette, M.J.; Brown, T.J. MT1-MMP is the critical determinant of matrix degradation and invasion by ovarian cancer cells. *Br. J. Cancer* **2007**, *97*, 358–367. [CrossRef] [PubMed]
81. Barbolina, M.V.; Adley, B.P.; Ariztia, E.V.; Liu, Y.; Stack, M.S. Microenvironmental Regulation of Membrane Type 1 Matrix Metalloproteinase Activity in Ovarian Carcinoma Cells via Collagen-induced EGR1 Expression. *J. Biol. Chem.* **2007**, *282*, 4924–4931. [CrossRef] [PubMed]
82. Barbolina, M.V.; Liu, Y.; Gurler, H.; Kim, M.; Kajdacsy-Balla, A.A.; Rooper, L.; Shepard, J.; Weiss, M.; Shea, L.D.; Penzes, P.; et al. Matrix Rigidity Activates Wnt Signaling through Down-regulation of Dickkopf-1 Protein. *J. Biol. Chem.* **2013**, *288*, 141–151. [CrossRef] [PubMed]
83. Moss, N.M.; Liu, Y.; Johnson, J.J.; Debiase, P.; Jones, J.; Hudson, L.G.; Munshi, H.G.; Stack, M.S. Epidermal growth factor receptor-mediated membrane type 1 matrix metalloproteinase endocytosis regulates the transition between invasive versus expansive growth of ovarian carcinoma cells in three-dimensional collagen. *Mol. Cancer Res.* **2009**, *7*, 809–820. [CrossRef] [PubMed]
84. Yoo, E.; Kim, J.H.; Kim, M.J.; Yu, J.S.; Chung, J.J.; Yoo, H.S.; Kim, K.W. Greater and lesser omenta: Normal anatomy and pathologic processes. *Radiographics* **2007**, *27*, 707–720. [CrossRef] [PubMed]
85. Meza-Perez, S.; Randall, T.D. Immunological Functions of the Omentum. *Trends Immunol.* **2017**, *38*, 526–536. [CrossRef] [PubMed]
86. Platell, C.; Cooper, D.; Papadimitriou, J.M.; Hall, J.C. The omentum. *World J. Gastroenterol.* **2000**, *6*, 169–176. [PubMed]

87. Nieman, K.M.; Kenny, H.A.; Penicka, C.V.; Ladanyi, A.; Buell-Gutbrod, R.; Zillhardt, M.R.; Romero, I.L.; Carey, M.S.; Mills, G.B.; Hotamisligil, G.S.; et al. Adipocytes promote ovarian cancer metastasis and provide energy for rapid tumor growth. *Nat. Med.* **2011**, *17*, 1498–1503. [CrossRef] [PubMed]
88. Khan, S.M.; Funk, H.M.; Thiolloy, S.; Lotan, T.L.; Hickson, J.; Prins, G.S.; Drew, A.F.; Rinker-Schaeffer, C.W. In vitro metastatic colonization of human ovarian cancer cells to the omentum. *Clin. Exp. Metastasis* **2010**, *27*, 185–196. [CrossRef] [PubMed]
89. Arie, A.B.; McNally, L.; Kapp, D.S.; Teng, N.N. The omentum and omentectomy in epithelial ovarian cancer: A reappraisal: Part II–The role of omentectomy in the staging and treatment of apparent early stage epithelial ovarian cancer. *Gynecol. Oncol.* **2013**, *131*, 784–790. [CrossRef] [PubMed]
90. Sato, M.; Kawana, K.; Adachi, K.; Fujimoto, A.; Yoshida, M.; Nakamura, H.; Nishida, H.; Inoue, T.; Taguchi, A.; Ogishima, J.; et al. Detachment from the primary site and suspension in ascites as the initial step in metabolic reprogramming and metastasis to the omentum in ovarian cancer. *Oncol. Lett.* **2018**, *15*, 1357–1361. [CrossRef] [PubMed]
91. Yung, M.M.; Tang, H.W.; Cai, P.C.; Leung, T.H.; Ngu, S.F.; Chan, K.K.; Xu, D.; Yang, H.; Ngan, H.Y.; Chan, D.W. GRO-alpha and IL-8 enhance ovarian cancer metastatic potential via the CXCR2-mediated TAK1/NFkappaB signaling cascade. *Theranostics* **2018**, *8*, 1270–1285. [CrossRef] [PubMed]
92. Shimotsuma, M.; Shields, J.W.; Simpson-Morgan, M.W.; Sakuyama, A.; Shirasu, M.; Hagiwara, A.; Takahashi, T. Morpho-physiological function and role of omental milky spots as omentum-associated lymphoid tissue (OALT) in the peritoneal cavity. *Lymphology* **1993**, *26*, 90–101. [PubMed]
93. Clark, R.; Krishnan, V.; Schoof, M.; Rodriguez, I.; Theriault, B.; Chekmareva, M.; Rinker-Schaeffer, C. Milky spots promote ovarian cancer metastatic colonization of peritoneal adipose in experimental models. *Am. J. Pathol.* **2013**, *183*, 576–591. [CrossRef] [PubMed]
94. Ladanyi, A.; Mukherjee, A.; Kenny, H.A.; Johnson, A.; Mitra, A.K.; Sundaresan, S.; Nieman, K.M.; Pascual, G.; Benitah, S.A.; Montag, A.; et al. Adipocyte-induced CD36 expression drives ovarian cancer progression and metastasis. *Oncogene* **2018**, *37*, 2285–2301. [CrossRef] [PubMed]
95. Mitra, A.K.; Chiang, C.Y.; Tiwari, P.; Tomar, S.; Watters, K.M.; Peter, M.E.; Lengyel, E. Microenvironment-induced downregulation of miR-193b drives ovarian cancer metastasis. *Oncogene* **2015**, *34*, 5923–5932. [CrossRef] [PubMed]
96. Lau, T.S.; Chan, L.K.; Wong, E.C.; Hui, C.W.; Sneddon, K.; Cheung, T.H.; Yim, S.F.; Lee, J.H.; Yeung, C.S.; Chung, T.K.; et al. A loop of cancer-stroma-cancer interaction promotes peritoneal metastasis of ovarian cancer via TNFalpha-TGFalpha-EGFR. *Oncogene* **2017**, *36*, 3576–3587. [CrossRef] [PubMed]
97. Balkwill, F. Cancer and the chemokine network. *Nat. Rev. Cancer* **2004**, *4*, 540–550. [CrossRef] [PubMed]
98. Proudfoot, A.E. Chemokine receptors: Multifaceted therapeutic targets. *Nat. Rev. Immunol.* **2002**, *2*, 106–115. [CrossRef] [PubMed]
99. Scotton, C.J.; Wilson, J.L.; Scott, K.; Stamp, G.; Wilbanks, G.D.; Fricker, S.; Bridger, G.; Balkwill, F.R. Multiple actions of the chemokine CXCL12 on epithelial tumor cells in human ovarian cancer. *Cancer Res.* **2002**, *62*, 5930–5938. [PubMed]
100. Barbolina, M.V.; Kim, M.; Liu, Y.; Shepard, J.; Belmadani, A.; Miller, R.J.; Shea, L.D.; Stack, M.S. Microenvironmental regulation of chemokine (C-X-C-motif) receptor 4 in ovarian carcinoma. *Mol. Cancer Res.* **2010**, *8*, 653–664. [CrossRef] [PubMed]
101. Righi, E.; Kashiwagi, S.; Yuan, J.; Santosuosso, M.; Leblanc, P.; Ingraham, R.; Forbes, B.; Edelblute, B.; Collette, B.; Xing, D.; et al. CXCL12/CXCR4 Blockade Induces Multimodal Antitumor Effects that Prolong Survival in an Immunocompetent Mouse Model of Ovarian Cancer. *Cancer Res.* **2011**, *71*, 5522–5534. [CrossRef] [PubMed]
102. Kim, M.; Rooper, L.; Xie, J.; Rayahin, J.; Burdette, J.E.; Kajdacsy-Balla, A.; Barbolina, M.V. The Lymphotactin Receptor Is Expressed in Epithelial Ovarian Carcinoma and Contributes to Cell Migration and Proliferation. *Mol. Cancer Res.* **2012**. [CrossRef] [PubMed]
103. Holcroft, J.; Kressel, H.Y.; Prager, R.; Trunkey, D.; Jacobs, R.P. An experience with a LeVeen ascites shunt. *Arch. Surg.* **1976**, *111*, 302–303. [CrossRef] [PubMed]
104. Reinhardt, G.F.; Stanley, M.M. Peritoneovenous shunting for ascites. *Surg. Gynecol. Obstet.* **1977**, *145*, 419–424. [PubMed]

105. White, M.A.; Agle, S.C.; Padia, R.K.; Zervos, E.E. Denver peritoneovenous shunts for the management of malignant ascites: A review of the literature in the post LeVeen Era. *Am. Surg.* **2011**, *77*, 1070–1075. [PubMed]
106. Souter, R.G.; Wells, C.; Tarin, D.; Kettlewell, M.G. Surgical and pathologic complications associated with peritoneovenous shunts in management of malignant ascites. *Cancer* **1985**, *55*, 1973–1978. [CrossRef]
107. Tarin, D.; Price, J.E.; Kettlewell, M.G.; Souter, R.G.; Vass, A.C.; Crossley, B. Clinicopathological observations on metastasis in man studied in patients treated with peritoneovenous shunts. *Br. Med. J. (Clin. Res. Ed.)* **1984**, *288*, 749–751. [CrossRef]
108. Tarin, D.; Price, J.E.; Kettlewell, M.G.W.; Souter, R.G.; Vass, A.C.R.; Crossley, B. Mechanisms of Human-Tumor Metastasis Studied in Patients with Peritoneovenous Shunts. *Cancer Res.* **1984**, *44*, 3584–3592. [PubMed]
109. Tarin, D.; Vass, A.C.; Kettlewell, M.G.; Price, J.E. Absence of metastatic sequelae during long-term treatment of malignant ascites by peritoneo-venous shunting. A clinico-pathological report. *Invasion Metastasis* **1984**, *4*, 1–12. [PubMed]
110. Matsuo, K.; Carter, C.M.; Ahn, E.H.; Prather, C.P.; Eno, M.L.; Im, D.D.; Rosenshein, N.B. Inferior vena cava filter placement and risk of hematogenous distant metastasis in ovarian cancer. *Am. J. Clin. Oncol.* **2013**, *36*, 362–367. [CrossRef] [PubMed]
111. Kamran, P.; Sereti, K.I.; Zhao, P.; Ali, S.R.; Weissman, I.L.; Ardehali, R. Parabiosis in mice: A detailed protocol. *J. Vis. Exp.* **2013**, *80*, e50556. [CrossRef] [PubMed]
112. Figueras, A.; Alsina-Sanchis, E.; Lahiguera, A.; Abreu, M.; Muinelo-Romay, L.; Moreno-Bueno, G.; Casanovas, O.; Graupera, M.; Matias-Guiu, X.; Vidal, A.; et al. A Role for CXCR4 in Peritoneal and Hematogenous Ovarian Cancer Dissemination. *Mol. Cancer Ther.* **2018**, *17*, 532–543. [CrossRef] [PubMed]
113. Berek, J.S.; Kehoe, S.T.; Kumar, L.; Friedlander, M. Cancer of the ovary, fallopian tube, and peritoneum. *Int J. Gynaecol. Obstet.* **2018**, *143* (Suppl. 2), 59–78. [CrossRef] [PubMed]
114. Tammela, T.; Alitalo, K. Lymphangiogenesis: Molecular mechanisms and future promise. *Cell* **2010**, *140*, 460–476. [CrossRef] [PubMed]
115. Kuerti, S.; Oliveira-Ferrer, L.; Milde-Langosch, K.; Schmalfeldt, B.; Legler, K.; Woelber, L.; Prieske, K.; Mahner, S.; Trillsch, F. VEGF-C expression attributes the risk for lymphatic metastases to ovarian cancer patients. *Oncotarget* **2017**, *8*, 43218–43227. [CrossRef] [PubMed]
116. Ma, M.; Yu, N. Ubiquitin-specific protease 7 expression is a prognostic factor in epithelial ovarian cancer and correlates with lymph node metastasis. *OncoTargets Ther.* **2016**, *9*, 1559–1569.
117. Kang, Y.; Pu, T.; Cai, Q.; Hong, S.; Zhang, M.; Li, G.; Zhu, Z.; Xu, C. Identification of lymphatic metastasis-associated genes in a metastatic ovarian cancer cell line. *Mol. Med. Rep.* **2015**, *12*, 2741–2748. [CrossRef] [PubMed]
118. Li, Y.; Hermanson, D.L.; Moriarity, B.S.; Kaufman, D.S. Human iPSC-Derived Natural Killer Cells Engineered with Chimeric Antigen Receptors Enhance Anti-tumor Activity. *Cell Stem Cell* **2018**, *23*, 181–192. [CrossRef] [PubMed]
119. Golfier, S.; Kopitz, C.; Kahnert, A.; Heisler, I.; Schatz, C.A.; Stelte-Ludwig, B.; Mayer-Bartschmid, A.; Unterschemmann, K.; Bruder, S.; Linden, L.; et al. Anetumab ravtansine: A novel mesothelin-targeting antibody-drug conjugate cures tumors with heterogeneous target expression favored by bystander effect. *Mol. Cancer Ther.* **2014**, *13*, 1537–1548. [CrossRef] [PubMed]
120. Bell-McGuinn, K.M.; Matthews, C.M.; Ho, S.N.; Barve, M.; Gilbert, L.; Penson, R.T.; Lengyel, E.; Palaparthy, R.; Gilder, K.; Vassos, A.; et al. A phase II, single-arm study of the anti-alpha5beta1 integrin antibody volociximab as monotherapy in patients with platinum-resistant advanced epithelial ovarian or primary peritoneal cancer. *Gynecol. Oncol.* **2011**, *121*, 273–279. [CrossRef] [PubMed]
121. Delmonte, A.; Del Conte, G.; Sessa, C.; Perotti, A.; Fasolo, A.; Williams, E.; Smith, K.; Dall'O', E.; Colombo, N.; Gianni, L. Results from a phase 1/2 study of volociximab in combination with liposomal doxorubicin in relapsed advanced epithelial ovarian and primary peritoneal carcinoma. *J. Clin. Oncol.* **2008**, *26*, 16527. [CrossRef]
122. Reyes, H.D.; Thiel, K.W.; Carlson, M.J.; Meng, X.; Yang, S.; Stephan, J.M.; Leslie, K.K. Comprehensive profiling of EGFR/HER receptors for personalized treatment of gynecologic cancers. *Mol. Diagn. Ther.* **2014**, *18*, 137–151. [CrossRef] [PubMed]
123. Citri, A.; Yarden, Y. EGF-ERBB signalling: Towards the systems level. *Nat. Rev. Mol. Cell Biol.* **2006**, *7*, 505–516. [CrossRef] [PubMed]

124. Killock, D. Lung cancer: A new generation of EGFR inhibition. *Nat. Rev. Clin. Oncol.* **2015**, *12*, 373. [CrossRef] [PubMed]
125. Singh, D.; Attri, B.K.; Gill, R.K.; Bariwal, J. Review on EGFR Inhibitors: Critical Updates. *Mini Rev. Med. Chem.* **2016**, *16*, 1134–1166. [CrossRef] [PubMed]
126. Morrison, J.; Thoma, C.; Goodall, R.J.; Lyons, T.J.; Gaitskell, K.; Wiggans, A.J.; Bryant, A. Epidermal growth factor receptor blockers for the treatment of ovarian cancer. *Cochrane Database Syst. Rev.* **2018**, *10*, CD007927. [CrossRef] [PubMed]
127. Monk, B.J.; Minion, L.E.; Coleman, R.L. Anti-angiogenic agents in ovarian cancer: Past, present, and future. *Ann. Oncol.* **2016**, *27* (Suppl. 1), i33–i39. [CrossRef] [PubMed]
128. Coward, J.I.; Middleton, K.; Murphy, F. New perspectives on targeted therapy in ovarian cancer. *Int. J. Womens Health* **2015**, *7*, 189–203. [CrossRef] [PubMed]

cancers

MDPI

Review

Lysophospholipid Signaling in the Epithelial Ovarian Cancer Tumor Microenvironment

Yan Xu

Department of Obstetrics and Gynecology, Indiana University School of Medicine, 1044W. Walnut Street R4-W037, Indianapolis, IN 46202, USA; xu2@iu.edu

Received: 4 June 2018; Accepted: 5 July 2018; Published: 9 July 2018

Abstract: As one of the important cancer hallmarks, metabolism reprogramming, including lipid metabolism alterations, occurs in tumor cells and the tumor microenvironment (TME). It plays an important role in tumorigenesis, progression, and metastasis. Lipids, and several lysophospholipids in particular, are elevated in the blood, ascites, and/or epithelial ovarian cancer (EOC) tissues, making them not only useful biomarkers, but also potential therapeutic targets. While the roles and signaling of these lipids in tumor cells are extensively studied, there is a significant gap in our understanding of their regulations and functions in the context of the microenvironment. This review focuses on the recent study development in several oncolipids, including lysophosphatidic acid and sphingosine-1-phosphate, with emphasis on TME in ovarian cancer.

Keywords: lipids; lysophospholipids (LPLs); lysophosphatidic acid (LPA); sphingosine-1-phosphate; tumor microenvironment (TME); epithelial ovarian cancer (EOC)

1. Introduction

The tumor microenvironment (TME) for epithelial ovarian cancer (EOC) is rather unique. It is mainly confined within the peritoneal cavity and frequently associated with ascetic fluid [1–6]. The TME consists of many stromal cell types including: tumor-associated macrophages (TAMs), T cells (e.g., regulatory T cells), tumor associated fibroblasts (TAFs), mesothelial cells, adipocytes, endothelial cells (ECs), myeloid-derived suppressor cells (MDSCs), pericytes, platelets, extracellular matrix components (EMCs), and cell-free factors [1–6].

The presence of ascetic fluid provides a mobile, easy access, and more dynamic environment for tumor–stromal interactions. In addition to tumor and stromal cells, EOC ascites is rich in cell-free inflammatory cytokines, chemokines, matrix metalloproteinases, integrins, and other secreted molecules, including bioactive lipid factors. These factors are generated by and mutually function in both tumor and stromal cells via autocrine/paracrine mechanisms. They may exist in either extracellular vesicles (EVs) or in "free" forms, including bond forms to proteins or other molecules [7]. EVs are membrane surrounded structures released by cells in an evolutionally conserved manner, but their release, contents, and/or up-take may be abnormally regulated in cancer. The major populations include microvesicles (MVs, 100–1000 nm), exosomes (30–100 nm), and apoptotic bodies [8]. Exosomes have emerged as new as diagnostic markers, as well as cell-to-cell communication vehicles, with therapeutic applications [4,9,10]. The compositions of exosomes from different cell types are complex, containing ~200 lipids, >3000 proteins, ~1600 mRNA and ~800 microRNAs [11–13].

Metabolic reprogramming is one of the major cancer hallmarks [14] that is critical for cancer cells to adapt to stress from TME and the increased nutritional requirements during their growth. These modifications occur through cross-talk between tumor and stroma cells in TME in a dynamic network that connects different molecular processes, such as energy production, inflammatory response, and drug resistance [15]. In particular, primary EOC is characterized by abnormal lipid metabolism and energy disorders. In addition, recurrent EOC patients have been shown to have increased amino acid and lipid metabolism compared with primary EOC patients [16].

Compared to other major living cell components, including DNA, RNA (all composed by four bases), proteins (all composed of 20 amino acids), and carbohydrates (all have the basic element CH_2O with ring, chain, and branched structures), lipids are very diverse in both their respective structures and functions. These diverse compounds are grouped into classes; glycerophospholipids (PLs) (including lysophospholipids (LPLs)) sphingolipids, sterol lipids, prenol lipids, saccharolipids, and polyketides [17]. The functional involvement of many of these lipids in EOC, PLs, LPLs, sphingophospholipids (SPLs), fatty acids, cholesterol, vitamins, and triglycerides (TGs) in particular, have been covered by many reviews [3,18–23]. This review will focus on recent development of signaling LPLs with an emphasis on TME in EOC.

2. LPLs

Compared to PLs, which have two fatty acid chains, LPLs only have one fatty acid chain and thus have reduced hydrophobicity (Figure 1). In addition, many LPLs are either negatively or positively charged, increasing their polarity and solubility in water. With these chemical properties, LPLs are synthesized and/or secreted extracellularly and many of them function as signaling molecules through their specific membrane receptors. In addition, several of these LPLs have tumor promoting activities and are thus termed as "oncolipids" [24]. They are accumulated in the TME. LPA is a prototype of the LPL signaling molecules. It exhibits pleiomorphic functions in almost all cell lineages tested. Since our early report on LPA's effect in EOC [25,26], more than 1000 papers have been published reporting the roles and signaling mechanisms of LPA in various cancer types [24,27].

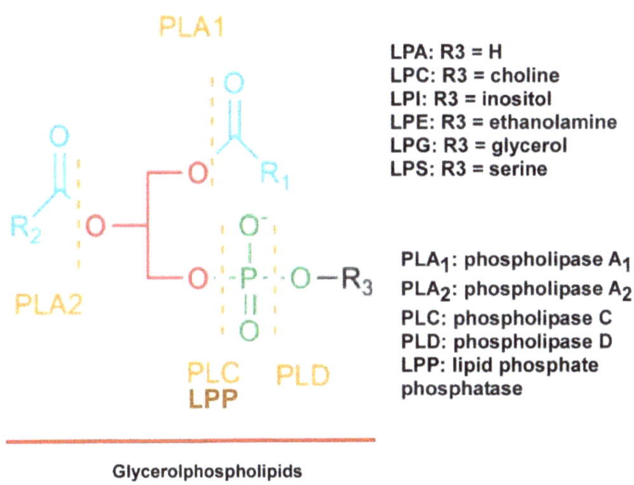

Figure 1. Structures of glycerophospholipids (PLs), lysophospholipids (LPLs only have R_1 or R_2) and the action sites of phospholipases.

2.1. LPA

Lysophosphatidic acids (LPAs) are a group of compounds with various fatty acid side chains, differing in their length (commonly 14–22 carbons) and number of double bonds (commonly from zero to six in most tissues). In addition, the chemical linkages of the fatty acid chain to the glycerol backbone can be differently grouped as acyl-, alkyl-, and alkenyl-LPAs [28]. We have originally identified LPA as an EOC growth factor, termed ovarian cancer activating factor [26,29]. Numerous papers have now been published to show that LPA induces a broad range of tumor promoting activities in EOC and LPA is a therapeutic target for EOC [30–32].

2.1.1. Increased LPA Levels in EOC

We have initially reported LPA as a potential marker for EOC [33]. Blinded [34] and numerous independent studies have confirmed that LPA is elevated in the blood from EOC patients, when compare to those with benign diseases and/or healthy controls [35–42]. The mean values of physiological and pathologic concentrations of plasma LPA in healthy women and EOC patients are 0.6–0.9 μM and 2.0–22 μM, respectively [33,34]. The mean values of benign and malignant ascites acyl-LPA concentrations are 2.9 μM and 19–95 μM, respectively [28,43]. The mean non-acyl (both alkyl-, and alkenyl)-LPA levels are 3.7 ± 1.7 and 0.9 ± 0.7 μM for benign and malignant ascites, respectively [28] (Table 1). The concentrations of LPA and other lipids mentioned below were measured in cell-free plasma or ascites. In most reports, it is unclear whether these lipids are associated with MVs, as mentioned in the Introduction, in a protein bound, or a "free" form. In an attempt to test this, we have separated different MVs and the cell- and vesicle-free (S4) portion of ascites via step-wise centrifugation. We found that human EOC ascites S4 portion potently promotes proliferation, migration, and invasion of human EOC cells in a PLA_2-dependent manner, suggesting that LPA, and maybe other LPLs, may present in a protein bound or a "free" form [44]. However, this issue needs to be more systematically investigated and lipid association with different extracellular components may also be cell-type and context dependent.

The challenges to move LPA or other lipid molecules as cancer markers into clinics are several fold. First, as metabolites, these molecules have quick turnover times by their producing and degradation enzymes. In addition, their levels are likely to be affected by other physiological/pathologic conditions—such as diet, smoking, and drinking—which have not been completely investigated. Secondly, due to their chemical properties, the different extraction, storage and detection conditions/methods used in different labs significantly affect the levels detected. However, there are no standard procedures established. Finally, the best detecting method for these lipid markers is electrospray-mass spectrometry (ESI-MS), which is not a routine setting in clinics currently. These challenges have made cross-examination and validation of these markers difficult. Studies to standardize the procedures of collection, extraction, storage, and measurements of lipid marker are critical. Nevertheless, technique advancements are emerging. In particular, mass spectrometry will likely become a routine technique in regular clinic settings in the near future.

2.1.2. LPA Production and Regulation

LPA is produced from secreted enzymes from lysophosphatidylcholine (LPC) by autotaxin (ATX), as well as phospholipase A_2 (PLA_2) by either providing the substrate LPC for ATX, or directly produce LPA from phosphatidic acid (PA) [49–53]. LPA is degraded outside cells by a family of three enzymes called the lipid phosphate phosphatases (LPPs) (Figure 1). Imbalanced expression and/or activity levels of ATX, PLA_2s, and/or LPPs are involved in EOC [54] (see Figure 1 in [54]). The ATX/LPA axis has received increasing interest as a target in cancers, fibrotic diseases, autoimmune diseases, arthritis, chronic hepatitis, obesity, and impaired glucose homeostasis [55]. At least one of the synthetic ATX selective inhibitors is in clinical trials for idiopathic pulmonary disease [55].

EOC cells may produce LPA upon stimulations [50–52]. However, the tissues and cells in EOC TME are likely to be the major source of elevated LPA in EOC. The cell types involved in LPA production include, but are not limited to, platelets, adipocytes, mesothelial cells, and immune cells.

Platelet activation generated LPA used to be considered the major source of plasma LPA [56,57]. Paraneoplastic thrombocytosis has been recognized as a prevalent phenomenon in patients with ovarian cancer since the early 1970s [58]. Cancer patients have a ~4-fold increased risk of venous thromboembolism compared with the general population and this is associated with significant morbidity and mortality [59,60]. The preventative and therapeutic significance of blocking thrombopoietic factors (cytokines and lipids) have been noted as an interesting direction in EOC research [60–62]. Tumor cells hijack platelet functions by activating them though platelet aggregation [63]. Activated platelets may help tumor cells survive immune surveillance by acting as protective "cloaks" against immune destruction. However, the major tumor cell beneficiary activities are likely mediated by the factors secreted from platelets after activation. These factors include cytokines (such as interleukin-6 (IL-6)), TGF-β, and lipid factors, which mediate the inflammatory, proliferative, and proangiogenic activities of platelets to promote tumor growth, tissue invasion, and metastasis [63]. Increased platelet counts and platelet activation associated with EOC are likely to be important contributors for the elevated LPA levels in EOC TME. In addition, LPA promotes platelet aggregation [64,65] and blocking platelet function leads to inhibition of metastasis of breast cancer through decreased LPA signaling. ATX is detected in platelet α-granules. Functionally active ATX is eventually released following tumor cell-induced platelet aggregation, thereby promoting metastasis [66].

EOC cells preferentially metastasize to omentum, the adipose tissue, which secretes many chemotactic cytokines and growth factors, including LPA [67–69]. Adipocytes promote ovarian cancer metastasis and provide energy for rapid tumor growth [70]. Since the identification of ATX as the major LPA producing enzyme [21,71] and studies conducted using mouse ATX knockout models, it has become clear that adipose tissue are at least one of the major tissues in EOC TME that produces LPA [72,73]. Approximately 40% of body ATX is produced by adipocytes, and this is further increased by inflammation [74,75]. The use of ATX inhibitors seems an attractive strategy to produce novel medicinal agents, for example anticancer agents [55].

EOC cells that metastasize within the peritoneal cavity wall and the organs enclosed are coved by a layer of peritoneal mesothelial cells. We have shown that human peritoneal mesothelial cells constitutively produce LPA, which accounts for a significant portion of the chemotactic activity of the conditioned medium from peritoneal mesothelial cells to ovarian cancer cells [76]. The calcium-independent phospholipase A_2 (iPLA_2), and cytosolic PLA_2 (cPLA_2) are involved in this production and LPA's tumor promoting activities [76].

Although many types of immune and endothelial cells are involved in EOC TME, their contributions to LPA production and/or degradation are less known. Steady-state ATX is expressed by only a few tissues, including high endothelial venules in lymph nodes, but inflammatory signals (enriched in EOC TME) can upregulate ATX expression in different tissues [77]. In addition, when ECs are con-cultured with EOC cells, coherent and non-cell line specific changes in fatty acids, glycerophospholipids, and carbohydrates, induced by endothelial cell contact are observed over time [78]. Wong and Reinartz et al. have reported that macrophage-derived phospholipase PLA_2G7, which may produce extracellular LPA, is involved in EOC and associated with early relapse of EOC. It is a secreted enzyme that may produce LPA and arachidonic acid [79,80].

Mice with homozygous deletion of LPP1 (a LPA degradation enzyme) in stromal cells result in elevated levels and decreased turnover of LPA in vivo. In turn, enhanced tumor seeding in the LPP1 KO mice compared to wild type was observed [81].

Taken together, the host cells play an important role in producing and degradation LPA, which may be present in cell free forms in either exosome and/or EV-free forms [44].

2.1.3. Major Cellular Functions and Signaling Mechanisms of LPA in EOC

LPA stimulates almost every aspects of tumor promoting activities, including cell proliferation or differentiation, prevents apoptosis induced by stress or stimuli, induces platelet aggregation stimulates cell morphology changes, cell adhesion, cell migration, and cell invasion. It also stimulates tumorigenesis and metastasis in vivo [25,26,30–32,50,51,76,82–99].

LPA regulates many pro-tumorigenic and pro-inflammatory factors, including vascular endothelial growth factor (VEGF), matrix metalloproteinases (MMPs), urokinase plasminogen activator, IL-6, IL-8, CXC motif chemokine ligand 12/CXC receptor 4, COX2, cyclin D1, Hippo–YAP, and growth-regulated oncogene alpha. These regulations are at the transcriptional, translational, and epigenetic levels [19,34,44,51,76,82–84,86,88–91,94,95,100–105]. LPA induces loss of junctional β-catenin, stimulates clustering of β1 integrins, and enhances the conformationally active population of surface β1 integrins. Furthermore, LPA treatment initiates nuclear translocation of β-catenin and transcriptional activation of Wnt/β-catenin target genes, resulting in gain of mesenchymal marker expression [106]. Gglycodelin, a glycoprotein, is over-expressed in various malignancies, including EOC, and its expression correlates with the diagnosis and prognosis of cancer patients. The expression of glycodelin can be regulated by stromal cells and LPA [107]. While LPA's function and signaling in EOC has been rather extensively reviewed [19,100,108], several notions and recent developments are specifically noted here.

Firstly, LPA is a confirmed mitogen in many cell types. However, MTT dye reduction is not a good method to measure this effect. LPA affected MTT dye reduction with an unknown mechanism in EOC cells [109], making it an unreliable indicator for cell number changes. In addition, MTT dye reduction may not be sensitive enough to detect DNA replication as [^{3}H]tymidine incorporation [25,26,29].

Secondly, the most potent roles of LPA in EOC and other cancer cells are likely to be cell migration and invasion. This action is mediated by LPARs and G_i and $G_{12/13}$ [30,51,76,83,89,95,104]. This is correlated to LPA's in vivo effects, where LPA mainly stimulates metastasis, instead of primary tumor growth [89,95]. In comparison, EGF and other growth factors are likely to be more effective in cell proliferation than LPA, but the latter is more effective in induction of cell migration and invasion [30,51,89,90,93,95].

Thirdly, LPA has been recently shown to be involved in cancer stem cells (CSC) in EOC [84,110]. Seo et al. have shown that EOC CSC produces LPA, which augments CSC characteristics such as sphere-forming ability, resistance to anticancer drugs, tumorigenic potential in xenograft transplantation, and high expression of CSC-associated genes, including OCT4, SOX2, and aldehyde dehydrogenase 1 (ALDH1). These actions are mediated by $LPAR_1$. ATX is highly secreted from ovarian CSCs. Inhibition or knockdown of ATX markedly attenuates the LPA-producing, tumorigenic, and drug resistance potentials of CSCs. In addition, clinicopathological analysis shows a significant survival disadvantage of patients with positive staining of ATX. In addition, LPA is involved in the crosstalk between CSC in TME. EOC cells secrete LPA that activates the expression and secretion of CXCL12 by mesenchymal stem cells (MSCs), enhancing the resistance of OVCA cells to hyperthermia [23].

Fourthly, the majority of LPA signaling is mediated by its six G protein couples receptors (GPCRs) [100,111]. Among them, $LPAR_{1–3}$ belong to the endothelial differentiation gene (Edg) family of GPCR and $LPAR_{4–6}$ belong to the purinergic P2Y family of GPCRs [111,112]. While $LPAR_{1–3}$ in general mediate LPA's tumor promoting activities [76,113,114], limited reports showed that $LPAR_1$ may represent a negative regulatory LPA receptor inducing apoptosis in ovarian cancer cells [96]. At least three compounds blocking these receptors have passed phase I and phase II clinical trials [115]. Compared to $LPAR_{1–3}$, $LPAR_{4–6}$ are less studied. Both pro- and anticancer effects mediated by $LPA_{4–6}$ in various cancers have been reported, with the majority of them reporting anti-cancer effects [116–118].

Finally, LPA has also been identified as a ligand for the nuclear receptor peroxisome proliferator-activated receptor gamma (PPARγ) [119,120]. However, the LPA-PPARγ studies are mainly limited to the vascular and metabolic processes [121,122]. The roles of PPARγ-mediated LPA effects in cancer are essentially unknown until recently. Emerging evidence, however, suggest that this is an important missing opportunity in cancer research. We have recently shown that LPA dose- and time-dependently upregulated SOX9 in EOC cells. This upregulation is mediated by PPARγ. SOX9 was involved in cellular activities related to Cancer Stem Cells (CSC), including anokis-resistance, regulation of CSC marker CD44, and spheroid-formation [85]. In addition, we have shown that LPA effectively upregulates ZIP4 (a zinc transporter) expression via by PPARγ and LPA's promoting effects in CSC-related activities in HGSOC cells is at least partially mediated by ZIP4 in an extracellular zinc-independent manner [84]. These findings emphasize the importance of targeting by PPARγ for LPA's tumorigenic actions.

2.1.4. LPA in the Immune System

The ATX-LPA axis has emerged as a novel regulator of lymphocyte homing and inflammation. LPARs are expressed by T cells and LPA enhances the motility of human and mouse T cells in vitro, although generally not in a direct manner [77]. Cancer cells must evade the immune system during metastasis. LPA facilitates this important process by inhibiting $CD8^{+}$ T cell activation [75]. LPA also regulates macrophage differentiation and T cell motility [123,124]. Although EOC cells mainly express the $LPAR_{1-3}$, $LPAR_6$ is the main LPA receptor on TAM and tumor-associated T cells [123].

ATX is expressed by lymphoid organ high endothelial venule. LPARs receptors are expressed by NK cells, mast cells, eosinophils, and B cells [77]. In addition, tumor-associated macrophages (TAMs) produce LPA [123]. However, how LPA signaling in stroma cells, and in immune cells in EOC in particular, remains to be further investigated.

In summary, LPA, a simple molecule that mediates a plethora of biological effects, may be targeted at its levels of production by ATX or PLA_2s, LPA receptors, including PPARγ, or through LPA degradation by lipid phosphate phosphatases (LPPs). The targeting strategy should take TME into serious consideration. Drugs for these applications have been and will soon be entering clinical practice [27].

2.2. LPC

Compared to LPA, plasma LPC levels are usually 10 to 100-fold higher and are in the 100–200 µM range in human subjects [34] (Table 1). LPC levels have been shown to be significantly elevated in the plasma of ovarian cancer patients [34,125,126]. On the other hand, others and we have shown that patients with malignant cancer diseases have attenuated LPC plasma levels [127–131]. Moreover, different phospholipase A_2 enzymes, which mainly convert phosphatidylcholine (PC) to LPC have been shown to be functionally involved in EOC and/or as markers for various cancer types [7,44,50,51,91,126,132,133].

LPC is present at the highest concentrations among LPLs. Its role in signing is still debatable. Although both of its tumor-promoting and suppressing activities have been reported in various cancers [96,129,134,135], specific attention should be paid that LPC is not present in a free form in most physiological and pathologic conditions. It binds to albumin and other carrier proteins [136,137]. The bound form of LPC may not have many of the effects reported previously [91,137]. In particular, when high concentrations (>20 µM) of free LPC are used, it may have non-specific and detergent-like effects, which are unlikely to be physiologically or pathologically significant. LPC may function as a component of cell membrane and carrier to deliver choline to tissues and LPC is the precursor/substrate for ATX to produce LPA. However, its levels are hardly rate-limiting. This fact is also pertinent to developing ATX inhibitors. Those ATX substrate analog inhibitors are difficult to work in vivo, due to the competitive high concentrations of LPC present in the human blood and/or other tissues.

Table 1. Concentrations (in µM) of major lysophospholipids (LPLs) involved in epithelial ovarian cancer (EOC).

Lipid	EOC or BC Plasma	Healthy Control Plasma	EOC Ascites	Benign Ascites
Acyl-LPA	2–22 [33,34]	0.6–0.9 [33,34]	19–95 [28,43]	2.9 [28,43]
Alkyl-, and alkenyl-LPA			3.7 ± 1.7 [28]	0.9 ± 0.7 [28]
LPI	to 3.0 [28,35]	0–1.5 [28,35]	14.7 ± 9.7 [28,35]	2.9 ± 2.0 [28,35]
LPC	120 ± 0.30 [45] 117–153 [34]	128 ± 46 [45] 122 [34]		
S1P	0.52 ± 0.12 [137]	0.58 ± 0.18 [45–48]	sub-µM to low µM [45–48]	sub-µM to low µM [45–48]

BC: breast cancer.

2.3. Lysophosphatidylinositol (LPI) and Other LPLs

While phosphatidylinositol (PI) is the substrate of PI3K, one of the most pertinent signaling pathways in cancer [138], LPI as a signaling molecule is much less studied. We have shown that the plasma and ascites levels of LPI in EOC are elevated. In healthy controls, the plasma levels of LPI are in the range of 0–1.5 μM in healthy subjects, which are increased to 1.1–3.0 μM in patients with EOC [28,35]. The means and SDs levels of LPI in non-malignant ascites vs. malignant EOC are 2.9 ± 2.0 μM and 14.7 ± 9.7 μM respectively (Table 1). However, in our lab, unlike LPA and sphingosine-1-phosphate (S1P), neither positive nor negative effects of LPI have been detected in EOC cells. A very recent report has shown that LPC and LPI regulate gene expression, including adhesion molecules, cytokines, and chemokines, as well as those involved in cholesterol biosynthesis (by LPC), or gene transcripts critical for the metabolism of glucose, lipids, and amino acids (by LPI) in human aortic endothelial cells (HAECs). Moreover, LPC and LPI share the ability to transdifferentiate HAECs into innate immune cells [139].

Although a specific receptor of LPI has been reported [140], they warrant further validation for their rules and signaling in cancers. We and others also detected several other LPLs in EOC plasma and/or ascites, including lysophosphatidylethanolamine (LPE), lysophosphatidylglycerol (LPG), lysophosphatidylserine (LPS), lyso-platelet activating factor (lyso-PAF), and PAF [28,35,45,141]. However, the role and signaling of these LPLs in EOC are much less studied.

2.4. Sphingosine-1-Phosphate (S1P)

2.4.1. S1P Levels and Production

S1P is the orthologue of LPA with a different backbone (a sphingoid base vs. a glycerol backbone). The physiological and pathologic concentrations of S1P is approximately one order of magnitude lower than that of LPA and is usually present in sub-μM to low μM range [45–48] (Table 1).

In contrast to LPA, S1P is mainly produced intracellularly by two sphingosine kinases (SphK1 and SphK2; see Figure 1 in [142] and Figure 2 in [143]). S1P may be irreversibly degraded by S1P lyase (SPL) or dephosphorylated by S1P phosphatases (SPPs). Since the S1P lyase level in the blood was much lower than that in tissues and erythrocytes, as well as platelets lack SPL and SPP activity when they mature, higher S1P levels in the blood and lower amounts in tissues are present [144].

Intracellularly produced S1P is exported out of the cell either by the specific transporter Spinster 2 (Spns2) or by several members of the ABC transporter family [145]. This autocrine and/or paracrine action of S1P is known as "inside-out signaling". In the last few years, it has become evident that S1P also exerts intracellular functions by targeting different molecules, including the PPARγ family factors [145].

2.4.2. S1P Functions and Signaling Mechanisms in EOC

Over the past two decades, increasing evidence demonstrates a strong link between S1P and both normal physiology and progression of different diseases, including cancer and inflammation. S1P may affect survival, proliferation, angiogenesis, and metastatic spread of cancer [144,146].

LPA and S1P share structural similarity. In addition, $LPAR_{1-3}$ and $S1PR_{1-5}$ belong to the same edg-receptor family [147]. Moreover, S1P has been shown to have many similar tumor promoting activities as LPA, and is considered as a cancer treatment target [148,149], which has been reviewed extensively [32,144,146,150,151] (Figure 2). However, there are several major differences between LPA and S1P. Most of all, while LPA displays, in most cases, tumor promoting activities; S1P is multi-facet at several levels, which is emphasized as follows.

Firstly, S1P has strong concentration dependent differential effects. As mentioned above, the physiological/pathological concentrations of S1P are in general lower than those of LPA [46–48]. The effects of S1P in EOC cells tested are highly concentration-dependent [152–157]. While lower concentrations of S1P (≤1 μM) are usually stimulatory, higher concentrations (10–30 μM) of S1P are inhibitory. The S1P effects are also dependent on cell culture conditions. For example, S1P (10 μM) induced cell death when cells were in suspension but stimulated cell growth when cells were attached. The calcium-dependent induction of cell death by S1P is apparently associated with its inhibitory effect on cell attachment and cell adhesion [152]. N-cadherin, γ- and β-catenins, FAK, and integrin β1 are among the proteins affected by S1P and/or LPA [152,156].

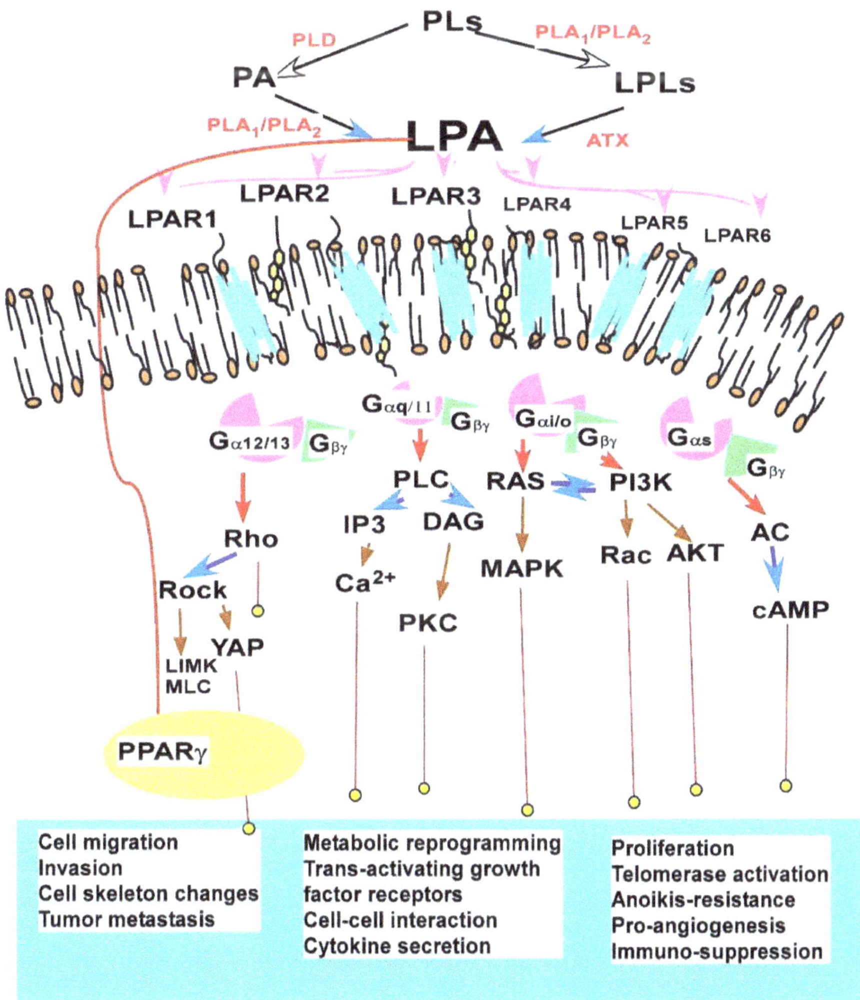

Figure 2. Diagram illustration of LPA receptors, signaling pathways and functions shown in EOC. LPA stimulates almost every aspect of tumor promoting activities [24,25,29–31,44,45,70,77–94]. This figure illuminates updated information related to LPA receptors ($LPAR_1$ to $LPAR_6$, and PPARγ), signaling pathways, and functions shown in EOC, modified from a previous review article by Yung et al. [94]. In particular, the nuclear receptor for LPA, PPARγ is included. While Gα12/13, Gαq, and Gαi mediate tumor promoting activities in most cases, Gαs is likely to be a negative regulator counter-reacting Gαi actions. Abbreviations: autotaxin (ATX); LIM kinase (LIMK); lysophospholipids (LPLs); myosin light chain (MLC) phosphatidic acid (PA); phospholipids (PLs); phospholipase D (PLD); phospholipase A_1 (PLA_1); and phospholipase A_2 (PLA_2).

Based on chemically measured S1P concentrations in biological fluids and the binding affinities of S1P to its receptors (in nM to low μM range [158]), the effects of low concentrations of S1P ($\leq$1 μM) may be more pathophysiologically relevant. The effects of high concentrations of S1P (10–30 μM) may be more artificial and/or non-specific.

Secondly, SphK1 and SphK2 have distinct cellular locations, regulations and functions. In general, SphK1 is tumor promoting and SphK2 is suppressive; SphK1 is upregulated in cancer, while SphK2 is downregulated [151,159,160]. Numerous tumor promoting agonists including TNF-α and other inflammatory signaling molecules, such as IL-1β, IFN-γ, IgE, and C5a, stimulate cytosolic SphK1, which translocates to the plasma membrane and uses sphingosine as a substrate to generate S1P. Elevated SphK1 has been shown in EOC cells and functionally involved in drug-resistance and other tumor promoting activities [161,162]. In contrast, SphK2 is located in cytosol or in the nucleus [144]. S1P produced by SphK2 inhibits histone deacetylases (HDACs), which modulates the dynamic balance of histone acetylation and influences the epigenetic regulation of specific target genes [163]. The two SphKs are also likely to have cooperative roles as evidence by knockout mice. Double-knockout animals were embryonic lethal, due to the incomplete maturation of the vascular system and brain, although mice deficient in either SphK1 or SphK2 had no obvious abnormalities [151].

Thirdly, different and opposing effects are mediated by different S1PRs. S1P receptors have been identified so far and named $S1PR_{1-5}$ (formerly referred to as endothelial differentiation gene (Edg1, 5, 3, 6, 8) [147]. Following receptor activation, multiple signaling cascades are activated, which are very similar to or opposing to those stimulated by LPA [164,165]. Among the five S1PRs, $S1PR_1$/$S1PR_3$ and $S1PR_2$ receptors may mediate opposing effects [149,151,153,154,157,159,160]. $S1PR_1$ and $S1PR_3$ mediate S1P's tumor promoting activities, such as cell migration and invasion via activation of Rac. Blockage of SphK1, but not SphK2, or $S1PR_{1/3}$ could attenuate ovarian cancer angiogenesis and inhibit angiogenic factor expression in a mouse model [159]. $S1PR_1$ is upregulated in ovarian cancer tissues and cell lines, which is negatively regulated by miR-148a in EOC cells [166]. On the other hand, $S1PR_2$ generally mediates the inhibitory effect via Rho-mediated inhibition of Rac [160]. $S1PR_2$ is also involved in negative regulation of tumor angiogenesis and tumor growth in vivo via RhoC activation [167], although one study has shown that the growth of SKOV3 cells could be decreased by $S1PR_2$ inhibition in vitro and in vivo [168]. In addition, $S1PR_2$ has an inhibitor role in macrophage recruitment during inflammation [169]. Goetzl et al. reported that both $S1PR_2$ and $S1PR_3$ are expressed higher in ovarian surface epithelial cells than in ovarian cancer cells [170].

Finally, S1P may have profound regulator effects on inflammation and in the immune system. The SphKs/S1P/$S1PR_1$ axis plays an important role in the immune regulation. It is involved in the mature vascular system; pathological angiogenesis; immune cell egress from tissue compartments; hematopoietic, vascular, and stem cell survival; and cytokine production. In particular, S1P induces STAT3 activation in tumor-associated myeloid derived suppressing cells (MDSCs) [151]. In addition, the roles of LPA and S1P on angiogenesis are likely to be different. S1P may have a direct proangiogenic role on ECs [159]. S1P and its receptors are involved in vessel morphogenesis and angiogenesis during embryonic development and in the adult organism both under normal and pathological conditions [171,172]. On the other hand, LPA's role on ECs may be indirect and mediated by its effect on tumor and/or TME cells via releasing proangiogenic factors, such as IL-8 [92,94,173].

SphK1 is highly expressed in the tumor stroma of high grade serous ovarian cancer (HGSOC) and is required for the differentiation and tumor promoting function of cancer-associated fibroblasts (CAFs) [174]. While increasing S1P catabolism or inhibiting S1P biosynthesis could become a new way to treat cancer, some studies found that the inhibition of S1P raised secondary malignancy [151,175].

A biospecific monoclonal antibody to S1P (S1P mAb) has been developed and investigated for its role in tumorigenesis. The anti-S1P mAb substantially reduced tumor progression and in some cases eliminated measurable tumors in murine xenograft and allograft models. Tumor growth inhibition was attributed to antiangiogenic and antitumorigenic effects of the antibody [176]. The anti-S1P mAb blocked EC migration and resulting capillary formation, inhibited blood vessel formation induced by

VEGF and bFGF, and arrested tumor-associated angiogenesis [176]. In this study, SKOV3 cells were used for ovarian cancer, but they are not cells from HGSOC, which accounts for about 70% of EOC cases, with less than 30% of patients currently surviving more than five years after diagnosis with little improvement in overall survival over the past 40 years [177–179]. Hence, the therapeutic significance of targeting S1P in EOC warrants further studies.

In summary, the role of S1P in the pathogenesis of ovarian cancer remains unclear and controversial and more studies are clearly required. Due to the multi-faceted nature of S1P's roles and signaling, targeting S1P signaling may be a double-edged sword.

2.5. *Sphingosylphosphorylcholine (SPC)*

SPC is an orthologue of LPC with a different backbone (a sphingoid base vs. a glycerol backbone). The levels of SPC in EOC vs. non-malignant ascites is low: 71.5 ± 50.8 nM vs. 17.9 ± 10.1 nM, respectively [28]. The levels of plasma SPC are also at nM range [45]. SPC is a potential calcium-release inducer in EOC cells [25,26,29]. SPC also shows other cellular activities in EOC cells, including regulation of IL-8 expression in EOC cells [94]. However, high concentrations (at μM level) of SPC is very toxic to cells. SPC induces dendritic cells (DC) chemotaxis and stimulates the production of IL-12 from DC [180,181]. However, the real physiological or pathological roles of SPC in EOC are still very elusive.

3. Conclusions

The reciprocal interplay of cancer cells and TME is an indispensable prerequisite for tumor growth and progression. Ovarian cancer, the most lethal of all gynecological malignancies, is characterized by a unique TME that enables specific and efficient metastatic mechanisms/routes, impairs immune surveillance, and mediates therapy resistance. More specifically, detached cancer cells—as well as large numbers of T cells, TAMs, and other host cells—cooperate with resident host cells to support tumor progression and immune evasion. The presence of the peritoneal fluid (ascites) enables more efficient tumor-stromal cell interactions and the transcoelomic spread of tumor cells to other pelvic and peritoneal organs. In particular, this fluid is rich in tumor-promoting soluble factors including elevated LPLs, either in EV or non-EV forms. Several important future directions and unresolved questions include, but are not limited to: development of standard and uniform methods for lipid extraction and analyses; further characterization of LPL regulation (both production and degradation) and their signaling mechanisms; development of strategies for cancer-specific targeting those tumor promoting lipids; and conducting more studies on their extracellular associations in order to better develop markers and targeting. Overall, it is critical to take TME into consideration to develop the next generation of therapeutic strategies.

Funding: This work is supported in part by the National Institutes of Health (RO1 155145 to Y.X.); and the Mary Fendrich-Hulman Charitable Trust Fund to Y.X.

Acknowledgments: Y.X. would like to thank Kevin McClelland for editing the manuscript.

Conflicts of Interest: The author declares no conflict of interest.

References

1. Westermann, A.M.; Beijnen, J.H.; Moolenaar, W.H.; Rodenhuis, S. Growth factors in human ovarian cancer. *Cancer Treat. Rev.* **1997**, *23*, 113–131. [CrossRef]
2. Amsterdam, A. Novel role of growth factors in ovary function. *Harefuah* **2010**, *149*, 789–793. [PubMed]
3. Thibault, B.; Castells, M.; Delord, J.P.; Couderc, B. Ovarian cancer microenvironment: Implications for cancer dissemination and chemoresistance acquisition. *Cancer Metastasis Rev.* **2014**, *33*, 17–39. [CrossRef] [PubMed]
4. Luo, Z.; Wang, Q.; Lau, W.B.; Lau, B.; Xu, L.; Zhao, L.; Yang, H.; Feng, M.; Xuan, Y.; Yang, Y.; et al. Tumor microenvironment: The culprit for ovarian cancer metastasis? *Cancer Lett.* **2016**, *377*, 174–182. [CrossRef] [PubMed]

5. Hodeib, M.; Serna-Gallegos, T.; Tewari, K.S. A review of HER2-targeted therapy in breast and ovarian cancer: Lessons from antiquity—Cleopatra and Penelope. *Future Oncol.* **2015**, *11*, 3113–3131. [CrossRef] [PubMed]
6. Ntanasis-Stathopoulos, I.; Fotopoulos, G.; Tzanninis, I.G.; Kotteas, E.A. The emerging role of tyrosine kinase inhibitors in ovarian cancer treatment: A systematic review. *Cancer Investig.* **2016**, *34*, 313–339. [CrossRef] [PubMed]
7. Cai, H.; Chiorean, E.G.; Chiorean, M.V.; Rex, D.K.; Robb, B.W.; Hahn, N.M.; Liu, Z.; Loehrer, P.J.; Harrison, M.L.; Xu, Y. Elevated phospholipase A2 activities in plasma samples from multiple cancers. *PLoS ONE* **2013**, *8*, e57081. [CrossRef] [PubMed]
8. Pap, E.; Pallinger, E.; Pasztoi, M.; Falus, A. Highlights of a new type of intercellular communication: Microvesicle-based information transfer. *Inflamm. Res.* **2009**, *58*, 1–8. [CrossRef] [PubMed]
9. He, C.; Zheng, S.; Luo, Y.; Wang, B. Exosome theranostics: Biology and translational medicine. *Theranostics* **2018**, *8*, 237–255. [CrossRef] [PubMed]
10. Taylor, D.D.; Gercel-Taylor, C. MicroRNA signatures of tumor-derived exosomes as diagnostic biomarkers of ovarian cancer. *Gynecol. Oncol.* **2008**, *110*, 13–21. [CrossRef] [PubMed]
11. Beach, A.; Zhang, H.G.; Ratajczak, M.Z.; Kakar, S.S. Exosomes: An overview of biogenesis, composition and role in ovarian cancer. *J. Ovarian Res.* **2014**, *7*, 14. [CrossRef] [PubMed]
12. Saleem, S.N.; Abdel-Mageed, A.B. Tumor-derived exosomes in oncogenic reprogramming and cancer progression. *Cell. Mol. Life Sci.* **2015**, *72*, 1–10. [CrossRef] [PubMed]
13. Cheng, L.; Wu, S.; Zhang, K.; Qing, Y.; Xu, T. A comprehensive overview of exosomes in ovarian cancer: Emerging biomarkers and therapeutic strategies. *J. Ovarian Res.* **2017**, *10*, 73. [CrossRef] [PubMed]
14. Hanahan, D.; Weinberg, R.A. Hallmarks of cancer: The next generation. *Cell* **2011**, *144*, 646–674. [CrossRef] [PubMed]
15. Amoroso, M.R.; Matassa, D.S.; Agliarulo, I.; Avolio, R.; Maddalena, F.; Condelli, V.; Landriscina, M.; Esposito, F. Stress-adaptive response in ovarian cancer drug resistance: Role of trap1 in oxidative metabolism-driven inflammation. *Adv. Protein Chem. Struct. Biol.* **2017**, *108*, 163–198. [PubMed]
16. Ke, C.; Li, A.; Hou, Y.; Sun, M.; Yang, K.; Cheng, J.; Wang, J.; Ge, T.; Zhang, F.; Li, Q.; et al. Metabolic phenotyping for monitoring ovarian cancer patients. *Sci. Rep.* **2016**, *6*, 23334. [CrossRef] [PubMed]
17. Fahy, E.; Subramaniam, S.; Murphy, R.C.; Nishijima, M.; Raetz, C.R.; Shimizu, T.; Spener, F.; van Meer, G.; Wakelam, M.J.; Dennis, E.A. Update of the lipid maps comprehensive classification system for lipids. *J. Lipid Res.* **2009**, *50*, S9–S14. [CrossRef] [PubMed]
18. Tania, M.; Khan, M.A.; Song, Y. Association of lipid metabolism with ovarian cancer. *Curr. Oncol.* **2010**, *17*, 6–11. [PubMed]
19. Tsujiuchi, T.; Araki, M.; Hirane, M.; Dong, Y.; Fukushima, N. Lysophosphatidic acid receptors in cancer pathobiology. *Histol. Histopathol.* **2014**, *29*, 313–321. [PubMed]
20. Pua, T.L.; Wang, F.Q.; Fishman, D.A. Roles of LPA in ovarian cancer development and progression. *Future Oncol.* **2009**, *5*, 1659–1673. [CrossRef] [PubMed]
21. Tokumura, A. Physiological and pathophysiological roles of lysophosphatidic acids produced by secretory lysophospholipase d in body fluids. *Biochim. Biophys. Acta* **2002**, *1582*, 18–25. [CrossRef]
22. Turkoglu, O.; Zeb, A.; Graham, S.; Szyperski, T.; Szender, J.B.; Odunsi, K.; Bahado-Singh, R. Metabolomics of biomarker discovery in ovarian cancer: A systematic review of the current literature. *Metabolomics* **2016**, *12*. [CrossRef] [PubMed]
23. Varas-Godoy, M.; Rice, G.; Illanes, S.E. The crosstalk between ovarian cancer stem cell niche and the tumor microenvironment. *Stem Cells Int.* **2017**, *2017*, 5263974. [CrossRef] [PubMed]
24. Ray, U.; Roy Chowdhury, S.; Vasudevan, M.; Bankar, K.; Roychoudhury, S.; Roy, S.S. Gene regulatory networking reveals the molecular cue to lysophosphatidic acid-induced metabolic adaptations in ovarian cancer cells. *Mol. Oncol.* **2017**, *11*, 491–516. [CrossRef] [PubMed]
25. Xu, Y.; Fang, X.J.; Casey, G.; Mills, G.B. Lysophospholipids activate ovarian and breast cancer cells. *Biochem. J.* **1995**, *309*, 933–940. [CrossRef] [PubMed]
26. Xu, Y.; Gaudette, D.C.; Boynton, J.D.; Frankel, A.; Fang, X.J.; Sharma, A.; Hurteau, J.; Casey, G.; Goodbody, A.; Mellors, A.; et al. Characterization of an ovarian cancer activating factor in ascites from ovarian cancer patients. *Clin. Cancer Res.* **1995**, *1*, 1223–1232. [PubMed]

27. Benesch, M.G.K.; MacIntyre, I.T.K.; McMullen, T.P.W.; Brindley, D.N. Coming of age for autotaxin and lysophosphatidate signaling: Clinical applications for preventing, detecting and targeting tumor-promoting inflammation. *Cancers* **2018**, *10*, 73. [CrossRef] [PubMed]
28. Xiao, Y.J.; Schwartz, B.; Washington, M.; Kennedy, A.; Webster, K.; Belinson, J.; Xu, Y. Electrospray ionization mass spectrometry analysis of lysophospholipids in human ascitic fluids: Comparison of the lysophospholipid contents in malignant vs. nonmalignant ascitic fluids. *Anal. Biochem.* **2001**, *290*, 302–313. [CrossRef] [PubMed]
29. Xu, Y.; Casey, G.; Mills, G.B. Effect of lysophospholipids on signaling in the human jurkat T cell line. *J. Cell. Physiol.* **1995**, *163*, 441–450. [CrossRef] [PubMed]
30. Sengupta, S.; Wang, Z.; Tipps, R.; Xu, Y. Biology of lpa in health and disease. *Semin. Cell Dev. Biol.* **2004**, *15*, 503–512. [CrossRef] [PubMed]
31. Mills, G.B.; Moolenaar, W.H. The emerging role of lysophosphatidic acid in cancer. *Nat. Rev. Cancer* **2003**, *3*, 582–591. [CrossRef] [PubMed]
32. Murph, M.; Mills, G.B. Targeting the lipids LPA and S1p and their signalling pathways to inhibit tumour progression. *Expert. Rev. Mol. Med.* **2007**, *9*, 1–18. [CrossRef] [PubMed]
33. Xu, Y.; Shen, Z.; Wiper, D.W.; Wu, M.; Morton, R.E.; Elson, P.; Kennedy, A.W.; Belinson, J.; Markman, M.; Casey, G. Lysophosphatidic acid as a potential biomarker for ovarian and other gynecologic cancers. *JAMA* **1998**, *280*, 719–723. [CrossRef] [PubMed]
34. Sutphen, R.; Xu, Y.; Wilbanks, G.D.; Fiorica, J.; Grendys, E.C., Jr.; LaPolla, J.P.; Arango, H.; Hoffman, M.S.; Martino, M.; Wakeley, K.; et al. Lysophospholipids are potential biomarkers of ovarian cancer. *Cancer Epidemiol. Biomarkers Prev.* **2004**, *13*, 1185–1191. [PubMed]
35. Xiao, Y.; Chen, Y.; Kennedy, A.W.; Belinson, J.; Xu, Y. Evaluation of plasma lysophospholipids for diagnostic significance using electrospray ionization mass spectrometry (ESI-MS) analyses. *Ann. N. Y. Acad. Sci.* **2000**, *905*, 242–259. [CrossRef] [PubMed]
36. Sedlakova, I.; Vavrova, J.; Tosner, J.; Hanousek, L. Lysophosphatidic acid in ovarian cancer patients. *Ceska Gynekol.* **2006**, *71*, 312–317. [CrossRef] [PubMed]
37. Meleh, M.; Pozlep, B.; Mlakar, A.; Meden-Vrtovec, H.; Zupancic-Kralj, L. Determination of serum lysophosphatidic acid as a potential biomarker for ovarian cancer. *J. Chromatogr. B Analyt. Technol. Biomed. Life Sci.* **2007**, *858*, 287–291. [CrossRef] [PubMed]
38. Sedlakova, I.; Vavrova, J.; Tosner, J.; Hanousek, L. Lysophosphatidic acid: An ovarian cancer marker. *Eur. J. Gynaecol. Oncol.* **2008**, *29*, 511–514. [PubMed]
39. Nakamura, K.; Igarashi, K.; Ohkawa, R.; Yokota, H.; Masuda, A.; Nakagawa, S.; Yano, T.; Ikeda, H.; Aoki, J.; Yatomi, Y. Serum autotaxin is not a useful biomarker for ovarian cancer. *Lipids* **2012**, *47*, 927–930. [CrossRef] [PubMed]
40. Lu, Z.; Chen, Y.; Hu, Z.; Hu, C. Diagnostic value of total plasma lysophosphatidic acid in ovarian cancer: A meta-analysis. *Int. J. Gynecol. Cancer* **2015**, *25*, 18–23. [CrossRef] [PubMed]
41. Zhang, Y.J.; Cao, L.Y.; Fu, Z.Z.; Wang, Y.J.; Wang, G.X.; Gu, T. Clinical significance of plasma lysophosphatidic acid levels in the differential diagnosis of ovarian cancer. *J. Cancer Res. Ther.* **2015**, *11*, 375–380. [CrossRef] [PubMed]
42. Li, Y.Y.; Zhang, W.C.; Zhang, J.L.; Zheng, C.J.; Zhu, H.; Yu, H.M.; Fan, L.M. Plasma levels of lysophosphatidic acid in ovarian cancer versus controls: A meta-analysis. *Lipids Health Dis.* **2015**, *14*, 72. [CrossRef] [PubMed]
43. Westermann, A.M.; Havik, E.; Postma, F.R.; Beijnen, J.H.; Dalesio, O.; Moolenaar, W.H.; Rodenhuis, S. Malignant effusions contain lysophosphatidic acid (LPA)-like activity. *Ann. Oncol.* **1998**, *9*, 437–442. [CrossRef] [PubMed]
44. Cai, Q.; Zhao, Z.; Antalis, C.; Yan, L.; Del Priore, G.; Hamed, A.H.; Stehman, F.B.; Schilder, J.M.; Xu, Y. Elevated and secreted phospholipase A(2) activities as new potential therapeutic targets in human epithelial ovarian cancer. *FASEB J.* **2012**, *26*, 3306–3320. [CrossRef] [PubMed]
45. Murph, M.; Tanaka, T.; Pang, J.; Felix, E.; Liu, S.; Trost, R.; Godwin, A.K.; Newman, R.; Mills, G. Liquid chromatography mass spectrometry for quantifying plasma lysophospholipids: Potential biomarkers for cancer diagnosis. *Methods Enzymol.* **2007**, *433*, 1–25. [PubMed]
46. Zhao, Z.; Xu, Y. An extremely simple method for extraction of lysophospholipids and phospholipids from blood samples. *J. Lipid Res.* **2010**, *51*, 652–659. [CrossRef] [PubMed]

47. Michels, M.; Japtok, L.; Alisjahbana, B.; Wisaksana, R.; Sumardi, U.; Puspita, M.; Kleuser, B.; de Mast, Q.; van der Ven, A.J. Decreased plasma levels of the endothelial protective sphingosine-1-phosphate are associated with dengue-induced plasma leakage. *J. Infect.* **2015**, *71*, 480–487. [CrossRef] [PubMed]
48. Ramanathan, R.; Raza, A.; Sturgill, J.; Lyon, D.; Young, J.; Hait, N.C.; Takabe, K. Paradoxical association of postoperative plasma sphingosine-1-phosphate with breast cancer aggressiveness and chemotherapy. *Mediators Inflamm.* **2017**, *2017*, 5984819. [CrossRef] [PubMed]
49. Ma, L.; Uchida, H.; Nagai, J.; Inoue, M.; Aoki, J.; Ueda, H. Evidence for de novo synthesis of lysophosphatidic acid in the spinal cord through phospholipase A2 and autotaxin in nerve injury-induced neuropathic pain. *J. Pharmacol. Exp. Ther.* **2010**, *333*, 540–546. [CrossRef] [PubMed]
50. Eder, A.M.; Sasagawa, T.; Mao, M.; Aoki, J.; Mills, G.B. Constitutive and lysophosphatidic acid (LPA)-induced LPA production: Role of phospholipase D and phospholipase a2. *Clin. Cancer Res.* **2000**, *6*, 2482–2491. [PubMed]
51. Sengupta, S.; Xiao, Y.J.; Xu, Y. A novel laminin-induced LPA autocrine loop in the migration of ovarian cancer cells. *FASEB J.* **2003**, *17*, 1570–1572. [CrossRef] [PubMed]
52. Shen, Z.; Belinson, J.; Morton, R.E.; Xu, Y.; Xu, Y. Phorbol 12-myristate 13-acetate stimulates lysophosphatidic acid secretion from ovarian and cervical cancer cells but not from breast or leukemia cells. *Gynecol. Oncol.* **1998**, *71*, 364–368. [CrossRef] [PubMed]
53. Aoki, J. Mechanisms of lysophosphatidic acid production. *Semin. Cell. Dev. Biol.* **2004**, *15*, 477–489. [CrossRef] [PubMed]
54. Benesch, M.G.; Tang, X.; Venkatraman, G.; Bekele, R.T.; Brindley, D.N. Recent advances in targeting the autotaxin-lysophosphatidate-lipid phosphate phosphatase axis in vivo. *J. Biomed. Res.* **2016**, *30*, 272–284. [PubMed]
55. Nikolaou, A.; Kokotou, M.G.; Limnios, D.; Psarra, A.; Kokotos, G. Autotaxin inhibitors: A patent review (2012–2016). *Expert. Opin. Ther. Pat.* **2017**, *27*, 815–829. [CrossRef] [PubMed]
56. Gaits, F.; Fourcade, O.; Le Balle, F.; Gueguen, G.; Gaige, B.; Gassama-Diagne, A.; Fauvel, J.; Salles, J.P.; Mauco, G.; Simon, M.F.; et al. Lysophosphatidic acid as a phospholipid mediator: Pathways of synthesis. *FEBS Lett.* **1997**, *410*, 54–58. [CrossRef]
57. Pages, C.; Simon, M.F.; Valet, P.; Saulnier-Blache, J.S. Lysophosphatidic acid synthesis and release. *Prostaglandins Other Lipid Mediat.* **2001**, *64*, 1–10. [CrossRef]
58. Roszkowski, I.; Niewiarowska, M.; Czerwinska, J.; Bar-Pratkowska, J.; Obrebski, T. Problems of surgical treatment of a patient with blood platelet disorders. *Ginekol. Pol.* **1971**, *42*, 1499–1501. [PubMed]
59. Hisada, Y.; Geddings, J.E.; Ay, C.; Mackman, N. Venous thrombosis and cancer: From mouse models to clinical trials. *J. Thromb. Haemost.* **2015**, *13*, 1372–1382. [CrossRef] [PubMed]
60. Menczer, J. Preoperative elevated platelet count and thrombocytosis in gynecologic malignancies. *Arch. Gynecol. Obstet.* **2017**, *295*, 9–15. [CrossRef] [PubMed]
61. Swier, N.; Versteeg, H.H. Reciprocal links between venous thromboembolism, coagulation factors and ovarian cancer progression. *Thromb. Res.* **2017**, *150*, 8–18. [CrossRef] [PubMed]
62. Zhou, Q.; Huang, F.; He, Z.; Zuo, M.Z. Clinicopathological and prognostic significance of platelet count in patients with ovarian cancer. *Climacteric* **2018**, *21*, 60–68. [CrossRef] [PubMed]
63. Lin, R.J.; Afshar-Kharghan, V.; Schafer, A.I. Paraneoplastic thrombocytosis: The secrets of tumor self-promotion. *Blood* **2014**, *124*, 184–187. [CrossRef] [PubMed]
64. Nugent, D.; Belinson, J.L.; Xu, Y. The synergistic interactions of oleoyl-lysophosphatidic acid in platelet aggregation. *Med. Sci. Res.* **1999**, *27*, 435–441.
65. Nugent, D.; Xu, Y. Sphingosine-1-phosphate: Characterization of its inhibition of platelet aggregation. *Platelets* **2000**, *11*, 226–232. [PubMed]
66. Leblanc, R.; Houssin, A.; Peyruchaud, O. Platelets, autotaxin and lysophosphatidic acid signaling: Win-win factors for cancer metastasis. *Br. J. Pharmacol.* **2018**. [CrossRef] [PubMed]
67. Krishnan, V.; Clark, R.; Chekmareva, M.; Johnson, A.; George, S.; Shaw, P.; Seewaldt, V.; Rinker-Schaeffer, C. In vivo and ex vivo approaches to study ovarian cancer metastatic colonization of milky spot structures in peritoneal adipose. *J. Vis. Exp.* **2015**, e52721. [CrossRef] [PubMed]
68. Feist, P.E.; Loughran, E.A.; Stack, M.S.; Hummon, A.B. Quantitative proteomic analysis of murine white adipose tissue for peritoneal cancer metastasis. *Anal. Bioanal. Chem.* **2018**, *410*, 1583–1594. [CrossRef] [PubMed]

69. Cai, Q.; Yan, L.; Xu, Y. Anoikis resistance is a critical feature of highly aggressive ovarian cancer cells. *Oncogene* **2015**, *34*, 3315–3324. [CrossRef] [PubMed]
70. Nieman, K.M.; Kenny, H.A.; Penicka, C.V.; Ladanyi, A.; Buell-Gutbrod, R.; Zillhardt, M.R.; Romero, I.L.; Carey, M.S.; Mills, G.B.; Hotamisligil, G.S.; et al. Adipocytes promote ovarian cancer metastasis and provide energy for rapid tumor growth. *Nat. Med.* **2011**, *17*, 1498–1503. [CrossRef] [PubMed]
71. Umezu-Goto, M.; Kishi, Y.; Taira, A.; Hama, K.; Dohmae, N.; Takio, K.; Yamori, T.; Mills, G.B.; Inoue, K.; Aoki, J.; et al. Autotaxin has lysophospholipase d activity leading to tumor cell growth and motility by lysophosphatidic acid production. *J. Cell. Biol.* **2002**, *158*, 227–233. [CrossRef] [PubMed]
72. Benesch, M.G.; Zhao, Y.Y.; Curtis, J.M.; McMullen, T.P.; Brindley, D.N. Regulation of autotaxin expression and secretion by lysophosphatidate and sphingosine 1-phosphate. *J. Lipid Res.* **2015**, *56*, 1134–1144. [CrossRef] [PubMed]
73. Volden, P.A.; Skor, M.N.; Johnson, M.B.; Singh, P.; Patel, F.N.; McClintock, M.K.; Brady, M.J.; Conzen, S.D. Mammary adipose tissue-derived lysophospholipids promote estrogen receptor-negative mammary epithelial cell proliferation. *Cancer Prev. Res.* **2016**, *9*, 367–378. [CrossRef] [PubMed]
74. Dusaulcy, R.; Rancoule, C.; Gres, S.; Wanecq, E.; Colom, A.; Guigne, C.; van Meeteren, L.A.; Moolenaar, W.H.; Valet, P.; Saulnier-Blache, J.S. Adipose-specific disruption of autotaxin enhances nutritional fattening and reduces plasma lysophosphatidic acid. *J. Lipid Res.* **2011**, *52*, 1247–1255. [CrossRef] [PubMed]
75. Benesch, M.G.K.; Yang, Z.; Tang, X.; Meng, G.; Brindley, D.N. Lysophosphatidate signaling: The tumor microenvironment's new nemesis. *Trends Cancer* **2017**, *3*, 748–752. [CrossRef] [PubMed]
76. Ren, J.; Xiao, Y.J.; Singh, L.S.; Zhao, X.; Zhao, Z.; Feng, L.; Rose, T.M.; Prestwich, G.D.; Xu, Y. Lysophosphatidic acid is constitutively produced by human peritoneal mesothelial cells and enhances adhesion, migration, and invasion of ovarian cancer cells. *Cancer Res.* **2006**, *66*, 3006–3014. [CrossRef] [PubMed]
77. Knowlden, S.; Georas, S.N. The autotaxin-LPA axis emerges as a novel regulator of lymphocyte homing and inflammation. *J. Immunol.* **2014**, *192*, 851–857. [CrossRef] [PubMed]
78. Halama, A.; Guerrouahen, B.S.; Pasquier, J.; Satheesh, N.J.; Suhre, K.; Rafii, A. Nesting of colon and ovarian cancer cells in the endothelial niche is associated with alterations in glycan and lipid metabolism. *Sci. Rep.* **2017**, *7*, 39999. [CrossRef] [PubMed]
79. Wong, J.L.; Obermajer, N.; Odunsi, K.; Edwards, R.P.; Kalinski, P. Synergistic COX2 induction by IFNgamma and TNFalpha self-limits type-1 immunity in the human tumor microenvironment. *Cancer Immunol. Res.* **2016**, *4*, 303–311. [CrossRef] [PubMed]
80. Reinartz, S.; Finkernagel, F.; Adhikary, T.; Rohnalter, V.; Schumann, T.; Schober, Y.; Nockher, W.A.; Nist, A.; Stiewe, T.; Jansen, J.M.; et al. A transcriptome-based global map of signaling pathways in the ovarian cancer microenvironment associated with clinical outcome. *Genome Biol.* **2016**, *17*, 108. [CrossRef] [PubMed]
81. Nakayama, J.; Raines, T.A.; Lynch, K.R.; Slack-Davis, J.K. Decreased peritoneal ovarian cancer growth in mice lacking expression of lipid phosphate phosphohydrolase 1. *PLoS ONE* **2015**, *10*, e0120071. [CrossRef] [PubMed]
82. Baudhuin, L.M.; Cristina, K.L.; Lu, J.; Xu, Y. AKT activation induced by lysophosphatidic acid and sphingosine-1-phosphate requires both mitogen-activated protein kinase kinase and p38 mitogen-activated protein kinase and is cell-line specific. *Mol. Pharmacol.* **2002**, *62*, 660–671. [CrossRef] [PubMed]
83. Cai, H.; Xu, Y. The role of lpa and yap signaling in long-term migration of human ovarian cancer cells. *Cell Commun. Signal.* **2013**, *11*, 31. [CrossRef] [PubMed]
84. Fan, Q.; Cai, Q.; Li, P.; Wang, W.; Wang, J.; Gerry, E.; Wang, T.L.; Shih, I.M.; Nephew, K.P.; Xu, Y. The novel zip4 regulation and its role in ovarian cancer. *Oncotarget* **2017**, *8*, 90090–90107. [CrossRef] [PubMed]
85. Fan, Q.; Cai, Q.; Xu, Y. *LPA Regulates sox9 in Ovarian Cancer Cells*; Gavin Publishers: Lisle, IL, USA, 2017.
86. Fang, X.; Yu, S.; Bast, R.C.; Liu, S.; Xu, H.J.; Hu, S.X.; LaPushin, R.; Claret, F.X.; Aggarwal, B.B.; Lu, Y.; et al. Mechanisms for lysophosphatidic acid-induced cytokine production in ovarian cancer cells. *J. Biol. Chem.* **2004**, *279*, 9653–9661. [CrossRef] [PubMed]
87. Ha, J.H.; Ward, J.D.; Radhakrishnan, R.; Jayaraman, M.; Song, Y.S.; Dhanasekaran, D.N. Lysophosphatidic acid stimulates epithelial to mesenchymal transition marker slug/snail2 in ovarian cancer cells via galphai2, src, and hif1alpha signaling nexus. *Oncotarget* **2016**, *7*, 37664. [CrossRef] [PubMed]
88. Jiang, Y.; Berk, M.; Singh, L.S.; Tan, H.; Yin, L.; Powell, C.T.; Xu, Y. Kiss1 suppresses metastasis in human ovarian cancer via inhibition of protein kinase c alpha. *Clin. Exp. Metastasis* **2005**, *22*, 369–376. [CrossRef] [PubMed]

89. Kim, K.S.; Sengupta, S.; Berk, M.; Kwak, Y.G.; Escobar, P.F.; Belinson, J.; Mok, S.C.; Xu, Y. Hypoxia enhances lysophosphatidic acid responsiveness in ovarian cancer cells and lysophosphatidic acid induces ovarian tumor metastasis in vivo. *Cancer Res.* **2006**, *66*, 7983–7990. [CrossRef] [PubMed]
90. Li, H.; Wang, D.; Zhang, H.; Kirmani, K.; Zhao, Z.; Steinmetz, R.; Xu, Y. Lysophosphatidic acid stimulates cell migration, invasion, and colony formation as well as tumorigenesis/metastasis of mouse ovarian cancer in immunocompetent mice. *Mol. Cancer Ther.* **2009**, *8*, 1692–1701. [CrossRef] [PubMed]
91. Li, H.; Zhao, Z.; Wei, G.; Yan, L.; Wang, D.; Zhang, H.; Sandusky, G.E.; Turk, J.; Xu, Y. Group via phospholipase a2 in both host and tumor cells is involved in ovarian cancer development. *FASEB J.* **2010**, *24*, 4103–4116. [CrossRef] [PubMed]
92. Lu, J.; Xiao Yj, Y.J.; Baudhuin, L.M.; Hong, G.; Xu, Y. Role of ether-linked lysophosphatidic acids in ovarian cancer cells. *J. Lipid Res.* **2002**, *43*, 463–476. [PubMed]
93. Ren, J.; Li, Y.; Zhang, Y.L.; Zhou, X.H.; Zhang, L.; Yang, Y.; Li, Y. Effect of inhibitors of phospholipase A(2); on the metastasis potentials of human ovarian cancer cells. *Xi Bao Yu Fen Zi Mian Yi Xue Za Zhi* **2010**, *26*, 992–995. (In Chinese) [PubMed]
94. Schwartz, B.M.; Hong, G.; Morrison, B.H.; Wu, W.; Baudhuin, L.M.; Xiao, Y.J.; Mok, S.C.; Xu, Y. Lysophospholipids increase interleukin-8 expression in ovarian cancer cells. *Gynecol. Oncol.* **2001**, *81*, 291–300. [CrossRef] [PubMed]
95. Sengupta, S.; Kim, K.S.; Berk, M.P.; Oates, R.; Escobar, P.; Belinson, J.; Li, W.; Lindner, D.J.; Williams, B.; Xu, Y. Lysophosphatidic acid downregulates tissue inhibitor of metalloproteinases, which are negatively involved in lysophosphatidic acid-induced cell invasion. *Oncogene* **2007**, *26*, 2894–2901. [CrossRef] [PubMed]
96. Fang, X.; Gaudette, D.; Furui, T.; Mao, M.; Estrella, V.; Eder, A.; Pustilnik, T.; Sasagawa, T.; Lapushin, R.; Yu, S.; et al. Lysophospholipid growth factors in the initiation, progression, metastases, and management of ovarian cancer. *Ann. N. Y. Acad. Sci.* **2000**, *905*, 188–208. [CrossRef] [PubMed]
97. Fang, X.; Schummer, M.; Mao, M.; Yu, S.; Tabassam, F.H.; Swaby, R.; Hasegawa, Y.; Tanyi, J.L.; LaPushin, R.; Eder, A.; et al. Lysophosphatidic acid is a bioactive mediator in ovarian cancer. *Biochim. Biophys. Acta* **2002**, *1582*, 257–264. [CrossRef]
98. Mills, G.B.; Eder, A.; Fang, X.; Hasegawa, Y.; Mao, M.; Lu, Y.; Tanyi, J.; Tabassam, F.H.; Wiener, J.; Lapushin, R.; et al. Critical role of lysophospholipids in the pathophysiology, diagnosis, and management of ovarian cancer. *Cancer Treat. Res.* **2002**, *107*, 259–283. [PubMed]
99. Yung, Y.C.; Stoddard, N.C.; Chun, J. Lpa receptor signaling: Pharmacology, physiology, and pathophysiology. *J. Lipid Res.* **2014**, *55*, 1192–1214. [CrossRef] [PubMed]
100. Jesionowska, A.; Cecerska-Heryc, E.; Matoszka, N.; Dolegowska, B. Lysophosphatidic acid signaling in ovarian cancer. *J. Recept Signal. Transduct. Res.* **2015**, *35*, 578–584. [CrossRef] [PubMed]
101. Wang, G.L.; Wen, Z.Q.; Xu, W.P.; Wang, Z.Y.; Du, X.L.; Wang, F. Inhibition of lysophosphatidic acid receptor-2 expression by rna interference decreases lysophosphatidic acid-induced urokinase plasminogen activator activation, cell invasion, and migration in ovarian cancer SKOV-3 cells. *Croat Med. J.* **2008**, *49*, 175–181. [CrossRef] [PubMed]
102. Yang, K.; Zheng, D.; Deng, X.; Bai, L.; Xu, Y.; Cong, Y.S. Lysophosphatidic acid activates telomerase in ovarian cancer cells through hypoxia-inducible factor-1alpha and the PI3K pathway. *J. Cell Biochem.* **2008**, *105*, 1194–1201. [CrossRef] [PubMed]
103. Bai, C.Q.; Yao, Y.W.; Liu, C.H.; Zhang, H.; Xu, X.B.; Zeng, J.L.; Liang, W.J.; Yang, W.; Song, Y. Diagnostic and prognostic significance of lysophosphatidic acid in malignant pleural effusions. *J. Thorac. Dis.* **2014**, *6*, 483–490. [PubMed]
104. Fan, Q.; Cai, Q.; Xu, Y. Foxm1 is a downstream target of lpa and yap oncogenic signaling pathways in high grade serous ovarian cancer. *Oncotarget* **2015**, *6*, 27688–27699. [CrossRef] [PubMed]
105. Fishman, D.A.; Liu, Y.; Ellerbroek, S.M.; Stack, M.S. Lysophosphatidic acid promotes matrix metalloproteinase (MMP) activation and mmp-dependent invasion in ovarian cancer cells. *Cancer Res.* **2001**, *61*, 3194–3199. [PubMed]
106. Burkhalter, R.J.; Westfall, S.D.; Liu, Y.; Stack, M.S. Lysophosphatidic acid initiates epithelial to mesenchymal transition and induces beta-catenin-mediated transcription in epithelial ovarian carcinoma. *J. Biol. Chem.* **2015**, *290*, 22143–22154. [CrossRef] [PubMed]
107. Cui, J.; Liu, Y.; Wang, X. The roles of glycodelin in cancer development and progression. *Front. Immunol.* **2017**, *8*, 1685. [CrossRef] [PubMed]

108. Xu, Y.; Xiao, Y.J.; Zhu, K.; Baudhuin, L.M.; Lu, J.; Hong, G.; Kim, K.S.; Cristina, K.L.; Song, L.; F, S.W.; et al. Unfolding the pathophysiological role of bioactive lysophospholipids. *Curr. Drug Targets Immune Endocr. Metabol. Disord.* **2003**, *3*, 23–32. [CrossRef] [PubMed]
109. Li, H.; Xu, Y. Indiana University School of Medicine, Indianapolis, IN, USA. Unpublished observation, 2007.
110. Seo, E.J.; Kwon, Y.W.; Jang, I.H.; Kim, D.K.; Lee, S.I.; Choi, E.J.; Kim, K.H.; Suh, D.S.; Lee, J.H.; Choi, K.U.; et al. Autotaxin regulates maintenance of ovarian cancer stem cells through lysophosphatidic acid-mediated autocrine mechanism. *Stem Cells* **2016**, *34*, 551–564. [CrossRef] [PubMed]
111. Bar-Shavit, R.; Maoz, M.; Kancharla, A.; Nag, J.K.; Agranovich, D.; Grisaru-Granovsky, S.; Uziely, B. G protein-coupled receptors in cancer. *Int. J. Mol. Sci.* **2016**, *17*, 1320. [CrossRef] [PubMed]
112. Taniguchi, R.; Inoue, A.; Sayama, M.; Uwamizu, A.; Yamashita, K.; Hirata, K.; Yoshida, M.; Tanaka, Y.; Kato, H.E.; Nakada-Nakura, Y.; et al. Structural insights into ligand recognition by the lysophosphatidic acid receptor LPA6. *Nature* **2017**, *548*, 356–360. [CrossRef] [PubMed]
113. Hope, J.M.; Wang, F.Q.; Whyte, J.S.; Ariztia, E.V.; Abdalla, W.; Long, K.; Fishman, D.A. LPA receptor 2 mediates LPA-induced endometrial cancer invasion. *Gynecol. Oncol.* **2009**, *112*, 215–223. [CrossRef] [PubMed]
114. Lin, S.; Wang, D.; Iyer, S.; Ghaleb, A.M.; Shim, H.; Yang, V.W.; Chun, J.; Yun, C.C. The absence of lpa2 attenuates tumor formation in an experimental model of colitis-associated cancer. *Gastroenterology* **2009**, *136*, 1711–1720. [CrossRef] [PubMed]
115. Stoddard, N.C.; Chun, J. Promising pharmacological directions in the world of lysophosphatidic acid signaling. *Biomol. Ther.* **2015**, *23*, 1–11. [CrossRef] [PubMed]
116. Takahashi, K.; Fukushima, K.; Onishi, Y.; Inui, K.; Node, Y.; Fukushima, N.; Honoki, K.; Tsujiuchi, T. Lysophosphatidic acid (LPA) signaling via LPA4 and LPA6 negatively regulates cell motile activities of colon cancer cells. *Biochem. Biophys. Res. Commun.* **2017**, *483*, 652–657. [CrossRef] [PubMed]
117. Ishii, S.; Hirane, M.; Fukushima, K.; Tomimatsu, A.; Fukushima, N.; Tsujiuchi, T. Diverse effects of LPA4, LPA5 and LPA6 on the activation of tumor progression in pancreatic cancer cells. *Biochem. Biophys. Res. Commun.* **2015**, *461*, 59–64. [CrossRef] [PubMed]
118. Takahashi, K.; Fukushima, K.; Otagaki, S.; Ishimoto, K.; Minami, K.; Fukushima, N.; Honoki, K.; Tsujiuchi, T. Effects of LPA1 and LPA6 on the regulation of colony formation activity in colon cancer cells treated with anticancer drugs. *J. Recept. Signal. Transduct. Res.* **2018**, *38*, 71–75. [CrossRef] [PubMed]
119. McIntyre, T.M.; Pontsler, A.V.; Silva, A.R.; St Hilaire, A.; Xu, Y.; Hinshaw, J.C.; Zimmerman, G.A.; Hama, K.; Aoki, J.; Arai, H.; et al. Identification of an intracellular receptor for lysophosphatidic acid (LPA): LPA is a transcellular ppargamma agonist. *Proc. Natl. Acad. Sci. USA* **2003**, *100*, 131–136. [CrossRef] [PubMed]
120. Tsukahara, T.; Tsukahara, R.; Yasuda, S.; Makarova, N.; Valentine, W.J.; Allison, P.; Yuan, H.; Baker, D.L.; Li, Z.; Bittman, R.; et al. Different residues mediate recognition of 1-*O*-oleyllysophosphatidic acid and rosiglitazone in the ligand binding domain of peroxisome proliferator-activated receptor gamma. *J. Biol. Chem.* **2006**, *281*, 3398–3407. [CrossRef] [PubMed]
121. Tsukahara, T. Ppar gamma networks in cell signaling: Update and impact of cyclic phosphatidic acid. *J. Lipids* **2013**, *2013*, 246597. [CrossRef] [PubMed]
122. Tsukahara, T.; Haniu, H.; Matsuda, Y. Effect of alkyl glycerophosphate on the activation of peroxisome proliferator-activated receptor gamma and glucose uptake in C2C12 cells. *Biochem. Biophys. Res. Commun.* **2013**, *433*, 281–285. [CrossRef] [PubMed]
123. Worzfeld, T.; Finkernagel, F.; Reinartz, S.; Konzer, A.; Adhikary, T.; Nist, A.; Stiewe, T.; Wagner, U.; Looso, M.; Graumann, J.; et al. Proteotranscriptomics reveal signaling networks in the ovarian cancer microenvironment. *Mol. Cell. Proteom.* **2018**, *17*, 270–289. [CrossRef] [PubMed]
124. Knowlden, S.A.; Capece, T.; Popovic, M.; Chapman, T.J.; Rezaee, F.; Kim, M.; Georas, S.N. Regulation of T cell motility in vitro and in vivo by LPA and LPA2. *PLoS ONE* **2014**, *9*, e101655. [CrossRef] [PubMed]
125. Okita, M.; Gaudette, D.C.; Mills, G.B.; Holub, B.J. Elevated levels and altered fatty acid composition of plasma lysophosphatidylcholine(lysopc) in ovarian cancer patients. *Int. J. Cancer* **1997**, *71*, 31–34. [CrossRef]
126. Zhang, Y.; Liu, Y.; Li, L.; Wei, J.; Xiong, S.; Zhao, Z. High resolution mass spectrometry coupled with multivariate data analysis revealing plasma lipidomic alteration in ovarian cancer in asian women. *Talanta* **2016**, *150*, 88–96. [CrossRef] [PubMed]

127. Zhao, Z.; Xiao, Y.; Elson, P.; Tan, H.; Plummer, S.J.; Berk, M.; Aung, P.P.; Lavery, I.C.; Achkar, J.P.; Li, L.; et al. Plasma lysophosphatidylcholine levels: Potential biomarkers for colorectal cancer. *J. Clin. Oncol.* **2007**, *25*, 2696–2701. [CrossRef] [PubMed]
128. Zhao, Z.; Xu, Y. Measurement of endogenous lysophosphatidic acid by ESI-MS/MS in plasma samples requires pre-separation of lysophosphatidylcholine. *J. Chromatogr. B Analyt. Technol. Biomed. Life Sci.* **2009**, *877*, 3739–3742. [CrossRef] [PubMed]
129. Ross, T.; Jakubzig, B.; Grundmann, M.; Massing, U.; Kostenis, E.; Schlesinger, M.; Bendas, G. The molecular mechanism by which saturated lysophosphatidylcholine attenuates the metastatic capacity of melanoma cells. *FEBS Open Bio* **2016**, *6*, 1297–1309. [CrossRef] [PubMed]
130. Kuhn, T.; Floegel, A.; Sookthai, D.; Johnson, T.; Rolle-Kampczyk, U.; Otto, W.; von Bergen, M.; Boeing, H.; Kaaks, R. Higher plasma levels of lysophosphatidylcholine 18:0 are related to a lower risk of common cancers in a prospective metabolomics study. *BMC Med.* **2016**, *14*, 13. [CrossRef] [PubMed]
131. Goto, T.; Terada, N.; Inoue, T.; Kobayashi, T.; Nakayama, K.; Okada, Y.; Yoshikawa, T.; Miyazaki, Y.; Uegaki, M.; Utsunomiya, N.; et al. Decreased expression of lysophosphatidylcholine (16:0/OH) in high resolution imaging mass spectrometry independently predicts biochemical recurrence after surgical treatment for prostate cancer. *Prostate* **2015**, *75*, 1821–1830. [CrossRef] [PubMed]
132. Song, Y.; Wilkins, P.; Hu, W.; Murthy, K.S.; Chen, J.; Lee, Z.; Oyesanya, R.; Wu, J.; Barbour, S.E.; Fang, X. Inhibition of calcium-independent phospholipase A2 suppresses proliferation and tumorigenicity of ovarian carcinoma cells. *Biochem. J.* **2007**, *406*, 427–436. [CrossRef] [PubMed]
133. Li, H.; Zhao, Z.; Antalis, C.; Zhao, Z.; Emerson, R.; Wei, G.; Zhang, S.; Zhang, Z.Y.; Xu, Y. Combination therapy of an inhibitor of group via phospholipase A2 with paclitaxel is highly effective in blocking ovarian cancer development. *Am. J. Pathol.* **2011**, *179*, 452–461. [CrossRef] [PubMed]
134. Carneiro, A.B.; Iaciura, B.M.; Nohara, L.L.; Lopes, C.D.; Veas, E.M.; Mariano, V.S.; Bozza, P.T.; Lopes, U.G.; Atella, G.C.; Almeida, I.C.; et al. Lysophosphatidylcholine triggers TLR2- and TLR4-mediated signaling pathways but counteracts LPS-induced no synthesis in peritoneal macrophages by inhibiting NF-kappab translocation and MAPK/ERK phosphorylation. *PLoS ONE* **2013**, *8*, e76233. [CrossRef] [PubMed]
135. Li, X.; Fang, P.; Li, Y.; Kuo, Y.M.; Andrews, A.J.; Nanayakkara, G.; Johnson, C.; Fu, H.; Shan, H.; Du, F.; et al. Mitochondrial reactive oxygen species mediate lysophosphatidylcholine-induced endothelial cell activation. *Arterioscler. Thromb. Vasc. Biol.* **2016**, *36*, 1090–1100. [CrossRef] [PubMed]
136. de Bony, J.; Dufourcq, J.; Clin, B. Lipid-protein interactions: NMR study of melittin and its binding to lysophosphatidylcholine. *Biochim. Biophys. Acta* **1979**, *552*, 531–534. [CrossRef]
137. Kim, Y.L.; Im, Y.J.; Ha, N.C.; Im, D.S. Albumin inhibits cytotoxic activity of lysophosphatidylcholine by direct binding. *Prostaglandins Other Lipid Mediat.* **2007**, *83*, 130–138. [CrossRef] [PubMed]
138. Sanchez-Vega, F.; Mina, M.; Armenia, J.; Chatila, W.K.; Luna, A.; La, K.C.; Dimitriadoy, S.; Liu, D.L.; Kantheti, H.S.; Saghafinia, S.; et al. Oncogenic signaling pathways in the cancer genome atlas. *Cell* **2018**, *173*, 321–337.e10. [CrossRef] [PubMed]
139. Li, X.; Wang, L.; Fang, P.; Sun, Y.; Jiang, X.; Wang, H.; Yang, X.F. Lysophospholipids induce innate immune transdifferentiation of endothelial cells, resulting in prolonged endothelial activation. *J. Biol. Chem.* **2018**. [CrossRef] [PubMed]
140. Hurst, K.; Badgley, C.; Ellsworth, T.; Bell, S.; Friend, L.; Prince, B.; Welch, J.; Cowan, Z.; Williamson, R.; Lyon, C.; et al. A putative lysophosphatidylinositol receptor GPR55 modulates hippocampal synaptic plasticity. *Hippocampus* **2017**, *27*, 985–998. [CrossRef] [PubMed]
141. Zhao, Z.; Cai, Q.; Xu, Y. The lipidomic analyses in low and highly aggressive ovarian cancer cell lines. *Lipids* **2016**, *51*, 179–187. [CrossRef] [PubMed]
142. Cannavo, A.; Liccardo, D.; Komici, K.; Corbi, G.; de Lucia, C.; Femminella, G.D.; Elia, A.; Bencivenga, L.; Ferrara, N.; Koch, W.J.; et al. Sphingosine kinases and sphingosine 1-phosphate receptors: Signaling and actions in the cardiovascular system. *Front. Pharmacol.* **2017**, *8*, 556. [CrossRef] [PubMed]
143. Hatoum, D.; Haddadi, N.; Lin, Y.; Nassif, N.T.; McGowan, E.M. Mammalian sphingosine kinase (SPHK) isoenzymes and isoform expression: Challenges for sphk as an oncotarget. *Oncotarget* **2017**, *8*, 36898–36929. [CrossRef] [PubMed]
144. Rodriguez, Y.I.; Campos, L.E.; Castro, M.G.; Aladhami, A.; Oskeritzian, C.A.; Alvarez, S.E. Sphingosine-1 phosphate: A new modulator of immune plasticity in the tumor microenvironment. *Front. Oncol.* **2016**, *6*, 218. [CrossRef] [PubMed]

145. Nagahashi, M.; Takabe, K.; Terracina, K.P.; Soma, D.; Hirose, Y.; Kobayashi, T.; Matsuda, Y.; Wakai, T. Sphingosine-1-phosphate transporters as targets for cancer therapy. *Biomed. Res. Int.* **2014**, *2014*, 651727. [CrossRef] [PubMed]
146. Kunkel, G.T.; Maceyka, M.; Milstien, S.; Spiegel, S. Targeting the sphingosine-1-phosphate axis in cancer, inflammation and beyond. *Nat. Rev. Drug Discov.* **2013**, *12*, 688–702. [CrossRef] [PubMed]
147. Kostenis, E. Novel clusters of receptors for sphingosine-1-phosphate, sphingosylphosphorylcholine, and (lyso)-phosphatidic acid: New receptors for "old" ligands. *J. Cell. Biochem.* **2004**, *92*, 923–936. [CrossRef] [PubMed]
148. Kang, Y.C.; Kim, K.M.; Lee, K.S.; Namkoong, S.; Lee, S.J.; Han, J.A.; Jeoung, D.; Ha, K.S.; Kwon, Y.G.; Kim, Y.M. Serum bioactive lysophospholipids prevent trail-induced apoptosis via PI3K/AKT-dependent cflip expression and bad phosphorylation. *Cell Death Differ.* **2004**, *11*, 1287–1298. [CrossRef] [PubMed]
149. Park, K.S.; Kim, M.K.; Lee, H.Y.; Kim, S.D.; Lee, S.Y.; Kim, J.M.; Ryu, S.H.; Bae, Y.S. S1p stimulates chemotactic migration and invasion in ovcar3 ovarian cancer cells. *Biochem. Biophys Res. Commun.* **2007**, *356*, 239–244. [CrossRef] [PubMed]
150. Dai, L.; Xia, P.; Di, W. Sphingosine 1-phosphate: A potential molecular target for ovarian cancer therapy? *Cancer Investig.* **2014**, *32*, 71–80. [CrossRef] [PubMed]
151. Jin, L.; Liu, W.R.; Tian, M.X.; Fan, J.; Shi, Y.H. The SPHKS/S1P/S1PR1 axis in immunity and cancer: More ore to be mined. *World J. Surg. Oncol.* **2016**, *14*, 131. [CrossRef] [PubMed]
152. Hong, G.; Baudhuin, L.M.; Xu, Y. Sphingosine-1-phosphate modulates growth and adhesion of ovarian cancer cells. *FEBS Lett.* **1999**, *460*, 513–518. [CrossRef]
153. Baudhuin, L.M.; Jiang, Y.; Zaslavsky, A.; Ishii, I.; Chun, J.; Xu, Y. S1p3-mediated AKT activation and cross-talk with platelet-derived growth factor receptor (PDGFR). *FASEB J.* **2004**, *18*, 341–343. [CrossRef] [PubMed]
154. Wang, D.; Zhao, Z.; Caperell-Grant, A.; Yang, G.; Mok, S.C.; Liu, J.; Bigsby, R.M.; Xu, Y. S1p differentially regulates migration of human ovarian cancer and human ovarian surface epithelial cells. *Mol. Cancer Ther.* **2008**, *7*, 1993–2002. [CrossRef] [PubMed]
155. Devine, K.M.; Smicun, Y.; Hope, J.M.; Fishman, D.A. S1p induced changes in epithelial ovarian cancer proteolysis, invasion, and attachment are mediated by GI and RAC. *Gynecol. Oncol.* **2008**, *110*, 237–245. [CrossRef] [PubMed]
156. Smicun, Y.; Gil, O.; Devine, K.; Fishman, D.A. S1p and LPA have an attachment-dependent regulatory effect on invasion of epithelial ovarian cancer cells. *Gynecol. Oncol.* **2007**, *107*, 298–309. [CrossRef] [PubMed]
157. Smicun, Y.; Reierstad, S.; Wang, F.Q.; Lee, C.; Fishman, D.A. S1p regulation of ovarian carcinoma invasiveness. *Gynecol. Oncol.* **2006**, *103*, 952–959. [CrossRef] [PubMed]
158. Troupiotis-Tsailaki, A.; Zachmann, J.; Gonzalez-Gil, I.; Gonzalez, A.; Ortega-Gutierrez, S.; Lopez-Rodriguez, M.L.; Pardo, L.; Govaerts, C. Ligand chain length drives activation of lipid g protein-coupled receptors. *Sci. Rep.* **2017**, *7*, 2020. [CrossRef] [PubMed]
159. Dai, L.; Liu, Y.; Xie, L.; Wu, X.; Qiu, L.; Di, W. Sphingosine kinase 1/sphingosine-1-phosphate (S1p)/S1p receptor axis is involved in ovarian cancer angiogenesis. *Oncotarget* **2017**, *8*, 74947–74961. [CrossRef] [PubMed]
160. Fyrst, H.; Saba, J.D. An update on sphingosine-1-phosphate and other sphingolipid mediators. *Nat. Chem. Biol.* **2010**, *6*, 489–497. [CrossRef] [PubMed]
161. Illuzzi, G.; Bernacchioni, C.; Aureli, M.; Prioni, S.; Frera, G.; Donati, C.; Valsecchi, M.; Chigorno, V.; Bruni, P.; Sonnino, S.; et al. Sphingosine kinase mediates resistance to the synthetic retinoid *N*-(4-hydroxyphenyl)retinamide in human ovarian cancer cells. *J. Biol. Chem.* **2010**, *285*, 18594–18602. [CrossRef] [PubMed]
162. Snider, A.J.; Orr Gandy, K.A.; Obeid, L.M. Sphingosine kinase: Role in regulation of bioactive sphingolipid mediators in inflammation. *Biochimie* **2010**, *92*, 707–715. [CrossRef] [PubMed]
163. Hait, N.C.; Allegood, J.; Maceyka, M.; Strub, G.M.; Harikumar, K.B.; Singh, S.K.; Luo, C.; Marmorstein, R.; Kordula, T.; Milstien, S.; et al. Regulation of histone acetylation in the nucleus by sphingosine-1-phosphate. *Science* **2009**, *325*, 1254–1257. [CrossRef] [PubMed]
164. Patmanathan, S.N.; Wang, W.; Yap, L.F.; Herr, D.R.; Paterson, I.C. Mechanisms of sphingosine 1-phosphate receptor signalling in cancer. *Cell Signal.* **2017**, *34*, 66–75. [CrossRef] [PubMed]

165. Fan, Q.; Cheng, Y.; Chang, H.M.; Deguchi, M.; Hsueh, A.J.; Leung, P.C.K. Sphingosine-1-phosphate promotes ovarian cancer cell proliferation by disrupting hippo signaling. *Oncotarget* **2017**, *8*, 27166–27176. [CrossRef] [PubMed]
166. Wen, Z.; Zhao, S.; Liu, S.; Liu, Y.; Li, X.; Li, S. MicroRNA-148a inhibits migration and invasion of ovarian cancer cells via targeting sphingosine-1-phosphate receptor 1. *Mol. Med. Rep.* **2015**, *12*, 3775–3780. [CrossRef] [PubMed]
167. Du, W.; Takuwa, N.; Yoshioka, K.; Okamoto, Y.; Gonda, K.; Sugihara, K.; Fukamizu, A.; Asano, M.; Takuwa, Y. S1p(2), the G protein-coupled receptor for sphingosine-1-phosphate, negatively regulates tumor angiogenesis and tumor growth in vivo in mice. *Cancer Res.* **2010**, *70*, 772–781. [CrossRef] [PubMed]
168. Dai, L.; Liu, Y.X.; Xie, L.; Di, W. Effect of S1PR2 inhibition on epithelial ovarian cancer SKOV3 cell proliferation in vitro and in vivo. *Zhonghua Fu Chan Ke Za Zhi* **2018**, *53*, 106–110. (In Chinese) [PubMed]
169. Michaud, J.; Im, D.S.; Hla, T. Inhibitory role of sphingosine 1-phosphate receptor 2 in macrophage recruitment during inflammation. *J. Immunol.* **2010**, *184*, 1475–1483. [CrossRef] [PubMed]
170. Goetzl, E.J.; Dolezalova, H.; Kong, Y.; Hu, Y.L.; Jaffe, R.B.; Kalli, K.R.; Conover, C.A. Distinctive expression and functions of the type 4 endothelial differentiation gene-encoded G protein-coupled receptor for lysophosphatidic acid in ovarian cancer. *Cancer Res.* **1999**, *59*, 5370–5375. [PubMed]
171. Argraves, K.M.; Wilkerson, B.A.; Argraves, W.S. Sphingosine-1-phosphate signaling in vasculogenesis and angiogenesis. *World J. Biol. Chem.* **2010**, *1*, 291–297. [CrossRef] [PubMed]
172. Lucke, S.; Levkau, B. Endothelial functions of sphingosine-1-phosphate. *Cell. Physiol. Biochem.* **2010**, *26*, 87–96. [CrossRef] [PubMed]
173. Kim, K.S.; Ren, J.; Jiang, Y.; Ebrahem, Q.; Tipps, R.; Cristina, K.; Xiao, Y.J.; Qiao, J.; Taylor, K.L.; Lum, H.; et al. GPR4 plays a critical role in endothelial cell function and mediates the effects of sphingosylphosphorylcholine. *FASEB J.* **2005**, *19*, 819–821. [CrossRef] [PubMed]
174. Beach, J.A.; Aspuria, P.J.; Cheon, D.J.; Lawrenson, K.; Agadjanian, H.; Walsh, C.S.; Karlan, B.Y.; Orsulic, S. Sphingosine kinase 1 is required for TGF-beta mediated fibroblastto- myofibroblast differentiation in ovarian cancer. *Oncotarget* **2016**, *7*, 4167–4182. [CrossRef] [PubMed]
175. Shida, D.; Takabe, K.; Kapitonov, D.; Milstien, S.; Spiegel, S. Targeting SPHK1 as a new strategy against cancer. *Curr. Drug Targets* **2008**, *9*, 662–673. [CrossRef] [PubMed]
176. Visentin, B.; Vekich, J.A.; Sibbald, B.J.; Cavalli, A.L.; Moreno, K.M.; Matteo, R.G.; Garland, W.A.; Lu, Y.; Yu, S.; Hall, H.S.; et al. Validation of an anti-sphingosine-1-phosphate antibody as a potential therapeutic in reducing growth, invasion, and angiogenesis in multiple tumor lineages. *Cancer Cell* **2006**, *9*, 225–238. [CrossRef] [PubMed]
177. Siegel, R.L.; Miller, K.D.; Jemal, A. Cancer statistics, 2018. *CA Cancer J. Clin.* **2018**, *68*, 7–30. [CrossRef] [PubMed]
178. Domcke, S.; Sinha, R.; Levine, D.A.; Sander, C.; Schultz, N. Evaluating cell lines as tumour models by comparison of genomic profiles. *Nat. Commun.* **2013**, *4*, 2126. [CrossRef] [PubMed]
179. Beaufort, C.M.; Helmijr, J.C.; Piskorz, A.M.; Hoogstraat, M.; Ruigrok-Ritstier, K.; Besselink, N.; Murtaza, M.; van, I.W.F.; Heine, A.A.; Smid, M.; et al. Ovarian cancer cell line panel (OCCP): Clinical importance of in vitro morphological subtypes. *PLoS ONE* **2014**, *9*, e103988. [CrossRef] [PubMed]
180. Lee, H.Y.; Shin, E.H.; Bae, Y.S. Sphingosylphosphorylcholine stimulates human monocyte-derived dendritic cell chemotaxis. *Acta Pharmacol. Sin.* **2006**, *27*, 1359–1366. [CrossRef] [PubMed]
181. Ceballos, A.; Sabatte, J.; Nahmod, K.; Martinez, D.; Salamone, G.; Vermeulen, M.; Maggini, J.; Salomon, H.; Geffner, J. Sphingosylphosphorylcholine activates dendritic cells, stimulating the production of interleukin-12. *Immunology* **2007**, *121*, 328–336. [CrossRef] [PubMed]

cancers

MDPI

Review

Oxidative Phosphorylation: A Target for Novel Therapeutic Strategies Against Ovarian Cancer

Amruta P. Nayak [1,2], Arvinder Kapur [2], Lisa Barroilhet [2] and Manish S. Patankar [2,*]

1 Indian Institute of Science Education and Research, Pune 411008, India; amruta.pn@students.iiserpune.ac.in

2 Department of Obstetrics and Gynecology, University of Wisconsin-Madison, Madison, WI 54911, USA; akaur@wisc.edu (A.K.); barroilhet@wisc.edu (L.B.)

* Correspondence: patankar@wisc.edu; Tel.: +1-608-262-8871

Received: 25 July 2018; Accepted: 17 September 2018; Published: 18 September 2018

Abstract: Aerobic glycolysis is an important metabolic adaptation of cancer cells. There is growing evidence that oxidative phosphorylation is also an active metabolic pathway in many tumors, including in high grade serous ovarian cancer. Metastasized ovarian tumors use fatty acids for their energy needs. There is also evidence of ovarian cancer stem cells privileging oxidative phosphorylation (OXPHOS) for their metabolic needs. Metformin and thiazolidinediones such as rosiglitazone restrict tumor growth by inhibiting specific steps in the mitochondrial electron transport chain. These observations suggest that strategies to interfere with oxidative phosphorylation should be considered for the treatment of ovarian tumors. Here, we review the literature that supports this hypothesis and describe potential agents and critical control points in the oxidative phosphorylation pathway that can be targeted using small molecule agents. In this review, we also discuss potential barriers that can reduce the efficacy of the inhibitors of oxidative phosphorylation.

Keywords: high grade serous ovarian cancer; metabolism; mitochondria; oxidative phosphorylation; oxidative stress; biguanides; atovaquone; plumbagin; thiazolidinediones; ubiquinone; Nrf-2

1. Introduction

Metabolic adaptations allow tumors to maintain a highly proliferative state. Evidence in support of such adaptations was obtained nearly a century ago by Otto Warburg, and Carl and Gert Cori and their colleagues when they demonstrated an increased uptake of glucose by tumors as compared to normal tissues [1,2]. Warburg further demonstrated that even when sufficient oxygen was available, tumors used glycolysis to metabolize glucose to lactic acid [3,4]. In this respect, glucose metabolism in tumor cells resembles that occurring under anaerobic conditions. However, because glucose was being metabolized to lactate in the presence of oxygen, Warburg coined the term "aerobic glycolysis" to accurately describe this metabolic process in tumors [3]. Although glucose breakdown through oxidative phosphorylation (OXPHOS) yields maximum number of ATP, curtailing the metabolism to glycolysis provides the necessary biomolecule precursors needed by the tumors to maintain a high level of proliferation [5–8]. Several key enzymes in the glycolytic pathway and tricarboxylic acid cycle, (pyruvate kinase M2, pyruvate dehydrogenase kinase, isocitrate dehydrogenase, succinate dehydrogenase, lactate dehydrogenase and others (representative articles include [9–17]) are targets for anti-cancer drugs.

A rapidly growing body of evidence is demonstrating that an adaptation to aerobic glycolysis does not entail a complete shutdown of oxidative phosphorylation (OXPHOS) in tumors. Active electron transport occurs in cancer cells that trigger tumor recurrence and in cancer stem cells [18–20]. Here, we review evidence supporting the importance of OXPHOS in high grade serous ovarian cancer (HGSOC), discuss small molecule inhibitors of OXPHOS, their mechanism of action, and potential barriers to the use of such agents for the treatment of HGSOC.

2. Oxphos As Target for Hgsoc Therapy

Ovarian cancer is classified into type I and II diseases [21–23]. Clear cell cancer and low grade endometrioid are major types of ovarian tumors classified as Type I malignancies with mutations in ARID1A (AT-rich interactive domain-containing protein 1A), K-Ras (Kirsten rat sarcoma) and PTEN (phosphatase and tensin homolog). High grade serous ovarian cancer (HGSOC), the predominant subtype, is classified as Type II disease and is characterized by mutations in p53 and copy number variations [24–29]. In the majority of the patients, HGSOC is detected at an advanced stage when the tumor has progressed to sites beyond the ovaries. While cytoreductive surgery and chemotherapy with platinum and taxanes are initially effective, they fail to prevent recurrence of HGSOC. Recurrent HGSOC responds poorly to most established and experimental therapies. While PARP (ADP ribose polymerase) inhibitors have extended overall survival [30–33], there remains a need for additional novel therapeutic approaches to treat HGSOC. In this review, we make the case that OXPHOS be considered as a druggable pathway while developing novel therapies against HGSOC.

In normal cells, glucose is metabolized through glycolysis, tricarboxylic acid cycle and OXPHOS to produce 34–38 molecules of ATP per molecule of glucose (Figure 1). In cancer and other highly proliferative and activated cells (immune cells, for example), the end product of glycolysis, pyruvate, is not transferred to the mitochondria and consumed in the tricarboxylic acid cycle, but instead is converted to lactate (Figure 1). This conversion to lactate allows the cells to regenerate NAD (Nicotinamide Adenosine Dinucleotide) needed to drive the conversion of glyceraldehyde-3-phosphate to 1,3-bisphosphoglycerate in glycolysis.

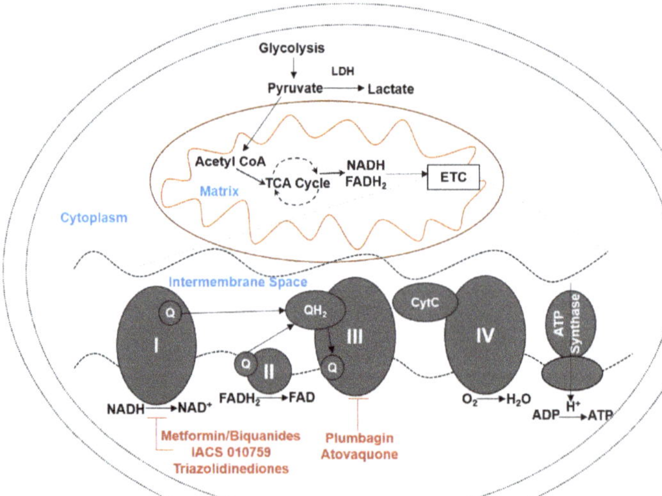

Figure 1. Oxidative phosphorylation. Aerobic glycolysis metabolizes glucose to lactic acid. Oxidative phosphorylation (OXPHOS) occurs in mitochondria and leads to efficient generation of ATP. OXPHOS is an active pathway in tumors and cancer stem cells. Several inhibitors or the various subunits of the mitochondrial electron transport complexes can serve as candidates for tumor therapy. Prominent drug candidates for HGSOC are shown. CytC, Cytochrome C, ETC, Electron Transport Chain, LDH, Lactate Dehydrogenase, Q, unbiquinone, QH2, Ubiquinol, TCA, Tricarboxylic Acid Cycle, $FADH_2$, Flavine Adenine Dinucleotide, NADH, Nicotinamide Adenine Dinucleotide.

An active glycolytic pathway supplies the biochemical precursors required for protein, nucleotide and lipid synthesis. This is an important reason why cancer cells limit glucose metabolism to glycolysis even when there is sufficient availability of oxygen. Ovarian tumors also show metabolic adaptation to aerobic glycolysis that allows them to maintain an increased proliferative capacity and survive under anchorage dependent conditions [34–36]. Adaptation of the ovarian cancer cells to aerobic glycolysis is supported by the increased expression of the glycolytic enzymes pyruvate kinase isoform M2 (PKM2), hexokinase II, and lactate dehydrogenase A (LDHA). PKM2 catalyzes the conversion of phosphoenolpyruvate to pyruvate and regulates the flux of acetyl coA available to enter the tricarboxylic aid cycle. Elevated expression of PKM2 correlates with decreased progression free survival in HGSOC although overall survival is not affected [37]. Hexokinase II is also upregulated in HGSOC [38,39]. Elevated hexokinase II expression contributes to chemoresistance in ovarian tumors [40].

The metabolic shift to aerobic glycolysis provides the precursors for synthesis of proteins, nucleotides and lipids, at the expense of ATP. To compensate, cancer cells overexpress glucose transporters. As a result, there is an increased uptake of glucose that is catabolized through aerobic glycolysis. Serous ovarian tumors express high levels of glucose transporters GLUT1, GLUT3 and GLUT4, as compared to healthy and benign ovarian tissues [41,42]. Increased uptake of glucose through the upregulation of glucose transporters (GLUT) is a hallmark of HGSOC allowing their imaging by 18-fluoro-deoxyglucose positron emission tomography (FDG-PET) [43,44].

The molecular mechanisms responsible for the metabolic reprogramming in cancer are under investigation. One mechanism is through the regulation of pyruvate kinase M2 isoform (PKM2) activity through its interactions with growth factor receptors [45]. There is emerging evidence that mutated *BRCA1* [46] and mutated p53 (reviewed in [47]), two genes that are most frequently mutated in HGSOC, also contribute to the shift to aerobic glycolysis. In HGSOC, the sulfatase, h-Sulf-1 is downregulated [48]. Loss of h-sulf-1 in HGSOC increases glycolytic activity through the phosphorylation of pyruvate dehydrogenase causing a decrease in availability of pyruvate in the tricarboxylic acid cycle [49]. All of these results clearly demonstrate that glycolysis is a major metabolic adaptation occurring in HGSOC.

2.1. Relevance of OXPHOS in Solid Tumors

While aerobic glycolysis is an important adaptation in HGSOC, OXPHOS is also an active pathway in cancer cells [50–56]. Initial data in support of this observation was gained from experiments with the tetracycline-inducible K-Ras (G12D) mouse model for pancreatic cancer [57]. Withdrawal of doxycycline caused regression of the pancreatic ductal carcinomas [55]. However, tumor recurrence was observed in the mice 4–5 months after doxycycline withdrawal. Tumor relapse was attributed to cancer stem cells surviving the ablation of mutant K-Ras. These surviving cancer stem cells had increased mitochondrial biogenesis with higher OXPHOS activity but impaired glycolysis. The relapsing tumors were responsive to the OXPHOS inhibitor, oligomycin [55].

Tumor cells that utilize aerobic glycolysis, coexist with cancer cells with active OXPHOS. A recent report by Yu et al [58] developed a model to predict the predominant metabolic pathway utilized by normal and cancer cells. Glycolysis is indicated by high expression of HIF-1α (Hypoxia inducible factor-1α) and low levels of phospho-5′ AMP-activated protein kinase (pAMPK), whereas OXPHOS-reliant tumors have low levels of HIF-1α and high levels of pAMPK. Some cancer cells express high levels of both HIF-1α and pAMPK indicating active glycolysis as well as OXPHOS.

There are, however, some indicators that active mitochondrial metabolism may have a favorable outcome. The Bioenergetic Cellular (BEC) index, a ratio β-F1ATPase (F1 portion of adenosine triphosphate synthase) to HSP60 (Heat Shock Protein 60) and GAPDH (Glyceraldehyde 3-Phosphate dehydrogenase) expression, predicts the metabolic state of a cell [59]. A higher BEC is an indicator of active OXPHOS. In one study, thirty six of 55 HGSOC patients had a BEC of less than 2.65 [60]. Progression free survival was higher in patients with <2.65 compared to the 19 patients with higher BEC (9.8 versus 5.3 months). However, the BEC does not account for metabolic heterogeneity and therefore these observations do not rule out the presence of tumor foci with active OXPHOS. There is evidence that agents targeting OXPHOS can be used to target cancer initiating cells, chemoresistant tumors as well as non-tumor cells from the tumor microenvironment.

2.2. Reliance of Ovarian Cancer Stem Cells on OXPHOS

Perhaps the largest impact of OXPHOS is in the survival and proliferation of cancer initiating stem cells. Tumor initiating cells isolated from tumorigenic murine ovarian surface epithelial (MOSE) cells showed increased expression of glucose transporters and an overactive glycolytic pathway [61,62]. However, these tumor initiating cells also had a higher capacity than non-tumor initiating tumor-forming MOSE cells for mitochondrial oxygen consumption. The tumor initiating MOSE cells also exhibited higher survivability when cultured in media that did not contain glucose but was supplemented with glutamine and fatty acids. The observation that the tumor initiating cells are better able to survive on glutamine-supplemented media suggests that they are less dependent on glycolysis and, through the entry of glutamine in the tricarboxylic acid cycle are able to generate sufficient NADH (Nicotinamide Adenine Dinucleotide) and FADH2 (Flavine Adenine Dinucleotide). The higher mitochondrial capacity facilitates production of sufficient levels of ATP that drive their proliferation.

$CD44^+/CD117^+$ cancer stem cells isolated from the peritoneal fluid of HGSOC patients that had the ability to form tumors in mice, showed decreased levels of pyruvate dehydrogenase kinase (PDHK1) and increased expression of isocitrate dehydrogenase (IDH2) [18]. This observation is in stark contrast to the non-cancer stem cells ($CD44^+/CD117^-$) from HGSOC patients where the PDHK1 was upregulated and IDH2 was significantly lower. PDHK1 negatively regulates pyruvate dehydrogenase and as a result controls the amount of acetyl-CoA (Coenzyme A) available for the tricarboxylic acid cycle. The decrease in PDHK1 and increase in IDH2 in the $CD44^+/CD117^+$ ovarian cancer stem cell population are indicators of enhanced tricarboxylic acid cycle. The $CD44^+/CD117^+$ cancer stem cells produced higher levels of oxygen radicals and had enhanced OXPHOS than the non-stem cell ($CD44^+/CD117^-$) population. $RAG2^{-/-}$ mice implanted with HGSOC tumors when maintained on a diet supplemented with the glycolysis inhibitor, 2-deoxyglucose, instead of glucose, showed a decrease in tumor size. However, the surviving tumors from these mice were enriched in $CD44^+/CD117^+$ cancer stem cells [18].

2.3. OXPHOS in Chemoresistant HGSOC

There is also evidence that OXPHOS is an important pathway to target in chemoresistant tumors. Tumor necrosis factor receptor-associated protein 1 (TRAP1) is a mitochondrial chaperone from the Hsp90 family [63]. Increase in TRAP1 expression elevates aerobic glycolysis in ovarian cancer cell lines [64]. HGSOC lines with lower expression of TRAP1 or silencing of this gene increased the oxygen consumption rate and decreased extracellular acidification (a measure of aerobic glycolysis) [65]. Low expression of TRAP1 results in higher reliance on OXPHOS and is associated with resistance to platinum [64]. Chemoresistant ovarian cancer cells show increased OXPHOS activity and survive under limiting glucose levels or when the resistant tumors were implanted in mice that were fed 2-deoxyglucose [34].

2.4. OXPHOS and the Tumor Microenvironment

Tumor cells produce high levels of lactic acid in the tumor microenvironment. While the lactic acid can be transported by cancer cells through monocarboxylic acid transporters (MCTs) and used to promote tumor proliferation, the effect of the acidic environment on the non-malignant cells is also an important factor to consider when determining the metabolic profile of the tumor. For example, the lactic acid in the tumor microenvironment can regulate the activity of immune cells infiltrating the tumor microenvironment (reviewed in [66]).

Such crosstalk in the tumor microenvironment is not unidirectional as the tumor cells can also be affected by the metabolic status of the fibroblasts from the microenvironment. For example, Ras (glycine at position 12 mutated to valine) mutations alter metabolism in cancer cells and increase the release of oxygen radicals [67]. These radicals induce oxidative stress in the intratumoral stroma, forcing a catabolic state that produces lactate, ketones, glutamine and fatty acids that serve as fuel to the cancer.

Stromal cells from the tumor microenvironment express low levels of caveolin 1 and high MCT4 allowing them to expel lactate into the tumor microenvironment [68–70]. In a recent study, patients with *BRCA1* mutated breast cancers were treated with the anti-oxidant, *N*-acetyl cysteine [71]. Pathological examination showed a reversal in the expression of caveolin 1 and MCT4 by the stromal cells suggesting that neutralization of the oxygen radicals can inhibit the symbiotic relationship between the cancer cells and the stromal cells in the microenvironment. Additionally, this study also observed a decrease in ki67 stained cancer cells. Since the mitochondria are the major source for oxygen radicals, it can be argued that the stromal cells from the tumor microenvironment are OXPHOS-active and the oxygen radicals generated by these cells promote the proliferation of cancer cells. Therefore, inhibitors of OXPHOS can not only be successful because of their direct cytotoxic effects on cancer cells but also through the potential modulation of metabolism in stromal and other non-cancer cells from the tumor microenvironment.

3. OXPHOS Provides Multiple Targets for Drug Development

High energy electrons from NADH and $FADH_2$ are harvested in OXPHOS and transferred to molecular oxygen. Four multiprotein complexes located in the inner membrane of the mitochondria (Complexes I-IV) are required for electron transport (Figure 1). Electrons from NADH and $FADH_2$ are extracted in complex I and complex II, respectively. Electrons from complex I and II are delivered to complex III via the electron carrier, ubiquinone (coenzyme 10, CoQ10). The quinone head group of CoQ10 participates in two electron redox reactions. Addition of one electron to CoQ10 yields semiquinone and further reduction of this intermediate leads to formation of ubiquinol. The electron transfer from NADH/$FADH_2$ to ubiquinone occurs at the ubiquinone and ubiquinol binding sites, Q_0 and $Q_{i,}$ in the mitochondrial complexes I-III. Electron transport from complex III to complex IV is aided by cytochrome C (Cyt C). In complex IV, the electrons are delivered to molecular oxygen to form water.

The transfer of electrons is coupled with pumping of protons from the matrix to the intermembrane of the mitochondria. This transfer of protons leads to the maintenance of a proton gradient ($\Delta\Psi_{pion}$). The proton efflux from the matrix helps maintain a negative charge in the matrix and contributes to an electrical potential gradient ($\Delta\Psi_m$). The electromotive force generated through proton transport provides the energy necessary for the fifth mitochondrial complex, the ATP synthase, to convert ADP to ATP.

Complex chemical reactions and biochemical control points are required to regulate OXPHOS. From the standpoint of cancer drug discovery, this situation provides opportunities for the development of novel therapeutic strategies. Agents that interfere with electron transport, maintenance of the proton gradient ($\Delta\Psi_{pion}$ and $\Delta\Psi_m$) and transfer of electrons to oxygen and ATP synthesis can be developed as cancer therapeutics. While small molecules are likely the preferred agents to target OXPHOS, efforts are also underway to develop peptides that can specifically target this mitochondrial metabolic pathway (reviewed in Reference [72]). In the subsequent sections, we will discuss small molecule agents that interfere with OXPHOS.

4. OXPHOS Inhibitors

4.1. Complex I Inhibiting Biguanides

Metformin and proguanil are biguanides with complex I inhibitory activities (Figure 1). Regular use of metformin reduces risk of ovarian cancer (OR 0.61, 95% CI 0.3–1.25) [73]. Nearly 70% of HGSOC patients using metformin survived for 5 years. In comparison, only 47% of HGSOC patients who were not on metformin survived for 5-years or more [74]. Romero and colleagues have analyzed the positive benefits of metformin use in HGSOC patients with diabetes. Approximately 51% of the diabetic patients who regularly used metformin had progression-free survival at 5-years post initial diagnosis of the cancer. In contrast, 23% of the nondiabetic metformin users and only 8% of non-diabetic non-metformin users had progression free survival at 5-years postdiagnosis [75]. The overall survival at 5-years post initial diagnosis of HGSOC was reported to be 63%, 37% and 23%, respectively, for these three cohorts [75].

Metformin inhibits complex I and thereby reduces ATP production. As a result of decreased ATP levels, AMPK is activated in cancer cells along with inhibition of mTORC1 (mammalian Target of Rapamycin Complex). Millimolar concentrations of metformin are required to inhibit complex I activity and there remains an active question of whether such high levels of metformin can be achieved in solid tumors. Proguanil inhibits complex I activity in the malarial parasite and is therefore administered in conjunction with atovaquone, a complex III inhibitor. However, proguanil has limited effect on human complex I and is therefore not suitable for cancer therapy [76,77]. Another bigiuanide, phenformin triggers lactic acidosis and therefore has major clinical toxicity.

A novel agent, IACS-0107059, that likely mimics the biguanide functional group has been investigated as therapy for acute myeloid leukemia [78–82]. This compound blocks complex I at subnanomolar-nanomolar range and inhibits proliferation of HGSOC cells. Clinical trials are currently underway to test this compound against AML and solid tumors.

Eight clinical trials are currently posted in *clinicaltrials.gov* to test the effect of metformin in ovarian cancer patients. The majority of these trials are evaluating the combination of metformin with chemotherapy and are currently recruiting patients. Results from one trial (NCT01579812) showed that metformin was well tolerated. Ex vivo evaluation of the tumors showed significant decrease in viable cancer stem cell population. Additional studies are needed to determine if these positive benefits are due to the OXPHOS inhibitory effects of metformin. As demonstrated by Yu et al. [58], monitoring pAMPK and HIF-1α levels in the metformin clinical trials can potentially be used as biomarkers for the status of OXPHOS versus aerobic glycolysis in tumors providing insight into the metabolic adaptations occurring in the tumors in response to this biguanide.

4.2. Oxidative Stress Inducers

Oxidative stress results from an imbalance between the processes responsible for generation and sequestration of reactive oxygen radicals (ROS) [83]. Since the transfer of electrons to molecular oxygen is an integral step of the electron transport chain, the OXPHOS pathway is a major generator of oxygen radicals (Figure 2). A rapid increase in intracellular levels of oxygen radicals causes cellular damage and cell death. Atovaquone, a Food and Drug Administration (FDA) approved anti-malarial agent that inhibits complex III activity, is being repurposed for treatment of solid tumors [84–86].

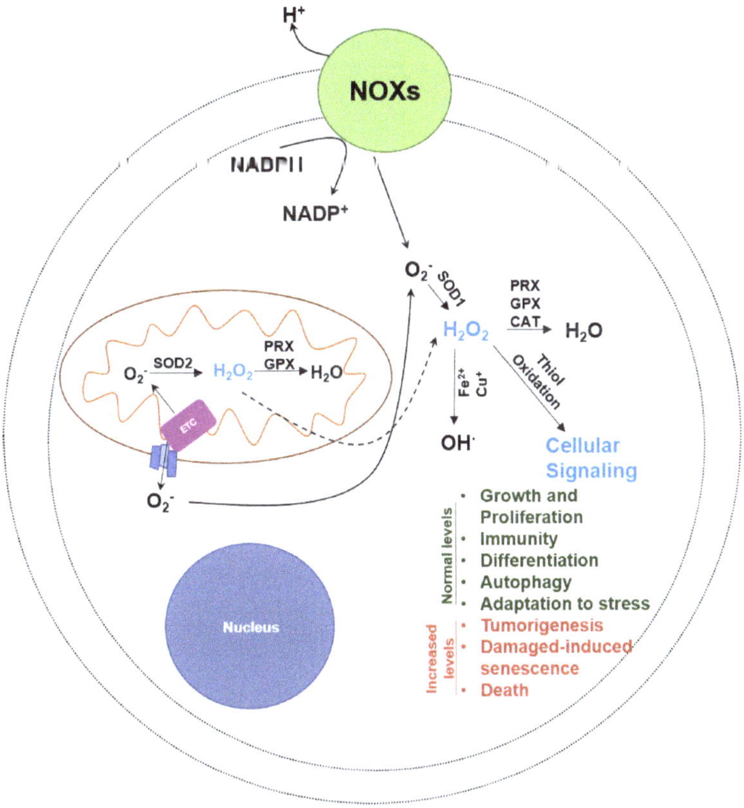

Figure 2. Uncontrolled oxidative stress is cytotoxic to cancer cells. OXPHOS is a major producer of oxygen radicals. While oxygen radicals have positive benefits in cells, a rapid and uncontrolled rise in hydroxyl radicals can lead to cancer cell death. PRX, peroxiredoxin, GPX, glutathione peroxidase, CAT, catalase, SOD, superoxide dismutase, and NOX, NADPH oxidase.

Unpublished results from our group are demonstrating that atovaquone should be investigated for treatment of HGSOC. The naphthoquinone unit of atovaquone engages in redox reactions and interferes with electron transport mediated by ubiquinone [84–86].

There are several naturally occurring and synthetic molecules that contain the quinone, naphthoquinone or anthroquinone head groups. Plumbagin and juglone are examples of such compounds. Treatment of HGSOC cells with plumbagin results in an immediate increase in intracellular oxygen radical flux [87]. Plumbagin also inhibits oxygen consumption rate, decreases ATP synthesis and increases the redox ratio (NADH/FAD) and extracellular acidification rate (ECAR) [87]. These results indicate that plumbagin is likely to be an inhibitor of mitochondrial electron transport.

Molecules that interfere with ubiquinone-mediated electron transport induce severe oxidative stress. There are at three major types of reactive oxygen species (superoxide anion O_2^-, hydrogen peroxide H_2O_2 and hydroxyl radicals $OH^{\cdot}$) that are formed due to incomplete transfer of electrons to molecular oxygen (Figure 2).

4.3. Superoxide Anion (O_2^-)

The primary source of superoxide anions is the electron transport chain in the mitochondria [88,89]. Leakage of electrons travelling through the multiple complexes in the electron transport chain results in one-electron reduction of oxygen to produce the superoxide anions. The second major producer of the superoxide anion are the NADPH oxidases (NOXs) which are transmembrane enzymes present at the different membranes in the cell (Figure 2) [90]. Superoxide anions are restricted in the cellular damage they can cause because they typically only react with peptide epitopes located near the iron sulfur complexes and hence do not cause indiscriminate cellular damage [91].

4.4. Hydrogen Peroxide (H_2O_2)

Reduction of the superoxide anions by superoxide dismutases (SODs) yields hydrogen peroxide (H_2O_2). H_2O_2, existing at nanomolar concentration in the cell, is the main ROS signaling molecule of the cell. It functions by oxidizing the thiolate anion (Cys-S^-) of a cysteine to its sulfenic form (Cys-SO^-). Oxidation of the cysteine thiol affects the formation of Inter- and Intramolecular disulfide bonds and has serious consequences on the biological properties of proteins [92]. This oxidation is reversed by the enzymatic action of thioredoxin (TRX) and glutaredoxin (GRX), which themselves are reduced back by thioredoxin reductase (TR). These set of enzymes essentially constitute the main group of molecules executing the redox signaling in the cell [93]. At abnormally high concentrations of H_2O_2 (as those observed during oxidative stress), the sulfenic form (Cys-SO^-) is irreversibly oxidized to higher oxidized states of sulfinic (Cys-$SO_2{}^-$) and sulfonic (Cys-$SO_3{}^-$), which cannot be repaired through redox control and hence can cause significant cellular damage.

4.5. Hydroxyl Radical ($OH^{\cdot}$)

The hydroxyl radical ($OH^{\cdot}$) is the most reactive of the three ROS molecules described in this section. H_2O_2 reacts with metal cations (Fe^{2+}, Cu^+) present in the cytosol in a reaction called Fenton reaction, to produce $OH^{\cdot}$. Additionally, nitric oxide synthases (NOS) also produce $OH^{\cdot}$ along with $NO_2{}^{\cdot}$ under limiting concentration of cofactors and co-substrates. $OH^{\cdot}$ reacts indiscriminately with various substrates such as lipids, proteins and DNA and leads to genomic instability [94]. Presence of $OH^{\cdot}$ is abnormal and therefore, an indicator of high oxidative stress in the cell.

Oxidative stress induces pleiotropic effects in the cells. These include, but are not restricted to, activation of p53, inhibition of NFκB (Nuclear Factor kappa-B), activation of protein kinases and other signaling molecules, and decrease in the expression of survivin. Molecules such as plumbagin that increase intracellular oxygen radicals, are often thought to mediate pleiotropic effects that culminate in cancer cell death. However, it is important to consider that such molecules may also be specific in their ability to compete with ubiquinone and inhibit electron transport in the mitochondria and can therefore serve as important OXPHOS-targeting agents.

5. Barriers to Using OXPHOS Inhibitors for HGSOC Therapy

5.1. Potential Toxicity of OXPHOS Inhibitors

Our studies with plumbagin clearly show that inhibition of electron transport results in a rapid increase in harmful oxygen radicals that cause significant cellular damage [87]. With OXPHOS serving as a major mechanism for energy generation, there is significant risk that inhibitors of this pathway may damage healthy tissues. Toxicity of the OXPHOS inhibitors is therefore a major concern that may curtail their use for the treatment of HGSOC and other tumors. It should be noted, however, that plumbagin did not produce major toxicity in pre-clinical studies in mouse models [95,96]. Additionally, metformin is generally a safe drug with minimum toxicity. While proguanil is effective inhibitor of the OXPHOS pathway in the malarial and other parasites, its specific activity as OXPHOS inhibitor is reduced in human cells [76,77]. This experience with atovaquone and proguanil suggests that rational drug development approaches can be applied to develop inhibitors that have higher potency against human OXPHOS complexes. Given the higher susceptibility of cancer cells to oxidative stress, well-designed and more potent inhibitors can potentially be used at lower concentrations to produce optimum activity in tumors while reducing toxicity in healthy tissues. Additionally, the OXPHOS inhibitors can also be functionalized with folate and other tumor targeting moieties to facilitate selective delivery of these agents to the tumor, thereby achieving higher efficacy with lower toxicity.

5.2. Anti-Oxidant Mechanisms and Chemoresistance

There are elaborate antioxidant mechanisms to maintain steady state levels of oxygen radicals in all cells. Superoxide dismutase, catalase, peroxiredoxins, glutathione, glutathione reductase, thioredoxins and others form the network of anti-oxidant mechanisms that control oxidative stress. This network is controlled by a master regulatory transcription factor, Nrf-2. Cancer cells respond to inhibition of complex III by atovaquone, (unpublished observation) and plumbagin [87] by increasing the expression of Nrf-2. Therefore, the oxidative stress triggered by these agents is relatively short lived and therefore attenuates their cytolytic activity. The rise in Nrf-2 should therefore be considered as a chemoresistance mechanism to oxidative stress-inducing OXPHOS inhibitors. Combining these OXPHOS inhibitors with Nrf-2 modulators such as brusatol, results in a synergistic increase in inhibition of cancer cell proliferation [87]. Agents that enhance Nrf-2 activity are being developed to control oxidative damage in neurologic diseases. Similar efforts are needed to develop Nrf-2 inhibitors to enhance oxidative stress in HGSOC and other tumors.

Use of Nrf-2 inhibitors for cancer treatment also raises the possibility that such approaches may inhibit the natural protection against oxygen radicals in healthy tissues. Rational drug design, targeted delivery and specific drug formulations will be required to maximize the effect of Nrf-2 inhibitors in cancer cells while attenuating the side-effects of such drugs in healthy tissues.

5.3. Mitochondrial Adaptations to Oxidative Stress

The extensive use of atovaquone has led to the realization that some malarial parasites have developed resistance to this drug through mutations in cytochrome B (Cyt B), an essential component of complex III [97–100]. Cyt B is encoded by the mitochondrial genome. The mitochondrial population with mutated Cyt B is likely to be exposed to minimum oxidative damage in response to atovaquone and will therefore show enrichment through successive mitochondrial replications. A similar situation can also be envisioned in cancer where oxidative stress-inducing OXPHOS inhibitors may result in an increase in the drug-resistant mitochondrial pool. Studies are needed to determine the contributions of such mitochondrial adaptations to chemoresistance against OXPHOS inhibitors.

6. Conclusions

Inhibition of glucose metabolism will result in significantly curtailing the ability of cancer cells to proliferate and modulate the tumor microenvironment through the release of lactic acid and other intermediates. OXPHOS pathway in tumors, cancer stem cells and the stromal and immune cells in the tumor microenvironment is recognized as a target for development of novel anti-cancer therapies. The multimeric complexes of the OXPHOS pathway are targets for small molecule inhibitors that can inhibit metabolism as well as induce oxidative damage and cancer cell death. OXPHOS inhibitors can be paired with immunologic and other therapies. While the development of novel OXPHOS inhibitors will be necessary, it will be important that these agents are specifically targeted to the cancer or the tumor microenvironment in order to reduce toxicity in healthy tissues. Finally, the use of OXPHOS inhibitors may result in Nrf-2 activation and mitochondrial adaptations that may pose as pathways that are typically not considered to contribute towards chemoresistance. Studies on atovaquone, proguanil, metformin have provided the foundation that will support the development of additional and more potent OXPHOS inhibitors for the treatment of HGSOC and other tumors.

Author Contributions: This manuscript was conceptualized by A.P.N. and M.S.P. who also compiled the majority of the text. A.P.N. and A.K. collected the data necessary to interpret the literature and propose the central concept of this review. L.B. participated in discussions and helped write the text and provided her clinical expertise.

Funding: This research was funded by the Department of Obstetrics and Gynecology and funds from the Wisconsin Ovarian Cancer Alliance (WOCA) to M.S.P. and L.B.

Acknowledgments: Support from the University of Wisconsin Carbone Cancer Center Support Grant P30 CA014520 that supports the Flow Lab and the Small Molecule Screening Facility is acknowledged.

Conflicts of Interest: Authors declare no conflict of interest.

References

1. Warburg, O.; Wind, F.; Negelein, E. The metabolism of tumors in the body. *J. Gen. Physiol.* **1927**, *8*, 519–530. [CrossRef] [PubMed]
2. Cori, C.F.; Cori, G.T. The Carbohydrate metabolism of tumors: I. The free sugar, lactic acid, and glycogen content of malignant tumors. *J. Biol. Chem.* **1925**, *64*, 11–22.
3. Warburg, O. On the origin of cancer cells. *Science* **1956**, *123*, 309–314. [CrossRef] [PubMed]
4. Warburg, O. On respiratory impairment in cancer cells. *Science* **1956**, *124*, 269–270. [PubMed]
5. Hatzivassiliou, G.; Zhao, F.; Bauer, D.E.; Andreadis, C.; Shaw, A.N.; Dhanak, D.; Hingorani, S.R.; Tuveson, D.A.; Thompson, C.B. ATP citrate lyase inhibition can suppress tumor cell growth. *Cancer Cell* **2005**, *8*, 311–321. [CrossRef] [PubMed]
6. Menendez, J.A.; Lupu, R. Fatty acid synthase and the lipogenic phenotype in cancer pathogenesis. *Nat. Rev. Cancer* **2007**, *7*, 763–777. [CrossRef] [PubMed]
7. Christofk, H.R.; Vander Heiden, M.G.; Harris, M.H.; Ramanathan, A.; Gerszten, R.E.; Wei, R.; Fleming, M.D.; Schreiber, S.L.; Cantley, L.C. The M2 splice isoform of pyruvate kinase is important for cancer metabolism and tumour growth. *Nature* **2008**, *452*, 230–233. [CrossRef] [PubMed]
8. Christofk, H.R.; Vander Heiden, M.G.; Wu, N.; Asara, J.M.; Cantley, L.C. Pyruvate kinase M2 is a phosphotyrosine-binding protein. *Nature* **2008**, *452*, 181–186. [CrossRef] [PubMed]
9. Boxer, M.B.; Jiang, J.K.; Vander Heiden, M.G.; Shen, M.; Skoumbourdis, A.P.; Southall, N.; Veith, H.; Leister, W.; Austin, C.P.; Park, H.W.; et al. Evaluation of substituted N,N′diarylsulfonamides as activators of the tumor cell specific M2 isoform of pyruvate kinase. *J. Med. Chem.* **2010**, *53*, 1048–1055. [CrossRef] [PubMed]
10. Anastasiou, D.; Yu, Y.; Israelsen, W.J.; Jiang, J.K.; Boxer, M.B.; Hong, B.S.; Tempel, W.; Dimov, S.; Shen, M.; Jha, A.; et al. Pyruvate kinase M2 activators promote tetramer formation and suppress tumorigenesis. *Nat. Chem. Biol.* **2012**, *8*, 839–847. [CrossRef] [PubMed]
11. Ning, X.; Qi, H.; Li, R.; Jin, Y.; McNutt, M.A.; Yin, Y. Synthesis and antitumor activity of novel 2, 3-didithiocarbamate substituted naphthoquinones as inhibitors of pyruvate kinase M2 isoform. *J. Enzyme Inhib. Med. Chem.* **2018**, *33*, 126–129. [CrossRef] [PubMed]
12. Sugden, M.C.; Holness, M.J. Therapeutic potential of the mammalian pyruvate dehydrogenase kinases in the prevention of hyperglycaemia. *Curr. Drug Targets Immune Endocr. Metabol. Disord.* **2002**, *2*, 151–165. [CrossRef] [PubMed]
13. Zhang, W.; Zhang, S.L.; Hu, X.; Tam, K.Y. Phenyl butyrate inhibits pyruvate dehydrogenase kinase 1 and contributes to its anti-cancer effect. *Eur. J. Pharm. Sci.* **2017**, *110*, 93–100. [CrossRef] [PubMed]
14. Zhang, S.L.; Hu, X.; Zhang, W.; Yao, H.; Tam, K.Y. Development of pyruvate dehydrogenase kinase inhibitors in medicinal chemistry with particular emphasis as anticancer agents. *Drug Discov. Today* **2015**, *20*, 1112–1119. [CrossRef] [PubMed]
15. Abdel-Magid, A.F. Inhibitors of Mutant IDH for the Treatment of Cancer. *ACS Med. Chem. Lett.* **2014**, *5*, 1184–1185. [CrossRef] [PubMed]
16. Wang, H.; Huwaimel, B.; Verma, K.; Miller, J.; Germain, T.M.; Kinarivala, N.; Pappas, D.; Brookes, P.S.; Trippier, P.C. Synthesis and Antineoplastic Evaluation of Mitochondrial Complex II (Succinate Dehydrogenase) Inhibitors Derived from Atpenin A5. *ChemMedChem* **2017**, *12*, 1033–1044. [CrossRef] [PubMed]
17. Rai, G.; Brimacombe, K.R.; Mott, B.T.; Urban, D.J.; Hu, X.; Yang, S.M.; Lee, T.D.; Cheff, D.M.; Kouznetsova, J.; Benavides, G.A.; et al. Discovery and Optimization of Potent, Cell-Active Pyrazole-Based Inhibitors of Lactate Dehydrogenase (LDH). *J. Med. Chem.* **2017**, *60*, 9184–9204. [CrossRef] [PubMed]
18. Pasto, A.; Bellio, C.; Pilotto, G.; Ciminale, V.; Silic-Benussi, M.; Guzzo, G.; Rasola, A.; Frasson, C.; Nardo, G.; Zulato, E.; et al. Cancer stem cells from epithelial ovarian cancer patients privilege oxidative phosphorylation, and resist glucose deprivation. *Oncotarget* **2014**, *5*, 4305–4319. [CrossRef] [PubMed]

19. De Luca, A.; Fiorillo, M.; Peiris-Pages, M.; Ozsvari, B.; Smith, D.L.; Sanchez-Alvarez, R.; Martinez-Outschoorn, U.E.; Cappello, A.R.; Pezzi, V.; Lisanti, M.P.; et al. Mitochondrial biogenesis is required for the anchorage-independent survival and propagation of stem-like cancer cells. *Oncotarget* **2015**, *6*, 14777–14795. [CrossRef] [PubMed]
20. Lamb, R.; Bonuccelli, G.; Ozsvari, B.; Peiris-Pages, M.; Fiorillo, M.; Smith, D.L.; Bevilacqua, G.; Mazzanti, C.M.; McDonnell, L.A.; Naccarato, A.G.; et al. Mitochondrial mass, a new metabolic biomarker for stem-like cancer cells: Understanding WNT/FGF-driven anabolic signaling. *Oncotarget* **2015**, *6*, 30453–30471. [CrossRef] [PubMed]
21. Kurman, R.J.; Shih Ie, M. Molecular pathogenesis and extraovarian origin of epithelial ovarian cancer–shifting the paradigm. *Hum. Pathol.* **2011**, *42*, 918–931. [CrossRef] [PubMed]
22. Kurman, R.J.; Shih Ie, M. The origin and pathogenesis of epithelial ovarian cancer: A proposed unifying theory. *Am. J. Surg. Pathol.* **2010**, *34*, 433–443. [CrossRef] [PubMed]
23. Kurman, R.J.; Shih Ie, M. Pathogenesis of ovarian cancer: lessons from morphology and molecular biology and their clinical implications. *Int. J. Gynecol. Pathol.* **2008**, *27*, 151–160. [CrossRef] [PubMed]
24. Network, C.G.A.R. Integrated genomic analyses of ovarian carcinoma. *Nature* **2011**, *474*, 609–615. [CrossRef]
25. Kuhn, E.; Kurman, R.J.; Vang, R.; Sehdev, A.S.; Han, G.; Soslow, R.; Wang, T.L.; Shih Ie, M. TP53 mutations in serous tubal intraepithelial carcinoma and concurrent pelvic high-grade serous carcinoma-evidence supporting the clonal relationship of the two lesions. *J. Pathol.* **2012**, *226*, 421–426. [CrossRef] [PubMed]
26. Vang, R.; Levine, D.A.; Soslow, R.A.; Zaloudek, C.; Shih Ie, M.; Kurman, R.J. Molecular Alterations of TP53 are a Defining Feature of Ovarian High-Grade Serous Carcinoma: A Rereview of Cases Lacking TP53 Mutations in The Cancer Genome Atlas Ovarian Study. *Int. J. Gynecol. Pathol.* **2016**, *35*, 48–55. [CrossRef] [PubMed]
27. Kindelberger, D.W.; Lee, Y.; Miron, A.; Hirsch, M.S.; Feltmate, C.; Medeiros, F.; Callahan, M.J.; Garner, E.O.; Gordon, R.W.; Birch, C.; et al. Intraepithelial carcinoma of the fimbria and pelvic serous carcinoma: Evidence for a causal relationship. *Am. J. Surg. Pathol.* **2007**, *31*, 161–169. [CrossRef] [PubMed]
28. Perets, R.; Wyant, G.A.; Muto, K.W.; Bijron, J.G.; Poole, B.B.; Chin, K.T.; Chen, J.Y.; Ohman, A.W.; Stepule, C.D.; Kwak, S.; et al. Transformation of the fallopian tube secretory epithelium leads to high-grade serous ovarian cancer in Brca;Tp53;Pten models. *Cancer Cell* **2013**, *24*, 751–765. [CrossRef] [PubMed]
29. Przybycin, C.G.; Kurman, R.J.; Ronnett, B.M.; Shih Ie, M.; Vang, R. Are all pelvic (nonuterine) serous carcinomas of tubal origin? *Am. J. Surg. Pathol.* **2010**, *34*, 1407–1416. [CrossRef] [PubMed]
30. Matulonis, U.A.; Harter, P.; Gourley, C.; Friedlander, M.; Vergote, I.; Rustin, G.; Scott, C.; Meier, W.; Shapira-Frommer, R.; Safra, T.; et al. Olaparib maintenance therapy in patients with platinum-sensitive, relapsed serous ovarian cancer and a BRCA mutation: Overall survival adjusted for postprogression poly(adenosine diphosphate ribose) polymerase inhibitor therapy. *Cancer* **2016**, *122*, 1844–1852. [CrossRef] [PubMed]
31. Ledermann, J.A.; Harter, P.; Gourley, C.; Friedlander, M.; Vergote, I.; Rustin, G.; Scott, C.; Meier, W.; Shapira-Frommer, R.; Safra, T.; et al. Overall survival in patients with platinum-sensitive recurrent serous ovarian cancer receiving olaparib maintenance monotherapy: An updated analysis from a randomised, placebo-controlled, double-blind, phase 2 trial. *Lancet Oncol.* **2016**, *17*, 1579–1589. [CrossRef]
32. Ledermann, J.; Harter, P.; Gourley, C.; Friedlander, M.; Vergote, I.; Rustin, G.; Scott, C.L.; Meier, W.; Shapira-Frommer, R.; Safra, T.; et al. Olaparib maintenance therapy in patients with platinum-sensitive relapsed serous ovarian cancer: A preplanned retrospective analysis of outcomes by BRCA status in a randomised phase 2 trial. *Lancet Oncol.* **2014**, *15*, 852–861. [CrossRef]
33. Ledermann, J.; Harter, P.; Gourley, C.; Friedlander, M.; Vergote, I.; Rustin, G.; Scott, C.; Meier, W.; Shapira-Frommer, R.; Safra, T.; et al. Olaparib maintenance therapy in platinum-sensitive relapsed ovarian cancer. *N. Engl. J. Med.* **2012**, *366*, 1382–1392. [CrossRef] [PubMed]
34. Dar, S.; Chhina, J.; Mert, I.; Chitale, D.; Buekers, T.; Kaur, H.; Giri, S.; Munkarah, A.; Rattan, R. Bioenergetic Adaptations in Chemoresistant Ovarian Cancer Cells. *Sci. Rep.* **2017**, *7*, 8760. [CrossRef] [PubMed]
35. Anderson, A.S.; Roberts, P.C.; Frisard, M.I.; McMillan, R.P.; Brown, T.J.; Lawless, M.H.; Hulver, M.W.; Schmelz, E.M. Metabolic changes during ovarian cancer progression as targets for sphingosine treatment. *Exp. Cell Res.* **2013**, *319*, 1431–1442. [CrossRef] [PubMed]

36. Dier, U.; Shin, D.H.; Hemachandra, L.P.; Uusitalo, L.M.; Hempel, N. Bioenergetic analysis of ovarian cancer cell lines: Profiling of histological subtypes and identification of a mitochondria-defective cell line. *PLoS ONE* **2014**, *9*, e98479. [CrossRef] [PubMed]
37. Chao, T.K.; Huang, T.S.; Liao, Y.P.; Huang, R.L.; Su, P.H.; Shen, H.Y.; Lai, H.C.; Wang, Y.C. Pyruvate kinase M2 is a poor prognostic marker of and a therapeutic target in ovarian cancer. *PLoS ONE* **2017**, *12*, e0182166. [CrossRef] [PubMed]
38. Jin, Z.; Gu, J.; Xin, X.; Li, Y.; Wang, H. Expression of hexokinase 2 in epithelial ovarian tumors and its clinical significance in serous ovarian cancer. *Eur. J. Gynaecol. Oncol.* **2014**, *35*, 519–524. [PubMed]
39. Xintaropoulou, C.; Ward, C.; Wise, A.; Queckborner, S.; Turnbull, A.; Michie, C.O.; Williams, A.R.W.; Rye, T.; Gourley, C.; Langdon, S.P. Expression of glycolytic enzymes in ovarian cancers and evaluation of the glycolytic pathway as a strategy for ovarian cancer treatment. *BMC Cancer* **2018**, *18*, 636. [CrossRef] [PubMed]
40. Suh, D.H.; Kim, M.A.; Kim, H.; Kim, M.K.; Kim, H.S.; Chung, H.H.; Kim, Y.B.; Song, Y.S. Association of overexpression of hexokinase II with chemoresistance in epithelial ovarian cancer. *Clin. Exp. Med.* **2014**, *14*, 345–353. [CrossRef] [PubMed]
41. Tsukioka, M.; Matsumoto, Y.; Noriyuki, M.; Yoshida, C.; Nobeyama, H.; Yoshida, H.; Yasui, T.; Sumi, T.; Honda, K.; Ishiko, O. Expression of glucose transporters in epithelial ovarian carcinoma: Correlation with clinical characteristics and tumor angiogenesis. *Oncol. Rep.* **2007**, *18*, 361–367. [CrossRef] [PubMed]
42. Kalir, T.; Wang, B.Y.; Goldfischer, M.; Haber, R.S.; Reder, I.; Demopoulos, R.; Cohen, C.J.; Burstein, D.E. Immunohistochemical staining of GLUT1 in benign, borderline, and malignant ovarian epithelia. *Cancer* **2002**, *94*, 1078–1082. [CrossRef] [PubMed]
43. Cho, H.; Lee, Y.S.; Kim, J.; Chung, J.Y.; Kim, J.H. Overexpression of glucose transporter-1 (GLUT-1) predicts poor prognosis in epithelial ovarian cancer. *Cancer Investig.* **2013**, *31*, 607–615. [CrossRef] [PubMed]
44. Kurokawa, T.; Yoshida, Y.; Kawahara, K.; Tsuchida, T.; Okazawa, H.; Fujibayashi, Y.; Yonekura, Y.; Kotsuji, F. Expression of GLUT-1 glucose transfer, cellular proliferation activity and grade of tumor correlate with [F-18]-fluorodeoxyglucose uptake by positron emission tomography in epithelial tumors of the ovary. *Int. J. Cancer* **2004**, *109*, 926–932. [CrossRef] [PubMed]
45. Hitosugi, T.; Kang, S.; Vander Heiden, M.G.; Chung, T.-W.; Elf, S.; Lythgoe, K.; Dong, S.; Lonial, S.; Wang, X.; Chen, G.Z.; et al. Tyrosine Phosphorylation Inhibits PKM2 to Promote the Warburg Effect and Tumor Growth. *Sci. Signal.* **2009**, *2*, ra73. [CrossRef] [PubMed]
46. Privat, M.; Radosevic-Robin, N.; Aubel, C.; Cayre, A.; Penault-Llorca, F.; Marceau, G.; Sapin, V.; Bignon, Y.J.; Morvan, D. BRCA1 induces major energetic metabolism reprogramming in breast cancer cells. *PLoS ONE* **2014**, *9*, e102438. [CrossRef] [PubMed]
47. Simabuco, F.M.; Morale, M.G.; Pavan, I.C.B.; Morelli, A.P.; Silva, F.R.; Tamura, R.E. p53 and metabolism: From mechanism to therapeutics. *Oncotarget* **2018**, *9*, 23780–23823. [CrossRef] [PubMed]
48. Lai, J.; Chien, J.; Staub, J.; Avula, R.; Greene, E.L.; Matthews, T.A.; Smith, D.I.; Kaufmann, S.H.; Roberts, L.R.; Shridhar, V. Loss of HSulf-1 up-regulates heparin-binding growth factor signaling in cancer. *J. Biol. Chem.* **2003**, *278*, 23107–23117. [CrossRef] [PubMed]
49. Mondal, S.; Roy, D.; Camacho-Pereira, J.; Khurana, A.; Chini, E.; Yang, L.; Baddour, J.; Stilles, K.; Padmabandu, S.; Leung, S.; et al. HSulf-1 deficiency dictates a metabolic reprograming of glycolysis and TCA cycle in ovarian cancer. *Oncotarget* **2015**, *6*, 33705–33719. [CrossRef] [PubMed]
50. Hayashi, J.; Takemitsu, M.; Nonaka, I. Recovery of the missing tumorigenicity in mitochondrial DNA-less HeLa cells by introduction of mitochondrial DNA from normal human cells. *Somat. Cell Mol. Genet.* **1992**, *18*, 123–129. [CrossRef] [PubMed]
51. Cavalli, L.R.; Varella-Garcia, M.; Liang, B.C. Diminished tumorigenic phenotype after depletion of mitochondrial DNA. *Cell Growth Differ.* **1997**, *8*, 1189–1198. [PubMed]
52. Morais, R.; Zinkewich-Peotti, K.; Parent, M.; Wang, H.; Babai, F.; Zollinger, M. Tumor-forming ability in athymic nude mice of human cell lines devoid of mitochondrial DNA. *Cancer Res.* **1994**, *54*, 3889–3896. [PubMed]
53. Lim, H.Y.; Ho, Q.S.; Low, J.; Choolani, M.; Wong, K.P. Respiratory competent mitochondria in human ovarian and peritoneal cancer. *Mitochondrion* **2011**, *11*, 437–443. [CrossRef] [PubMed]

54. Tan, A.S.; Baty, J.W.; Dong, L.F.; Bezawork-Geleta, A.; Endaya, B.; Goodwin, J.; Bajzikova, M.; Kovarova, J.; Peterka, M.; Yan, B.; et al. Mitochondrial genome acquisition restores respiratory function and tumorigenic potential of cancer cells without mitochondrial DNA. *Cell Metab.* **2015**, *21*, 81–94. [CrossRef] [PubMed]
55. Viale, A.; Pettazzoni, P.; Lyssiotis, C.A.; Ying, H.; Sanchez, N.; Marchesini, M.; Carugo, A.; Green, T.; Seth, S.; Giuliani, V.; et al. Oncogene ablation-resistant pancreatic cancer cells depend on mitochondrial function. *Nature* **2014**, *514*, 628–632. [CrossRef] [PubMed]
56. Viale, A.; Corti, D.; Draetta, G.F. Tumors and mitochondrial respiration: A neglected connection. *Cancer Res.* **2015**, *75*, 3685–3686. [CrossRef] [PubMed]
57. Ying, H.; Kimmelman, A.C.; Lyssiotis, C.A.; Hua, S.; Chu, G.C.; Fletcher-Sananikone, E.; Locasale, J.W.; Son, J.; Zhang, H.; Coloff, J.L.; et al. Oncogenic Kras maintains pancreatic tumors through regulation of anabolic glucose metabolism. *Cell* **2012**, *149*, 656–670. [CrossRef] [PubMed]
58. Yu, L.; Lu, M.; Jia, D.; Ma, J.; Ben-Jacob, E.; Levine, H.; Kaipparettu, B.A.; Onuchic, J.N. Modeling the Genetic Regulation of Cancer Metabolism: Interplay between Glycolysis and Oxidative Phosphorylation. *Cancer Res.* **2017**, *77*, 1564–1574. [CrossRef] [PubMed]
59. Cuezva, J.M.; Krajewska, M.; de Heredia, M.L.; Krajewski, S.; Santamaria, G.; Kim, H.; Zapata, J.M.; Marusawa, H.; Chamorro, M.; Reed, J.C. The bioenergetic signature of cancer: A marker of tumor progression. *Cancer Res.* **2002**, *62*, 6674–6681. [PubMed]
60. Hjerpe, E.; Egyhazi Brage, S.; Carlson, J.; Frostvik Stolt, M.; Schedvins, K.; Johansson, H.; Shoshan, M.; Avall-Lundqvist, E. Metabolic markers GAPDH, PKM2, ATP5B and BEC-index in advanced serous ovarian cancer. *BMC Clin. Pathol.* **2013**, *13*, 30. [CrossRef] [PubMed]
61. Anderson, A.S.; Roberts, P.C.; Frisard, M.I.; Hulver, M.W.; Schmelz, E.M. Ovarian tumor-initiating cells display a flexible metabolism. *Exp. Cell Res.* **2014**, *328*, 44–57. [CrossRef] [PubMed]
62. Roberts, P.C.; Mottillo, E.P.; Baxa, A.C.; Heng, H.H.; Doyon-Reale, N.; Gregoire, L.; Lancaster, W.D.; Rabah, R.; Schmelz, E.M. Sequential molecular and cellular events during neoplastic progression: A mouse syngeneic ovarian cancer model. *Neoplasia* **2005**, *7*, 944–956. [CrossRef] [PubMed]
63. Song, H.Y.; Dunbar, J.D.; Zhang, Y.X.; Guo, D.; Donner, D.B. Identification of a protein with homology to hsp90 that binds the type 1 tumor necrosis factor receptor. *J. Biol. Chem.* **1995**, *270*, 3574–3581. [CrossRef] [PubMed]
64. Matassa, D.S.; Amoroso, M.R.; Lu, H.; Avolio, R.; Arzeni, D.; Procaccini, C.; Faicchia, D.; Maddalena, F.; Simeon, V.; Agliarulo, I.; et al. Oxidative metabolism drives inflammation-induced platinum resistance in human ovarian cancer. *Cell Death Differ.* **2016**, *23*, 1542–1554. [CrossRef] [PubMed]
65. Matassa, D.S.; Agliarulo, I.; Avolio, R.; Landriscina, M.; Esposito, F. TRAP1 regulation of cancer metabolism: Dual role as oncogene or tumor suppressor. *Genes* **2018**, *9*, 195. [CrossRef] [PubMed]
66. Morrot, A.; da Fonseca, L.M.; Salustiano, E.J.; Gentile, L.B.; Conde, L.; Filardy, A.A.; Franklim, T.N.; da Costa, K.M.; Freire-de-Lima, C.G.; Freire-de-Lima, L. Metabolic Symbiosis and Immunomodulation: How Tumor Cell-Derived Lactate May Disturb Innate and Adaptive Immune Responses. *Front. Oncol.* **2018**, *8*, 81. [CrossRef] [PubMed]
67. Martinez-Outschoorn, U.E.; Curry, J.M.; Ko, Y.H.; Lin, Z.; Tuluc, M.; Cognetti, D.; Birbe, R.C.; Pribitkin, E.; Bombonati, A.; Pestell, R.G.; et al. Oncogenes and inflammation rewire host energy metabolism in the tumor microenvironment: RAS and NFkappaB target stromal MCT4. *Cell Cycle* **2013**, *12*, 2580–2597. [CrossRef] [PubMed]
68. Whitaker-Menezes, D.; Martinez-Outschoorn, U.E.; Lin, Z.; Ertel, A.; Flomenberg, N.; Witkiewicz, A.K.; Birbe, R.C.; Howell, A.; Pavlides, S.; Gandara, R.; et al. Evidence for a stromal-epithelial "lactate shuttle" in human tumors: MCT4 is a marker of oxidative stress in cancer-associated fibroblasts. *Cell Cycle* **2011**, *10*, 1772–1783. [CrossRef] [PubMed]
69. Witkiewicz, A.K.; Dasgupta, A.; Nguyen, K.H.; Liu, C.; Kovatich, A.J.; Schwartz, G.F.; Pestell, R.G.; Sotgia, F.; Rui, H.; Lisanti, M.P. Stromal caveolin-1 levels predict early DCIS progression to invasive breast cancer. *Cancer Biol. Ther.* **2009**, *8*, 1071–1079. [CrossRef] [PubMed]
70. Sotgia, F.; Martinez-Outschoorn, U.E.; Lisanti, M.P. Cancer metabolism: New validated targets for drug discovery. *Oncotarget* **2013**, *4*, 1309–1316. [CrossRef] [PubMed]

71. Monti, D.; Sotgia, F.; Whitaker-Menezes, D.; Tuluc, M.; Birbe, R.; Berger, A.; Lazar, M.; Cotzia, P.; Draganova-Tacheva, R.; Lin, Z.; et al. Pilot study demonstrating metabolic and anti-proliferative effects of in vivo anti-oxidant supplementation with N-Acetylcysteine in Breast Cancer. *Semin. Oncol.* **2017**, *44*, 226–232. [CrossRef] [PubMed]
72. Constance, J.E.; Lim, C.S. Targeting malignant mitochondria with therapeutic peptides. *Ther. Deliv.* **2012**, *3*, 961–979. [CrossRef] [PubMed]
73. Bodmer, M.; Becker, C.; Meier, C.; Jick, S.S.; Meier, C.R. Use of metformin and the risk of ovarian cancer: A case-control analysis. *Gynecol. Oncol.* **2011**, *123*, 200–204. [CrossRef] [PubMed]
74. Kumar, S.; Meuter, A.; Thapa, P.; Langstraat, C.; Giri, S.; Chien, J.; Rattan, R.; Cliby, W.; Shridhar, V. Metformin intake is associated with better survival in ovarian cancer: A case-control study. *Cancer* **2013**, *119*, 555–562. [CrossRef] [PubMed]
75. Romero, I.L.; McCormick, A.; McEwen, K.A.; Park, S.; Karrison, T.; Yamada, S.D.; Pannain, S.; Lengyel, E. Relationship of type II diabetes and metformin use to ovarian cancer progression, survival, and chemosensitivity. *Obstet. Gynecol.* **2012**, *119*, 61–67. [CrossRef] [PubMed]
76. Bakhru, A.; Buckanovich, R.J.; Griggs, J.J. The impact of diabetes on survival in women with ovarian cancer. *Gynecol. Oncol.* **2011**, *121*, 106–111. [CrossRef] [PubMed]
77. Bridges, H.R.; Sirvio, V.A.; Agip, A.N.; Hirst, J. Molecular features of biguanides required for targeting of mitochondrial respiratory complex I and activation of AMP-kinase. *BMC Biol.* **2016**, *14*, 65. [CrossRef] [PubMed]
78. Bridges, H.R.; Jones, A.J.; Pollak, M.N.; Hirst, J. Effects of metformin and other biguanides on oxidative phosphorylation in mitochondria. *Biochem. J.* **2014**, *462*, 475–487. [CrossRef] [PubMed]
79. Baran, N.; Han, L.; Herbrich, S.; Sweeney, S.R.; Lodi, A.; Gay, J.; Feng, N.; Molina, J.R.; Kaminski, M.; Guzman, M.L.; et al. Novel Complex I inhibitor IACS-010759 Targets Leukemia Initiating Cells (LICs) in AML Patients. *Clin. Lymphoma Myeloma Leuk.* **2017**, *17*, S297. [CrossRef]
80. Baran, N.; Molina, J.; Cavazos, A.; Harutyunyan, K.; Feng, N.; Gay, J.; Piya, S.; Shanmuga Velandy, S.; Jabbour, E.J.; Andreeff, M.; et al. Mitochondrial Complex I Inhibition with Iacs-010759 in T-ALL Preclinical Models. *Blood* **2016**, *128*, 4028.
81. Lissanu Deribe, Y.; Sun, Y.; Terranova, C.; Khan, F.; Martinez-Ledesma, J.; Gay, J.; Gao, G.; Mullinax, R.A.; Khor, T.; Feng, N.; et al. Mutations in the SWI/SNF complex induce a targetable dependence on oxidative phosphorylation in lung cancer. *Nat. Med.* **2018**, *24*, 1047–1057. [CrossRef] [PubMed]
82. Molina, J.R.; Sun, Y.; Protopopova, M.; Gera, S.; Bandi, M.; Bristow, C.; McAfoos, T.; Morlacchi, P.; Ackroyd, J.; Agip, A.A.; et al. An inhibitor of oxidative phosphorylation exploits cancer vulnerability. *Nat. Med.* **2018**, *24*, 1036–1046. [CrossRef] [PubMed]
83. Vangapandu, H.V.; Alston, B.; Morse, J.; Ayres, M.L.; Wierda, W.G.; Keating, M.J.; Marszalek, J.R.; Gandhi, V. Biological and metabolic effects of IACS-010759, an OxPhos inhibitor, on chronic lymphocytic leukemia cells. *Oncotarget* **2018**, *9*, 24980–24991. [CrossRef] [PubMed]
84. Sies, H. *Oxidative Stress: Introductory Remarks*; Academic Press: London, UK, 1985.
85. Lv, Z.; Yan, X.; Lu, L.; Su, C.; He, Y. Atovaquone enhances doxorubicin's efficacy via inhibiting mitochondrial respiration and STAT3 in aggressive thyroid cancer. *J. Bioenerg. Biomembr.* **2018**, *50*, 263–270. [CrossRef] [PubMed]
86. Fiorillo, M.; Lamb, R.; Tanowitz, H.B.; Mutti, L.; Krstic-Demonacos, M.; Cappello, A.R.; Martinez-Outschoorn, U.E.; Sotgia, F.; Lisanti, M.P. Repurposing atovaquone: Targeting mitochondrial complex III and OXPHOS to eradicate cancer stem cells. *Oncotarget* **2016**, *7*, 34084–34099. [CrossRef] [PubMed]
87. Ashton, T.M.; Fokas, E.; Kunz-Schughart, L.A.; Folkes, L.K.; Anbalagan, S.; Huether, M.; Kelly, C.J.; Pirovano, G.; Buffa, F.M.; Hammond, E.M.; et al. The anti-malarial atovaquone increases radiosensitivity by alleviating tumour hypoxia. *Nat. Commun.* **2016**, *7*, 12308. [CrossRef] [PubMed]
88. Kapur, A.; Beres, T.; Rathi, K.; Nayak, A.P.; Czarnecki, A.; Felder, M.; Gillette, A.; Ericksen, S.S.; Sampene, E.; Skala, M.C.; et al. Oxidative stress via inhibition of the mitochondrial electron transport and Nrf-2-mediated anti-oxidative response regulate the cytotoxic activity of plumbagin. *Sci. Rep.* **2018**, *8*, 1073. [CrossRef] [PubMed]

89. Goncalves, R.L.; Quinlan, C.L.; Perevoshchikova, I.V.; Hey-Mogensen, M.; Brand, M.D. Sites of superoxide and hydrogen peroxide production by muscle mitochondria assessed ex vivo under conditions mimicking rest and exercise. *J. Biol. Chem.* **2015**, *290*, 209–227. [CrossRef] [PubMed]
90. Murphy, M.P. How mitochondria produce reactive oxygen species. *Biochem. J.* **2009**, *417*, 1–13. [CrossRef] [PubMed]
91. Bedard, K.; Krause, K.H. The NOX family of ROS-generating NADPH oxidases: Physiology and pathophysiology. *Physiol. Rev.* **2007**, *87*, 245–313. [CrossRef] [PubMed]
92. Chen, Y.; Azad, M.B.; Gibson, S.B. Superoxide is the major reactive oxygen species regulating autophagy. *Cell Death Differ.* **2009**, *16*, 1040–1052. [CrossRef] [PubMed]
93. Rhee, S.G. Cell signaling. H_2O_2, a necessary evil for cell signaling. *Science* **2006**, *312*, 1882–1883. [CrossRef] [PubMed]
94. Winterbourn, C.C.; Hampton, M.B. Thiol chemistry and specificity in redox signaling. *Free Radic. Biol. Med.* **2008**, *45*, 549–561. [CrossRef] [PubMed]
95. Dizdaroglu, M.; Jaruga, P. Mechanisms of free radical-induced damage to DNA. *Free Radic. Res.* **2012**, *46*, 382–419. [CrossRef] [PubMed]
96. Hafeez, B.B.; Zhong, W.; Mustafa, A.; Fischer, J.W.; Witkowsky, O.; Verma, A.K. Plumbagin inhibits prostate cancer development in TRAMP mice via targeting PKCepsilon, Stat3 and neuroendocrine markers. *Carcinogenesis* **2012**, *33*, 2586–2592. [CrossRef] [PubMed]
97. Hafeez, B.B.; Fischer, J.W.; Singh, A.; Zhong, W.; Mustafa, A.; Meske, L.; Sheikhani, M.O.; Verma, A.K. Plumbagin Inhibits Prostate Carcinogenesis in Intact and Castrated PTEN Knockout Mice via Targeting PKCepsilon, Stat3, and Epithelial-to-Mesenchymal Transition Markers. *Cancer Prev. Res.* **2015**, *8*, 375–386. [CrossRef] [PubMed]
98. Srivastava, I.K.; Morrisey, J.M.; Darrouzet, E.; Daldal, F.; Vaidya, A.B. Resistance mutations reveal the atovaquone-binding domain of cytochrome b in malaria parasites. *Mol. Microbiol.* **1999**, *33*, 704–711. [CrossRef] [PubMed]
99. Plucinski, M.M.; Huber, C.S.; Akinyi, S.; Dalton, W.; Eschete, M.; Grady, K.; Silva-Flannery, L.; Mathison, B.A.; Udhayakumar, V.; Arguin, P.M.; et al. Novel Mutation in Cytochrome B of Plasmodium falciparum in One of Two Atovaquone-Proguanil Treatment Failures in Travelers Returning From Same Site in Nigeria. *Open Forum Infect. Dis.* **2014**, *1*, ofu059. [CrossRef] [PubMed]
100. Gil, J.P.; Nogueira, F.; Stromberg-Norklit, J.; Lindberg, J.; Carrolo, M.; Casimiro, C.; Lopes, D.; Arez, A.P.; Cravo, P.V.; Rosario, V.E. Detection of atovaquone and Malarone resistance conferring mutations in Plasmodium falciparum cytochrome b gene (cytb). *Mol. Cell. Probes* **2003**, *17*, 85–89. [CrossRef]

cancers

MDPI

Review

Organotypic 3D Models of the Ovarian Cancer Tumor Microenvironment

Karen M. Watters [1], Preety Bajwa [2] and Hilary A. Kenny [2,*]

[1] Department of Chemistry, University of Chicago, Chicago, IL 60637, USA; kmwatters@uchicago.edu
[2] Department of Obstetrics and Gynecology, Section of Gynecologic Oncology, University of Chicago, Chicago, IL 60637, USA; preetybjw25@gmail.com
* Correspondence: hkenny@bsd.uchicago.edu

Received: 19 June 2018; Accepted: 8 August 2018; Published: 9 August 2018

Abstract: Ovarian cancer progression involves multifaceted and variable tumor microenvironments (TMEs), from the in situ carcinoma in the fallopian tube or ovary to dissemination into the peritoneal cavity as single cells or spheroids and attachment to the mesothelial-lined surfaces of the omentum, bowel, and abdominal wall. The TME comprises the tumor vasculature and lymphatics (including endothelial cells and pericytes), in addition to mesothelial cells, fibroblasts, immune cells, adipocytes and extracellular matrix (ECM) proteins. When generating 3D models of the ovarian cancer TME, researchers must incorporate the most relevant stromal components depending on the TME in question (e.g., early or late disease). Such complexity cannot be captured by monolayer 2D culture systems. Moreover, immortalized stromal cell lines, such as mesothelial or fibroblast cell lines, do not always behave the same as primary cells whose response in functional assays may vary from donor to donor; 3D models with primary stromal cells may have more physiological relevance than those using stromal cell lines. In the current review, we discuss the latest developments in organotypic 3D models of the ovarian cancer early metastatic microenvironment. Organotypic culture models comprise two or more interacting cell types from a particular tissue. We focus on organotypic 3D models that include at least one type of primary stromal cell type in an ECM background, such as collagen or fibronectin, plus ovarian cancer cells. We provide an overview of the two most comprehensive current models—a 3D model of the omental mesothelium and a microfluidic model. We describe the cellular and non-cellular components of the models, the incorporation of mechanical forces, and how the models have been adapted and utilized in functional assays. Finally, we review a number of 3D models that do not incorporate primary stromal cells and summarize how integration of current models may be the next essential step in tackling the complexity of the different ovarian cancer TMEs.

Keywords: ovarian cancer; tumor microenvironment; 3D models

---

1. Introduction

From tumor initiation to metastasis, intricate and reciprocal interactions between ovarian cancer cells and the stromal components of their surrounding milieu create complex and fluctuating tumor microenvironments (TMEs) [1,2]. Stromal components of the TME include the tumor vasculature and lymphatics (including endothelial cells and pericytes), mesothelial cells, fibroblasts, immune cells, and extracellular matrix (ECM) proteins. Mechanical forces such as sheer stress caused by increased peritoneal fluid flow also contribute to this environment, inducing changes in cell morphology and gene expression [3,4]. All of these elements are associated with specific facets of tumorigenesis and metastasis, and their involvement cannot be accurately captured by traditional 2D cell culture systems. Cancer cell cultures in 3D microenvironments are far more representative of disease than traditional 2D systems. Three-dimensional systems provide 1) conditions which are structurally similar to the in vivo environment and are amenable to changes in oxygen and growth factor gradients (e.g., cell spheroids) [5] and 2) cell–cell and cell–ECM communication (e.g., scaffold-based models) [6,7].

Ovarian cancer progression involves detachment of cancer cells from the in situ carcinoma in the fallopian tube or the primary ovarian tumor, dissemination into the peritoneal cavity as single cells or spheroids, and attachment to the mesothelial-lined surfaces of the omentum, bowel, and abdominal wall [8,9] (Figure 1). Ovarian cancer complexity and heterogeneity has meant that development of in vitro 3D ovarian cancer TME models to recapitulate in vivo pathophysiological features has been challenging. Our group has previously published comprehensive reviews of the different 3D culture methods used to study the ovarian cancer TME [10–12]. In the current review, we focus on the latest organotypic 3D models that utilize primary stromal cells, in particular a 3D model of the omental mesothelium and a microfluidic model. We provide an overview of these models, both of which are used to study the early steps of ovarian cancer metastasis, describing the cellular and non-cellular components, the consideration of mechanical forces, and their utilization. We discuss the challenges and limitations associated with the current models and put forward the essential steps to establish an archetype model that will faithfully recreate the in vivo scenario.

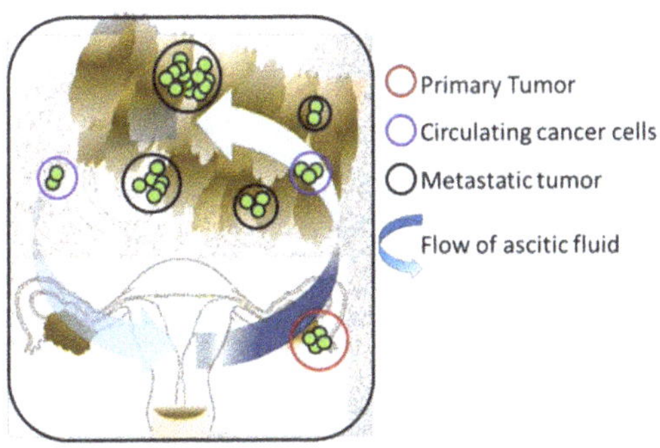

Figure 1. Pathogenesis of the ovarian cancer disease.

2. Three-Dimensional Modelling of Early Metastasis TME Interactions in Ovarian Cancer

Organotypic models refer to 3D models, usually containing ECM, that are comprised of two or more cell types to mimic the complex interactions within a tissue. For this review, we focus on organotypic models of the ovarian cancer TME that comprise a 3D culture containing at least one primary stromal cell type in an ECM background, such as collagen or fibronectin, plus ovarian cancer cells.

Ovarian cancer cells have a special predilection for the peritoneum and the omentum as sites of metastasis [8]. The outer lining of these sites consists of a single layer of mesothelial cells with an underlying ECM. During the metastatic process, microscopic non-invasive omental metastases proliferate on top of this layer of mesothelial cells. As the metastases increase in size, the cancer cells induce pro-tumorigenic changes in the stromal cells of the microenvironment, including an increase in the number of fibroblasts and a more rigid basement membrane [13]. Tumor cells then invade the omental adipose tissue. In 1985, Niedbala et al. were the first to establish an organotypic culture of the ovarian cancer TME and investigate the mechanism through which ovarian cancer cells infiltrate the mesothelial cell layer and attach to the ECM [14]. Human primary mesothelial cells (HPMCs) were grown in a monolayer on ECM derived from bovine corneal endothelial cells, onto which ovarian cancer cells derived from patient ascites were seeded. A current version of this organotypic model of the ovarian cancer TME was developed by Kenny et al. This model allows examination of the role that the ECM, HPMCs and fibroblasts play in the initial adhesion, migration, invasion and proliferation of ovarian cancer cells during early metastasis to the mesothelium [13]; it is referred to in this review as the "mesothelium model". Adding a different element, other models recreate the dynamic mechanical forces that act upon ovarian cancer spheroids in the peritoneal cavity using microfluidic devices [15,16].

3. Three-Dimensional Organotypic Model of Human Mesothelium

The 3D organotypic mesothelium model was created to elucidate the role of specific cellular and non-cellular components, namely fibroblasts, HPMCs, and different ECM proteins, in early ovarian cancer metastasis to the omentum [13]. Two key factors set this mesothelium model apart from other 3D cultures: (1) prior to construction of the model, the authors analyzed hematoxylin and eosin (H&E) stains of normal omental biopsies to form the best picture of the physiological framework of normal omentum and (2) the authors included two types of primary stromal cells, HPMCs and fibroblasts. Using primary HPMCs and fibroblasts at early passages extracted from fresh biopsies of omentum obtained during surgery, the authors recreated the omental ovarian cancer TME in vitro (Figure 2A). Primary human omental fibroblasts were embedded in ECM and overlaid with a layer of HPMCs (1:5 ratio of fibroblasts and HPMCs). With the addition of ovarian cancer cells or immortalized ovarian surface epithelial cells, this highly reproducible construction was used to determine the role of each of the TME components, including different ECM proteins, during ovarian cancer adhesion and invasion. Results from the model showed that both HPMCs and fibroblasts play key roles in these processes. Customization of the model with different ECM proteins revealed that ovarian cancer cell adhesion and invasion is greatest in the presence of collagen, compared with vitronectin, fibronectin, or laminin. This tool was also shared as a JoVE video article to improve dissemination of the protocol and as a resource for other scientists in the ovarian cancer field [17].

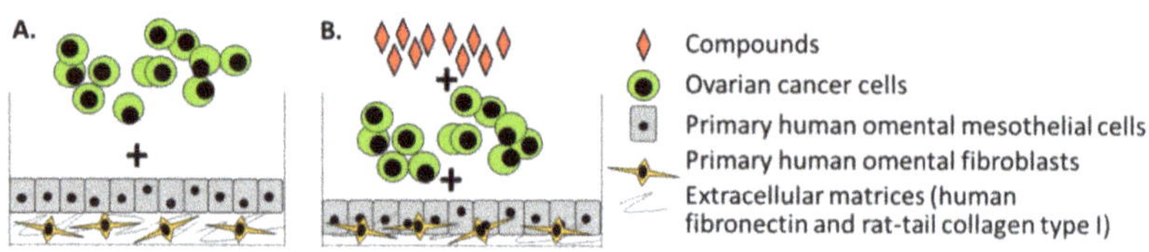

Figure 2. Three-dimensional organotypic model of human mesothelium. (**A**) Layered model for investigating ovarian cancer adhesion, migration, invasion and proliferation in a metastatic microenvironment. In this layered model, the extracellular matrix (ECM) and fibroblasts are cultured together prior to the sequential addition of mesothelial cells and cancer cells. (**B**) Model for high-throughput screening (HTS) to identify compounds that inhibit ovarian cancer adhesion/invasion or proliferation. In this HTS model, ECM, fibroblasts and mesothelial cells are plated simultaneously, followed by the addition of cancer cells and compounds.

This modular mesothelium model has been used in numerous publications that further illuminate the mechanisms involved in early ovarian cancer metastasis [18–24]. Kenny and colleagues proceeded to show that ovarian cancer cells recruit HPMCs to establish metastatic colonies by inducing an upregulation in the levels of fibronectin 1 (FN1) mRNA and protein in HPMCs which promote cell adhesion [18]. They also demonstrated that adhering ovarian cancer cells express matrix-metalloproteinase, which cleaves matrix proteins into smaller fragments, thereby facilitating invasion [19,20]. Other functional assays using the model and an antibody against the urokinase plasminogen activator (uPA) receptor (u-PAR) revealed that targeting the uPA/u-PAR proteolytic system reduced metastasis and induced apoptosis of ovarian cancer cells [21]. Building on this, Mitra et al. used the mesothelium model to identify miRNAs involved in omental colonization, demonstrating that upregulation of uPA in ovarian cancer cells is due to downregulation in miR-193b levels, which is in turn due to ovarian cancer cell interaction with HPMCs on the surface of the omentum [22]. More recently, a study by Caroline Ford's group used this model to expand on the synergistic role of Wnt receptors ROR1 and ROR2 in early ovarian cancer metastasis, specifically their role in ovarian cancer cell adhesion to the omentum [23].

The ability to customize the mesothelium model led to its reshaping and utilization in high throughput screening (HTS) assays. Through optimization of parameters such as incubation time, plating sequence, number of ovarian cancer cells, HPMCs, fibroblasts, and ECM, the model was adapted for use in reproducible 384- and 1536- multi-well HTS assays [25] (Figure 2B). Fully automated 3D HTS assays were carried out to screen small molecule inhibitors that could potentially target ovarian cancer adhesion/invasion or proliferation [25,26]. The effect of oncology drugs from three small molecule compound libraries, the National Center for Advancing Translational Sciences (NCATS) Mechanism Interrogation PlatE oncology collection, the Prestwick library, and the Library of Pharmacologically Active Compounds (LOPAC^{1280}) on ovarian cancer adhesion/invasion or proliferation was investigated. These assays were followed by confirmatory, counter, and secondary biological assays utilizing the 3D organotypic model of human mesothelium to identify lead compounds. Ultimately, inhibitory activity of the lead compounds on ovarian cancer metastasis was validated in different in vivo xenograft models. A key takeaway from the HTS assays was that many of the compounds screened were active in cancer cells on plastic (>90%), but only a few compounds were effective in the 3D HTS platform (<1%) which directly translated to in vivo activity in xenograft mouse models. Differences in drug response between 2D cultures and 3D organotypic models have also been demonstrated in studies of skin melanoma. These studies reported that treatment with tumor necrosis factor-related apoptosis-inducing ligand (TRAIL) in combination with either UVB or cisplatin killed melanoma cells in 2D cultures, but only the TRAIL plus cisplatin combination was effective in their layered 3D organotypic skin melanoma spheroid model [27–29]. These differences further highlight the value of 3D organotypic models that can accurately represent the complexity of the ovarian cancer TME.

This mesothelium model is a first step at recapitulating the metastatic microenvironment of ovarian cancer, but it still lacks other in vivo factors such as vasculature, adipocytes, and host immune cells. However, it represents a significantly more complex experimental system than ovarian cancer cells grown in monolayer to analyze the complex mechanisms of tumorigenesis and to potentially identify new therapeutics. Omental cells from different patients in the 3D organotypic cultures reveal a broader picture of donor-to-donor variability in terms of drug response, cellular function and cell signaling.

4. Three-Dimensional Organotypic Model of Cancer Cells Circulating in Ascites

Peritoneal dissemination of ovarian cancer spheroids and their interactions with omental mesothelial cells are not static processes. Hydrodynamic forces generated by increased production of fluid in the peritoneal cavity must be considered in addition to the 3D culture itself. To recreate this aspect of the ovarian cancer TME, Li et al. developed a 3D microfluidic-based platform in which living cells are infused into micrometer-sized chambers [15]. These platforms enable accurate control of the cellular microenvironment, allowing a continuous release of growth factors or nutrients. In their device, Li et al. plated mesothelial cells on fibronectin, and added fluorescently labelled ovarian cancer spheroids under continuous fluidic flow to mimic the flow of peritoneal fluid induced by ovarian cancer in the clinical setting (Figure 3).

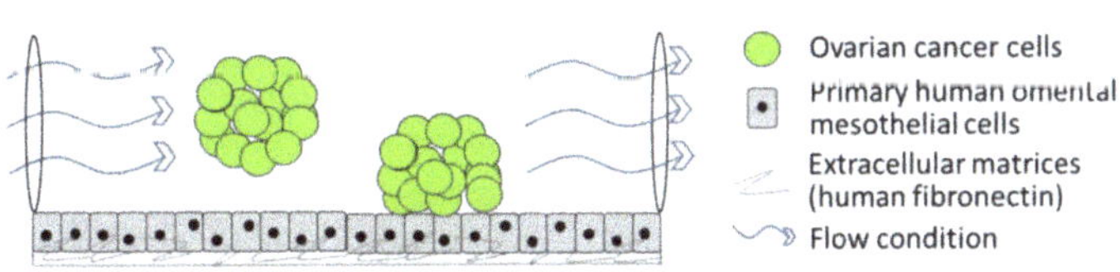

Figure 3. Three-dimensional organotypic model of cancer cells circulating in ascites.

A 2018 publication by Carroll et al. added another layer of complexity by investigating the interactions between alternatively activated macrophages (AAMs), mesothelial cells, and ovarian cancer cells in dynamic flow experiments of ovarian cancer cell adhesion [16]. The authors first determined, under static 3D conditions, that AAM-secreted macrophage inflammatory protein-1 induced expression of P-selectin in mesothelial cell lines, which in turn increased ovarian cancer cell adhesion to the mesothelial cells. Using a parallel-plate flow chamber, which simulates fluid sheer stress on cells, the authors went on to demonstrate that these increased levels of P-selectin in mesothelial cell lines led to increased rolling of ovarian cancer cells.

Compared with experiments under static conditions, experiments performed under flow conditions provided valuable insights into features of transcoelomic metastasis that cannot be reproduced in standard static cultures, such as increased adhesion under flow conditions [16]. Although these microfluidic 3D models contained only one stromal cell type, their modularity means that they can be customized to include other stromal cells or ECM components for use in functional assays such as adhesion, invasion, and proliferation.

5. Other 3D Models of the Ovarian Cancer TME

5.1. Organoids

Identification of precursor lesions in the fallopian tube fimbria of ovarian cancer patients and BRCA mutation carriers point towards the fimbria as the likely site of origin of high-grade serous ovarian cancer [30–32], but the fallopian tube TME has not been well explored. Organoids are in vitro cellular clusters (3D) derived from primary tissue that use ECM hydrogels to self-assemble with architecture, histology, and genetic features resembling the original tissue [33]. Kessler et al. re-constructed the microenvironmental milieu with growth factors and Matrigel to successfully culture fallopian tube organoids from fallopian tube epithelial stem cells [34]. By supplementing this culture with a selection of growth factors, the authors determined that both Notch and Wnt regulate stemness and differentiation in fallopian tube organoids.

Fallopian tube organoid in vitro models have also been generated from induced pluripotent stem cells (iPSCs). Yucer et al. guided differentiation of iPSC lines into fallopian tube epithelium precursor cells through exposure to BMP4 and WNT4 followed by follistatin, an activin-binding protein that bio-neutralizes members of the TGF-β superfamily [35]. When spheroids of these differentiated cells were grown on Matrigel and supplemented with estrogen, progesterone, and crucially, conditioned media from primary fallopian tube epithelial cells, they self-organized into luminal structures representative of the fallopian tube architecture with ciliated and secretory components.

Organoids do not contain any stromal components, but can be incorporated into organotypic culture systems to study the interactions between the organoid cells and the cells of their microenvironment. Taking this concept a step further, one could envision a model in which transformed fallopian tube epithelial cells [36] are propagated in organoids and integrated into an organotypic model to investigate the early ovarian cancer TME.

5.2. Explant Cultures

While not a model in the sense that models are constructed, explants of omentum, ovary or fallopian tube pieces cultured in the presence of ovarian cancer cells represent another form of 3D culture. In particular, mouse omentum, ovarian and fallopian tube organ pieces can be cultured for up to two weeks [37–39]. Human omentum and fallopian tube explants have been cultured for up to five days with ovarian cancer cells [19,40], and ultimately revealed that ovarian cancer cells could metastasize to the fallopian tube. In addition, these explant cultures can be used to test the effect of different drugs or treatments on ovarian cancer adhesion, migration, invasion and proliferation by targeting either the cancer or stromal cells.

5.3. Cell Line Spheroids

For most researchers studying the ovarian cancer TME, access to patient tissue to obtain primary cells will be the limiting factor. A number of valuable 3D models that do not include primary cells have been published and utilized in functional assays. These non-organotypic endeavors to recapitulate the primary ovarian cancer TME in vitro include ovarian cancer cell spheroid cultures on synthetic matrices [41], on ECM [9], in low-adherent plastics, in hanging-drops, or in spinner flasks [42–44]. While these 3D systems lack a primary stromal cell component, their multi-component concept is more faithful to the TME than cells grown in a monolayer on plastic. Utilization of such systems in numerous studies elucidating the mechanisms of drug resistance demonstrate that they can be used as predictive preclinical models [41,45–48].

6. Challenges and Future Perspectives

6.1. Picturing the Prototype Ovarian Cancer TME Model

Developing an ideal model for the ovarian cancer TME is not straightforward. Multiple TMEs with varying components mean that a minimum of four models are likely required: in situ carcinoma in the fallopian tube; dissemination in the peritoneal cavity; early metastatic attachment to the mesothelial-lined surfaces of the omentum, bowel, and abdominal wall; and late chemoresistant metastases. Each complete model will first require the comprehensive characterization (e.g., by immunohistochemistry) of the associated stromal cells and ECM components, the growth factor and metabolite milieu, and, if applicable, the flow rate. Once the components of each TME have been characterized, the primary cells and ECM will need to be isolated, followed by reconstruction of the tissue of interest, with the aid of a bioprinter or synthetic matrices that can be degraded by cells once they form their own ECM architecture. Functionality of the model will then have to be verified. One option for this may be to confirm that the in vitro secreted proteins are analogous to those of the in vivo secretome, for example in terms of drug response or activation of immune cells. Each of these phases of model development is a significant undertaking, and the current models do not come close to the in vivo scenario in terms of the variety of cell types that are involved in each TME.

6.2. Future Directions

Multiple potential sites of origin and the continuously changing microenvironments at each stage of the disease demand the development of more diverse (i.e., fallopian tube, ovary, peritoneum) and complex 3D models of the ovarian cancer TME. Each phase of progression has a distinct TME with specific components; for example, models of chemoresistance would include cancer-associated fibroblasts [49], which are not included in the models of early metastasis discussed here. Each of the models presented has its own advantages and limitations, leading us to propose that integration of these models will be a first step towards a more accurate model.

Currently, the mesothelium model is the only 3D organotypic model of the ovarian cancer TME that is utilized by multiple research groups [13]. The mesothelium model was designed to mimic the tissue organization of the mesothelium that lines the human omentum and peritoneum. It recapitulates the initial adhesion, migration, invasion and proliferation of ovarian cancer cells on the mesothelium lining. This platform has been modified to investigate the individual and cooperative role of different cell types in the TME on ovarian cancer progression. It has evolved and been adapted for HTS of over 100,000 small molecule compounds which could potentially identify new therapeutics for prevention of ovarian cancer metastasis.

The organotypic models discussed here are restrained from reaching their full potential due to the limitations of working with primary tissue, including access to the tissue and the lifespan of the 3D models, the absence of other essential primary features such as vasculature [50,51], and the inclusion of artificial ECM components. Vascularization appears to be next obvious step in advancing the organotypic models towards the in vivo scenario. An elegant model of a vascularized TME was recently published by Magdeldin et al., in which the authors created a 3D model of the tumor stroma using colorectal cancer cell spheroids, collagen hydrogels, the basement membrane protein laminin, human dermal fibroblasts, and human umbilical vein endothelial cells (HUVECs) [52]. Customization of the stromal composition revealed that laminin was critical for regulating vascular network formation, while the addition of the cancer cells to the model disrupted the interconnectivity of the network. Jeon et al. reported on an organ-specific 3D microfluidic model to study human breast cancer cell extravasation during metastasis [53]. In their microfluidic model, primary bone marrow-derived mesenchymal stem cells (hBM-MSCs), osteo-differentiated primary hBM-MSCs, and primary GFP-HUVECs were embedded in a fibrin gel in the microfluidic device. The endothelial cells formed the vasculature, and the other cells contributed to a microenvironment that mimicked bone, a frequent site of metastasis in advanced breast cancer. Addition of breast cancer cells to this modular model enabled the authors to investigate the roles of the different components in extravasation.

Matrices incorporated into the models presented here are purified from other human, rat, or mouse sources. Scaffold properties [54], including the concentration of ECM proteins, can affect the stiffness of the artificial matrix; therefore, the accessibility of drugs in in vitro screening must also be considered and optimized. Incorporation of perictyes and endothelial cells, as well as ECM from patient-matched mesothelium or prolonged cultures where the microenvironmental cells secrete and organize their own ECM, could clarify key mechanisms of metastasis, chemoresistance and recurrence. Bioprinting has emerged as a very promising approach to in vitro 3D cancer models owing to its ability to create complex 3D architectures [55].

Ultimately, 3D organotypic models of ovarian cancer aim to recapitulate but systematically simplify the in vivo human microenvironment. Our hope is that by increasing the physiological relevance of 3D organotypic microenvironment models of tumor initiation, primary tumor growth, circulating tumor multi-cellular aggregates, different metastatic sites, and chemoresistant ovarian cancer, the clinical significance of ovarian cancer research will be improved. If we want to offer personalized medicine for ovarian cancer patients, we will also need to successfully establish ovarian cancer organoids for biobanking, as observed with the establishment of organoid cultures in breast, bladder and colorectal cancers [56–58]. By recreating the different TMEs in vitro, we can clarify the role of the TME in the transformation of the original epithelial stem cells into metastatic and chemoresistant cancer cells to ultimately prevent and effectively treat ovarian cancer.

Author Contributions: K.M.W., P.B. and H.A.K. performed the literature review and wrote, reviewed, and approved the final version of the manuscript.

Funding: This research was funded by the Bears Care Foundation (H.A.K.) and Liz Tilberis Award (H.A.K.).

Conflicts of Interest: The authors declare that they have no conflict of interest.

References

1. Hanahan, D.; Coussens, L.M. Accessories to the crime: Functions of cell recruited to the tumor microenvironment. *Cancer Cell* **2012**, *21*, 309–322. [CrossRef] [PubMed]
2. Hanahan, D.; Weinberg, R.A. Hallmarks of cancer: The next generation. *Cell* **2011**, *144*, 646–674. [CrossRef] [PubMed]
3. Rutkowski, J.M.; Swartz, M.A. A driving force for change: Interstitial flow as a morphoregulator. *Trends Cell Biol.* **2007**, *17*, 44–50. [CrossRef] [PubMed]
4. Chien, S.; Li, S.; Shyy, Y.J. Effects of mechanical forces on signal transduction and gene expression in endothelial cells. *Hypertension* **1998**, *31*, 162–169. [CrossRef] [PubMed]
5. Achilli, T.M.; Meyer, J.; Morgan, J.R. Advances in the formation, use and understanding of multi-cellular spheroids. *Expert Opin. Biol. Ther.* **2012**, *12*, 1347–1360. [CrossRef] [PubMed]
6. Griffith, L.G.; Swartz, M.A. Capturing complex 3d tissue physiology in vitro. *Nat. Rev. Mol. Cell Biol.* **2006**, *7*, 211–224. [CrossRef] [PubMed]
7. Langhans, S.A. Three-dimensional in vitro cell culture models in drug discovery and drug repositioning. *Front. Pharmacol.* **2018**, *9*, 6. [CrossRef] [PubMed]

8. Lengyel, E. Ovarian cancer development and metastasis. *Am. J. Pathol.* **2010**, *177*, 1053–1064. [CrossRef] [PubMed]
9. Klymenko, Y.; Kim, O.; Loughran, E.; Yang, J.; Lombard, R.; Alber, M.; Stack, M.S. Cadherin composition and multicellular aggregate invasion in organotypic models of epithelial ovarian cancer intraperitoneal metastasis. *Oncogene* **2017**, *36*, 5840–5851. [CrossRef] [PubMed]
10. Kenny, H.A.; Dogan, S.; Zillhardt, M.; Mitra, A.B.; Yamada, S.D.; Krausz, T.; Lengyel, E. Organotypic models of metastasis: A three dimensional culture mimicking the human peritoneum and omentum for the study of the early steps of ovarian cancer metastasis. *Cancer Treat. Res.* **2009**, *149*, 335–351. [PubMed]
11. White, E.A.; Kenny, H.A.; Lengyel, E. Three-dimensional modeling of ovarian cancer. *Adv. Drug Del. Rev.* **2014**, *79–80*, 184–192. [CrossRef] [PubMed]
12. Lengyel, E.; Burdette, J.E.; Kenny, H.A.; Matei, D.; Pilrose, J.; Haluska, P.; Nephew, K.P.; Hales, D.B.; Stack, M.S. Epithelial ovarian cancer experimental models. *Oncogene* **2014**, *33*, 3619–3633. [CrossRef] [PubMed]
13. Kenny, H.A.; Krausz, T.; Yamada, S.D.; Lengyel, E. Use of a novel 3d culture model to elucidate the role of mesothelial cells, fibroblasts and extra-cellular matrices on adhesion and invasion of ovarian cancer cells to the omentum. *Int. J. Cancer* **2007**, *121*, 1463–1472. [CrossRef] [PubMed]
14. Niedbala, M.J.; Crickard, K.; Bernacki, R. Interactions of human ovarian tumor cells with human mesothelial cells grown on extracellular matrix. *Exp. Cell Res.* **1985**, *160*, 499–513. [CrossRef]
15. Li, S.S.; Ip, C.K.; Tang, M.Y.; Sy, S.K.; Yung, S.; Chan, T.M.; Yang, M.; Shum, H.C.; Wong, A.S. Modeling ovarian cancer multicellular spheroid behavior in a dynamic 3d peritoneal microdevice. *J. Vis. Expe.* **2017**. [CrossRef] [PubMed]
16. Carroll, M.J.; Fogg, K.C.; Patel, H.A.; Krause, H.B.; Mancha, A.S.; Patankar, M.S.; Weisman, P.S.; Barroilhet, L.; Kreeger, P.K. Alternatively-activated macrophages upregulate mesothelial expression of p-selectin to enhance adhesion of ovarian cancer cells. *Cancer Res.* **2018**, *78*, 3560–3573. [CrossRef] [PubMed]
17. Peters, P.N.; Schryver, E.M.; Lengyel, E.; Kenny, H. Modeling the early steps of ovarian cancer dissemination in an organotypic culture of the human peritoneal cavity. *J. Vis. Exp.* **2015**, e53541. [CrossRef] [PubMed]
18. Kenny, H.A.; Chiang, C.Y.; White, E.A.; Schryver, E.M.; Habis, M.; Romero, I.L.; Ladanyi, A.; Penicka, C.V.; George, J.; Matlin, K.; et al. Mesothelial cells promote early ovarian cancer metastasis through fibronectin secretion. *J. Clin. Invest.* **2014**, *124*, 4614–4628. [CrossRef] [PubMed]
19. Kenny, H.A.; Kaur, S.; Coussens, L.M.; Lengyel, E. The initial steps of ovarian cancer cell metastasis are mediated by mmp-2 cleavage of vitronectin and fibronectin. *J. Clin. Invest.* **2008**, *118*, 1367–1379. [CrossRef] [PubMed]
20. Kenny, H.A.; Lengyel, E. Mmp-2 functions as an early response protein in ovarian cancer metastasis. *Cell Cycle* **2009**, *8*, 683–688. [CrossRef] [PubMed]
21. Kenny, H.A.; Leonhardt, P.; Ladanyi, A.; Yamada, S.D.; Montag, A.; Im, H.K.; Jagadeeswaran, S.; Shaw, D.E.; Mazar, A.P.; Lengyel, E. Targeting the urokinase plasminogen activator receptor inhibits ovarian cancer metastasis. *Clin. Cancer Res.* **2011**, *17*, 459–471. [CrossRef] [PubMed]
22. Mitra, A.K.; Chiang, C.Y.; Tiwari, P.; Tomar, S.; Watters, K.M.; Peter, M.E.; Lengyel, E. Microenvironment-induced downregulation of mir-193b drives ovarian cancer metastasis. *Oncogene* **2015**, *34*, 5923–5932. [CrossRef] [PubMed]
23. Henry, C.; Hacker, N.; Ford, C. Silencing ror1 and ror2 inhibits invasion and adhesion in an organotypic model of ovarian cancer metastasis. *Oncotarget* **2017**, *8*, 112727–112738. [CrossRef] [PubMed]
24. Tomar, S.; Plotnik, J.P.; Haley, J.; Scantland, J.; Dasari, S.; Sheikh, Z.; Emerson, R.; Lenz, D.; Hollenhorst, P.C.; Mitra, A.K. Ets1 induction by the microenvironment promotes ovarian cancer metastasis through focal adhesion kinase. *Cancer Lett.* **2018**, *414*, 190–204. [CrossRef] [PubMed]
25. Kenny, H.A.; Lal-Nag, M.; White, E.A.; Shen, M.; Chiang, C.Y.; Mitra, A.K.; Zhang, Y.; Curtis, M.; Schryver, E.M.; Bettis, S.; et al. Quantitative high throughput screening using a primary human three-dimensional organotypic culture predicts in vivo efficacy. *Nat. Commun.* **2015**, *6*, 6220. [CrossRef] [PubMed]
26. Lal-Nag, M.; McGee, L.; Guha, R.; Lengyel, E.; Kenny, H.A.; Ferrer, M. A high-throughput screening model of the tumor microenvironment for ovarian cancer cell growth. *SLAS Discov.* **2017**, *22*, 494–506. [CrossRef] [PubMed]

27. Vorsmann, H.; Groeber, F.; Walles, H.; Busch, S.; Beissert, S.; Walczak, H.; Kulms, D. Development of a human three-dimensional organotypic skin-melanoma spheroid model for in vitro drug testing. *Cell Death Dis.* **2013**, *4*, e719. [CrossRef] [PubMed]
28. Thayaparasingham, B.; Kunz, A.; Peters, N.; Kulms, D. Sensitization of melanoma cells to trail by uvb-induced and nf-kappab-mediated downregulation of xiap. *Oncogene* **2009**, *28*, 345–362. [CrossRef] [PubMed]
29. Hornle, M.; Peters, N.; Thayaparasingham, B.; Vorsmann, H.; Kashkar, H.; Kulms, D. Caspase-3 cleaves xiap in a positive feedback loop to sensitize melanoma cells to trail-induced apoptosis. *Oncogene* **2011**, *30*, 575–587. [CrossRef] [PubMed]
30. Callahan, M.J.; Crum, C.P.; Medeiros, F.; Kindelberger, D.W.; Elvin, J.A.; Garber, J.E.; Feltmate, C.M.; Berkowitz, R.S.; Muto, M.G. Primary fallopian tube malignancies in brca-positive women undergoing surgery for ovarian cancer risk reduction. *J. Clin. Oncol.* **2007**, *25*, 3985–3990. [CrossRef] [PubMed]
31. Kindelberger, D.W.; Lee, Y.; Miron, A.; Hirsch, M.S.; Feltmate, C.; Medeiros, F.; Callahan, M.J.; Garner, E.O.; Gordon, R.W.; Birch, C.; et al. Intraepithelial carcinoma of the fimbria and pelvic serous carcinoma: Evidence for a causal relationship. *Am. J. Surg. Pathol.* **2007**, *31*, 161–169. [CrossRef] [PubMed]
32. Piek, J.M.; van Diest, P.J.; Zweemer, R.P.; Jansen, J.W.; Poort-Keesom, R.J.; Menko, F.H.; Gille, J.J.; Jongsma, A.P.; Pals, G.; Kenemans, P.; et al. Dysplastic changes in prophylactically removed fallopian tubes of women predisposed to developing ovarian cancer. *J. Pathol.* **2001**, *195*, 451–456. [CrossRef] [PubMed]
33. Fatehullah, A.; Tan, S.H.; Barker, N. Organoids as an in vitro model of human development and disease. *Nat. Cell Biol.* **2016**, *18*, 246–254. [CrossRef] [PubMed]
34. Kessler, M.; Hoffmann, K.; Brinkmann, V.; Thieck, O.; Jackisch, S.; Toelle, B.; Berger, H.; Mollenkopf, H.J.; Mangler, M.; Sehouli, J.; et al. The notch and wnt pathways regulate stemness and differentiation in human fallopian tube organoids. *Nat. Commun.* **2015**, *6*, 8989. [CrossRef] [PubMed]
35. Yucer, N.; Holzapfel, M.; Jenkins Vogel, T.; Lenaeus, L.; Ornelas, L.; Laury, A.; Sareen, D.; Barrett, R.; Karlan, B.Y.; Svendsen, C.N. Directed differentiation of human induced pluripotent stem cells into fallopian tube epithelium. *Sci. Rep.* **2017**, *7*, 10741. [CrossRef] [PubMed]
36. Karst, A.M.; Levanon, K.; Drapkin, R. Modeling high-grade serous ovarian carcinogenesis from the fallopian tube. *Proc. Nat. Acad. Sci. USA* **2011**, *108*, 7547–7552. [CrossRef] [PubMed]
37. King, S.M.; Quartuccio, S.; Hillard, T.S.; Inoue, K.; Burdette, J.E. Alginate hydrogels for three-dimensional organ culture of ovaries and oviducts. *J. Vis. Exp.* **2011**, *52*, 1–6. [CrossRef] [PubMed]
38. Clark, R.; Krishnan, V.; Schoof, M.; Rodriguez, I.; Theriault, B.; Chekmareva, M.; Rinker-Schaeffer, C. Milky spots promote ovarian cancer metastatic colonization of peritoneal adipose in experimental models. *Ame. J. Pathol.* **2013**, *183*, 576–591. [CrossRef] [PubMed]
39. Krishnan, V.; Clark, R.; Chekmareva, M.; Johnson, A.; George, S.; Shaw, P.; Seewaldt, V.; Rinker-Schaeffer, C. In vivo and ex vivo approaches to study ovarian cancer metastatic colonization of milky spot structures in peritoneal adipose. *J. Vis. Exp.* **2015**, e52721. [CrossRef] [PubMed]
40. Eckert, M.A.; Pan, S.; Hernandez, K.M.; Loth, R.M.; Andrade, J.; Volchenboum, S.L.; Faber, P.; Montag, A.; Lastra, R.; Peter, M.E.; et al. Genomics of ovarian cancer progression reveals diverse metastatic trajectories including intraepithelial metastasis to the fallopian tube. *Cancer Discov.* **2016**, *6*, 1342–1351. [CrossRef] [PubMed]
41. Yang, Z.; Zhao, X. A 3d model of ovarian cancer cell lines on peptide nanofiber scaffold to explore the cell-scaffold interaction and chemotherapeutic resistance of anticancer drugs. *Int. J. Nanomedicine* **2011**, *6*, 303–310. [CrossRef] [PubMed]
42. Letourneau, I.J.; Quinn, M.C.; Wang, L.L.; Portelance, L.; Caceres, K.Y.; Cyr, L.; Delvoye, N.; Meunier, L.; de Ladurantaye, M.; Shen, Z.; et al. Derivation and characterization of matched cell lines from primary and recurrent serous ovarian cancer. *BMC Cancer* **2012**, *12*, 379. [CrossRef] [PubMed]
43. Zietarska, M.; Maugard, C.; Filali-Mouhim, A.; Alam-Fahmy, M.; Tonin, P.N.; Provencher, D.; Mes-Masson, A.M. Molecular description of a 3d in vitro model for the study of epithelial ovarian cancer (eoc). *Mol. Carcinog.* **2007**, *46*, 872–885. [CrossRef] [PubMed]
44. Lawrenson, K.; Benjamin, E.; Turmaine, M.; Jacobs, I.; Gayther, S.; Dafou, D. In vitro three-dimensional modeling of human ovarian surface epithelial cells. *Cell Prolif.* **2009**, *42*, 385–393. [CrossRef] [PubMed]

45. Loessner, D.; Stok, K.S.; Lutolf, M.P.; Hutmacher, D.W.; Clements, J.A.; Rizzi, S.C. Bioengineered 3d platform to explore cell-ecm interactions and drug resistance of epithelial ovarian cancer cells. *Biomaterials* **2010**, *31*, 8494–8506. [CrossRef] [PubMed]
46. Chen, J.; Wang, J.; Chen, D.; Yang, J.; Yang, C.; Zhang, Y.; Zhang, H.; Dou, J. Evaluation of characteristics of cd44+cd117+ ovarian cancer stem cells in three dimensional basement membrane extract scaffold versus two dimensional monocultures. *BMC Cell Biol.* **2013**, *14*, 7. [CrossRef] [PubMed]
47. Dong, Y.; Stephens, C.; Walpole, C.; Swedberg, J.E.; Boyle, G.M.; Parsons, P.G.; McGuckin, M.A.; Harris, J.M.; Clements, J.A. Paclitaxel resistance and multicellular spheroid formation are induced by kallikrein-related peptidase 4 in serous ovarian cancer cells in an ascites mimicking microenvironment. *PLoS ONE* **2013**, *8*, e57056. [CrossRef] [PubMed]
48. Lee, J.M.; Mhawech-Fauceglia, P.; Lee, N.; Parsanian, L.C.; Lin, Y.G.; Gayther, S.A.; Lawrenson, K. A three-dimensional microenvironment alters protein expression and chemosensitivity of epithelial ovarian cancer cells in vitro. *Lab. Invest.* **2013**, *93*, 528–542. [PubMed]
49. Schauer, I.G.; Sood, A.K.; Mok, S.; Liu, J. Cancer-associated fibroblasts and their putative role in potentiating the initiation and development of epithelial ovarian cancer. *Neoplasia* **2011**, *13*, 393–405. [CrossRef] [PubMed]
50. Butler, J.M.; Kobayashi, H.; Rafii, S. Instructive role of the vascular niche in promoting tumour growth and tissue repair by angiocrine factors. *Nat. Rev. Cancer* **2010**, *10*, 138–146. [CrossRef] [PubMed]
51. Monk, B.J.; Minion, L.E.; Coleman, R.L. Anti-angiogenic agents in ovarian cancer: Past, present, and future. *Ann. Oncol.* **2016**, *27*, i33–i39. [CrossRef] [PubMed]
52. Magdeldin, T.; Lopez-Davila, V.; Pape, J.; Cameron, G.W.; Emberton, M.; Loizidou, M.; Cheema, U. Engineering a vascularised 3d in vitro model of cancer progression. *Sci. Rep.* **2017**, *7*, 44045. [CrossRef] [PubMed]
53. Jeon, J.S.; Bersini, S.; Gilardi, M.; Dubini, G.; Charest, J.L.; Moretti, M.; Kamm, R.D. Human 3d vascularized organotypic microfluidic assays to study breast cancer cell extravasation. *Proc. Natl. Acad. Sci. USA* **2015**, *112*, 214–219. [CrossRef] [PubMed]
54. Zheng, L.; Hu, X.; Huang, Y.; Xu, G.; Yang, J.; Li, L. In vivo bioengineered ovarian tumors based on collagen, matrigel, alginate and agarose hydrogels: A comparative study. *Biomed. Mater.* **2015**, *10*, 015016. [CrossRef] [PubMed]
55. Murphy, S.V.; Atala, A. 3d bioprinting of tissues and organs. *Nat. Biotechnol.* **2014**, *32*, 773–785. [CrossRef] [PubMed]
56. Sachs, N.; de Ligt, J.; Kopper, O.; Gogola, E.; Bounova, G.; Weeber, F.; Balgobind, A.V.; Wind, K.; Gracanin, A.; Begthel, H.; et al. A living biobank of breast cancer organoids captures disease heterogeneity. *Cell* **2018**, *172*, 373–386. [CrossRef] [PubMed]
57. Lee, S.H.; Hu, W.; Matulay, J.T.; Silva, M.V.; Owczarek, T.B.; Kim, K.; Chua, C.W.; Barlow, L.J.; Kandoth, C.; Williams, A.B.; et al. Tumor evolution and drug response in patient-derived organoid models of bladder cancer. *Cell* **2018**, *173*, 515–528.e517. [CrossRef] [PubMed]
58. Cristobal, A.; van den Toorn, H.W.P.; van de Wetering, M.; Clevers, H.; Heck, A.J.R.; Mohammed, S. Personalized proteome profiles of healthy and tumor human colon organoids reveal both individual diversity and basic features of colorectal cancer. *Cell Rep.* **2017**, *18*, 263–274. [CrossRef] [PubMed]

cancers

MDPI

Review

The Role of Inflammation and Inflammatory Mediators in the Development, Progression, Metastasis, and Chemoresistance of Epithelial Ovarian Cancer

Sudha S. Savant [1,†], Shruthi Sriramkumar [2,†] and Heather M. O'Hagan [1,3,*]

1 Medical Sciences, Indiana University School of Medicine, Bloomington, IN 47405, USA; ssavant@iu.edu
2 Cell, Molecular and Cancer Biology Graduate Program, Indiana University, Bloomington, IN 47405, USA; ssriramk@iu.edu
3 Indiana University Melvin and Bren Simon Cancer Center, Indianapolis, IN 46202, USA
* Correspondence: hmohagan@indiana.edu; Tel.: +1-812-855-3035
† These authors contributed equally to this work.

Received: 14 June 2018; Accepted: 24 July 2018; Published: 30 July 2018

Abstract: Inflammation plays a role in the initiation and development of many types of cancers, including epithelial ovarian cancer (EOC) and high grade serous ovarian cancer (HGSC), a type of EOC. There are connections between EOC and both peritoneal and ovulation-induced inflammation. Additionally, EOCs have an inflammatory component that contributes to their progression. At sites of inflammation, epithelial cells are exposed to increased levels of inflammatory mediators such as reactive oxygen species, cytokines, prostaglandins, and growth factors that contribute to increased cell division, and genetic and epigenetic changes. These exposure-induced changes promote excessive cell proliferation, increased survival, malignant transformation, and cancer development. Furthermore, the pro-inflammatory tumor microenvironment environment (TME) contributes to EOC metastasis and chemoresistance. In this review we will discuss the roles inflammation and inflammatory mediators play in the development, progression, metastasis, and chemoresistance of EOC.

Keywords: inflammation; epithelial ovarian cancer; cytokines; reactive oxygen species; growth factors

1. Inflammation and EOC

Inflammation is part of the immune response that protects against foreign pathogens and aids in healing. Inflammation is elicited in response to cellular damage either by infection, exposure to foreign particles (pollutants or irritants), or an increase in cellular stress [1]. The ultimate goal of the inflammatory response is to restore tissue homeostasis, either by destruction or healing of the damaged tissue. The acute or immediate inflammatory response involves modification of the vasculature surrounding the site of stress or damage to increase blood flow. This alteration is then followed by activation of innate immune cells already present in the tissue, including macrophages, dendritic cells (DC), and mast cells, and an increase in infiltration of additional innate immune cells into the affected tissue. At sites of inflammation there are high levels of reactive oxygen species (ROS), cytokines, chemokines, and growth factors that are produced by the immune cells and other cells in the tissue. Acute inflammation is essential for tissue homeostasis and to protect against normal exposure to pathogens. However, in certain cases the body is unable to resolve this response or is subjected to repeated stimulation resulting in chronic inflammation.

Ovarian cancer (OC) is the fifth leading cause of cancer-related deaths in women in the United States [2] and can originate in the germ cells, sex-cord stroma, the fallopian tube (FT), or ovary epithelium. Epithelial ovarian cancer (EOC) which originates from the ovary or fallopian tube epithelium, accounts for 85–90% of all OCs. Chronic inflammation is an important risk factor associated for EOC and high grade serous ovarian cancer (HGSC), the most malignant subtype of EOC. Chronic inflammation results in activation of signaling pathways, transcription factors, and the innate and adaptive immune responses [3,4]. In this review we primarily focus on inflammation as a risk factor for invasive EOC, but have also included supportive evidence from other OC subtypes, studies that do not define the subtype of OC, and other tumor types as indicated.

1.1. Signaling Pathways and Transcription Factors

Several signaling pathways and transcription factors involved in the inflammatory response also play critical roles in EOC. Here we briefly introduce relevant pathways that will be linked to OC formation in later sections. Cytokines produced during inflammation bind to and activate toll like receptors (TLRs) on cell surfaces, which results in activation of the signaling pathways involving mitogen-activated protein kinases (MAPKs) p38 and JNK (c-Jun N-terminal kinase) and transcription factors including nuclear factor kappa-light-chain-enhancer of activated B cells (NF-κB) and the signal transducer and activator of transcription (STATs). The MAPK pathway regulates cellular processes like proliferation, differentiation, growth, migration, and cell death by upregulating the expression of transcription factors like AP-1, c-Jun, FOS and by activating NF-κB and STATs, that either by themselves or along with AP-1 or c-Jun regulate expression of pro-survival and pro-growth genes. NF-κB and AP-1 also regulate production of cytokines like IL-6 [5–7].

During inflammation these transcription factors play an important role to maintain tissue homeostasis. However, in case of chronic inflammation, the signaling pathways are continuously stimulated, which can contribute to tumorigenesis.

1.2. Innate Immune Response

Inflammation activates the innate immune response, which signals macrophages and DCs to secrete chemoattractants like Interleukin-8 (IL-8), monocyte chemotactic protein-1 (MCP-1), and various other inflammatory mediators. These chemoattractants in turn result in recruitment of neutrophils, lymphocytes, and natural killer (NK) cells to the site of damage. All of these cells then secrete cytokines like IL-1, IL-3, IL-6, IL-8, tumor necrosis factor alpha (TNF-α), interferon (IFN) α, and colony-stimulating factors (CSF) like granulocyte macrophage CSF (GM-CSF). The cytokines bind to transmembrane receptors on the cell surfaces of other cells to activate transcription factors that regulate gene expression downstream of the cytokine activated pathway. This creates a pro-inflammatory environment resulting in recruitment of other immune cells, migration of endothelial cells, and proliferation of fibroblasts. Activation of macrophages and NK cells results in the production of high levels of ROS and reactive nitrogen species (RNS), which are used by these cells to kill foreign pathogens, but also end up damaging neighboring normal cells [8]. The lymphocytes also secrete growth factors like platelet derived growth factor (PDGF), transforming growth factor beta (TGF-β), and fibroblast growth factor (FGF), which facilitate wound healing. Overall the acute immune response is a rapid response that typically only lasts a few days. It results in removal of the pathogen, release of proteolytic enzymes to destroy damaged tissue, or stimulation of the proliferation of fibroblasts and epithelial cells to repair the tissue [1].

1.3. Adaptive Immune Response

If the infection is not resolved by the innate immune response, the adaptive immune response is activated, which is less inflammatory in nature. The adaptive immune response also provides longstanding protection against specific pathogens and/or antigens. B cells and T cells are the effector cells of the adaptive immune system that are derived from lymphocytes when they are presented with specific antigens by the antigen presenting cells (APC). T cells respond to the APCs by producing IL-2, which induces expression of transcription factors that facilitate T cells to differentiate into T regulatory (Tregs) and T effector (Teff) cells. There are two major classes of T effector cells;

$CD8^+$ cytotoxic T cells and $CD4^+$ T helper (Th) cells. Th cells are further differentiated into Th1, Th2, or Th17 depending on the ILs secreted and the transcription factors expressed. IFN-y activates STAT1 to induce formation of Th1 and IL-6, and TGF-β can induce Th17 cell formation. Th1 and Th17 secrete ILs and activate macrophages and B cells to create a pro-inflammatory microenvironment (ME) that can be protumorigenic depending on the context. Tregs are immunosuppressive cells that turn off the immune response [1,9,10].

2. Inflammation as a Risk Factor for EOC

Amongst other factors such as hereditary, environmental, and lifestyle, inflammation emerges as an important risk factor for EOC. EOC arises either in the epithelial layer surrounding the ovary or in the epithelium of the distal FT, which could then spread to the ovary. A significant portion of HGSC is thought to originate in the FT, in part because removal of the FT significantly reduces OC risk [11]. Interestingly, while surgical specimens from mutation carriers rarely had premalignant ovarian epithelial changes, early lesions called serous tubal intraepithelial carcinomas (STICs) were found in the FTs of 5–10% of the patients. Copy number and mutational analysis suggest that STICs shed cells with metastatic potential that then colonize the ovary to form HGSC. STICs are mostly found in the fimbriae, the distal end of the FT that shares a ME with the ovary. During a woman's lifetime, the repeated secretion of ROS, cytokines, and other growth factors by the ovaries and immune cells creates a chronic inflammatory ME in the peritoneum that in turn potentiates the initiation of normal cells to malignant ones in the FT and the ovary, supports tumor progression, metastasis, and development of resistance to chemotherapy.

During ovulation, infection and other causes of inflammation ovary and FT tissue is damaged and undergoes repair. We will briefly discuss how each of these processes evoke or involve an inflammatory response that can persist, leading to a cytokine and growth factor rich environment in the peritoneum and contribute to EOC.

2.1. Ovulation

The process of ovulation itself is comparable to that of inflammation as described in the early 20th century. The development of the follicle to its rupture and release of the egg results in recruitment of activated immune cells to the ovary and production of enormous amounts of chemokines, cytokines, and growth factors. Ovulation is initiated by a surge of Luteinizing hormones (LH) that results in increased blood flow to the ovarian follicles. Before release of the egg, the surge of LH hormone recruits neutrophils and macrophages to the graafian follicles [12–14]. Macrophages in the theca have been shown to support growth of follicles [15]. During ovulation macrophages secrete growth factors like hepatocyte growth factor (HGF), TGF-β, and epidermal growth factor (EGF), which stimulate cellular proliferation and follicle growth. Simultaneously the macrophages also secrete ROS, TNF-α, and IL1β, which stimulate local apoptosis resulting in rupture of the follicle, which bathes the ovarian surface and fimbriae with follicular fluid. Exposure of FT cells to follicular fluid results in altered expression of genes associated with inflammation, including increased expression of IL8 and cyclooxygenase-2 (COX-2) [16]. Quiescent fibroblasts are present in the thecal layer surrounding the follicles. Exposure to growth factors stimulates their proliferation and they then secrete prostaglandins, collagenases, and plasminogen activator. In the corpus luteum, after the follicle is released, the macrophages secrete prostaglandins, ROS, and TNF-α, which stimulate apoptosis of the corpus luteum cells. Therefore, ovulation results in the cyclic exposure of FT and ovarian epithelial cells to high levels of ROS, cytokines, and growth factors [17] Although the other causes of inflammation discussed below are important and result in increased overall risk for EOC, the process of ovulation itself occurs often in the lifetime of the majority of women and may be the most important inflammation-related risk factor for EOC. This hypothesis is corroborated by the laying hen model, which is commonly used to study ovarian cancer [18]. In this model, hens develop spontaneous EOC, likely due to their high ovulation rate, thus linking ovulation directly as an increased risk factor for EOC. Delayed onset of menarche and early onset of menopause have been shown to be inversely related to the risk of OC, likely due to the reduction in number of ovulation cycles in a woman's lifetime [19,20]. Further, ovulation has also been connected to EOC because contraceptive pills, pregnancy, and breastfeeding reduce the risk of OC. These factors reduce, halt, or delay overall ovulation cycles, respectively, which in turn reduces overall exposure to inflammation of the ovary and FT. The associations of parity and oral contraceptive use with invasive EOC were recently confirmed in a large, prospective study using the European Prospective Investigation into Cancer and Nutrition (EPIC) cohort that found only limited heterogeneity in the risk between reproductive factors and EOC subtypes [21]. Hysterectomy, tube ligation, and removal of ovaries are also protective against development of OC [22,23].

2.2. Infection

Pelvic inflammatory disorder (PID) is the infection of the female reproductive organs like cervix, uterus, FTs, and ovaries. It is a significant risk factor for OC and is caused by various bacteria and virus such as *Chlamydia trachomatis*, *Mycoplasma genitalium*, *Neisseria gonorrhoeae*, human papilloma virus, and cytomegalovirus [24,25]. Infection by these microbes results in DNA damage and production of ROS and induces a pro-inflammatory response, which involves secretion of cytokines and migration of immune cells [24]. PID is generally resolved with antibiotics within 48–72 hours of detection. However, repeated infection and unresolved inflammation can lead to chronic inflammation that is a risk factor for EOC.

2.3. Other Sources of Inflammation

The other causes of inflammation in the ovaries and/or FTs are endometriosis, obesity, Polycystic Ovarian Syndrome (PCOS), and talc exposure. Endometriosis is defined as presence of stroma and endometrial gland tissues in the pelvic peritoneum, rectovaginal septum, and ovaries [26]. Retrograde menstruation is the most commonly accepted theory for endometriosis. Retrograde menstruation results in aberrant accumulation of red blood cells (RBCs) and tissue, which can trigger an inflammatory response, activating the macrophages in the peritoneal cavity [27,28]. The macrophages lyse the RBCs, resulting in an increase in iron accumulation in the endometric implants and peritoneal fluid. The accumulated iron can catalyze formation of free radicals like RNS and ROS in the peritoneum and results in increased oxidative stress (OS). OS can activate NF-κB, in macrophages resulting in secretion of growth factors, cytokines, and IFNs. Around one third of women are affected by mild endometriosis, which resolves on its own over time. For the remaining cases, endometriosis results in chronic pain and inflammation, which can be resolved by excision of affected tissue or the outgrowth. However, in 45% of these cases, the endometriosis reoccurs resulting in repeated bouts of chronic inflammation [29,30].

Obese women have higher risks of EOC and HGSC and pro-inflammatory cytokines are associated with higher body mass index (BMI) levels. Adipose tissues secrete the cytokines TNF-α, IL-6, IL-8, and MCP-1, which can induce an inflammatory reaction in the peritoneum [31]. Continuous secretion of these cytokines leads to a state of chronic inflammation, which includes activation of macrophages and recruitment of NK cells and results in high levels of OS. Once the tumor has been initiated, the continuous secretion of cytokines by adipose tissue or omentum can facilitate migration of cancer cells to the omentum, promoting metastasis of the tumor into the peritoneum [30]. High levels (>10 mg/L) of C-reactive protein (CRP), a marker of global inflammation, are associated with an increased risk of EOC [32,33]. IL-6 itself is not a risk factor for EOC but in obese women IL-6 and CRP may be associated with increased EOC risk [33].

PCOS also contributes to inflammation in women and may increase risk of EOC [34]. PCOS is a hormonal disorder occurring in reproductive aged women during which ovaries may develop numerous small collections of fluid and fail to release eggs properly. Obesity, hyperandrogenism, and increased insulin resistance further characterize PCOS. Increased C-Reactive protein (CRP) and MCP-1 levels, indicative of low-level chronic inflammation, are elevated in women with PCOS [35–38].

Simultaneously chemokines like IL-18, IL-6, and TNF-α are also increased in circulation in women with PCOS [39–42]. The increase in inflammatory mediators correlates positively with BMI, suggesting that increased obesity in women with PCOS may be the source of inflammation. Increased DNA damage and OS is observed in women with PCOS, which may also increase risk for EOC [43]. Evidence linking PCOS directly to EOC is limited due to small study sizes, PCOS being associated with other EOC risk factors such as obesity, and PCOS possibly being only associated with one subtype of EOC, borderline serous [44].

Talc is a silicate mineral and exposure to it can cause inflammation of the ovaries and poses a risk hazard for development of EOC [45]. It has been proposed that talc from talcum powder used for dusting and from condoms and vaginal diaphragms can migrate up to the ovaries via retrograde flow of fluids and mucous and get lodged in the ovaries. Tubal ligation, which is protective for EOC, is thought to block the transport off talc from the lower genital tract. Talc behaves as a foreign particle, triggering an inflammatory response [46,47]. The talc attracts macrophages, which try to phagocytose it. The macrophages then send chemotactic signals to other immune response mediators and initiate a wound healing process. Since talc is not degradable by the body, it inhibits the wound healing process, resulting in chronic inflammation.

2.4. NSAIDS and Reduced Risk of EOC

Further connecting inflammation to EOC are several studies that demonstrate that intake of non-steroidal anti-inflammatory drugs (NSAIDs), specifically of aspirin, correlates inversely with risk of OC and endometrial cancer [48–52]. In vitro studies with OC cell lines and NSAIDS show that NSAIDs and COX-2 inhibitors facilitate apoptosis, however this effect is not dependent on COX-2 and may be due to upregulation of p21, a protein important for cell cycle arrest [53]. Another study by Arango et al., demonstrates that acetylsalicylic acid or aspirin resulted in increased apoptosis via downregulation of Bcl2 in an endometrial cancer cell line [54]. A third study has shown that a selective COX-2 inhibitor, JTE-522, can inhibit proliferation and increase apoptosis of endometrial cancer cells by increasing levels of p53 and p21 and decreasing phosphorylation of retinoblastoma (Rb) protein, which results in its activation; all of which results in cell cycle arrest [55,56]. Simultaneously, there was an increase in caspase-3 activity, which is indicative of increased apoptosis. Another mechanism by which aspirin could facilitate its chemopreventive nature is by inhibiting oxidative induced DNA damage [57]. COX-1 is also expressed in normal ovaries of the laying hen, with expression increasing in post-ovulatory follicles suggesting its importance for or a role in ovulation. With the onset of OC, COX-1 expression is increased [58] and COX-1 inhibition and NSAIDs have shown to decrease proliferation of ascites in the laying hen OC model [59]. Further, when 0.1% aspirin was included in their diet for one year, although the onset of OC was not different, the progression of cancer was slower when compared to hens fed with regular diet [60].

As discussed, inflammation results in secretion of ROS, growth factors, cytokines, and chemokines into the shared environment surrounding the ovary and distal FT. Exposure of normal tissue to these inflammatory mediators results in activation of downstream signaling that can promote the transformation of normal cells or survival of already transformed cells. Once EOC has already formed further exposure of cancer cells to these inflammatory mediators also results in activation of downstream signaling within the cancer cell and in the surrounding tissue, creating an inflammatory environment that can further promote EOC (Figure 1). We will discuss in more detail how key inflammatory mediators contribute to EOC initiation, progression, metastasis, and chemoresistance.

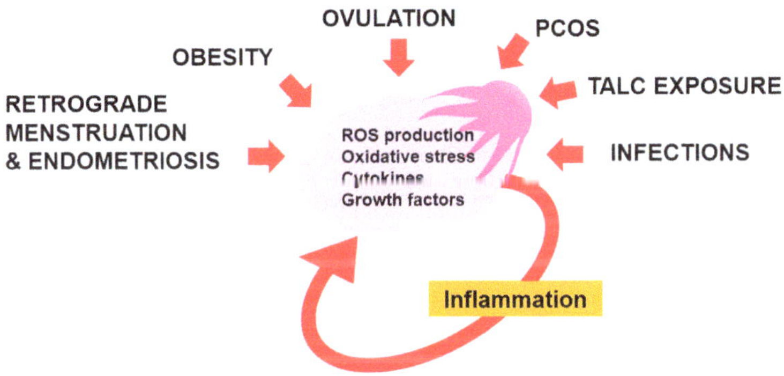

Figure 1. Sources of inflammation in the ovary and fimbriae. Ovulation, retrograde menstruation, endometriosis, infections, exposure to talc, Polycystic Ovarian Syndrome (PCOS), and obesity result in exposure of the ovary and fimbriae to reactive oxygen species (ROS), oxidative stress, cytokines, and growth factors, generating an inflammatory response that leads to additional production of ROS and cytokines in the ovary. Unresolved, chronic inflammation is a critical risk factor for tumor initiation.

3. Inflammation and EOC Initiation and Progression

Tumorigenesis is a multistep process that requires cells to gain the ability to evade apoptosis and antigrowth signals, proliferate independently of stimuli, develop a support system (angiogenesis), and have the capacity to invade and metastasize. Tumorigenesis is initiated by the transformation of a normal cell to a malignant one. The deregulation of the above mentioned processes in the malignant cell could potentiate its progression to cancer.

One mechanism of cancer initiation is genomic instability due to DNA damage [61] and EOCs exhibit a high number of chromosomal aberrations and genomic instability [62]. The most common gene mutations in HGSCs include *BRCA*, *TP53*, and genes in involved in mismatch repair and the DNA damage response [63]. A pro-inflammatory ME can also contribute to genetic instability and therefore play a role in EOC cancer initiation. A pro-inflammatory ME, which is continuously supplemented by ROS, cytokines, and growth factors, can cause DNA damage in epithelial ovarian and FT cells, switch on antiapoptotic pathways, and initiate transformation of normal cells. When cells transformed either by oncogenic alterations or by exposure to inflammation are in a pro-inflammatory ME they are able to turn on pro-survival signaling pathways rather than the senescence pathways that are normally induced by oncogene expression in normal cells. For example, disruption of the RAS pathway results in activated NF-κB signaling and upregulation of its downstream targets including cytokines like IL-1β, IL-6, and IL-8. These cytokines are upregulated in EOC patients and their increased levels correlate with decreased survival [64–71]. The inflammatory mediators like cytokines, chemokines, growth factors, and prostaglandins secreted by the transformed epithelial cells further promote a pro-inflammatory environment, which can reprogram the surrounding cells to form the TME. The TME is mainly composed of endothelial cells, cancer associated fibroblasts (CAFs), adipocytes, tumor associated macrophages (TAMs), regulatory T-cells, pericytes, infiltrated immune cells such as neutrophils, lymphocytes, and various other cells that further secrete growth factors and cytokines which potentiate tumor progression (Figure 2, Table 1). Furthermore, OC-initiating cells (OCICs) have been identified in tumors and ascites that exhibit stem cell like properties and are capable of forming tumors [65,66,72]. Cytokines can promote self-renewal of CD133$^+$ OCICs to potentiate tumor progression [73].

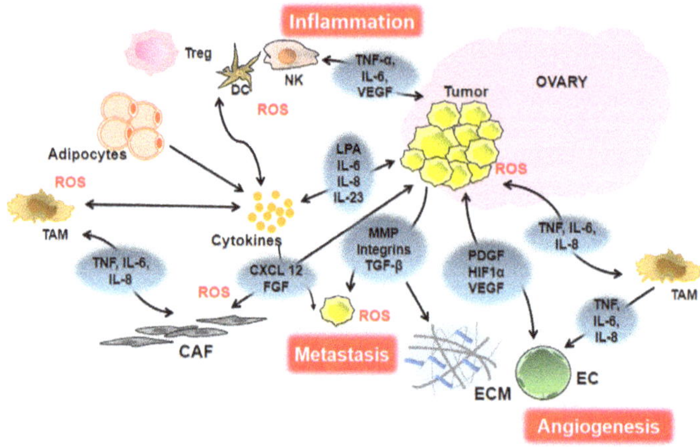

Figure 2. Inflammatory mediators contributing to EOC progression, metastasis, and angiogenesis. EOC cells produce ROS, chemokines, cytokines, and growth factors that can: (1) Lead to recruitment of immune cells like Dentric cells (DC), Natural killer cells (NK), Tumor associated macrophages (TAMs), and T-regulatory (Treg) cells into the TME, which generate additional cytokines, ROS, and growth factors, resulting in chronic inflammation. (2) Stimulate the tumor cells themselves, the TAMs, and the surrounding fibroblasts (also known as cancer associated fibroblasts or CAFs) to proliferate and secrete growth factors like TGF-β and FGF that stimulate production of integrins and Matrix Metalloproteins (MMPs), resulting in migration of the tumor cell via degradation of the extra cellular matrix (ECM). (3) Stimulate endothelial cells (EC) to produce growth factors like PDGF and EGF and factors like VEGF that stimulate angiogenesis. The double arrows indicate that the cells are a source of the factor as well as stimulated by it.

The innate immune response can prevent tumorigenesis by recognizing and eliminating transformed cells. However, chronic inflammation can contribute to the ability of premalignant cells to evade apoptosis, escape the immune surveillance, and continue to grow, resulting in tumor formation. As mentioned, EOC can originate from either distal FT or ovarian epithelial cells. Since both the ovary and fimbria are exposed to the same ME, exposures reviewed here are relevant to initiation in either tissue. [74]. In this section we will review the role of OS and some specific pro-inflammatory mediators and signaling pathways in the initiation and progression of EOC.

3.1. ROS and Oxidative Stress

ROS plays an important role in the normal female reproductive cycle, from affecting maturation of the oocyte to ovulation, apoptosis of cells in corpus luteum, and embryo development [75]. Ovulation results in increased levels of DNA damage in the FT epithelium that is likely a result of the ROS generated during ovulation or the ovulation-associated increase in numbers of infiltrating macrophages in the FT [17]. Additionally, during infection and inflammatory responses immune and damaged cells produce ROS resulting in continuous exposure of the ovaries, FTs, and peritoneal cavity to ROS [76–78]. ROS exposure could potentially lead to epithelial cells in the ovary and FT undergoing transformative changes, as has been demonstrated for ovarian surface epithelium cells grown in 3D culture [79]. Elevated ROS and RNS levels beyond the level that cells can neutralize results in OS. Increased OS results in DNA damage, activation of signaling cascades, and epigenetic alterations.

DNA damage in a cell results in stimulation of DNA damage repair pathways. These repair pathways can be inactivated or be erroneous, which results in increased genotoxic stress and mutated DNA. Secretory tubal epithelial cells in the FT, a cell of origin for HGSC, are particularly susceptible to genotoxic injury with persistent DNA damage that could lead to mutation and STIC formation [80].

Mutations in tumor oncogenes and suppressors result in overexpression, constitutive activation of the protein, loss of expression, or expression of nonfunctional proteins, resulting in a transformed cell. Follicular fluid may have transformative properties as it has been demonstrated that bathing fimbriae with follicular fluid containing high levels of ROS results in increased levels of DNA damage. Bathing fimbriae that have loss of p53 and Rb with this follicular fluid results in evasion of apoptosis and cells with persistent DNA damage [81].

ROS can activate pro-survival intracellular tyrosine phosphorylation signaling cascades, mainly regulated by the MAPKs and redox sensitive kinases. Activation of c-Jun, JNK, ERK (extracellular signal-regulated kinase), and p38-MAPK signaling cascades results in upregulation of cell cycle proteins that enhances proliferation. Activation of JNK can also activate NF-κB, which can suppress apoptosis. The MAPK pathway inhibits apoptosis and regulates differentiation. When activated in transformed cells these pathways are important for tumor initiation. ROS affects redox sensitive factors like thioreoxin, which is also found elevated in OC cell lines [82]. Thioredoxin is involved in redox regulation of transcription factors such as NF-κB, NRF2, forkhead box class O (FOXO) proteins, reducing factor-1 (ref-1), and hypoxia inducible factor (HIF-1α), thereby increasing their binding to the DNA. Most of these transcription factors promote tumor growth and progression by regulating expression of genes that affect cell survival and growth [83,84]. For example, FOXO, NRF2, and ref-1 transcription factors upregulate transcription of anti-oxidant proteins that scavenge free radicals and allow survival of damaged or transformed cells [85]. HIF-1α upregulates the antiapoptotic factor, bcl-2 as well as vascular endothelial growth factor (VEGF), a factor important for angiogenesis.

OS has also been shown to facilitate epigenetic mechanisms in many cancers, including EOC [86]. Innate immune-mediated inflammation drives epigenetic silencing of tumor suppressor genes (TSGs) [87]. At sites of inflammation high levels of OS result in oxidative DNA damage that is recognized by the mismatch repair proteins mutS homolog MSH2 and MSH6. MSH2 and MSH6 then recruit epigenetic silencing proteins, including DNA methyltransferase 1 (DNMT1) to the sites of damage [88]. In an in vivo model of inflammation-driven colon tumorigenesis this early recruitment to sites of oxidative DNA damage results in permanent methylation of TSGs in tumors that form at the sites of inflammation [89]. While such a mechanism has not been directly proven in EOC models, Sapoznik et al. have demonstrated that exposure to follicular fluid or inflammation can induce Activation-Induced Cytidine Deaminase (AIDS) in fallopian tube epithelial cells, which results in epigenetic and genetic changes, increase in DNA damage and genotoxic stress and may be a contributing factor to EOC [90].

3.2. TNF-α

The cytokine TNF-α plays an important role in the process of ovulation and in removal of damaged corpus luteum. TNF-α ligand and its receptors, TNFRI and TNFRII are upregulated in ovarian tumors compared to normal ovarian tissue and high levels of TNF-α are found in ascites from OC patients [91–93]. OC cells have also been shown to secrete high levels of TNF-α as compared to normal ovarian epithelial cells resulting in autocrine upregulation of TNF-α mRNA and in expression of other pro-inflammatory cytokines, chemokines, and angiogenic factors like IL-6, M-CSF, CXCL2, CCL2, and VEGF [93,94]. Kellie et al. have shown using mouse models that TNF-α stimulates IL-17 production via TNFRI resulting in myeloid cell recruitment to the ovarian TME and increased tumor growth [95]. TNF-α, also upregulates AIDS transcript levels which can contribute to genotoxic stress [90].

3.3. IL-6

The cytokine IL-6 has been associated with poor survival in OC and is emerging as a potential therapeutic target for EOC [67,68,96,97]. IL-6 is normally produced by ovarian epithelial and OC cells. Macrophage migration inhibitory factor (MIF), EGF, and Transglutaminase secreted by OC cells can stimulate IL-6 production via activation of NF-κB [98–100]. IL-6 increases proliferation of OC cells by facilitating their exit from G1 into S phase of the cell cycle and by activation of the MAPK-ERK-Akt (protein kinase B) growth promoting signaling pathway [101]. ERK activation can promote formation of ascites by increasing the migration of tumor cells [70]. IL-6 production by M2 macrophages present in ascites in later stages of EOC can also stimulate cancer cell proliferation via STAT3 activation [102]. High levels of IL-6 can result in immune suppression by downregulation of IL-2, which stimulates Teff cell production [103]. IL-6 also stimulates production of Metallomatrix proteins (MMPs) in OC cells, which increases their invasive properties and promotes tumorigenesis [101,104].

3.4. IL-8

IL-8 a member of C-X-C chemokine family is present in the preovulatory follicle [105] where it may play a role in increasing leukocyte infiltration [106]. It is also elevated in ovarian cysts and in OC patients compared to healthy controls [107,108]. IL-8 has been found to be present in significantly higher levels in the ascites of patients with OC in comparison to patients with benign gynecological disorders [109]. Increased IL-8 expression has been associated with poor prognosis in OC patients [107]. Treatment of EOC cells with IL-8 results in their increased proliferation, which is accompanied by an increase in cyclins B1 and D1 and is dependent on phosphorylation of Akt and ERK [110]. Cyclins B1 and D1 are important for cell cycle progression, and an increase in their expression leads to increased cell growth. On the other hand, two independent studies have demonstrated that IL-8 inhibits EOC growth by increasing neutrophil infiltration [111,112].

3.5. Lyophosphotidic Acid (LPA)

LPA is a phospholipid that binds to and activates the endothelial differentiation gene (Edg) family of receptors. LPA is present in ovarian follicular fluid and it stimulates IL-6 and IL-8 production in the corpus luteum [113,114]. OC cells have been shown to produce LPA, which functions like a growth factor [115–119]. Plasma and ascites of OC patients have elevated levels of LPA that contribute to OC progression via upregulation of COX-2 and MMP2 [115,120,121]. LPA can bind to LPA_2 receptor and induce expression of IL-6 and IL-8 via activation of NF-κB and AP-1 in OC cell lines [122]. It can induce ROS dependent Akt and ERK phosphorylation and inhibition of LPA can increase apoptosis of EOC cells [123]. ERK phosphorylation can induce phosphorylation of HIF-1α, which then can upregulate VEGF and promote tumorigenesis. Another group demonstrated that stimulating EOC cells with ether-linked LPA resulted in their increased proliferation and survival by increased synthesis of DNA and activation of Akt via PI3K, which contributes to tumor progression [124].

3.6. Prostaglandins and COX-1 and COX-2

Prostaglandins are secreted in the ovary, FT, and uterus. They are important for maturation of the oocyte and facilitate the movement of the FT so that the mature oocyte can move from the ovary to the uterus. In the uterus prostaglandins help regulate and maintain uterine blood flow. COX-1 and COX-2 are enzymes that catalyze the production of prostaglandins from arachidonic acid and are overexpressed in OC patients [22,125,126]. High COX levels positively correlate with increased cell proliferation, angiogenesis, and malignancy in ovarian tumors [126,127]. COX-1 and COX-2 are normally involved in the acute inflammatory response but can become dysregulated in chronic inflammatory or TMEs. Obermajer et al. have demonstrated that prostaglandins produced by COX-2 can stimulate production of CXCR4 and its ligand Stromal cell derived factor 1 (SDF1) CXCL12 in myeloid derived suppressor cells (MDSC), which stimulates them to migrate towards OC ascites [128]. MDSCs inhibit the proliferation and differentiation of T cells, resulting in overall immune suppression, which allows the tumor cells to escape immune surveillance and continue to grow. Genetically engineered mouse models of EOC; one harboring the *p53* and *Rb* deletion and other the $KRAS^{G12D}$ mutation and *Pten* deletion, demonstrate increased COX-1 levels, thus suggestive that COX-1 could be used as a potential biomarker and therapeutic target for EOC [129]. Further when

COX-1 was inhibited in EOC cells, it led to reduction in prostacyclin (a type of prostaglandin) synthesis and reduced tumor growth by enhanced apoptosis [130].

4. Inflammation and EOC Angiogenesis

Angiogenesis is required for the growth of both primary and metastatic tumors [131]. The process of angiogenesis is a complex multi step process reviewed previously [132]. It is regulated by a balance between pro-angiogenic and antiangiogenic factors. Hypoxic and ischemic areas are present at sites of inflammation and also in tumors mainly due to obstruction of local blood vessels, differences in pace of growth of blood vessels and growth of the tumor and/or infiltration of immune cells. Macrophages accumulate at hypoxic sites and alter their gene expression profiles in response to the hypoxic conditions. One of the important genes for angiogenesis that is upregulated by hypoxia is VEGF [133,134]. The rate-limiting step in angiogenesis is VEGF signaling in endothelial cells (ECs) [135]. VEGF functions via tyrosine kinase receptors VEGF-1 and VEGF-2 and promotes migration, survival, proliferation of ECs, and formation of new blood vessels [136–138]. Many of the inflammatory mediators discussed so far are also involved in promoting angiogenesis in EOC as detailed below (Figure 2, Table 1).

4.1. TNF-α

TNF-α creates a pro-inflammatory TME and has also been associated with promoting angiogenesis. It has been hypothesized that TNF-α induces the production of soluble factors that promote tumor angiogenesis. Culture supernatants from TNF-α expressing cells induce the growth of mouse lung endothelial cells in vitro while culture supernatants from TNF-α lacking cells do not exert the same effect [94]. In pituitary adenomas TNF-α is known to induce VEGF that in turn induces CXCL12 [139,140]. VEGF and CXCL12 synergistically induce angiogenesis in EOC [141]. Mice injected with OC cells lacking TNF-α have reduced vascular density in their tumors and reduced formation of blood vessels in the peritoneal deposits. These mice also did not have accumulation of ascetic fluid suggesting the importance of TNF-α in angiogenesis and EOC progression [94].

4.2. IL-6

In physiological conditions, IL-6 is involved in angiogenesis in the ovary during the development of ovarian follicles [142]. IL-6 induces the phosphorylation of STAT3 and MAPK in ovarian endothelial cells thereby enhancing their migratory ability, a key step in angiogenesis [143]. As explained before, OC cells also secrete increased amounts of IL-6. Some OC cells also secrete an alternative splice variant of IL-6Rα, the soluble form sIL-6R, which consists of only the ectodomain of the transmembrane receptor. By a process called trans-signaling, the sIL-6R-IL-6 complex initiates signaling in cells in the ME that do not express the transmembrane receptor facilitating angiogenesis [144].

4.3. IL-8

Several studies have clearly established the role of IL-8 in promoting angiogenesis. Hu et al., demonstrated that IL-8 plays a role in angiogenesis using a rat sponge model [145]. IL-8 was also able to induce angiogenesis in the rat cornea, which is normally avascular [146]. As explained in the previous section, there are several sources of IL-8 in ovarian TME. Overexpression of IL-8 in A2780 (non-IL-8 expressing) OC cells has been shown to increase the expression of VEGF, MMP-2, and MMP-9; while depletion of IL-8 in SKOV3 (IL-8 expressing) cells has been shown to reduce VEGF, MMP-2, and MMP-9 [110]. The process of angiogenesis involves degradation of extracellular matrix components and proliferation and migration of endothelial cells. MMPs are a family of endopeptidases that breakdown components of extracellular matrix and have been implicated in angiogenesis [147]. Because of the importance of VEGF and MMPs in angiogenesis these findings suggest that IL-8 in the ovarian TME will promote the formation of new blood vessels in EOC. Targeting IL-8 using mouse models reduces EOC growth and decreases angiogenesis [112].

Table 1. Role of inflammatory mediators in different stages of tumor progression.

Inflammatory Mediators	Secreting Cell Type	Stages in Tumor Progression			
		Initiation and Progression	Angiogenesis	Metastasis	Chemoresistance
TNF-α ligands, TNFRI, TNFRII	OC cells, infiltrating monocytes, macrophages	↑ autocrine production of TNF-α and IL-6, M-CSF, CXCL2, CCL2 [93,94] and AIDS mRNA level [90]	↑ VEGF, VEGF↑ CXCL12 and promotes angiogenesis [139–141]	↑ TGF-α secretion by stromal fibroblasts which promote peritoneal metastasis [148] Enhances migration of OC cells towards CXCL12 [149,150]	
IL-6	Ovarian epithelial cells, OC cells, M2 macrophages, mesothelial cells, TAMS, ascites	↑Proliferation by promoting G1 to S transition and MAPK-ERK- Akt activation and STAT3 activation [101,102] ↓IL-2, resulting in immune suppression [103]	Induces STAT3 and MAPK phosphorylation which enhances migration of endothelial cells [143] sIL-6R-IL-6 facilitates angiogenesis in cells lacking IL-6 receptor [144]	Stimulates production of MMPs in OCs which ↑ invasion and migration [101,104] ↑ IL-6 in ascites enhances invasion via JAK-STAT signaling [151]	↓ Caspase- 3 cleavage and makes OC cells resistant to cisplatin and paclitaxel [152] ↑Expression of MDR1, GSTpi, Bcl-2, Bcl-xL, and XIAP [152]
IL-8	Pre-ovulatory follicles, OC cells, ascites	↑ Proliferation by ↑ cyclin B1 and cyclin D1 via pAkt [110]	↑ Expression of VEGF, MMP-2, MMP-9 promoting angiogenesis [110]	Activates TAK1/ NF-κB via CXCR2 [153]	Blocks TRAIL induced apoptosis to promote resistance [154]
LPA	Follicular fluid, corpus luteum, OC cells, ascites	↑ IL-6 and IL-8 via NF-κB and AP-1 [113,114,122] ↑COX-2 AND MMP2 [115,120,121] ↑ phosphorylation of Akt and ERK resulting in increased cell cycle [123,124]	↑ Expression of VEGF via Myc and Sp-1 [155]	↑ urokinase, which results in degradation of basemembrane protein to promote metastasis [156,157]	
Prostaglandins, COX-1 and COX-2	Ovary, FT, uterus, MDSCs	↑ CXCR4 and SDF1 in MDSCs resulting in immune suppression [128]	↑ Bcl-2 and blood vessel formation [158,159]		↑ Bcl-2, thus inhibiting apoptosis in lung, colon, breast and prostate cancers [158,159]
TGF-β and EGF	OC cells, CAFs			TGF-β ↑ VCAN, which activates NF-κB and ↑MM-9 [160]	↑ EGF protects cells from cisplatin-induced apoptosis [161]. Inhibiting TGF-β sensitizes resistant cells [162]

4.4. LPA

In addition to playing a role in initiation, and progression, LPA has also been implicated in angiogenesis in OC. LPA has been shown to induce transcriptional activation of VEGF in EOC cell lines [163]. Transcriptional activation of VEGF primarily occurs through HIF-1α under oxygen limiting conditions in Hep3B hepatocellular carcinoma cells [164]. LPA mediated induction of VEGF expression has been shown to be independent of HIF-1α in EOC cell lines. Transition metal cobalt treatment also leads to stabilization of HIF1α similar to hypoxia. Combination treatment of EOC cells with cobalt and LPA additively increased VEGF production suggesting the effect of two different pathways [155]. LPA activates c-Myc and Sp-1, which induce VEGF expression through consensus binding sites in the VEGF promoter that have been implicated in HIFα independent induction of VEGF [155].

5. Inflammation and EOC Metastasis

Tumor metastasis is the major cause of mortality in most cancers, including EOC. Most EOC patients are diagnosed at an advanced stage when the cancer has already metastasized [165]. Dissemination of cancer cells to distant sites is a complex multi-step process called the invasion-metastasis cascade and is reviewed in detail in previous papers [166–168]. Briefly, some major steps in metastasis are—invasion through the basement membrane, intravasation into the lymphatics and circulation, survival of disseminating cancer cells in circulation, extravasation into surrounding tissues, colonization, and finally, formation of micro and macro metastases. However, unlike other epithelial malignancies, EOC has a different pattern of metastasis. EOC cells directly shed from the primary tumor into the peritoneal space and disseminate to organs in the peritoneal cavity. One of the prerequisites for cancer cells to metastasize is to undergo a process called epithelial to mesenchymal transition (EMT) where they lose their ability to attach to the basement membrane and acquire a mesenchymal phenotype and characteristics. Several recent evidences have indicated that the TME aids tumor cells to acquire these properties facilitating the metastatic cascade. An example of the ME promoting metastasis is the presence of STICs in the distal part of the FT, which shares its ME with ovary. Yang-Hartwich et al. have demonstrated that granulosa cells in the ovary secrete SDF-1 (stromal cell-derived factor 1) [169]. SDF-1 functions as a chemoattractant and recruits malignant FT cells to the ovary suggesting that the ovary is a primary site of metastasis, not the primary tumor site. Russo et al. demonstrated that loss of PTEN (phosphatase and tensin homolog) by the malignant FT cells and upregulation of WNT4 (wingless-related MMTV integration site 4) is crucial for initial metastasis to the ovary thereby supporting the tubal origin of EOC and the ovary as the primary site of metastasis [170]. The cells that make up the TME also secrete various inflammatory mediators, which facilitate progression and metastasis of OC cells (Figure 2, Table 1). These factors enable tumor metastasis by deregulating signal transduction pathways. Examples include the PI3-Akt and RAS-ERK pathways, which control migration and invasion through downstream effectors like Rho family GTPases, extracellular proteases, integrins, matrix associated proteins like focal adhesion kinases (FAK), and transcription factors like ETS2 and AP-1 [171–173]. Robinson-Smith et al. demonstrated that peritoneal inflammation correlated with dissemination of cancer cells from the ovaries in SCID mice. Augmenting the inflammatory response using thioglycolate accelerated ascites formation and metastasis while suppressing the inflammation using acetyl salicyclic acid impeded ascites formation and reduced metastasis. This inflammation-induced metastasis of OC cells was found to be primarily mediated by macrophages and not neutrophils or NK cells [174]. As explained in one of the previous sections a pro-inflammatory environment can be created in the peritoneum due to secretion of cytokines like IL-6 and TNF-α by adipose cells [31]. Omentum, the primary site of metastasis of OC, is largely composed of adipose cells. In addition to adipocytes, omentum also consists of blood and lymph vessels, immune cells, and stromal cells [175]. Adipocytes have been shown to increase migration, invasion, and proliferation of EOC cells. Upregulation of SUSD2 a secreted tumor suppressor by adipocytes by guadecitabine treatment reduced EOC migration and invasion. This finding suggests that epigenetic changes in the stromal cells in addition to EOC cells can facilitate EOC metastasis [176]. Omentum has aggregates of immune cells around the vasculature commonly referred to as milky spots [177]. Melanoma, lung carcinoma, ovarian carcinoma, and mammary carcinoma cell lines have been shown to specifically metastasize to the immune cell aggregates in the omentum when injected intraperitonealy into C57BL/6 mice [178]. These milky spots in the omentum have also been shown to facilitate metastatic colonization of the OC cells. Clark et al. have suggested that both adipocytes and milky spots have specific and important roles in metastatic colonization of OC cells [179]. These evidences imply that omentum potentially provides a good niche for the growth of ovarian cancer cells. Here we will specifically discuss how inflammatory mediators promote tumor metastasis in EOC.

5.1. ROS

EOC cells produce a large amount of ROS [180]. Loss of E-cadherin is one of the characteristic features of tumor cells with increased ability to migrate and invade. Wang et al. demonstrated that ROS leads to HIFα mediated activation of lysl oxidase. Lysl oxidase was shown to inversely correlate with E-cadherin expression promoting migration and invasion in EOC cells [181]. Tumor cells treated with sub-lethal doses of H_2O_2 failed to attach to the extracellular matrix components fibronectin and laminin and had increased metastatic colonization of lung, thereby establishing a role for ROS in tumor cell metastasis [182].

5.2. TNF-α

TNF-α provides a good example of how interactions between cancer and stroma aid in OC metastasis. Ascitic fluid and OCs contain a large number infiltrating macrophages in part because OCs constitutively produce M-CSF, which functions as a chemoattractant for monocytes [183]. These infiltrating monocytes produce many cytokines one of which is TNF-α [184,185]. OC cells also have elevated TNF-α expression that is regulated by DNA hypomethylation and chromatin remodeling of the TNF-α promoter. Increased TNF-α produced by OC cells and macrophages stimulates increased expression of TGF-α in stromal fibroblasts. TGF-α secreting stromal fibroblasts promote peritoneal metastasis of OC via EGF receptor signaling [148].

Furthermore, in EOC cells and clinical biopsies TNF-α expression correlates with one of the most commonly expressed cytokine receptors CXCR4. TNF-α stimulation of EOC cells enhanced their migration toward the only CXCR4 ligand, CXCL12. Stimulation of EOC cells by CXCL12 induced mRNA and protein expression of TNF-α. Therefore, a positive feedback loop has been suggested where in CXCL12 induced TNF-α potentially acts on the cancer cells and induces CXCR4 expression thereby enhancing tumor cell migration [149,150].

5.3. IL-6

IL-6 has also been implicated in metastasis of OC. Elevated levels of IL-6 found in serum and peritoneal fluid of EOC and OC patients have many sources [186–188]. Mesothelial cells in the peritoneum, TAMs, and EOC cells all secrete IL-6 [67]. M2 polarized macrophages in the ovarian TME induce proliferation and invasion of EOC cells by secretion of IL-6 [189]. Increased IL-6 present in ascites from OC patients enhanced the invasive ability of OC cells via the JAK-STAT signaling pathway. Canonically IL-6 signaling occurs by binding of the ligand to its transmembrane receptor IL-6Rα. The effect of IL-6 on invasion of OC cells correlated with their IL-6R expression [151]. Because through trans-signaling, the sIL-6R–IL-6 complex initiates signaling in cells that do not express the transmembrane receptor [144], we hypothesize that IL-6 produced by macrophages could also promote invasion of OC cells similar to the mechanism of induction of angiogenesis.

5.4. IL-8

Increased proliferation, anchorage independent growth, and angiogenic potential are some prerequisites for cells to metastasize. IL-8 increases the proliferation of OC cells and upregulates VEGF and MMP2 and 9 via activation of NF-κB, which results in enhanced invasive phenotype of OC cells. IL-8 has been shown to activate TAK1/NF-κB signaling via CXCR2, thereby facilitating the seeding and growth of OC cells in the peritoneal cavity during metastasis [153].

5.5. LPA

LPA promotes proliferation, survival, and metastasis of EOC cells by inducing the expression of c-Myc, VEGF, IL-8, MMPs and COX-2 [163,190–193]. LPA acts through its receptors LPAR1-3, which are members of G-protein coupled receptor superfamily. Invasive EOC cells have significantly higher expression of LPAR1 in comparison to non-invasive cell lines and LPA induces EOC cell invasion specifically through LPAR1 and not through LPAR2 or LPAR3 [194]. It can also induce secretion of urokinase in EOC cells, which has been shown to play a role in metastasis and its high levels correlate with advanced OC and poor survival in patients. LPA has been shown to increase promoter activity, mRNA levels, protein levels, and enzyme activity of Urokinase plasminogen activator (uPA) possibly via the edg-4 LPA receptor [156]. uPA is involved in converting plasminogen to plasmin, which facilitates the degradation of basement membrane and extracellular membrane proteins like fibronection aiding in metastasis [157].

5.6. TGF-β

TGF-β initiates signaling by dimerization of serine/threonine kinase receptors. The dimerization of receptors results in their phosphorylation, which then relays signals downstream via SMAD dependent and SMAD independent pathways. Phosphorylation by the TGF-β receptor causes R-SMADs to bind to Co-SMAD and translocate to the nucleus, where they activate transcription of genes that promote invasion, migration. Bone morphogenic proteins (BMPs) are cytokines that belong to TGF-β family and have been associated with progression of many different cancer types. Their mechanism of promoting tumor progression depends on the TME in which the cancer grows and their mode of metastatic spread [195]. Specifically, BMP-2 overexpression has been associated with poor prognosis in OC [196]. Additionally, TGF-β could potentially modify the TME to promote tumorigenesis. Veriscan (VCAN), an extracellular matrix associated protein, was upregulated by TGF-β through TGF-β receptor II (TGFBR2) and SMAD signaling making the EOC cells more aggressive. Increased VCAN expression enhanced motility and invasion of EOC cells by activating NF-κB signaling, increased expression of MMP-9, and hyaluronidase mediated motility receptor [160]. CAFs have higher expression of TGF-β receptors in comparison to normal ovarian fibroblasts and EOC cells suggesting that CAFs within the TME are more responsive to TGF-β then the other cell types [160].

6. Inflammation and EOC Chemoresistance

The standard treatment for EOC patients is cytoreductive surgery followed by platinum/taxane-based chemotherapy [197]. The main obstacle in treatment of EOC patients is development of chemoresistance. Resistance to chemotherapy can be either intrinsic or acquired. Inherent gene expression patterns harbored by chemo-naïve tumor cells contribute to intrinsic resistance. Acquired resistance is a consequence of different alterations induced after exposure to chemotherapeutic agents [198]. Different mechanisms, including increased drug efflux, decreased uptake of the drug, inactivation of the drug, increased DNA repair, and reduced apoptotic response, have been implicated in development of platinum resistance [199]. Several recent studies have demonstrated that the TME contributes to both intrinsic and acquired resistance. One type of intrinsic drug resistance influenced by the TME is referred to as environment mediated drug resistance (EMDR). In EMDR, factors and cells present in the TME activate diverse signaling events, transiently protecting the tumor cells from undergoing apoptosis in response to chemotherapeutic agents [200,201]. Another type of drug resistance induced by cytokines, chemokines, and growth factors secreted by fibroblast cells in the tumor stroma is called soluble factor mediated drug resistance (SFM-DR). A good example of SFM-DR is IL-6 mediated drug resistance in multiple myeloma. IL-6 is important for growth of multiple myeloma cells. IL-6 activates STAT3 signaling in these cells and protects them from Fas mediated apoptosis by upregulating antiapoptotic protein Bcl-X_L [202]. Myeloma cells that produced IL-6 in an autocrine manner were found to be resistant to dexamethasone induced apoptosis while non-IL-6 producing cells were sensitive [203]. Cell adhesion mediated drug resistance (CAM-DR) occurs due to adhesion of tumor cells to extracellular matrix components like laminin, collagen, and fibronectin or due to fibroblasts present in the tumor stroma [204]. An example of this type of resistance is when drug sensitive myeloma cells were adhered to an extracellular matrix component fibronectin, they exhibited a reversible drug resistant phenotype which was not due reduced drug accumulation or increase in antiapoptotic proteins like Bcl-X_L [201]. Here we will discuss specific inflammatory mediators and their role in OC chemoresistance (Figure 3).

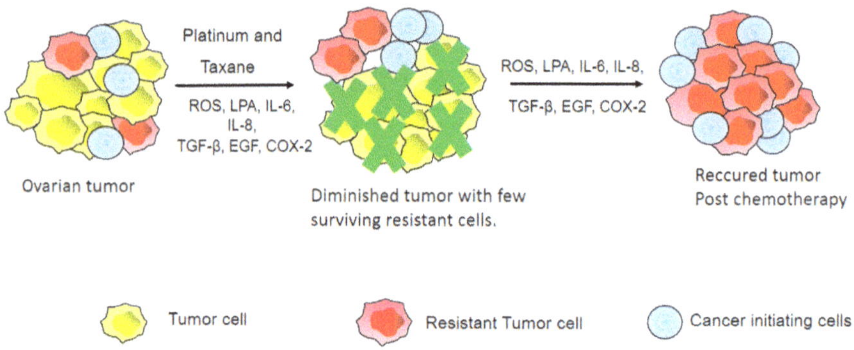

Figure 3. Inflammatory mediators contribute to chemoresistance of EOC. A combination of platinum and taxane drugs is currently used as chemotherapy for OC. ROS, Lyophosphotidic Acid (LPA), cytokines, and growth factors like TGF-β and EGF increase tumor cell survival by upregulating antiapoptotic genes, by stimulating stemness and proliferation of cancer initiating cells, by increasing repair of damaged DNA, or by increasing efflux of the drug. The resistant tumor cells and the cancer initiating cells can then proliferate under the influence of growth factors and cytokines resulting in a recurrent chemoresistant tumor.

6.1. ROS

ROS are abundant in the pro-inflammatory TME. Malignant EOC tissues have been shown to have 96% higher ROS levels than normal controls [205]. OC stem like cells or OCICs are more drug resistant and responsible for relapse of chemoresistant tumors [66]. OCICs produce ROS and superoxide. This ROS induces the expression of peroxisome proliferator-activated receptor-gamma coactivator (PCG)-1α, which regulates mitochondrial biogenesis and is required for expression of detoxifying enzymes [206,207]. PCG1α increases the aldehyde dehydrogenase (ALDH) activity and expression of multidrug resistance gene (MDR1). MDR1 is an ATP dependent transporter that has been associated with efflux of platinum based drugs from OC cells contributing to platinum resistance. Scavenging ROS reduced expression of PCG1α and drug resistant related genes thereby linking ROS to development of chemoresistance [207].

6.2. IL-6

IL-6 in the OC TME is associated with increased chemoresistance. Wang et al. demonstrated that autocrine production of IL-6 by EOC cells makes them resistant to cisplatin and paclitaxel by causing decreased proteolytic cleavage of capase-3. Paclitaxel resistant EOC cells have increased expression of IL-6 and one of its downstream effectors STAT3 [208,209]. IL-6 producing OC cells also had increased expression of multidrug resistant genes MDR1 and GSTpi and anti-apoptotic genes

Bcl-2, Bcl-xL, and XIAP, suggesting that IL6 promotes drug resistance by increasing drug efflux and reducing apoptosis [152].

6.3. IL-8

IL-8 blocks TRAIL-induced apoptosis and reduces caspase cleavage in EOC cell lines by decreasing the expression of death receptor (DR) 4 [210]. TRAIL is a cell death inducing ligand that belongs to the TNF superfamily and has been shown to induce apoptosis specifically in tumor cells and not in nontransformed cells [211,212]. Combination of TRAIL and the chemotherapeutic drugs—cisplatin, doxorubicin, and paclitaxel has been shown to induce apoptosis in chemoresistant EOC cell lines by causing increased caspase and PARP cleavage [154]. This finding suggests that IL8 may contribute to chemoresistance by blocking TRAIL.

6.4. LPA

LPA has been shown to contribute to platinum resistance by preventing cells from undergoing cisplatin-induced apoptosis without affecting their proliferation rate. The mechanism of how LPA inhibits apoptosis in EOC cells in response to cisplatin is not yet clearly understood [161].

6.5. TGF-β and EGF

Recurrent OC show significantly higher expression of TGF-β1 and TGF-β3 in comparison to primary tumors and normal ovary tissue [213]. Inhibition of TGF-β by the inhibitor LY2109761 sensitizes resistant SKOV3 cells to cisplatin suggesting that TGF-β contributes to the development of platinum resistance in EOC cells [162]. Cisplatin resistant A2780P cells had hypomethylation and upregulation of TGFBR2 confirming the involvement of the pathway in acquisition of platinum resistance [214]. An elevated level of EGF receptor (EGFR) has also been associated with poor prognosis in OC patients [215]. EGF has been shown to stimulate the growth of EOC cells expressing EGFR and alters their cell cycle distribution [216]. EGF similar to LPA has been shown to protect EOC cells from undergoing cisplatin induced apoptosis [161].

6.6. COX-2

In addition to being associated with tumor initiation and progression, COX-2 has also been associated with chemoresistance. Ferrandina et al. reported that a statistically significant higher percentage of primary OC patients unresponsive to platinum-containing chemotherapy were positive for COX-2 than responsive patients (84.6% versus 34.6%, respectively) [217]. The percentage of positive COX-2 staining per tumor area in COX-2 positive patients ranged from 15 to 45%. The results from this study suggest that COX-2 levels may influence the response of patients to different chemotherapy regimens, but the sample size of this study was small and the results need to be confirmed in a larger group of patients. Furthermore, this association needs to be corroborated biochemically [217]. In both patients groups undergoing cytoreductive surgery and explorative laparotomy, COX-2 expression was higher in nonresponders [218]. Using lung, colon, and prostate cancer models, COX-2 has been shown to induce Bcl-2 and promote tumor growth by facilitating the formation of new blood vessels [158,159]. These findings suggest that COX-2 may contribute to chemoresistance by inhibiting apoptosis and promoting angiogenesis in OC as well.

7. Treatment Strategies Targeting Inflammatory Mediators in EOC

As discussed, development of resistance to available chemotherapeutic drugs remains the major obstacle in management of OC patients. While several immunotherapies have been developed to improve the antitumor response of T-cells and/or modulate the immune response, here we will discuss EOC treatment strategies that specifically target the inflammatory mediators that have been reviewed above.

A monoclonal antibody directed at VEGF, bevacizumab, has been widely studied and is a promising target in EOC [219]. Bevacizumab is a recombinant humanized monoclonal antibody and has been approved by the FDA for treatment of metastatic breast, non-small cell lung, and colorectal cancer. Phase II clinical studies have shown that it is active in treatment of recurrent OC patients [220]. OCEANS trial was a randomized phase III clinical trial that evaluated the safety and efficacy of bevacizumab in combination with gemcitabine and carboplatin (GC) in comparison with GC alone in recurrent platinum sensitive ovarian, primary peritoneal, or FT cancer. This trial demonstrated that bevacizumab was able to prolong the PFS in platinum-sensitive recurrent EOC patients [221]. In addition to OCEANS, GOG218, and ICON7 have also shown that bevacizumab prolongs the PFS in OC patients confirming the promise this therapeutic target holds for management of OC [222,223].

We have discussed some mechanisms by which the pro-inflammatory cytokine TNF-α promotes OC metastasis and angiogenesis making it a good target for development of therapeutic agents. The safety profile and biological activity of a monoclonal anti-TNF-α antibody, Infliximab was assessed in a clinical study consisting of patients with advanced solid tumors, including OC. Infliximab did not have any toxic effects and was well tolerated by these patients. Reduced plasma levels of IL-6 and CCL12 in these patients was observed 24 h and 48 h after administration of Infliximab, while neutralization of TNF-α was detected after an hour indicating some biological activity [224]. This response warrants further study of Infliximab as a therapeutic agent for treatment of OC.

IL-6/STAT3 signaling has been implicated at different stages of OC progression and is a promising target although most agents are still in preclinical or early clinical trial stages. Siltuximab, an anti-IL-6 antibody, suppresses IL-6-induced STAT3 phosphorylation and nuclear translocation in OC cell lines. Siltuximab treatment also reduced the level of pro-survival proteins like Bcl-X_L and Survivin, which are downstream of STAT3. Siltuximab was able to sensitize paclitaxel resistant OC cell lines, but did not show the same effect in vivo [225]. sc144 is a novel small molecule inhibitor has shown significant promise in preclinical studies. sc144 binds gp130, which is a signal transducer in STAT3 signaling. It causes phosphorylation of gp130 leading to its deglycosylation. This abrogates downstream STAT3 phosphorylation and nuclear translocation inhibiting transcription of downstream genes. sc144 has increased potency in EOC cells in comparison to normal epithelial cells and slows down the growth of tumors in xenograft models of EOC [226]. A phase I clinical trial combining carboplatin, the monoclonal antibody Tocilizumab, which blocks IL-6R, and immune enhancer INF-α showed good promise. The EOC patients who received the highest dose of Tocilizumab had increased serum levels of IL-6 and sIL-6R and also showed longer median overall survival [227].

We have discussed the role of TGF-β in EOC tumor progression substantiating it as a good therapeutic target. A preclinical study of LY2109761 (TGFβRI and TGFβRII kinase inhibitor) in combination with cisplatin was conducted by Gao et al. This inhibitor significantly increased apoptosis in cisplatin resistant cells. Combining LY2109761 with cisplatin had antiproliferative effects and increased the rate of apoptosis in parental and cisplatin resistant xenograft models [162]. In triple negative breast cancer, LY2157299 a TGF-β1 receptor kinase inhibitor, prevented recurrence of tumors in xenograft models after treatment with paclitaxel [228]. Early phase clinical trials of LY2157299 in patients with advanced or metastasized pancreatic cancer have been completed. Early phase trials in triple negative metastatic breast cancer, unresectable hepatocellular carcinoma, and metastatic castration resistant prostate cancer are underway [229].

EGF has also been associated with chemoresistance in EOC. Cetuximab, a chimerized monoclonal antibody that targets EGFR, was tested in combination with carboplatin in patients with recurrent platinum sensitive OC. Cetuximab showed modest activity in these patients [230]. Panitumumab, a human monoclonal antibody specific to EGFR, in combination with carboplatin did not improve efficacy or progression free survival in platinum sensitive EOC patients [231].

8. Conclusions and Future Perspectives

Several studies in the last decade have associated increased inflammation and inflammatory mediators with increased EOC risk and reduced survival in EOC patients. We have presented published evidence suggesting that inflammation and inflammatory mediators promote ovarian tumorigenesis. However the mechanisms by which the process of inflammation culminates in ovarian tumor initiation need to be further understood. Such links have been established in colon and pancreatic cancer. Understanding these mechanisms is important for developing ways to target inflammatory mediators and reduce OC risk. Furthermore, epidemiological studies of NSAIDs and early clinical trials targeting IL-6 and TNF-α have shown significant promise, thus suggesting that targeting inflammatory mediators as treatment for OC warrants future research.

Author Contributions: S.S.S. and S.S wrote the manuscript. S.S.S., S.S. and H.M.O. edited the manuscript.

Funding: This research was funded by National Institute of Environmental Health Sciences grant number R01ES023183 and Ovarian Cancer Research Fund grant number 458788.

Acknowledgments: We acknowledge the Leo and Anne Albert Charitable Trust for its ongoing support of ovarian cancer research.

Conflicts of Interest: The authors declare no conflicts of interest.

References

1. Maiuri, A.R.; O'Hagan, H.M. Interplay Between Inflammation and Epigenetic Changes in Cancer. *Prog. Mol. Biol. Transl. Sci.* **2016**, *144*, 69–117. [PubMed]
2. Siegel, R.L.; Miller, K.D.; Jemal, A. Cancer statistics, 2018. *CA Cancer J. Clin.* **2018**, *68*, 7–30. [CrossRef] [PubMed]
3. Grivennikov, S.I.; Greten, F.R.; Karin, M. Immunity, inflammation, and cancer. *Cell* **2010**, *140*, 883–899. [CrossRef] [PubMed]
4. Clendenen, T.V.; Lundin, E.; Zeleniuch-Jacquotte, A.; Koenig, K.L.; Berrino, F.; Lukanova, A.; Lokshin, A.E.; Idahl, A.; Ohlson, N.; Hallmans, G.; et al. Circulating inflammation markers and risk of epithelial ovarian cancer. *Cancer Epidemiol. Biomark. Prev.* **2011**, *20*, 799–810. [CrossRef] [PubMed]
5. Chou, C.H.; Wei, L.H.; Kuo, M.L.; Huang, Y.J.; Lai, K.P.; Chen, C.A.; Hsieh, C.Y. Up-regulation of interleukin-6 in human ovarian cancer cell via a Gi/PI3K-Akt/NF-κB pathway by lysophosphatidic acid, an ovarian cancer-activating factor. *Carcinogenesis* **2005**, *26*, 45–52. [CrossRef] [PubMed]
6. Dhillon, A.S.; Hagan, S.; Rath, O.; Kolch, W. MAP kinase signalling pathways in cancer. *Oncogene* **2007**, *26*, 3279–3290. [CrossRef] [PubMed]
7. Schulze-Osthoff, K.; Ferrari, D.; Riehemann, K.; Wesselborg, S. Regulation of NF-κ B activation by MAP kinase cascades. *Immunobiology* **1997**, *198*, 35–49. [CrossRef]
8. Coussens, L.M.; Werb, Z. Inflammation and cancer. *Nature* **2002**, *420*, 860–867. [CrossRef] [PubMed]
9. Gajewski, T.F.; Schreiber, H.; Fu, Y.X. Innate and adaptive immune cells in the tumor microenvironment. *Nat. Immunol.* **2013**, *14*, 1014–1022. [CrossRef] [PubMed]
10. Wilson, C.B.; Rowell, E.; Sekimata, M. Epigenetic control of T-helper-cell differentiation. *Nat. Rev. Immunol.* **2009**, *9*, 91–105. [CrossRef] [PubMed]
11. Tone, A.A.; Salvador, S.; Finlayson, S.J.; Tinker, A.V.; Kwon, J.S.; Lee, C.H.; Cohen, T.; Ehlen, T.; Lee, M.; Carey, M.S.; et al. The role of the fallopian tube in ovarian cancer. *Clin. Adv. Hematol. Oncol.* **2012**, *10*, 296–306. [PubMed]
12. Petrovska, M.; Dimitrov, D.G.; Michael, S.D. Quantitative changes in macrophage distribution in normal mouse ovary over the course of the oestrous cycle examined with an image analysis system. *Am. J. Reprod. Immunol.* **1996**, *36*, 175–183. [CrossRef] [PubMed]
13. Takaya, R.; Fukaya, T.; Sasano, H.; Suzuki, T.; Tamura, M.; Yajima, A. Macrophages in normal cycling human ovaries; immunohistochemical localization and characterization. *Hum. Reprod.* **1997**, *12*, 1508–1512. [CrossRef] [PubMed]
14. Wu, R.; Van der Hoek, K.H.; Ryan, N.K.; Norman, R.J.; Robker, R.L. Macrophage contributions to ovarian function. *Hum. Reprod. Update* **2004**, *10*, 119–133. [CrossRef] [PubMed]

15. Tingen, C.M.; Kiesewetter, S.E.; Jozefik, J.; Thomas, C.; Tagler, D.; Shea, L.; Woodruff, T.K. A macrophage and theca cell-enriched stromal cell population influences growth and survival of immature murine follicles in vitro. *Reproduction* **2011**, *141*, 809–820. [CrossRef] [PubMed]
16. Lau, A.; Kollara, A.; St John, E.; Tone, A.A.; Virtanen, C.; Greenblatt, E.M.; King, W.A.; Brown, T.J. Altered expression of inflammation-associated genes in oviductal cells following follicular fluid exposure: Implications for ovarian carcinogenesis. *Exp. Biol. Med.* **2014**, *239*, 24–32. [CrossRef] [PubMed]
17. King, S.M.; Hilliard, T.S.; Wu, L.Y.; Jaffe, R.C.; Fazleabas, A.T.; Burdette, J.E. The impact of ovulation on fallopian tube epithelial cells: Evaluating three hypotheses connecting ovulation and serous ovarian cancer. *Endocr. Relat. Cancer* **2011**, *18*, 627–642. [CrossRef] [PubMed]
18. Johnson, P.A.; Giles, J.R. The hen as a model of ovarian cancer. *Nat. Rev. Cancer* **2013**, *13*, 432–436. [CrossRef] [PubMed]
19. Gong, T.T.; Wu, Q.J.; Vogtmann, E.; Lin, B.; Wang, Y.L. Age at menarche and risk of ovarian cancer: A meta-analysis of epidemiological studies. *Int. J. Cancer* **2013**, *132*, 2894–2900. [CrossRef] [PubMed]
20. Chiaffarino, F.; Pelucchi, C.; Parazzini, F.; Negri, E.; Franceschi, S.; Talamini, R.; Conti, E.; Montella, M.; La Vecchia, C. Reproductive and hormonal factors and ovarian cancer. *Ann. Oncol.* **2001**, *12*, 337–341. [CrossRef] [PubMed]
21. Fortner, R.T.; Ose, J.; Merritt, M.A.; Schock, H.; Tjonneland, A.; Hansen, L.; Overvad, K.; Dossus, L.; Clavel-Chapelon, F.; Baglietto, L.; et al. Reproductive and hormone-related risk factors for epithelial ovarian cancer by histologic pathways, invasiveness and histologic subtypes: Results from the EPIC cohort. *Int. J. Cancer* **2015**, *137*, 1196–1208. [CrossRef] [PubMed]
22. Espey, L.L. Ovulation as an inflammatory reaction—A hypothesis. *Biol. Reprod.* **1980**, *22*, 73–106. [CrossRef] [PubMed]
23. Machelon, V.; Emilie, D. Production of ovarian cytokines and their role in ovulation in the mammalian ovary. *Eur. Cytokine Netw.* **1997**, *8*, 137–143. [PubMed]
24. Chumduri, C.; Gurumurthy, R.K.; Zadora, P.K.; Mi, Y.; Meyer, T.F. Chlamydia infection promotes host DNA damage and proliferation but impairs the DNA damage response. *Cell. Host Microbe* **2013**, *13*, 746–758. [CrossRef] [PubMed]
25. Ingerslev, K.; Hogdall, E.; Schnack, T.H.; Skovrider-Ruminski, W.; Hogdall, C.; Blaakaer, J. The potential role of infectious agents and pelvic inflammatory disease in ovarian carcinogenesis. *Infect. Agent Cancer* **2017**, *12*, 25. [CrossRef] [PubMed]
26. Burney, R.O.; Giudice, L.C. Pathogenesis and pathophysiology of endometriosis. *Fertil. Steril.* **2012**, *98*, 511–519. [CrossRef] [PubMed]
27. Vercellini, P.; Crosignani, P.; Somigliana, E.; Viganò, P.; Buggio, L.; Bolis, G.; Fedele, L. The 'incessant menstruation' hypothesis: A mechanistic ovarian cancer model with implications for prevention. *Hum. Reprod.* **2011**, *26*, 2262–2273. [CrossRef] [PubMed]
28. Burghaus, S.; Haberle, L.; Schrauder, M.G.; Heusinger, K.; Thiel, F.C.; Hein, A.; Wachter, D.; Strehl, J.; Hartmann, A.; Ekici, A.B.; et al. Endometriosis as a risk factor for ovarian or endometrial cancer—Results of a hospital-based case-control study. *BMC Cancer* **2015**, *15*, 751. [CrossRef] [PubMed]
29. Sayasneh, A.; Tsivos, D.; Crawford, R. Endometriosis and ovarian cancer: A systematic review. *ISRN Obstet. Gynecol.* **2011**, *2011*, 140310. [CrossRef] [PubMed]
30. Rizi, B.S.; Nagrath, D. Linking omentum and ovarian cancer: NO. *Oncoscience* **2015**, *2*, 797–798. [PubMed]
31. Gunderson, C.C.; Ding, K.; Dvorak, J.; Moore, K.N.; McMeekin, D.S.; Benbrook, D.M. The pro-inflammatory effect of obesity on high grade serous ovarian cancer. *Gynecol. Oncol.* **2016**, *143*, 40–45. [CrossRef] [PubMed]
32. Poole, E.M.; Lee, I.M.; Ridker, P.M.; Buring, J.E.; Hankinson, S.E.; Tworoger, S.S. A prospective study of circulating C-reactive protein, interleukin-6, and tumor necrosis factor alpha receptor 2 levels and risk of ovarian cancer. *Am. J. Epidemiol.* **2013**, *178*, 1256–1264. [CrossRef] [PubMed]
33. Ose, J.; Schock, H.; Tjonneland, A.; Hansen, L.; Overvad, K.; Dossus, L.; Clavel-Chapelon, F.; Baglietto, L.; Boeing, H.; Trichopolou, A.; et al. Inflammatory Markers and Risk of Epithelial Ovarian Cancer by Tumor Subtypes: The EPIC Cohort. *Cancer Epidemiol. Biomark. Prev.* **2015**, *24*, 951–961. [CrossRef] [PubMed]
34. Duleba, A.J.; Dokras, A. Is PCOS an inflammatory process? *Fertil. Steril.* **2012**, *97*, 7–12. [CrossRef] [PubMed]
35. Kelly, C.C.; Lyall, H.; Petrie, J.R.; Gould, G.W.; Connell, J.M.; Sattar, N. Low grade chronic inflammation in women with polycystic ovarian syndrome. *J. Clin. Endocrinol. Metab.* **2001**, *86*, 2453–2455. [CrossRef] [PubMed]

36. Escobar-Morreale, H.F.; Luque-Ramirez, M.; Gonzalez, F. Circulating inflammatory markers in polycystic ovary syndrome: A systematic review and metaanalysis. *Fertil. Steril.* **2011**, *95*, 1048–1058.e1041-1042. [CrossRef] [PubMed]
37. Gonzalez, F.; Rote, N.S.; Minium, J.; Kirwan, J.P. Evidence of proatherogenic inflammation in polycystic ovary syndrome. *Metabolism* **2009**, *58*, 954–962. [CrossRef] [PubMed]
38. Glintborg, D.; Andersen, M.; Richelsen, B.; Bruun, J.M. Plasma monocyte chemoattractant protein-1 (MCP-1) and macrophage inflammatory protein-1α are increased in patients with polycystic ovary syndrome (PCOS) and associated with adiposity, but unaffected by pioglitazone treatment. *Clin. Endocrinol.* **2009**, *71*, 652–658. [CrossRef] [PubMed]
39. Escobar-Morreale, H.F.; Botella-Carretero, J.I.; Villuendas, G.; Sancho, J.; San Millan, J.L. Serum interleukin-18 concentrations are increased in the polycystic ovary syndrome: Relationship to insulin resistance and to obesity. *J. Clin. Endocrinol. Metab.* **2004**, *89*, 806–811. [CrossRef] [PubMed]
40. Yang, Y.; Qiao, J.; Li, R.; Li, M.Z. Is interleukin-18 associated with polycystic ovary syndrome? *Reprod. Biol. Endocrinol.* **2011**, *9*, 7. [CrossRef] [PubMed]
41. Tarkun, I.; Cetinarslan, B.; Turemen, E.; Canturk, Z.; Biyikli, M. Association between Circulating Tumor Necrosis Factor-α, Interleukin-6, and Insulin Resistance in Normal-Weight Women with Polycystic Ovary Syndrome. *Metab. Syndr. Relat. Disord.* **2006**, *4*, 122–128. [CrossRef] [PubMed]
42. Vgontzas, A.N.; Trakada, G.; Bixler, E.O.; Lin, H.M.; Pejovic, S.; Zoumakis, E.; Chrousos, G.P.; Legro, R.S. Plasma interleukin 6 levels are elevated in polycystic ovary syndrome independently of obesity or sleep apnea. *Metabolism* **2006**, *55*, 1076–1082. [CrossRef] [PubMed]
43. Dinger, Y.; Akcay, T.; Erdem, T.; Ilker Saygili, E.; Gundogdu, S. DNA damage, DNA susceptibility to oxidation and glutathione level in women with polycystic ovary syndrome. *Scand. J. Clin Lab. Investig.* **2005**, *65*, 721–728.
44. Harris, H.R.; Terry, K.L. Polycystic ovary syndrome and risk of endometrial, ovarian, and breast cancer: A systematic review. *Fertil. Res. Pract.* **2016**, *2*, 14. [CrossRef] [PubMed]
45. Heller, D.S.; Westhoff, C.; Gordon, R.E.; Katz, N. The relationship between perineal cosmetic talc usage and ovarian talc particle burden. *Am. J. Obstet. Gynecol.* **1996**, *174*, 1507–1510. [CrossRef]
46. Henderson, W.J.; Hamilton, T.C.; Griffiths, K. Talc in normal and malignant ovarian tissue. *Lancet* **1979**, *1*, 499. [CrossRef]
47. Muscat, J.E.; Huncharek, M.S. Perineal talc use and ovarian cancer: A critical review. *Eur J. Cancer Prev.* **2008**, *17*, 139–146. [CrossRef] [PubMed]
48. Brasky, T.M.; Moysich, K.B.; Cohn, D.E.; White, E. Non-steroidal anti-inflammatory drugs and endometrial cancer risk in the VITamins And Lifestyle (VITAL) cohort. *Gynecol. Oncol.* **2013**, *128*, 113–119. [CrossRef] [PubMed]
49. Prizment, A.E.; Folsom, A.R.; Anderson, K.E. Nonsteroidal anti-inflammatory drugs and risk for ovarian and endometrial cancers in the Iowa Women's Health Study. *Cancer Epidemiol. Biomark. Prev.* **2010**, *19*, 435–442. [CrossRef] [PubMed]
50. Fairfield, K.M.; Hunter, D.J.; Fuchs, C.S.; Colditz, G.A.; Hankinson, S.E. Aspirin, other NSAIDs, and ovarian cancer risk (United States). *Cancer Causes Control.* **2002**, *13*, 535–542. [CrossRef] [PubMed]
51. Trabert, B.; Poole, E.M.; White, E.; Visvanathan, K.; Adami, H.O.; Anderson, G.L.; Brasky, T.M.; Brinton, L.A.; Fortner, R.T.; Gaudet, M.; et al. Analgesic Use and Ovarian Cancer Risk: An Analysis in the Ovarian Cancer Cohort Consortium. *J. Natl. Cancer Inst.* **2018**. [CrossRef] [PubMed]
52. Peres, L.C.; Camacho, F.; Abbott, S.E.; Alberg, A.J.; Bandera, E.V.; Barnholtz-Sloan, J.; Bondy, M.; Cote, M.L.; Crankshaw, S.; Funkhouser, E.; et al. Analgesic medication use and risk of epithelial ovarian cancer in African American women. *Br. J. Cancer* **2016**, *114*, 819–825. [CrossRef] [PubMed]
53. Rodriguez-Burford, C.; Barnes, M.N.; Oelschlager, D.K.; Myers, R.B.; Talley, L.I.; Partridge, E.E.; Grizzle, W.E. Effects of nonsteroidal anti-inflammatory agents (NSAIDs) on ovarian carcinoma cell lines: Preclinical evaluation of NSAIDs as chemopreventive agents. *Clin. Cancer Res.* **2002**, *8*, 202–209. [PubMed]
54. Arango, H.A.; Icely, S.; Roberts, W.S.; Cavanagh, D.; Becker, J.L. Aspirin effects on endometrial cancer cell growth. *Obstet. Gynecol.* **2001**, *97*, 423–427. [PubMed]
55. Gao, J.; Niwa, K.; Sun, W.; Takemura, M.; Lian, Z.; Onogi, K.; Seishima, M.; Mori, H.; Tamaya, T. Non-steroidal anti-inflammatory drugs inhibit cellular proliferation and upregulate cyclooxygenase-2 protein expression in endometrial cancer cells. *Cancer Sci.* **2004**, *95*, 901–907. [CrossRef] [PubMed]

56. Li, H.L.; Zhang, H.W.; Chen, D.D.; Zhong, L.; Ren, X.D.; St-Tu, R. JTE-522, a selective COX-2 inhibitor, inhibits cell proliferation and induces apoptosis in RL95-2 cells. *Acta Pharmacol Sin.* **2002**, *23*, 631–637. [PubMed]
57. Hsu, C.S.; Li, Y. Aspirin potently inhibits oxidative DNA strand breaks: Implications for cancer chemoprevention. *Biochem. Biophys. Res. Commun.* **2002**, *293*, 705–709. [PubMed]
58. Hales, D.B.; Zhuge, Y.; Lagman, J.A.; Ansenberger, K.; Mahon, C.; Barua, A.; Luborsky, J.L.; Bahr, J.M. Cyclooxygenases expression and distribution in the normal ovary and their role in ovarian cancer in the domestic hen (Gallus domesticus). *Endocrine* **2008**, *33*, 235–244. [CrossRef] [PubMed]
59. Urick, M.E.; Giles, J.R.; Johnson, P.A. VEGF expression and the effect of NSAIDs on ascites cell proliferation in the hen model of ovarian cancer. *Gynecol. Oncol.* **2008**, *110*, 418–424. [CrossRef] [PubMed]
60. Urick, M.E.; Giles, J.R.; Johnson, P.A. Dietary aspirin decreases the stage of ovarian cancer in the hen. *Gynecol. Oncol.* **2009**, *112*, 166–170. [CrossRef] [PubMed]
61. Hanahan, D.; Weinberg, R.A. Hallmarks of cancer: The next generation. *Cell* **2011**, *144*, 646–674. [CrossRef] [PubMed]
62. Cancer Genome Atlas Research, N. Integrated genomic analyses of ovarian carcinoma. *Nature* **2011**, *474*, 609–615.
63. Toss, A.; Tomasello, C.; Razzaboni, E.; Contu, G.; Grandi, G.; Cagnacci, A.; Schilder, R.J.; Cortesi, L. Hereditary ovarian cancer: Not only BRCA 1 and 2 genes. *Biomed. Res. Int.* **2015**, *2015*, 341723. [CrossRef] [PubMed]
64. Liu, J.; Yang, G.; Thompson-Lanza, J.A.; Glassman, A.; Hayes, K.; Patterson, A.; Marquez, R.T.; Auersperg, N.; Yu, Y.; Hahn, W.C.; et al. A genetically defined model for human ovarian cancer. *Cancer Res.* **2004**, *64*, 1655–1663. [CrossRef] [PubMed]
65. Szotek, P.P.; Pieretti-Vanmarcke, R.; Masiakos, P.T.; Dinulescu, D.M.; Connolly, D.; Foster, R.; Dombkowski, D.; Preffer, F.; Maclaughlin, D.T.; Donahoe, P.K. Ovarian cancer side population defines cells with stem cell-like characteristics and Mullerian Inhibiting Substance responsiveness. *Proc. Natl. Acad. Sci. USA* **2006**, *103*, 11154–11159. [CrossRef] [PubMed]
66. Zhang, S.; Balch, C.; Chan, M.W.; Lai, H.C.; Matei, D.; Schilder, J.M.; Yan, P.S.; Huang, T.H.; Nephew, K.P. Identification and characterization of ovarian cancer-initiating cells from primary human tumors. *Cancer Res.* **2008**, *68*, 4311–4320. [CrossRef] [PubMed]
67. Lane, D.; Matte, I.; Rancourt, C.; Piche, A. Prognostic significance of IL-6 and IL-8 ascites levels in ovarian cancer patients. *BMC Cancer* **2011**, *11*, 210. [CrossRef] [PubMed]
68. Masoumi-Moghaddam, S.; Amini, A.; Wei, A.Q.; Robertson, G.; Morris, D.L. Intratumoral interleukin-6 predicts ascites formation in patients with epithelial ovarian cancer: A potential tool for close monitoring. *J. Ovarian Res.* **2015**, *8*, 58. [CrossRef] [PubMed]
69. Rath, K.S.; Funk, H.M.; Bowling, M.C.; Richards, W.E.; Drew, A.F. Expression of soluble interleukin-6 receptor in malignant ovarian tissue. *Am. J. Obstet. Gynecol.* **2010**, *203*, 230.e1–230.e8. [CrossRef] [PubMed]
70. Lo, C.W.; Chen, M.W.; Hsiao, M.; Wang, S.; Chen, C.A.; Hsiao, S.M.; Chang, J.S.; Lai, T.C.; Rose-John, S.; Kuo, M.L.; et al. IL-6 trans-signaling in formation and progression of malignant ascites in ovarian cancer. *Cancer Res.* **2011**, *71*, 424–434. [CrossRef] [PubMed]
71. Dalal, V.; Kumar, R.; Kumar, S.; Sharma, A.; Kumar, L.; Sharma, J.B.; Roy, K.K.; Singh, N.; Vanamail, P. Biomarker potential of IL-6 and VEGF-A in ascitic fluid of epithelial ovarian cancer patients. *Clin. Chim. Acta* **2018**, *482*, 27–32. [CrossRef] [PubMed]
72. Bapat, S.A.; Mali, A.M.; Koppikar, C.B.; Kurrey, N.K. Stem and progenitor-like cells contribute to the aggressive behavior of human epithelial ovarian cancer. *Cancer Res.* **2005**, *65*, 3025–3029. [CrossRef] [PubMed]
73. Wang, D.; Xiang, T.; Zhao, Z.; Lin, K.; Yin, P.; Jiang, L.; Liang, Z.; Zhu, B. Autocrine interleukin-23 promotes self-renewal of CD133+ ovarian cancer stem-like cells. *Oncotarget* **2016**, *7*, 76006–76020. [CrossRef] [PubMed]
74. Maccio, A.; Madeddu, C. Inflammation and ovarian cancer. *Cytokine* **2012**, *58*, 133–147. [CrossRef] [PubMed]
75. Agarwal, A.; Gupta, S.; Sharma, R.K. Role of oxidative stress in female reproduction. *Reprod. Biol. Endocrinol.* **2005**, *3*, 28. [CrossRef] [PubMed]
76. Shkolnik, K.; Tadmor, A.; Ben-Dor, S.; Nevo, N.; Galiani, D.; Dekel, N. Reactive oxygen species are indispensable in ovulation. *Proc. Natl. Acad. Sci. USA* **2011**, *108*, 1462–1467. [CrossRef] [PubMed]
77. Waris, G.; Ahsan, H. Reactive oxygen species: Role in the development of cancer and various chronic conditions. *J. Carcinog.* **2006**, *5*, 14. [CrossRef] [PubMed]
78. Liou, G.Y.; Storz, P. Reactive oxygen species in cancer. *Free Radic. Res.* **2010**, *44*, 479–496. [CrossRef] [PubMed]

79. King, S.M.; Quartuccio, S.M.; Vanderhyden, B.C.; Burdette, J.E. Early transformative changes in normal ovarian surface epithelium induced by oxidative stress require Akt upregulation, DNA damage and epithelial-stromal interaction. *Carcinogenesis* **2013**, *34*, 1125–1133. [CrossRef] [PubMed]
80. Levanon, K.; Ng, V.; Piao, H.Y.; Zhang, Y.; Chang, M.C.; Roh, M.H.; Kindelberger, D.W.; Hirsch, M.S.; Crum, C.P.; Marto, J.A.; et al. Primary ex vivo cultures of human fallopian tube epithelium as a model for serous ovarian carcinogenesis. *Oncogene* **2010**, *29*, 1103–1113. [CrossRef] [PubMed]
81. Huang, H.S.; Chu, S.C.; Hsu, C.F.; Chen, P.C.; Ding, D.C.; Chang, M.Y.; Chu, T.Y. Mutagenic, surviving and tumorigenic effects of follicular fluid in the context of p53 loss: Initiation of fimbria carcinogenesis. *Carcinogenesis* **2015**, *36*, 1419–1428. [CrossRef] [PubMed]
82. Kalinina, E.V.; Chernov, N.N.; Saprin, A.N.; Kotova, Y.N.; Gavrilova, Y.A.; Chermnykh, N.S.; Shcherbak, N.P. Expression of genes for thioredoxin 1 and thioredoxin 2 in multidrug resistance ovarian carcinoma cells SKVLB. *Bull. Exp. Biol. Med.* **2007**, *144*, 301–303. [CrossRef] [PubMed]
83. Ohno, T.; Hirota, K.; Nakamura, H.; Masutani, H.; Sasada, T.; Yodoi, J. Thioredoxin and Its Involvement in the Redox Regulation of Transcription Factors, NF-κB and AP-1. In *Oxygen Homeostasis and Its Dynamics*; Springer: Tokyo, Japan, 1998; pp. 450–456.
84. Van der Wijst, M.G.; Huisman, C.; Mposhi, A.; Roelfes, G.; Rots, M.G. Targeting Nrf2 in healthy and malignant ovarian epithelial cells: Protection versus promotion. *Mol. Oncol.* **2015**, *9*, 1259–1273. [CrossRef] [PubMed]
85. Klotz, L.O.; Sanchez-Ramos, C.; Prieto-Arroyo, I.; Urbanek, P.; Steinbrenner, H.; Monsalve, M. Redox regulation of FoxO transcription factors. *Redox Biol* **2015**, *6*, 51–72. [CrossRef] [PubMed]
86. Ozdemir, F.; Altinisik, J.; Karateke, A.; Coksuer, H.; Buyru, N. Methylation of tumor suppressor genes in ovarian cancer. *Exp. Ther. Med.* **2012**, *4*, 1092–1096. [CrossRef] [PubMed]
87. Niwa, T.; Tsukamoto, T.; Toyoda, T.; Mori, A.; Tanaka, H.; Maekita, T.; Ichinose, M.; Tatematsu, M.; Ushijima, T. Inflammatory processes triggered by Helicobacter pylori infection cause aberrant DNA methylation in gastric epithelial cells. *Cancer Res.* **2010**, *70*, 1430–1440. [CrossRef] [PubMed]
88. Ding, N.; Bonham, E.M.; Hannon, B.E.; Amick, T.R.; Baylin, S.B.; O'Hagan, H.M. Mismatch repair proteins recruit DNA methyltransferase 1 to sites of oxidative DNA damage. *J. Mol. Cell. Biol.* **2016**, *8*, 244–254. [CrossRef] [PubMed]
89. Maiuri, A.R.; Peng, M.; Sriramkumar, S.; Kamplain, C.M.; DeStefano Shields, C.E.; Sears, C.L.; O'Hagan, H.M. Mismatch Repair Proteins Initiate Epigenetic Alterations during Inflammation-Driven Tumorigenesis. *Cancer Res.* **2017**, *77*, 3467–3478. [CrossRef] [PubMed]
90. Sapoznik, S.; Bahar-Shany, K.; Brand, H.; Pinto, Y.; Gabay, O.; Glick-Saar, E.; Dor, C.; Zadok, O.; Barshack, I.; Zundelevich, A.; et al. Activation-Induced Cytidine Deaminase Links Ovulation-Induced Inflammation and Serous Carcinogenesis. *Neoplasia* **2016**, *18*, 90–99. [CrossRef] [PubMed]
91. Gupta, M.; Babic, A.; Beck, A.H.; Terry, K. TNF-α expression, risk factors, and inflammatory exposures in ovarian cancer: Evidence for an inflammatory pathway of ovarian carcinogenesis? *Hum. Pathol.* **2016**, *54*, 82–91. [CrossRef] [PubMed]
92. Moradi, M.M.; Carson, L.F.; Weinberg, B.; Haney, A.F.; Twiggs, L.B.; Ramakrishnan, S. Serum and ascitic fluid levels of interleukin-1, interleukin-6, and tumor necrosis factor-α in patients with ovarian epithelial cancer. *Cancer* **1993**, *72*, 2433–2440. [CrossRef]
93. Szlosarek, P.W.; Grimshaw, M.J.; Kulbe, H.; Wilson, J.L.; Wilbanks, G.D.; Burke, F.; Balkwill, F.R. Expression and regulation of tumor necrosis factor alpha in normal and malignant ovarian epithelium. *Mol. Cancer Ther.* **2006**, *5*, 382–390. [CrossRef] [PubMed]
94. Kulbe, H.; Thompson, R.; Wilson, J.L.; Robinson, S.; Hagemann, T.; Fatah, R.; Gould, D.; Ayhan, A.; Balkwill, F. The inflammatory cytokine tumor necrosis factor-α generates an autocrine tumor-promoting network in epithelial ovarian cancer cells. *Cancer Res.* **2007**, *67*, 585–592. [CrossRef] [PubMed]
95. Charles, K.A.; Kulbe, H.; Soper, R.; Escorcio-Correia, M.; Lawrence, T.; Schultheis, A.; Chakravarty, P.; Thompson, R.G.; Kollias, G.; Smyth, J.F.; et al. The tumor-promoting actions of TNF-α involve TNFR1 and IL-17 in ovarian cancer in mice and humans. *J. Clin. Investig.* **2009**, *119*, 3011–3023. [CrossRef] [PubMed]
96. Berek, J.S.; Chung, C.; Kaldi, K.; Watson, J.M.; Knox, R.M.; Martinez-Maza, O. Serum interleukin-6 levels correlate with disease status in patients with epithelial ovarian cancer. *Am. J. Obstet. Gynecol.* **1991**, *164*, 1038–1042; discussion 1042–1033. [CrossRef]

97. Scambia, G.; Testa, U.; Benedetti Panici, P.; Foti, E.; Martucci, R.; Gadducci, A.; Perillo, A.; Facchini, V.; Peschle, C.; Mancuso, S. Prognostic significance of interleukin 6 serum levels in patients with ovarian cancer. *Br. J. Cancer* **1995**, *71*, 354. [CrossRef] [PubMed]
98. Alberti, C.; Pinciroli, P.; Valeri, B.; Ferri, R.; Ditto, A.; Umezawa, K.; Sensi, M.; Canevari, S.; Tomassetti, A. Ligand-dependent EGFR activation induces the co-expression of IL-6 and PAI-1 via the NFkB pathway in advanced-stage epithelial ovarian cancer. *Oncogene* **2012**, *31*, 4139–4149. [CrossRef] [PubMed]
99. Hagemann, T.; Robinson, S.C.; Thompson, R.G.; Charles, K.; Kulbe, H.; Balkwill, F.R. Ovarian cancer cell-derived migration inhibitory factor enhances tumor growth, progression, and angiogenesis. *Mol. Cancer Ther.* **2007**, *6*, 1993–2002. [CrossRef] [PubMed]
100. Oh, K.; Moon, H.G.; Lee, D.S.; Yoo, Y.B. Tissue transglutaminase-interleukin-6 axis facilitates peritoneal tumor spreading and metastasis of human ovarian cancer cells. *Lab. Anim. Res.* **2015**, *31*, 188–197. [CrossRef] [PubMed]
101. Wang, Y.; Li, L.; Guo, X.; Jin, X.; Sun, W.; Zhang, X.; Xu, R.C. Interleukin-6 signaling regulates anchorage-independent growth, proliferation, adhesion and invasion in human ovarian cancer cells. *Cytokine* **2012**, *59*, 228–236. [CrossRef] [PubMed]
102. Takaishi, K.; Komohara, Y.; Tashiro, H.; Ohtake, H.; Nakagawa, T.; Katabuchi, H.; Takeya, M. Involvement of M2-polarized macrophages in the ascites from advanced epithelial ovarian carcinoma in tumor progression via Stat3 activation. *Cancer Sci.* **2010**, *101*, 2128–2136. [CrossRef] [PubMed]
103. Maccio, A.; Lai, P.; Santona, M.C.; Pagliara, L.; Melis, G.B.; Mantovani, G. High serum levels of soluble IL-2 receptor, cytokines, and C reactive protein correlate with impairment of T cell response in patients with advanced epithelial ovarian cancer. *Gynecol. Oncol.* **1998**, *69*, 248–252. [CrossRef] [PubMed]
104. Rabinovich, A.; Medina, L.; Piura, B.; Segal, S.; Huleihel, M. Regulation of ovarian carcinoma SKOV-3 cell proliferation and secretion of MMPs by autocrine IL-6. *Anticancer Res.* **2007**, *27*, 267–272. [PubMed]
105. Runesson, E.; Bostrom, E.K.; Janson, P.O.; Brannstrom, M. The human preovulatory follicle is a source of the chemotactic cytokine interleukin-8. *Mol. Hum. Reprod* **1996**, *2*, 245–250. [CrossRef] [PubMed]
106. Brannstrom, M.; Mayrhofer, G.; Robertson, S.A. Localization of leukocyte subsets in the rat ovary during the periovulatory period. *Biol. Reprod.* **1993**, *48*, 277–286. [CrossRef] [PubMed]
107. Kassim, S.K.; El-Salahy, E.M.; Fayed, S.T.; Helal, S.A.; Helal, T.; Azzam Eel, D.; Khalifa, A. Vascular endothelial growth factor and interleukin-8 are associated with poor prognosis in epithelial ovarian cancer patients. *Clin. Biochem.* **2004**, *37*, 363–369. [CrossRef] [PubMed]
108. Ivarsson, K.; Runesson, E.; Sundfeldt, K.; Haeger, M.; Hedin, L.; Janson, P.O.; Brannstrom, M. The chemotactic cytokine interleukin-8—A cyst fluid marker for malignant epithelial ovarian cancer? *Gynecol. Oncol.* **1998**, *71*, 420–423. [CrossRef] [PubMed]
109. Radke, J.; Schmidt, D.; Bohme, M.; Schmidt, U.; Weise, W.; Morenz, J. Cytokine level in malignant ascites and peripheral blood of patients with advanced ovarian carcinoma. *Geburtshilfe Frauenheilkd.* **1996**, *56*, 83–87. [CrossRef] [PubMed]
110. Wang, Y.; Xu, R.C.; Zhang, X.L.; Niu, X.L.; Qu, Y.; Li, L.Z.; Meng, X.Y. Interleukin-8 secretion by ovarian cancer cells increases anchorage-independent growth, proliferation, angiogenic potential, adhesion and invasion. *Cytokine* **2012**, *59*, 145–155. [CrossRef] [PubMed]
111. Hirose, K.; Hakozaki, M.; Nyunoya, Y.; Kobayashi, Y.; Matsushita, K.; Takenouchi, T.; Mikata, A.; Mukaida, N.; Matsushima, K. Chemokine gene transfection into tumour cells reduced tumorigenicity in nude mice in association with neutrophilic infiltration. *Br. J. Cancer* **1995**, *72*, 708–714. [CrossRef] [PubMed]
112. Lee, L.F.; Hellendall, R.P.; Wang, Y.; Haskill, J.S.; Mukaida, N.; Matsushima, K.; Ting, J.P. IL-8 reduced tumorigenicity of human ovarian cancer in vivo due to neutrophil infiltration. *J. Immunol.* **2000**, *164*, 2769–2775. [CrossRef] [PubMed]
113. Tokumura, A.; Miyake, M.; Nishioka, Y.; Yamano, S.; Aono, T.; Fukuzawa, K. Production of lysophosphatidic acids by lysophospholipase D in human follicular fluids of In vitro fertilization patients. *Biol. Reprod.* **1999**, *61*, 195–199. [CrossRef] [PubMed]
114. Chen, S.U.; Chou, C.H.; Lee, H.; Ho, C.H.; Lin, C.W.; Yang, Y.S. Lysophosphatidic acid up-regulates expression of interleukin-8 and -6 in granulosa-lutein cells through its receptors and nuclear factor-κB dependent pathways: Implications for angiogenesis of corpus luteum and ovarian hyperstimulation syndrome. *J. Clin. Endocrinol. Metab.* **2008**, *93*, 935–943. [CrossRef] [PubMed]

115. Fang, X.; Gaudette, D.; Furui, T.; Mao, M.; Estrella, V.; Eder, A.; Pustilnik, T.; Sasagawa, T.; Lapushin, R.; Yu, S.; et al. Lysophospholipid growth factors in the initiation, progression, metastases, and management of ovarian cancer. *Ann. N. Y. Acad. Sci.* **2000**, *905*, 188–208. [CrossRef] [PubMed]
116. Xu, Y.; Shen, Z.; Wiper, D.W.; Wu, M.; Morton, R.E.; Elson, P.; Kennedy, A.W.; Belinson, J.; Markman, M.; Casey, G. Lysophosphatidic acid as a potential biomarker for ovarian and other gynecologic cancers. *JAMA* **1998**, *280*, 719–723. [CrossRef] [PubMed]
117. Xiao, Y.J.; Schwartz, B.; Washington, M.; Kennedy, A.; Webster, K.; Belinson, J.; Xu, Y. Electrospray ionization mass spectrometry analysis of lysophospholipids in human ascitic fluids: Comparison of the lysophospholipid contents in malignant vs nonmalignant ascitic fluids. *Anal. Biochem.* **2001**, *290*, 302–313. [CrossRef] [PubMed]
118. Mills, G.B.; Eder, A.; Fang, X.; Hasegawa, Y.; Mao, M.; Lu, Y.; Tanyi, J.; Tabassam, F.H.; Wiener, J.; Lapushin, R.; et al. Critical role of lysophospholipids in the pathophysiology, diagnosis, and management of ovarian cancer. *Cancer Treat. Res.* **2002**, *107*, 259–283. [PubMed]
119. Westermann, A.M.; Havik, E.; Postma, F.R.; Beijnen, J.H.; Dalesio, O.; Moolenaar, W.H.; Rodenhuis, S. Malignant effusions contain lysophosphatidic acid (LPA)-like activity. *Ann. Oncol.* **1998**, *9*, 437–442. [CrossRef] [PubMed]
120. Xu, Y.; Gaudette, D.C.; Boynton, J.D.; Frankel, A.; Fang, X.J.; Sharma, A.; Hurteau, J.; Casey, G.; Goodbody, A.; Mellors, A.; et al. Characterization of an ovarian cancer activating factor in ascites from ovarian cancer patients. *Clin. Cancer Res.* **1995**, *1*, 1223–1232. [PubMed]
121. Symowicz, J.; Adley, B.P.; Woo, M.M.; Auersperg, N.; Hudson, L.G.; Stack, M.S. Cyclooxygenase-2 functions as a downstream mediator of lysophosphatidic acid to promote aggressive behavior in ovarian carcinoma cells. *Cancer Res.* **2005**, *65*, 2234–2242. [CrossRef] [PubMed]
122. Fang, X.; Yu, S.; Bast, R.C.; Liu, S.; Xu, H.J.; Hu, S.X.; LaPushin, R.; Claret, F.X.; Aggarwal, B.B.; Lu, Y.; et al. Mechanisms for lysophosphatidic acid-induced cytokine production in ovarian cancer cells. *J. Biol. Chem.* **2004**, *279*, 9653–9661. [CrossRef] [PubMed]
123. Saunders, J.A.; Rogers, L.C.; Klomsiri, C.; Poole, L.B.; Daniel, L.W. Reactive oxygen species mediate lysophosphatidic acid induced signaling in ovarian cancer cells. *Free Radic. Biol. Med.* **2010**, *49*, 2058–2067. [CrossRef] [PubMed]
124. Lu, J.; Xiao Yj, Y.J.; Baudhuin, L.M.; Hong, G.; Xu, Y. Role of ether-linked lysophosphatidic acids in ovarian cancer cells. *J. Lipid Res.* **2002**, *43*, 463–476. [PubMed]
125. Ali-Fehmi, R.; Morris, R.T.; Bandyopadhyay, S.; Che, M.; Schimp, V.; Malone, J.M., Jr.; Munkarah, A.R. Expression of cyclooxygenase-2 in advanced stage ovarian serous carcinoma: Correlation with tumor cell proliferation, apoptosis, angiogenesis, and survival. *Am. J. Obstet. Gynecol.* **2005**, *192*, 819–825. [CrossRef] [PubMed]
126. Rask, K.; Zhu, Y.; Wang, W.; Hedin, L.; Sundfeldt, K. Ovarian epithelial cancer: A role for PGE2-synthesis and signalling in malignant transformation and progression. *Mol. Cancer* **2006**, *5*, 62. [CrossRef] [PubMed]
127. Heinonen, P.K.; Metsa-Ketela, T. Prostaglandin and thromboxane production in ovarian cancer tissue. *Gynecol. Obstet. Investig.* **1984**, *18*, 225–229. [CrossRef] [PubMed]
128. Obermajer, N.; Muthuswamy, R.; Odunsi, K.; Edwards, R.P.; Kalinski, P. PGE(2)-induced CXCL12 production and CXCR4 expression controls the accumulation of human MDSCs in ovarian cancer environment. *Cancer Res.* **2011**, *71*, 7463–7470. [CrossRef] [PubMed]
129. Daikoku, T.; Tranguch, S.; Trofimova, I.N.; Dinulescu, D.M.; Jacks, T.; Nikitin, A.Y.; Connolly, D.C.; Dey, S.K. Cyclooxygenase-1 is overexpressed in multiple genetically engineered mouse models of epithelial ovarian cancer. *Cancer Res.* **2006**, *66*, 2527–2531. [CrossRef] [PubMed]
130. Li, W.; Cai, J.H.; Zhang, J.; Tang, Y.X.; Wan, L. Effects of cyclooxygenase inhibitors in combination with taxol on expression of cyclin D1 and Ki-67 in a xenograft model of ovarian carcinoma. *Int. J. Mol. Sci.* **2012**, *13*, 9741–9753. [CrossRef] [PubMed]
131. Paweletz, N.; Knierim, M. Tumor-related angiogenesis. *Crit. Rev. Oncol. Hematol.* **1989**, *9*, 197–242. [CrossRef]
132. Potente, M.; Gerhardt, H.; Carmeliet, P. Basic and therapeutic aspects of angiogenesis. *Cell* **2011**, *146*, 873–887. [CrossRef] [PubMed]
133. Burke, B.; Giannoudis, A.; Corke, K.P.; Gill, D.; Wells, M.; Ziegler-Heitbrock, L.; Lewis, C.E. Hypoxia-induced gene expression in human macrophages: Implications for ischemic tissues and hypoxia-regulated gene therapy. *Am. J. Pathol.* **2003**, *163*, 1233–1243. [CrossRef]

134. Murdoch, C.; Muthana, M.; Lewis, C.E. Hypoxia regulates macrophage functions in inflammation. *J. Immunol.* **2005**, *175*, 6257–6263. [CrossRef] [PubMed]
135. Ferrara, N.; Gerber, H.P.; LeCouter, J. The biology of VEGF and its receptors. *Nat. Med.* **2003**, *9*, 669–676. [CrossRef] [PubMed]
136. Wheeler-Jones, C.; Abu-Ghazaleh, R.; Cospedal, R.; Houliston, R.A.; Martin, J.; Zachary, I. Vascular endothelial growth factor stimulates prostacyclin production and activation of cytosolic phospholipase A2 in endothelial cells via p42/p44 mitogen-activated protein kinase. *FEBS Lett.* **1997**, *420*, 28–32. [CrossRef]
137. Gerber, H.P.; McMurtrey, A.; Kowalski, J.; Yan, M.; Keyt, B.A.; Dixit, V.; Ferrara, N. Vascular endothelial growth factor regulates endothelial cell survival through the phosphatidylinositol 3′-kinase/Akt signal transduction pathway. Requirement for Flk-1/KDR activation. *J. Biol. Chem.* **1998**, *273*, 30336–30343. [CrossRef] [PubMed]
138. Gerber, H.P.; Dixit, V.; Ferrara, N. Vascular endothelial growth factor induces expression of the antiapoptotic proteins Bcl-2 and A1 in vascular endothelial cells. *J. Biol. Chem.* **1998**, *273*, 13313–13316. [CrossRef] [PubMed]
139. Xiao, Z.; Liu, Q.; Mao, F.; Wu, J.; Lei, T. TNF-α-induced VEGF and MMP-9 expression promotes hemorrhagic transformation in pituitary adenomas. *Int. J. Mol. Sci.* **2011**, *12*, 4165–4179. [CrossRef] [PubMed]
140. Liang, Z.; Brooks, J.; Willard, M.; Liang, K.; Yoon, Y.; Kang, S.; Shim, H. CXCR4/CXCL12 axis promotes VEGF-mediated tumor angiogenesis through Akt signaling pathway. *Biochem. Biophys. Res. Commun.* **2007**, *359*, 716–722. [CrossRef] [PubMed]
141. Kryczek, I.; Lange, A.; Mottram, P.; Alvarez, X.; Cheng, P.; Hogan, M.; Moons, L.; Wei, S.; Zou, L.; Machelon, V.; et al. CXCL12 and vascular endothelial growth factor synergistically induce neoangiogenesis in human ovarian cancers. *Cancer Res.* **2005**, *65*, 465–472. [PubMed]
142. Motro, B.; Itin, A.; Sachs, L.; Keshet, E. Pattern of interleukin 6 gene expression in vivo suggests a role for this cytokine in angiogenesis. *Proc. Natl. Acad. Sci. USA* **1990**, *87*, 3092–3096. [CrossRef] [PubMed]
143. Nilsson, M.B.; Langley, R.R.; Fidler, I.J. Interleukin-6, secreted by human ovarian carcinoma cells, is a potent proangiogenic cytokine. *Cancer Res.* **2005**, *65*, 10794–10800. [CrossRef] [PubMed]
144. Rose-John, S.; Scheller, J.; Elson, G.; Jones, S.A. Interleukin-6 biology is coordinated by membrane-bound and soluble receptors: Role in inflammation and cancer. *J. Leukocyte Biol.* **2006**, *80*, 227–236. [CrossRef] [PubMed]
145. Hu, D.E.; Hori, Y.; Fan, T.P. Interleukin-8 stimulates angiogenesis in rats. *Inflammation* **1993**, *17*, 135–143. [CrossRef] [PubMed]
146. Koch, A.E.; Polverini, P.J.; Kunkel, S.L.; Harlow, L.A.; DiPietro, L.A.; Elner, V.M.; Elner, S.G.; Strieter, R.M. Interleukin-8 as a macrophage-derived mediator of angiogenesis. *Science* **1992**, *258*, 1798–1801. [CrossRef] [PubMed]
147. Stetler-Stevenson, W.G. Matrix metalloproteinases in angiogenesis: A moving target for therapeutic intervention. *J. Clin. Investig.* **1999**, *103*, 1237–1241. [CrossRef] [PubMed]
148. Lau, T.S.; Chan, L.K.; Wong, E.C.; Hui, C.W.; Sneddon, K.; Cheung, T.H.; Yim, S.F.; Lee, J.H.; Yeung, C.S.; Chung, T.K.; et al. A loop of cancer-stroma-cancer interaction promotes peritoneal metastasis of ovarian cancer via TNFα-TGFα-EGFR. *Oncogene* **2017**, *36*, 3576–3587. [CrossRef] [PubMed]
149. Scotton, C.J.; Wilson, J.L.; Scott, K.; Stamp, G.; Wilbanks, G.D.; Fricker, S.; Bridger, G.; Balkwill, F.R. Multiple actions of the chemokine CXCL12 on epithelial tumor cells in human ovarian cancer. *Cancer Res.* **2002**, *62*, 5930–5938. [PubMed]
150. Kulbe, H.; Hagemann, T.; Szlosarek, P.W.; Balkwill, F.R.; Wilson, J.L. The inflammatory cytokine tumor necrosis factor-α regulates chemokine receptor expression on ovarian cancer cells. *Cancer Res.* **2005**, *65*, 10355–10362. [CrossRef] [PubMed]
151. Kim, S.; Gwak, H.; Kim, H.S.; Kim, B.; Dhanasekaran, D.N.; Song, Y.S. Malignant ascites enhances migratory and invasive properties of ovarian cancer cells with membrane bound IL-6R in vitro. *Oncotarget* **2016**, *7*, 83148–83159. [CrossRef] [PubMed]
152. Wang, Y.; Niu, X.L.; Qu, Y.; Wu, J.; Zhu, Y.Q.; Sun, W.J.; Li, L.Z. Autocrine production of interleukin-6 confers cisplatin and paclitaxel resistance in ovarian cancer cells. *Cancer Lett.* **2010**, *295*, 110–123. [CrossRef] [PubMed]

153. Yung, M.M.; Tang, H.W.; Cai, P.C.; Leung, T.H.; Ngu, S.F.; Chan, K.K.; Xu, D.; Yang, H.; Ngan, H.Y.; Chan, D.W. GRO-α and IL-8 enhance ovarian cancer metastatic potential via the CXCR2-mediated TAK1/NFκB signaling cascade. *Theranostics* **2018**, *8*, 1270–1285. [CrossRef] [PubMed]
154. Cuello, M.; Ettenberg, S.A.; Nau, M.M.; Lipkowitz, S. Synergistic induction of apoptosis by the combination of trail and chemotherapy in chemoresistant ovarian cancer cells. *Gynecol. Oncol.* **2001**, *81*, 380–390. [CrossRef] [PubMed]
155. Song, Y.; Wu, J.; Oyesanya, R.A.; Lee, Z.; Mukherjee, A.; Fang, X. Sp-1 and c-Myc mediate lysophosphatidic acid-induced expression of vascular endothelial growth factor in ovarian cancer cells via a hypoxia-inducible factor-1-independent mechanism. *Clin. Cancer Res.* **2009**, *15*, 492–501. [CrossRef] [PubMed]
156. Pustilnik, T.B.; Estrella, V.; Wiener, J.R.; Mao, M.; Eder, A.; Watt, M.A.; Bast, R.C., Jr.; Mills, G.B. Lysophosphatidic acid induces urokinase secretion by ovarian cancer cells. *Clin. Cancer Res.* **1999**, *5*, 3704–3710. [PubMed]
157. Duffy, M.J. Proteases as prognostic markers in cancer. *Clin. Cancer Res.* **1996**, *2*, 613–618. [PubMed]
158. Masferrer, J.L.; Leahy, K.M.; Koki, A.T.; Zweifel, B.S.; Settle, S.L.; Woerner, B.M.; Edwards, D.A.; Flickinger, A.G.; Moore, R.J.; Seibert, K. Antiangiogenic and antitumor activities of cyclooxygenase-2 inhibitors. *Cancer Res.* **2000**, *60*, 1306–1311. [PubMed]
159. Liu, X.H.; Yao, S.; Kirschenbaum, A.; Levine, A.C. NS398, a selective cyclooxygenase-2 inhibitor, induces apoptosis and down-regulates bcl-2 expression in LNCaP cells. *Cancer Res.* **1998**, *58*, 4245–4249. [PubMed]
160. Yeung, T.L.; Leung, C.S.; Wong, K.K.; Samimi, G.; Thompson, M.S.; Liu, J.; Zaid, T.M.; Ghosh, S.; Birrer, M.J.; Mok, S.C. TGF-beta modulates ovarian cancer invasion by upregulating CAF-derived versican in the tumor microenvironment. *Cancer Res.* **2013**, *73*, 5016–5028. [CrossRef] [PubMed]
161. Frankel, A.; Mills, G.B. Peptide and lipid growth factors decrease cis-diamminedichloroplatinum-induced cell death in human ovarian cancer cells. *Clin. Cancer Res.* **1996**, *2*, 1307–1313. [PubMed]
162. Gao, Y.; Shan, N.; Zhao, C.; Wang, Y.; Xu, F.; Li, J.; Yu, X.; Gao, L.; Yi, Z. LY2109761 enhances cisplatin antitumor activity in ovarian cancer cells. *Int J. Clin. Exp. Pathol.* **2015**, *8*, 4923–4932. [PubMed]
163. Hu, Y.L.; Tee, M.K.; Goetzl, E.J.; Auersperg, N.; Mills, G.B.; Ferrara, N.; Jaffe, R.B. Lysophosphatidic acid induction of vascular endothelial growth factor expression in human ovarian cancer cells. *J. Natl. Cancer Inst.* **2001**, *93*, 762–768. [CrossRef] [PubMed]
164. Forsythe, J.A.; Jiang, B.H.; Iyer, N.V.; Agani, F.; Leung, S.W.; Koos, R.D.; Semenza, G.L. Activation of vascular endothelial growth factor gene transcription by hypoxia-inducible factor 1. *Mol. Cell. Biol.* **1996**, *16*, 4604–4613. [CrossRef] [PubMed]
165. Tan, D.S.; Agarwal, R.; Kaye, S.B. Mechanisms of transcoelomic metastasis in ovarian cancer. *Lancet Oncol.* **2006**, *7*, 925–934. [CrossRef]
166. Fidler, I.J. The pathogenesis of cancer metastasis: The 'seed and soil' hypothesis revisited. *Nat. Rev. Cancer* **2003**, *3*, 453–458. [CrossRef] [PubMed]
167. Gupta, G.P.; Massague, J. Cancer metastasis: Building a framework. *Cell* **2006**, *127*, 679–695. [CrossRef] [PubMed]
168. Talmadge, J.E.; Fidler, I.J. AACR centennial series: The biology of cancer metastasis: Historical perspective. *Cancer Res.* **2010**, *70*, 5649–5669. [CrossRef] [PubMed]
169. Yang-Hartwich, Y.; Gurrea-Soteras, M.; Sumi, N.; Joo, W.D.; Holmberg, J.C.; Craveiro, V.; Alvero, A.B.; Mor, G. Ovulation and extra-ovarian origin of ovarian cancer. *Sci. Rep.* **2014**, *4*, 6116. [CrossRef] [PubMed]
170. Russo, A.; Czarnecki, A.A.; Dean, M.; Modi, D.A.; Lantvit, D.D.; Hardy, L.; Baligod, S.; Davis, D.A.; Wei, J.J.; Burdette, J.E. PTEN loss in the fallopian tube induces hyperplasia and ovarian tumor formation. *Oncogene* **2018**, *37*, 1976–1990. [CrossRef] [PubMed]
171. Cain, R.J.; Ridley, A.J. Phosphoinositide 3-kinases in cell migration. *Biol. Cell* **2009**, *101*, 13–29. [CrossRef] [PubMed]
172. Raftopoulou, M.; Hall, A. Cell migration: Rho GTPases lead the way. *Dev. Biol.* **2004**, *265*, 23–32. [CrossRef] [PubMed]
173. Devreotes, P.; Horwitz, A.R. Signaling networks that regulate cell migration. *Cold Spring Harb Perspect. Biol.* **2015**, *7*, a005959. [CrossRef] [PubMed]
174. Robinson-Smith, T.M.; Isaacsohn, I.; Mercer, C.A.; Zhou, M.; Van Rooijen, N.; Husseinzadeh, N.; McFarland-Mancini, M.M.; Drew, A.F. Macrophages mediate inflammation-enhanced metastasis of ovarian tumors in mice. *Cancer Res.* **2007**, *67*, 5708–5716. [CrossRef] [PubMed]

175. Wilkosz, S.; Ireland, G.; Khwaja, N.; Walker, M.; Butt, R.; de Giorgio-Miller, A.; Herrick, S.E. A comparative study of the structure of human and murine greater omentum. *Anat. Embryol.* **2005**, *209*, 251–261. [CrossRef] [PubMed]
176. Tang, J.; Pulliam, N.; Ozes, A.; Buechlein, A.; Ding, N.; Keer, H.; Rusch, D.; O'Hagan, H.; Stack, M.S.; Nephew, K.P. Epigenetic Targeting of Adipocytes Inhibits High-Grade Serous Ovarian Cancer Cell Migration and Invasion. *Mol. Cancer Res.* **2018**. [CrossRef] [PubMed]
177. Meza-Perez, S.; Randall, T.D. Immunological Functions of the Omentum. *Trends Immunol.* **2017**, *38*, 526–536. [CrossRef] [PubMed]
178. Gerber, S.A.; Rybalko, V.Y.; Bigelow, C.E.; Lugade, A.A.; Foster, T.H.; Frelinger, J.G.; Lord, E.M. Preferential attachment of peritoneal tumor metastases to omental immune aggregates and possible role of a unique vascular microenvironment in metastatic survival and growth. *Am. J. Pathol.* **2006**, *169*, 1739–1752. [CrossRef] [PubMed]
179. Clark, R.; Krishnan, V.; Schoof, M.; Rodriguez, I.; Theriault, B.; Chekmareva, M.; Rinker-Schaeffer, C. Milky spots promote ovarian cancer metastatic colonization of peritoneal adipose in experimental models. *Am. J. Pathol.* **2013**, *183*, 576–591. [CrossRef] [PubMed]
180. Szatrowski, T.P.; Nathan, C.F. Production of large amounts of hydrogen peroxide by human tumor cells. *Cancer Res.* **1991**, *51*, 794–798. [PubMed]
181. Wang, Y.; Ma, J.; Shen, H.; Wang, C.; Sun, Y.; Howell, S.B.; Lin, X. Reactive oxygen species promote ovarian cancer progression via the HIF-1α/LOX/E-cadherin pathway. *Oncol. Rep.* **2014**, *32*, 2150–2158. [CrossRef] [PubMed]
182. Kundu, N.; Zhang, S.; Fulton, A.M. Sublethal oxidative stress inhibits tumor cell adhesion and enhances experimental metastasis of murine mammary carcinoma. *Clin. Exp. Metast.* **1995**, *13*, 16–22. [CrossRef]
183. Haskill, S.; Becker, S.; Fowler, W.; Walton, L. Mononuclear-cell infiltration in ovarian cancer. I. Inflammatory-cell infiltrates from tumour and ascites material. *Br. J. Cancer* **1982**, *45*, 728–736. [CrossRef] [PubMed]
184. Wang, J.M.; Griffin, J.D.; Rambaldi, A.; Chen, Z.G.; Mantovani, A. Induction of monocyte migration by recombinant macrophage colony-stimulating factor. *J. Immunol.* **1988**, *141*, 575–579. [PubMed]
185. Ramakrishnan, S.; Xu, F.J.; Brandt, S.J.; Niedel, J.E.; Bast, R.C., Jr.; Brown, E.L. Constitutive production of macrophage colony-stimulating factor by human ovarian and breast cancer cell lines. *J. Clin. Investig.* **1989**, *83*, 921–926. [CrossRef] [PubMed]
186. Plante, M.; Rubin, S.C.; Wong, G.Y.; Federici, M.G.; Finstad, C.L.; Gastl, G.A. Interleukin-6 level in serum and ascites as a prognostic factor in patients with epithelial ovarian cancer. *Cancer* **1994**, *73*, 1882–1888. [CrossRef]
187. Scambia, G.; Testa, U.; Panici, P.B.; Martucci, R.; Foti, E.; Petrini, M.; Amoroso, M.; Masciullo, V.; Peschle, C.; Mancuso, S. Interleukin-6 serum levels in patients with gynecological tumors. *Int. J. Cancer* **1994**, *57*, 318–323. [CrossRef] [PubMed]
188. Tempfer, C.; Zeisler, H.; Sliutz, G.; Haeusler, G.; Hanzal, E.; Kainz, C. Serum evaluation of interleukin 6 in ovarian cancer patients. *Gynecol. Oncol.* **1997**, *66*, 27–30. [CrossRef] [PubMed]
189. Isobe, A.; Sawada, K.; Kinose, Y.; Ohyagi-Hara, C.; Nakatsuka, E.; Makino, H.; Ogura, T.; Mizuno, T.; Suzuki, N.; Morii, E.; et al. Interleukin 6 receptor is an independent prognostic factor and a potential therapeutic target of ovarian cancer. *PLoS ONE* **2015**, *10*, e0118080. [CrossRef] [PubMed]
190. Fishman, D.A.; Liu, Y.; Ellerbroek, S.M.; Stack, M.S. Lysophosphatidic acid promotes matrix metalloproteinase (MMP) activation and MMP-dependent invasion in ovarian cancer cells. *Cancer Res.* **2001**, *61*, 3194–3199. [PubMed]
191. Reiser, C.O.; Lanz, T.; Hofmann, F.; Hofer, G.; Rupprecht, H.D.; Goppelt-Struebe, M. Lysophosphatidic acid-mediated signal-transduction pathways involved in the induction of the early-response genes prostaglandin G/H synthase-2 and Egr-1: A critical role for the mitogen-activated protein kinase p38 and for Rho proteins. *Biochem. J.* **1998**, *330 (Pt 3)*, 1107–1114. [CrossRef] [PubMed]
192. Moolenaar, W.H.; Kruijer, W.; Tilly, B.C.; Verlaan, I.; Bierman, A.J.; de Laat, S.W. Growth factor-like action of phosphatidic acid. *Nature* **1986**, *323*, 171–173. [CrossRef] [PubMed]
193. Schwartz, B.M.; Hong, G.; Morrison, B.H.; Wu, W.; Baudhuin, L.M.; Xiao, Y.J.; Mok, S.C.; Xu, Y. Lysophospholipids increase interleukin-8 expression in ovarian cancer cells. *Gynecol. Oncol.* **2001**, *81*, 291–300. [CrossRef] [PubMed]

194. Yu, X.; Zhang, Y.; Chen, H. LPA receptor 1 mediates LPA-induced ovarian cancer metastasis: An in vitro and in vivo study. *BMC Cancer* **2016**, *16*, 846. [CrossRef] [PubMed]
195. Loizzi, V.; Del Vecchio, V.; Gargano, G.; De Liso, M.; Kardashi, A.; Naglieri, E.; Resta, L.; Cicinelli, E.; Cormio, G. Biological Pathways Involved in Tumor Angiogenesis and Bevacizumab Based Anti-Angiogenic Therapy with Special References to Ovarian Cancer. *Int. J. Mol. Sci.* **2017**, *18*, 1967. [CrossRef] [PubMed]
196. Le Page, C.; Puiffe, M.L.; Meunier, L.; Zietarska, M.; de Ladurantaye, M.; Tonin, P.N.; Provencher, D.; Mes-Masson, A.M. BMP-2 signaling in ovarian cancer and its association with poor prognosis. *J. Ovarian Res.* **2009**, *2*, 4. [CrossRef] [PubMed]
197. Ozols, R.F. Treatment goals in ovarian cancer. *Int. J. Gynecol. Cancer* **2005**, *15* (Suppl. 1), 3–11. [CrossRef] [PubMed]
198. Koti, M.; Siu, A.; Clement, I.; Bidarimath, M.; Turashvili, G.; Edwards, A.; Rahimi, K.; Mes-Masson, A.M.; Squire, J.A. A distinct pre-existing inflammatory tumour microenvironment is associated with chemotherapy resistance in high-grade serous epithelial ovarian cancer. *Br. J. Cancer* **2015**, *112*, 1215–1222. [CrossRef] [PubMed]
199. Stewart, D.J. Mechanisms of resistance to cisplatin and carboplatin. *Crit Rev. Oncol Hematol* **2007**, *63*, 12–31. [CrossRef] [PubMed]
200. Damiano, J.S.; Cress, A.E.; Hazlehurst, L.A.; Shtil, A.A.; Dalton, W.S. Cell adhesion mediated drug resistance (CAM-DR): Role of integrins and resistance to apoptosis in human myeloma cell lines. *Blood* **1999**, *93*, 1658–1667. [PubMed]
201. Meads, M.B.; Hazlehurst, L.A.; Dalton, W.S. The bone marrow microenvironment as a tumor sanctuary and contributor to drug resistance. *Clin. Cancer Res.* **2008**, *14*, 2519–2526. [CrossRef] [PubMed]
202. Catlett-Falcone, R.; Landowski, T.H.; Oshiro, M.M.; Turkson, J.; Levitzki, A.; Savino, R.; Ciliberto, G.; Moscinski, L.; Fernandez-Luna, J.L.; Nunez, G.; et al. Constitutive activation of Stat3 signaling confers resistance to apoptosis in human U266 myeloma cells. *Immunity* **1999**, *10*, 105–115. [CrossRef]
203. Frassanito, M.A.; Cusmai, A.; Iodice, G.; Dammacco, F. Autocrine interleukin-6 production and highly malignant multiple myeloma: Relation with resistance to drug-induced apoptosis. *Blood* **2001**, *97*, 483–489. [CrossRef] [PubMed]
204. Meads, M.B.; Gatenby, R.A.; Dalton, W.S. Environment-mediated drug resistance: A major contributor to minimal residual disease. *Nat. Rev. Cancer* **2009**, *9*, 665–674. [CrossRef] [PubMed]
205. Cohen, S.; Mehrabi, S.; Yao, X.; Millingen, S.; Aikhionbare, F.O. Reactive Oxygen Species and Serous Epithelial Ovarian Adenocarcinoma. *Cancer Res. J.* **2016**, *4*, 106–114. [CrossRef] [PubMed]
206. St-Pierre, J.; Drori, S.; Uldry, M.; Silvaggi, J.M.; Rhee, J.; Jager, S.; Handschin, C.; Zheng, K.; Lin, J.; Yang, W.; et al. Suppression of reactive oxygen species and neurodegeneration by the PGC-1 transcriptional coactivators. *Cell* **2006**, *127*, 397–408. [CrossRef] [PubMed]
207. Kim, B.; Jung, J.W.; Jung, J.; Han, Y.; Suh, D.H.; Kim, H.S.; Dhanasekaran, D.N.; Song, Y.S. PGC1α induced by reactive oxygen species contributes to chemoresistance of ovarian cancer cells. *Oncotarget* **2017**, *8*, 60299–60311. [PubMed]
208. Duan, Z.; Feller, A.J.; Penson, R.T.; Chabner, B.A.; Seiden, M.V. Discovery of differentially expressed genes associated with paclitaxel resistance using cDNA array technology: Analysis of interleukin (IL) 6, IL-8, and monocyte chemotactic protein 1 in the paclitaxel-resistant phenotype. *Clin. Cancer Res.* **1999**, *5*, 3445–3453. [PubMed]
209. Duan, Z.; Foster, R.; Bell, D.A.; Mahoney, J.; Wolak, K.; Vaidya, A.; Hampel, C.; Lee, H.; Seiden, M.V. Signal transducers and activators of transcription 3 pathway activation in drug-resistant ovarian cancer. *Clin. Cancer Res.* **2006**, *12*, 5055–5063. [CrossRef] [PubMed]
210. Abdollahi, T.; Robertson, N.M.; Abdollahi, A.; Litwack, G. Identification of interleukin 8 as an inhibitor of tumor necrosis factor-related apoptosis-inducing ligand-induced apoptosis in the ovarian carcinoma cell line OVCAR3. *Cancer Res.* **2003**, *63*, 4521–4526. [PubMed]
211. Wiley, S.R.; Schooley, K.; Smolak, P.J.; Din, W.S.; Huang, C.P.; Nicholl, J.K.; Sutherland, G.R.; Smith, T.D.; Rauch, C.; Smith, C.A.; et al. Identification and characterization of a new member of the TNF family that induces apoptosis. *Immunity* **1995**, *3*, 673–682. [CrossRef]
212. Walczak, H.; Miller, R.E.; Ariail, K.; Gliniak, B.; Griffith, T.S.; Kubin, M.; Chin, W.; Jones, J.; Woodward, A.; Le, T.; et al. Tumoricidal activity of tumor necrosis factor-related apoptosis-inducing ligand in vivo. *Nat. Med.* **1999**, *5*, 157–163. [CrossRef] [PubMed]

213. Bristow, R.E.; Baldwin, R.L.; Yamada, S.D.; Korc, M.; Karlan, B.Y. Altered expression of transforming growth factor-beta ligands and receptors in primary and recurrent ovarian carcinoma. *Cancer* **1999**, *85*, 658–668. [CrossRef]
214. Li, M.; Balch, C.; Montgomery, J.S.; Jeong, M.; Chung, J.H.; Yan, P.; Huang, T.H.; Kim, S.; Nephew, K.P. Integrated analysis of DNA methylation and gene expression reveals specific signaling pathways associated with platinum resistance in ovarian cancer. *BMC Med. Genom.* **2009**, *2*, 34. [CrossRef] [PubMed]
215. Psyrri, A.; Kassar, M.; Yu, Z.; Bamias, A.; Weinberger, P.M.; Markakis, S.; Kowalski, D.; Camp, R.L.; Rimm, D.L.; Dimopoulos, M.A. Effect of epidermal growth factor receptor expression level on survival in patients with epithelial ovarian cancer. *Clin. Cancer Res.* **2005**, *11*, 8637–8643. [CrossRef] [PubMed]
216. Crew, A.J.; Langdon, S.P.; Miller, E.P.; Miller, W.R. Mitogenic effects of epidermal growth factor and transforming growth factor-α on EGF-receptor positive human ovarian carcinoma cell lines. *Eur. J. Cancer* **1992**, *28*, 337–341. [CrossRef]
217. Ferrandina, G.; Ranelletti, F.O.; Martinelli, E.; Paglia, A.; Zannoni, G.F.; Scambia, G. Cyclo-oxygenase-2 (Cox-2) expression and resistance to platinum versus platinum/paclitaxel containing chemotherapy in advanced ovarian cancer. *BMC Cancer* **2006**, *6*, 182. [CrossRef] [PubMed]
218. Ferrandina, G.; Lauriola, L.; Zannoni, G.F.; Fagotti, A.; Fanfani, F.; Legge, F.; Maggiano, N.; Gessi, M.; Mancuso, S.; Ranelletti, F.O.; et al. Increased cyclooxygenase-2 (COX-2) expression is associated with chemotherapy resistance and outcome in ovarian cancer patients. *Ann. Oncol.* **2002**, *13*, 1205–1211. [CrossRef] [PubMed]
219. Kim, A.; Ueda, Y.; Naka, T.; Enomoto, T. Therapeutic strategies in epithelial ovarian cancer. *J. Exp. Clin. Cancer Res.* **2012**, *31*, 14. [CrossRef] [PubMed]
220. Ellis, L.M.; Hicklin, D.J. VEGF-targeted therapy: Mechanisms of anti-tumour activity. *Nat. Rev. Cancer* **2008**, *8*, 579–591. [CrossRef] [PubMed]
221. Aghajanian, C.; Blank, S.V.; Goff, B.A.; Judson, P.L.; Teneriello, M.G.; Husain, A.; Sovak, M.A.; Yi, J.; Nycum, L.R. OCEANS: A randomized, double-blind, placebo-controlled phase III trial of chemotherapy with or without bevacizumab in patients with platinum-sensitive recurrent epithelial ovarian, primary peritoneal, or fallopian tube cancer. *J. Clin. Oncol.* **2012**, *30*, 2039–2045. [CrossRef] [PubMed]
222. Burger, R.A.; Brady, M.F.; Bookman, M.A.; Fleming, G.F.; Monk, B.J.; Huang, H.; Mannel, R.S.; Homesley, H.D.; Fowler, J.; Greer, B.E.; et al. Incorporation of bevacizumab in the primary treatment of ovarian cancer. *N. Engl. J. Med.* **2011**, *365*, 2473–2483. [CrossRef] [PubMed]
223. Perren, T.J.; Swart, A.M.; Pfisterer, J.; Ledermann, J.A.; Pujade-Lauraine, E.; Kristensen, G.; Carey, M.S.; Beale, P.; Cervantes, A.; Kurzeder, C.; et al. A phase 3 trial of bevacizumab in ovarian cancer. *N. Engl. J. Med.* **2011**, *365*, 2484–2496. [CrossRef] [PubMed]
224. Brown, E.R.; Charles, K.A.; Hoare, S.A.; Rye, R.L.; Jodrell, D.I.; Aird, R.E.; Vora, R.; Prabhakar, U.; Nakada, M.; Corringham, R.E.; et al. A clinical study assessing the tolerability and biological effects of infliximab, a TNF-α inhibitor, in patients with advanced cancer. *Ann. Oncol.* **2008**, *19*, 1340–1346. [CrossRef] [PubMed]
225. Guo, Y.; Nemeth, J.; O'Brien, C.; Susa, M.; Liu, X.; Zhang, Z.; Choy, E.; Mankin, H.; Hornicek, F.; Duan, Z. Effects of siltuximab on the IL-6-induced signaling pathway in ovarian cancer. *Clin. Cancer Res.* **2010**, *16*, 5759–5769. [CrossRef] [PubMed]
226. Xu, S.; Grande, F.; Garofalo, A.; Neamati, N. Discovery of a novel orally active small-molecule gp130 inhibitor for the treatment of ovarian cancer. *Mol. Cancer Ther.* **2013**, *12*, 937–949. [CrossRef] [PubMed]
227. Dijkgraaf, E.M.; Santegoets, S.J.; Reyners, A.K.; Goedemans, R.; Wouters, M.C.; Kenter, G.G.; van Erkel, A.R.; van Poelgeest, M.I.; Nijman, H.W.; van der Hoeven, J.J.; et al. A phase I trial combining carboplatin/doxorubicin with tocilizumab, an anti-IL-6R monoclonal antibody, and interferon-α2b in patients with recurrent epithelial ovarian cancer. *Ann. Oncol.* **2015**, *26*, 2141–2149. [CrossRef] [PubMed]
228. Bhola, N.E.; Balko, J.M.; Dugger, T.C.; Kuba, M.G.; Sanchez, V.; Sanders, M.; Stanford, J.; Cook, R.S.; Arteaga, C.L. TGF-beta inhibition enhances chemotherapy action against triple-negative breast cancer. *J. Clin. Investig.* **2013**, *123*, 1348–1358. [CrossRef] [PubMed]
229. Alsina-Sanchis, E.; Figueras, A.; Lahiguera, A.; Gil-Martin, M.; Pardo, B.; Piulats, J.M.; Marti, L.; Ponce, J.; Matias-Guiu, X.; Vidal, A.; et al. TGFbeta Controls Ovarian Cancer Cell Proliferation. *Int. J. Mol. Sci.* **2017**, *18*, 1658. [CrossRef] [PubMed]

230. Secord, A.A.; Blessing, J.A.; Armstrong, D.K.; Rodgers, W.H.; Miner, Z.; Barnes, M.N.; Lewandowski, G.; Mannel, R.S.; Gynecologic Oncology, G. Phase II trial of cetuximab and carboplatin in relapsed platinum-sensitive ovarian cancer and evaluation of epidermal growth factor receptor expression: A Gynecologic Oncology Group study. *Gynecol. Oncol.* **2008**, *108*, 493–499. [CrossRef] [PubMed]
231. Chekerov, R.; Klare, P.; Krabisch, P.; Potenberg, J.; Heinrich, G.; Mueller, L.; Kurbacher, C.M.; Grischke, E.-M.; Braicu, E.I.; Wimberger, P.; et al. Panitumumab in platinum-sensitive epithelial ovarian cancer patients with KRAS wild-type: The PROVE-study, a phase II randomized multicenter study of the North-Eastern German Society of Gynaecologic Oncology. *J. Clin. Oncol.* **2017**, *35* (Suppl. 1), 5558.

cancers

MDPI

Review

The Endometriotic Tumor Microenvironment in Ovarian Cancer

Jillian R. Hufgard Wendel, Xiyin Wang and Shannon M. Hawkins *

Department of Obstetrics and Gynecology, Indiana University School of Medicine, Indianapolis, IN 46202, USA; jhufgard@iu.edu (J.R.H.W.); xw49@iu.edu (X.W.)

* Correspondence: shhawkin@iu.edu; Tel.: +1-317-274-8225

Received: 29 June 2018; Accepted: 2 August 2018; Published: 7 August 2018

Abstract: Women with endometriosis are at increased risk of developing ovarian cancer, specifically ovarian endometrioid, low-grade serous, and clear-cell adenocarcinoma. An important clinical caveat to the association of endometriosis with ovarian cancer is the improved prognosis for women with endometriosis at time of ovarian cancer staging. Whether endometriosis-associated ovarian cancers develop from the molecular transformation of endometriosis or develop because of the endometriotic tumor microenvironment remain unknown. Additionally, how the presence of endometriosis improves prognosis is also undefined, but likely relies on the endometriotic microenvironment. The unique tumor microenvironment of endometriosis is composed of epithelial, stromal, and immune cells, which adapt to survive in hypoxic conditions with high levels of iron, estrogen, and inflammatory cytokines and chemokines. Understanding the unique molecular features of the endometriotic tumor microenvironment may lead to impactful precision therapies and/or modalities for prevention. A challenge to this important study is the rarity of well-characterized clinical samples and the limited model systems. In this review, we will describe the unique molecular features of endometriosis-associated ovarian cancers, the endometriotic tumor microenvironment, and available model systems for endometriosis-associated ovarian cancers. Continued research on these unique ovarian cancers may lead to improved prevention and treatment options.

Keywords: ovarian cancer; endometriosis; tumor microenvironment; miRNA molecules; genes; hypoxia; inflammation; model systems

---

1. Introduction

Endometriosis is a debilitating disease that is estimated to affect up to 5 million U.S. women and girls. Endometriosis results in considerable morbidity, including pelvic pain, multiple operations, infertility, and negative effects on psychosocial quality of life [1–5]. Unfortunately, endometriosis is also a significant risk factor for development of ovarian cancer [6]. The presence of endometriosis increases the risk of ovarian endometrioid, low-grade serous, and clear-cell adenocarcinoma by up to 8.9-fold but not high-grade serous adenocarcinoma [7–12]. Thus, ovarian endometrioid, low-grade serous, and clear-cell adenocarcinomas are considered endometriosis-associated ovarian cancers. Ovarian cancer is considered a top-five cancer killer in U.S. women, claiming more than 14,000 lives in 2015 [13]. Therefore, 5 million U.S. women and girls with endometriosis are at risk for developing deadly ovarian cancer. Fortunately, ovarian endometrioid and clear-cell adenocarcinoma represent roughly 20% of all ovarian cancers and account for less than 10% of deaths [14–16]. Clinically, studies suggest that co-occurrence of endometriosis with ovarian cancer is associated with an improved prognosis [17–20]. Important factors in this improved prognosis include discovery at early age and early stage disease in women with endometriosis at time of ovarian cancer staging [21–24], but may also represent the unique biology from the endometriotic tumor microenvironment. This review will focus on the contributions of the endometriotic tumor microenvironment to ovarian cancer biology.

2. Unique Molecular Features of Endometriosis-Associated Ovarian Cancer

Each histotype of epithelial ovarian cancer is thought to arise from a distinct precursor lesion. For example, endometriosis is thought to give rise to both ovarian endometrioid and clear-cell adenocarcinomas [25]. Recently, sophisticated proteomic tracing studies suggest that ovarian endometrioid adenocarcinomas arise from secretory cells of endometriosis or the endometrium, while ovarian clear-cell adenocarcinomas arise from ciliated cells. Importantly, it is hypothesized that the unique cellular environment dictates the development of ciliated or secretory cells, which then gain mutations to become malignant [26]. Recently, next-generation sequencing studies showed mutations in cancer-driver genes (i.e., AT-rich interaction domain 1A (*ARID1A*), Phosphatidylinositol-4, 5-bisphosphate 3-kinase catalytic subunit alpha (*PIK3CA*), and Kirsten rat sarcoma viral oncogene homolog (*KRAS*)) in deep infiltrating endometriotic lesions, supporting the idea that the endometriotic microenvironment facilitates mutations [27]. Because deep infiltrating endometriotic lesions do not pose a risk of malignant transformation, the unique contributions of driver mutations in these particular endometriotic lesions are still relatively unknown [27]. Interestingly, these mutations in cancer-driver genes were only present in glandular epithelium and not underlying stroma [27]. These data support the idea that both epithelium and stromal populations of deep infiltrating endometriosis do not represent similar clonal populations. Further, this data may represent the idea that unique stromal populations are recruited to the area [28,29]. Detailed studies of unique genetic contributions of both epithelial and/or stromal compartments in malignant transformation are needed.

Studies examining endometriotic lesions and ovarian cancer from the same patient have shown concordant mutations in *ARID1A*, phosphatase and tensin homolog (*PTEN*), *PIK3CA*, and *KRAS*, suggesting that mutations in endometriosis cause a predisposition to ovarian cancer [30–33]. Mutations in *KRAS* and *ARID1A* have been discovered in endometriosis, including ovarian endometriosis and deep infiltrating endometriosis [27,34]. Loss of ARID1A is higher in atypical endometriosis and non-atypical endometriosis adjacent to ovarian cancer than non-atypical endometriotic distal lesions [30,32,35–39]. In general, both endometrioid and clear cell ovarian cancer with or without endometriosis have common high frequency mutations in *ARID1A*, *PIK3CA*, catenin betat 1 (*CTNNB1*), *PTEN*, and *KRAS* [33,40–45]. In terms of unique molecular features, 29% of low-grade ovarian endometrioid adenocarcinomas with concurrent endometriosis contained mutations in *KRAS* compared to 3% of low-grade endometrioid adenocarcinomas lacking endometriosis [33]. Importantly, Ishikawa et al. showed high frequency of *ARID1A* mutations and one patient with both *ARID1A* and *KRAS* mutations in endometriosis-associated ovarian cancers [43]. The contributions of both ARID1A and KRAS warrant further study in terms of endometriosis, the endometriotic tumor microenvironment, and endometriosis-associated ovarian cancer.

In terms of low-grade serous tumors, an A to T substitution in *BRAF* has been identified in 36–68% of low-grade serous ovarian cancers and is associated with improved prognosis [46–48]. Additionally, increased expression of B-raf proto-oncogene, serine/threonine kinase (BRAF) was also noted in eutopic and ectopic endometrium of women with endometriosis when compared to control endometrium [49]. The contributions of BRAF to endometriosis and endometriosis-associated ovarian cancers, specifically, low-grade serous ovarian cancers are understudied.

In addition to mutational changes, epigenetic changes play a role in both endometriosis and endometriosis-associated ovarian cancers. Methylation changes in both endometriosis and endometriosis associated ovarian cancer have recently been reviewed [50,51]. Along those lines, endometriosis tissues have decrease expression of ten-eleven translocation genes (*TET1*, *TET2*, and *TET3*), which convert 5-methylcytosine to 5-hydroxymethlcytosine and play a role in changes in levels of 5-hydroxymethylcytosine marks in endometriosis tissues and blood [52]. Unfortunately, the authors did not assess 5-hydroxymethlcytosine marks in specific genes. Further studies are needed in endometriosis-associated ovarian cancer to examine changes in these and other alternative DNA marks. MicroRNA (miRNA) molecules, which are also considered epigenetic changes, are dysregulated in endometriosis (reviewed in [53]). While dysregulated miRNAs in epithelial ovarian cancers have been recently reviewed [54,55], dysregulated miRNA molecules in endometriosis-associated ovarian cancers have not been individually reviewed. Given that miRNA molecules can be secreted from cells, we have included miRNA molecules under endometriotic tumor microenvironment (below).

A challenge to studies on the endometriotic tumor microenvironment is the rarity of clinical samples of ovarian cancer with concurrent endometriosis and the rigor of details provided for patient characterization. Given over 22,000 women will be diagnosed with ovarian cancer in 2016 [13], only 10% will be endometrioid and roughly 10% will be clear-cell [14–16]. Additionally, a majority of women with endometriosis-associated ovarian cancers do not have endometriosis at time of staging. Roughly 30% of ovarian endometrioid or clear-cell adenocarcinomas will have concurrent endometriosis, further narrowing the number of tumors to study with concurrent endometriosis [56–59]. Many studies do not describe the patient population in terms of absence or presence of endometriosis, leaving readers to believe that the women may not have endometriosis, which may not be accurate. Efforts for data harmonization for rare tumors may improve reproducibility. Using well-characterized samples, Banz et al. used transcriptome microarray analysis to evaluate normal ovary, endometriomas, and endometrioid ovarian cancer with and without endometriosis [60]. The results showed a small group of cytokines dysregulated in ovarian cancers with endometriosis, consistent with the inflammatory milieu of endometriosis [60]. Additionally, Zhang et al. showed a unique gene signature in ovarian endometrioid adenocarcinoma with concurrent endometriosis compared to ovarian endometrioid adenocarcinoma without concurrent endometriosis [61]. Highly dysregulated signaling pathways included nuclear factor kappa B (NFkB), transforming growth factor beta (TGFβ), and KRAS signaling [61]. Most likely there are contributions from genetics and epigenetics that may be mediated from the endometriotic tumor microenvironment [62]. However, further studies are needed to examine how endometriosis affects ovarian cancer.

3. The Unique Endometriotic Tumor Microenvironment

While the pathogenesis of endometriosis is still largely poorly understood, the most accepted theory is the implantation theory following retrograde menstruation (reviewed in [63]). Most menstruating women have retrograde menstruation [64], but only 10% have endometriosis [1–3], suggesting that unique conditions occur in women with endometriosis. The endometriotic microenvironment contains multiple cell types—endometrial epithelial cells, stromal fibroblasts, endothelial cells, and immune cells—as well as inflammatory mediators, metabolic waste products such as iron from the breakdown of red blood cells, steroid hormones, and small RNA molecules. Thus, it is not surprising that the conditions found in endometriosis are also advantageous to the growth and development of ovarian cancer. However, very little is known about how these stressful conditions directly affect ovarian cancer. In this section, we will describe these important factors within the scope of endometriosis and how these important factors pertain to ovarian cancer. Figure 1 summarizes graphically key players in the endometriotic tumor microenvironment as it pertains to ovarian cancer.

3.1. Hypoxia and Endothelial Cells

Hypoxia is thought to be critical to the survival and invasion of endometriotic cells through multiple mechanisms including autophagy [65–68], TGFβ signaling [69], and signal transducer and activator of transcription 3 (STAT3) signaling [70–72]. In endometriosis, hypoxia stabilizes hypoxia inducible factor-1α (HIF1A) which downregulates dual-specificity phosphatase-2 (DUSP2) directly and indirectly through miR-20a [73]. Ultimately, this downregulation leads to increased angiogenesis and proliferation through activation of extracellular signal-regulated kinase (ERK) signaling cascades [73,74]. As such, molecular immunohistochemistry shows a high correlation between precursor endometriosis lesions and matched clear-cell adenocarcinomas for expression of HIF1A and phosphorylated mechanistic target of rapamycin kinase (P-mTOR) [75]. Importantly, vascular endothelial growth factor (VEGF), leptin (LEP), cysteine rich angiongenic inducer 61

(CYR61), and osteopontin (SPP1) work together in response to hypoxia to establish a local vascular network within the endometriotic lesion [74]. In addition to neoangiogenesis mediated through HIF1A, as endometriotic lesions undergo hypoxia and inflammation from repeated menstrual cycles, the expression of tissue factor increases. Tissue factor is a critical protein for extrinsic coagulation cascade, leading to hypercoagulation. Clinically, women with clear-cell ovarian cancer have more frequent venous thromboembolism [76]. Hypoxia may also lead to cellular proliferation through estrogen receptor, leptin, and prostaglandin modulation [77]. These studies suggest that the hypoxic microenvironment of endometriosis plays a role in not only the potentiation of endometriosis by promoting cell proliferation and nutrient availability through vascularization but may also play roles in outcomes for women with clear-cell ovarian cancer. The increased expression of HIF1A in endometriosis may represent a novel therapeutic target for endometriosis or ovarian cancer [78].

3.2. Fibroblasts and Extracellular Matrix Components

Endometriosis is pathologically complex, containing endometrial epithelial and stromal fibroblasts outside the uterine cavity, alongside invading hemosiderin-laden macrophages [79]. The endometriotic extracellular matrix (ECM) plays a significant role in paracrine/autocrine signaling between epithelial and stromal cells [80–83]. Studies have shown unique functional properties of primary cultures of human endometrial stromal fibroblasts from women with endometriosis compared to cultures from women without endometriosis. Specifically, fibroblast cultures from women with endometriosis have a deficiency in decidualization, the differentiation process by which the uterus prepares for pregnancy [84]. Additionally, these fibroblasts from women with endometriosis have increased ERK signaling, high proliferative potential from progesterone resistance, and acquire an inflammatory phenotype [85–89]. While the importance of stromal-epithelial crosstalk is noted in embryo implantation in the uterus [80], the role of similar crosstalk in endometriosis or epithelial ovarian cancers is still understudied but may represent a key component of the endometriotic tumor microenvironment.

To examine the tumor microenvironment in ovarian cancer, Zhang et al. used computer-aided image analysis and showed that the number of cancer-associated fibroblasts, as indicated by cells positive for smooth muscle antigen, was higher in epithelial ovarian cancers compared to benign adnexal masses. Unfortunately, the specific histology of ovarian cancers and the pathology of the benign adnexal masses were not described in these studies. Large numbers of similarly staining cancer-associated fibroblasts were also found in omental metastatic lesions [90]. Co-culture of cancer-associated fibroblast with ovarian cancer cell lines (SKOV3, CAOV3) led to increased invasion and migration when compared to ovarian cancer cell lines grown in co-culture with normal fibroblasts [90]. One of the main questions regarding cancer-associated fibroblasts is how and why they are becoming activated to benefit tumor cells. Mitra et al. proposed that ovarian cancer cells reprogram fibroblasts into cancer-associated fibroblasts through miRNA expression changes [91]. Specifically, cancer-associated fibroblasts have a significant downregulation of miR-31 and miR-214 and upregulation of miR-155. C-C motif ligand 5 (CCL5), a chemokine known to be highly upregulated in ovarian cancers, is a direct target of miR-214. Similarly, endometriomas have high expression of chemokines and dysregulated miRNA expression [92]. Advancements in the understanding of the role of non-epithelial ovarian cancer cells in ovarian cancer may lead to better treatments which block tumor promotion brought on by tumor adjacent cells.

3.3. Immune Cells and Inflammatory Mediators

Dysregulated inflammation plays a key role in endometriosis-associated pathology [93]. For example, Capobianco and Rovere-Querini provide an in-depth review of the role of macrophages in endometriosis, showing a relationship between components of the endometriotic microenvironment such as high iron, hypoxia, and angiogenesis with macrophage recruitment and activation [94]. Additionally, a syngeneic mouse model of endometriosis showed that endometriotic lesions failed to grow without macrophages, and if macrophages were removed after implantation, angiogenesis was halted, blocking the progression of the endometriotic lesion [95]. Further, Canet et al. suggest that retainment of a specific macrophage population in endometriomas, the cell division cycle 42 (CDC42)-positive population, protects endometriomas from malignant transformation [96]. Similarly, platelet factor 4 (PF4) also known as chemokine (C-X-C Motif) ligand 4 (CXCL4) is highly expressed on macrophages in endometriomas, but not on tumor-associated macrophages of clear cell ovarian cancers [97]. Thus, specific details of the macrophage population in endometriosis and ovarian cancer are important and require further study.

Transcriptomic work on endometriomas showed that the inflammatory cytokine transforming growth factor beta 1 (TGFβ1), regulates other inflammatory mediators relevant to endometriosis, including tumor necrosis factor alpha (TNFα) and interleukin-6 (IL6) [92]. These inflammatory mediators are highly elevated in peritoneal fluid from women with endometriosis [98–101]. The acute and chronic inflammation of endometriosis is a response to the invading tissue, leading to the release of regulated on activation normal T cell expressed and secreted (RANTES), monocyte chemotactic protein-1 (MCP1), and interleukin-8 (IL8), which act as chemoattractants recruiting more macrophages to the area [102]. In terms of the endometriotic tumor microenvironment, the promotion of tumor invasion via macrophages may be dependent on TNFα [103], which is elevated in women with endometriosis [98,99]. Along the same lines, work using an estrogen receptor beta (ERβ)-overexpressing syngeneic mouse model of endometriosis suggests that non-genomic effects of ERβ play a role in the TNFα-mediated dysregulation of endometriosis progression [104]. Encouragingly, treatment of a syngeneic mouse model of endometriosis with a long-acting TNFα-blocking agent decreased endometriotic implant size [105]. However, treatment of women with rectovaginal nodules with infliximab, a TNFα monoclonal antibody, had no improved clinical effect over placebo [106]. Understanding the immune response to misplaced endometrial tissue will be a large factor in understanding the onset and progression of endometriosis and lead to a better understanding of how endometriosis creates a unique and potentially tumor-promoting microenvironment.

3.4. Altered Metabolism

Endometriotic cysts contain blood. When blood is metabolized, heme and iron are released into the microenvironment [107]. Because of this, endometriotic cysts contain higher iron levels than other benign ovarian cysts [108]. Consequently, an iron-rich microenvironment can lead to increased proliferation, DNA synthesis, and adhesion, and promote chronic inflammation, allowing for the spread of endometriosis [107]. High iron also leads to excessive oxidative stress, which creates a microenvironment conducive to the induction of mutations and has been linked to cancer development in the liver and lung [107,109]. Shigetomi et al. outlines how endometriotic cells under oxidative stress from excess iron are able to bypass cell cycle checkpoints after DNA damage by overexpressing hepatocyte nuclear factor-1 beta (HNF1B), which activates forkhead box transcription factors and alters miRNA expression promoting cell survival [110]. Due to the excess iron exposure, endometriotic cysts have higher expression of lactose dehydogenase, lipid peroxidase, and 8-hydroxy-2′-deoxyguanosine. High expression of these markers of oxidative stress link endometriosis, high iron, and higher frequencies of gene mutations [108]. These data corroborate the hypothesis that endometriosis produces a high iron microenvironment that may lead to increased DNA damage through oxidative stress, but also promotes cell survival, leading to a highly mutated subpopulation of cells that continue to grow [111].

Alongside high iron levels, endometriotic peritoneal fluid has elevated lactate. Further, endometriotic lesions express high levels of glycolysis genes compared to eutopic endometrium [112]. Increased expression of HNF1α in the endometriotic peritoneum leads to the conversion of glucose to lactate in a process known as the "Warburg Effect," known for its promotion of cell survival in stressful microenvironments [113]. Lipidomics has also been pursued for understanding the metabolomic profile of the endometriotic microenvironment. Lipid profiling studies on endometrial aspirates have shown a reduction of saturated diacylglycerols and triacylglycerols in endometriosis patients compared to healthy controls [114]. In fact, this study generated a panel of 123 metabolites which were differentially expressed in endometriosis women and correctly identified 86% of samples to either the endometriosis or control group [114]. A similar study on endometrial biopsies used five lipid metabolites as biomarkers and were able to predict endometriosis with 75% specificity and 90.5% sensitivity [115]. A true model of the endometriotic tumor microenvironment should include increased iron levels, higher levels of glycolysis-associated proteins, and endometriosis-associated lipidomic profiles.

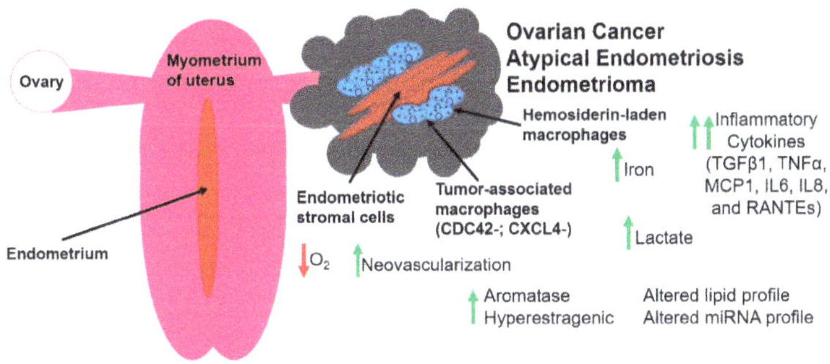

Figure 1. Composition of the endometriotic tumor microenvironment. Endometriosis represents a pathologically benign disease. Endometriosis may be classified into endometriomas, superficial peritoneal disease, or deep infiltrating endometriosis (invasion > 5 mm). Although deep infiltrating endometriosis is invading, typically into the muscularis layer of the bowel, it is clinically not associated with ovarian cancer. Endometriomas are epithelial lined cysts of the ovary, which can be filled with a brown cyst fluid, and thus the name "chocolate cysts." Endometriomas can be associated with ovarian cancer, with atypical endometriomas having a higher risk of malignant transformation. Atypical endometriomas are characterized by epithelial cells with enlarged hyperchromatic and pleomorphic nuclei, with cellular crowding and high nuclear-to-cytoplasmic ratio. The altered endometriotic tumor microenvironment may lead to malignant transformation or propagation of proliferative potential [107]. RANTES: regulated on activation normal T cell expressed and secreted; MCP1: monocyte chemotactic protein-1; IL: interleukin; TGFβ1: transforming growth factor beta 1; TNFα: tumor necrosis factor alpha; CDC42: cell division cycle 42; CXCL4: chemokine (C-X-C motif) ligand 4.

3.5. Steroid Hormones

Endometriosis is an estrogen-responsive disorder with lesion-level hyperestrogenism. Specifically, endometriotic tissue differs from eutopic endometrial tissue by the high expression of aromatase (CYP19A1) and 17β-hydroxysteroid-dehydrogenase (17β-HSD) type 1 and the absence of 17β-HSD type 2 [107,116]. Aromatase converts androstenedione or testosterone to estrone and estradiol at the level of the endometriotic microenvironment. High levels of estradiol have been linked to IL8 and RANTES production, which facilitate proliferation, inflammation, and feedback to increased expression of aromatase [107,117]. Aromatase activity is also stimulated through prostaglandin E_2, an inflammatory product of cyclooxygenase 1 and 2 (COX1/2), found in endometriotic lesions in high levels [118]. Inhibitors of prostaglandin E2 receptor show promising effects in a xenograft model of endometriosis [119]. At the endometriotic lesion level, there is significant feed forward production and maintenance of estrogen, associated with pro-tumorigenic qualities. Medical management of endometriosis with oral contraceptives lowers overall steroid hormone levels. This may explain why the protection from combined oral contraceptive therapy on ovarian cancer risk is more robust for women with endometriosis (odds ratio 0.21 (0.08–0.58), $p = 0.003$) compared to non-endometriosis population (odds ratio 0.47 (0.37–0.61, $p < 0.001$)) [120]. Thus, the role of steroid hormones on endometriosis-associated ovarian cancers needs further study.

3.6. Small RNA Molecules

Small RNA molecules are non-coding RNA molecules that can play an important role in the post-transcriptional regulation of gene expression. Multiple groups of small RNAs have been identified, such as microRNAs (miRNAs), small nucleolar RNA (snoRNAs), small interfering RNAs (siRNAs), and Piwi-interacting RNA (piRNAs) [121]. The most studied type of small RNA molecules in endometriosis-associated ovarian cancers are miRNAs. In general, miRNAs regulate gene expression by mRNA cleavage and translational repression [122,123]. Studies have shown that miRNAs are frequently dysregulated in endometriosis and endometriosis-associated ovarian cancers (reviewed in [53–55]). Compilation of dysregulated miRNAs in ovarian endometrioid and clear-cell adenocarcinomas, as well as endometriosis (Supplemental Table S1) shows dysregulated miRNA molecules for each tissue type [53,55,124–133]. Figure 2 shows the number of miRNAs dysregulated in ovarian clear-cell and endometrioid adenocarcinomas, and endometriosis tissues. Supplemental Table S1 details the specific miRNA molecules in the each unique and overlapping group. MiR-126 was found downregulated in all three groups. While the function of miR-126 is still unknown, miR-126 was significantly downregulated in endometriosis compared with eutopic endometrium [134]. Additionally, downregulation of miR-126 induced non-ovarian cancer cell proliferation, migration, and invasion, mediated through numerous validated targets, such as PI3K, KRAS, and VEGF. Reduced levels of miR-126 were a significant predictor of poor survival of cancer patients, although women with ovarian cancer were not included in the study [135]. Thus, miR-126 may play a role in endometriosis and ovarian cancer, even though these functional studies did not have ovarian cancer samples with concurrent endometriosis.

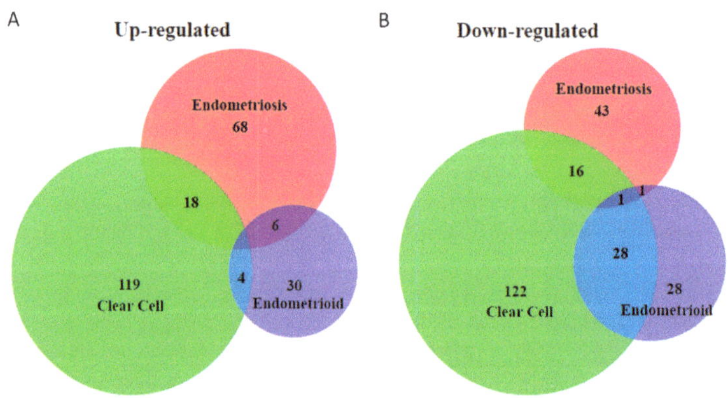

Figure 2. Venn diagram of overlap of number of miRNAs differentially expressed in endometriosis and ovarian clear-cell and endometrioid adenocarcinoma. The miRNAs differentially expressed are depicted in three overlapping circles. The numbers indicate the miRNA counts that are unique or in common between the groups. (**A**) Upregulated miRNAs; (**B**) downregulated miRNAs. Supplemental Table S1 details the miRNAs in each group above.

MiR-30a, miR-30c, miR-31, miR-532-5p, and miR-885-5p were upregulated in clear cell ovarian cancer by multiple studies [124–127,131,132]. MiR-30 was found to be 5-fold overexpressed in ovarian clear-cell adenocarcinoma [132]. Sestito et al. showed that overexpression of miR-30a delayed tumor formation in xenograft tumors, and overexpression of miR-30a sensitized ovarian cancer cells to chemotherapy [136]. Downregulation of miR-532 was associated with poor survival in women with ovarian cancer, and overexpression of miR-532 suppressed the proliferative and invasive capacity of the ovarian cancer cell lines, ES2 and SKOV-3, and inhibited tumor growth in vivo [137]. Endometrioid ovarian cancer had the shortest list of dysregulated miRNAs (Figure 2 and Supplemental Table S1). MiR-200 family miRNAs (miR-200a, -200b, -200c, -141, and -429) were upregulated in ovarian cancer and may play crucial roles in ovarian cancer metastasis, diagnosis, and treatment [126,129,130,138].

4. Model Systems for Studying Rare Ovarian Cancers

Multiple model systems have been employed to study endometriosis and endometriosis-associated ovarian cancers (reviewed in [139,140]). This review will highlight the tumor microenvironment of the genetically engineered mouse models of endometriosis-associated ovarian cancers. We have chosen to focus on spontaneous models instead of transplant models (reviewed in [140]). Because there has yet to be a comprehensive mouse model that replicates ovarian cancer with endometriosis, this review will also focus on the role of immortalized cell lines, xenograft models, co-culture systems, and three-dimensional (3D) models.

4.1. Genetically Engineered Mouse Models

4.1.1. Candidate Genes in Genetically Engineered Mouse Models

High-grade serous ovarian cancer is a genomically complex disease [141] and although neither endometrioid nor clear-cell ovarian cancer have been as extensively profiled, they are likely complex as well. For the study of genetically engineered mouse models, fortunately, both endometrioid and clear cell ovarian cancer have high frequency mutations in only a handful of genes: *ARID1A*, *PIK3CA*, *CTNNB1*, *PTEN*, and *KRAS* [33,40–42,44,45]. Use of traditional *Cre* recombinase technology with candidate-gene floxed alleles has had mixed results in terms of single gene knockout developing endometriosis-associated ovarian cancers. Table 1 lists the promoters driving *Cre* recombinase, and Table 2 details the brief rationale behind the use of specific genes in these mouse models. Table 3 lists these genes with combinations of tissue-specific promoters driving *Cre* recombinase. Despite the promising allele targets and the tissue-specific promoters driving *Cre* recombinase, there are no genetic mouse models of endometriosis and concurrent ovarian cancer. Investigators have created genetically engineered mouse models, which developed ovarian low-grade serous, clear-cell, or endometrioid adenocarcinoma (Table 3). However, none of these models have concurrent endometriosis. This suggests that different genetic combinations are required to model concurrent endometriosis and ovarian cancer. The discussion below highlights the role of the microenvironment of each model, and how this microenvironment may be playing a role in ovarian cancer development. Even though the presented models do not completely represent the endometriotic tumor microenvironment, they are still useful for understanding development of endometrioid or clear-cell ovarian cancer.

Table 1. *Cre* recombinase promoters and site of effects.

Cre	Gene Promoter	Location of Expression	Ref.
Adenovirus (Ad)	Cytomegalovirus	Injection site	[142]
Amhr2	Anti-Mullërian hormone receptor type 2	Oviduct: stroma Uterus: stroma and smooth muscle cells Ovary: granulosa cells and ovarian surface epithelium	[143,144]
Cyp19	Cytochrome P450 family 19	Granulosa cells of antral follicles and luteal cells	[145]
Ovgp1	Oviductal glycoprotein 1	Non-ciliated oviductal epithelial cells	[146]
Pax8	Paired box gene 8	Fallopian tube, cervix, uterus, and endometrium	[147]
Pgr	Progesterone receptor	Oviduct: epithelium Uterus: epithelium, stroma, myometrium Ovary: time-limited granulosa cells	[148]

Table 2. Genes important in mouse models of endometriosis-associated ovarian cancer.

Mouse Allele	Gene Name and Mouse Ref	Effect of *Cre* Recombination	Endometriosis-Associated Ovarian Cancer Implications and Ref.
Arid1a$^{f/f}$	AT-rich interactive domain 1A	ARID1A loss	46–95% of clear-cell and 30% of endometrioid tumors have loss of ARID1A [30,43–45]
Apc$^{f/f}$	Adenomatous polyposis coli	Overexpression of β-catenin	Mutations in *APC* lead to activation of β-catenin which is frequently activated in endometrioisis-associated ovarian cancers [149]
Ctnnb1$^{f/f}$	Catenin beta-1	Overexpression of β-catenin	16–54% of endometriod ovarian cancers have mutations in β-catenin, leading to nuclear localization, and activation of wingless integration site (WNT) signaling [150–153]
Kras$^{lsl\text{-}G12D}$	Kirsten rat sarcoma	Expression of oncogenic *Kras*	29% of low-grade endometrioid ovarian tumors with concurrent endometriosis [33]
MUC1$^{+/-}$	Mucin 1	Expression human MUC1 in mouse	Expressed in endometrium and endometriosis; potential biomarker for endometriosis or ovarian cancer [154]
Pik3ca^{H1047R}	Phosphatidylinositol-4,5-bisphosphate 3-kinase catalytic subunit alpha	Mutation in *Pik3ca* kinase domain	20% of clear-cell and 20% of endometrioid ovarian cancers with mutations [155]
Pten$^{f/f}$	Phosphatase and tensin homolog	PTEN loss and activation of AKT	20% of clear-cell and 20% of endometrioid cancers [156]

Table 3. Mouse models with implications in endometriosis and endometriosis-associated ovarian cancers.

Genotype	Phenotype	Penetrance	Details	Ref.
Arid1a$^{f/f}$;*Ad*^{Cre} (Ovarian bursa)	No cancer	0/29 with adnexal masses 0/42 with adnexal masses	No endometriosis	[157,158]
Arid1a$^{f/f}$;*Amhr2*^{Cre}	No cancer	0/20 with adnexal masses	No endometriosis	[159]
Arid1a$^{f/f}$;*Pgr*^{Cre}	No cancer	0/20 with adnexal masses	No endometriosis	[160]
Pten$^{f/f}$;*Ad*^{Cre} (Ovarian bursa)	No cancer	0/5 with adnexal masses 0/63 with adnexal masses	No endometriosis	[158,161]
Pten$^{f/f}$;*Ad*^{Cre} (Infundibulum to ovarian bursa)	Low penetrance endometrioid ovarian cancer at 26 weeks	8/13 with ovarian endometriosis like lesions 1/13 with ovarian cancer by 26 weeks	Endometriosis-like lesions of ovary (lacked stromal component)	[162]
Pten$^{f/f}$;*Cyp19*^{Cre}	No cancer	0/4 with adnexal masses	No endometriosis	[163]
Pten$^{f/f}$;*Amhr2*^{Cre}	Granulosa cell tumor	5/70 with ovarian cancers by 7 months	No endometriosis	[164]
Pten$^{f/f}$;*Apc*$^{f/f}$;*Ovgp1*^{Cre}	Endometrioid ovarian carcinoma	10/15 with ovarian cancers	Metastatic lesions	[146]
Pten$^{f/f}$;*Pax8*^{Cre}	Endometrioid oviductal adenocarcinoma	3/4 with oviductal cancers by 7 months	Oviductal tumors metastasized to ovary	[147]

Table 3. *Cont.*

Genotype	Phenotype	Penetrance	Details	Ref.
Pik3ca^{H1047R};*Ad*^{Cre} (Ovarian bursa)	No cancer	0/6 with adnexal masses	4/5 ovarian surface epithelium hyperplasia (microscopic)	[157]
Kras^{G12D};*Ad*^{Cre} (Infundibulum to ovarian bursa)	15/15 endometriosis-like lesions of ovary	15/15 with endometriosis-like lesions of ovary	Endometriosis-like lesions of ovary (lacked stromal component)	[162]
Kras^{G12D};*Ad*^{Cre} (Uterotubal injection to ovarian bursa)	7/15 with peritoneal endometriosis	7/15 with peritoneal endometriosis	Peritoneal endometriosis	[162]
Kras^{G12D};*Ad*^{Cre} (IP injection)	No cancer	0/13 with adnexal masses	No endometriosis	[162]
Kras^{G12D};*Amhr2*^{Cre}	No cancer	0/4 with adnexal masses	No endometriosis Abnormal follicles	[145,163]
Kras^{G12D};*Cyp19*^{Cre}	No cancer	0/4 with adnexal masses	No endometriosis Abnormal follicles	[145,163]
Kras^{G12D};*Pgr*^{Cre}	No cancer	0/3 with adnexal masses	No endometriosis	[163]
Ctnnb1$^{f/+}$;*Amhr2*^{Cre}	Endometrioid ovarian carcinoma	5/6 with ovarian cancer by 6 months	No endometriosis	[165]
Arid1a$^{f/f}$;*Pik3ca*^{H1047R};*Ad*^{Cre} (Ovarian bursa)	Poorly differentiated clear-cell ovarian carcinoma	23/30 with ovarian cancer by 7 weeks	77% penetrance No endometriosis Aggressive metastatic tumors	[157]
Arid1a$^{f/f}$;*Pten*$^{f/f}$;*Ad*^{Cre} (Ovarian bursa)	5/13 endometrioid ovarian carcinoma 8/13 undifferentiated adenocarcinoma	13/22 with ovarian cancer by 9 months	59% penetrance No endometriosis Aggressive undifferentiated tumors	[158]
Apc$^{f/f}$;*Pgr*^{Cre}	Endometrioid ovarian carcinoma	12/43 with ovarian cancer	No endometriosis 16% endometrioid ovarian cysts	[166]
Pten$^{f/f}$;*Apc*$^{f/f}$;*Ad*^{Cre} (Ovarian bursa)	Endometrioid ovarian carcinoma	29/29 with ovarian cancer	100% penetrance No endometriosis Aggressive metastatic tumors	[161]
Pten$^{f/f}$;*Apc*$^{f/f}$;*Pik3ca*^{H1047R};*Ad*^{Cre} (Ovarian bursa)	Endometrioid ovarian carcinoma	11/11 with ovarian cancer	No endometriosis Aggressive metastatic tumors	[167]
Kras^{G12D};*Pten*$^{f/f}$;*Ad*^{Cre} (Infundibulum to ovarian bursa)	Endometrioid ovarian carcinoma	9/9 with ovarian cancer by 12 weeks	100% penetrance Aggressive metastatic disease No endometriosis	[162]
MUC1$^{+/-}$;*Kras*^{G12D};*Ad*^{Cre} (Ovarian bursa)	Endometriosis-like lesions of ovary	No ovarian cancer	endometriosis-like lesions of ovary	[168]
Ctnnb1$^{f/+}$;*Pten*$^{f/f}$;*Amhr2*^{Cre}	Endometrioid ovarian carcinoma	5/5 with ovarian cancer by 6 weeks	No endometriosis	[165]
Kras^{G12D};*Pten*$^{f/f}$;*Amhr2*^{Cre}	Low grade ovarian serous papillary adenocarcinomas	100% with ovarian tumors by 10 weeks	No endometriosis	[143,163]
Kras^{G12D};*Pten*$^{f/f}$;*Pgr*^{Cre}	No cancer	0/3 with adnexal masses	No endometriosis	[163]
Kras^{G12D};*Pten*$^{f/f}$;*Cyp19*^{Cre}	No cancer	0/3 with adnexal masses	No endometriosis	[163]

4.1.2. Endometriosis

The only genetically engineered mouse model to spontaneously develop endometriosis with a single gene change is a highly innovative mouse model developed by Dinulescu et al. [162]. Using an oncogenic *KRAS* knock-in allele mouse (*Kras*^{G12D}), peritoneal endometriosis developed after injection of adenovirus-driven *Cre* (Ad^{Cre}) through the uterotubal junction to infect the ovarian bursa. This true peritoneal endometriosis model contained glandular epithelium and stromal components validated by molecular immunohistochemistry to cytokeratin 7, 8, and 20, estrogen receptor, progesterone receptor, smooth muscle actin, and CD10 [162]. Conversely, when Ad^{Cre} was injected through the infundibulum to the ovarian bursa, the model develops ovarian endometriosis-like lesions without the stromal component [162]. A transplantation experiment hints that the peritoneal endometriosis is uterine or tubal in origin while the ovarian endometriosis-like lesions are ovarian surface epithelium derived [162]. While long-term follow up showed no development of ovarian cancer, future studies into the molecular lineage using secretory or ciliary markers may allow better definition of cell of origin [26,62]. A similar mouse model adds human mucin 1 (*MUC1*) to oncogenic *Kras*^{G12D} with Ad^{Cre} intrabursal injection [168]. This mouse model similarly exhibits endometriosis-like lesions of the ovary. Importantly, these mice developed an immune response to MUC1 with high numbers of CD4+ Foxp3+ regulatory T cells in para-aortic lymph nodes compared to uninjected mice without lesions [168]. Models which recapitulate the immune response are needed to study the endometriotic tumor microenvironment.

Because mice do not normally menstruate, modeling retrograde menstruation requires significant manipulation. In homologous mouse models of endometriosis, endometrium from an estrogen-primed donor mouse is injected into a syngeneic estrogen-treated recipient mouse. However, homologous mouse models such as these grow poorly without exogenous estrogen [162]. A variation is the menstrual mouse model. In this model, the donor mouse undergoes significant hormonal manipulation followed by a stimulation of the uterus leading to decidualization. Hormone withdrawal leads to degeneration of the endometrium with leukocyte invasion, similar to menstruation in women [169–171]. Donor sloughed endometrium is then placed into recipient syngeneic mouse. Using this approach, Cheng et al. placed oncogenic *Kras*^{G12V} endometrial tissue into the subcuticular ventral abdomen of syngeneic mice without exogenous hormonal stimulation or matrix [172]. These lesions contained glandular epithelium, stroma, immune cells, extracellular matrix, and blood vessels with both estrogen receptor alpha and beta expression [172]. Similarly, Greaves et al. used a similar approach with endometrial tissue from a menstrual model of wild type mice. Using hormonally stimulated receptor mice, injection of tissue intraperitoneal with this non-genetically modified endometrial tissue leads to peritoneal endometriosis [173]. Again, these tissues were histologically and molecularly similar to human endometriosis [173]. Hormonal levels (i.e., endogenous versus exogenous high levels), tissue placement (i.e., subcuticular versus intraperitoneal), and genetic changes important to endometriosis-associated ovarian cancers (i.e., oncogenic *KRAS*, loss of function *ARID1A*) must be considered when using these menstrual endometriosis models. Additionally, genetically engineered mouse models that are unable to undergo decidualization such as *Pgr*^{Cre};*Arid1a*$^{f/f}$ mice [160] do not allow such studies.

4.1.3. Clear Cell Ovarian Cancer

Poorly differentiated clear-cell ovarian carcinoma develops at 7.5 weeks post-injection in *Ad*^{Cre};*Arid1a*$^{f/f}$;*Pik3ca*^{*H1047R} female mice with 77% penetrance and with 57% of injected mice having peritoneal metastasis [157]. Similar deletion of *ARID1A* alone or with knock-in of *Pik3ca* mutations showed ovarian surface epithelium hyperplasia but no endometriosis [157,158]. Although clear cell features are present two weeks post-injection, endometriotic-like lesions are not described [157]. Microarray analysis, comparing primary ovarian tumors to contralateral un-injected ovary, found almost 600 genes dysregulated with significant enrichment in immune system function [157]. Consistent with an endometriotic tumor microenvironment, IL6 signaling was found to be increased in the primary tumors, peritoneal metastases, body fluids, and ascites [157]. IL6 signaling and tumor cell growth was blocked with IL6 neutralizing antibodies. While IL6 expression was also implicated in normal ovarian surface epithelium hyperplasia with *ARID1A* deletion or *Pik3ca* mutation alone, the combination further enhanced IL6 production [157]. Cross-species, global gene expression profiling showed similar dysregulated genes in this mouse model compared to ovarian clear-cell adenocarcinoma from women [174]. Together these data suggest that the deletion of *ARID1A* and mutation in *Pik3ca*^{*H1057R} results in increased IL6 expression leading to the ovarian surface epithelial hyperplasia and eventually clear cell ovarian cancer. These tumor cells perpetuate IL6 production, creating a positive feedback loop of increased IL6 and increased cell (normal and cancerous) proliferation [157,174]. This interaction highlights how the tumor and its microenvironment can interact with one another to generate a more tumor-promoting environment.

4.1.4. Endometrioid

ARID1A, PIK3CA, CTNNB1, PTEN, and *KRAS* [33,40–42,44,45] are commonly mutated in both endometrioid and clear cell ovarian cancers from women. However, manipulation of these genes in mice typically results in endometrioid but not clear cell ovarian cancer. On injection of adenovirus-driven Cre (*Ad*^{Cre}) into the ovarian bursa through the infundibulum of *Pten*$^{f/f}$;*Kras*^{G12D} female mice generated female mice with 100% penetrance of highly aggressive and metastatic endometrioid ovarian cancer at 12 weeks. Interestingly, this mouse model has ovarian endometriosis-like lesions with either addition of oncogenic *Kras*^{G12D} or deletion of *Pten* alone, but only results in endometrioid ovarian cancer when both *Pten* and *Kras*^{G12D} are simultaneously mutated [162].

A mouse model targeting both *Pten* and *Apc* resulted in endometrioid ovarian cancer with high penetrance and metastatic disease [161]. Unfortunately, this conditional knockout (*Ad*^{Cre}) did not result in endometriosis, which may be due to the early (6-week post-injection) tumor development [161]. Another model of endometrioid carcinoma in mice utilized a double conditional knockout of *Pten* and *Arid1a* and intrabursal *Ad*^{Cre} injection to show a progression of ovarian surface epithelium hyperplasia, endometrioid carcinoma, and finally poorly differentiated carcinoma [158]. The well-differentiated endometrioid carcinoma was confined to the ovaries, suggesting the place of origin, while the undifferentiated tumors had metastasized into the peritoneal cavity [158]. Guan et al. hypothesizes that *ARID1A* plays a role in both tumor initiation and progression but requires the collaborative second hit of *Pten* to produce tumors [158]. Although the hyperplasia was not linked to endometriosis in these mice, it does speak to an environment of uncontrolled cellular proliferation giving rise to endometrioid ovarian cancer when left untreated.

High nuclear β-catenin levels have uniquely been found in endometrioid ovarian cancer from women, where this nuclear accumulation leads to activation of the WNT pathway [149]. Gain-of-function deletion of exon 3 of *Ctnnb1* leads to stable β-catenin expression in mice [175]. *Amhr2*^{Cre}*Ctnnb1*$^{f/f}$ female mice have aggressive endometrioid ovarian cancers with 100% penetrance by 6 months. Addition of *Pten* deletion to this model allows for tumors that are even more aggressive by 6 weeks [165]. Similar to deletion of exon 3 of *Ctnnb1*, deletion of *Apc* leads to stable β-catenin and WNT signaling activation [149]. Only with deletion of *Pten* did mice develop ovarian tumors [161]. To model the progression of type I tumors to the more aggressive type II tumors, Wu et al. (2013) added *Pik3ca*$^{E545K/+}$ to *Apc*$^{f/f}$ *Pten*$^{f/f}$ mice with Ad^{Cre} and showed peritoneal and lung metastasis [167].

While these models used *Amhr2*^{Cre} or Ad^{Cre} to focus genetic changes in the ovarian surface epithelium, other studies have created conditional genetic changes in the oviduct. When *Apc* and *Pten* were concurrently deleted in the fallopian tube using *Ovgp1*^{Cre}, endometrioid tumors of the ovaries developed in 10 of 15 mice, with 50% of those resulting in metastasis to the lungs or omentum [146]. Deletion of *Pten* in the fallopian tube by *Pax8*^{Cre} also resulted in endometrioid tumors. Specifically, 75% of female mice developed primary tumors in the fallopian tube by 7 months, and 75% of tumor-burdened mice had metastasis to the ovaries [147]. Deletion of *Apc* with *Pgr*^{Cre} female mice revealed tumors in both the oviduct and ovaries. Specifically, 25 of 40 female mice developed endometrioid oviductal tumors, one of 43 developed granulosa cell tumors, and 12 of 43 developed endometrioid ovarian tumors. While these female mice had simple ovarian cysts, the authors did not specifically denominate them as endometriosis [166]. Taken together, these mouse models suggest that the oviduct and/or the ovary may be involved in endometrioid cancer development in the mouse.

4.1.5. Low-Grade Serous Ovarian Cancer

Addition of oncogenic *Kras* (*Kras*^{G12D}) with either *Amhr2*^{Cre} or *Cyp19*^{Cre} resulted in ovaries with abnormal follicles, which were non-tumorigenic but also non-mitotic and non-apoptotic [145]. Deletion of *Pten* using *Amhr2*^{Cre} did result in increased proliferation and increased cell survival of ovarian surface epithelium [163]. However, the loss of the tumor suppressor *Pten* alone is not tumorigenic in somatic cells of the ovary. When *Pten* is deleted in the context of oncogenic *Kras* with *Amhr2*^{Cre}, there is development of low-grade serous papillary cystadenocarcinoma [163]. Although no endometriosis was noted, these mice were shown to have ovarian surface epithelium hyperplasia and abnormal follicle-derived ovarian lesions. Mullany et al. continued work on the *Kras*^{G12D};*Pten*$^{f/f}$;*Amhr2*^{Cre} mice and showed that ovarian surface epithelium cells, removed from mutant mice prior to tumor formation, developed into tumors when grown in soft agar [143]. This key result suggests that *Kras* and *Pte//* play a significant role in the development of tumors in the ovarian surface epithelium, and the genetic mutations are the primary driver, since tumor formation occurred even outside of the ovarian microenvironment [143].

4.2. *Other Models*

4.2.1. Immortalized Cell Lines

Immortalized human ovarian cancer cell lines have been widely used for studying molecular mechanisms of ovarian cancer. Ovarian cancer cell lines are used to study cancer biology, connecting genetic and epigenetic alterations to cancer development, progression, and drug response. Importantly, ovarian cancer cell lines have been developed from different histological and molecular subtypes of ovarian cancer. Unfortunately, molecular characterization has revealed that common ovarian cancer cell lines (i.e., SKOV3, HEYA8) do not molecularly represent the histology of tumor of origin. The number of cell lines derived from either endometrioid or clear cell ovarian cancers is more limited than high-grade serous cell lines. However, molecular profiling, including attention to gene mutations common in these endometriosis-associated ovarian cancers (i.e., *ARID1A*, *PIK3CA*, *CTNNB1*, *PTEN*, and *KRAS*) and mutations common in high-grade serous (i.e., TP53), have allowed better molecular and biological distinction [176–182]. Table 4 shows the common endometrioid and clear-cell ovarian cancer cell lines, including lines that were not derived from endometriosis-associated ovarian cancers, but which may molecularly represent non-high grade serous cell lines. Even fewer cell endometriotic cell lines exist, with 12Z cells being the only widely shared epithelial-like endometriosis immortalized cell line [183]. For rigor and reproducibility, additional well-characterized endometriotic cell lines and possibly ovarian cancer cell lines derived from women with endometriosis need to be created.

Table 4. Endometriosis and endometriosis-associated ovarian cancer cell lines.

Cell Line	Original Derivation	Putative Histotype by Molecular Studies	Genetic Mutations	Genetic Gains	Ref.
11Z	Red peritoneal endometriotic lesion	Benign	Unknown	Unknown	[183]
12Z	Red peritoneal endometriotic lesion	Benign (epithelial-like)	Unknown	Unknown	[183]
EEC16	Benign endometriotic lesion (epithelial-like)	Benign	Unknown	Unknown	[184]
EMosis-CC/TERT	Benign endometriotic lesion (epithelial-like)	Benign	Unknown	Unknown	[185]
22B	Red peritoneal endometriotic lesion (Stromal/fibroblast-like)	Benign	Unknown	Unknown	[183]
Hs 832(C).T (CRL-7566)	Benign endometriotic ovarian cyst	Benign	Unknown	Unknown	ATCC
OVTOKO	Clear-cell (spleen metastasis)	Clear-cell	None	*ERRB2, HNF1B, MET, PPM1D, STAT3, TP53, YAP1, ZNF217, CDKN2A, CDKN2B*	[177–179,18[illegible]]
OVMANA	Clear-cell (primary tumor)	Clear-cell	*BRCA2, PIK3CA, ARID1A*	*ARID1A, MET, PPM1D, TP53, ZNF217*	[178,179,182,186]
TOV21G	Clear-cell (primary tumor)	Clear-cell	*KRAS, PTEN, PIK3CA, CTNNB1, ARID1A, TPX2*		[178–182,187]
RMG-1	Clear-cell (ascites)	Clear-cell	*TP53* *	*ERBB2*	[178,179,182,188]
RMG-2	Clear-cell	Clear-cell	*PPP2R1A, ARID1A*	*ERBB2, HNF1B, MET, PIK3CA, PPM1D, STAT3, ZNF217, CDKN2A, CDKN2B*	[179]
OCC1	Clear-cell	Clear-cell			[189]
JHOC-5	Clear-cell (pelvic metastasis)	Clear-cell		*ARID1A, ERBB2, HNF1B, MET, PIK3CA, PPM1D, STAT2, ZNF217, CDKN2A, CDKN2B*	[178,179,182,190]
JHOC-7	Clear-cell	Clear-cell	*PIK3CA*	*ARID1A, HNF1B, PIK3CA, PPM1D, STAT3, ZNF217*	[179]

Table 4. *Cont.*

Cell Line	Original Derivation	Putative Histotype by Molecular Studies	Genetic Mutations	Genetic Gains	Ref.
JHOC-9	Clear-cell	Clear-cell	*PTEN, ARID1A*	*HNF1B, ZNF217*	[179]
ES2	Poorly differentiated clear-cell (primary tumor)	Endometrioid/Clear-cell	*BRAF, TP53, APC, MYC*		[178–182,191]
OVISE	Clear-cell (pelvic metastasis)	Endometrioid/Clear-cell	*ARID1A*		[177–179,182]
OVSAYO	Clear-cell	Serous	*TP53*		[179]
TOV112D	Endometrioid (primary tumor)	Endometrioid	*CTNNB1, TP53*		[179–182,187]
OVK18	Endometrioid (ascites)	Endometrioid	*TP53, PTEN, KRAS, ARID1A*		[178,182,192]
SNU-251	Endometrioid	Endometrioid	*BRCA1*		[193]
2008	Endometrioid	Atypical non-serous	*TP53*		[179]
IGROV1	Endometrioid with serous/clear cell (primary tumor)	Endometrioid/Clear-cell	*PTEN, TP53, ARID1A, BRCA1, BRCA2, PIK3CA, TPX2*		[178–180,182,194]
59M	Endometrioid with clear cell (ascites)	Endometrioid/Clear-cell	*TP53*	*MYC*	[178,180,182,193,195]
COV362	Endometrioid (pleural effusion)	Serous	*TP53, BRCA1, RB1 *, EGFR, APC*	*MYC*	[178,180,182,196]
A2780	Unknown adenocarcinoma	Endometrioid	*PTEN, ARID1A, PIK3CA, BRAF*		[178–182,197]
HEYA8	Moderately differentiated papillary serous (peritoneal metastasis)	Unlikely serous	*KRAS, BRAF*		[178,179,182,198]
SKOV3	Well differentiated, adenocarcinoma (ascites)	Endometrioid/Clear-cell	*PIK3CA, ARID1A*	*ERBB2*	[178–182,199]

* Homozygous deletion.

4.2.2. Xenograft Models

Implantation of immortalized human cell lines typically requires immunocompromised mice. A Japanese group created telomerase transformed endometriosis epithelial cell lines and confirmed cellular growth, steroid hormone response, and lack of malignant transformation in nude mice [185]. Further, these cells have been used in xenograft models to study treatment effects of small molecular inhibitors in endometriosis [104,200]. However, limited distribution outside Japan has restricted the use of these cells for studies of endometriosis-associated ovarian cancers. A similarly developed endometriotic epithelial cell line (EEC16) does not grow in SCID mice [184].

In terms of the endometriotic tumor microenvironment, Komiyama et al. placed normal endometrium of women without endometriosis into SCID mice. RMG-1 cells, a clear-cell ovarian cancer cell line, were grown in mice then transplanted into mice with or without endometrial implants. Although the tumors weighed less when grown with endometrium, proliferation was significantly higher in mice with transplanted endometrium. Additionally, these tumors expressed high levels of TGFβ and IL6. Addition of normal human endometrium changed the xenograft model to a more endometriotic microenvironment [201].

4.2.3. Three Dimensional (3D) and Co-Culture Models

Immortalized cell lines in monolayer two-dimensional (2D) culture fail to recapitulate the complexity of tumor tissue. Tumors are three-dimensional (3D) structures, surrounded by other cell types and a unique extracellular matrix (ECM) that is biologically optimized for growth of each cell type [202]. To recapitulate this for in vitro model systems, immortalized cell lines can be grown in Matrigel, ultra-low-adhesive plates, or a hanging drop. Using these methods, many immortalized cell lines will form 3D spheroids. Three-dimensional spheroid models can be highly instructive towards the understanding of current drug resistance and new therapeutics because they better mimic the way 3D tumors or de novo spheroids interact with the surrounding microenvironment. Specifically, the architecture of spheroids results in non-heterogeneity of nutrient and drug penetration, which can cause differential responses to varying layers of the spheroid. For example, Lee et al. compared 31 ovarian cancer cell lines in both 2D monolayer and 3D spheroids to primary tumors. Three-dimensional spheroids showed slower rates of proliferation and decreased drug sensitivity than the same cells grown in 2D [202]. Additionally, these 3D spheroids mimicked histological characteristics of primary tumors. Although the authors did not perform genome-wide transcriptomic analysis, candidate biomarkers such as mucin 16, cell surface associated (CA125), Wilms Tumor 1 (WT1), estrogen receptor, Paired box gene 8 (PAX8), and β-catenin were examined by IHC on a tissue microarray composed of 2D and 3D samples. The expression of these biomarkers correlated well with expression in primary tumors [202]. These data suggest that 3D spheroid models alter the microenvironment in a potentially more biological way compared to other in vitro systems. Additionally, Lal-Nag et al. used high-throughput screening to test multiple oncological drugs against the HEYA8 cell line. The cells responded differently to various drugs if they were grown in monolayer, in the process of forming spheroids, or already in pre-formed spheroids. This work establishes that the dimensionality of ovarian cancer cells plays a role in how they respond to their environment [203]. Similarly, Chowwanadisai et al. created cisplatin-resistant ovarian cancer spheroids by treating cells with sub-threshold doses of cisplatin, which resulted in a mesenchymal-enriched gene expression signature [204]. While molecular changes within spheroids may play a role in chemotherapy resistance, size of spheroids, similar to remaining disease after debulking surgery, plays a role in response. Tanenbaum et al. [205] showed that small spheroids treated with either short-term high-dose or prolonged low-dose cisplatin underwent significant shrinkage. Importantly, large spheroids preferentially responded to short-term high doses of cisplatin [205]. The investigators did not explore if the remaining cells became chemotherapy-resistant [205]. Although immortalized cell lines from endometrioid or clear cell ovarian cancers have not been extensively tested in 3D culture, we anticipate that they would behave similarly.

In addition to single cell types within 3D spheroids, co-culture systems can be useful. For example, endometrial epithelial cells are inhibited at a rate of 65–80% when grown in co-culture with endometrial stromal cells, highlighting the need for complex co-culture models [206]. Additionally, co-culture models of epithelial and stromal endometriosis cells show that stromal cells are responsible for metabolism of iron. The authors hypothesize that storage of iron by stromal cells is protective against malignant transformation of epithelial cells. Specifically, a lack of stromal cells and an abundance of epithelial cells, which cannot metabolize iron, leads to oxidative damage and oncogenic change [207]. This hypothesis fits with data from Anglesio et al. showing tumorigenic mutations in *KRAS* in epithelial cells of endometriosis but not stromal cells [27]. Similarly, co-culture of macrophages with endometriotic epithelial or endometriotic stromal cells leads to an increase in invasion that is more robust in epithelial than stromal cells [208]. Three-dimensional organoids made from endometrium and decidua have been developed simultaneously by two independent laboratories and represent promising models for in vitro study [209,210]. Development of additional endometriotic tumor microenvironment models are needed to study ovarian cancer cells within spheroids, 3D organoids, or co-culture systems.

5. Future of Precision Therapy for/or Prevention of Ovarian Cancer

Endometriosis is a known risk factor for ovarian cancer [41]. However, early treatment of endometriosis represents a known prevention strategy for ovarian cancer. For example, a woman on oral contraceptive therapy has a more robust protection against ovarian cancer if she has endometriosis than if she does not [120]. While treatment of endometriosis with contraceptives is effective, women desiring fertility do not enjoy the side of effects of contraception, and when medical management is stopped, 73% of women have return of symptoms. Additionally, surgical treatment of endometriosis with removal of one or both ovaries results in significant decrease in ovarian cancer risk. However, 55% of women undergoing local resection of endometriosis will have at least one more surgery over the course of seven years [211,212]. Morbidity associated with multiple operations makes selection of timing for endometriosis surgery important in pre-menopausal women. New treatments for endometriosis are needed. Importantly, discovery of new treatments for endometriosis should be a priority for ovarian cancer funding agencies as these therapies may lead to prevention of ovarian cancer.

In terms of therapy highlighting the importance of the molecular signaling between cells within tumors, Mok et al. used a systems biology approach to study individual cell types. Machine learning with large databases of drugs and molecular effects highlighted an FDA-approved drug for potential targeted treatment. While this study used high-grade serous ovarian tumors, it brings forward the importance of non-epithelial ovarian cancer cells in cancer treatment [213]. Importantly, the study focused on TGFβ signaling pathways [213]. Endometriosis also has dysregulated TGFβ signaling pathways [92]. Similar treatment of endometriosis may prevent ovarian cancer.

6. Conclusions

The Gynecologic Cancers Steering Committee of the National Institutes of Health (NIH) proposed strategic priorities for ovarian cancer. These research priorities focus on discovery of biomarkers, identification of cancer subsets to drive treatment recommendations, immunotherapy, combination therapies, and manipulation of the host-tumor microenvironment. While these priorities are not specific for a particular histotype, they are highly applicable to both the more common high-grade serous and less common endometriosis-associated ovarian cancers and warrant further study in endometriosis-associated ovarian cancer models.

Supplementary Materials: The following are available online at http://www.mdpi.com/2072-6694/10/8/261/s1, Table S1: Dysregulated miRNAs in ovarian endometrioid and clear-cell adenocarcinoma and endometriosis.

Author Contributions: J.R.H.W., X.W. and S.M.H. constructed the outline of the review, synthesized and referenced the works included, and wrote the manuscript.

Funding: This work was supported by NIH/NCI R03CA19127, Rivkin Center for Ovarian Cancer Pilot Award.

Conflicts of Interest: The authors declare no conflict of interest.

References

1. Bulun, S.E. Endometriosis. *N. Engl. J. Med.* **2009**, *360*, 268–279. [CrossRef] [PubMed]
2. Giudice, L.C.; Kao, L.C. Endometriosis. *Lancet* **2004**, *364*, 1789–1799. [CrossRef]
3. Buck Louis, G.M.; Hediger, M.L.; Peterson, C.M.; Croughan, M.; Sundaram, R.; Stanford, J.; Chen, Z.; Fujimoto, V.Y.; Varner, M.W.; Trumble, A.; et al. Incidence of endometriosis by study population and diagnostic method: The endo study. *Fertil. Steril.* **2011**, *96*, 360–365. [CrossRef] [PubMed]
4. DiVasta, A.D.; Vitonis, A.F.; Laufer, M.R.; Missmer, S.A. Spectrum of symptoms in women diagnosed with endometriosis during adolescence vs. adulthood. *Am. J. Obstet. Gynecol.* **2018**, *218*, 324.e1–324.e11. [CrossRef] [PubMed]
5. Aerts, L.; Grangier, L.; Streuli, I.; Dallenbach, P.; Marci, R.; Wenger, J.M.; Pluchino, N. Psychosocial impact of endometriosis: From co-morbidity to intervention. *Best Pract. Res. Clin. Obstet. Gynaecol.* **2018**. [CrossRef] [PubMed]
6. Pearce, C.L.; Stram, D.O.; Ness, R.B.; Stram, D.A.; Roman, L.D.; Templeman, C.; Lee, A.W.; Menon, U.; Fasching, P.A.; McAlpine, J.N.; et al. Population distribution of lifetime risk of ovarian cancer in the united states. *Cancer Epidemiol. Biomark. Prev.* **2015**, *24*, 671–676. [CrossRef] [PubMed]
7. Pearce, C.L.; Templeman, C.; Rossing, M.A.; Lee, A.; Near, A.M.; Webb, P.M.; Nagle, C.M.; Doherty, J.A.; Cushing-Haugen, K.L.; Wicklund, K.G.; et al. Association between endometriosis and risk of histological subtypes of ovarian cancer: A pooled analysis of case-control studies. *Lancet Oncol.* **2012**, *13*, 385–394. [CrossRef]
8. Nagle, C.M.; Olsen, C.M.; Webb, P.M.; Jordan, S.J.; Whiteman, D.C.; Green, A.C.; Australian Cancer Study Group; Australian Ovarian Cancer Study Group. Endometrioid and clear cell ovarian cancers: A comparative analysis of risk factors. *Eur. J. Cancer* **2008**, *44*, 2477–2484. [CrossRef] [PubMed]
9. Rossing, M.A.; Cushing-Haugen, K.L.; Wicklund, K.G.; Doherty, J.A.; Weiss, N.S. Risk of epithelial ovarian cancer in relation to benign ovarian conditions and ovarian surgery. *Cancer Causes Control* **2008**, *19*, 1357–1364. [CrossRef] [PubMed]
10. Brinton, L.A.; Gridley, G.; Persson, I.; Baron, J.; Bergqvist, A. Cancer risk after a hospital discharge diagnosis of endometriosis. *Am. J. Obstet. Gynecol.* **1997**, *176*, 572–579. [CrossRef]
11. Kobayashi, H.; Sumimoto, K.; Moniwa, N.; Imai, M.; Takakura, K.; Kuromaki, T.; Morioka, E.; Arisawa, K.; Terao, T. Risk of devloping ovarian cancer among women with ovarian endometrioma: A cohort study in shizuoka, Japan. *Int. J. Cancer* **2007**, *17*, 37–43. [CrossRef] [PubMed]
12. Somigliana, E.; Vigano, P.; Parazzini, F.; Stoppelli, S.; Giambattista, E.; Vercellini, P. Association between endometriosis and cancer: A comprehensive review and a critical analysis of clinical and epidemiological evidence. *Gynecol. Oncol.* **2006**, *101*, 331–341. [CrossRef] [PubMed]
13. Siegel, R.L.; Miller, K.D.; Jemal, A. Cancer statistics, 2018. *CA Cancer J. Clin.* **2018**, *68*, 7–30. [CrossRef] [PubMed]
14. Kurman, R.J.; Shih Ie, M. The dualistic model of ovarian carcinogenesis: Revisited, revised, and expanded. *Am. J. Pathol.* **2016**, *186*, 733–747. [CrossRef] [PubMed]
15. Reid, B.M.; Permuth, J.B.; Sellers, T.A. Epidemiology of ovarian cancer: A review. *Cancer Biol. Med.* **2017**, *14*, 9–32. [PubMed]
16. Prat, J. Pathology of borderline and invasive cancers. *Best Pract. Res. Clin. Obstet. Gynaecol.* **2017**, *41*, 15–30. [CrossRef] [PubMed]
17. Lu, J.; Tao, X.; Zhou, J.; Lu, Y.; Wang, Z.; Liu, H.; Xu, C. Improved clinical outcomes of patients with ovarian carcinoma arising in endometriosis. *Oncotarget* **2017**, *8*, 5843–5852. [CrossRef] [PubMed]
18. Dinkelspiel, H.E.; Matrai, C.; Pauk, S.; Pierre-Louis, A.; Chiu, Y.L.; Gupta, D.; Caputo, T.; Ellenson, L.H.; Holcomb, K. Does the presence of endometriosis affect prognosis of ovarian cancer? *Cancer Investig.* **2016**, *34*, 148–154. [CrossRef] [PubMed]
19. McMeekin, D.S.; Burger, R.A.; Manetta, A.; DiSaia, P.; Berman, M.L. Endometrioid adenocarcinoma of the ovary and its relationship to endometriosis. *Gynecol. Oncol.* **1995**, *59*, 81–86. [CrossRef] [PubMed]

20. Melin, A.; Lundholm, C.; Malki, N.; Swahn, M.L.; Sparen, P.; Bergqvist, A. Endometriosis as a prognostic factor for cancer survival. *Int. J. Cancer* **2011**, *129*, 948–955. [CrossRef] [PubMed]
21. Quirk, J.T.; Natarajan, N.; Mettlin, C.J. Age-specific ovarian cancer incidence rate patterns in the united states. *Gynecol. Oncol.* **2005**, *99*, 248–250. [CrossRef] [PubMed]
22. Gershenson, D.M. Treatment of ovarian cancer in young women. *Clin. Obstet. Gynecol.* **2012**, *55*, 65–74. [CrossRef] [PubMed]
23. Storey, D.J.; Rush, R.; Stewart, M.; Rye, T.; Al-Nafussi, A.; Williams, A.R.; Smyth, J.F.; Gabra, H. Endometrioid epithelial ovarian cancer: 20 years of prospectively collected data from a single center. *Cancer* **2008**, *112*, 2211–2220. [CrossRef] [PubMed]
24. Noli, S.; Cipriani, S.; Scarfone, G.; Villa, A.; Grossi, E.; Monti, E.; Vercellini, P.; Parazzini, F. Long term survival of ovarian endometriosis associated clear cell and endometrioid ovarian cancers. *Int. J. Gynecol. Cancer* **2013**, *23*, 244–248. [CrossRef] [PubMed]
25. Ahmed, A.A.; Becker, C.M.; Bast, R.C., Jr. The origin of ovarian cancer. *BJOG* **2012**, *119*, 134–136. [CrossRef] [PubMed]
26. Cochrane, D.R.; Tessier-Cloutier, B.; Lawrence, K.M.; Nazeran, T.; Karnezis, A.N.; Salamanca, C.; Cheng, A.S.; McAlpine, J.N.; Hoang, L.N.; Gilks, C.B.; et al. Clear cell and endometrioid carcinomas: Are their differences attributable to distinct cells of origin? *J. Pathol.* **2017**, *243*, 26–36. [CrossRef] [PubMed]
27. Anglesio, M.S.; Papadopoulos, N.; Ayhan, A.; Nazeran, T.M.; Noe, M.; Horlings, H.M.; Lum, A.; Jones, S.; Senz, J.; Seckin, T.; et al. Cancer-associated mutations in endometriosis without cancer. *N. Engl. J. Med.* **2017**, *376*, 1835–1848. [CrossRef] [PubMed]
28. Chui, M.H.; Wang, T.L.; Shih, I.M. Endometriosis: Benign, malignant, or something in between? *Oncotarget* **2017**, *8*, 78263–78264. [CrossRef] [PubMed]
29. Lac, V.; Huntsman, D.G. Distinct developmental trajectories of endometriotic epithelium and stroma: Implications for the origins of endometriosis. *J. Pathol.* **2018**. [CrossRef] [PubMed]
30. Wiegand, K.C.; Shah, S.P.; Al-Agha, O.M.; Zhao, Y.; Tse, K.; Zeng, T.; Senz, J.; McConechy, M.K.; Anglesio, M.S.; Kalloger, S.E.; et al. Arid1a mutations in endometriosis-associated ovarian carcinomas. *N. Engl. J. Med.* **2010**, *363*, 1532–1543. [CrossRef] [PubMed]
31. Sato, N.; Tsunoda, H.; Nishida, M.; Morishita, Y.; Takimoto, Y.; Kubo, T.; Noguchi, M. Loss of heterozygosity on 10q23.3 and mutation of the tumor suppressor gene Pten in benign endometrial cyst of the ovary: Possible sequence progression from benign endometrial cyst to endometrioid carcinoma and clear cell carcinoma of the ovary. *Cancer Res.* **2000**, *60*, 7052–7056. [PubMed]
32. Yamamoto, S.; Tsuda, H.; Takano, M.; Tamai, S.; Matsubara, O. Loss of arid1a protein expression occurs as an early event in ovarian clear-cell carcinoma development and frequently coexists with PIK3CA mutations. *Mod. Pathol.* **2012**, *25*, 615–624. [CrossRef] [PubMed]
33. Stewart, C.J.; Leung, Y.; Walsh, M.D.; Walters, R.J.; Young, J.P.; Buchanan, D.D. Kras mutations in ovarian low-grade endometrioid adenocarcinoma: Association with concurrent endometriosis. *Hum. Pathol.* **2012**, *43*, 1177–1183. [CrossRef] [PubMed]
34. Zou, Y.; Zhou, J.Y.; Guo, J.B.; Wang, L.Q.; Luo, Y.; Zhang, Z.Y.; Liu, F.Y.; Tan, J.; Wang, F.; Huang, O.P. The presence of kras, ppp2r1a and arid1a mutations in 101 Chinese samples with ovarian endometriosis. *Mutat. Res.* **2018**, *809*, 1–5. [CrossRef] [PubMed]
35. Wiegand, K.C.; Lee, A.F.; Al-Agha, O.M.; Chow, C.; Kalloger, S.E.; Scott, D.W.; Steidl, C.; Wiseman, S.M.; Gascoyne, R.D.; Gilks, B.; et al. Loss of baf250a (arid1a) is frequent in high-grade endometrial carcinomas. *J. Pathol.* **2011**, *224*, 328–333. [CrossRef] [PubMed]
36. Yamamoto, S.; Tsuda, H.; Takano, M.; Tamai, S.; Matsubara, O. Pik3ca mutations and loss of arid1a protein expression are early events in the development of cystic ovarian clear cell adenocarcinoma. *Virchows Arch.* **2012**, *460*, 77–87. [CrossRef] [PubMed]
37. Samartzis, E.P.; Samartzis, N.; Noske, A.; Fedier, A.; Caduff, R.; Dedes, K.J.; Fink, D.; Imesch, P. Loss of arid1a/baf250a-expression in endometriosis: A biomarker for risk of carcinogenic transformation? *Mod. Pathol.* **2012**, *25*, 885–892. [CrossRef] [PubMed]
38. Ayhan, A.; Mao, T.L.; Seckin, T.; Wu, C.H.; Guan, B.; Ogawa, H.; Futagami, M.; Mizukami, H.; Yokoyama, Y.; Kurman, R.J.; et al. Loss of arid1a expression is an early molecular event in tumor progression from ovarian endometriotic cyst to clear cell and endometrioid carcinoma. *Int. J. Gynecol. Cancer* **2012**, *22*, 1310–1315. [CrossRef] [PubMed]

39. Xiao, W.; Awadallah, A.; Xin, W. Loss of arid1a/baf250a expression in ovarian endometriosis and clear cell carcinoma. *Int. J. Clin. Exp. Pathol.* **2012**, *5*, 642–650. [PubMed]
40. Anglesio, M.S.; Yong, P.J. Endometriosis-associated ovarian cancers. *Clin. Obstet. Gynecol.* **2017**, *60*, 711–727. [CrossRef] [PubMed]
41. Vercellini, P.; Viganò, P.; Buggio, L.; Makieva, S.; Scarfone, G.; Cribiu, F.M.; Parazzini, F.; Somigliana, E. Perimenopausal management of ovarian endometriosis and associated cancer risk: When is medical or surgical treatment indicated? *Best Pract. Res. Clin. Obstet. Gynaecol.* **2018**. [CrossRef] [PubMed]
42. Friedlander, M.L.; Russell, K.; Millis, S.; Gatalica, Z.; Bender, R.; Voss, A. Molecular profiling of clear cell ovarian cancers: Identifying potential treatment targets for clinical trials. *Int. J. Gynecol. Cancer* **2016**, *26*, 648–654. [CrossRef] [PubMed]
43. Ishikawa, M.; Nakayama, K.; Nakamura, K.; Ono, R.; Sanuki, K.; Yamashita, H.; Ishibashi, T.; Minamoto, T.; Iida, K.; Razia, S.; et al. Affinity-purified DNA-based mutation profiles of endometriosis-related ovarian neoplasms in Japanese patients. *Oncotarget* **2018**, *9*, 14754–14763. [CrossRef] [PubMed]
44. Jones, S.; Wang, T.L.; Shih Ie, M.; Mao, T.L.; Nakayama, K.; Roden, R.; Glas, R.; Slamon, D.; Diaz, L.A., Jr.; Vogelstein, B.; et al. Frequent mutations of chromatin remodeling gene arid1a in ovarian clear cell carcinoma. *Science* **2010**, *330*, 228–231. [CrossRef] [PubMed]
45. Shibuya, Y.; Tokunaga, H.; Saito, S.; Shimokawa, K.; Katsuoka, F.; Bin, L.; Kojima, K.; Nagasaki, M.; Yamamoto, M.; Yaegashi, N.; et al. Identification of somatic genetic alterations in ovarian clear cell carcinoma with next generation sequencing. *Genes Chromosom. Cancer* **2018**, *57*, 51–60. [CrossRef] [PubMed]
46. Grisham, R.N.; Iyer, G.; Garg, K.; DeLair, D.; Hyman, D.M.; Zhou, Q.; Iasonos, A.; Berger, M.F.; Dao, F.; Spriggs, D.R. Braf mutation is associated with early stage disease and improved outcome in patients with low-grade serous ovarian cancer. *Cancer* **2013**, *119*, 548–554. [CrossRef] [PubMed]
47. Sieben, N.L.; Macropoulos, P.; Roemen, G.M.; Kolkman-Uljee, S.M.; Jan Fleuren, G.; Houmadi, R.; Diss, T.; Warren, B.; Al Adnani, M.; De Goeij, A.P.; et al. In ovarian neoplasms, braf, but not kras, mutations are restricted to low-grade serous tumours. *J. Pathol.* **2004**, *202*, 336–340. [CrossRef] [PubMed]
48. Singer, G.; Oldt, R., 3rd; Cohen, Y.; Wang, B.G.; Sidransky, D.; Kurman, R.J.; Shih Ie, M. Mutations in BRAF and KRAS characterize the development of low-grade ovarian serous carcinoma. *J. Natl. Cancer Inst.* **2003**, *95*, 484–486. [CrossRef] [PubMed]
49. Lv, X.; Wang, D.; Ma, Y.; Long, Z. Analysis of the oncogene braf mutation and the correlation of the expression of wild-type BRAF and CREB1 in endometriosis. *Int. J. Mol. Med.* **2018**, *41*, 1349–1356. [CrossRef] [PubMed]
50. Saare, M.; Krigul, K.L.; Laisk-Podar, T.; Ponandai-Srinivasan, S.; Rahmioglu, N.; Lalit Kumar, P.G.; Zondervan, K.; Salumets, A.; Peters, M. DNA methylation alterations-potential cause of endometriosis pathogenesis or a reflection of tissue heterogeneity? *Biol. Reprod.* **2018**. [CrossRef] [PubMed]
51. He, J.; Chang, W.; Feng, C.; Cui, M.; Xu, T. Endometriosis malignant transformation: Epigenetics as a probable mechanism in ovarian tumorigenesis. *Int. J. Genom.* **2018**, *2018*, 1465348. [CrossRef] [PubMed]
52. Roca, F.J.; Loomans, H.A.; Wittman, A.T.; Creighton, C.J.; Hawkins, S.M. Ten-eleven translocation genes are downregulated in endometriosis. *Curr. Mol. Med.* **2016**, *16*, 288–298. [CrossRef] [PubMed]
53. Nothnick, W.B. MicroRNAs and endometriosis: Distinguishing drivers from passengers in disease pathogenesis. *Semin. Reprod. Med.* **2017**, *35*, 173–180. [CrossRef] [PubMed]
54. Logan, M.; Hawkins, S.M. Role of microRNAs in cancers of the female reproductive tract: Insights from recent clinical and experimental discovery studies. *Clin. Sci. (Lond. Engl. 1979)* **2015**, *128*, 153–180. [CrossRef] [PubMed]
55. Wang, X.; Ivan, M.; Hawkins, S.M. The role of microRNA molecules and microRNA-regulating machinery in the pathogenesis and progression of epithelial ovarian cancer. *Gynecol. Oncol.* **2017**, *147*, 481–487. [CrossRef] [PubMed]
56. Machado-Linde, F.; Sanchez-Ferrer, M.L.; Cascales, P.; Torroba, A.; Orozco, R.; Silva Sanchez, Y.; Nieto, A.; Fiol, G. Prevalence of endometriosis in epithelial ovarian cancer. Analysis of the associated clinical features and study on molecular mechanisms involved in the possible causality. *Eur. J. Gynaecol. Oncol.* **2015**, *36*, 21–24. [PubMed]
57. Stamp, J.P.; Gilks, C.B.; Wesseling, M.; Eshragh, S.; Ceballos, K.; Anglesio, M.S.; Kwon, J.S.; Tone, A.; Huntsman, D.G.; Carey, M.S. Baf250a expression in atypical endometriosis and endometriosis-associated ovarian cancer. *Int. J. Gynecol. Cancer* **2016**, *26*, 825–832. [CrossRef] [PubMed]

58. Van Gorp, T.; Amant, F.; Neven, P.; Vergote, I.; Moerman, P. Endometriosis and the development of malignant tumours of the pelvis. A review of literature. *Best Pract. Res. Clin. Obstet. Gynaecol.* **2004**, *18*, 349–371. [CrossRef] [PubMed]
59. Fukunaga, M.; Nomura, K.; Ishikawa, E.; Ushigome, S. Ovarian atypical endometriosis: Its close association with malignant epithelial tumours. *Histopathology* **1997**, *30*, 249–255. [CrossRef] [PubMed]
60. Banz, C.; Ungethuem, U.; Kuban, R.J.; Diedrich, K.; Lengyel, E.; Hornung, D. The molecular signature of endometriosis-associated endometrioid ovarian cancer differs significantly from endometriosis-independent endometrioid ovarian cancer. *Fertil. Steril.* **2010**, *94*, 1212–1217. [CrossRef] [PubMed]
61. Zhang, C.; Wang, X.; Anaya, Y.; Parodi, L.; Chen, L.; Anderson, M.L.; Hawkins, S.M. Distinct molecular pathways in ovarian endometrioid adenocarcinoma with concurrent endometriosis. *Int. J. Cancer* **2018**, in press.
62. Kolin, D.L.; Dinulescu, D.M.; Crum, C.P. Origin of clear cell carcinoma: Nature or nurture? *J. Pathol.* **2018**, *244*, 131–134. [CrossRef] [PubMed]
63. Klemmt, P.A.B.; Starzinski-Powitz, A. Molecular and cellular pathogenesis of endometriosis. *Curr. Womens Health Rev.* **2018**, *14*, 106–116. [CrossRef] [PubMed]
64. Liu, D.T.; Hitchcock, A. Endometriosis: Its association with retrograde menstruation, dysmenorrhoea and tubal pathology. *Br. J. Obstet. Gynaecol.* **1986**, *93*, 859–862. [CrossRef] [PubMed]
65. Liu, H.; Du, Y.; Zhang, Z.; Lv, L.; Xiong, W.; Zhang, L.; Li, N.; He, H.; Li, Q.; Liu, Y. Autophagy contributes to hypoxia-induced epithelial to mesenchymal transition of endometrial epithelial cells in endometriosis. *Biol. Reprod.* **2018**. [CrossRef] [PubMed]
66. Allavena, G.; Carrarelli, P.; Del Bello, B.; Luisi, S.; Petraglia, F.; Maellaro, E. Autophagy is upregulated in ovarian endometriosis: A possible interplay with p53 and heme oxygenase-1. *Fertil. Steril.* **2015**, *103*, 1244–1251.e1. [CrossRef] [PubMed]
67. Liu, H.; Zhang, Z.; Xiong, W.; Zhang, L.; Xiong, Y.; Li, N.; He, H.; Du, Y.; Liu, Y. Hypoxia-inducible factor-1alpha promotes endometrial stromal cells migration and invasion by upregulating autophagy in endometriosis. *Reproduction* **2017**, *153*, 809–820. [CrossRef] [PubMed]
68. Xu, T.X.; Zhao, S.Z.; Dong, M.; Yu, X.R. Hypoxia responsive miR-210 promotes cell survival and autophagy of endometriotic cells in hypoxia. *Eur. Rev. Med. Pharmacol. Sci.* **2016**, *20*, 399–406. [PubMed]
69. Lin, X.; Dai, Y.; Xu, W.; Shi, L.; Jin, X.; Li, C.; Zhou, F.; Pan, Y.; Zhang, Y.; Lin, X.; et al. Hypoxia promotes ectopic adhesion ability of endometrial stromal cells via tgf-beta1/smad signaling in endometriosis. *Endocrinology* **2018**, *159*, 1630–1641. [CrossRef] [PubMed]
70. Yoo, J.Y.; Kim, T.H.; Fazleabas, A.T.; Palomino, W.A.; Ahn, S.H.; Tayade, C.; Schammel, D.P.; Young, S.L.; Jeong, J.W.; Lessey, B.A. Kras activation and over-expression of sirt1/bcl6 contributes to the pathogenesis of endometriosis and progesterone resistance. *Sci. Rep.* **2017**, *7*, 6765. [CrossRef] [PubMed]
71. Hsiao, K.Y.; Chang, N.; Tsai, J.L.; Lin, S.C.; Tsai, S.J.; Wu, M.H. Hypoxia-inhibited DUSP2 expression promotes IL-6/STAT3 signaling in endometriosis. *Am. J. Reprod. Immunol.* **2017**, *78*. [CrossRef] [PubMed]
72. Kim, B.G.; Yoo, J.Y.; Kim, T.H.; Shin, J.H.; Langenheim, J.F.; Ferguson, S.D.; Fazleabas, A.T.; Young, S.L.; Lessey, B.A.; Jeong, J.W. Aberrant activation of signal transducer and activator of transcription-3 (stat3) signaling in endometriosis. *Hum. Reprod. (Oxf. Engl.)* **2015**, *30*, 1069–1078. [CrossRef] [PubMed]
73. Lin, S.C.; Wang, C.C.; Wu, M.H.; Yang, S.H.; Li, Y.H.; Tsai, S.J. Hypoxia-induced microRNA-20a expression increases ERK phosphorylation and angiogenic gene expression in endometriotic stromal cells. *J. Clin. Endocrinol. Metab.* **2012**, *97*, E1515–E1523. [CrossRef] [PubMed]
74. Hsiao, K.Y.; Chang, N.; Lin, S.C.; Li, Y.H.; Wu, M.H. Inhibition of dual specificity phosphatase-2 by hypoxia promotes interleukin-8-mediated angiogenesis in endometriosis. *Hum. Reprod. (Oxf. Engl.)* **2014**, *29*, 2747–2755. [CrossRef] [PubMed]
75. Kato, M.; Yamamoto, S.; Takano, M.; Matsubara, O.; Furuya, K. Aberrant expression of the mammalian target of rapamycin, hypoxia-inducible factor-1alpha, and glucose transporter 1 in the development of ovarian clear-cell adenocarcinoma. *Int. J. Gynecol. Pathol.* **2012**, *31*, 254–263. [CrossRef] [PubMed]
76. Duska, L.R.; Garrett, L.; Henretta, M.; Ferriss, J.S.; Lee, L.; Horowitz, N. When 'never-events' occur despite adherence to clinical guidelines: The case of venous thromboembolism in clear cell cancer of the ovary compared with other epithelial histologic subtypes. *Gynecol. Oncol.* **2010**, *116*, 374–377. [CrossRef] [PubMed]
77. Hsiao, K.Y.; Lin, S.C.; Wu, M.H.; Tsai, S.J. Pathological functions of hypoxia in endometriosis. *Front. Biosci. (Elite Ed.)* **2015**, *7*, 309–321. [CrossRef] [PubMed]

78. Zhan, L.; Wang, W.; Zhang, Y.; Song, E.; Fan, Y.; Wei, B. Hypoxia-inducible factor-1alpha: A promising therapeutic target in endometriosis. *Biochimie* **2016**, *123*, 130–137. [CrossRef] [PubMed]
79. Meserve, E.E.; Crum, C.P. Benign conditions of the ovary. In *Diagnostic Gynecologic and Obstetric Pathology*, 3rd ed.; Crum, C.P., Haefner, H.K., Peters, W.A., III, Eds.; Elsevier, Inc.: Philadephia, PA, USA, 2018; pp. 761–799.
80. Hantak, A.M.; Bagchi, I.C.; Bagchi, M.K. Role of uterine stromal-epithelial crosstalk in embryo implantation. *Int. J. Dev. Biol.* **2014**, *58*, 139–146. [CrossRef] [PubMed]
81. Kurita, T.; Medina, R.; Schabel, A.B.; Young, P.; Gama, P.; Parekh, T.V.; Brody, J.; Cunha, G.R.; Osteen, K.G.; Bruner-Tran, K.L.; et al. The activation function-1 domain of estrogen receptor alpha in uterine stromal cells is required for mouse but not human uterine epithelial response to estrogen. *Differentiation* **2005**, *73*, 313–322. [CrossRef] [PubMed]
82. Osteen, K.G.; Rodgers, W.H.; Gaire, M.; Hargrove, J.T.; Gorstein, F.; Matrisian, L.M. Stromal-epithelial interaction mediates steroidal regulation of metalloproteinase expression in human endometrium. *Proc. Natl. Acad. Sci. USA* **1994**, *91*, 10129–10133. [CrossRef] [PubMed]
83. Valdez, J.; Cook, C.D.; Ahrens, C.C.; Wang, A.J.; Brown, A.; Kumar, M.; Stockdale, L.; Rothenberg, D.; Renggli, K.; Gordon, E.; et al. On-demand dissolution of modular, synthetic extracellular matrix reveals local epithelial-stromal communication networks. *Biomaterials* **2017**, *130*, 90–103. [CrossRef] [PubMed]
84. Aghajanova, L.; Hamilton, A.; Kwintkiewicz, J.; Vo, K.C.; Giudice, L.C. Steroidogenic enzyme and key decidualization marker dysregulation in endometrial stromal cells from women with versus without endometriosis. *Biol. Reprod.* **2009**, *80*, 105–114. [CrossRef] [PubMed]
85. Velarde, M.C.; Aghajanova, L.; Nezhat, C.R.; Giudice, L.C. Increased mitogen-activated protein kinase kinase/extracellularly regulated kinase activity in human endometrial stromal fibroblasts of women with endometriosis reduces 3′,5′-cyclic adenosine 5′-monophosphate inhibition of cyclin D1. *Endocrinology* **2009**, *150*, 4701–4712. [CrossRef] [PubMed]
86. Aghajanova, L.; Horcajadas, J.A.; Weeks, J.L.; Esteban, F.J.; Nezhat, C.N.; Conti, M.; Giudice, L.C. The protein kinase a pathway-regulated transcriptome of endometrial stromal fibroblasts reveals compromised differentiation and persistent proliferative potential in endometriosis. *Endocrinology* **2010**, *151*, 1341–1355. [CrossRef] [PubMed]
87. Aghajanova, L.; Tatsumi, K.; Horcajadas, J.A.; Zamah, A.M.; Esteban, F.J.; Herndon, C.N.; Conti, M.; Giudice, L.C. Unique transcriptome, pathways, and networks in the human endometrial fibroblast response to progesterone in endometriosis. *Biol. Reprod.* **2011**, *84*, 801–815. [CrossRef] [PubMed]
88. Aghajanova, L.; Velarde, M.C.; Giudice, L.C. The progesterone receptor coactivator hic-5 is involved in the pathophysiology of endometriosis. *Endocrinology* **2009**, *150*, 3863–3870. [CrossRef] [PubMed]
89. Barragan, F.; Irwin, J.C.; Balayan, S.; Erikson, D.W.; Chen, J.C.; Houshdaran, S.; Piltonen, T.T.; Spitzer, T.L.; George, A.; Rabban, J.T.; et al. Human endometrial fibroblasts derived from mesenchymal progenitors inherit progesterone resistance and acquire an inflammatory phenotype in the endometrial niche in endometriosis. *Biol. Reprod.* **2016**, *94*, 118. [CrossRef] [PubMed]
90. Zhang, Y.; Tang, H.; Cai, J.; Zhang, T.; Guo, J.; Feng, D.; Wang, Z. Ovarian cancer-associated fibroblasts contribute to epithelial ovarian carcinoma metastasis by promoting angiogenesis, lymphangiogenesis and tumor cell invasion. *Cancer Lett.* **2011**, *303*, 47–55. [CrossRef] [PubMed]
91. Mitra, A.K.; Zillhardt, M.; Hua, Y.; Tiwari, P.; Murmann, A.E.; Peter, M.E.; Lengyel, E. MicroRNAs reprogram normal fibroblasts into cancer-associated fibroblasts in ovarian cancer. *Cancer Discov.* **2012**, *2*, 1100–1108. [CrossRef] [PubMed]
92. Hawkins, S.M.; Creighton, C.J.; Han, D.Y.; Zariff, A.; Anderson, M.L.; Gunaratne, P.H.; Matzuk, M.M. Functional microRNA involved in endometriosis. *Mol. Endocrinol.* **2011**, *25*, 821–832. [CrossRef] [PubMed]
93. Lessey, B.A.; Kim, J.J. Endometrial receptivity in the eutopic endometrium of women with endometriosis: It is affected, and let me show you why. *Fertil. Steril.* **2017**, *108*, 19–27. [CrossRef] [PubMed]
94. Capobianco, A.; Rovere-Querini, P. Endometriosis, a disease of the macrophage. *Front. Immunol.* **2013**, *4*, 9. [CrossRef] [PubMed]
95. Bacci, M.; Capobianco, A.; Monno, A.; Cottone, L.; Di Puppo, F.; Camisa, B.; Mariani, M.; Brignole, C.; Ponzoni, M.; Ferrari, S.; et al. Macrophages are alternatively activated in patients with endometriosis and required for growth and vascularization of lesions in a mouse model of disease. *Am. J. Pathol.* **2009**, *175*, 547–556. [CrossRef] [PubMed]

96. Canet, B.; Pons, C.; Espinosa, I.; Prat, J. Cdc42-positive macrophages may prevent malignant transformation of ovarian endometriosis. *Hum. Pathol.* **2012**, *43*, 720–725. [CrossRef] [PubMed]
97. Furuya, M.; Tanaka, R.; Miyagi, E.; Kami, D.; Nagahama, K.; Miyagi, Y.; Nagashima, Y.; Hirahara, F.; Inayama, Y.; Aoki, I. Impaired CXCL4 expression in tumor-associated macrophages (tams) of ovarian cancers arising in endometriosis. *Cancer Biol. Ther.* **2012**, *13*, 671–680. [CrossRef] [PubMed]
98. Eisermann, J.; Gast, M.J.; Pineda, J.; Odem, R.R.; Collins, J.L. Tumor necrosis factor in peritoneal fluid of women undergoing laparoscopic surgery. *Fertil. Steril.* **1988**, *50*, 573–579. [CrossRef]
99. Calhaz-Jorge, C.; Costa, A.P.; Barata, M.; Santos, M.C.; Melo, A.; Palma-Carlos, M.L. Tumour necrosis factor alpha concentrations in the peritoneal fluid of infertile women with minimal or mild endometriosis are lower in patients with red lesions only than in patients without red lesions. *Hum. Reprod. (Oxf. Engl.)* **2000**, *15*, 1256–1260. [CrossRef]
100. Jin, C.H.; Yi, K.W.; Ha, Y.R.; Shin, J.H.; Park, H.T.; Kim, T.; Hur, J.Y. Chemerin expression in the peritoneal fluid, serum, and ovarian endometrioma of women with endometriosis. *Am. J. Reprod. Immunol.* **2015**, *74*, 379–386. [CrossRef] [PubMed]
101. Young, V.J.; Brown, J.K.; Saunders, P.T.; Duncan, W.C.; Horne, A.W. The peritoneum is both a source and target of TGF-Beta in women with endometriosis. *PLoS ONE* **2014**, *9*, e106773. [CrossRef] [PubMed]
102. Worley, M.J.; Welch, W.R.; Berkowitz, R.S.; Ng, S.W. Endometriosis-associated ovarian cancer: A review of pathogenesis. *Int. J. Mol. Sci.* **2013**, *14*, 5367–5379. [CrossRef] [PubMed]
103. Lewis, C.E.; Pollard, J.W. Distinct role of macrophages in different tumor microenvironments. *Cancer Res.* **2006**, *66*, 605–612. [CrossRef] [PubMed]
104. Han, S.J.; Jung, S.Y.; Wu, S.P.; Hawkins, S.M.; Park, M.J.; Kyo, S.; Qin, J.; Lydon, J.P.; Tsai, S.Y.; Tsai, M.J.; et al. Estrogen receptor beta modulates apoptosis complexes and the inflammasome to drive the pathogenesis of endometriosis. *Cell* **2015**, *163*, 960–974. [CrossRef] [PubMed]
105. Altan, Z.M.; Denis, D.; Kagan, D.; Grund, E.M.; Palmer, S.S.; Nataraja, S.G. A long-acting tumor necrosis factor alpha-binding protein demonstrates activity in both in vitro and in vivo models of endometriosis. *J. Pharmacol. Exp. Ther.* **2010**, *334*, 460–466. [CrossRef] [PubMed]
106. Koninckx, P.R.; Craessaerts, M.; Timmerman, D.; Cornillie, F.; Kennedy, S. Anti-TNF-alpha treatment for deep endometriosis-associated pain: A randomized placebo-controlled trial. *Hum. Reprod. (Oxf. Engl.)* **2008**, *23*, 2017–2023. [CrossRef] [PubMed]
107. Wei, J.J.; William, J.; Bulun, S. Endometriosis and ovarian cancer: A review of clinical, pathologic, and molecular aspects. *Int. J. Gynecol. Pathol.* **2011**, *30*, 553–568. [CrossRef] [PubMed]
108. Yamaguchi, K.; Mandai, M.; Toyokuni, S.; Hamanishi, J.; Higuchi, T.; Takakura, K.; Fujii, S. Contents of endometriotic cysts, especially the high concentration of free iron, are a possible cause of carcinogenesis in the cysts through the iron-induced persistent oxidative stress. *Clin. Cancer Res.* **2008**, *14*, 32–40. [CrossRef] [PubMed]
109. Mandai, M.; Matsumura, N.; Baba, T.; Yamaguchi, K.; Hamanishi, J.; Konishi, I. Ovarian clear cell carcinoma as a stress-responsive cancer: Influence of the microenvironment on the carcinogenesis and cancer phenotype. *Cancer Lett.* **2011**, *310*, 129–133. [CrossRef] [PubMed]
110. Shigetomi, H.; Higashiura, Y.; Kajihara, H.; Kobayashi, H. A potential link of oxidative stress and cell cycle regulation for development of endometriosis. *Gynecol. Endocrinol.* **2012**, *28*, 897–902. [CrossRef] [PubMed]
111. Vercellini, P.; Crosignani, P.; Somigliana, E.; Vigano, P.; Buggio, L.; Bolis, G.; Fedele, L. The 'incessant menstruation' hypothesis: A mechanistic ovarian cancer model with implications for prevention. *Hum. Reprod. (Oxf. Engl.)* **2011**, *26*, 2262–2273. [CrossRef] [PubMed]
112. Young, V.J.; Brown, J.K.; Maybin, J.; Saunders, P.T.; Duncan, W.C.; Horne, A.W. Transforming growth factor-β induced warburg-like metabolic reprogramming may underpin the development of peritoneal endometriosis. *J. Clin. Endocrinol. Metab.* **2014**, *99*, 3450–3459. [CrossRef] [PubMed]
113. Young, V.J.; Ahmad, S.F.; Brown, J.K.; Duncan, W.C.; Horne, A.W. Id2 mediates the transforming growth factor-β1-induced warburg-like effect seen in the peritoneum of women with endometriosis. *MHR Basic Sci. Reprod. Med.* **2016**, *22*, 648–654. [CrossRef] [PubMed]
114. Dominguez, F.; Ferrando, M.; Diaz-Gimeno, P.; Quintana, F.; Fernandez, G.; Castells, I.; Simon, C. Lipidomic profiling of endometrial fluid in women with ovarian endometriosisdagger. *Biol. Reprod.* **2017**, *96*, 772–779. [CrossRef] [PubMed]

115. Li, J.; Gao, Y.; Guan, L.; Zhang, H.; Sun, J.; Gong, X.; Li, D.; Chen, P.; Ma, Z.; Liang, X.; et al. Discovery of phosphatidic acid, phosphatidylcholine, and phosphatidylserine as biomarkers for early diagnosis of endometriosis. *Front. Physiol* **2018**, *9*, 14. [CrossRef] [PubMed]
116. Cheng, Y.H.; Imir, A.; Fenkci, V.; Yilmaz, M.B.; Bulun, S.E. Stromal cells of endometriosis fail to produce paracrine factors that induce epithelial 17beta hydroxysteroid dehydrogenase type 2 gene and its transcriptional regulator sp1: A mechanism for defective estradiol metabolism. *Am. J. Obstet. Gynecol.* **2007**, *196*, 391.e1–391.e7; discussion 391.e7–391.e8. [CrossRef] [PubMed]
117. Khorram, O.; Taylor, R.N.; Ryan, I.P.; Schall, T.J.; Landers, D.V. Peritoneal fluid concentrations of the cytokine rantes correlate with the severity of endometriosis. *Am. J. Obstet. Gynecol.* **1993**, *169*, 1545–1549. [CrossRef]
118. Tsai, S.J.; Wu, M.H.; Lin, C.C.; Sun, H.S.; Chen, H.M. Regulation of steroidogenic acute regulatory protein expression and progesterone production in endometriotic stromal cells. *J. Clin. Endocrinol. Metab.* **2001**, *86*, 5765–5773. [CrossRef] [PubMed]
119. Arosh, J.A.; Lee, J.; Balasubbramanian, D.; Stanley, J.A.; Long, C.R.; Meagher, M.W.; Osteen, K.G.; Bruner-Tran, K.L.; Burghardt, R.C.; Starzinski-Powitz, A.; et al. Molecular and preclinical basis to inhibit PGE2 receptors EP2 and EP4 as a novel nonsteroidal therapy for endometriosis. *Proc. Natl. Acad. Sci. USA* **2015**, *112*, 9716–9721. [CrossRef] [PubMed]
120. Modugno, F.; Ness, R.B.; Allen, G.O.; Schildkraut, J.M.; Davis, F.G.; Goodman, M.T. Oral contraceptive use, reproductive history, and risk of epithelial ovarian cancer in women with and without endometriosis. *Am. J. Obstet. Gynecol.* **2004**, *191*, 733–740. [CrossRef] [PubMed]
121. Hawkins, S.M.; Buchold, G.M.; Matzuk, M.M. Minireview: The roles of small rna pathways in reproductive medicine. *Mol. Endocrinol.* **2011**, *25*, 1257–1279. [CrossRef] [PubMed]
122. Bartel, D.P. MicroRNAs: Genomics, biogenesis, mechanism, and function. *Cell* **2004**, *116*, 281–297. [CrossRef]
123. Bartel, D.P.; Chen, C.Z. Micromanagers of gene expression: The potentially widespread influence of metazoan microRNAs. *Nat. Rev.* **2004**, *5*, 396–400. [CrossRef] [PubMed]
124. Yanaihara, N.; Noguchi, Y.; Saito, M.; Takenaka, M.; Takakura, S.; Yamada, K.; Okamoto, A. microRNA gene expression signature driven by miR-9 overexpression in ovarian clear cell carcinoma. *PLoS ONE* **2016**, *11*, e0162584. [CrossRef] [PubMed]
125. Zhang, X.; Guo, G.; Wang, G.; Zhao, J.; Wang, B.; Yu, X.; Ding, Y. Profile of differentially expressed miRNAs in high-grade serous carcinoma and clear cell ovarian carcinoma, and the expression of miR-510 in ovarian carcinoma. *Mol. Med. Rep.* **2015**, *12*, 8021–8031. [CrossRef] [PubMed]
126. Vilming Elgaaen, B.; Olstad, O.K.; Haug, K.B.; Brusletto, B.; Sandvik, L.; Staff, A.C.; Gautvik, K.M.; Davidson, B. Global miRNA expression analysis of serous and clear cell ovarian carcinomas identifies differentially expressed miRNAs including miR-200c-3p as a prognostic marker. *BMC Cancer* **2014**, *14*, 80. [CrossRef] [PubMed]
127. Wyman, S.K.; Parkin, R.K.; Mitchell, P.S.; Fritz, B.R.; O'Briant, K.; Godwin, A.K.; Urban, N.; Drescher, C.W.; Knudsen, B.S.; Tewari, M. Repertoire of microRNAs in epithelial ovarian cancer as determined by next generation sequencing of small RNA cDNA libraries. *PLoS ONE* **2009**, *4*, e5311. [CrossRef] [PubMed]
128. Wu, R.L.; Ali, S.; Bandyopadhyay, S.; Alosh, B.; Hayek, K.; Daaboul, M.F.; Winer, I.; Sarkar, F.H.; Ali-Fehmi, R. Comparative analysis of differentially expressed miRNAs and their downstream mRNAs in ovarian cancer and its associated endometriosis. *J Cancer Sci. Ther.* **2015**, *7*, 258–265. [PubMed]
129. Iorio, M.V.; Visone, R.; Di Leva, G.; Donati, V.; Petrocca, F.; Casalini, P.; Taccioli, C.; Volinia, S.; Liu, C.G.; Alder, H.; et al. microRNA signatures in human ovarian cancer. *Cancer Res.* **2007**, *67*, 8699–8707. [CrossRef] [PubMed]
130. Braicu, O.L.; Budisan, L.; Buiga, R.; Jurj, A.; Achimas-Cadariu, P.; Pop, L.A.; Braicu, C.; Irimie, A.; Berindan-Neagoe, I. miRNA expression profiling in formalin-fixed paraffin-embedded endometriosis and ovarian cancer samples. *Onco Targets Ther.* **2017**, *10*, 4225–4238. [CrossRef] [PubMed]
131. Zhao, H.; Ding, Y.; Tie, B.; Sun, Z.F.; Jiang, J.Y.; Zhao, J.; Lin, X.; Cui, S. miRNA expression pattern associated with prognosis in elderly patients with advanced OPSC and OCC. *Int. J. Oncol.* **2013**, *43*, 839–849. [CrossRef] [PubMed]
132. Calura, E.; Fruscio, R.; Paracchini, L.; Bignotti, E.; Ravaggi, A.; Martini, P.; Sales, G.; Beltrame, L.; Clivio, L.; Ceppi, L.; et al. miRNA landscape in stage i epithelial ovarian cancer defines the histotype specificities. *Clin. Cancer Res.* **2013**, *19*, 4114–4123. [CrossRef] [PubMed]

133. Suryawanshi, S.; Vlad, A.M.; Lin, H.M.; Mantia-Smaldone, G.; Laskey, R.; Lee, M.; Lin, Y.; Donnellan, N.; Klein-Patel, M.; Lee, T.; et al. Plasma microRNAs as novel biomarkers for endometriosis and endometriosis-associated ovarian cancer. *Clin. Cancer Res.* **2013**, *19*, 1213–1224. [CrossRef] [PubMed]
134. Liu, S.; Gao, S.; Wang, X.Y.; Wang, D.B. Expression of miR-126 and Crk in endometriosis: miR-126 may affect the progression of endometriosis by regulating Crk expression. *Arch. Gynecol. Obstet.* **2012**, *285*, 1065–1072. [CrossRef] [PubMed]
135. Ebrahimi, F.; Gopalan, V.; Smith, R.A.; Lam, A.K. miR-126 in human cancers: Clinical roles and current perspectives. *Exp. Mol. Pathol.* **2014**, *96*, 98–107. [CrossRef] [PubMed]
136. Sestito, R.; Cianfrocca, R.; Rosano, L.; Tocci, P.; Semprucci, E.; Di Castro, V.; Caprara, V.; Ferrandina, G.; Sacconi, A.; Blandino, G.; et al. miR-30a inhibits endothelin a receptor and chemoresistance in ovarian carcinoma. *Oncotarget* **2016**, *7*, 4009–4023. [CrossRef] [PubMed]
137. Bai, L.; Wang, H.; Wang, A.H.; Zhang, L.Y.; Bai, J. microRNA-532 and microRNA-3064 inhibit cell proliferation and invasion by acting as direct regulators of human telomerase reverse transcriptase in ovarian cancer. *PLoS ONE* **2017**, *12*, e0173912. [CrossRef] [PubMed]
138. Humphries, B.; Yang, C. The microRNA-200 family: Small molecules with novel roles in cancer development, progression and therapy. *Oncotarget* **2015**, *6*, 6472–6498. [CrossRef] [PubMed]
139. Greaves, E.; Critchley, H.O.D.; Horne, A.W.; Saunders, P.T.K. Relevant human tissue resources and laboratory models for use in endometriosis research. *Acta. Obstet. Gynecol. Scand.* **2017**, *96*, 644–658. [CrossRef] [PubMed]
140. King, C.M.; Barbara, C.; Prentice, A.; Brenton, J.D.; Charnock-Jones, D.S. Models of endometriosis and their utility in studying progression to ovarian clear cell carcinoma. *J. Pathol.* **2016**, *238*, 185–196. [CrossRef] [PubMed]
141. Cancer Genome Atlas Research Network. Integrated genomic analyses of ovarian carcinoma. *Nature* **2011**, *474*, 609–615. [CrossRef] [PubMed]
142. Hardy, S.; Kitamura, M.; HarrisStansil, T.; Dai, Y.M.; Phipps, M.L. Construction of adenovirus vectors through Cre-lox recombination. *J. Virol.* **1997**, *71*, 1842–1849. [PubMed]
143. Mullany, L.K.; Fan, H.Y.; Liu, Z.; White, L.D.; Marshall, A.; Gunaratne, P.; Anderson, M.L.; Creighton, C.J.; Xin, L.; Deavers, M.; et al. Molecular and functional characteristics of ovarian surface epithelial cells transformed by KrasG12D and loss of Pten in a mouse model in vivo. *Oncogene* **2011**, *30*, 3522–3536. [CrossRef] [PubMed]
144. Jamin, S.P.; Arango, N.A.; Mishina, Y.; Hanks, M.C.; Behringer, R.R. Requirement of bmpr1a for mullerian duct regression during male sexual development. *Nat. Genet.* **2002**, *32*, 408–410. [CrossRef] [PubMed]
145. Fan, H.Y.; Shimada, M.; Liu, Z.; Cahill, N.; Noma, N.; Wu, Y.; Gossen, J.; Richards, J.S. Selective expression of KrasG12D in granulosa cells of the mouse ovary causes defects in follicle development and ovulation. *Development* **2008**, *135*, 2127–2137. [CrossRef] [PubMed]
146. Wu, R.; Zhai, Y.; Kuick, R.; Karnezis, A.N.; Garcia, P.; Naseem, A.; Hu, T.C.; Fearon, E.R.; Cho, K.R. Impact of oviductal versus ovarian epithelial cell of origin on ovarian endometrioid carcinoma phenotype in the mouse. *J. Pathol.* **2016**, *240*, 341–351. [CrossRef] [PubMed]
147. Russo, A.; Czarnecki, A.A.; Dean, M.; Modi, D.A.; Lantvit, D.D.; Hardy, L.; Baligod, S.; Davis, D.A.; Wei, J.J.; Burdette, J.E. Pten loss in the fallopian tube induces hyperplasia and ovarian tumor formation. *Oncogene* **2018**, *37*, 1976–1990. [CrossRef] [PubMed]
148. Soyal, S.M.; Mukherjee, A.; Lee, K.Y.S.; Li, J.; Li, H.; DeMayo, F.J.; Lydon, J.P. Cre-mediated recombination in cell lineages that express the progesterone receptor. *Genesis* **2005**, *41*, 58–66. [CrossRef] [PubMed]
149. Wu, R.; Zhai, Y.; Fearon, E.R.; Cho, K.R. Diverse mechanisms of beta-catenin deregulation in ovarian endometrioid adenocarcinomas. *Cancer Res.* **2001**, *61*, 8247–8255. [PubMed]
150. Saegusa, M.; Okayasu, I. Frequent nuclear beta-catenin accumulation and associated mutations in endometrioid-type endometrial and ovarian carcinomas with squamous differentiation. *J. Pathol.* **2001**, *194*, 59–67. [CrossRef] [PubMed]
151. Palacios, J.; Gamallo, C. Mutations in the beta-catenin gene (CTNNB1) in endometrioid ovarian carcinomas. *Cancer Res.* **1998**, *58*, 1344–1347. [PubMed]
152. Gamallo, C.; Palacios, J.; Moreno, G.; Calvo de Mora, J.; Suarez, A.; Armas, A. Beta-catenin expression pattern in stage I and II ovarian carcinomas: Relationship with beta-catenin gene mutations, clinicopathological features, and clinical outcome. *Am. J. Pathol.* **1999**, *155*, 527–536. [CrossRef]

153. Wright, K.; Wilson, P.; Morland, S.; Campbell, I.; Walsh, M.; Hurst, T.; Ward, B.; Cummings, M.; Chenevix-Trench, G. Beta-catenin mutation and expression analysis in ovarian cancer: Exon 3 mutations and nuclear translocation in 16% of endometrioid tumours. *Int. J. Cancer* **1999**, *82*, 625–629. [CrossRef]
154. Dharmaraj, N.; Chapela, P.J.; Morgado, M.; Hawkins, S.M.; Lessey, B.A.; Young, S.L.; Carson, D.D. Expression of the transmembrane mucins, MUC1, MUC4 and MUC16, in normal endometrium and in endometriosis. *Hum. Reprod. (Oxf. Engl.)* **2014**, *29*, 1730–1738. [CrossRef] [PubMed]
155. Campbell, I.G.; Russell, S.E.; Choong, D.Y.; Montgomery, K.G.; Ciavarella, M.L.; Hooi, C.S.; Cristiano, B.E.; Pearson, R.B.; Phillips, W.A. Mutation of the PIK3CA gene in ovarian and breast cancer. *Cancer Res.* **2004**, *64*, 7678–7681. [CrossRef] [PubMed]
156. Kurman, R.J.; Shih, I.-M. Molecular pathogenesis and extraovarian origin of epithelial ovarian cancer—Shifting the paradigm. *Hum. Pathol.* **2011**, *42*, 918–931. [CrossRef] [PubMed]
157. Chandler, R.L.; Damrauer, J.S.; Raab, J.R.; Schisler, J.C.; Wilkerson, M.D.; Didion, J.P.; Starmer, J.; Serber, D.; Yee, D.; Xiong, J. Coexistent ARID1A–PIK3CA mutations promote ovarian clear-cell tumorigenesis through pro-tumorigenic inflammatory cytokine signalling. *Nat. Commun.* **2015**, *6*, 6118. [CrossRef] [PubMed]
158. Guan, B.; Rahmanto, Y.S.; Wu, R.C.; Wang, Y.; Wang, Z.; Wang, T.L.; Shih Ie, M. Roles of deletion of ARID1A, a tumor suppressor, in mouse ovarian tumorigenesis. *J. Natl. Cancer Inst.* **2014**, *106*. [CrossRef] [PubMed]
159. Wang, X.; Khatri, S.; Broaddus, R.; Wang, Z.; Hawkins, S.M. Deletion of ARID1A in reproductive tract mesenchymal cells reduces fertility in female mice. *Biol. Reprod.* **2016**, *94*, 93. [CrossRef] [PubMed]
160. Kim, T.H.; Yoo, J.Y.; Wang, Z.; Lydon, J.P.; Khatri, S.; Hawkins, S.M.; Leach, R.E.; Fazleabas, A.T.; Young, S.L.; Lessey, B.A.; et al. Arid1a is essential for endometrial function during early pregnancy. *PLoS Genet.* **2015**, *11*, e1005537. [CrossRef] [PubMed]
161. Wu, R.; Hendrix-Lucas, N.; Kuick, R.; Zhai, Y.; Schwartz, D.R.; Akyol, A.; Hanash, S.; Misek, D.E.; Katabuchi, H.; Williams, B.O.; et al. Mouse model of human ovarian endometrioid adenocarcinoma based on somatic defects in the Wnt/beta-catenin and PI3K/Pten signaling pathways. *Cancer Cell* **2007**, *11*, 321–333. [CrossRef] [PubMed]
162. Dinulescu, D.M.; Ince, T.A.; Quade, B.J.; Shafer, S.A.; Crowley, D.; Jacks, T. Role of K-ras and Pten in the development of mouse models of endometriosis and endometrioid ovarian cancer. *Nat. Med.* **2005**, *11*, 63–70. [CrossRef] [PubMed]
163. Fan, H.-Y.; Liu, Z.; Paquet, M.; Wang, J.; Lydon, J.P.; DeMayo, F.J.; Richards, J.S. Cell type–specific targeted mutations of Kras and Pten document proliferation arrest in granulosa cells versus oncogenic insult to ovarian surface epithelial cells. *Cancer Res.* **2009**, *69*, 6463–6472. [CrossRef] [PubMed]
164. Lague, M.N.; Paquet, M.; Fan, H.Y.; Kaartinen, M.J.; Chu, S.; Jamin, S.P.; Behringer, R.R.; Fuller, P.J.; Mitchell, A.; Dore, M.; et al. Synergistic effects of Pten loss and Wnt/CTNNB1 signaling pathway activation in ovarian granulosa cell tumor development and progression. *Carcinogenesis* **2008**, *29*, 2062–2072. [CrossRef] [PubMed]
165. Tanwar, P.S.; Zhang, L.; Kaneko-Tarui, T.; Curley, M.D.; Taketo, M.M.; Rani, P.; Roberts, D.J.; Teixeira, J.M. Mammalian target of rapamycin is a therapeutic target for murine ovarian endometrioid adenocarcinomas with dysregulated Wnt/β-catenin and Pten. *PLoS ONE* **2011**, *6*, e20715. [CrossRef] [PubMed]
166. Van der Horst, P.H.; van der Zee, M.; Heijmans-Antonissen, C.; Jia, Y.; DeMayo, F.J.; Lydon, J.P.; van Deurzen, C.H.; Ewing, P.C.; Burger, C.W.; Blok, L.J. A mouse model for endometrioid ovarian cancer arising from the distal oviduct. *Int. J. Cancer* **2014**, *135*, 1028–1037. [CrossRef] [PubMed]
167. Wu, R.; Baker, S.J.; Hu, T.C.; Norman, K.M.; Fearon, E.R.; Cho, K.R. Type I to type II ovarian carcinoma progression: Mutant Trp53 or PIK3CA confers a more aggressive tumor phenotype in a mouse model of ovarian cancer. *Am. J. Pathol.* **2013**, *182*, 1391–1399. [CrossRef] [PubMed]
168. Budiu, R.A.; Diaconu, I.; Chrissluis, R.; Dricu, A.; Edwards, R.P.; Vlad, A.M. A conditional mouse model for human muc1-positive endometriosis shows the presence of anti-muc1 antibodies and foxp3+ regulatory t cells. *Dis. Model Mech.* **2009**, *2*, 593–603. [CrossRef] [PubMed]
169. Cousins, F.L.; Murray, A.; Esnal, A.; Gibson, D.A.; Critchley, H.O.; Saunders, P.T. Evidence from a mouse model that epithelial cell migration and mesenchymal-epithelial transition contribute to rapid restoration of uterine tissue integrity during menstruation. *PLoS ONE* **2014**, *9*, e86378. [CrossRef] [PubMed]
170. Brasted, M.; White, C.A.; Kennedy, T.G.; Salamonsen, L.A. Mimicking the events of menstruation in the murine uterus. *Biol. Reprod.* **2003**, *69*, 1273–1280. [CrossRef] [PubMed]

171. Cheng, C.W.; Bielby, H.; Licence, D.; Smith, S.K.; Print, C.G.; Charnock-Jones, D.S. Quantitative cellular and molecular analysis of the effect of progesterone withdrawal in a murine model of decidualization. *Biol. Reprod.* **2007**, *76*, 871–883. [CrossRef] [PubMed]
172. Cheng, C.W.; Licence, D.; Cook, E.; Luo, F.; Arends, M.J.; Smith, S.K.; Print, C.G.; Charnock-Jones, D.S. Activation of mutated K-ras in donor endometrial epithelium and stroma promotes lesion growth in an intact immunocompetent murine model of endometriosis. *J. Pathol.* **2011**, *224*, 261–269. [CrossRef] [PubMed]
173. Greaves, E.; Cousins, F.L.; Murray, A.; Esnal-Zufiaurre, A.; Fassbender, A.; Horne, A.W.; Saunders, P.T. A novel mouse model of endometriosis mimics human phenotype and reveals insights into the inflammatory contribution of shed endometrium. *Am. J. Pathol.* **2014**, *184*, 1930–1939. [CrossRef] [PubMed]
174. Chandler, R.L.; Raab, J.R.; Vernon, M.; Magnuson, T.; Schisler, J.C. Global gene expression profiling of a mouse model of ovarian clear cell carcinoma caused by ARID1A and PIK3CA mutations implicates a role for inflammatory cytokine signaling. *Genom. Data* **2015**, *5*, 329–332. [CrossRef] [PubMed]
175. Harada, N.; Tamai, Y.; Ishikawa, T.; Sauer, B.; Takaku, K.; Oshima, M.; Taketo, M.M. Intestinal polyposis in mice with a dominant stable mutation of the beta-catenin gene. *EMBO J.* **1999**, *18*, 5931–5942. [CrossRef] [PubMed]
176. Uehara, S.; Abe, H.; Hoshiai, H.; Yajima, A.; Suzuki, M. Establishment and characterization of ovarian endometrioid carcinoma cell line. *Gynecol. Oncol.* **1984**, *17*, 314–325. [CrossRef]
177. Gorai, I.; Nakazawa, T.; Miyagi, E.; Hirahara, F.; Nagashima, Y.; Minaguchi, H. Establishment and characterization of two human ovarian clear cell adenocarcinoma lines from metastatic lesions with different properties. *Gynecol. Oncol.* **1995**, *57*, 33–46. [CrossRef] [PubMed]
178. Domcke, S.; Sinha, R.; Levine, D.A.; Sander, C.; Schultz, N. Evaluating cell lines as tumour models by comparison of genomic profiles. *Nat. Commun.* **2013**, *4*, 2126. [CrossRef] [PubMed]
179. Anglesio, M.S.; Wiegand, K.C.; Melnyk, N.; Chow, C.; Salamanca, C.; Prentice, L.M.; Senz, J.; Yang, W.; Spillman, M.A.; Cochrane, D.R.; et al. Type-specific cell line models for type-specific ovarian cancer research. *PLoS ONE* **2013**, *8*, e72162. [CrossRef]
180. Beaufort, C.M.; Helmijr, J.C.; Piskorz, A.M.; Hoogstraat, M.; Ruigrok-Ritstier, K.; Besselink, N.; Murtaza, M.; van, I.W.F.; Heine, A.A.; Smid, M.; et al. Ovarian cancer cell line panel (OCCP): Clinical importance of in vitro morphological subtypes. *PLoS ONE* **2014**, *9*, e103988. [CrossRef] [PubMed]
181. Ince, T.A.; Sousa, A.D.; Jones, M.A.; Harrell, J.C.; Agoston, E.S.; Krohn, M.; Selfors, L.M.; Liu, W.; Chen, K.; Yong, M.; et al. Characterization of twenty-five ovarian tumour cell lines that phenocopy primary tumours. *Nat. Commun.* **2015**, *6*, 7419. [CrossRef] [PubMed]
182. Blayney, J.K.; Davison, T.; McCabe, N.; Walker, S.; Keating, K.; Delaney, T.; Greenan, C.; Williams, A.R.; McCluggage, W.G.; Capes-Davis, A.; et al. Prior knowledge transfer across transcriptional data sets and technologies using compositional statistics yields new mislabelled ovarian cell line. *Nucleic Acids Res.* **2016**, *44*, e137. [CrossRef] [PubMed]
183. Zeitvogel, A.; Baumann, R.; Starzinski-Powitz, A. Identification of an invasive, n-cadherin-expressing epithelial cell type in endometriosis using a new cell culture model. *Am. J. Pathol.* **2001**, *159*, 1839–1852. [CrossRef]
184. Brueggmann, D.; Templeman, C.; Starzinski-Powitz, A.; Rao, N.P.; Gayther, S.A.; Lawrenson, K. Novel three-dimensional in vitro models of ovarian endometriosis. *J. Ovarian Res.* **2014**, *7*, 17. [CrossRef] [PubMed]
185. Bono, Y.; Kyo, S.; Takakura, M.; Maida, Y.; Mizumoto, Y.; Nakamura, M.; Nomura, K.; Kiyono, T.; Inoue, M. Creation of immortalised epithelial cells from ovarian endometrioma. *Br. J. Cancer* **2012**, *106*, 1205–1213. [CrossRef] [PubMed]
186. Ohta, I.; Gorai, I.; Miyamoto, Y.; Yang, J.; Zheng, J.H.; Kawata, N.; Hirahara, F.; Shirotake, S. Cyclophosphamide and 5-fluorouracil act synergistically in ovarian clear cell adenocarcinoma cells. *Cancer Lett.* **2001**, *162*, 39–48. [CrossRef]
187. Provencher, D.M.; Lounis, H.; Champoux, L.; Tetrault, M.; Manderson, E.N.; Wang, J.C.; Eydoux, P.; Savoie, R.; Tonin, P.N.; Mes-Masson, A.M. Characterization of four novel epithelial ovarian cancer cell lines. *In Vitro Cell Dev. Biol. Anim.* **2000**, *36*, 357–361. [CrossRef]
188. Nozawa, S.; Tsukazaki, K.; Sakayori, M.; Jeng, C.H.; Iizuka, R. Establishment of a human ovarian clear cell carcinoma cell line (RMG-I) and its single cell cloning-with special reference to the stem cell of the tumor. *Hum. Cell* **1988**, *1*, 426–435. [PubMed]

189. Wong, W.S.; Wong, Y.F.; Ng, Y.T.; Huang, P.D.; Chew, E.C.; Ho, T.H.; Chang, M.Z. Establishment and characterization of a new human cell line derived from ovarian clear cell carcinoma. *Gynecol. Oncol.* **1990**, *38*, 37–45. [CrossRef]
190. Yamada, K.; Tachibana, T.; Hashimoto, H.; Suzuki, K.; Yanagida, S.; Endoh, H.; Kimura, E.; Yasuda, M.; Tanaka, T.; Ishikawa, H. Establishment and characterization of cell lines derived from serous adenocarcinoma (JHOS-2) and clear cell adenocarcinoma (JHOC-5, JHOC-6) of human ovary. *Hum. Cell* **1999**, *12*, 131–138. [PubMed]
191. Lau, D.H.; Lewis, A.D.; Ehsan, M.N.; Sikic, B.I. Multifactorial mechanisms associated with broad cross-resistance of ovarian carcinoma cells selected by cyanomorpholino doxorubicin. *Cancer Res.* **1991**, *51*, 5181–5187. [PubMed]
192. Yanagibashi, T.; Gorai, I.; Nakazawa, T.; Miyagi, E.; Hirahara, F.; Kitamura, H.; Minaguchi, H. Complexity of expression of the intermediate filaments of six new human ovarian carcinoma cell lines: New expression of cytokeratin 20. *Br. J. Cancer* **1997**, *76*, 829–835. [CrossRef] [PubMed]
193. Stordal, B.; Timms, K.; Farrelly, A.; Gallagher, D.; Busschots, S.; Renaud, M.; Thery, J.; Williams, D.; Potter, J.; Tran, T.; et al. Brca1/2 mutation analysis in 41 ovarian cell lines reveals only one functionally deleterious brca1 mutation. *Mol. Oncol.* **2013**, *7*, 567–579. [CrossRef] [PubMed]
194. Benard, J.; Da Silva, J.; De Blois, M.C.; Boyer, P.; Duvillard, P.; Chiric, E.; Riou, G. Characterization of a human ovarian adenocarcinoma line, IGROV1, in tissue culture and in nude mice. *Cancer Res.* **1985**, *45*, 4970–4979. [PubMed]
195. Hills, C.A.; Kelland, L.R.; Abel, G.; Siracky, J.; Wilson, A.P.; Harrap, K.R. Biological properties of ten human ovarian carcinoma cell lines: Calibration in vitro against four platinum complexes. *Br. J. Cancer* **1989**, *59*, 527–534. [CrossRef] [PubMed]
196. Van den Berg-Bakker, C.A.; Hagemeijer, A.; Franken-Postma, E.M.; Smit, V.T.; Kuppen, P.J.; van Ravenswaay Claasen, H.H.; Cornelisse, C.J.; Schrier, P.I. Establishment and characterization of 7 ovarian carcinoma cell lines and one granulosa tumor cell line: Growth features and cytogenetics. *Int. J. Cancer* **1993**, *53*, 613–620. [CrossRef] [PubMed]
197. Eva, A.; Robbins, K.C.; Andersen, P.R.; Srinivasan, A.; Tronick, S.R.; Reddy, E.P.; Ellmore, N.W.; Galen, A.T.; Lautenberger, J.A.; Papas, T.S.; et al. Cellular genes analogous to retroviral onc genes are transcribed in human tumour cells. *Nature* **1982**, *295*, 116–119. [CrossRef] [PubMed]
198. Buick, R.N.; Pullano, R.; Trent, J.M. Comparative properties of five human ovarian adenocarcinoma cell lines. *Cancer Res.* **1985**, *45*, 3668–3676. [PubMed]
199. Fogh, J. MSKCC. Available online: http://www.mskcc.org (accessed on 14 June 2018).
200. Han, S.J.; Hawkins, S.M.; Begum, K.; Jung, S.Y.; Kovanci, E.; Qin, J.; Lydon, J.P.; DeMayo, F.J.; O'Malley, B.W. A new isoform of steroid receptor coactivator-1 is crucial for pathogenic progression of endometriosis. *Nat. Med.* **2012**, *18*, 1102–1111. [CrossRef] [PubMed]
201. Komiyama, S.; Aoki, D.; Katsuki, Y.; Nozawa, S. Proliferative activity of early ovarian clear cell adenocarcinoma depends on association with endometriosis. *Eur. J. Obstet. Gynecol. Reprod. Biol.* **2006**, *127*, 130–136. [CrossRef] [PubMed]
202. Lee, J.M.; Mhawech-Fauceglia, P.; Lee, N.; Parsanian, L.C.; Lin, Y.G.; Gayther, S.A.; Lawrenson, K. A three-dimensional microenvironment alters protein expression and chemosensitivity of epithelial ovarian cancer cells in vitro. *Lab. Investig.* **2013**, *93*, 528–542. [PubMed]
203. Lal-Nag, M.; McGee, L.; Guha, R.; Lengyel, E.; Kenny, H.A.; Ferrer, M. A high-throughput screening model of the tumor microenvironment for ovarian cancer cell growth. *SLAS Discov.* **2017**, *22*, 494–506. [CrossRef] [PubMed]
204. Chowanadisai, W.; Messerli, S.M.; Miller, D.H.; Medina, J.E.; Hamilton, J.W.; Messerli, M.A.; Brodsky, A.S. Cisplatin resistant spheroids model clinically relevant survival mechanisms in ovarian tumors. *PLoS ONE* **2016**, *11*, e0151089. [CrossRef] [PubMed]
205. Tanenbaum, L.M.; Mantzavinou, A.; Subramanyam, K.S.; Del Carmen, M.G.; Cima, M.J. Ovarian cancer spheroid shrinkage following continuous exposure to cisplatin is a function of spheroid diameter. *Gynecol. Oncol.* **2017**, *146*, 161–169. [CrossRef] [PubMed]
206. Arnold, J.T.; Kaufman, D.G.; Seppala, M.; Lessey, B.A. Endometrial stromal cells regulate epithelial cell growth in vitro: A new co-culture model. *Hum. Reprod. (Oxf. Engl.)* **2001**, *16*, 836–845. [CrossRef]

207. Mori, M.; Ito, F.; Shi, L.; Wang, Y.; Ishida, C.; Hattori, Y.; Niwa, M.; Hirayama, T.; Nagasawa, H.; Iwase, A.; et al. Ovarian endometriosis-associated stromal cells reveal persistently high affinity for iron. *Redox Biol.* **2015**, *6*, 578–586. [CrossRef] [PubMed]
208. Chan, R.W.S.; Lee, C.L.; Ng, E.H.Y.; Yeung, W.S.B. Co-culture with macrophages enhances the clonogenic and invasion activity of endometriotic stromal cells. *Cell Prolif.* **2017**, *50*. [CrossRef] [PubMed]
209. Turco, M.Y.; Gardner, L.; Hughes, J.; Cindrova-Davies, T.; Gomez, M.J.; Farrell, L.; Hollinshead, M.; Marsh, S.G.E.; Brosens, J.J.; Critchley, H.O.; et al. Long-term, hormone-responsive organoid cultures of human endometrium in a chemically defined medium. *Nat. Cell Biol.* **2017**, *19*, 568–577. [CrossRef] [PubMed]
210. Boretto, M.; Cox, B.; Noben, M.; Hendriks, N.; Fassbender, A.; Roose, H.; Amant, F.; Timmerman, D.; Tomassetti, C.; Vanhie, A.; et al. Development of organoids from mouse and human endometrium showing endometrial epithelium physiology and long-term expandability. *Development* **2017**, *144*, 1775–1786. [CrossRef] [PubMed]
211. Practice bulletin No. 114: Management of endometriosis. *Obstet. Gynecol.* **2010**, *116*, 223–236.
212. Shakiba, K.; Bena, J.F.; McGill, K.M.; Minger, J.; Falcone, T. Surgical treatment of endometriosis: A 7-year follow-up on the requirement for further surgery. *Obstet. Gynecol.* **2008**, *111*, 1285–1292. [CrossRef] [PubMed]
213. Yeung, T.L.; Sheng, J.; Leung, C.S.; Li, F.; Kim, J.; Ho, S.Y.; Matzuk, M.M.; Lu, K.H.; Wong, S.T.C.; Mok, S.C. Systematic identification of druggable epithelial-stromal crosstalk signaling networks in ovarian cancer. *J. Natl. Cancer Inst.* **2018**. [CrossRef] [PubMed]

cancers

MDPI

Review

Can Stemness and Chemoresistance Be Therapeutically Targeted via Signaling Pathways in Ovarian Cancer?

Lynn Roy [1,2] and Karen D. Cowden Dahl [1,2,3,4,*]

1 Harper Cancer Research Institute, South Bend, IN 46617, USA; lmroy@iupui.edu
2 Department of Biochemistry and Molecular Biology, Indiana University School of Medicine-South Bend, South Bend, IN 46617, USA
3 Department of Chemistry and Biochemistry, University of Notre Dame, Notre Dame, IN 46617, USA
4 Indiana University Melvin and Bren Simon Cancer Center, Indianapolis, IN 46202, USA
* Correspondence: kcowdend@iupui.edu; Tel.: +1-574-631-1675

Received: 15 June 2018; Accepted: 17 July 2018; Published: 24 July 2018

Abstract: Ovarian cancer is the most lethal gynecological malignancy. Poor overall survival, particularly for patients with high grade serous (HGS) ovarian cancer, is often attributed to late stage at diagnosis and relapse following chemotherapy. HGS ovarian cancer is a heterogenous disease in that few genes are consistently mutated between patients. Additionally, HGS ovarian cancer is characterized by high genomic instability. For these reasons, personalized approaches may be necessary for effective treatment and cure. Understanding the molecular mechanisms that contribute to tumor metastasis and chemoresistance are essential to improve survival rates. One favored model for tumor metastasis and chemoresistance is the cancer stem cell (CSC) model. CSCs are cells with enhanced self-renewal properties that are enriched following chemotherapy. Elimination of this cell population is thought to be a mechanism to increase therapeutic response. Therefore, accurate identification of stem cell populations that are most clinically relevant is necessary. While many CSC identifiers (ALDH, OCT4, CD133, and side population) have been established, it is still not clear which population(s) will be most beneficial to target in patients. Therefore, there is a critical need to characterize CSCs with reliable markers and find their weaknesses that will make the CSCs amenable to therapy. Many signaling pathways are implicated for their roles in CSC initiation and maintenance. Therapeutically targeting pathways needed for CSC initiation or maintenance may be an effective way of treating HGS ovarian cancer patients. In conclusion, the prognosis for HGS ovarian cancer may be improved by combining CSC phenotyping with targeted therapies for pathways involved in CSC maintenance.

Keywords: ovarian cancer; cancer stem cells; signaling; chemoresistance; metastasis

1. Introduction

In the United States, ovarian cancer is the fifth leading cause of cancer death in women [1]. The American Cancer Society (ACS) estimates that this year approximately 22,240 women will be newly diagnosed with ovarian cancer, and ~14,075 women will die as a result of the disease, making it the most lethal gynecologic malignancy (ACS Facts and Figures 2018). The vagueness of symptoms (bloating, abdominal/pelvic pain, difficulty eating/feeling of fullness, and frequent urination) and the lack of early detection methods contribute to the majority of patients (70–75%) receiving diagnoses in advanced stages (stage III or stage IV) when the cancer has metastasized throughout the peritoneal cavity [1,2]. The five-year survival rate for women with advanced-stage ovarian cancer is ~25% [3,4].

There are several major ovarian cancer subtypes. Additionally, there is mutational and gene expression heterogeneity within each subgenre. Mutational and gene expression heterogeneity is also found in different subpopulations within a single tumor. Patients with the same pathological diagnosis, such as high grade serous (HGS) carcinoma, often vary greatly with respect to gene expression and specific genetic mutations [3,5,6]. The lack of consistent mutations or mis-expressed genes makes developing novel targeted therapeutics difficult. The current standard of care is a "one size fits all" approach consisting of aggressive debulking surgery to resect visible tumor followed by platinum and taxane combination chemotherapy [1,7–9]. Residual tumor implants measuring less than 1 cm are considered indicative of optimal debulking [1]. Debulking surgery performed by a gynecological oncologist improves the chance of survival; however, many patients are not treated by gynecological oncologists [1,7,8]. Therefore, in some cases, chemotherapy prior to surgery is equally effective as primary debulking [4]. Chemotherapy treatment is initially effective in 70–80% of patients [2,10,11]. However, recurrence of the disease will occur in the majority of patients (80–90%) within 5 years, and the tumors often acquire resistance to the chemotherapeutics [1,9,11]. The presence of microscopic tumors left behind during surgical debulking and the limitations of current chemotherapeutics contribute to the likelihood of relapse. The presence or enrichment of cancer stem cells (CSCs), which are defined as tumor cells that survive and/or accumulate after chemotherapy, have activation of self-renewing signaling pathways, and exhibit increased tumor-initiating properties, may contribute to relapse [11–13]. We will discuss how CSC properties contribute to chemoresistance and how investigating these properties may lead to novel therapeutics to eliminate ovarian cancer and prevent relapse.

2. Histologic Types of Ovarian Cancer

Ovarian tumors are divided into three types: epithelial (60%), germ cell (30%), and specialized stromal cells tumors (8%) [3,14]. Epithelial tumors comprise the majority of malignant ovarian tumors (80–90%) [10,14]. Within the epithelial tumors there are four major subtypes: serous, endometrioid, clear cell, and mucinous [5,15,16]. Serous tumors are the most common of the epithelial subtypes and comprise two-thirds of all cases [2,3,5,15]. Historically, serous ovarian cancer is classified according to three different three-tiered systems based on morphology/histology. The three systems are the FIGO (the International Federation of Gynecology and Obstetrics) system based on architectural features, the World Health Organization system based on architectural and cytological features, and the Shimizu/Silverberg system based on architectural features, degree of atypical cytological features, and mitotic index, with the most common system being the FIGO system [17]. Within the FIGO system, serous ovarian carcinomas are classified as low grade (Grade 1), intermediate grade (Grade 2), and high grade (Grade 3) [16]. Historically, low grade and high grade serous ovarian tumors were considered to be different grades of the same tumor [5]. However, molecular and genetic studies suggest that it is likely low grade and HGS tumors are distinct diseases with different genetic mutations and different prognoses [5,15,18]. A newer two-tier system combines the current histopathological classification system with molecular genetic findings and clinical features. In this system, ovarian tumors are designated as Type I or Type II [17,19] (Figure 1).

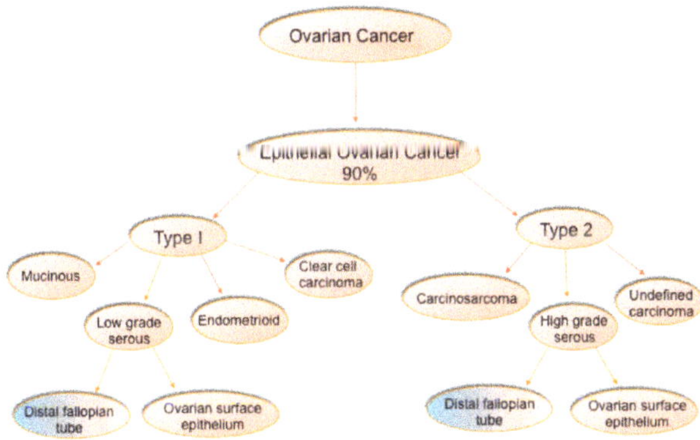

Figure 1. Classification of the Epithelial Ovarian Cancer histological subtype according to the two tier system. Type I tumors include endometroid, clear cell carcinoma, mucinous, and low grade serous. Type II tumors are mostly comprised of high grade serous but also include carcinosarcoma and undefined carcinomas [5,15,18,20].

Low grade serous, mucinous, endometrioid, and clear cell carcinomas fall within the Type I classification [5]. These tumors arise from endometrial tissue, fallopian tube tissue, germ cells, and transitional epithelium [5,14,15,18,21,22]. Type I tumors grow more slowly (are indolent) and are considered to be more genetically stable [5,14,20]. Type II tumors typically have a higher disease volume throughout the peritoneal cavity and a higher incidence of ascites than Type I tumors [20]. They appear to follow a stepwise pattern from a benign precursor to a malignancy with genetic changes in specific cell signaling pathways [2]. Type I tumors are predominantly of non-serous type [10]. Low grade serous ovarian cancer accounts for approximately 5–10% of all serous ovarian cancers [2,10,16]. The most common pathway disrupted in low grade serous ovarian cancer is the mitogen-activated protein kinase (MAPK) pathway [5,6,16,17]. Specifically, activating mutations in BRAF and KRAS are common [2,10,23]. An active MAPK pathway is found in 80% of low grade serous tumors as well as in 78% of their putative precursor lesions (borderline tumors) [16]. Other genes/pathways that are commonly altered in Type I tumors include PTEN, PI3K, ARID1A, Wnt/β-catenin, and ERRB2 [2,6,15,18,20,24,25] (Figure 2).

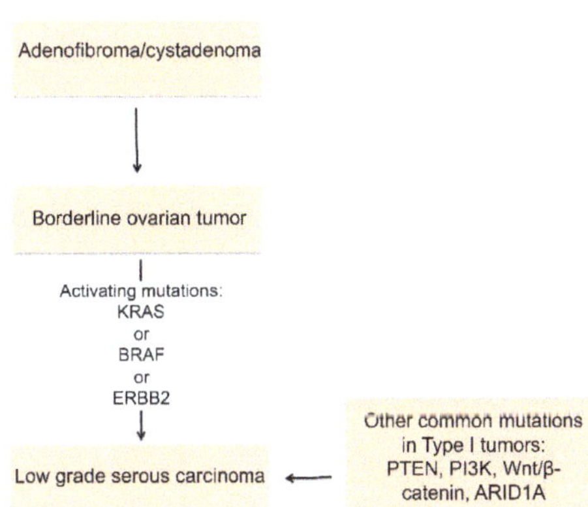

Figure 2. Pathway for Type I tumor formation. Type I tumors appear to form in a stepwise manner from benign precursor lesions. Progression from a borderline ovarian tumors to low grade serous carcinoma commonly includes activating mutations in one of the following members of the MAPK pathway: KRAS, BRAF, or ERBB2.

Prognosis for early-stage tumors is good with a >80% 5-year survival rate with chemotherapy [9]. When dividing all ovarian tumors between stages, Type I tumors are heavily represented in stage I/II (clear cell, 26%; endometrioid, 27%; mucinous, 8%). Only about 36% of early stage tumors are serous [18]. Treatment options for Type I ovarian tumors are identical to those used for Type II tumors and include debulking surgery followed by chemotherapy [17,18]. The response of Type I tumors to chemotherapy is poor due to the relative insensitivity to drug regimens and lack of targeted therapies [7,26]. Low grade serous ovarian tumors have a low response rate to platinum-based therapies with 4% showing a complete response, none with a partial response, 88% with stable disease, and 8% with progression [27]. Another study demonstrated that low grade serous tumors are less responsive than HGS tumors to both paclitaxel (69% vs. 14%) and carboplatin (50% vs. 17%) [27,28]. Type I tumors account for only 10% of ovarian cancer deaths [20]. The poor response of Type I tumors to therapy and the chemoresistance that arises in Type II tumors highlight the need for novel treatment strategies.

HGS tumors comprise 75% of all Type II tumors [3]. HGS neoplasms are typically aggressive and develop rapidly (high mitotic activity) [5,18,20]. Previously, it was thought that HGS ovarian cancer was derived from the ovarian surface epithelium or from cortical inclusion cysts [18,29]. Recent molecular and mouse studies suggest that these tumors likely arise from the epithelium of the distal fallopian tube and that serous tubal intra-epithelial carcinoma (STIC) lesions are the precursors to HGS ovarian cancer [29–31]. One study examined histological sections from fallopian tubes of ovarian cancer patients for evidence of STIC lesions. STIC lesions were identified in 61% of the fallopian tubes from HGS patients with 92% of the lesions being in the fimbriated end of the fallopian tube [32]. Kroeger et al. compiled a list of 15 studies showing that approximately 50–60% of HGS tumors are associated with STIC lesions in the fimbriated end of the fallopian tube [3]. Furthermore, in a molecular profiling analysis, HGS tumors with and without STIC lesions exhibited molecular profiles similar to fallopian tube epithelium [29]. To establish if HGS ovarian cancer can be recapitulated in the mouse, transgenic mouse models have been developed. Dicer and PTEN were conditionally deleted in the reproductive tract using anti-Müllerian hormone receptor type 2-directed Cre (*Amhr2-Cre*) [33]. These mice exhibited abnormal proliferation in the stromal compartment of the fallopian tube [33]. Primary and metastatic tumors that developed in the mice were histologically serous carcinoma, and they shared a similar gene expression profile with human HGS tumors [33]. In another model, Pax8-Cre was used to drive the deletion of *Brca/Pten/Tp53* in the fallopian tube. These mice developed STIC lesions and serous carcinomas [31]. Interestingly, loss of PTEN alone in the fallopian tube (via Pax-8-Cre) was sufficient to generate endometrioid and serous borderline tumors [34]. This raises the possibility of fallopian tube origins for some Type I tumors and non-HGS tumors. While it is possible that a portion of HGS tumors arise from the ovarian surface epithelium, it is likely that a major site of origin for HGS tumors is the fallopian tube [30,35].

Unlike Type I tumors, there is a significant amount of genetic instability within the Type II subgroup, and few genes are consistently mutated [5,14]. The main exception is that in Type II tumors, TP53 mutations are common (both inactivating and gain of function) [36,37]. TP53 mutations are rare in Type I tumors [6]. Type II tumors often exhibit active DNA damage repair mechanisms (e.g., PARP) [3,20]. Overexpression of oncogenes ERRB2 (20–67%) and AKT (12–30%) also occur in some cases [6]. Other common mutations in Type II tumors are BRCA1 or BRCA2. Epithelial ovarian cancer is sporadic in 90% of cases with the remaining 10% being hereditary [2]. In 90–95% of hereditary Type II ovarian tumors, there are germline mutations in BRCA1 or BRCA2 [2]. Importantly, BRCA1 and BRCA2 are often mutated or inactivated in spontaneous ovarian cancer. BRCA1 and BRCA2 mutations are detected in around 5–9% and 3–4% of spontaneous ovarian cancer, respectively [38–42]. Loss of BRCA function through other means, particularly promoter methylation, is common in ovarian cancer (particularly when mutations are not present) [43,44]. Therefore, the p53 and BRCA1/2 pathways are highly implicated in development of HGS ovarian cancer.

Most Type II tumors are found in advanced stages of the disease, which leads to a poor overall prognosis. While Type II tumors respond well to chemotherapy (70–80%) initially, almost all patients relapse and Type II tumors result in 90% of all deaths from ovarian cancer [20]. The advanced stage of disease and development of chemoresistance with Type II tumors results in high mortality. A contributing factor to tumor metastasis and chemoresistance is the presence or enrichment of tumor-initiating/cancer stem cells (CSCs) [45]. Devising new treatments that eliminate this cell demographic is of particular interest for HGS ovarian cancer.

3. Definition of Ovarian Cancer Stem Cells

Heterogeneity is a common feature in ovarian cancer tumors. Different models are proposed to explain tumor heterogeneity. In the stochastic or clonal model, tumors arise from a group of homogeneous cells (clonal). Tumor heterogeneity then occurs through random (stochastic) events within this population. Any of the cells within this population can be tumor initiating provided they possess the necessary genetic mutations, epigenetic changes, and a receptive microenvironment [46–50]. The second model (CSC model) recapitulates the stem cell hierarchy found in development of tissues like the hematopoietic system. In this model, tumors are made of groups of heterogeneous cells that all arise from precursor cells with stem-like properties. These "stem-like" precursors differentiate and/or acquire different mutations that lead to diverse activation of pathways. The resultant cells have unique phenotypes and a hierarchical pattern of inheritance from the initiating CSCs [47,49–52] (Figure 3).

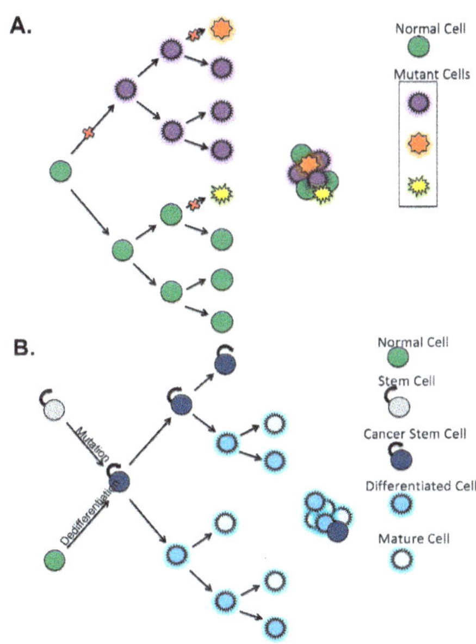

Figure 3. Models of tumor development and heterogeneity. (**A**) The clonal evolution model for tumor initiation. A genetic event occurs in a cell giving rise to a mutant cell population. Any cell is capable of becoming a tumor cell if there is an initiating genetic event. Tumor heterogeneity is due to propagation of cells carrying mutations that are the result of multiple genetic events. (**B**) The cancer stem cell model for tumor initiation. Either a normal stem cell has a genetic event resulting in a cancer stem cell capable of indefinite self-renewal and/or differentiation or a differentiated cell has a genetic event that activates a stem like program within the cell resulting in a cancer stem cell. Tumor cells have a hierarchical inheritance pattern from their cancer stem cell but develop different phenotypes as they acquire further mutations as they differentiate resulting in tumor heterogeneity.

Normal stem cells divide asymmetrically, allowing for self-renewal. One daughter cell retains all the characteristics and programing of the parent cell while the other daughter cell differentiates or acquires new properties [53]. To maintain their undifferentiated state and the ability to self-renew, stem cells reside in a "stem cell niche" comprising various stromal cells, vascular support, and soluble factors that provide a permissive environment [49,54]. CSCs display self-renewal characteristics and retain the ability to produce cells that are highly proliferative and invasive [47,53]. Other characteristics of CSCs include significant DNA repair capability and resistance to therapy [49,53]. In fact, ovarian CSCs ($CD133^+$ and $Sca1^+$) persisted following chemotherapy in a mouse model of ovarian cancer and in cells treated with carboplatin in vitro [45,55]. Moreover, these cells express stem cell markers and maintain tumor initiating potential [45]. Additionally, in vitro studies demonstrated that treatment of ovarian cancer cells with chemotherapy enriches the stem cell pool [56–58]. These studies imply that CSCs are protected from chemotherapy and may be initiators of tumor relapse.

4. Stem Cell Identification in Ovarian Cancer

In 2005, Bapat et al. described the first example of a putative ovarian CSC. A single cell was taken from the ascites of an ovarian cancer patient. Once propagated, the cell was able to form anchorage-independent spheroids in culture and was able to seed tumors in mice via serial transplantation over several generations, illustrating the stem-like capabilities of the cell [59]. Since this initial study, many other investigations have been conducted to identify and validate ovarian CSCS. Identification of CSCs relies on the presence of markers (cell surface and intracellular) that are unique to this particular subset of tumor cells [46,47,50]. In ovarian cancer, a variety of markers are used to denote the presence of CSCs. Cells isolated based on these markers can be tested for "stemness" in vitro via spheroid forming assays, resistance to chemotherapeutics, and in vivo with limiting dilution assays (LDAs) to examine the tumorgenicity of the sample [52]. In the LDA, mice are injected with a defined number of cells from a mixed population of cells or cells isolated that express the stem cell markers. The population that is more stem-like will initiate tumors from significantly fewer cells [60]. Table 1 contains a list of some putative ovarian CSC markers.

Table 1. Putative Ovarian Cancer Stem Cell Markers.

Marker	Type of Protein	Suspected Role in Stem Cells	References
CD24	Cell surface transmembrane glycoprotein	Stem gene expression, tumor initiation, chemoresistance, stem cell maintenance	[46,53,61,62]
CD44	Cell surface transmembrane glycoprotein (hyaluronic acid receptor)	Chemoresistance, tumor initiation, stem gene expression, spheroid formation	[13,46,53,61–67]
cKit/CD117	Tyrosine kinase receptor	Chemoresistance, stem cell maintenance, tumor initiation	[11,53,59,61,68,69]
PROM1/CD133	Cell surface transmembrane glycoprotein	Tumor initiation, chemoresistance, spheroid formation, high cell proliferation	[13,46,53,61,62,70–76]
ALDH1	Cytosolic aldehyde dehydrogenase enzyme	Tumor initiation, chemoresistance, spheroid formation	[46,53,61,75,77,78]
ROR1	Tyrosine kinase receptor	Spheroid formation, tumor initiation, proliferation	[79,80]
SOX2	Transcription factor	Stem cell maintenance, self-renewal	[8,81–84]
NANOG	Transcription factor	Stem cell maintenance, self-renewal, chemoresistance	[8,53,61,66,81–83]
POU5F1/OCT4	Transcription factor	Tumor initiation, chemoresistance	[8,53,61,81–83]
MYC	Transcription factor	Tumor initiation, chemoresistance	[85,86]
EpCAM	Cell surface membrane glycoprotein	Tumor initiation, spheroid formation, proliferation	[13,46,53,61,62]
MDR1/ABCB1	ATP binding cassette transporter	Chemoresistance	[46,49,53,61,66,87–91]
ABCG2	ATP binding cassette transporter	Chemoresistance	[46,49,53,61,87,88,90,91]

4.1. Side Population

One way in which ovarian CSCs are identified is by their ability to efflux DNA-binding dyes such as Hoechst 33342 and Rhodamine 123 resulting in a side population (SP) using flow cytometry. The ability to efflux these dyes identifies a CSC population that overexpress ATP binding cassette transporters such as MDR1/ABCB1 and ABCG2 that can efflux chemotherapeutic agents [46,49,61,87,88]. This SP demonstrates stem cell properties including the ability to repopulate tumors in an LDA and resistance to chemotherapy. Expression of ABCB1 and ABCG2 correlates with resistance to cisplatin and paclitaxel in ovarian cancer cell lines (2008, KF28, TU-OM-1, OVCAR3, SKOV3) and in cells from patient and mouse ascites [89–91]. However, the SP of cells is heterogeneous and can display different combinations of other stem cell markers, so it may be unknown which cells within this population is most "stem-like" or which population(s) are reconstituting the tumor [53].

4.2. Cell Surface Markers

Cell surface makers are essential in the identification of CSCs for multiple tumor types. When Bapat et al. first described ovarian CSCs, CD117 was demonstrated to be a cell surface marker for the ovarian CSCs [59]. Human serous ovarian cancer patient-derived xenografts (PDXs) showed that $CD117^+$ cells isolated from the xenografts were able to recapitulate a tumor with only 10,000 cells; this was a 100-fold increase in tumor initiating capability compared with the $CD117^-$ cells [68]. $CD117^+$ cells were also successful at generating tumors when serially transplanted [68]. Other ovarian CSC surface markers include CD24, CD44, EpCAM, and CD133 [13,46,53,61,62]. One of the most commonly reported ovarian CSC markers is CD133. CD133 expression correlates with poor prognosis in ovarian cancer and increased chemoresistance [70–72]. In cell lines, CD133 promotes a number of stem characteristics. $CD133^+$ and $CD133^-$ cells were single cell isolated and expanded from A2780 and PEO1 cell lines [73]. The $CD133^-$ cells only produced $CD133^-$ cells while $CD133^+$ cells divided asymmetrically to produce both $CD133^+$ and $CD133^-$ cells, suggesting that the $CD133^+$ cells retain stem cell properties [73]. $CD133^+$ cells exhibit increased resistance to cisplatin and were more tumorigenic in xenograft and serial transplantation studies [73,74]. Another one of the common CSC markers is CD44. CD44 is the hyaluronate receptor and is important in adhesion. In ovarian cancer, CD44 correlates with chemoresistance and tumor progression [63–65]. One function of CD44 is to activate Stat3 [66]. CD44 is commonly used as a stem cell marker in combination with CD117, MyD88, E-cadherin/CD34, and CD24/EpCAM. Each of these $CD44^+$ cell populations has been demonstrated to have stem-like properties (reviewed in Klemba et al.) [67]. In conclusion, there are multiple surface markers used to identify CSCs in ovarian cancer. Some investigations use these surface markers alone or in combination with other markers. However, we are still uncertain if there is a definitive ovarian CSC marker/population, if multiple CSC populations co-exist, or if CSC identity varies by patient.

4.3. ALDH Activity

In addition to cell surface markers, CSCs often are identified using the expression of the enzyme aldehyde dehydrogenase 1 (ALDH1) and its activity. The enzymatic activity of ALDH1 is used to identify and define CSCs in cancer types including breast, colon, liver, and ovarian [46]. Several studies suggest that ALDH1 expression correlates with poor prognosis. In one study of ovarian cancer patients, ALDH1A1 expression was found in 72.9% of tumors, and this expression correlated with decreased progression-free survival (6.05 vs. 13.81 months) [77]. A second study demonstrated that patients with high ALDH1 expression (by immunohistochemistry in >50% of the tumor section) exhibited poorer prognosis [78]. Cell lines with high ALDH1 exhibited increased chemoresistance and tumorgenicity [78]. Silva et al. examined 13 primary human ovarian tumors and 5 ascites samples for various putative CSC markers. ALDH1 was expressed in all cases [75]. Ovarian cancer cell lines were then examined for these CSC markers. Each of the cell lines examined (A2008, SKOV3, HEY-1, A2780, OVCAR8, OVCAR3, and OVCAR432) had a subpopulation of cells with ALDH1 expression [75]. Conversely, knockdown of ALDH1A1 in an orthotopic mouse model (from both taxane- and platinum-resistant cell lines) sensitized the tumors to treatment, resulting in reduced tumor growth [77]. The expression and activity of ALDH1 alone or in combination with cell surface stem cell markers is a popular and accepted method for identifying ovarian CSCs.

4.4. Transcription Factors

Pluripotency transcription factors necessary for normal stem cell maintenance are commonly expressed in ovarian CSCs [53,81–83]. In addition to being markers for ovarian CSCs, transcription factors such as OCT4, SOX2, and NANOG are expressed during development and are essential for normal stem cell maintenance and proliferation [62,66,84,92–95]. Aberrant expression of stem cell genes in differentiated cells, progenitor cells, or stem cell populations can lead to enhanced self-renewal and proliferative capability [96]. Expression of stem cell transcription factors not only provides evidence for the CSC model of tumor development, it also explains in part how stem cell properties of self-renewal and asymmetric division are maintained in CSCs. By comparing normal stem cell populations to CSCs we can gain insight into tumor initiation and regulation of the CSC phenotype. In embryonic stem cells (ESCs) the pluripotency transcription factors form a protein interaction network [83]. Many of these interactions are critical for stem cell functions. In addition, expression of pluripotency factors and protein–protein interactions are retained in CSCs. Among these factors is ARID3B. ARID3B and its paralog ARID3A are expressed in ESCs in a complex with NANOG, OCT4, and NAC1 [83]. ARID3B is overexpressed in serous ovarian cancer and its expression in the nucleus correlates with relapse following chemotherapy [58,97]. ARID3B increases expression of stem cell markers [76]. In particular, ARID3B induces expression of the stem cell marker Prom1 (CD133) [58]. ARID3B additionally increases the pool of CD133^{+} cells, suggesting that it has a role in promoting a stem cell phenotype [58,76]. In fact, ARID3A and ARID3B co-localize with CD133 in ovarian cancer tumor sections. Additionally, ARID3B is enriched in ovarian cancer ascites sorted for CD133^{+} cells (Figure 4). These data suggest that ARID3B^{+} cells are found in a stem cell niche (Figure 4). Future studies on pluripotency factors common in ovarian CSCs including OCT4, MYC, and ARID3B will provide clarity for how cancer stemness is maintained [85,86].

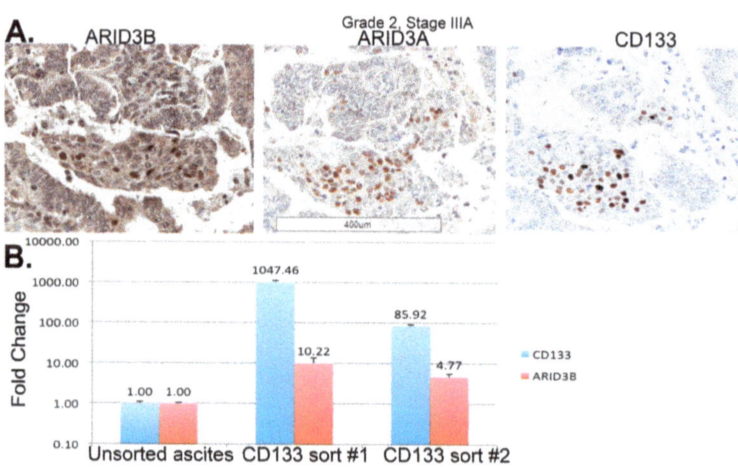

Figure 4. ARID3B expression correlates with CD133-stem cell niche. (**A**) IHC shows nuclear ARID3A and ARID3B co-localize with CD133+ regions in serial HGSOC sections. (**B**) HGSOC patient ascites was sorted for CD133+ cells. RT-qPCR was conducted for Prom1(CD133) and ARID3B on unsorted and independent sorts [98].

Different stem cell markers may confer different selective advantages to different pools of "CSCs". Patients may have more than one pool of stem cells and different patients may have CSCs with different phenotypes. An example is included in Figure 5. To enrich for CSCs, OVCA429 and Kuramochi cells were untreated or treated with cisplatin and paclitaxel and then cultured on nonadherent plates in stem cell media [56]. Flow cytometry was performed for CD117 (gene = CKIT) and CD133. OVCA429 cells have a clear CD117^{+}CD133^{-} population of CSCs that is enriched following chemotherapy treatment. Following chemotherapy treatment, multiple cell populations are expanded in Kuramochi cells including CD133^{+}/CD117^{-}, CD133^{+}/CD117^{+}, and CD117^{+}/CD133^{-}. These experiments suggest that different stem cell pools may be more prevalent in an individual cell type or patient tumor. Importantly, each of the CSC markers may have its own each unique function. The kinase activity of CD117 may provide a survival advantage over CD117^{-} cells [69]. However, CD133^{+} cells may have an adhesion or metastatic advantage over cells lacking CD133 [76]. Although we can detect cell-to-cell variation in the expression of markers, we do not know if these different CSC lineages arise from common progenitors. CSC lineage tracing to define the hierarchy of cells in a stem cell population has not been conducted for all putative ovarian CSC subtypes. Additionally, LDAs need to be conducted to verify stem cell potential for each putative ovarian CSC population. In order for studies of CSCs to be translational, we will need to define how the different CSC populations pertain to patient prognosis, relapse, and response to therapy. Moving forward, we need to establish the clinical significance of different ovarian CSC marker profiles [47,52,53,61,99]. Comparing survival and relapse potential for patients based on these different marker profiles is essential for us to develop effective treatments for the clinically relevant ovarian CSC populations.

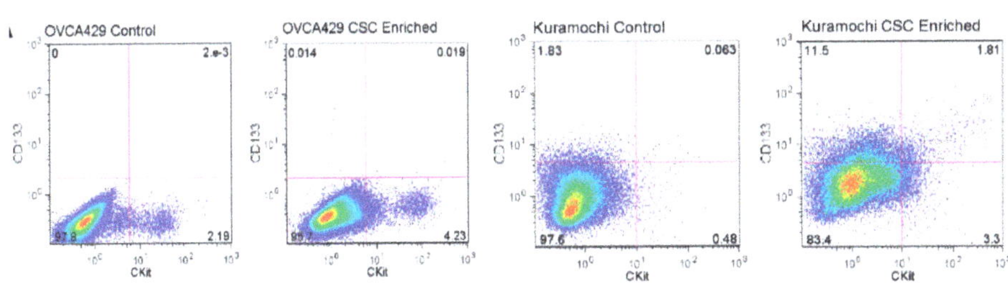

Figure 5. Flow cytometry for the stem cell markers CD117 and CD133 on ovarian cancer cells before and after CSC enrichment. Untreated OVCA429 and Kuramochi cells or cells enriched for CSCs (by treatment with cisplatin and paclitaxel followed by culturing CSCs in stem cell media on ultra low adhesion plates) [56] were stained for stem cell markers CD117 (cKIT is the gene that encodes CD117) (X-axis) and CD133 (Y-axis).

5. Pathways That Promote Stemness and Chemoresistance in HGSOC

We chose to focus on the major pathways that drive both stemness and chemoresistance in HGS ovarian cancer. These properties of highly metastatic HGS ovarian cancer are inextricably linked. Understanding the pathways that are most pertinent to metastatic HGS ovarian cancer will provide us with putative targets to develop efficacious therapeutic agents. As there are numerous pathways involved in stemness and chemoresistance, we will highlight the ones that have a clear role in ovarian cancer and are potentially targetable.

5.1. PI3K/PTEN/AKT Signaling

Aberrant PI3K/PTEN/AKT signaling often results from genomic alterations in many cancers including clear cell ovarian cancer. In HGS carcinoma, there are few mutations in the components of the PI3K/PTEN/AKT pathway, but by immunohistochemistry (IHC) about half of the HGS tumors have evidence of pathway activation [100,101]. A meta-analysis of the literature reports that both univariate and multivariate analysis show that high expression of activated AKT (pAKT) is associated with poor progression-free survival and poor overall survival [102]. Due to mutations in many parts of the PI3K/PTEN/AKT pathway, activated AKT signaling is highly relevant for ovarian cancer development and progression.

The PI3K/PTEN/AKT pathway is also implicated in ovarian CSCs. PI3K/PTEN/AKT signaling regulates enrichment of CSCs, maintenance of a CSC phenotype, and chemoresistance [103–106]. Spheroids derived from SKOV3 and HO8910 cell lines expressed elevated phosphorylated AKT1 and decreased expression of PTEN [103]. The spheroids exhibited increased resistance to paclitaxel [103]. Conversely, inhibiting AKT1 activation decreased spheroid formation and migration [104]. Knockdown of AKT1 via siRNA resulted in the loss of CSC marker expression (OCT4, SOX2, ALDH1, and ABCG2) as well as loss of spheroid formation and paclitaxel resistance [104]. These studies demonstrate the importance of the PI3K/PTEN/AKT pathway in CSC formation, maintenance, and chemoresistance to paclitaxel.

The PI3K/PTEN/AKT pathway also regulates cisplatin resistance in ovarian cancer. In cisplatin-resistant A2780 cells (A2780-CP), AKT regulates the expression of PPM1D [105]. PPM1D inhibits the DNA damage and apoptotic response after DNA damage occurs [105]. Downregulation of AKT activity results in loss of PPM1D stability and increases its degradation [105]. Loss of PPM1D increases the response of the A2780-CP cells to cisplatin [105].

The PI3K/PTEN/AKT signaling pathway promotes the enrichment of ovarian CSC populations and regulates ovarian CSC chemoresistance, thus making it an ideal target for therapeutics to eliminate ovarian CSCs. There are currently PI3K/PTEN/AKT inhibitors such as BKM120, Everdimus, and Perifosine that are being used to treat cancer patients [100]. Future efforts to stratify patients that are likely to benefit from PI3K/PTEN/AKT inhibition will be needed for this therapy to be effective in ovarian cancer patients.

5.2. *Jak2/STAT3*

Proliferation, survival, and differentiation are all regulated by the Jak2/STAT3 pathway in several solid tumors [107]. In ovarian cancer, the Jak/STAT pathway is constitutively active in most cases [108]. Jak/STAT is implicated for having a key role in the development of HGS ovarian cancer. Activation of STAT3 via phosphorylation at Tyr705 and the loss of the STAT3 inhibitor PIAS3 may serve as a tumor-initiating event in the distal fallopian tube for the formation of HGS ovarian cancer [109]. Phosphorylated STAT3 is expressed in 86% of ovarian tumors examined (from different histotypes) and constitutive pSTAT3 expression is expressed in 63% of the HGS tumors examined [110]. Phosphorylated, nuclear STAT3 is associated with poor prognosis [110]. In tissue microarrays (TMAs), patients whose tumors had high nuclear pSTAT3 staining (>10% nuclei stained) had poorer survival rates than women with low nuclear pSTAT3 staining (<10% nuclei stained) [110]. These patient findings implicate the Jak/STAT pathway as being highly important for ovarian cancer initiation and progression.

The Jak/STAT pathway also regulates ovarian CSCs. $CD24^+$ ovarian CSCs require Jak2/STAT3 signaling for growth and metastasis [111]. Primary tumors generated in the *Apc*$^-$; *Pten*$^-$; *Trp53*$^-$ (transgenic mouse model in which APC, PTEN, and Trp53 are conditionally deleted in the ovarian surface epithelium) were collected, dissociated, and sorted via fluorescence-activated cell sorting (FACS) using stem cell markers [111]. LDAs confirmed that the $CD24^+$ cells isolated were a CSC population [111]. This population of cells expressed elevated pSTAT3 and stem cell marker NANOG, which is required for stem cell renewal [111]. $CD24^+$ cells were injected into mice and the mice were then treated with cisplatin or with cisplatin+TG101209, a Jak2 inhibitor [111]. The mice treated with cisplatin+TG101209 showed significantly increased survival and almost no metastases (1 out of 14) [111].

Other studies show a role for the Jak/STAT pathway in ovarian CSC maintenance and chemoresistance. Abubaker et al. collected tumor cells from patient ascites or the HEY8 ovarian cancer cell line and treated them with paclitaxel [108]. Treatment with paclitaxel induced the expression of CSC markers CD117, OCT4, and EpCAM in ascites and HEY8 cells [108]. In both the paclitaxel-treated ascites and HEY8 cells, the Jak2/STAT3 pathway was activated [108]. This suggests that the Jak2/STAT3 pathway regulates the expression of stem-like genes necessary for CSC maintenance. Moreover, paclitaxel-treated cells were also treated with the Jak2-specific small molecule inhibitor (CYT387), which resulted in inhibition of the Jak2/STAT3 pathway activation, loss of stem cell marker expression, and increased sensitivity of the cells to paclitaxel treatment [108]. When paclitaxel-treated and paclitaxel+CYT387-treated cells were injected into mice, the mice injected with the paclitaxel+CYT387-treated cells showed a reduced tumor burden and enhanced sensitivity to paclitaxel [108]. These studies demonstrate that in models of ovarian cancer, Jak2 inhibitors are effective at reducing stem cell characteristics and inhibiting tumor growth. These inhibitors also increase survival and response to therapy. Because the Jak/STAT pathway promotes stemness and chemoresistance in the CSC population, it is a viable target for therapies aimed at reducing ovarian CSC populations.

5.3. NFκB

The NFκB pathway plays a role in normal cellular processes such as survival, proliferation, and apoptosis. In cancer the NFκB pathway is implicated in invasion and metastasis. However, the pathway is also involved in CSC maintenance [112]. In ovarian cancer, both the canonical and noncanonical NFκB pathways are active. A $CD44^+$ ovarian CSC population isolated from patient ascites exhibited constitutive NFκB pathway activation via a luciferase reporter assay, formed spheroids in culture, and formed tumors when injected into mice [13]. Another study showed that $CD44^+$ CSCs from SKOV3 cells (that also express NANOG, SOX2, and OCT4) exhibited increased expression of NFκB pathway members RelA, RelB, and IKKα [113]. Inhibition of the NFκB pathway with a dominant-negative form of IκBα resulted in a decrease in the $CD44^+$ CSC population with a reduction from 65.3% $CD44^+$ cells to just 27.7% [113]. These data suggest that NFκB signaling regulates expression of stemness genes.

The NFκB pathway is also involved in ovarian CSC chemoresistance. $CD44^+$ ovarian CSCs from patient ascites have constitutively active NFkB [13]. When treated with TNFα, the $CD44^+$ cells showed increased NFκB activity and cytokine production as well as resistance to TNFα-induced apoptosis [13]. The resistance to apoptotic pathway activation suggests a mechanism for ovarian CSC survival when treated with chemotherapeutics. Treatment of ovarian CSCs with Eriocalyxin B (EriB) inhibits the NFκB pathway and induces cell death in ovarian CSCs [114]. EriB inhibited the TNFα-induced NFκB activity and cytokine production and sensitized the cells to TNFα- and FasL-induced cell death [114]. This suggests that inhibition of the canonical NFκB pathway could sensitize ovarian CSCs to therapy [114].

While many studies focused on the canonical NFκB pathway, the noncanonical pathway is also active in promoting stemness and chemoresistance in ovarian cancer. RelB in particular is important for ovarian CSC regulation. RelB is overexpressed in ovarian CSC populations including $CD44^+$ SKOV3 cells and $ALDH^+/CD133^+$ OV90 and ACI23 cell lines [113,115]. In the OV90 and ACI23 cells, ALDH1 activity and expression of RelB both increase with carboplatin treatment [115]. This suggests a role for the noncanonical NFκB pathway and RelB in promoting stemness and chemoresistance. Knockdown of RelB with shRNA reduced the number of $ALDH^+/CD133^+$ CSCs in vitro in both cell lines and in xenografts by 50% [115]. The RelB knockdown decreased expression of other stem cell markers (NANOG and CD44) and increased sensitivity to carboplatin [115]. In addition, ACI23 and OV90 cells, when stably transfected with inducible shRNA for RelB, showed reduced spheroid formation and reduced tumorgenicity [115]. The noncanonical pathway through RelB promotes tumor growth as well as the expression of stemness genes [115]. RelB also regulates chemoresistance in ovarian CSCs [115]. Thus, both the canonical and noncanonical NFκB pathways are excellent targets for therapeutics to reduce the CSC population.

5.4. Notch

Notch signaling has a role in multiple cellular processes. Notch is a critical component in regulating progenitor cell maintenance, differentiation, cell proliferation, and apoptosis. Notch is also important for cell–cell communication [116,117]. In HGS ovarian cancer, Notch3 expression is amplified/overexpressed [118]. By analyzing 31 fresh HGS ovarian cancer samples, Notch3 amplification correlated with protein expression [118]. Notch3 was overexpressed more often in high grade tumors (66%) than in low grade tumors (33%) [118]. Further, according to The Cancer Genome Atlas (TCGA), Notch3 is amplified in 17% of HGS tumors. The most highly expressed Notch3 ligand in ovarian serous carcinoma is Jagged 1, which is predominantly expressed in the mesothelial cells within the tumor microenvironment, suggesting a role for Notch3/Jagged 1 signaling in cell adhesion and proliferation [119].

In the majority of patients with recurrent HGS ovarian cancer, Notch3 is overexpressed [120]. Tumors from patients with either primary disease or recurrent disease were examined for Notch3 overexpression and survival [120]. In the group with primary disease, there was no difference in survival between those with Notch3 overexpression and those without [120]. Those in the group with recurrent disease did show a difference. Those expressing high Notch3 levels had decreased overall survival (22 vs. 37 months) and decreased progression-free survival (3 vs. 8 months) suggesting that Notch3 expression is a factor in the recurrence of ovarian cancer as well as a prognostic indicator in recurrent disease [120].

Chemoresistance is a hallmark of CSCs and disease recurrence/relapse, and Notch3 expression affects the expression of stemness factors as well as chemoresistance. The transcription factor OCT4 promotes self-renewal of ovarian CSCs while SOX2 is required for their maintenance [84,92]. Overexpression of Notch3 in ovarian cancer cell lines (IOSE-80pc and MPSC1) enhances expression of stem cell markers (NANOG, OCT4, and SOX2) and increases expression of the ABCB1 transporter protein [120]. The ABCB1 transporter increases chemoresistance in these ovarian CSCs and NANOG promotes the epithelial to mesenchymal transition (EMT) in ovarian cancer [121]. To demonstrate the role of Notch3 on chemoresistance, Nocth3 was knocked down in OVCAR3 cells using shRNA resulting in reduced IC_{50} compared to control cells [120]. These studies all implicate Notch3 signaling in ovarian CSC chemoresistance.

Other Notch signaling molecules are also implicated in stemness and chemoresistance including Jagged 1 and downstream signaling molecules. Downregulation of Jagged 1 in SKOV3TRip2 cells via siRNA increased sensitivity of cells to docetaxel [122]. In ovarian cancer cells isolated for the SP, Notch pathway genes (FPTG, ST3GAL6, and ADAM19), stem cell markers NANOG and OCT4, and three ABC transporter genes (ABCG2 [both lines], ABCC4 [SKOV3 only], and ABCB1 [A224 only]) were induced [95]. Collectively, the data suggest that Notch signaling is involved in promoting stemness and chemoresistance, and expression of Notch3 in particular may serve as a prognostic indicator for patients with recurrent disease. Notch signaling is an attractive target for therapeutics aimed at ovarian CSCs. Currently, there are experimental γ-secretase inhibitors, γ-secretase modifiers, Notch soluble decoys, and negative regulatory region monoclonal antibodies that are already being developed [116].

5.5. Wnt

Wnt signaling is particularly important during development where it regulates cell fate determination during embryogenesis including the cardiovascular system, central nervous system, and craniofacial development [116,123]. In adults, Wnt signaling is critical for self-renewal in tissues (e.g., bone growth plate, hair follicles, colon, etc.) [116,124,125]. The major processes regulated by noncanonical Wnt signaling include cell polarity and motility; however, Wnt also plays a role in maintaining stem cells, quiescence, and chemoresistance [126]. Wnt signaling is complex and many components of Wnt signaling are implicated in ovarian CSCs and chemoresistance (Figure 6).

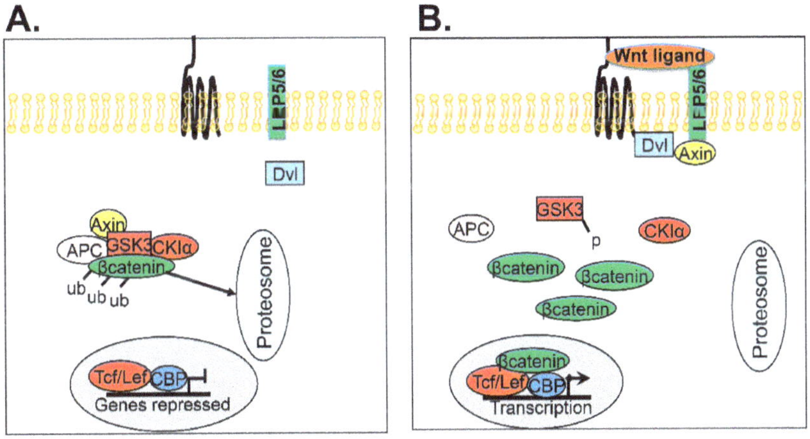

Figure 6. Wnt Signaling Cascade. (**A**) Basal state without the presence of Wnt ligand activation. β-catenin is ubiquitinated and sent to the proteosome for destruction. (**B**) Activation of the Wnt pathway via binding of a Wnt ligand to the Frizzled receptor and LRP5/6 resulting in recruitment of Disheveled (Dvl) and axin to the cell membrane. β-catenin is released from the destruction complex and translocates to the nucleus to act as a co-transcription factor.

With regards to ovarian cancer, Wnt signaling is involved in normal development of the ovarian and fallopian tube stem cells. Wnt signaling also has functions in tumor development. LGR5 is a stem cell marker for ovarian stem cells and LGR6 is a stem cell marker for the fallopian tube, and expression of either one is a sign of elevated Wnt signaling [127–129]. LGR5 and LGR6 are expressed in HGS tumors [127]. LGR5$^+$ cell-driven lineage tracing was performed in mice, illustrating the importance of LGR5 and Wnt signaling in embryonic and adult ovarian stem cells for homeostasis and regenerative repair and self-renewal [130]. Since the fimbria of the fallopian tube are implicated as a site of origin in HGS tumors, fallopian tube stems cells also must be examined [129]. Using a Tcf-eGFP reporter and confocal microscopy on fallopian tube organoid cultures, active Wnt signaling was needed for the expression of stem cell factors to support organoid growth [129]. Understanding how abnormal regulation of Wnt signaling drives initiation or maintenance of ovarian CSCs is critical.

Disregulation of Wnt signaling is frequently involved in the development of cancer [123,131]. In ovarian cancer, aberrant Wnt signaling differs by histotype. Wnt signaling stabilization and subsequent nuclear translocation of β-catenin leads to activation of Wnt target genes including those involved in stemness. β-catenin is frequently mutated at GSK3β phosphorylation sites that allow β-catenin to be ubiquinated and degraded in the absence of Wnt signaling (54%) resulting in nuclear localization in approximately 70% of cases of low grade endometrioid ovarian carcinomas [132]. Activating mutations of proteins in the Wnt pathway are rare in serous ovarian carcinomas [132]. However, there is evidence of nuclear β-catenin in HGS [132]. With regards to the noncanonical Wnt pathway, Wnt5A was highly expressed in a collection of 583 ovarian tumors and it is found in the ascites [126,132]. Receptor tyrosine kinase-like orphan receptor 1 (ROR1) (a pseudokinase and receptor for Wnt5A) is expressed in ovarian cancer and is correlated with poor outcomes [79]. Survival analysis showed that patients with high expression of ROR1 had significantly reduced progression-free survival and overall survival [79]. Cells isolated from ROR1$^+$ patient-derived xenografts exhibited stem-like qualities including ALDH1 expression, ability to form spheroids, and increased tumorgenicity [80]. These data suggest that ROR1 is a potential CSC marker for ovarian cancer and that noncanonical Wnt signaling is a component of ovarian cancer stemness.

In ovarian CSCs, Wnt signaling helps promote both stemness and chemoresistance. The CSC marker/receptor tyrosine kinase, CD117, is upregulated in ovarian CSCs. Many factors contribute to acquisition of CD117 expression including the hypoxic microenvironment of the stem cell niche [106]. CD117 leads to activation of AKT and the phosphorylation of GSK3β and nuclear expression of β-catenin [106]. β-catenin activity induces expression of ABCG2, a drug transporter which increases cisplatin and paclitaxel resistance [106]. Therefore, the hypoxic niche supports stemness by activation of Wnt target genes.

Wnt signaling in ovarian cancer CSCs is complex. Collectively, the patient studies combined with cell culture and animal models suggest that multiple Wnt signaling pathways contribute to stemness and chemoresistance in ovarian cancer. A number of potential molecules in the Wnt pathways may be viable targets for therapeutic intervention. Wnt inhibitors such as compounds that target Disheveled (NSC668036 and FJ9), Frizzled receptor antibody, Thiazolidinedione (target β-catenin reverse transport), and Sulindac (unknown action but potentially effects β-catenin proteasomal degradation) are being examined for use in cancer treatment [116]. Deciphering the cross-talk between Wnt and other pathways in addition to more sophisticated assessment of the contribution of particular Wnt molecules and pathways will enable development of future Wnt-targeted drugs that can be used in ovarian cancer treatment.

5.6. Hedgehog

During embryogenesis, Hedgehog signaling (Hh) regulates tissue polarity as well as patterning and stem cell maintenance [116]. In cancer, the Hh pathway is dysregulated in one of two ways: (1) constitutive expression of endogenous ligand (e.g., Sonic hedgehog [Shh]) or (2) mutations of proteins within the pathway (Patched, SMO, SUFU) [133]. We will explore the ways Hedgehog signaling has emerged as an important regulator of proliferation, chemoresistance, and stemness in ovarian cancer [133,134].

Overexpression of Gli1 (a transcription factor activated by Hh signaling) as well as PTCH (Hh receptor) is correlated with poor prognosis and survival in patients [133]. Eighty cases of epithelial ovarian tumor were examined by IHC [133]. All cases expressed PTCH, though PTCH was highly expressed in 34.1% of cases [133]. Gli1 expression varied by histotype of the tumor with high Gli1 expression being most common in serous tumors [133]. High expression of either Gli1 or PTCH correlated with poor survival compared to those patients with low expression [133]. These data suggest that Gli1 and/or PTCH expression may be prognostic indicators for ovarian cancer patients. Gli1 antagonists such as HPI 1–4 that are currently being developed as well as drugs targeting PTCH may be useful therapies for ovarian cancer patients with activated Hh signaling.

In ovarian cancer, Gli1 appears to be a critical contributor. Gli1 is a regulator of proliferation and tumor growth in ovarian cancer. Gli1 is elevated in several ovarian cancer cell lines (OVCAR5, OV-202, and OV-167) compared with normal ovarian surface epithelium [135]. Inhibition of the Hh pathway with cyclopamine resulted in Gli1 decreasing in a dose-dependent manner (60–80%) [135]. The decrease in Gli1 mRNA and protein correlated with a decrease in proliferation in all three cancer lines [135]. In addition to the in vitro results, a mouse xenograft model using OVCAR5 cells found that cyclopamine significantly inhibited tumor growth [135]. In agreement with these findings, exogenous expression of Gli1 in ovarian cancer cell lines SKOV3, OVCAR3, and OVCA433 increased cell proliferation 2-fold and increased invasiveness 200–500% over control; whereas knockdown of Gli1 with siRNA suppressed proliferation and invasiveness (40–60%) [133]. These studies suggest that Gli1 is an important regulator of proliferation and tumor growth in ovarian cancer.

The Hh pathway regulates stemness in ovarian cancer. In one study, ES2, SKOV3, and TOV112D cells were treated with recombinant Shh and Ihh, both Hh pathway agonists [134]. In all three cell lines, spheroid formation increased significantly [134]. When treated with cyclopamine, there was significant impairment of spheroid formation [134]. This demonstrates a role for the Hh pathway in maintaining stemness in ovarian cancer.

Gli1 also is implicated in chemoresistance in ovarian cancer cells. Gli1 has an interesting role in the DNA damage response following cisplatin treatment [136]. In cisplatin-resistant A2780 cells

(A2780-CP), cells with anti-Gli1 shRNA or a scrambled shRNA were treated with cisplatin and then DNA repair was assessed [136]. After 12 h the control cells had repaired 78% of the DNA adducts compared to 33% in cells treated with anti-Gli1 shRNA [136]. In addition to impairing the cell's ability to repair the cisplatin adducts, pretreatment with the anti-Gli1 shRNA sensitized the cells to cisplatin resulting, in a shift of the IC_{50} from 30 μM to 5 μM [136]. This suggests that Gli1 regulates DNA adduct repair and sensitivity to cisplatin in ovarian cancer. Additionally, Gli1, SMO, and PTCH are overexpressed in borderline and malignant ovarian cancer [137]. Moreover, Gli1 and SMO were highly overexpressed in platinum-resistant ovarian cancer [137]. Both cell culture and patient studies suggest an important role for Gli1 and Hh signaling in ovarian cancer chemoresistance.

While Hh signaling is studied in regard to other cancer types, Hh signaling in ovarian cancer is relatively understudied. Current findings suggest that Gli1 has an important role in ovarian cancer stemness, tumorigenicity, and chemoresistance. Further studies on the role of Hh signaling in ovarian cancer will allow for personalized medicine approaches for those patients with active Hh. Future therapy options could include the Hh inhibitor GDC-0449 that is currently in clinical trials for use in ovarian cancer [138].

5.7. Developing Therapeutics Targeting Ovarian Cancer Stem Cells

There are multiple pathways involved in promoting a stem cell phenotype and chemoresistance in ovarian cancer. Each pathway has the potential to be therapeutically targeted. However, a major challenge is defining which population of cells needs to be targeted with pathway inhibitors.

If a therapeutic goal is to eliminate the CSC population, more studies are needed to define CSC populations, markers, and critical pathways that are required for stem cell maintenance (Table 2: Summary of targetable genes).

Table 2. Summary of targetable genes.

Pathway	Gene	Potential Therapeutics in Trials
PI3K/PTEN/AKT	AKT1 PTEN PPMID	BKM120, Everdimus, Perifosine
Jak/STAT	STAT3 JAK2	
NFκB	RelA RelB IKK IκBα TNFα	
Notch	Notch3 Jagged1	γ-secretase inhibitors, γ-secretase modifiers, Notch soluble decoys, negative regulatory region monoclonal antibodies
Wnt	β-catenin Wnt5A Disheveled Frizzled	NSC668036, FJ9, Frizzled receptor antibodies, Thiazoldinedone, Suldinac
Hedgehog	Patched Gli1	HPI 1, HPI 2, HPI 3, HPI 4, GDC-0449

6. Future Studies

Ovarian CSCs in HGS ovarian cancer are an attractive target for therapeutics in order to prevent relapse following chemotherapy. Prior to targeting these insidious cells, a number of issues should be considered. One complication in treating patients with HGS ovarian cancer is the amount of heterogeneity found within the tumors. Additionally, HGS is characterized by genomic instability rather than specific driving mutations. This level of heterogeneity makes identifying drug targets that help a wide population of HGS ovarian cancer patients difficult. More phenotypic, genetic, and epigenetic studies of patient CSCs need to be conducted to assess which CSC populations are the most critical ones to target. Hierarchical lineage tracing efforts will allow us to decipher if different CSC populations arise from a common progenitor cell. Detailing the mechanisms that are required for CSC maintenance is critical. Delineating the role of the microenvironment in CSC maintenance is also important. Do these varying marker profiles denote differing niches for the CSCs and, therefore, different survival and renewal pathways that are active in different populations of CSCs? Are different CSC subpopulations present at different times during cancer progression? These questions underscore the need for personalized medicine in the treatment of ovarian cancer. Three potential targets for new therapeutics include stem cell markers, stem cell signaling pathways needed for renewal and/or survival, and the stem cell niche. Careful studies examining the contribution of CSC subpopulations and signaling pathways to CSC survival and maintenance will lead to directed therapeutic target design.

Author Contributions: L.R. contributed to the conceptualization, data curation, writing—original draft preparation, and writing—review and editing. K.D.C.D. contributed to conceptualization and writing—review and editing.

Funding: This research received no external funding.

Acknowledgments: We would like to acknowledge Elizabeth Fisher for her help in gathering sources, Paige Dausinas for proofreading, and Jeffrey Kurkewich for the flow cytometry data. We would also like to acknowledge the Department of Defense Ovarian Cancer Research Program (W81XWH-15-1-0071) and the Indiana CTSI Research Enhancement Grant for their support of KDCD.

Conflicts of Interest: The authors declare no conflict of interest.

References

1. Schorge, J.O.; McCann, C.; Del Carmen, M.G. Surgical debulking of ovarian cancer: What difference does it make? *Rev. Obstet. Gynecol.* **2010**, *3*, 111–117. [PubMed]
2. Lalwani, N.; Prasad, S.R.; Vikram, R.; Shanbhogue, A.K.; Huettner, P.C.; Fasih, N. Histologic, molecular, and cytogenetic features of ovarian cancers: Implications for diagnosis and treatment. *Radiographics* **2011**, *31*, 625–646. [CrossRef] [PubMed]
3. Kroeger, P.T., Jr.; Drapkin, R. Pathogenesis and heterogeneity of ovarian cancer. *Curr. Opin. Obstet. Gynecol.* **2017**, *29*, 26–34. [CrossRef] [PubMed]
4. Kehoe, S.; Hook, J.; Nankivell, M.; Jayson, G.C.; Kitchener, H.; Lopes, T.; Luesley, D.; Perren, T.; Bannoo, S.; Mascarenhas, M.; et al. Primary chemotherapy versus primary surgery for newly diagnosed advanced ovarian cancer (chorus): An open-label, randomised, controlled, non-inferiority trial. *Lancet* **2015**, *386*, 249–257. [CrossRef]
5. Rojas, V.; Hirshfield, K.M.; Ganesan, S.; Rodriguez-Rodriguez, L. Molecular characterization of epithelial ovarian cancer: Implications for diagnosis and treatment. *Int. J. Mol. Sci.* **2016**, *17*, 2113. [CrossRef] [PubMed]
6. Koshiyama, M.; Matsumura, N.; Konishi, I. Recent concepts of ovarian carcinogenesis: Type I and type II. *Biomed. Res. Int.* **2014**, *2014*. [CrossRef] [PubMed]
7. Kommoss, S.; Gilks, C.B.; du Bois, A.; Kommoss, F. Ovarian carcinoma diagnosis: The clinical impact of 15 years of change. *Br. J. Cancer* **2016**, *115*, 993–999. [CrossRef] [PubMed]
8. Ottevanger, P.B. Ovarian cancer stem cells more questions than answers. *Semin. Cancer Biol.* **2017**, *44*, 67–71. [CrossRef] [PubMed]
9. Ozols, R.F. Treatment goals in ovarian cancer. *Int. J. Gynecol. Cancer* **2005**, *15*, 3–11. [CrossRef] [PubMed]
10. Gilks, C.B.; Prat, J. Ovarian carcinoma pathology and genetics: Recent advances. *Hum. Pathol.* **2009**, *40*, 1213–1223. [CrossRef] [PubMed]
11. Zhang, S.; Balch, C.; Chan, M.W.; Lai, H.C.; Matei, D.; Schilder, J.M.; Yan, P.S.; Huang, T.H.; Nephew, K.P. Identification and characterization of ovarian cancer-initiating cells from primary human tumors. *Cancer Res.* **2008**, *68*, 4311–4320. [CrossRef] [PubMed]

12. Steg, A.D.; Bevis, K.S.; Katre, A.A.; Ziebarth, A.; Dobbin, Z.C.; Alvarez, R.D.; Zhang, K.; Conner, M.; Landen, C.N. Stem cell pathways contribute to clinical chemoresistance in ovarian cancer. *Clin. Cancer Res.* **2012**, *18*, 869–881. [CrossRef] [PubMed]
13. Alvero, A.B.; Chen, R.; Fu, H.H.; Montagna, M.; Schwartz, P.E.; Rutherford, T.; Silasi, D.A.; Steffensen, K.D.; Waldstrom, M.; Visintin, I.; et al. Molecular phenotyping of human ovarian cancer stem cells unravels the mechanisms for repair and chemoresistance. *Cell Cycle* **2009**, *8*, 158–166. [CrossRef] [PubMed]
14. Devouassoux-Shisheboran, M.; Genestie, C. Pathobiology of ovarian carcinomas. *Chin. J. Cancer* **2015**, *34*, 50–55. [CrossRef] [PubMed]
15. Ramalingam, P. Morphologic, immunophenotypic, and molecular features of epithelial ovarian cancer. *Oncology* **2016**, *30*, 166–176. [PubMed]
16. Kaldawy, A.; Segev, Y.; Lavie, O.; Auslender, R.; Sopik, V.; Narod, S.A. Low-grade serous ovarian cancer: A review. *Gynecol. Oncol.* **2016**, *143*, 433–438. [CrossRef] [PubMed]
17. Diaz-Padilla, I.; Malpica, A.L.; Minig, L.; Chiva, L.M.; Gershenson, D.M.; Gonzalez-Martin, A. Ovarian low-grade serous carcinoma: A comprehensive update. *Gynecol. Oncol.* **2012**, *126*, 279–285. [CrossRef] [PubMed]
18. McCluggage, W.G. Morphological subtypes of ovarian carcinoma: A review with emphasis on new developments and pathogenesis. *Pathology* **2011**, *43*, 420–432. [CrossRef] [PubMed]
19. Malpica, A.; Deavers, M.T.; Lu, K.; Bodurka, D.C.; Atkinson, E.N.; Gershenson, D.M.; Silva, E.G. Grading ovarian serous carcinoma using a two-tier system. *Am. J. Surg. Pathol.* **2004**, *28*, 496–504. [CrossRef] [PubMed]
20. Kurman, R.J.; Shih, I.M. The dualistic model of ovarian carcinogenesis: Revisited, revised, and expanded. *Am. J. Pathol.* **2016**, *186*, 733–747. [CrossRef] [PubMed]
21. Moss, E.L.; Evans, T.; Pearmain, P.; Askew, S.; Singh, K.; Chan, K.K.; Ganesan, R.; Hirschowitz, L. Should all cases of high-grade serous ovarian, tubal, and primary peritoneal carcinomas be reclassified as tubo-ovarian serous carcinoma? *Int. J. Gynecol. Cancer* **2015**, *25*, 1201–1207. [CrossRef] [PubMed]
22. Soslow, R.A. Histologic subtypes of ovarian carcinoma: An overview. *Int. J. Gynecol. Pathol.* **2008**, *27*, 161–174. [CrossRef] [PubMed]
23. O'Neill, C.J.; McBride, H.A.; Connolly, L.E.; Deavers, M.T.; Malpica, A.; McCluggage, W.G. High-grade ovarian serous carcinoma exhibits significantly higher p16 expression than low-grade serous carcinoma and serous borderline tumour. *Histopathology* **2007**, *50*, 773–779. [CrossRef] [PubMed]
24. Jones, S.; Wang, T.L.; Shih, I.M.; Mao, T.L.; Nakayama, K.; Roden, R.; Glas, R.; Slamon, D.; Diaz, L.A., Jr.; Vogelstein, B.; et al. Frequent mutations of chromatin remodeling gene arid1a in ovarian clear cell carcinoma. *Science* **2010**, *330*, 228–231. [CrossRef] [PubMed]
25. Huang, H.N.; Lin, M.C.; Huang, W.C.; Chiang, Y.C.; Kuo, K.T. Loss of arid1a expression and its relationship with PI3K-Akt pathway alterations and ZNF217 amplification in ovarian clear cell carcinoma. *Mod. Pathol.* **2014**, *27*, 983–990. [CrossRef] [PubMed]
26. Gershenson, D.M.; Sun, C.C.; Bodurka, D.; Coleman, R.L.; Lu, K.H.; Sood, A.K.; Deavers, M.; Malpica, A.L.; Kavanagh, J.J. Recurrent low-grade serous ovarian carcinoma is relatively chemoresistant. *Gynecol. Oncol.* **2009**, *114*, 48–52. [CrossRef] [PubMed]
27. Schmeler, K.M.; Sun, C.C.; Bodurka, D.C.; Deavers, M.T.; Malpica, A.; Coleman, R.L.; Ramirez, P.T.; Gershenson, D.M. Neoadjuvant chemotherapy for low-grade serous carcinoma of the ovary or peritoneum. *Gynecol. Oncol.* **2008**, *108*, 510–514. [CrossRef] [PubMed]
28. Santillan, A.; Kim, Y.W.; Zahurak, M.L.; Gardner, G.J.; Giuntoli, R.L., 2nd; Shih, I.M.; Bristow, R.E. Differences of chemoresistance assay between invasive micropapillary/low-grade serous ovarian carcinoma and high-grade serous ovarian carcinoma. *Int. J. Gynecol. Cancer* **2007**, *17*, 601–606. [CrossRef] [PubMed]
29. Ducie, J.; Dao, F.; Considine, M.; Olvera, N.; Shaw, P.A.; Kurman, R.J.; Shih, I.M.; Soslow, R.A.; Cope, L.; Levine, D.A. Molecular analysis of high-grade serous ovarian carcinoma with and without associated serous tubal intra-epithelial carcinoma. *Nat. Commun.* **2017**, *8*, 990. [CrossRef] [PubMed]
30. Kessler, M.; Fotopoulou, C.; Meyer, T. The molecular fingerprint of high grade serous ovarian cancer reflects its fallopian tube origin. *Int. J. Mol. Sci.* **2013**, *14*, 6571–6596. [CrossRef] [PubMed]
31. Perets, R.; Wyant, G.A.; Muto, K.W.; Bijron, J.G.; Poole, B.B.; Chin, K.T.; Chen, J.Y.; Ohman, A.W.; Stepule, C.D.; Kwak, S.; et al. Transformation of the fallopian tube secretory epithelium leads to high-grade serous ovarian cancer in BRCA; TP53; PTEN models. *Cancer Cell* **2013**, *24*, 751–765. [CrossRef] [PubMed]

32. Przybycin, C.G.; Kurman, R.J.; Ronnett, B.M.; Shih, I.M.; Vang, R. Are all pelvic (nonuterine) serous carcinomas of tubal origin? *Am. J. Surg. Pathol.* **2010**, *34*, 1407–1416. [CrossRef] [PubMed]
33. Kim, J.; Coffey, D.M.; Creighton, C.J.; Yu, Z.; Hawkins, S.M.; Matzuk, M.M. High-grade serous ovarian cancer arises from fallopian tube in a mouse model. *Proc. Natl. Acad. Sci. USA* **2012**, *109*, 3921–3926. [CrossRef] [PubMed]
34. Russo, A.; Czarnecki, A.A.; Dean, M.; Modi, D.A.; Lantvit, D.D.; Hardy, L.; Baligod, S.; Davis, D.A.; Wei, J.J.; Burdette, J.E. Pten loss in the fallopian tube induces hyperplasia and ovarian tumor formation. *Oncogene* **2018**, *37*, 1976–1990. [CrossRef] [PubMed]
35. Kim, J.; Coffey, D.M.; Ma, L.; Matzuk, M.M. The ovary is an alternative site of origin for high-grade serous ovarian cancer in mice. *Endocrinology* **2015**, *156*, 1975–1981. [CrossRef] [PubMed]
36. Ahmed, A.A.; Etemadmoghadam, D.; Temple, J.; Lynch, A.G.; Riad, M.; Sharma, R.; Stewart, C.; Fereday, S.; Caldas, C.; Defazio, A.; et al. Driver mutations in TP53 are ubiquitous in high grade serous carcinoma of the ovary. *J. Pathol.* **2010**, *221*, 49–56. [CrossRef] [PubMed]
37. Mullany, L.K.; Wong, K.K.; Marciano, D.C.; Katsonis, P.; King-Crane, E.R.; Ren, Y.A.; Lichtarge, O.; Richards, J.S. Specific TP53 mutants overrepresented in ovarian cancer impact CNV, TP53 activity, responses to nutlin-3a, and cell survival. *Neoplasia* **2015**, *17*, 789–803. [CrossRef] [PubMed]
38. Merajver, S.D.; Pham, T.M.; Caduff, R.F.; Chen, M.; Poy, E.L.; Cooney, K.A.; Weber, B.L.; Collins, F.S.; Johnston, C.; Frank, T.S. Somatic mutations in the BRCA1 gene in sporadic ovarian tumours. *Nat. Genet.* **1995**, *9*, 439–443. [CrossRef] [PubMed]
39. Berchuck, A.; Heron, K.A.; Carney, M.E.; Lancaster, J.M.; Fraser, E.G.; Vinson, V.L.; Deffenbaugh, A.M.; Miron, A.; Marks, J.R.; Futreal, P.A.; et al. Frequency of germline and somatic BRCA1 mutations in ovarian cancer. *Clin. Cancer Res.* **1998**, *4*, 2433–2437. [PubMed]
40. Hennessy, B.T.; Timms, K.M.; Carey, M.S.; Gutin, A.; Meyer, L.A.; Flake, D.D., 2nd; Abkevich, V.; Potter, J.; Pruss, D.; Glenn, P.; et al. Somatic mutations in BRCA1 and BRCA2 could expand the number of patients that benefit from poly (ADP ribose) polymerase inhibitors in ovarian cancer. *J. Clin. Oncol.* **2010**, *28*, 3570–3576. [CrossRef] [PubMed]
41. Foster, K.A.; Harrington, P.; Kerr, J.; Russell, P.; DiCioccio, R.A.; Scott, I.V.; Jacobs, I.; Chenevix-Trench, G.; Ponder, B.A.; Gayther, S.A. Somatic and germline mutations of the BRCA2 gene in sporadic ovarian cancer. *Cancer Res.* **1996**, *56*, 3622–3625. [PubMed]
42. Hilton, J.L.; Geisler, J.P.; Rathe, J.A.; Hattermann-Zogg, M.A.; DeYoung, B.; Buller, R.E. Inactivation of BRCA1 and BRCA2 in ovarian cancer. *J. Natl. Cancer Inst.* **2002**, *94*, 1396–1406. [CrossRef] [PubMed]
43. Geisler, J.P.; Hatterman-Zogg, M.A.; Rathe, J.A.; Buller, R.E. Frequency of BRCA1 dysfunction in ovarian cancer. *J. Natl. Cancer Inst.* **2002**, *94*, 61–67. [CrossRef] [PubMed]
44. Rzepecka, I.K.; Szafron, L.; Stys, A.; Bujko, M.; Plisiecka-Halasa, J.; Madry, R.; Osuch, B.; Markowska, J.; Bidzinski, M.; Kupryjanczyk, J. High frequency of allelic loss at the BRCA1 locus in ovarian cancers: Clinicopathologic and molecular associations. *Cancer Genet.* **2012**, *205*, 94–100. [CrossRef] [PubMed]
45. Kulkarni-Datar, K.; Orsulic, S.; Foster, R.; Rueda, B.R. Ovarian tumor initiating cell populations persist following paclitaxel and carboplatin chemotherapy treatment in vivo. *Cancer Lett.* **2013**, *339*, 237–246. [CrossRef] [PubMed]
46. Foster, R.; Buckanovich, R.J.; Rueda, B.R. Ovarian cancer stem cells: Working towards the root of stemness. *Cancer Lett.* **2013**, *338*, 147–157. [CrossRef] [PubMed]
47. Clevers, H. The cancer stem cell: Premises, promises and challenges. *Nat. Med.* **2011**, *17*, 313–319. [CrossRef] [PubMed]
48. Nowell, P.C. The clonal evolution of tumor cell populations. *Science* **1976**, *194*, 23–28. [CrossRef] [PubMed]
49. Prieto-Vila, M.; Takahashi, R.U.; Usuba, W.; Kohama, I.; Ochiya, T. Drug resistance driven by cancer stem cells and their niche. *Int. J. Mol. Sci.* **2017**, *18*, 2574. [CrossRef] [PubMed]
50. Abdullah, L.N.; Chow, E.K. Mechanisms of chemoresistance in cancer stem cells. *Clin. Transl. Med.* **2013**, *2*, 3. [CrossRef] [PubMed]
51. Burgos-Ojeda, D.; Rueda, B.R.; Buckanovich, R.J. Ovarian cancer stem cell markers: Prognostic and therapeutic implications. *Cancer Lett.* **2012**, *322*, 1–7. [CrossRef] [PubMed]
52. Visvader, J.E.; Lindeman, G.J. Cancer stem cells in solid tumours: Accumulating evidence and unresolved questions. *Nat. Rev. Cancer* **2008**, *8*, 755–768. [CrossRef] [PubMed]

53. Ahmed, N.; Abubaker, K.; Findlay, J.; Quinn, M. Cancerous ovarian stem cells: Obscure targets for therapy but relevant to chemoresistance. *J. Cell. Biochem.* **2013**, *114*, 21–34. [CrossRef] [PubMed]
54. Borovski, T.; De Sousa, E.M.F.; Vermeulen, L.; Medema, J.P. Cancer stem cell niche: The place to be. *Cancer Res.* **2011**, *71*, 634–639. [CrossRef] [PubMed]
55. Starbuck, K.; Al-Alem, L.; Eavarone, D.A.; Hernandez, S.F.; Bellio, C.; Prendergast, J.M.; Stein, J.; Dransfield, D.T.; Zarrella, B.; Growdon, W.B.; et al. Treatment of ovarian cancer by targeting the tumor stem cell-associated carbohydrate antigen, sialyl-thomsen-nouveau. *Oncotarget* **2018**, *9*, 23289–23305. [CrossRef] [PubMed]
56. Cole, J.M.; Joseph, S.; Sudhahar, C.G.; Cowden Dahl, K.D. Enrichment for chemoresistant ovarian cancer stem cells from human cell lines. *J. Vis. Exp.* **2014**, 51891. [CrossRef] [PubMed]
57. Ma, L.; Lai, D.; Liu, T.; Cheng, W.; Guo, L. Cancer stem-like cells can be isolated with drug selection in human ovarian cancer cell line SKOV3. *Acta Biochim. Biophys. Sin.* **2010**, *42*, 593–602. [CrossRef] [PubMed]
58. Roy, L.; Samyesudhas, S.J.; Carrasco, M.; Li, J.; Joseph, S.; Dahl, R.; Cowden Dahl, K.D. Arid3b increases ovarian tumor burden and is associated with a cancer stem cell gene signature. *Oncotarget* **2014**, *5*, 8355–8366. [CrossRef] [PubMed]
59. Bapat, S.A.; Mali, A.M.; Koppikar, C.B.; Kurrey, N.K. Stem and progenitor-like cells contribute to the aggressive behavior of human epithelial ovarian cancer. *Cancer Res.* **2005**, *65*, 3025–3029. [CrossRef] [PubMed]
60. Agro, L.; O'Brien, C. In vitro and in vivo limiting dilution assay for colorectal cancer. *Bio-protocol* **2015**, *5*, 1–11. [CrossRef] [PubMed]
61. Ahmed, N.; Abubaker, K.; Findlay, J.K. Ovarian cancer stem cells: Molecular concepts and relevance as therapeutic targets. *Mol. Aspects Med.* **2014**, *39*, 110–125. [CrossRef] [PubMed]
62. Gao, M.Q.; Choi, Y.P.; Kang, S.; Youn, J.H.; Cho, N.H. CD24^{+} cells from hierarchically organized ovarian cancer are enriched in cancer stem cells. *Oncogene* **2010**, *29*, 2672–2680. [CrossRef] [PubMed]
63. Zhao, L.; Gu, C.; Huang, K.; Zhang, Z.; Ye, M.; Fan, W.; Han, W.; Meng, Y. The prognostic value and clinicopathological significance of CD44 expression in ovarian cancer: A meta-analysis. *Arch. Gynecol. Obstet.* **2016**, *294*, 1019–1029. [CrossRef] [PubMed]
64. Elzarkaa, A.A.; Sabaa, B.E.; Abdelkhalik, D.; Mansour, H.; Melis, M.; Shaalan, W.; Farouk, M.; Malik, E.; Soliman, A.A. Clinical relevance of CD44 surface expression in advanced stage serous epithelial ovarian cancer: A prospective study. *J. Cancer Res. Clin. Oncol.* **2016**, *142*, 949–958. [CrossRef] [PubMed]
65. Bonneau, C.; Rouzier, R.; Geyl, C.; Cortez, A.; Castela, M.; Lis, R.; Darai, E.; Touboul, C. Predictive markers of chemoresistance in advanced stages epithelial ovarian carcinoma. *Gynecol. Oncol.* **2015**, *136*, 112–120. [CrossRef] [PubMed]
66. Bourguignon, L.Y.; Peyrollier, K.; Xia, W.; Gilad, E. Hyaluronan-CD44 interaction activates stem cell marker nanog, stat-3-mediated mdr1 gene expression, and ankyrin-regulated multidrug efflux in breast and ovarian tumor cells. *J. Biol. Chem.* **2008**, *283*, 17635–17651. [CrossRef] [PubMed]
67. Klemba, A.; Purzycka-Olewiecka, J.K.; Wcislo, G.; Czarnecka, A.M.; Lewicki, S.; Lesyng, B.; Szczylik, C.; Kieda, C. Surface markers of cancer stem-like cells of ovarian cancer and their clinical relevance. *Contemp. Oncol.* **2018**, *22*, 48–55. [CrossRef] [PubMed]
68. Luo, L.; Zeng, J.; Liang, B.; Zhao, Z.; Sun, L.; Cao, D.; Yang, J.; Shen, K. Ovarian cancer cells with the CD117 phenotype are highly tumorigenic and are related to chemotherapy outcome. *Exp. Mol. Pathol.* **2011**, *91*, 596–602. [CrossRef] [PubMed]
69. Shaw, T.J.; Vanderhyden, B.C. AKT mediates the pro-survival effects of kit in ovarian cancer cells and is a determinant of sensitivity to imatinib mesylate. *Gynecol. Oncol.* **2007**, *105*, 122–131. [CrossRef] [PubMed]
70. Ruscito, I.; Cacsire Castillo-Tong, D.; Vergote, I.; Ignat, I.; Stanske, M.; Vanderstichele, A.; Ganapathi, R.N.; Glajzer, J.; Kulbe, H.; Trillsch, F.; et al. Exploring the clonal evolution of CD133/aldehyde-dehydrogenase-1 (ALDH1)-positive cancer stem-like cells from primary to recurrent high-grade serous ovarian cancer (HGSOC). A study of the ovarian cancer therapy-innovative models prolong survival (OCTIPS) consortium. *Eur. J. Cancer* **2017**, *79*, 214–225. [PubMed]
71. Liang, J.; Yang, B.; Cao, Q.; Wu, X. Association of vasculogenic mimicry formation and CD133 expression with poor prognosis in ovarian cancer. *Gynecol. Obstet. Investig.* **2016**, *81*, 529–536. [CrossRef] [PubMed]
72. Zhou, Q.; Chen, A.; Song, H.; Tao, J.; Yang, H.; Zuo, M. Prognostic value of cancer stem cell marker CD133 in ovarian cancer: A meta-analysis. *Int. J. Clin. Exp. Med.* **2015**, *8*, 3080–3088. [PubMed]

73. Baba, T.; Convery, P.A.; Matsumura, N.; Whitaker, R.S.; Kondoh, E.; Perry, T.; Huang, Z.; Bentley, R.C.; Mori, S.; Fujii, S.; et al. Epigenetic regulation of CD133 and tumorigenicity of CD133+ ovarian cancer cells. *Oncogene* **2009**, *28*, 209–218. [CrossRef] [PubMed]
74. Curley, M.D.; Therrien, V.A.; Cummings, C.L.; Sergent, P.A.; Koulouris, C.R.; Friel, A.M.; Roberts, D.J.; Seiden, M.V.; Scadden, D.T.; Rueda, B.R.; et al. CD133 expression defines a tumor initiating cell population in primary human ovarian cancer. *Stem Cells* **2009**, *27*, 2875–2883. [CrossRef] [PubMed]
75. Silva, I.A.; Bai, S.; McLean, K.; Yang, K.; Griffith, K.; Thomas, D.; Ginestier, C.; Johnston, C.; Kueck, A.; Reynolds, R.K.; et al. Aldehyde dehydrogenase in combination with CD133 defines angiogenic ovarian cancer stem cells that portend poor patient survival. *Cancer Res.* **2011**, *71*, 3991–4001. [CrossRef] [PubMed]
76. Roy, L.; Bobbs, A.; Sattler, R.; Kurkewich, J.L.; Dausinas, P.B.; Nallathamby, P.; Cowden Dahl, K.D. CD133 promotes adhesion to the ovarian cancer metastatic niche. *Cancer Growth Metastasis* **2018**, *11*. [CrossRef] [PubMed]
77. Landen, C.N., Jr.; Goodman, B.; Katre, A.A.; Steg, A.D.; Nick, A.M.; Stone, R.L.; Miller, L.D.; Mejia, P.V.; Jennings, N.B.; Gershenson, D.M.; et al. Targeting aldehyde dehydrogenase cancer stem cells in ovarian cancer. *Mol. Cancer Ther.* **2010**, *9*, 3186–3199. [CrossRef] [PubMed]
78. Wang, Y.C.; Yo, Y.T.; Lee, H.Y.; Liao, Y.P.; Chao, T.K.; Su, P.H.; Lai, H.C. Aldh1-bright epithelial ovarian cancer cells are associated with CD44 expression, drug resistance, and poor clinical outcome. *Am. J. Pathol.* **2012**, *180*, 1159–1169. [CrossRef] [PubMed]
79. Zhang, H.; Qiu, J.; Ye, C.; Yang, D.; Gao, L.; Su, Y.; Tang, X.; Xu, N.; Zhang, D.; Xiong, L.; et al. Ror1 expression correlated with poor clinical outcome in human ovarian cancer. *Sci. Rep.* **2014**, *4*, 5811. [CrossRef] [PubMed]
80. Zhang, S.; Cui, B.; Lai, H.; Liu, G.; Ghia, E.M.; Widhopf, G.F., 2nd; Zhang, Z.; Wu, C.C.; Chen, L.; Wu, R.; et al. Ovarian cancer stem cells express ror1, which can be targeted for anti-cancer-stem-cell therapy. *Proc. Natl. Acad. Sci. USA* **2014**, *111*, 17266–17271. [CrossRef] [PubMed]
81. Zhao, W.; Ji, X.; Zhang, F.; Li, L.; Ma, L. Embryonic stem cell markers. *Molecules* **2012**, *17*, 6196–6236. [CrossRef] [PubMed]
82. Yu, Z.; Pestell, T.G.; Lisanti, M.P.; Pestell, R.G. Cancer stem cells. *Int. J. Biochem. Cell Biol.* **2012**, *44*, 2144–2151. [CrossRef] [PubMed]
83. Wang, J.; Rao, S.; Chu, J.; Shen, X.; Levasseur, D.N.; Theunissen, T.W.; Orkin, S.H. A protein interaction network for pluripotency of embryonic stem cells. *Nature* **2006**, *444*, 364–368. [CrossRef] [PubMed]
84. Wen, Y.; Hou, Y.; Huang, Z.; Cai, J.; Wang, Z. Sox2 is required to maintain cancer stem cells in ovarian cancer. *Cancer Sci.* **2017**, *108*, 719–731. [CrossRef] [PubMed]
85. Di, J.; Duiveman-de Boer, T.; Zusterzeel, P.L.; Figdor, C.G.; Massuger, L.F.; Torensma, R. The stem cell markers OCT4A, nanog and C-MYC are expressed in ascites cells and tumor tissue of ovarian cancer patients. *Cell. Oncol.* **2013**, *36*, 363–374. [CrossRef] [PubMed]
86. Ning, Y.X.; Luo, X.; Xu, M.; Feng, X.; Wang, J. Let-7d increases ovarian cancer cell sensitivity to a genistein analog by targeting c-MYC. *Oncotarget* **2017**, *8*, 74836–74845. [CrossRef] [PubMed]
87. Zhou, S.; Schuetz, J.D.; Bunting, K.D.; Colapietro, A.M.; Sampath, J.; Morris, J.J.; Lagutina, I.; Grosveld, G.C.; Osawa, M.; Nakauchi, H.; et al. The ABC transporter BCRP1/ABCG2 is expressed in a wide variety of stem cells and is a molecular determinant of the side-population phenotype. *Nat. Med.* **2001**, *7*, 1028–1034. [CrossRef] [PubMed]
88. Garson, K.; Vanderhyden, B.C. Epithelial ovarian cancer stem cells: Underlying complexity of a simple paradigm. *Reproduction* **2015**, *149*, R59–R70. [CrossRef] [PubMed]
89. Hu, L.; McArthur, C.; Jaffe, R.B. Ovarian cancer stem-like side-population cells are tumourigenic and chemoresistant. *Br. J. Cancer* **2010**, *102*, 1276–1283. [CrossRef] [PubMed]
90. Kobayashi, Y.; Seino, K.; Hosonuma, S.; Ohara, T.; Itamochi, H.; Isonishi, S.; Kita, T.; Wada, H.; Kojo, S.; Kiguchi, K. Side population is increased in paclitaxel-resistant ovarian cancer cell lines regardless of resistance to cisplatin. *Gynecol. Oncol.* **2011**, *121*, 390–394. [CrossRef] [PubMed]
91. Szotek, P.P.; Pieretti-Vanmarcke, R.; Masiakos, P.T.; Dinulescu, D.M.; Connolly, D.; Foster, R.; Dombkowski, D.; Preffer, F.; Maclaughlin, D.T.; Donahoe, P.K. Ovarian cancer side population defines cells with stem cell-like characteristics and mullerian inhibiting substance responsiveness. *Proc. Natl. Acad. Sci. USA* **2006**, *103*, 11154–11159. [CrossRef] [PubMed]

92. Karmakar, S.; Seshacharyulu, P.; Lakshmanan, I.; Vaz, A.P.; Chugh, S.; Sheinin, Y.M.; Mahapatra, S.; Batra, S.K.; Ponnusamy, M.P. HPAF1/PD2 interacts with OCT3/4 to promote self-renewal of ovarian cancer stem cells. *Oncotarget* **2017**, *8*, 14806–14820. [CrossRef] [PubMed]
93. Peng, S.; Maihle, N.J.; Huang, Y. Pluripotency factors LIN28 and OCT4 identify a sub-population of stem cell-like cells in ovarian cancer. *Oncogene* **2010**, *29*, 2153–2159. [CrossRef] [PubMed]
94. Varas-Godoy, M.; Rice, G.; Illanes, S.E. The crosstalk between ovarian cancer stem cell niche and the tumor microenvironment. *Stem Cells Int.* **2017**, *2017*. [CrossRef] [PubMed]
95. Vathipadiekal, V.; Saxena, D.; Mok, S.C.; Hauschka, P.V.; Ozbun, L.; Birrer, M.J. Identification of a potential ovarian cancer stem cell gene expression profile from advanced stage papillary serous ovarian cancer. *PLoS ONE* **2012**, *7*, e29079. [CrossRef] [PubMed]
96. Ponnusamy, M.P.; Batra, S.K. Ovarian cancer: Emerging concept on cancer stem cells. *J. Ovarian Res.* **2008**, *1*, 4. [CrossRef] [PubMed]
97. Cowden Dahl, K.D.; Dahl, R.; Kruichak, J.N.; Hudson, L.G. The epidermal growth factor receptor responsive MIR-125a represses mesenchymal morphology in ovarian cancer cells. *Neoplasia* **2009**, *11*, 1208–1215. [CrossRef] [PubMed]
98. Roy, L.; Bobbs, A.; Cowden Dahl, K. Indiana University School of Medicine-South Bend, University of Notre Dame. Unpublished data, 2016.
99. Visvader, J.E.; Lindeman, G.J. Cancer stem cells: Current status and evolving complexities. *Cell Stem Cell* **2012**, *10*, 717–728. [CrossRef] [PubMed]
100. Li, H.; Zeng, J.; Shen, K. PI3K/AKT/mTOR signaling pathway as a therapeutic target for ovarian cancer. *Arch. Gynecol. Obstet.* **2014**, *290*, 1067–1078. [CrossRef] [PubMed]
101. Altomare, D.A.; Wang, H.Q.; Skele, K.L.; De Rienzo, A.; Klein-Szanto, A.J.; Godwin, A.K.; Testa, J.R. AKT and mTOR phosphorylation is frequently detected in ovarian cancer and can be targeted to disrupt ovarian tumor cell growth. *Oncogene* **2004**, *23*, 5853–5857. [CrossRef] [PubMed]
102. Cai, J.; Xu, L.; Tang, H.; Yang, Q.; Yi, X.; Fang, Y.; Zhu, Y.; Wang, Z. The role of the PTEN/PI3K/AKT pathway on prognosis in epithelial ovarian cancer: A meta-analysis. *Oncologist* **2014**, *19*, 528–535. [CrossRef] [PubMed]
103. Luo, X.; Dong, Z.; Chen, Y.; Yang, L.; Lai, D. Enrichment of ovarian cancer stem-like cells is associated with epithelial to mesenchymal transition through an miRNA-activated AKT pathway. *Cell Prolif.* **2013**, *46*, 436–446. [CrossRef] [PubMed]
104. Seo, E.J.; Kwon, Y.W.; Jang, I.H.; Kim, D.K.; Lee, S.I.; Choi, E.J.; Kim, K.H.; Suh, D.S.; Lee, J.H.; Choi, K.U.; et al. Autotaxin regulates maintenance of ovarian cancer stem cells through lysophosphatidic acid-mediated autocrine mechanism. *Stem Cells* **2016**, *34*, 551–564. [CrossRef] [PubMed]
105. Ali, A.Y.; Kim, J.Y.; Pelletier, J.F.; Vanderhyden, B.C.; Bachvarov, D.R.; Tsang, B.K. AKT confers cisplatin chemoresistance in human gynecological carcinoma cells by modulating ppm1d stability. *Mol. Carcinog.* **2015**, *54*, 1301–1314. [CrossRef] [PubMed]
106. Chau, W.K.; Ip, C.K.; Mak, A.S.; Lai, H.C.; Wong, A.S. C-kit mediates chemoresistance and tumor-initiating capacity of ovarian cancer cells through activation of Wnt/BETA-catenin-ATP-binding cassette G2 signaling. *Oncogene* **2013**, *32*, 2767–2781. [CrossRef] [PubMed]
107. Yoshikawa, T.; Miyamoto, M.; Aoyama, T.; Soyama, H.; Goto, T.; Hirata, J.; Suzuki, A.; Nagaoka, I.; Tsuda, H.; Furuya, K.; et al. JAK2/STAT3 pathway as a therapeutic target in ovarian cancers. *Oncol. Lett.* **2018**, *15*, 5772–5780. [CrossRef] [PubMed]
108. Abubaker, K.; Luwor, R.B.; Zhu, H.; McNally, O.; Quinn, M.A.; Burns, C.J.; Thompson, E.W.; Findlay, J.K.; Ahmed, N. Inhibition of the JAK2/STAT3 pathway in ovarian cancer results in the loss of cancer stem cell-like characteristics and a reduced tumor burden. *BMC Cancer* **2014**, *14*, 317. [CrossRef] [PubMed]
109. Saini, U.; Suarez, A.A.; Naidu, S.; Wallbillich, J.J.; Bixel, K.; Wanner, R.A.; Bice, J.; Kladney, R.D.; Lester, J.; Karlan, B.Y.; et al. STAT3/PIAS3 levels serve as "early signature" genes in the development of high-grade serous carcinoma from the fallopian tube. *Cancer Res.* **2018**, *78*, 1739–1750. [CrossRef] [PubMed]
110. Rosen, D.G.; Mercado-Uribe, I.; Yang, G.; Bast, R.C., Jr.; Amin, H.M.; Lai, R.; Liu, J. The role of constitutively active signal transducer and activator of transcription 3 in ovarian tumorigenesis and prognosis. *Cancer* **2006**, *107*, 2730–2740. [CrossRef] [PubMed]

111. Burgos-Ojeda, D.; Wu, R.; McLean, K.; Chen, Y.C.; Talpaz, M.; Yoon, E.; Cho, K.R.; Buckanovich, R.J. CD24+ ovarian cancer cells are enriched for cancer-initiating cells and dependent on JAK2 signaling for growth and metastasis. *Mol. Cancer Ther.* **2015**, *14*, 1717–1727. [CrossRef] [PubMed]
112. Rinkenbaugh, A.L.; Baldwin, A.S. The NF-κB pathway and cancer stem cells. *Cells* **2016**, *5*, 16. [CrossRef] [PubMed]
113. Gonzalez-Torres, C.; Gaytan-Cervantes, J.; Vazquez-Santillan, K.; Mandujano-Tinoco, E.A.; Ceballos-Cancino, G.; Garcia-Venzor, A.; Zampedri, C.; Sanchez-Maldonado, P.; Mojica-Espinosa, R.; Jimenez-Hernandez, L.E.; et al. NF-κB participates in the stem cell phenotype of ovarian cancer cells. *Arch. Med. Res.* **2017**, *48*, 343–351. [CrossRef] [PubMed]
114. Leizer, A.L.; Alvero, A.B.; Fu, H.H.; Holmberg, J.C.; Cheng, Y.C.; Silasi, D.A.; Rutherford, T.; Mor, G. Regulation of inflammation by the NF-κB pathway in ovarian cancer stem cells. *Am. J. Reprod. Immunol.* **2011**, *65*, 438–447. [CrossRef] [PubMed]
115. House, C.D.; Jordan, E.; Hernandez, L.; Ozaki, M.; James, J.M.; Kim, M.; Kruhlak, M.J.; Batchelor, E.; Elloumi, F.; Cam, M.C.; et al. NF-κB promotes ovarian tumorigenesis via classical pathways that support proliferative cancer cells and alternative pathways that support ALDH(+) cancer stem-like cells. *Cancer Res.* **2017**, *77*, 6927–6940. [CrossRef] [PubMed]
116. Takebe, N.; Harris, P.J.; Warren, R.Q.; Ivy, S.P. Targeting cancer stem cells by inhibiting Wnt, notch, and hedgehog pathways. *Nat. Rev. Clin. Oncol.* **2011**, *8*, 97–106. [CrossRef] [PubMed]
117. Chen, X.; Thiaville, M.M.; Chen, L.; Stoeck, A.; Xuan, J.; Gao, M.; Shih, I.M.; Wang, T.L. Defining notch3 target genes in ovarian cancer. *Cancer Res.* **2012**, *72*, 2294–2303. [CrossRef] [PubMed]
118. Park, J.T.; Li, M.; Nakayama, K.; Mao, T.L.; Davidson, B.; Zhang, Z.; Kurman, R.J.; Eberhart, C.G.; Shih, I.M.; Wang, T.L. Notch3 gene amplification in ovarian cancer. *Cancer Res.* **2006**, *66*, 6312–6318. [CrossRef] [PubMed]
119. Choi, J.H.; Park, J.T.; Davidson, B.; Morin, P.J.; Shih, I.M.; Wang, T.L. Jagged-1 and notch3 juxtacrine loop regulates ovarian tumor growth and adhesion. *Cancer Res.* **2008**, *68*, 5716–5723. [CrossRef] [PubMed]
120. Park, J.T.; Chen, X.; Trope, C.G.; Davidson, B.; Shih, I.M.; Wang, T.L. Notch3 overexpression is related to the recurrence of ovarian cancer and confers resistance to carboplatin. *Am. J. Pathol.* **2010**, *177*, 1087–1094. [CrossRef] [PubMed]
121. Qin, S.; Li, Y.; Cao, X.; Du, J.; Huang, X. Nanog regulates epithelial-mesenchymal transition and chemoresistance in ovarian cancer. *Biosci. Rep.* **2017**, *37*. [CrossRef] [PubMed]
122. Steg, A.D.; Katre, A.A.; Goodman, B.; Han, H.D.; Nick, A.M.; Stone, R.L.; Coleman, R.L.; Alvarez, R.D.; Lopez-Berestein, G.; Sood, A.K.; et al. Targeting the notch ligand jagged1 in both tumor cells and stroma in ovarian cancer. *Clin. Cancer Res.* **2011**, *17*, 5674–5685. [CrossRef] [PubMed]
123. Grigoryan, T.; Wend, P.; Klaus, A.; Birchmeier, W. Deciphering the function of canonical Wnt signals in development and disease: Conditional loss- and gain-of-function mutations of beta-catenin in mice. *Genes Dev.* **2008**, *22*, 2308–2341. [CrossRef] [PubMed]
124. Gavert, N.; Ben-Ze'ev, A. Beta-catenin signaling in biological control and cancer. *J. Cell. Biochem.* **2007**, *102*, 820–828. [CrossRef] [PubMed]
125. Clevers, H. Wnt/beta-catenin signaling in development and disease. *Cell* **2006**, *127*, 469–480. [CrossRef] [PubMed]
126. Asem, M.S.; Buechler, S.; Wates, R.B.; Miller, D.L.; Stack, M.S. Wnt5a signaling in cancer. *Cancers* **2016**, *8*, 79. [CrossRef] [PubMed]
127. Schindler, A.J.; Watanabe, A.; Howell, S.B. LGR5 and LGR6 in stem cell biology and ovarian cancer. *Oncotarget* **2018**, *9*, 1346–1355. [CrossRef] [PubMed]
128. Flesken-Nikitin, A.; Hwang, C.I.; Cheng, C.Y.; Michurina, T.V.; Enikolopov, G.; Nikitin, A.Y. Ovarian surface epithelium at the junction area contains a cancer-prone stem cell niche. *Nature* **2013**, *495*, 241–245. [CrossRef] [PubMed]
129. Kessler, M.; Hoffmann, K.; Brinkmann, V.; Thieck, O.; Jackisch, S.; Toelle, B.; Berger, H.; Mollenkopf, H.J.; Mangler, M.; Sehouli, J.; et al. The notch and Wnt pathways regulate stemness and differentiation in human fallopian tube organoids. *Nat. Commun.* **2015**, *6*, 8989. [CrossRef] [PubMed]
130. Ng, A.; Tan, S.; Singh, G.; Rizk, P.; Swathi, Y.; Tan, T.Z.; Huang, R.Y.; Leushacke, M.; Barker, N. LGR5 marks stem/progenitor cells in ovary and tubal epithelia. *Nat. Cell Biol.* **2014**, *16*, 745–757. [CrossRef] [PubMed]

131. Anastas, J.N.; Moon, R.T. Wnt signalling pathways as therapeutic targets in cancer. *Nat. Rev. Cancer* **2013**, *13*, 11–26. [CrossRef] [PubMed]
132. Barbolina, M.V.; Burkhalter, R.J.; Stack, M.S. Diverse mechanisms for activation of Wnt signalling in the ovarian tumour microenvironment. *Biochem. J.* **2011**, *437*, 1–12. [CrossRef] [PubMed]
133. Liao, X.; Siu, M.K.; Au, C.W.; Wong, E.S.; Chan, H.Y.; Ip, P.P.; Ngan, H.Y.; Cheung, A.N. Aberrant activation of hedgehog signaling pathway in ovarian cancers: Effect on prognosis, cell invasion and differentiation. *Carcinogenesis* **2009**, *30*, 131–140. [CrossRef] [PubMed]
134. Ray, A.; Meng, E.; Reed, E.; Shevde, L.A.; Rocconi, R.P. Hedgehog signaling pathway regulates the growth of ovarian cancer spheroid forming cells. *Int. J. Oncol.* **2011**, *39*, 797–804. [PubMed]
135. Bhattacharya, R.; Kwon, J.; Ali, B.; Wang, E.; Patra, S.; Shridhar, V.; Mukherjee, P. Role of hedgehog signaling in ovarian cancer. *Clin. Cancer Res.* **2008**, *14*, 7659–7666. [CrossRef] [PubMed]
136. Kudo, K.; Gavin, E.; Das, S.; Amable, L.; Shevde, L.A.; Reed, E. Inhibition of gli1 results in altered c-JUN activation, inhibition of cisplatin-induced upregulation of ERCC1, XPD and XRCC1, and inhibition of platinum-DNA adduct repair. *Oncogene* **2012**, *31*, 4718–4724. [CrossRef] [PubMed]
137. Song, X.; Yan, L.; Lu, C.; Zhang, C.; Zhu, F.; Yang, J.; Jing, H.; Zhang, Y.; Qiao, J.; Guo, H. Activation of hedgehog signaling and its association with cisplatin resistance in ovarian epithelial tumors. *Oncol. Lett.* **2018**, *15*, 5569–5576. [CrossRef] [PubMed]
138. Merchant, A.A.; Matsui, W. Targeting hedgehog—A cancer stem cell pathway. *Clin. Cancer Res.* **2010**, *16*, 3130–3140. [CrossRef] [PubMed]

cancers

MDPI

Review

Targeting the Microenvironment in High Grade Serous Ovarian Cancer

Nkechiyere G. Nwani [1], Livia E. Sima [1], Wilberto Nieves-Neira [1,2] and Daniela Matei [1,2,*]

1 Department of Obstetrics and Gynecology, Northwestern University, Chicago, IL 60611, USA; Nnwani@northwestern.edu (N.G.N.); Livia.sima@northwestern.edu (L.E.S.); wilberto.nieves-neira@nm.org (W.N.-N.)

2 Robert H. Lurie Comprehensive Cancer Center, Chicago, IL 60611, USA

* Correspondence: daniela.matei@northwestern.edu

Received: 20 July 2018; Accepted: 6 August 2018; Published: 10 August 2018

Abstract: Cancer–stroma interactions play a key role in cancer progression and response to standard chemotherapy. Here, we provide a summary of the mechanisms by which the major cellular components of the ovarian cancer (OC) tumor microenvironment (TME) including cancer-associated fibroblasts (CAFs), myeloid, immune, endothelial, and mesothelial cells potentiate cancer progression. High-grade serous ovarian cancer (HGSOC) is characterized by a pro-inflammatory and angiogenic signature. This profile is correlated with clinical outcomes and can be a target for therapy. Accumulation of malignant ascites in the peritoneal cavity allows for secreted factors to fuel paracrine and autocrine circuits that augment cancer cell proliferation and invasiveness. Adhesion of cancer cells to the mesothelial matrix promotes peritoneal tumor dissemination and represents another attractive target to prevent metastasis. The immunosuppressed tumor milieu of HGSOC is permissive for tumor growth and can be modulated therapeutically. Results of emerging preclinical and clinical trials testing TME-modulating therapeutics for the treatment of OC are highlighted.

Keywords: high-grade serous ovarian cancer; tumor microenvironment; angiogenesis; immune response; metastasis; therapeutic targeting strategies

1. Introduction

High-grade serous ovarian cancer (HGSOC) comprises the majority of epithelial ovarian tumors, is associated with a p53-mutated signature and is characterized by initial sensitivity to platinum and a unique pattern of dissemination in the peritoneal space. The peritoneum consists of mesothelial cells that cover and protect the viscera. The sub-peritoneal stroma contains a collagen-based matrix, activated fibroblasts, blood vessels, and lymphatics. This unique milieu permits accumulation of factors secreted by both cancer and stromal cells and enables metastatic seeding and tumor proliferation. The immune component of the peritoneal milieu consists of monocytes/macrophages and cytotoxic T cells. Several studies have demonstrated an "activated" phenotype of the peritoneal environment associated with ovarian cancer (OC), as opposed to its quiescent state in benign conditions [1]. The pro-inflammatory signature associated with cancer favors angiogenesis and exerts chemotactic and protective effects on cancer cells. Chemokines, cytokines, and growth factors commonly secreted in the tumor microenvironment (TME) include the stromal cell-derived factor (SDF1), interleukin-6 (IL-6), interleukin (IL-8), monocyte chemoattractant protein 1 (MCP1), Chemokine (C-C motif) ligand 5 and 7 (CCL5 and CCL7), transforming growth factor-β1 TGF β1, tumor necrosis factor-α (TNFα), fibroblast growth factor (FGF), and others [1–4]. While tumor cells play a role in the secretion of factors that modulate angiogenesis, non-transformed tumor infiltrating cells such as fibroblasts, myeloid cells, immune cells, and endothelial precursors also play a crucial role modulating neo-vascularization [5]. OC metastasis commonly involves the omentum, an adipocyte-rich organ. Lipid transfer between adipocytes and cancer cells mediated by fatty acid binding protein 4 (FABP4), through a "symbiotic" process between cancer cells and the fatty microenvironment was described as a key regulator of peritoneal metastasis [6]. As the rich TME protects cancer cells from noxious stimuli promoting tumor growth (Figure 1), its disruption through targeted therapy could arrest cancer progression. Indeed, over the past decade, several classes of novel agents targeting the ovarian TME have been developed and tested clinically. The most active agents are antiangiogenic therapies, which have been recently approved by the Food and Drug Administraton FDA for OC. Other emerging strategies, particularly immunotherapy, are in various stages of development. Here, several targeted therapies directed against the main components of the TME will be reviewed.

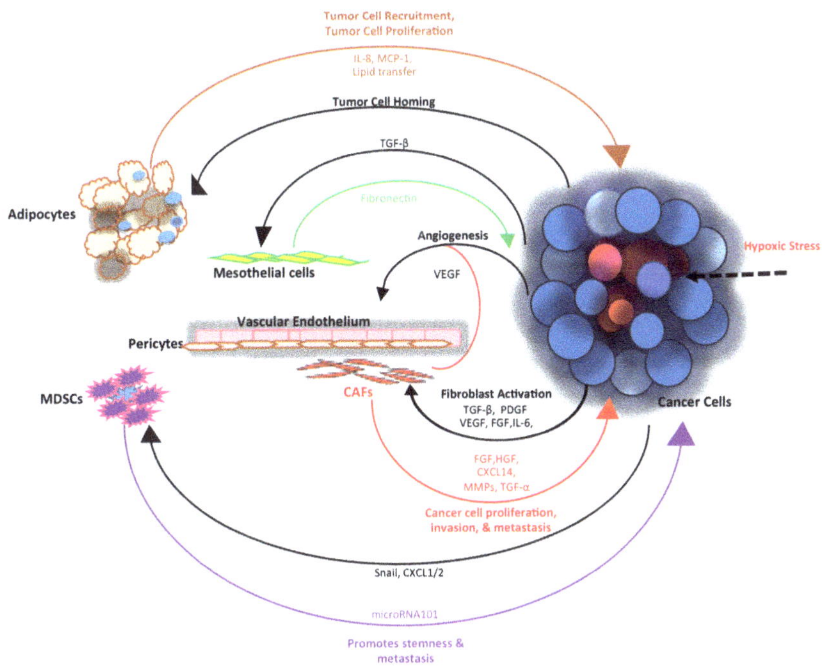

Figure 1. The interplay between cancer and stromal cells in the tumor microenvironent TME regulates tumor growth and metastasis: as tumors grow, hypoxic stress and low nutrient availability drives the release of tumor-secreted growth factors and cytokines that exert paracrine effects on the surrounding stroma. Sustained exposure to tumor-derived transforming growth factor-β (TGF-β), platelet-derived growth factor (PDGF), fibroblast growth factor (FGF) and vascular endothelial growth factor (VEGF) drives fibroblasts trans-differentiation into (cancer associated fibroblasts) CAFs. These factors also act upon endothelial cells, pericytes and immune cells to stimulate angiogenesis. CAF-derived FGF and hepatocyte growth factor (HGF) promote tumor cell proliferation, CAF-derived matrix metalloproteinases (MMPs) promote invasion while chemokine ligand 14 (CXCL14) and transforming growth factor-α (TGF-α) enhance metastasis. Ovarian cancer (OC) cell-derived TGF-β1 upregulates fibronectin secretion in mesothelial cells, which in turn enhances spheroid adhesion to the peritoneal wall. Adipocytes facilitate cells proliferation by providing energy dense lipids to the metastasized cancer cells. Cancer cells expressing Snail and chemokine (C-X-C motif) ligand 1/2 (CXCL1/2) recruit myeloid-derived suppressor cells (MDSCs) to the tumor site; conversely MDSC-secreted microRNA101 reprograms tumor cells to a stemness phenotype.

2. Fibroblasts

Fibroblasts represent the preeminent cellular component of connective tissues, the structural scaffold of many organs in the body. They are a heterogeneous population of mesenchymal-derived cells that maintain the composition of the extracellular matrix (ECM) [7,8]. As such, fibroblasts produce and deposit most of the proteins that comprise the ECM, including collagens, proteoglycans, tenascin, fibronectin, and laminin. Tissue homeostasis involves a tightly orchestrated balance of ECM synthesis and metabolism; in addition to ECM production, fibroblasts are also responsible for matrix metabolism. They produce several ECM-degrading matrix metalloproteinases (MMPs) and their inhibitors, tissue inhibitors of metalloproteinases (TIMPs) [9]. It has been observed that fibroblasts within the tumor milieu are phenotypically similar to activated fibroblasts associated with granulating tissue (wound healing) [10]. These cancer-associated fibroblasts (CAFs) function as tumor-promoting cells; playing important roles in tumor initiation and progression [11–13]. Although resident fibroblasts are a major source of CAFs, they can also arise from the trans-differentiation of other cell populations including epithelial cells, endothelial cells, pericytes, adipocytes and bone marrow-derived mesenchymal stem cells [14]. During tumorigenesis the trans-differentiation of the aforementioned cells into CAFs is driven by sustained exposure to tumor-derived factors including TGF-β, PDGF-BB, basic fibroblast growth factor (bFGF), vascular endothelial growth factor (VEGF), as well as microRNAs, reactive oxygen species (ROS), matrix metalloproteases (MMPs) and extracellular vesicles [15–19].

Current evidence suggests the mechanisms/downstream effectors that coordinate CAF activation vary and are contingent on CAF origin. For example, it was shown that SKOV3 cells stimulate normal fibroblasts conversion through TGF-β mediated induction of ROS and CLIC4, which led to the subsequent increase in the expression CAF markers αSMA and FAP. On the other hand, Jeon et al., demonstrated that cancer cell-derived lysophosphatidic acid induced TGF-β in adipose tissue-derived mesenchymal stem cells which then promoted their trans-differentiation into CAFs [18,20]. Likewise, expression of HOXA9, a differentiation related gene, was linked to paracrine secretion of TGF-β2 by OC cells, inducing adipose and mesenchymal stem cells to become CAFs [21]. It is unknown whether other stromal cells such as pericytes and endothelial could also contribute to the reactive stroma associated with HGSOC.

The role of fibroblasts in cancer progression is complex. Early studies provided evidence that fibroblasts possess anti-tumorigenic function by forming a restrictive stroma. However, the atypical cancer-stroma interactions promote fibroblasts to develop tumor-permissive properties [22–24]. Recent reports illustrate how the reciprocal cancer cell–fibroblast communication potentiates tumor growth and progression in OC models. For example, CAFs have been shown to suppress the immune response through miR141/200a-mediated expression of CAF-derived CXCL12. This chemokine promotes infiltration of immunosuppressive $CD25^+$ $FOXP3^+$ T lymphocytes in the HGSOC milieu, which in turn allows tumor growth [25]. CAFs have also been shown to drive tumor cell proliferation, migration and invasion by producing high amounts of mitogenic factors, hepatocyte growth factor (HGF) and FGF [26–28]. Additionally, CAF-secreted IL-8 and SDF-1 drive angiogenesis to facilitate oxygen and nutrients delivery to the tumor tissue [29,30]. Fibroblasts treated with SKOV3-derived extracellular vesicles acquired an activated phenotype; in turn these fibroblast enhanced tumor and endothelial cells proliferation [17]. In another study, OC cell-derived TNF-α induced TGF-α transcription in stromal fibroblasts. In turn, TGF-α secreted by these fibroblasts promoted metastasis via induction of EGFR signaling in cancer cells [31]. CAFs also produce metabolites that are essential to cancer cells' survival, such as lactate that is absorbed and utilized by oxidative phosphorylation in adjacent cancer cells [32]. The chemokine ligand 14 (CXCL14) is a CAFs secreted protein that is associated with a poor prognosis in OC. It was discovered that CXCL14 induced LINC00092 expression in OC cells, which resulted enhanced metastasis. LINC0009 interacted with 6-phosphofructo-2-kinase/fructose-2,6-biphosphatase 2 (PFKFB2) to induce a glycolytic phenotype in ovarian cancer cells. These interactions are necessary for maintaining the CAF-phenotype, thereby unearthing a positive feedback loop between CAF-cancer cells interactions that sustain a tumor-permissive microenvironment [33].

Cancer invasion and metastasis is also closely associated with MMPs secreted by CAFs and tumor cells and increased MMP expression has been associated with poor prognosis for various cancers [34]. In addition to modifying the ECM, MMPs can facilitate tumor growth and invasion by increasing the bioavailability of ECM tethered growth factors. For instance, CAF-secreted matrix metalloproteinase-13 (MMP-13) enhanced tumor cells invasion through proteolytic cleavage of matrix-bound VEGF and angiogenesis [35].

An additional factor involved in CAF-tumor cell cross-talk is the fibroblast activation protein (FAP). FAP is exclusively expressed on activated fibroblasts, and increased expression is associated with poor prognosis in many tumors [36]. In OC, FAP promoted HO-8910PM tumor cell proliferation, invasion and migration via interactions with integrin $\alpha3\beta1$ and urokinase-type plasminogen activator receptor (uPAR) signaling complex [37]. Moreover, elevated stromal FAP expression was a strong predictive marker of platinum resistance and relapse [38]. Due to the adverse effects of CAFs on cancer recurrence and patient survival, there has been extensive investment in developing strategies to effectively target CAFs.

3. Therapies Targeting Fibroblasts

FAP is overexpressed in many epithelial cancers including OC, and its expression is often associated with poor prognosis [36,38], cancer cell migration, invasion and immunosuppression [39–41]. As such, FAP has emerged as a potential therapeutic target to abate the tumor promoting effects of CAFs. The catalytic activity of FAP was shown to be necessary for tumor proliferation. However, inhibition of FAP enzymatic activity by small molecules has had little success in clinical trials [42,43]. In a transgenic mouse model, targeted depletion of FAP-expressing CAFs resulted in increased cancer cell death. Mechanistically, this effect was dependent on TNF-α and IFN-γ, which are known to be involved in $CD8^+$ T cell mediated cancer cell death [41]. Furthermore, pre-clinical studies using vaccines against FAP showed promising results for colon and lung cancer. Vaccines targeting FAP-expressing cells significantly suppressed tumor growth by eliciting $CD8^+$ or a combined $CD8^+$ and $CD4^+$-T cell response respectively [40,44].

TGF-β, a cytokine abundantly secreted by fibroblasts and detectable in ascites fluid, contributes to the development of a tumor-promoting microenvironment. Several TGF-β targeting agents have been evaluated in clinical trials. These include small molecule kinase, antisense oligonucleotides, and TGF-β-ligand traps [45,46]. In a mouse model of peritoneal metastasis, the TGF-β inhibitor A-83-01 improved overall survival [47,48]. Likewise, the transforming growth factor-β receptor 1 (TβRI) kinase inhibitor galunisertib inhibited tumor growth in a partly TME-dependent manner in various PDX tumors [49]. TGF-β inhibitors have also been shown to enhance the efficacy of conventional therapeutics. For example, combination treatment with TGF-β receptor inhibitor LY2109761 and cisplatin significantly blocked the growth of cisplatin-resistant ovarian xenografts [50]. Despite promising initial preclinical results, advancement of TGF-β signaling inhibitors to the clinical arena has been slow, marred by initial concerns over systemic (cardiac) toxicity, which fortunately appears to be limited in humans [51].

Several other tyrosine kinase inhibitors (TKI) have been employed to mitigate the pro-tumorigenic effects of growth factors secreted by fibroblasts in the tumor milieu, such as the platelet derived growth factor (PDGF) and fibroblast growth factor (FGF). PDGF-D over-expression was associated with lymph node metastasis and platinum resistance in ovarian cancer [52] and imatinib, a PDGFR inhibitor, was shown to inhibit OC cell growth [53]. While the precise effects of imatinib on ovarian stroma are not well defined, previous research demonstrated that this TKI suppressed angiogenesis in cervical tumors [54]. Dasatinib, another FDA approved TKI, which also targets the PDGF receptor has been shown to partially revert lung cancer-derived CAFs to a normal phenotype [55]. Clinical trials tested the PDGFR inhibitors imatinib and sorafenib in patients with recurrent platinum resistant OC and demonstrated modest clinical activity [56,57].

4. Angiogenesis

Angiogenesis is the process whereby new blood vessels sprout from the pre-existing vasculature. Angiogenesis is a tightly regulated and transient process observed in biological processes such as development, wound healing and reproduction [58]. However, pathological angiogenesis is a rate-limiting event in metastasis. As tumors increase in size (>1–2 mm^2), nutrient and oxygen availability are reduced and an angiogenic switch is activated; the newly formed blood vessels are able to deliver nutrients and oxygen necessary for cancer cell proliferation, facilitate waste expulsion, and also provide the primary route by which cancer cells migrate to secondary sites (metastasis) [59]. In fact, tumor vascularity serves as an indicator of metastatic potential for many cancers with highly vascularized tumors having greater incidence of metastasis and reduced survival [60,61]. In cancers, angiogenesis is driven by reduced levels of anti-angiogenic factors, and sustained overproduction of pro-angiogenic molecules by tumor and host cells [58]. Angiogenesis is triggered by growth factors such as VEGF, PDGF, (FGF), angiopoietin (Ang), as well as the chemokines IL-8 and interleukin-6 (IL-6) [59,62,63]. The association between HGSOC and an angiogenic signature was recognized more than two decades ago and has remained a staple in the study of this tumor's biology. VEGF is the most extensively studied angiogenic factor in pathological angiogenesis; it is overexpressed in HGSOC and secreted into malignant ascites [64–67]. Increased VEGF expression is associated with reduced survival rates in patients with OC [68–70]. In a cohort of 222 HGSOC specimens, high levels of VEGF-A were correlated with increased microvessel density and with infiltration by immune cells [71]. Interestingly, high levels of VEGF-A were associated with BRCA-mutated ovarian tumors [71]. Although cancer cells are a major source of angiogenic factors, non-neoplastic cells (immune cells, adipocytes, and CAFs) in the TME also produce the angiogenic factors required to sustain tumor growth and progression [72]. As such, there has been considerable focus on developing therapeutics to inhibit the angiogenic signaling as a means of mitigating cancer progression.

5. Anti-Angiogenic Therapy (AAT)

VEGF is the most extensively studied pro-angiogenic factor and therapies targeting this pathway use either inhibition of the ligand or of its receptor, vascular endothelial growth factor receptor (VEGFR). VEGF-A is a secreted glycoprotein that belongs a family of related growth factors that includes VEGF-B, VEGF-C, VEGF-D and VEGF-E and placental growth factor (PLGF), which have varying functions in angiogenesis [73]. The VEGF system functions as a mitogenic factor for endothelial cells, induces endothelial cell migration and differentiation, and protects immature endothelial cell against apoptosis [74,75]. VEGF exerts these functions by binding to the tyrosine kinase receptors VEGFR-1 (Flt-1) and VEGFR-2 (KDR/Flk-1) on the cell surface, causing them to dimerize and become activated [76]. Bevacizumab (Avastin, Roche, Basel, Switzerland), is a humanized monoclonal antibody against VEGF that binds and inactivates VEGF, thus inhibiting endothelial cell activation and proliferation. Bevacizumab was shown to reduce tumor growth and prolong survival in murine ovarian cancer models [77,78]. Clinical trials using bevacizumab as a single agent and in combination with other therapeutics have been successful and bevacizumab is currently FDA approved for use in the front-line setting, as well as in recurrent disease [79,80].

The first clinical trial to test the efficacy of bevacizumab in OC was performed by the Gynecologic Oncology Group (protocol GOG 170D) and tested the drug in 62 patients with recurrent, platinum-resistant disease. In this trial, 21% of patients exhibited objective clinical responses and 40.3% survived progression-free for at least 6 months. Median progression-free survival (PFS) and overall survival (OS) were 4.7 and 17 months respectively [81]. This initial success led to the development of combination therapies using bevacizumab with chemotherapy. In the ICON7 phase III trial, the efficacy of bevacizumab in combination with platinum and paclitaxel was tested in patients with advanced or metastatic epithelial ovarian cancer after cytoreductive surgery. Bevacizumab was continued for 12 additional cycles or until progression of disease. Progression-free survival at 42 months was increased from 22.4 months with chemotherapy alone to 24.1 months with combination treatment ($p = 0.04$). Interestingly, PFS and OS were most significantly increased in patients at high risk for progression. In this group, survival at 42 months was 28.8 months for patients receiving standard therapy vs. 36.6 months for patients receiving carboplatin/platinum and bevacizumab [82]. Similar results were observed in GOG protocol 218, where chemotherapy plus bevacizumab followed by bevacizumab maintenance improved PFS (but not OS) compared to platinum and paclitaxel alone after cytoreductive surgery [79]. In another randomized phase III clinical trial (AURELIA Trial), bevacizumab in combination with physician's choice chemotherapy was tested in women with recurrent platinum-resistant OC. The median PFS was 3.4 months for patients who received chemotherapy alone versus 6.7 months for patients treated with bevacizumab and chemotherapy [83]. These results summarized in Table 1 led to the approval and widespread clinical use of the first therapy targeting the ovarian cancer TME.

Table 1. Pivotal trials demonstrating Bevacizumab (Bev) clinical activity in OC.

Study	Course of Treatment	Target	TME Component	Patient Population	Phase Trial Size	Trial Endpoint	Clinical Outcome
ICON7	Chemo ± Bevac	VEGF-A	Endothelium	High risk ovarian cancer, stage IIIC or IV	Phase III N = 1528	PFS	At 42 months 22.4 vs. 24.1 months $p = 0.04$
GOG218	Chemo vs. Chemo + Bevac initiation vs. Chemo + Bevac Throughout	VEGF-A	Endothelium	New Diagnosed Stage III or IV OC	Phase III N = 1873	PFS, OS	Median PFS; 10.3 vs. 11.2 vs. 14.1 months; OS; *ns*
AURELIA	Chemo ± Bevac	VEGF-A	Endothelium	Recurrent OC PL-R	Phase III N = 361	PFS, OS	Median PFS; 3.4 vs. 6.7 months. OS; 13.3 vs. 16.6 months
OCEANS	Chemo ± Bevac	VEGF-A	Endothelium	Recurrent OC PL-S	Phase III N = 484	PFS	Median PFS 8.4 vs. 12.4 months
GOG213	Chemo ± Bevac	VEGF-A	Endothelium	Recurrent OC PL-S	Phase III N = 674	ORR	Median overall survival 37.3 vs. 42.2 months

Other modalities to block this pathway are in development. For example, aflibercept is a recombinant fusion protein of VEGFR1 and VEGFR 2 extracellular domain, which functions as a decoy receptor and inhibits VEGF-mediated signaling by trapping VEGF-A, VEGF-B, placental growth factor-1 (PlGF-1) and (PlGF-2). Aflibercept was shown to reduce ascites and decrease the peritoneal dissemination of OC xenograft models [53,84–86]. A phase II trial tested the efficacy of aflibercept in patients with advanced platinum-resistant OC and malignant ascites. Patients who required three or more previous paracenteses per month were given intravenous aflibercept 4 mg/kg every two weeks. The primary study endpoint was repeat paracentesis response rate (RPRR), and a response was defined as a minimum two-fold increase in time to repeat paracentesis compared with the baseline interval. Ten out of 16 patients treated achieved a response; RPRR was 62.5% (95% CI 35.4–84.8%). Median time to repeat paracentesis was 76.0 days (95% CI 64.0–178.0), 4.5 times longer than the baseline (16.8 days) and the median PFS was 59.5 days (95% CI 41.0–83.0) [87], demonstrating that targeting this growth factor in the TME leads to appreciable clinical benefits.

However, angiogenesis is a complex phenomenon tightly regulated by complementary and cross-talking pathways, which allows for the development of resistance [88]. Thus, inhibitors that concurrently block multiple receptors were tested in an effort to improve the efficacy of AAT. Cediranib (AZD2171, AstraZeneca) is a receptor tyrosine kinase inhibitor that inhibits vascular endothelial receptor 1–3 (VEGFR 1–3), platelet-derived growth factor-α and β (PDGFR-α and -β), and c-kit. A phase II clinical trial assessed the efficacy of cediranib in patients with recurrent gynecologic cancers who had received less than two lines of platinum-based chemotherapy. Of 46 patients treated, eight patients (17%) had partial responses (PR), six patients (13%) stable disease (SD), and there were no complete responses (CRs) [89]. In another phase II trial, the efficacy of single-agent cediranib was assessed in 74 patients with persistent/recurrent OC following one round of platinum-based chemotherapy. The patients were stratified into two groups; 39 platinum-sensitive (PL-S) and 35 platinum-resistant (PL-R), and the primary endpoint was objective response rate at 16 weeks. In the platinum sensitive (PL-S) group, 10 patients (26%) demonstrated partial responses (PR) and 20 (51%) had stable disease (SD). There were no confirmed PR in the platinum resistant (PL-R) group and 23 patients (66%) had SD. The median PFS was 7.2 months for PL-S and 3.7 months for PL-R groups, and the median OS was 27.7 and 11.9 months respectively [90]. Currently cediranib is being evaluated in combination with olaparib, a poly (ADP-ribose) polymerase PARP inhibitor in women with recurrent OC.

Nintedanib is another tyrosine kinase inhibitor for VEGFR-1-3, FGFR 1-3, PDGFR α and β. Nintedanib was tested as maintenance treatment after chemotherapy in a randomized trial. PFS at 36-weeks was 5.0% vs. 16.3% in placebo and nintedanib treated patients [91]. However, in a subsequent phase III trial (AGO-OVAR 12) nintedanib combined with platinum-based therapy did not induce a significant survival advantage after debulking surgery. The median PFS was 17.2 vs. 16.6 months for patients treated with nintedanib and placebo, respectively. A post-hoc analysis showed that nintedanib and platinum-based therapy combination improved PFS in non-high-risk patients [92]. Pazopanib (GW786034) is tyrosine kinase inhibitor for VEGFR-1, -2 and -3 PDGFR-α and -β and c-kit. An ongoing clinical phase II trial (MITO-11) is evaluating the safety and activity of pazopanib in combination with paclitaxel in patients with platinum-resistant or refractory OC. The median progression-free survival was 3.5 months in patients treated with weekly paclitaxel vs. 6.3 months in patients treated with weekly paclitaxel and pazopanib. The median overall survival was 14.8 months in paclitaxel treated vs. 18.7 months in patients treated with paclitaxel and pazopanib [93]. In all, these and other trials have convincingly demonstrated the activity of AAT in HGSOC, leading to the approval of bevacizumab for treatment in both the adjuvant and recurrent settings. New trials are evaluating the efficacy of anti-angiogenic drugs in combination with immune modulators or PARP inhibitors for treatment of gynecologic malignancies.

6. Interactions with the Mesothelial Matrix

In order to form secondary tumors, disseminated OC cell spheroids floating in the peritoneal cavity rely on their capacity to adhere to the mesothelial lining covering the peritoneal cavity and abdominal organs. During dissemination from the primary site, OC cells lose E-cadherin expression (Figure 2, upper left) and upregulate α5 integrin, which was proposed as a therapeutic target [94]. Secondary site invasion occurs upon displacement of the mesothelial monolayer cells (Figure 2, lower right), with cancer cells invading and submerging into the subjacent environment. The clearance of mesothelial cells is enabled by traction forces mediated by myosin and generated by the adhesion complex molecules, α5 integrin and talin-1, and is more efficiently accomplished by reprogrammed mesenchymal-like OC cells [95,96]. Other receptors that play a role in OC cell adhesion to mesothelium include CD44 and β1 integrin (Iβ1) [97]. OC cell-derived TGF-β1 upregulates fibronectin (FN) expression in mesotheial cells [98]. The adhesion of OC cells to the FN matrix secreted by mesothelial cells [98] is dependent upon α5β1 integrin clustering and talin recruitment to stabilize the adhesions (Figure 2) [95]. Integrin clustering is induced by secreted tissue transglutaminase (TG2), which forms a bridge connecting Iβ1 and FN together at the cell surface [99]. This event induces downstream RhoA activation and suppression of Src–p190RhoGAP signaling. A focus of our laboratory's work was to understand the role played by the TG2-Iβ1-FN ternary complex in the process of OC metastasis and to test it as a new therapeutic target. By using OC orthotopic and ip xenografts, we showed that TG2 knock-down blocked peritoneal dissemination of ovarian tumors through a mechanism dependent on β1-integrin mediated cell adhesion and signaling [100,101]. Our recent results also demonstrate that engagement of integrin β1 facilitated by TG2 activates β-catenin signaling and stemness associated pathways in in vivo and organoid models of HGSOC [102,103].

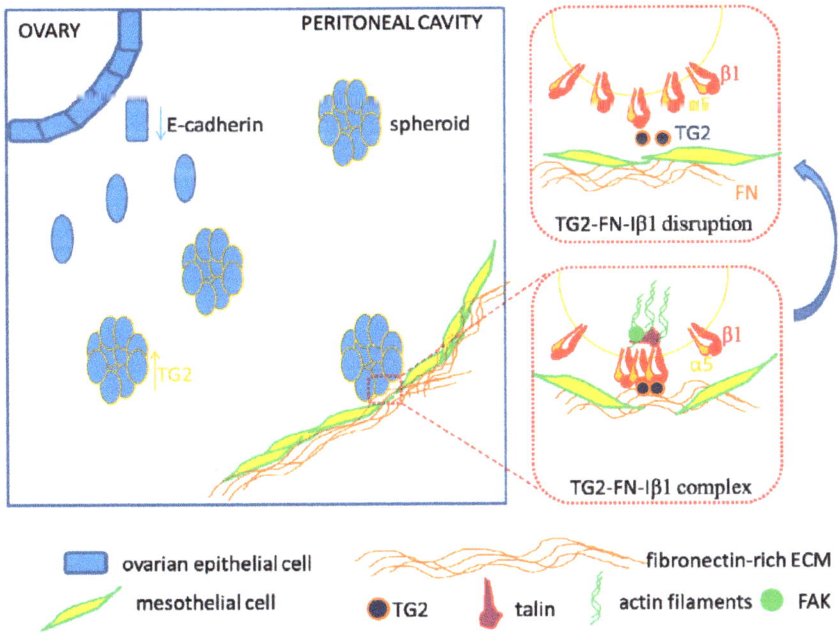

Figure 2. Ovarian cancer cells adhere to the mesothelial lining during tumor dissemination in the peritoneal cavity. Upon activation of EMT (epithelial-to-mesenchymal transition), cells progressively shed from the primary tumor into the peritoneal cavity (blue square). During the EMT process, there is a decrease in E-cadherin expression and increase in proteins associated to a mesenchymal phenotype, such as vimentin, tissue transglutaminase (TG2) and integrins. Cells that survive in the environment of the peritoneal cavity form spheroids. Spheroids attach to the fibronectin (FN) rich matrix secreted by the mesothelial cells, clear the subjacent monolayer and invade the underling tissue. These adhesion and invasion processes are mediated by interactions of integrin-β1 receptors with the FN fibrils in the ECM. Upon FN binding, α5β1 integrin receptors undergo clustering, which is enhanced by molecular bridges with TG2. Next, talin is recruited to the adhesion complex and provides the necessary traction force for the mesothelial monolayer displacement (red dotted bottom square). Also, "outside-in" signaling downstream of β1 integrin is activated, inducing focal adhesion kinase (FAK) phosphorylation. Therapeutic strategies targeting the TG2-FN-Iβ1 complex aim at interfering with the cell adhesion process and consequently preventing OC metastasis (red dotted top square).

7. Targeting Ovarian Cancer Cell Adhesion to the Peritoneal Matrix

Several strategies have been tested in an effort to block OC peritoneal dissemination. Treatment with blocking antibodies against integrins and the CD44 receptor were shown to inhibit OC cells adhesion to the mesothelial layer for short time intervals [104–106]. As α5β1 integrin is expressed on both OC cells as well as on the endothelial cells forming microvessels [107], it was expected that targeting this heterodimer (Figure 2, top square) will interfere with tumor growth and metastasis in many types of solid cancers, including OC [108]. Currently several drugs targeting integrins are under development (reviewed in [109]).

Volociximab, a chimeric antibody that binds α5β1 integrin with high affinity, was shown to block growth and dissemination of OC xenograft models [94]. However, the phase II clinical trial testing volociximab in patients with recurrent, platinum-resistant OC failed to demonstrate benefit although the drug was well tolerated [110]. Intetumumab (CNTO-95), a human αv-integrin specific monoclonal antibody that targets both αvβ3- and αvβ5-integrins showed anti-tumor and anti-angiogenic effects in xenografts models of breast cancer [111,112]. In a phase I clinical trial including patients with advanced solid tumors, one patient with ovarian carcinosarcoma had stable disease for six months [113]. Other integrin-blocking antibodies, such as etaracizumab, the humanized version of anti-αvβ3-integrin LM609 had minimal therapeutic benefit in other cancers [114]. Cilengitide is a stable cyclic pentapeptide containing an Arg-Gly-Asp (RGD) motif which allows selective binding to αvβ3 and αvβ5 integrins [115]. Cilengitide was tested in brain tumors and was found to not increase OS in glioblastoma patients during a phase III trial [116]. Given that αvβ3 integrin expression by tumor cells correlates with a favorable prognosis in OC patients [117], targeting this integrin might be a less appropriate strategy for OC. The initial disappointment with integrin targeting strategies may be related to their prior testing in the recurrent, advanced setting as single agents. Development of combination regimens and testing of these blocking antibodies in patients with low volume metastatic disease might overcome the lack of clinical success with this intervention.

FN is one of the most abundant ECM proteins in the omentum and peritoneum [118]. Adhesion of OC cells to FN via α5β1 integrin impacts "outside-in signaling" by inducing phosphorylation of focal adhesion kinase (FAK) either directly [119] or through c-Met [108]. This can further lead to activation of mitogenic pathways [120] which support tumor growth [121]. The β1 integrin–FN interaction is further enhanced by the bridging activity of TG2, a protein we discovered to be overexpressed in OC [122]. Previous work in our group has emphasized the importance of TG2 in the OC metastatic process, by providing evidence of its involvement in promoting OC cells' epithelial-to-mesenchymal transition through activation of non-canonical NF-κB [123], increasing cell proliferation by regulating β-catenin signaling [102], enhancing peritoneal dissemination [100], and increasing invasion by regulating MMP-2 [124]. As proof of principle that the TG2-FN-Iβ1 complex represents an interesting target in OC, we used a function-blocking antibody which targeted the FN binding domain of TG2, and showed that this antibody blocked OC spheroid proliferation and tumor initiating capacity by disrupting the interaction between OC stem cells and their niche [103].

To discover potent and selective TG2-FN inhibitors we used both virtual docking and high throughput screening strategies. Through an initial in silico docking approach, we identified a small molecule inhibitor capable of disrupting this complex and of blocking cancer cell adhesion to the FN matrix [125]. Subsequent efforts used an AlphaLISA-based assay adapted to high-throughput screening and applied to the ChemDiv library leading to the discovery and validation of several small molecules [126]. One hit selected from this screen (TG53) was validated in vitro to be an efficient inhibitor of OC cell adhesion to FN, migration and invasion. Future efforts focus on optimizing this compound through structure–activity relationship-based strategies to generate more selective, potent and drug-like compounds which block the TG2-FN protein–protein interaction and ultimately prevent OC metastasis.

8. Tumor Immune Response in Ovarian Cancer

Preclinical models and retrospective cohort analyses of human tumor specimens have demonstrated that the interaction between cancer cells and the host immune defense plays an important role harnessing tumor progression. There are several immune cell subsets relevant for tumor progression and response to immunotherapy [127]. These are classified in two categories: immune reactive and immune suppressive cells. The immune reactive cells include primarily cytotoxic T lymphocytes and activated $CD4^+$ T cells. The immune suppressive cells are myeloid lineage subpopulations known as myeloid-derived suppressor cells (MDSCs), tumor-associated macrophages (TAMs, especially M2 subtype), dendritic cells (DCs) and the lymphocyte subsets of T helper cells (Th2 subtype) and T regulatory cells (Tregs). A seminal study showed that the presence of $CD3^+$ tumor infiltrating lymphocytes (TILs) in OC is associated with increased survival [128]. The 5-year overall survival (OS) was 38% for patients whose tumors contained T cells compared to 4.5% for those whose tumors were devoid of T cells. Subsequently, a strong association between the presence of $CD8^+$ TILs and favorable clinical outcomes of HGSOC was recognized [129–131]. The $CD8^+$ to T regulatory (Tregs) cells ratio was also shown to correlate with increased survival of OC patients [130]. More recently, the presence of CD8$^+$ cells expressing the TNFR-family receptor CD137 (4-1BB) was reported as a prognostic marker associated with improved survival of OC patients [132]. A recent study evaluated the immune TME landscape in differentially growing metastases after several therapy cycles in an OC patient and reported heterogeneity in immune infiltrates that explained the evolution of tumor masses over nine years period [133]. This unique report revealed a correlation between the regressing or stable metastases and the presence of oligoclonal expanding T cells. Conversely, progressing tumors showed a lack of infiltration with anti-cancer lymphocytes. This study reinforces the importance of the tumor immune microenvironment to the outcome of OC disease. In all, these and other studies [134] strongly support the role of anti-tumor immunity as a key regulator in the evolution of the disease.

Enhancing the naturally occurring immune defense could therefore play an important role harnessing disease progression. Immunotherapy has demonstrated efficacy in various malignancies [135,136]. Several immune modulatory approaches (vaccines, IL2, CTLA-4 directed antibodies, adoptive transfer of activated T cells) have been tested in OC, with promising results in early interventions [137,138]. However, the impact of immunotherapy on the survival of OC patients remains unproven and predictive markers of positive outcomes remain undefined, highlighting the need to further optimize such strategies.

9. Immune Checkpoint Inhibitors

Recent advances have brought attention to the programmed cell death protein-1 (PD-1) mechanism used by cancer cells to evade immune surveillance, which can be effectively targeted by inhibitory antibodies [139]. This strategy demonstrated impressive clinical activity in several solid tumors (melanoma, lymphoma, renal, lung, and bladder cancer) leading to new FDA-approved interventions [140–142]. PD-1 signaling blocks T-cell activation keeping nascent T-cells in check and preventing immune responses against normal tissues. During cancer progression, this inhibitory pathway is activated by upregulating the expression of PD ligands (PD-L1 and PD-L2) on tumor and immune cells and permits evasion from immune surveillance [139]. The significance of the PD1 pathway to OC progression has been investigated; however, the emerging evidence is conflicting. On one hand, initial studies showed that the increased PD-L1 expression in ovarian tumors correlates with decreased intra-tumoral CD8$^+$ lymphocytes and worse patient survival [143]. Presence of dendritic cells expressing PD1 in the OC microenvironment was also found to be associated with decreased numbers of TILs and suppressed T cell activity [144], consistent with the concept that PD-L1 represents an escape mechanism. On the other hand, more recent studies using specific PD1 and PD-L1 detection antibodies provide evidence to the contrary. Two reports showed that expression of PD-L1 on immune cells in the tumor milieu, including on tumor associated macrophages (TAMs), is associated with increased total numbers of TILs and better survival in HGSOC [145,146]. It remains unresolved how expression of the PD1 pathway elements can be causally linked to a favorable prognosis in OC. It is possible that expression of PD-L1 reflects an active immune TME (defined by increased TILs density) able to attack and eliminate the tumor, or that PD-L1$^+$ TILs have a yet to be defined regulatory role in the immune response mechanism. Additional support for clinical interventions targeting this pathway includes that PD-1/PD-L1 blockade restored anti-tumor immunity in an OC xenograft model [147]. Two recent clinical trials tested PD-1 (pembrolizumab) and PD-L1 (avelumab) inhibitory antibodies in women with recurrent OC, reporting response rates of 11% (pembrolizumab) and 10% (avelumab), with 23% and 40% additional patients experiencing stable disease, respectively [148,149]. These early data suggest that immune checkpoint blockade in OC has defined, albeit modest activity.

Another emerging concept refers to the tumor neoantigen load as an important regulator of anti-tumor immune response and a marker for response to treatment [150,151]. Along these lines, a recent study showed that BRCA 1 and 2 mutated ovarian tumors are characterized by increased neoantigen load and that this correlates with increased number of TILs, increased expression of PD1 and PDL1, and is linked to improved clinical outcome [152]. These data support exploring

PD1 blockade in OC and continued investigation of the complex immune milieu associated with ovarian tumors. Therefore, identifying rational combinations to enhance the activity of PD1 blocking antibodies in OC and further analysis of the immune tumor milieu to identify predictive markers is necessary. Our group is exploring the combination of the PD1 inhibitor pembrolizumab and the DNA hypomethylating agent guadecitabine in women with recurrent platinum-resistant ovarian cancer (NCT02901899), testing the hypothesis that epigenomic priming will enhance the activity of immune checkpoint inhibitors.

10. Targeting Tumor Associated Macrophages (TAMs) and Myeloid-Derived Suppressor Cells (MDSCs)

Myeloid cells are frequently observed in the stroma of growing tumors [153]. The role of myeloid suppressor cells has been recognized first in late 1970s. In 2007, the term myeloid-derived suppressor cells (MDSCs) was coined for "bone marrow-derived cells of myeloid lineage comprising myeloid precursors and immature macrophages, granulocytes, and DCs, characterized by their high potential to suppress T cells" [154]. Immature myeloid suppressor cells were shown even earlier to accumulate in a variety of immune-related diseases, including cancer [155,156]. MDSC subsets were found to be responsible for immune suppression in 10 pre-clinical models of tumorigenesis [157]. In OC, macrophages are mainly found in ascites or infiltrate of the omentum. TAMs in the omentum were shown to harbor predominantly the M2 phenotype and to facilitate tumor progression [158,159]. Peritoneal TAMs support this process by secreting cytokines such as IL-6 and IL-8 [160]. In the ascites, M2 macrophage-like TAMs were found in the center of spheroids, where they participated in mechanisms supporting tumor cell proliferation and migration during OC metastasis [161]. The main signaling pathway involved in TAMs cross-talk to floating spheroid cancer cells was EGF–EGFR. TAMs promoted cancer cell invasiveness by activating the NF-κB and JNK signaling pathways [162]. Reversely, peritoneal macrophages were shown to adopt the M2 phenotype under the influence of OC cells expressing homeobox gene HOXA9 [163]. PD-L1 was primarily expressed by $CD68^+$ TAMs rather than tumor cells in HGSOC, and often colocalized with both cytotoxic T cells as well as T regulatory cells and was a positive prognostic marker [146].

The contribution of MDSCs defined as harboring $Lin^-CD45^+CD33^+$ markers combination was studied in a cohort of patients with HGSOC [164]. MDSCs comprised 37% of non-neoplastic cells in the TME and were responsible for inhibiting T-cell immunity, by blocking both T cell proliferation and effector function. Increased tumor MDSCs inversely correlated with $CD8^+$ TILs and overall survival in advanced OC [165]. Interestingly, the corresponding $Lin^-CD45^+CD33^+$ fraction in patients' blood did not have the same properties. MDSCs were shown to support metastasis and a cancer stem cell phenotype. Mechanistically, it was shown that tumor-resident MDSCs enhance stemness via microRNA101, which targets co-repressor gene C-terminal binding protein-2 (CtBP2) 3′-UTR region and interferes with its binding at NANOG, OCT4/3, and SOX2 promoters in primary OC cells [164]. Primary ovarian tumors expressing high levels of Snail were shown to recruit increased number of $CD33^+$ MDSCs through secretion of the CXCR2 ligands CXCL1/2 [166,167]. Therefore, blocking CXCR2 would represent a therapeutic approach for Snail-high OC tumors.

Targeting immature myeloid cells and their cross-talk with other immune cells and cancer cells is a potential strategy of combating tumor progression. Several classes of therapeutics targeting MDSCs or TAMs have been described and were recently reviewed [167]. They include agents which promote MDSCs apoptosis, antibodies that induce MDSCs and/or TAMs depletion, compounds that induce immature myeloid cells differentiation (such as retinoic acid, vitamin D3 or HDACi), inhibitors of immune suppression function (sildenafil, triterpenoids, inhibitors of COX-2, inducible nitric oxide), compounds which block recruitment (by targeting chemokines and chemokine receptors) or MDSCs proliferation, and lastly TAM reprogramming factors. Given that TAMs and MDSCs mediate resistance to immunotherapy targeting, this immune suppressive cell population could increase the success rate of checkpoint blockade inhibitors [168].

Several strategies have been tested in preclinical models, but progress towards clinical is still ongoing. For example, almetuzumab, which targets CD52 expressed by vascular leukocytes and Tie2$^+$ monocytes, was shown to have anti-myeloid and anti-angiogenic properties in OC models [169]. Anti-CD52 therapy decreased tumor growth in an OC murine model. Additionally, ovarian TAMs express high levels of folate receptor-2, which can be targeted by using methotrexate loaded G5-dendrimers (G5-MTX) [170]. Noteworthy, these G5-MTX nanoparticles were shown to overcome resistance to anti-VEGF-A therapy in OC preclinical models. Epigenetic modulators have also been shown to alter the myeloid population, triggering anti-tumor immune responses. For example, the bromodomain inhibitor JQ1 significantly reduced PD-L1 expression on TAMs and dendritic cells, induced increased T cell cytotoxic activity and suppressed OC tumor growth in preclinical models [171]. A combination of histone deacetylase inhibitors (HDACi) and DNA methyltransferase inhibitor (DNMTi) was shown to reduce TAMs and increase T and NK cell activation, delaying tumor progression in preclinical models [172]. The combination of DNMTi/HDACi also synergized with the immune checkpoint inhibitors. Clinical trials testing HDACi and DNMTi with anti-PD1 therapy in patients with recurrent OC are ongoing. Lastly, catumaxomab is a humanized antibody that targets three different cell types: tumor cells (via epithelial cell adhesion molecule (EpCAM) binding); T-cells (via CD3 binding); and accessory cells (macrophages, dendritic cells, and natural killer cells) via type I, IIa, and III Fcγ receptors (FcγR). Subsequently, catumaxomab induces several effects, including T-cell-mediated tumor lysis, antibody-dependent cell-mediated cytotoxicity, and phagocytosis via activation of NK cells and TAMs. Catumaxomab is administered intra-peritoneally and was shown to be clinically active in patients with malignant ascites, leading to its approval in Europe for the treatment of EpCAM$^+$ tumors associated with ascites, including HGSOC [173].

11. Conclusions

New targets at the interface between HGSOC cells and the TME have been characterized. Targeted treatments, alone or in combination with chemotherapy, are emerging and, in some situations, are already impacting clinical outcomes in women with HGSOC. Anti-angiogenic therapy in combination with chemotherapy has significantly improved the survival of women with advanced OC and has become part of the standard approach. In contrast, CAFs-directed strategies or therapeutics targeting cell adhesion to the matrix remain less impressive. Future development of combination and sequencing strategies based on a refined understanding of tumor biology and cross-talking pathways is critically needed. While immune interventions are still being optimized, early results suggest that combination strategies are needed to overcome the immune tolerant milieu of HGSOC. This could be due to silencing of tumor antigen and low tumor mutational burden, which render the ovarian tumors to be "cold", or to an infiltration of immunosuppressive cells. Therefore, current approaches investigate dual immune targeting or combinations with interventions that de-repress tumor antigens through epigenetic reprogramming or which increase the tumor mutational burden by inducing DNA damage. It is clear that in order to improve clinical outcomes in this fatal malignancy, interventions affecting both cancer cells and the stroma need to be implemented. Thus, we anticipate that clinical trials will continue to explore rationally designed combinations and/or sequences of therapies targeting vulnerabilities of both tumor cells and the TME.

Funding: This research was supported through funding from US Department of Veterans' Affairs, BX000792-06 and the Department of Defense W81XWH-17-1-0141 (to Daniela Matei), and the National Cancer Institute Training Grant T32 CA070085 (to Nkechiyere G. Nwani).

Conflicts of Interest: D.M. received consulting honoraria from Roche. The other authors declare no conflict of interest. The funders had no role in the design of the study; in the collection, analyses, or interpretation of data; in the writing of the manuscript, and in the decision to publish the results.

References

1. Freedman, R.S.; Deavers, M.; Liu, J.; Wang, E. Peritoneal inflammation—A microenvironment for epithelial ovarian cancer (eoc). *J. Transl. Med.* **2004**, *2*, 23. [CrossRef] [PubMed]
2. Said, N.; Socha, M.J.; Olearczyk, J.J.; Elmarakby, A.A.; Imig, J.D.; Motamed, K. Normalization of the ovarian cancer microenvironment by sparc. *Mol. Cancer Res.* **2007**, *5*, 1015–1030. [CrossRef] [PubMed]
3. Hurteau, J.; Rodriguez, G.C.; Whitaker, R.S.; Shah, S.; Mills, G.; Bast, R.C.; Berchuck, A. Transforming growth factor-beta inhibits proliferation of human ovarian cancer cells obtained from ascites. *Cancer* **1994**, *74*, 93–99. [CrossRef]
4. Gopinathan, G.; Milagre, C.; Pearce, O.M.; Reynolds, L.E.; Hodivala-Dilke, K.; Leinster, D.A.; Zhong, H.; Hollingsworth, R.E.; Thompson, R.; Whiteford, J.R.; et al. Interleukin-6 stimulates defective angiogenesis. *Cancer Res.* **2015**, *75*, 3098–3107. [CrossRef] [PubMed]
5. Nozawa, H.; Chiu, C.; Hanahan, D. Infiltrating neutrophils mediate the initial angiogenic switch in a mouse model of multistage carcinogenesis. *Proc. Natl. Acad. Sci. USA* **2006**, *103*, 12493–12498. [CrossRef] [PubMed]
6. Nieman, K.M.; Kenny, H.A.; Penicka, C.V.; Ladanyi, A.; Buell-Gutbrod, R.; Zillhardt, M.R.; Romero, I.L.; Carey, M.S.; Mills, G.B.; Hotamisligil, G.S.; et al. Adipocytes promote ovarian cancer metastasis and provide energy for rapid tumor growth. *Nat. Med.* **2011**, *17*, 1498–1503. [CrossRef] [PubMed]
7. Tarin, D.; Croft, C.B. Ultrastructural features of wound healing in mouse skin. *J. Anat.* **1969**, *105*, 189–190. [PubMed]
8. Chang, H.Y.; Chi, J.T.; Dudoit, S.; Bondre, C.; van de Rijn, M.; Botstein, D.; Brown, P.O. Diversity, topographic differentiation, and positional memory in human fibroblasts. *Proc. Natl. Acad. Sci. USA* **2002**, *99*, 12877–12882. [CrossRef] [PubMed]
9. McAnulty, R.J. Fibroblasts and myofibroblasts: Their source, function and role in disease. *Int. J. Biochem. Cell Biol.* **2007**, *39*, 666–671. [CrossRef] [PubMed]
10. Ryan, G.B.; Cliff, W.J.; Gabbiani, G.; Irle, C.; Montandon, D.; Statkov, P.R.; Majno, G. Myofibroblasts in human granulation tissue. *Hum. Pathol.* **1974**, *5*, 55–67. [CrossRef]
11. Kalluri, R.; Zeisberg, M. Fibroblasts in cancer. *Nat. Rev. Cancer* **2006**, *6*, 392–401. [CrossRef] [PubMed]
12. Bhowmick, N.A.; Chytil, A.; Plieth, D.; Gorska, A.E.; Dumont, N.; Shappell, S.; Washington, M.K.; Neilson, E.G.; Moses, H.L. Tgf-beta signaling in fibroblasts modulates the oncogenic potential of adjacent epithelia. *Science* **2004**, *303*, 848–851. [CrossRef] [PubMed]
13. Liao, D.; Luo, Y.; Markowitz, D.; Xiang, R.; Reisfeld, R.A. Cancer associated fibroblasts promote tumor growth and metastasis by modulating the tumor immune microenvironment in a 4t1 murine breast cancer model. *PLoS ONE* **2009**, *4*, e7965. [CrossRef] [PubMed]
14. Kendall, R.T.; Feghali-Bostwick, C.A. Fibroblasts in fibrosis: Novel roles and mediators. *Front. Pharmacol.* **2014**, *5*, 123. [CrossRef] [PubMed]
15. Hanahan, D.; Coussens, L.M. Accessories to the crime: Functions of cells recruited to the tumor microenvironment. *Cancer Cell* **2012**, *21*, 309–322. [CrossRef] [PubMed]
16. Madar, S.; Goldstein, I.; Rotter, V. 'Cancer associated fibroblasts'—More than meets the eye. *Trends Mol. Med.* **2013**, *19*, 447–453. [CrossRef] [PubMed]
17. Giusti, I.; Francesco, M.D.; Ascenzo, S.; Palmerini, M.G.; Macchiarelli, G.; Carta, G.; Dolo, V. Ovarian cancer-derived extracellular vesicles affect normal human fibroblast behavior. *Cancer Biol. Ther.* **2018**, *19*, 722–734. [CrossRef] [PubMed]
18. Yao, Q.; Qu, X.; Yang, Q.; Wei, M.; Kong, B. Clic4 mediates tgf-beta1-induced fibroblast-to-myofibroblast transdifferentiation in ovarian cancer. *Oncol. Rep.* **2009**, *22*, 541–548. [PubMed]
19. Fang, T.; Lv, H.; Lv, G.; Li, T.; Wang, C.; Han, Q.; Yu, L.; Su, B.; Guo, L.; Huang, S.; et al. Tumor-derived exosomal mir-1247-3p induces cancer-associated fibroblast activation to foster lung metastasis of liver cancer. *Nat. Commun.* **2018**, *9*, 191. [CrossRef] [PubMed]
20. Jeon, E.S.; Moon, H.J.; Lee, M.J.; Song, H.Y.; Kim, Y.M.; Cho, M.; Suh, D.S.; Yoon, M.S.; Chang, C.L.; Jung, J.S.; et al. Cancer-derived lysophosphatidic acid stimulates differentiation of human mesenchymal stem cells to myofibroblast-like cells. *Stem Cells* **2008**, *26*, 789–797. [CrossRef] [PubMed]
21. Ko, S.Y.; Barengo, N.; Ladanyi, A.; Lee, J.S.; Marini, F.; Lengyel, E.; Naora, H. Hoxa9 promotes ovarian cancer growth by stimulating cancer-associated fibroblasts. *J. Clin. Investig.* **2012**, *122*, 3603–3617. [CrossRef] [PubMed]

22. Dotto, G.P.; Weinberg, R.A.; Ariza, A. Malignant transformation of mouse primary keratinocytes by harvey sarcoma virus and its modulation by surrounding normal cells. *Proc. Natl. Acad. Sci. USA* **1988**, *85*, 6389–6393. [CrossRef] [PubMed]
23. Ozdemir, B.C.; Pentcheva-Hoang, T.; Carstens, J.L.; Zheng, X.F.; Wu, C.C.; Simpson, T.R.; Laklai, H.; Sugimoto, H.; Kahlert, C.; Novitskiy, S.V.; et al. Depletion of carcinoma-associated fibroblasts and fibrosis induces immunosuppression and accelerates pancreas cancer with reduced survival. *Cancer Cell* **2014**, *25*, 719–734. [CrossRef] [PubMed]
24. Cornil, I.; Theodorescu, D.; Man, S.; Herlyn, M.; Jambrosic, J.; Kerbel, R.S. Fibroblast cell-interactions with human-melanoma cells affect tumor-cell growth as a function of tumor progression. *Proc. Natl. Acad. Sci. USA* **1991**, *88*, 6028–6032. [CrossRef] [PubMed]
25. Givel, A.M.; Kieffer, Y.; Scholer-Dahirel, A.; Sirven, P.; Cardon, M.; Pelon, F.; Magagna, I.; Gentric, G.; Costa, A.; Bonneau, C.; et al. Mir200-regulated cxcl12beta promotes fibroblast heterogeneity and immunosuppression in ovarian cancers. *Nat. Commun.* **2018**, *9*, 1056. [CrossRef] [PubMed]
26. Wei, D.; Geng, F.; Liang, S.; Zhao, H.; Liu, M.; Wang, H. Caf-derived hgf promotes cell proliferation and drug resistance by up-regulating the c-met/pi3k/akt and grp78 signalling in ovarian cancer cells. *Biosci. Rep.* **2017**, *37*, BSR20160470.
27. Cirri, P.; Chiarugi, P. Cancer associated fibroblasts: The dark side of the coin. *Am. J. Cancer Res.* **2011**, *1*, 482–497. [PubMed]
28. Henriksson, M.L.; Edin, S.; Dahlin, A.M.; Oldenborg, P.A.; Oberg, A.; Van Guelpen, B.; Rutegard, J.; Stenling, R.; Palmqvist, R. Colorectal cancer cells activate adjacent fibroblasts resulting in fgf1/fgfr3 signaling and increased invasion. *Am. J. Pathol.* **2011**, *178*, 1387–1394. [CrossRef] [PubMed]
29. Matsuo, Y.; Ochi, N.; Sawai, H.; Yasuda, A.; Takahashi, H.; Funahashi, H.; Takeyama, H.; Tong, Z.; Guha, S. Cxcl8/il-8 and cxcl12/sdf-1alpha co-operatively promote invasiveness and angiogenesis in pancreatic cancer. *Int. J. Cancer* **2009**, *124*, 853–861. [CrossRef] [PubMed]
30. Orimo, A.; Gupta, P.B.; Sgroi, D.C.; Arenzana-Seisdedos, F.; Delaunay, T.; Naeem, R.; Carey, V.J.; Richardson, A.L.; Weinberg, R.A. Stromal fibroblasts present in invasive human breast carcinomas promote tumor growth and angiogenesis through elevated sdf-1/cxcl12 secretion. *Cell* **2005**, *121*, 335–348. [CrossRef] [PubMed]
31. Lau, T.S.; Chan, L.K.; Wong, E.C.; Hui, C.W.; Sneddon, K.; Cheung, T.H.; Yim, S.F.; Lee, J.H.; Yeung, C.S.; Chung, T.K.; et al. A loop of cancer-stroma-cancer interaction promotes peritoneal metastasis of ovarian cancer via tnfalpha-tgfalpha-egfr. *Oncogene* **2017**, *36*, 3576–3587. [CrossRef] [PubMed]
32. Mitchell, M.I.; Engelbrecht, A.M. Metabolic hijacking: A survival strategy cancer cells exploit? *Crit. Rev. Oncol. Hematol.* **2017**, *109*, 1–8. [CrossRef] [PubMed]
33. Zhao, L.J.; Ji, G.L.; Le, X.B.; Wang, C.; Xu, L.; Feng, M.; Zhang, Y.G.; Yang, H.L.; Xuan, Y.; Yang, Y.F.; et al. Long noncoding rna linc00092 acts in cancer-associated fibroblasts to drive glycolysis and progression of ovarian cancer. *Cancer Res.* **2017**, *77*, 1369–1382. [CrossRef] [PubMed]
34. Hadler-Olsen, E.; Winberg, J.O.; Uhlin-Hansen, L. Matrix metalloproteinases in cancer: Their value as diagnostic and prognostic markers and therapeutic targets. *Tumor Biol.* **2013**, *34*, 2041–2051. [CrossRef] [PubMed]
35. Lederle, W.; Hartenstein, B.; Meides, A.; Kunzelmann, H.; Werb, Z.; Angel, P.; Mueller, M.M. Mmp13 as a stromal mediator in controlling persistent angiogenesis in skin carcinoma. *Carcinogenesis* **2010**, *31*, 1175–1184. [CrossRef] [PubMed]
36. Garinchesa, P.; Old, L.J.; Rettig, W.J. Cell-surface glycoprotein of reactive stromal fibroblasts as a potential antibody target in human epithelial cancers. *Proc. Natl. Acad. Sci. USA* **1990**, *87*, 7235–7239. [CrossRef]
37. Yang, W.; Han, W.; Ye, S.; Liu, D.; Wu, J.; Liu, H.; Li, C.; Chen, H. Fibroblast activation protein-alpha promotes ovarian cancer cell proliferation and invasion via extracellular and intracellular signaling mechanisms. *Exp. Mol. Pathol.* **2013**, *95*, 105–110. [CrossRef] [PubMed]
38. Mhawech-Fauceglia, P.; Yan, L.; Sharifian, M.; Ren, X.; Liu, S.; Kim, G.; Gayther, S.A.; Pejovic, T.; Lawrenson, K. Stromal expression of fibroblast activation protein alpha (fap) predicts platinum resistance and shorter recurrence in patients with epithelial ovarian cancer. *Cancer Microenviron.* **2015**, *8*, 23–31. [CrossRef] [PubMed]

39. Cai, J.; Tang, H.; Xu, L.; Wang, X.; Yang, C.; Ruan, S.; Guo, J.; Hu, S.; Wang, Z. Fibroblasts in omentum activated by tumor cells promote ovarian cancer growth, adhesion and invasiveness. *Carcinogenesis* **2012**, *33*, 20–29. [CrossRef] [PubMed]
40. Zhang, Y.; Ertl, H.C.J. Depletion of fap(+) cells reduces immunosuppressive cells and improves metabolism and functions cd8(+)t cells within tumors. *Oncotarget* **2016**, *7*, 23282–23299. [CrossRef] [PubMed]
41. Kraman, M.; Bambrough, P.J.; Arnold, J.N.; Roberts, E.W.; Magiera, L.; Jones, J.O.; Gopinathan, A.; Tuveson, D.A.; Fearon, D.T. Suppression of antitumor immunity by stromal cells expressing fibroblast activation protein-alpha. *Science* **2010**, *330*, 827–830. [CrossRef] [PubMed]
42. Cheng, J.D.; Valianou, M.; Canutescu, A.A.; Jaffe, E.K.; Lee, H.O.; Wang, H.; Lai, J.H.; Bachovchin, W.W.; Weiner, L.M. Abrogation of fibroblast activation protein enzymatic activity attenuates tumor growth. *Mol. Cancer Ther.* **2005**, *4*, 351–360. [PubMed]
43. Kelly, T. Fibroblast activation protein-alpha and dipeptidyl peptidase iv (cd26): Cell-surface proteases that activate cell signaling and are potential targets for cancer therapy. *Drug Resist. Updat.* **2005**, *8*, 51–58. [CrossRef] [PubMed]
44. Wen, Y.; Wang, C.T.; Ma, T.T.; Li, Z.Y.; Zhou, L.N.; Mu, B.; Leng, F.; Shi, H.S.; Li, Y.O.; Wei, Y.Q. Immunotherapy targeting fibroblast activation protein inhibits tumor growth and increases survival in a murine colon cancer model. *Cancer Sci.* **2010**, *101*, 2325–2332. [CrossRef] [PubMed]
45. Fabregat, I.; Fernando, J.; Mainez, J.; Sancho, P. Tgf-beta signaling in cancer treatment. *Curr. Pharm. Des.* **2014**, *20*, 2934–2947. [CrossRef] [PubMed]
46. Akhurst, R.J. Targeting tgf-beta signaling for therapeutic gain. *Cold Spring Harb. Perspect. Biol.* **2017**, *9*, a022301. [CrossRef] [PubMed]
47. Abendstein, B.; Stadlmann, S.; Knabbe, C.; Buck, M.; Muller-Holzner, E.; Zeimet, A.G.; Marth, C.; Obrist, P.; Krugmann, J.; Offner, F.A. Regulation of transforming growth factor-beta secretion by human peritoneal mesothelial and ovarian carcinoma cells. *Cytokine* **2000**, *12*, 1115–1119. [CrossRef] [PubMed]
48. Yamamura, S.; Matsumura, N.; Mandai, M.; Huang, Z.; Oura, T.; Baba, T.; Hamanishi, J.; Yamaguchi, K.; Kang, H.S.; Okamoto, T.; et al. The activated transforming growth factor-beta signaling pathway in peritoneal metastases is a potential therapeutic target in ovarian cancer. *Int. J. Cancer* **2012**, *130*, 20–28. [CrossRef] [PubMed]
49. Maier, A.; Peille, A.L.; Vuaroqueaux, V.; Lahn, M. Anti-tumor activity of the tgf-beta receptor kinase inhibitor galunisertib (ly2157299 monohydrate) in patient-derived tumor xenografts. *Cell Oncol. (Dordr.)* **2015**, *38*, 131–144. [CrossRef] [PubMed]
50. Gao, Y.; Shan, N.; Zhao, C.; Wang, Y.; Xu, F.; Li, J.; Yu, X.; Gao, L.; Yi, Z. Ly2109761 enhances cisplatin antitumor activity in ovarian cancer cells. *Int. J. Clin. Exp. Pathol.* **2015**, *8*, 4923–4932. [PubMed]
51. Kovacs, R.J.; Maldonado, G.; Azaro, A.; Fernandez, M.S.; Romero, F.L.; Sepulveda-Sanchez, J.M.; Corretti, M.; Carducci, M.; Dolan, M.; Gueorguieva, I.; et al. Cardiac safety of tgf-beta receptor i kinase inhibitor ly2157299 monohydrate in cancer patients in a first-in-human dose study. *Cardiovasc. Toxicol.* **2015**, *15*, 309–323. [CrossRef] [PubMed]
52. Zhang, M.J.; Liu, T.B.; Xia, B.R.; Yang, C.Y.; Hou, S.Y.; Xie, W.L.; Lou, G. Platelet-derived growth factor d is a prognostic biomarker and is associated with platinum resistance in epithelial ovarian cancer. *Int. J. Gynecol. Cancer* **2018**, *28*, 323–331. [CrossRef] [PubMed]
53. Matei, D.; Chang, D.D.; Jeng, M.H. Imatinib mesylate (gleevec) inhibits ovarian cancer cell growth through a mechanism dependent on platelet-derived growth factor receptor alpha and akt inactivation. *Clin. Cancer Res.* **2004**, *10*, 681–690. [CrossRef] [PubMed]
54. Pietras, K.; Pahler, J.; Bergers, G.; Hanahan, D. Functions of paracrine pdgf signaling in the proangiogenic tumor stroma revealed by pharmacological targeting. *PLoS Med.* **2008**, *5*, e19. [CrossRef] [PubMed]
55. Haubeiss, S.; Schmid, J.O.; Murdter, T.E.; Sonnenberg, M.; Friedel, G.; van der Kuip, H.; Aulitzky, W.E. Dasatinib reverses cancer-associated fibroblasts (cafs) from primary lung carcinomas to a phenotype comparable to that of normal fibroblasts. *Mol. Cancer* **2010**, *9*, 1–8. [CrossRef] [PubMed]
56. Matei, D.; Emerson, R.E.; Schilder, J.; Menning, N.; Baldridge, L.A.; Johnson, C.S.; Breen, T.; McClean, J.; Stephens, D.; Whalen, C.; et al. Imatinib mesylate in combination with docetaxel for the treatment of patients with advanced, platinum-resistant ovarian cancer and primary peritoneal carcinomatosis: A hoosier oncology group trial. *Cancer* **2008**, *113*, 723–732. [CrossRef] [PubMed]

57. Matei, D.; Sill, M.W.; Lankes, H.A.; DeGeest, K.; Bristow, R.E.; Mutch, D.; Yamada, S.D.; Cohn, D.; Calvert, V.; Farley, J.; et al. Activity of sorafenib in recurrent ovarian cancer and primary peritoneal carcinomatosis: A gynecologic oncology group trial. *J. Clin. Oncol.* **2011**, *29*, 69–75. [CrossRef] [PubMed]
58. Otrock, Z.K.; Mahfouz, R.A.; Makarem, J.A.; Shamseddine, A.I. Understanding the biology of angiogenesis: Review of the most important molecular mechanisms. *Blood Cells Mol. Dis.* **2007**, *39*, 212–220. [CrossRef] [PubMed]
59. Folkman, J. Angiogenesis: An organizing principle for drug discovery? *Nat. Rev. Drug Discov.* **2007**, *6*, 273–286. [CrossRef] [PubMed]
60. Zetter, B.R. Angiogenesis and tumor metastasis. *Annu. Rev. Med.* **1998**, *49*, 407–424. [CrossRef] [PubMed]
61. Hollingsworth, H.C.; Kohn, E.C.; Steinberg, S.M.; Rothenberg, M.L.; Merino, M.J. Tumor angiogenesis in advanced stage ovarian carcinoma. *Am. J. Pathol.* **1995**, *147*, 33–41. [PubMed]
62. Hanahan, D.; Folkman, J. Patterns and emerging mechanisms of the angiogenic switch during tumorigenesis. *Cell* **1996**, *86*, 353–364. [CrossRef]
63. Ko, S.Y.; Naora, H. Therapeutic strategies for targeting the ovarian tumor stroma. *World J. Clin. Cases* **2014**, *2*, 194–200. [CrossRef] [PubMed]
64. Yoneda, J.; Kuniyasu, H.; Crispens, M.A.; Price, J.E.; Bucana, C.D.; Fidler, I.J. Expression of angiogenesis-related genes and progression of human ovarian carcinomas in nude mice. *J. Natl. Cancer Inst.* **1998**, *90*, 447–454. [CrossRef] [PubMed]
65. Barton, D.P.; Cai, A.; Wendt, K.; Young, M.; Gamero, A.; De Cesare, S. Angiogenic protein expression in advanced epithelial ovarian cancer. *Clin. Cancer Res.* **1997**, *3*, 1579–1586. [PubMed]
66. Kraft, A.; Weindel, K.; Ochs, A.; Marth, C.; Zmija, J.; Schumacher, P.; Unger, C.; Marme, D.; Gastl, G. Vascular endothelial growth factor in the sera and effusions of patients with malignant and nonmalignant disease. *Cancer* **1999**, *85*, 178–187. [CrossRef]
67. Mesiano, S.; Ferrara, N.; Jaffe, R.B. Role of vascular endothelial growth factor in ovarian cancer: Inhibition of ascites formation by immunoneutralization. *Am. J. Pathol.* **1998**, *153*, 1249–1256. [CrossRef]
68. Shen, G.H.; Ghazizadeh, M.; Kawanami, O.; Shimizu, H.; Jin, E.; Araki, T.; Sugisaki, Y. Prognostic significance of vascular endothelial growth factor expression in human ovarian carcinoma. *Br. J. Cancer* **2000**, *83*, 196–203. [CrossRef] [PubMed]
69. Duncan, T.J.; Al-Attar, A.; Rolland, P.; Scott, I.V.; Deen, S.; Liu, D.T.; Spendlove, I.; Durrant, L.G. Vascular endothelial growth factor expression in ovarian cancer: A model for targeted use of novel therapies? *Clin. Cancer Res.* **2008**, *14*, 3030–3035. [CrossRef] [PubMed]
70. Graybill, W.; Sood, A.K.; Monk, B.J.; Coleman, R.L. State of the science: Emerging therapeutic strategies for targeting angiogenesis in ovarian cancer. *Gynecol. Oncol.* **2015**, *138*, 223–226. [CrossRef] [PubMed]
71. Ruscito, I.; Cacsire Castillo-Tong, D.; Vergote, I.; Ignat, I.; Stanske, M.; Vanderstichele, A.; Glajzer, J.; Kulbe, H.; Trillsch, F.; Mustea, A.; et al. Characterisation of tumour microvessel density during progression of high-grade serous ovarian cancer: Clinico-pathological impact (an octips consortium study). *Br. J. Cancer* **2018**, *119*, 330–338. [CrossRef] [PubMed]
72. De Palma, M.; Biziato, D.; Petrova, T.V. Microenvironmental regulation of tumour angiogenesis. *Nat. Rev. Cancer* **2017**, *17*, 457–474. [CrossRef] [PubMed]
73. Holmes, D.I.; Zachary, I. The vascular endothelial growth factor (vegf) family: Angiogenic factors in health and disease. *Genome Biol.* **2005**, *6*, 209. [CrossRef] [PubMed]
74. Neufeld, G.; Cohen, T.; Gengrinovitch, S.; Poltorak, Z. Vascular endothelial growth factor (vegf) and its receptors. *FASEB J.* **1999**, *13*, 9–22. [CrossRef] [PubMed]
75. Gupta, K.; Kshirsagar, S.; Li, W.; Gui, L.; Ramakrishnan, S.; Gupta, P.; Law, P.Y.; Hebbel, R.P. Vegf prevents apoptosis of human microvascular endothelial cells via opposing effects on mapk/erk and sapk/jnk signaling. *Exp. Cell Res.* **1999**, *247*, 495–504. [CrossRef] [PubMed]
76. Cebe-Suarez, S.; Zehnder-Fjallman, A.; Ballmer-Hofer, K. The role of vegf receptors in angiogenesis; complex partnerships. *Cell. Mol. Life Sci.* **2006**, *63*, 601–615. [CrossRef] [PubMed]
77. Eskander, R.N.; Randall, L.M. Bevacizumab in the treatment of ovarian cancer. *Biologics* **2011**, *5*, 1–5. [PubMed]
78. Mabuchi, S.; Terai, Y.; Morishige, K.; Tanabe-Kimura, A.; Sasaki, H.; Kanemura, M.; Tsunetoh, S.; Tanaka, Y.; Sakata, M.; Burger, R.A.; et al. Maintenance treatment with bevacizumab prolongs survival in an in vivo ovarian cancer model. *Clin. Cancer Res.* **2008**, *14*, 7781–7789. [CrossRef] [PubMed]

79. Burger, R.A.; Brady, M.F.; Bookman, M.A.; Fleming, G.F.; Monk, B.J.; Huang, H.; Mannel, R.S.; Homesley, H.D.; Fowler, J.; Greer, B.E.; et al. Incorporation of bevacizumab in the primary treatment of ovarian cancer. *N. Engl. J. Med.* **2011**, *365*, 2473–2483. [CrossRef] [PubMed]
80. Coleman, R.L.; Brady, M.F.; Herzog, T.J.; Sabbatini, P.; Armstrong, D.K.; Walker, J.L.; Kim, B.G.; Fujiwara, K.; Tewari, K.S.; O'Malley, D.M.; et al. Bevacizumab and paclitaxel-carboplatin chemotherapy and secondary cytoreduction in recurrent, platinum-sensitive ovarian cancer (nrg oncology/gynecologic oncology group study gog-0213): A multicentre, open-label, randomised, phase 3 trial. *Lancet Oncol.* **2017**, *18*, 779–791. [CrossRef]
81. Burger, R.A.; Sill, M.W.; Monk, B.J.; Greer, B.E.; Sorosky, J.I. Phase II trial of bevacizumab in persistent or recurrent epithelial ovarian cancer or primary peritoneal cancer: A gynecologic oncology group study. *J. Clin. Oncol.* **2007**, *25*, 5165–5171. [CrossRef] [PubMed]
82. Perren, T.J.; Swart, A.M.; Pfisterer, J.; Ledermann, J.A.; Pujade-Lauraine, E.; Kristensen, G.; Carey, M.S.; Beale, P.; Cervantes, A.; Kurzeder, C.; et al. A phase 3 trial of bevacizumab in ovarian cancer. *N. Engl. J. Med.* **2011**, *365*, 2484–2496. [CrossRef] [PubMed]
83. Pujade-Lauraine, E.; Hilpert, F.; Weber, B.; Reuss, A.; Poveda, A.; Kristensen, G.; Sorio, R.; Vergote, I.; Witteveen, P.; Bamias, A.; et al. Bevacizumab combined with chemotherapy for platinum-resistant recurrent ovarian cancer: The aurelia open-label randomized phase iii trial. *J. Clin. Oncol.* **2014**, *32*, 1302–1308. [CrossRef] [PubMed]
84. Moroney, J.W.; Sood, A.K.; Coleman, R.L. Aflibercept in epithelial ovarian carcinoma. *Future Oncol.* **2009**, *5*, 591–600. [CrossRef] [PubMed]
85. Teng, L.S.; Jin, K.T.; He, K.F.; Zhang, J.; Wang, H.H.; Cao, J. Clinical applications of vegf-trap (aflibercept) in cancer treatment. *J. Chin. Med. Assoc.* **2010**, *73*, 449–456. [CrossRef]
86. Byrne, A.T.; Ross, L.; Holash, J.; Nakanishi, M.; Hu, L.; Hofmann, J.I.; Yancopoulos, G.D.; Jaffe, R.B. Vascular endothelial growth factor-trap decreases tumor burden, inhibits ascites, and causes dramatic vascular remodeling in an ovarian cancer model. *Clin. Cancer Res.* **2003**, *9*, 5721–5728. [PubMed]
87. Colombo, N.; Mangili, G.; Mammoliti, S.; Kalling, M.; Tholander, B.; Sternas, L.; Buzenet, G.; Chamberlain, D. A phase ii study of aflibercept in patients with advanced epithelial ovarian cancer and symptomatic malignant ascites. *Gynecol. Oncol.* **2012**, *125*, 42–47. [CrossRef] [PubMed]
88. Bergers, G.; Hanahan, D. Modes of resistance to anti-angiogenic therapy. *Nat. Rev. Cancer* **2008**, *8*, 592–603. [CrossRef] [PubMed]
89. Matulonis, U.A.; Berlin, S.; Ivy, P.; Tyburski, K.; Krasner, C.; Zarwan, C.; Berkenblit, A.; Campos, S.; Horowitz, N.; Cannistra, S.A.; et al. Cediranib, an oral inhibitor of vascular endothelial growth factor receptor kinases, is an active drug in recurrent epithelial ovarian, fallopian tube, and peritoneal cancer. *J. Clin. Oncol.* **2009**, *27*, 5601–5606. [CrossRef] [PubMed]
90. Hirte, H.; Lheureux, S.; Fleming, G.F.; Sugimoto, A.; Morgan, R.; Biagi, J.; Wang, L.; McGill, S.; Ivy, S.P.; Oza, A.M. A phase 2 study of cediranib in recurrent or persistent ovarian, peritoneal or fallopian tube cancer: A trial of the princess margaret, chicago and california phase ii consortia. *Gynecol. Oncol.* **2015**, *138*, 55–61. [CrossRef] [PubMed]
91. Ledermann, J.A.; Hackshaw, A.; Kaye, S.; Jayson, G.; Gabra, H.; McNeish, I.; Earl, H.; Perren, T.; Gore, M.; Persic, M.; et al. Randomized phase ii placebo-controlled trial of maintenance therapy using the oral triple angiokinase inhibitor bibf 1120 after chemotherapy for relapsed ovarian cancer. *J. Clin. Oncol.* **2011**, *29*, 3798–3804. [CrossRef] [PubMed]
92. Du Bois, A.; Kristensen, G.; Ray-Coquard, I.; Reuss, A.; Pignata, S.; Colombo, N.; Denison, U.; Vergote, I.; del Campo, J.M.; Ottevanger, P.; et al. Standard first-line chemotherapy with or without nintedanib for advanced ovarian cancer (ago-ovar 12): A randomised, double-blind, placebo-controlled phase 3 trial. *Lancet Oncol.* **2016**, *17*, 78–89. [CrossRef]
93. Pignata, S.; Lorusso, D.; Scambia, G.; Sambataro, D.; Tamberi, S.; Cinieri, S.; Mosconi, A.M.; Orditura, M.; Bartolini, S.; Arcangeli, V.; et al. Mito-11: A randomized multicenter phase ii trial testing the addition of pazopanib to weekly paclitaxel in platinum-resistant or -refractory advanced ovarian cancer (aoc). *J. Clin. Oncol.* **2014**, *32*.
94. Sawada, K.; Mitra, A.K.; Radjabi, A.R.; Bhaskar, V.; Kistner, E.O.; Tretiakova, M.; Jagadeeswaran, S.; Montag, A.; Becker, A.; Kenny, H.A.; et al. Loss of e-cadherin promotes ovarian cancer metastasis via alpha 5-integrin, which is a therapeutic target. *Cancer Res.* **2008**, *68*, 2329–2339. [CrossRef] [PubMed]

95. Iwanicki, M.P.; Davidowitz, R.A.; Ng, M.R.; Besser, A.; Muranen, T.; Merritt, M.; Danuser, G.; Ince, T.A.; Brugge, J.S. Ovarian cancer spheroids use myosin-generated force to clear the mesothelium. *Cancer Discov.* **2011**, *1*, 144–157. [CrossRef] [PubMed]
96. Davidowitz, R.A.; Selfors, L.M.; Iwanicki, M.P.; Elias, K.M.; Karst, A.; Piao, H.; Ince, T.A.; Drage, M.G.; Dering, J.; Konecny, G.E.; et al. Mesenchymal gene program-expressing ovarian cancer spheroids exhibit enhanced mesothelial clearance. *J. Clin. Investig.* **2014**, *124*, 2611–2625. [CrossRef] [PubMed]
97. Lessan, K.; Aguiar, D.J.; Oegema, T.; Siebenson, L.; Skubitz, A.P. Cd44 and beta1 integrin mediate ovarian carcinoma cell adhesion to peritoneal mesothelial cells. *Am. J. Pathol.* **1999**, *154*, 1525–1537. [CrossRef]
98. Kenny, H.A.; Chiang, C.Y.; White, E.A.; Schryver, E.M.; Habis, M.; Romero, I.L.; Ladanyi, A.; Penicka, C.V.; George, J.; Matlin, K.; et al. Mesothelial cells promote early ovarian cancer metastasis through fibronectin secretion. *J. Clin. Investig.* **2014**, *124*, 4614–4628. [CrossRef] [PubMed]
99. Janiak, A.; Zemskov, E.A.; Belkin, A.M. Cell surface transglutaminase promotes rhoa activation via integrin clustering and suppression of the src-p190rhogap signaling pathway. *Mol. Biol. Cell* **2006**, *17*, 1606–1619. [CrossRef] [PubMed]
100. Satpathy, M.; Cao, L.; Pincheira, R.; Emerson, R.; Bigsby, R.; Nakshatri, H.; Matei, D. Enhanced peritoneal ovarian tumor dissemination by tissue transglutaminase. *Cancer Res.* **2007**, *67*, 7194–7202. [CrossRef] [PubMed]
101. Shao, M.; Cao, L.; Shen, C.; Satpathy, M.; Chelladurai, B.; Bigsby, R.M.; Nakshatri, H.; Matei, D. Epithelial-to-mesenchymal transition and ovarian tumor progression induced by tissue transglutaminase. *Cancer Res.* **2009**, *69*, 9192–9201. [CrossRef] [PubMed]
102. Condello, S.; Cao, L.; Matei, D. Tissue transglutaminase regulates beta-catenin signaling through a c-src-dependent mechanism. *FASEB J.* **2013**, *27*, 3100–3112. [CrossRef] [PubMed]
103. Condello, S.; Sima, L.; Ivan, C.; Cardenas, H.; Schiltz, G.; Mishra, R.K.; Matei, D. Tissue tranglutaminase regulates interactions between ovarian cancer stem cells and the tumor niche. *Cancer Res.* **2018**, *78*, 2990–3001. [CrossRef] [PubMed]
104. Strobel, T.; Cannistra, S.A. Beta1-integrins partly mediate binding of ovarian cancer cells to peritoneal mesothelium in vitro. *Gynecol. Oncol.* **1999**, *73*, 362–367. [CrossRef] [PubMed]
105. Cannistra, S.A.; Kansas, G.S.; Niloff, J.; DeFranzo, B.; Kim, Y.; Ottensmeier, C. Binding of ovarian cancer cells to peritoneal mesothelium in vitro is partly mediated by cd44h. *Cancer Res.* **1993**, *53*, 3830–3838. [PubMed]
106. Strobel, T.; Swanson, L.; Cannistra, S.A. In vivo inhibition of cd44 limits intra-abdominal spread of a human ovarian cancer xenograft in nude mice: A novel role for cd44 in the process of peritoneal implantation. *Cancer Res.* **1997**, *57*, 1228–1232. [PubMed]
107. Slack-Davis, J.K.; Atkins, K.A.; Harrer, C.; Hershey, E.D.; Conaway, M. Vascular cell adhesion molecule-1 is a regulator of ovarian cancer peritoneal metastasis. *Cancer Res.* **2009**, *69*, 1469–1476. [CrossRef] [PubMed]
108. Mitra, A.K.; Sawada, K.; Tiwari, P.; Mui, K.; Gwin, K.; Lengyel, E. Ligand-independent activation of c-met by fibronectin and alpha(5)beta(1)-integrin regulates ovarian cancer invasion and metastasis. *Oncogene* **2011**, *30*, 1566–1576. [CrossRef] [PubMed]
109. Raab-Westphal, S.; Marshall, J.F.; Goodman, S.L. Integrins as therapeutic targets: Successes and cancers. *Cancers (Basel)* **2017**, *9*, 110. [CrossRef] [PubMed]
110. Bell-McGuinn, K.M.; Matthews, C.M.; Ho, S.N.; Barve, M.; Gilbert, L.; Penson, R.T.; Lengyel, E.; Palaparthy, R.; Gilder, K.; Vassos, A.; et al. A phase ii, single-arm study of the anti-alpha5beta1 integrin antibody volociximab as monotherapy in patients with platinum-resistant advanced epithelial ovarian or primary peritoneal cancer. *Gynecol. Oncol.* **2011**, *121*, 273–279. [CrossRef] [PubMed]
111. Trikha, M.; Zhou, Z.; Nemeth, J.A.; Chen, Q.; Sharp, C.; Emmell, E.; Giles-Komar, J.; Nakada, M.T. Cnto 95, a fully human monoclonal antibody that inhibits alphav integrins, has antitumor and antiangiogenic activity in vivo. *Int. J. Cancer* **2004**, *110*, 326–335. [CrossRef] [PubMed]
112. Chen, Q.; Manning, C.D.; Millar, H.; McCabe, F.L.; Ferrante, C.; Sharp, C.; Shahied-Arruda, L.; Doshi, P.; Nakada, M.T.; Anderson, G.M. Cnto 95, a fully human anti alphav integrin antibody, inhibits cell signaling, migration, invasion, and spontaneous metastasis of human breast cancer cells. *Clin. Exp. Metastasis* **2008**, *25*, 139–148. [CrossRef] [PubMed]
113. Mullamitha, S.A.; Ton, N.C.; Parker, G.J.; Jackson, A.; Julyan, P.J.; Roberts, C.; Buonaccorsi, G.A.; Watson, Y.; Davies, K.; Cheung, S.; et al. Phase i evaluation of a fully human anti-alphav integrin monoclonal antibody (cnto 95) in patients with advanced solid tumors. *Clin. Cancer Res.* **2007**, *13*, 2128–2135. [CrossRef] [PubMed]

114. Hersey, P.; Sosman, J.; O'Day, S.; Richards, J.; Bedikian, A.; Gonzalez, R.; Sharfman, W.; Weber, R.; Logan, T.; Buzoianu, M.; et al. A randomized phase 2 study of etaracizumab, a monoclonal antibody against integrin alpha(v)beta(3), + or - dacarbazine in patients with stage iv metastatic melanoma. *Cancer* **2010**, *116*, 1526–1534. [CrossRef] [PubMed]
115. Dolgos, H.; Freisleben, A.; Wimmer, E.; Scheible, H.; Kratzer, F.; Yamagata, T.; Gallemann, D.; Fluck, M. In vitro and in vivo drug disposition of cilengitide in animals and human. *Pharmacol. Res. Perspect.* **2016**, *4*, e00217. [CrossRef] [PubMed]
116. Soffietti, R.; Trevisan, E.; Ruda, R. What have we learned from trials on antiangiogenic agents in glioblastoma? *Expert Rev. Neurother.* **2014**, *14*, 1–3. [CrossRef] [PubMed]
117. Kaur, S.; Kenny, H.A.; Jagadeeswaran, S.; Zillhardt, M.R.; Montag, A.G.; Kistner, E.; Yamada, S.D.; Mitra, A.K.; Lengyel, E. {beta}3-integrin expression on tumor cells inhibits tumor progression, reduces metastasis, and is associated with a favorable prognosis in patients with ovarian cancer. *Am. J. Pathol.* **2009**, *175*, 2184–2196. [CrossRef] [PubMed]
118. Kenny, H.A.; Lengyel, E. Mmp-2 functions as an early response protein in ovarian cancer metastasis. *Cell Cycle* **2009**, *8*, 683–688. [CrossRef] [PubMed]
119. Schlaepfer, D.D.; Jones, K.C.; Hunter, T. Multiple grb2-mediated integrin-stimulated signaling pathways to erk2/mitogen-activated protein kinase: Summation of both c-src- and focal adhesion kinase-initiated tyrosine phosphorylation events. *Mol. Cell Biol.* **1998**, *18*, 2571–2585. [CrossRef] [PubMed]
120. Renshaw, M.W.; Price, L.S.; Schwartz, M.A. Focal adhesion kinase mediates the integrin signaling requirement for growth factor activation of map kinase. *J. Cell Biol.* **1999**, *147*, 611–618. [CrossRef] [PubMed]
121. Ward, K.K.; Tancioni, I.; Lawson, C.; Miller, N.L.; Jean, C.; Chen, X.L.; Uryu, S.; Kim, J.; Tarin, D.; Stupack, D.G.; et al. Inhibition of focal adhesion kinase (fak) activity prevents anchorage-independent ovarian carcinoma cell growth and tumor progression. *Clin. Exp. Metastasis* **2013**, *30*, 579–594. [CrossRef] [PubMed]
122. Matei, D.; Graeber, T.G.; Baldwin, R.L.; Karlan, B.Y.; Rao, J.; Chang, D.D. Gene expression in epithelial ovarian carcinoma. *Oncogene* **2002**, *21*, 6289–6298. [CrossRef] [PubMed]
123. Yakubov, B.; Chelladurai, B.; Schmitt, J.; Emerson, R.; Turchi, J.J.; Matei, D. Extracellular tissue transglutaminase activates noncanonical nf-kappab signaling and promotes metastasis in ovarian cancer. *Neoplasia* **2013**, *15*, 609–619. [CrossRef] [PubMed]
124. Satpathy, M.; Shao, M.; Emerson, R.; Donner, D.B.; Matei, D. Tissue transglutaminase regulates matrix metalloproteinase-2 in ovarian cancer by modulating camp-response element-binding protein activity. *J. Biol. Chem.* **2009**, *284*, 15390–15399. [CrossRef] [PubMed]
125. Khanna, M.; Chelladurai, B.; Gavini, A.; Li, L.; Shao, M.; Courtney, D.; Turchi, J.J.; Matei, D.; Meroueh, S. Targeting ovarian tumor cell adhesion mediated by tissue transglutaminase. *Mol. Cancer Ther.* **2011**, *10*, 626–636. [CrossRef] [PubMed]
126. Yakubov, B.; Chen, L.; Belkin, A.M.; Zhang, S.; Chelladurai, B.; Zhang, Z.Y.; Matei, D. Small molecule inhibitors target the tissue transglutaminase and fibronectin interaction. *PLoS ONE* **2014**, *9*, e89285. [CrossRef] [PubMed]
127. Burkholder, B.; Huang, R.Y.; Burgess, R.; Luo, S.; Jones, V.S.; Zhang, W.; Lv, Z.Q.; Gao, C.Y.; Wang, B.L.; Zhang, Y.M.; et al. Tumor-induced perturbations of cytokines and immune cell networks. *Biochim. Biophys. Acta.* **2014**, *1845*, 182–201. [CrossRef] [PubMed]
128. Zhang, L.; Conejo-Garcia, J.R.; Katsaros, D.; Gimotty, P.A.; Massobrio, M.; Regnani, G.; Makrigiannakis, A.; Gray, H.; Schlienger, K.; Liebman, M.N.; et al. Intratumoral t cells, recurrence, and survival in epithelial ovarian cancer. *Ne. Engl. J. Med.* **2003**, *348*, 203–213. [CrossRef] [PubMed]
129. Clarke, B.; Tinker, A.V.; Lee, C.H.; Subramanian, S.; van de Rijn, M.; Turbin, D.; Kalloger, S.; Han, G.; Ceballos, K.; Cadungog, M.G.; et al. Intraepithelial t cells and prognosis in ovarian carcinoma: Novel associations with stage, tumor type, and brca1 loss. *Mod. Pathol.* **2009**, *22*, 393–402. [CrossRef] [PubMed]
130. Sato, E.; Olson, S.H.; Ahn, J.; Bundy, B.; Nishikawa, H.; Qian, F.; Jungbluth, A.A.; Frosina, D.; Gnjatic, S.; Ambrosone, C.; et al. Intraepithelial cd8+ tumor-infiltrating lymphocytes and a high cd8+/regulatory t cell ratio are associated with favorable prognosis in ovarian cancer. *Proc. Natl. Acad. Sci. USA* **2005**, *102*, 18538–18543. [CrossRef] [PubMed]

131. Santoiemma, P.P.; Reyes, C.; Wang, L.P.; McLane, M.W.; Feldman, M.D.; Tanyi, J.L.; Powell, D.J., Jr. Systematic evaluation of multiple immune markers reveals prognostic factors in ovarian cancer. *Gynecol. Oncol.* **2016**, *143*, 120–127. [CrossRef] [PubMed]
132. Ye, Q.; Song, D.G.; Poussin, M.; Yamamoto, T.; Best, A.; Li, C.; Coukos, G.; Powell, D.J., Jr. Cd137 accurately identifies and enriches for naturally occurring tumor-reactive t cells in tumor. *Clin. Cancer Res.* **2014**, *20*, 44–55. [CrossRef] [PubMed]
133. Jimenez-Sanchez, A.; Memon, D.; Pourpe, S.; Veeraraghavan, H.; Li, Y.; Vargas, H.A.; Gill, M.B.; Park, K.J.; Zivanovic, O.; Konner, J.; et al. Heterogeneous tumor-immune microenvironments among differentially growing metastases in an ovarian cancer patient. *Cell* **2017**, *170*, 927–938. [CrossRef] [PubMed]
134. Bosmuller, H.C.; Wagner, P.; Peper, J.K.; Schuster, H.; Pham, D.L.; Greif, K.; Beschorner, C.; Rammensee, H.G.; Stevanovic, S.; Fend, F.; et al. Combined immunoscore of cd103 and cd3 identifies long-term survivors in high-grade serous ovarian cancer. *Int. J. Gynecol. Cancer* **2016**, *26*, 671–679. [CrossRef] [PubMed]
135. Garon, E.B.; Rizvi, N.A.; Hui, R.; Leighl, N.; Balmanoukian, A.S.; Eder, J.P.; Patnaik, A.; Aggarwal, C.; Gubens, M.; Horn, L.; et al. Pembrolizumab for the treatment of non-small-cell lung cancer. *N. Engl. J. Med.* **2015**, *372*, 2018–2028. [CrossRef] [PubMed]
136. Robert, C.; Schachter, J.; Long, G.V.; Arance, A.; Grob, J.J.; Mortier, L.; Daud, A.; Carlino, M.S.; McNeil, C.; Lotem, M.; et al. Pembrolizumab versus ipilimumab in advanced melanoma. *N. Engl. J. Med.* **2015**, *372*, 2521–2532. [CrossRef] [PubMed]
137. Ikarashi, H.; Fujita, K.; Takakuwa, K.; Kodama, S.; Tokunaga, A.; Takahashi, T.; Tanaka, K. Immunomodulation in patients with epithelial ovarian cancer after adoptive transfer of tumor-infiltrating lymphocytes. *Cancer Res.* **1994**, *54*, 190–196. [PubMed]
138. Kandalaft, L.E.; Chiang, C.L.; Tanyi, J.; Motz, G.; Balint, K.; Mick, R.; Coukos, G. A phase i vaccine trial using dendritic cells pulsed with autologous oxidized lysate for recurrent ovarian cancer. *J. Transl. Med.* **2013**, *11*, 149. [CrossRef] [PubMed]
139. Tumeh, P.C.; Harview, C.L.; Yearley, J.H.; Shintaku, I.P.; Taylor, E.J.; Robert, L.; Chmielowski, B.; Spasic, M.; Henry, G.; Ciobanu, V.; et al. Pd-1 blockade induces responses by inhibiting adaptive immune resistance. *Nature* **2014**, *515*, 568–571. [CrossRef] [PubMed]
140. Brahmer, J.R.; Tykodi, S.S.; Chow, L.Q.; Hwu, W.J.; Topalian, S.L.; Hwu, P.; Drake, C.G.; Camacho, L.H.; Kauh, J.; Odunsi, K.; et al. Safety and activity of anti-pd-l1 antibody in patients with advanced cancer. *N. Engl. J. Med.* **2012**, *366*, 2455–2465. [CrossRef] [PubMed]
141. Gettinger, S.N.; Horn, L.; Gandhi, L.; Spigel, D.R.; Antonia, S.J.; Rizvi, N.A.; Powderly, J.D.; Heist, R.S.; Carvajal, R.D.; Jackman, D.M.; et al. Overall survival and long-term safety of nivolumab (anti-programmed death 1 antibody, bms-936558, ono-4538) in patients with previously treated advanced non-small-cell lung cancer. *J. Clin. Oncol.* **2015**, *33*, 2004–2012. [CrossRef] [PubMed]
142. Mahoney, K.M.; Freeman, G.J.; McDermott, D.F. The next immune-checkpoint inhibitors: Pd-1/pd-l1 blockade in melanoma. *Clin. Ther.* **2015**, *37*, 764–782. [CrossRef] [PubMed]
143. Hamanishi, J.; Mandai, M.; Iwasaki, M.; Okazaki, T.; Tanaka, Y.; Yamaguchi, K.; Higuchi, T.; Yagi, H.; Takakura, K.; Minato, N.; et al. Programmed cell death 1 ligand 1 and tumor-infiltrating cd8+ t lymphocytes are prognostic factors of human ovarian cancer. *Proc. Natl. Acad. Sci. USA* **2007**, *104*, 3360–3365. [CrossRef] [PubMed]
144. Krempski, J.; Karyampudi, L.; Behrens, M.D.; Erskine, C.L.; Hartmann, L.; Dong, H.; Goode, E.L.; Kalli, K.R.; Knutson, K.L. Tumor-infiltrating programmed death receptor-1+ dendritic cells mediate immune suppression in ovarian cancer. *J. Immunol.* **2011**, *186*, 6905–6913. [CrossRef] [PubMed]
145. Darb-Esfahani, S.; Kunze, C.A.; Kulbe, H.; Sehouli, J.; Wienert, S.; Lindner, J.; Budczies, J.; Bockmayr, M.; Dietel, M.; Denkert, C.; et al. Prognostic impact of programmed cell death-1 (pd-1) and pd-ligand 1 (pd-l1) expression in cancer cells and tumor-infiltrating lymphocytes in ovarian high grade serous carcinoma. *Oncotarget* **2016**, *7*, 1486–1499. [CrossRef] [PubMed]
146. Webb, J.R.; Milne, K.; Kroeger, D.R.; Nelson, B.H. Pd-l1 expression is associated with tumor-infiltrating t cells and favorable prognosis in high-grade serous ovarian cancer. *Gynecol. Oncol.* **2016**, *141*, 293–302. [CrossRef] [PubMed]
147. Duraiswamy, J.; Freeman, G.J.; Coukos, G. Therapeutic pd-1 pathway blockade augments with other modalities of immunotherapy t-cell function to prevent immune decline in ovarian cancer. *Cancer Res.* **2013**, *73*, 6900–6912. [CrossRef] [PubMed]

148. Varga, A.; Pihapaul, S.A.; Ott, P.A.; Mehnert, J.M.; Bertonrigaud, D.; Johnson, E.A.; Cheng, J.D.; Yuan, S.; Rubin, E.H.; Matei, D.E. Antitumor activity and safety of pembrolizumab in patients (pts) with pd-l1 positive advanced ovarian cancer: Interim results from a phase ib study. *J. Clin. Oncol.* **2015**, *33*, 5510. [CrossRef]
149. Disis, M.L.; Patel, M.R.; Pant, S.; Infante, J.R.; Lockhart, A.C.; Kelly, K.; Beck, J.T.; Gordon, M.S.; Weiss, G.J.; Ejadi, S.; et al. Avelumab (msb0010718c), an anti-pd-l1 antibody, in patients with previously treated, recurrent or refractory ovarian cancer: A phase ib, open-label expansion trial. *J. Clin. Oncol.* **2015**, *33*, 5509. [CrossRef]
150. Lau, E. Mismatch repair deficiency predicts benefit of anti-pd-1 therapy. *Lancet Oncol.* **2015**, *16*, e319. [CrossRef]
151. Le, D.T.; Uram, J.N.; Wang, H.; Bartlett, B.R.; Kemberling, H.; Eyring, A.D.; Skora, A.D.; Luber, B.S.; Azad, N.S.; Laheru, D.; et al. Pd-1 blockade in tumors with mismatch-repair deficiency. *N. Engl. J. Med.* **2015**, *372*, 2509–2520. [CrossRef] [PubMed]
152. Strickland, K.C.; Howitt, B.E.; Shukla, S.A.; Rodig, S.; Ritterhouse, L.L.; Liu, J.F.; Garber, J.E.; Chowdhury, D.; Wu, C.J.; D'Andrea, A.D.; et al. Association and prognostic significance of brca1/2-mutation status with neoantigen load, number of tumor-infiltrating lymphocytes and expression of pd-1/pd-l1 in high grade serous ovarian cancer. *Oncotarget* **2016**, *7*, 13587–13598. [CrossRef] [PubMed]
153. Noy, R.; Pollard, J.W. Tumor-associated macrophages: From mechanisms to therapy. *Immunity* **2014**, *41*, 49–61. [CrossRef] [PubMed]
154. Rabinovich, G.A.; Gabrilovich, D.; Sotomayor, E.M. Immunosuppressive strategies that are mediated by tumor cells. *Annu. Rev. Immunol.* **2007**, *25*, 267–296. [CrossRef] [PubMed]
155. Bronte, V.; Serafini, P.; Apolloni, E.; Zanovello, P. Tumor-induced immune dysfunctions caused by myeloid suppressor cells. *J. Immunother.* **2001**, *24*, 431–446. [CrossRef] [PubMed]
156. Almand, B.; Clark, J.I.; Nikitina, E.; van Beynen, J.; English, N.R.; Knight, S.C.; Carbone, D.P.; Gabrilovich, D.I. Increased production of immature myeloid cells in cancer patients: A mechanism of immunosuppression in cancer. *J. Immunol.* **2001**, *166*, 678–689. [CrossRef] [PubMed]
157. Youn, J.I.; Nagaraj, S.; Collazo, M.; Gabrilovich, D.I. Subsets of myeloid-derived suppressor cells in tumor-bearing mice. *J. Immunol.* **2008**, *181*, 5791–5802. [CrossRef] [PubMed]
158. Zhang, M.; He, Y.; Sun, X.; Li, Q.; Wang, W.; Zhao, A.; Di, W. A high m1/m2 ratio of tumor-associated macrophages is associated with extended survival in ovarian cancer patients. *J. Ovarian Res.* **2014**, *7*, 19. [CrossRef] [PubMed]
159. Pollard, J.W. Tumour-educated macrophages promote tumour progression and metastasis. *Nat. Rev. Cancer* **2004**, *4*, 71–78. [CrossRef] [PubMed]
160. Wang, X.; Deavers, M.; Patenia, R.; Bassett, R.L., Jr.; Mueller, P.; Ma, Q.; Wang, E.; Freedman, R.S. Monocyte/macrophage and t-cell infiltrates in peritoneum of patients with ovarian cancer or benign pelvic disease. *J. Transl. Med.* **2006**, *4*, 30. [CrossRef] [PubMed]
161. Yin, M.; Li, X.; Tan, S.; Zhou, H.J.; Ji, W.; Bellone, S.; Xu, X.; Zhang, H.; Santin, A.D.; Lou, G.; et al. Tumor-associated macrophages drive spheroid formation during early transcoelomic metastasis of ovarian cancer. *J. Clin. Investig.* **2016**, *126*, 4157–4173. [CrossRef] [PubMed]
162. Hagemann, T.; Wilson, J.; Kulbe, H.; Li, N.F.; Leinster, D.A.; Charles, K.; Klemm, F.; Pukrop, T.; Binder, C.; Balkwill, F.R. Macrophages induce invasiveness of epithelial cancer cells via nf-kappa b and jnk. *J. Immunol.* **2005**, *175*, 1197–1205. [CrossRef] [PubMed]
163. Ko, S.Y.; Ladanyi, A.; Lengyel, E.; Naora, H. Expression of the homeobox gene hoxa9 in ovarian cancer induces peritoneal macrophages to acquire an m2 tumor-promoting phenotype. *Am. J. Pathol.* **2014**, *184*, 271–281. [CrossRef] [PubMed]
164. Cui, T.X.; Kryczek, I.; Zhao, L.; Zhao, E.; Kuick, R.; Roh, M.H.; Vatan, L.; Szeliga, W.; Mao, Y.; Thomas, D.G.; et al. Myeloid-derived suppressor cells enhance stemness of cancer cells by inducing microrna101 and suppressing the corepressor ctbp2. *Immunity* **2013**, *39*, 611–621. [CrossRef] [PubMed]
165. Horikawa, N.; Abiko, K.; Matsumura, N.; Hamanishi, J.; Baba, T.; Yamaguchi, K.; Yoshioka, Y.; Koshiyama, M.; Konishi, I. Expression of vascular endothelial growth factor in ovarian cancer inhibits tumor immunity through the accumulation of myeloid-derived suppressor cells. *Clin. Cancer Res.* **2017**, *23*, 587–599. [CrossRef] [PubMed]
166. Taki, M.; Abiko, K.; Baba, T.; Hamanishi, J.; Yamaguchi, K.; Murakami, R.; Yamanoi, K.; Horikawa, N.; Hosoe, Y.; Nakamura, E.; et al. Snail promotes ovarian cancer progression by recruiting myeloid-derived suppressor cells via cxcr2 ligand upregulation. *Nat. Commun.* **2018**, *9*, 1685. [CrossRef] [PubMed]

167. De Sanctis, F.; Bronte, V.; Ugel, S. Tumor-induced myeloid-derived suppressor cells. *Microbiol. Spectr.* **2016**, *4*. [CrossRef] [PubMed]
168. Engblom, C.; Pfirschke, C.; Pittet, M.J. The role of myeloid cells in cancer therapies. *Nat. Rev. Cancer* **2016**, *16*, 447–462. [CrossRef] [PubMed]
169. Pulaski, H.L.; Spahlinger, G.; Silva, I.A.; McLean, K.; Kueck, A.S.; Reynolds, R.K.; Coukos, G.; Conejo-Garcia, J.R.; Buckanovich, R.J. Identifying alemtuzumab as an anti-myeloid cell antiangiogenic therapy for the treatment of ovarian cancer. *J. Transl. Med.* **2009**, *7*, 49. [CrossRef] [PubMed]
170. Penn, C.A.; Yang, K.; Zong, H.; Lim, J.Y.; Cole, A.; Yang, D.; Baker, J.; Goonewardena, S.N.; Buckanovich, R.J. Therapeutic impact of nanoparticle therapy targeting tumor-associated macrophages. *Mol. Cancer Ther.* **2018**, *17*, 96–106. [CrossRef] [PubMed]
171. Zhu, H.; Bengsch, F.; Svoronos, N.; Rutkowski, M.R.; Bitler, B.G.; Allegrezza, M.J.; Yokoyama, Y.; Kossenkov, A.V.; Bradner, J.E.; Conejo-Garcia, J.R.; et al. Bet bromodomain inhibition promotes anti-tumor immunity by suppressing pd-l1 expression. *Cell Rep.* **2016**, *16*, 2829–2837. [CrossRef] [PubMed]
172. Stone, M.L.; Chiappinelli, K.B.; Li, H.; Murphy, L.M.; Travers, M.E.; Topper, M.J.; Mathios, D.; Lim, M.; Shih, I.M.; Wang, T.L.; et al. Epigenetic therapy activates type i interferon signaling in murine ovarian cancer to reduce immunosuppression and tumor burden. *Proc. Natl. Acad. Sci. USA* **2017**, *114*, E10981–E10990. [CrossRef] [PubMed]
173. Burges, A.; Wimberger, P.; Kumper, C.; Gorbounova, V.; Sommer, H.; Schmalfeldt, B.; Pfisterer, J.; Lichinitser, M.; Makhson, A.; Moiseyenko, V.; et al. Effective relief of malignant ascites in patients with advanced ovarian cancer by a trifunctional anti-epcam x anti-cd3 antibody: A phase i/ii study. *Clin. Cancer Res.* **2007**, *13*, 3899–3905. [CrossRef] [PubMed]

MDPI
St. Alban-Anlage 66
4052 Basel
Switzerland
Tel. +41 61 683 77 34
Fax +41 61 302 89 18
www.mdpi.com

Cancers Editorial Office
E-mail: cancers@mdpi.com
www.mdpi.com/journal/cancers

www.ingramcontent.com/pod-product-compliance
Lightning Source LLC
Chambersburg PA
CBHW041750220926
51847CB00014B/158
9783038975540